Methods in Enzymology

Volume 143
SULFUR AND SULFUR AMINO ACIDS

METHODS IN ENZYMOLOGY

EDITORS-IN-CHIEF

Sidney P. Colowick   Nathan O. Kaplan

Methods in Enzymology

Volume 143

Sulfur and Sulfur Amino Acids

EDITED BY

William B. Jakoby

NATIONAL INSTITUTES OF HEALTH
BETHESDA, MARYLAND

Owen W. Griffith

CORNELL UNIVERSITY MEDICAL COLLEGE
NEW YORK, NEW YORK

1987

ACADEMIC PRESS, INC.
Harcourt Brace Jovanovich, Publishers
Orlando   San Diego   New York   Austin
Boston   London   Sydney   Tokyo   Toronto

COPYRIGHT © 1987 BY ACADEMIC PRESS, INC.
ALL RIGHTS RESERVED.
NO PART OF THIS PUBLICATION MAY BE REPRODUCED OR
TRANSMITTED IN ANY FORM OR BY ANY MEANS, ELECTRONIC
OR MECHANICAL, INCLUDING PHOTOCOPY, RECORDING, OR
ANY INFORMATION STORAGE AND RETRIEVAL SYSTEM, WITHOUT
PERMISSION IN WRITING FROM THE PUBLISHER.

ACADEMIC PRESS, INC.
Orlando, Florida 32887

United Kingdom Edition published by
ACADEMIC PRESS INC. (LONDON) LTD.
24–28 Oval Road, London NW1 7DX

LIBRARY OF CONGRESS CATALOG CARD NUMBER: 54-9110

ISBN 0–12–182043–2 (alk. paper)

PRINTED IN THE UNITED STATES OF AMERICA

87 88 89 90   9 8 7 6 5 4 3 2 1

Table of Contents

Contributors to Volume 143 xiii
Preface xix
Volumes in Series xxi

Section I. Separation and Analysis

A. Inorganic Sulfur and Selenium Compounds

1.	Sulfate: Turbidimetric and Nephelometric Methods	Bo Sörbo	3
2.	Sulfate Determination: Ion Chromatography	Darryl M. Sullivan	7
3.	Sulfite Determination: Sulfite Oxidase	Hans-Otto Beutler	11
4.	Sulfite Determination: Fuchsin Method	Franz-Josef Leinweber and Kenneth J. Monty	15
5.	Sulfite and Thiosulfate: Using S-(2-Amino-2-carboxyethylsulfonyl)-L-cysteine	Toshihiko Ubuka	17
6.	Thiocyanate and Thiosulfate	John Westley	22
7.	Sulfane Sulfur	John L. Wood	25
8.	Sulfide Determination: Ion-Specific Electrode	Randolph T. Wedding	29
9.	Hydrogen Selenide and Methylselenol	Howard E. Ganther and R. J. Kraus	32
10.	Polarography of Sulfur Compounds	Howard Adler and John Westley	38

B. Thiols and Disulfides

11.	Spectrophotometric Assay of Thiols	Peter C. Jocelyn	44
12.	Fluorometric Assay of Thiols with Fluorobenzoxadiazoles	Kazuhiro Imai and Toshimasa Toyo'oka	67
13.	Thiol Labeling with Bromobimanes	Nechama S. Kosower and Edward M. Kosower	76

14. Determination of Low-Molecular-Weight Thiols Using Monobromobimane Fluorescent Labeling and High-Performance Liquid Chromatography	ROBERT C. FAHEY AND GERALD L. NEWTON	85
15. Purification of Thiols from Biological Samples	GERALD L. NEWTON AND ROBERT C. FAHEY	96
16. High-Performance Liquid Chromatography of Thiols and Disulfides: Dinitrophenol Derivatives	MARC W. FARISS AND DONALD J. REED	101
17. High-Performance Liquid Chromatography of Hepatic Thiols with Electrochemical Detection	EUGENE G. DEMASTER AND BETH REDFERN	110
18. Analysis for Disulfide Bonds in Peptides and Proteins	THEODORE W. THANNHAUSER, YASUO KONISHI, AND HAROLD A. SCHERAGA	115
19. Assay of Enzyme-Catalyzed Oxygen-Dependent Disulfide Bond Formation	MARK X. SLIWKOWSKI AND HAROLD E. SWAISGOOD	119
20. Cellular Protein-Mixed Disulfides	MARJORIE F. LOU, LAWRENCE L. POULSEN, AND DANIEL M. ZIEGLER	124
21. Measurement of Thiol–Disulfide Interchange Reactions and Thiol pK_a Values	JANETTE HOUK, RAJEEVA SINGH, AND GEORGE M. WHITESIDES	129

C. Other Organic Sulfur and Selenium Compounds

22. Cysteine and Cystine: High-Performance Liquid Chromatography of o-Phthalaldehyde Derivatives	JOHN D. H. COOPER AND D. C. TURNELL	141
23. Cystine: Binding Protein Assay	MARGARET SMITH, CLEMENT E. FURLONG, ALICE A. GREENE, AND JERRY A. SCHNEIDER	144
24. Selenocysteine	NOBUYOSHI ESAKI AND KENJI SODA	148
25. Cysteamine and Cystamine	MICHAEL W. DUFFEL, DANIEL J. LOGAN, AND DANIEL M. ZIEGLER	149
26. Cysteinesulfinic Acid, Hypotaurine, and Taurine: Reversed-Phase High-Performance Liquid Chromatography	MARTHA H. STIPANUK, LAWRENCE L. HIRSCHBERGER, AND JAMES DE LA ROSA	155
27. Cysteinesulfinic Acid: Fuchsin Method	FRANZ-JOSEF LEINWEBER AND KENNETH J. MONTY	160

28. Cysteinesulfinic Acid and Cysteic Acid: High-Performance Liquid Chromatography	K. Kuriyama and Y. Tanaka	164
29. Resolution of Cysteine and Methionine Enantiomers	Owen W. Griffith and Ernest B. Campbell	166
30. Isethionic Acid	Jack H. Fellman	172
31. 3-Mercaptopyruvate, 3-Mercaptolactate, and Mercaptoacetate	Bo Sörbo	178
32. Hypotaurine Aminotransferase Assay: [2-^{3}H]Hypotaurine	Jack H. Fellman	183
33. Penicillamine	Elisabeth M. Wolf-Heuss	186
34. S-Adenosylmethionine and Its Sulfur Metabolites	Richard K. Gordon, George A. Miura, Teresa Alonso, and Peter K. Chiang	191
35. Trimethylselenonium Ion	Howard E. Ganther, R. J. Kraus, and S. J. Foster	195
36. Sulfotransferase Assays	Sengoda G. Ramaswamy and William B. Jakoby	201
37. Arylsulfatases: Colorimetric and Fluorometric Assays	Alexander B. Roy	207

Section II. Preparative Methods

A. Preparation of Specific Metabolites

38. β-Sulfopyruvate	Owen W. Griffith and Catherine L. Weinstein	221
39. Alanine Sulfodisulfane	Jack H. Fellman	223
40. Thetin	H. Kondo and Makoto Ishimoto	227
41. Cysteine S-Conjugates	Patrick Hayden, Valentine H. Schaeffer, Gerald Larsen, and James L. Stevens	228
42. Chromogenic and Fluorigenic Substrates for Sulfurtransferases	Mary R. Burrous, John Lane, Aiko Westley, and John Westley	235
43. Selenocysteine	Hidehiko Tanaka and Kenji Soda	240

44.	Selenodjenkolic Acid	HIDEHIKO TANAKA AND KENJI SODA	243
45.	*Te*-Phenyltellurohomocysteine and *Te*-Phenyltellurocysteine	NOBUYOSHI ESAKI AND KENJI SODA	245

B. General Preparative Techniques

46.	Chemical Reduction of Disulfides	PETER C. JOCELYN	246
47.	Electrochemical Reduction of Disulfides	HAROLD KADIN	257
48.	Formation of Disulfides with Diamide	NECHAMA S. KOSOWER AND EDWARD M. KOSOWER	264
49.	Sulfinic Acids from Disulfides: L-[^{35}S]Cysteinesulfinic Acid	OWEN W. GRIFFITH AND CATHERINE L. WEINSTEIN	270
50.	Amino Acid Sulfones: *S*-Benzyl-DL-α-methylcysteine Sulfone	OWEN W. GRIFFITH	274
51.	Sulfonic Acids: L-Homocysteinesulfonic Acid	MICHAEL A. HAYWARD, ERNEST B. CAMPBELL, AND OWEN W. GRIFFITH	279
52.	Sulfoxides	CARL R. JOHNSON AND MARK P. WESTRICK	281
53.	Amino Acid Sulfoximines: α-Ethylmethionine Sulfoximine	OWEN W. GRIFFITH	286
54.	Preparation of Sulfur and Selenium Amino Acids with Microbial Pyridoxal Phosphate Enzymes	NOBUYOSHI ESAKI AND KENJI SODA	291

C. Nutritional Methods

55.	Construction of Assay Diets for Sulfur-Containing Amino Acids	DAVID H. BAKER	297
56.	Production of Selenium Deficiency in the Rat	RAYMOND F. BURK	307
57.	Intracellular Delivery of Cysteine	MARY E. ANDERSON AND ALTON MEISTER	313

Section III. Enzymes

A. Inorganic Sulfur and Sulfate

58.	Overview: Inorganic Sulfur and Sulfate Activation	JEROME A. SCHIFF AND TEKCHAND SAIDHA	329

59.	Sulfate-Activating Enzymes	Irwin H. Segel, Franco Renosto, and Peter A. Seubert	334
60.	Thiosulfate Reductase	Thomas R. Chauncey, Lawrence C. Uhteg, and John Westley	350
61.	Adenylylsulfate–Ammonia Adenylyltransferase from *Chlorella*	Heinz Fankhauser, Jerome A. Schiff, Leonard J. Garber, and Tekchand Saidha	354
62.	Sulfatases from *Helix pomatia*	Alexander B. Roy	361

B. Sulfur Amino Acids: Mammalian Systems

63.	Mammalian Sulfur Amino Acid Metabolism: An Overview	Owen W. Griffith	366
64.	Adenosylhomocysteinase (Bovine)	Peter K. Chiang	377
65.	Betaine–Homocysteine S-Methyltransferase (Human)	William E. Skiba, Marilyn S. Wells, John H. Mangum, and William M. Awad, Jr.	384
66.	Cystathionine β-Synthase (Human)	Jan P. Kraus	388
67.	Cysteine Dioxygenase	Kenji Yamaguchi and Yu Hosokawa	399
68.	Cysteinesulfinate Decarboxylase	Catherine L. Weinstein and Owen W. Griffith	404
69.	Cysteamine Dioxygenase	Russell B. Richerson and Daniel M. Ziegler	410
70.	Selenocysteine β-Lyase (Porcine)	Nobuyoshi Esaki and Kenji Soda	415

C. Sulfur Amino Acids: Plants

71.	Sulfur Amino Acids of Plants: An Overview	John Giovanelli	419
72.	1-Aminocyclopropane-1-carboxylate Synthase	Douglas O. Adams and Shang Fa Yang	426
73.	Adenosylhomocysteinase from Yellow Lupine	Andrzej Guranowski and Hieronim Jakubowski	430

74. S-Substituted L-Cysteine Sulfoxide Lyase from Shiitake Mushroom	Kyoden Yasumoto and Kimikazu Iwami	434
75. Cystine Lyase from Cabbage	Ivan K. Smith and David I. Hall	439
76. Cystathionine β-Lyase from Spinach	John Giovanelli	443
77. D-Cysteine Desulfhydrase from Spinach	Ahlert Schmidt	449

D. Sulfur Amino Acids: Microbial Systems

78. Microbial Sulfur Amino Acids: An Overview	Kenji Soda	453
79. L-Methionine γ-Lyase from *Pseudomonas putida* and *Aeromonas*	Nobuyoshi Esaki and Kenji Soda	459
80. O-Acetylhomoserine Sulfhydrylase from *Schizosaccharomyces pombe*	Shuzo Yamagata	465
81. O-Acetylhomoserine Sulfhydrylase from *Brevibacterium flavum*	Isamu Shiio and Hachiro Ozaki	470
82. O-Acetylserine Sulfhydrylase from *Bacillus sphaericus*	Toru Nagasawa and Hideaki Yamada	474
83. O-Acetyl-L-Serine–O-acetyl-L-Homoserine Sulfhydrylase from *Saccharomyces cerevisiae*	Shuzo Yamagata	478
84. Cystathionine β-Lyase from *Escherichia coli*	Jack R. Uren	483
85. Cystathionine γ-Lyase from *Streptomyces phaeochromogenes*	Toru Nagasawa, Hiroshi Kanzaki, and Hideaki Yamada	486
86. Selenocysteine β-Lyase from *Citrobacter freundii*	Nobuyoshi Esaki and Kenji Soda	493
87. Taurine Dehydrogenase	Hiroyuki Kondo and Makoto Ishimoto	496
88. ω-Amino Acid–Pyruvate Aminotransferase	Kazuo Yonaha, Seizen Toyama, and Kenji Soda	500

E. Enzymes Active with Disulfides

89. Sulfhydryl Oxidase from Milk	Harold E. Swaisgood and H. Robert Horton	504
90. Sulfhydryl Oxidase from Rat Skin	Lowell A. Goldsmith	510

91. Asparagusate Reductase		Hiroshi Yanagawa	516

Appendix: Index of Related Articles in Other Volumes of *Methods in Enzymology* 523

Author Index 529

Subject Index 553

Contributors to Volume 143

Article numbers are in parentheses following the names of contributors.
Affiliations listed are current.

DOUGLAS O. ADAMS (72), *Department of Viticulture and Enology, University of California, Davis, California 95616*

HOWARD ADLER (10), *Department of Biochemistry and Molecular Biology, The University of Chicago, Chicago, Illinois 60637*

TERESA ALONSO (34), *Department of Biochemistry, Walter Reed Army Institute of Research, Washington, D.C. 20307*

MARY E. ANDERSON (57), *Department of Biochemistry, Cornell University Medical College, New York, New York 10021*

WILLIAM M. AWAD, JR. (65), *Departments of Medicine and Biochemistry, University of Miami School of Medicine, Miami, Florida 33101*

DAVID H. BAKER (55), *Department of Animal Sciences and Division of Nutritional Sciences, University of Illinois, Urbana, Illinois 61801*

HANS-OTTO BEUTLER (3), *Boehringer-Mannheim GmbH, Biochemical Research Center, D-8132 Tutzing, Federal Republic of Germany*

RAYMOND F. BURK (56), *Department of Medicine, University of Texas Health Science Center, San Antonio, Texas 78284*

MARY R. BURROUS (42), *Department of Biochemistry and Molecular Biology, The University of Chicago, Chicago, Illinois 60637*

ERNEST B. CAMPBELL (29, 51), *Department of Biochemistry, Cornell University Medical College, New York, New York 10021*

THOMAS R. CHAUNCEY (60), *Department of Biochemistry and Molecular Biology, The University of Chicago, Chicago, Illinois 60637*

PETER K. CHIANG (34, 64), *Division of Biochemistry, Walter Reed Army Institute of Research, Washington, D.C. 20307*

JOHN D. H. COOPER (22), *Department of Biochemistry, Coventry and Warwickshire Hospital, Coventry CV1 4FH, England*

EUGENE G. DEMASTER (17), *Medical Research Laboratories, Veterans Administration Medical Center, Minneapolis, Minnesota 55417*

JAMES DE LA ROSA (26), *Division of Nutritional Sciences, Cornell University, Ithaca, New York 14853*

MICHAEL W. DUFFEL (25), *Department of Medicinal Chemistry and Natural Products, College of Pharmacy, University of Iowa, Iowa City, Iowa 52242*

NOBUYOSHI ESAKI (24, 45, 54, 70, 79, 86), *Institute for Chemical Research, Kyoto University, Uji, Kyoto-Fu 611, Japan*

ROBERT C. FAHEY (14, 15), *Department of Chemistry, University of California-San Diego, La Jolla, California 92093*

HEINZ FANKHAUSER (61), *CIBA-Geigy Ltd., CH-4002 Basel, Switzerland*

MARC W. FARISS (16), *Department of Pathology, Medical College of Virginia, Richmond, Virginia 23298*

JACK H. FELLMAN (30, 32, 39), *Department of Biochemistry, The Oregon Health Sciences University, Portland, Oregon 97201*

S. J. FOSTER (35), *Department of Nutritional Sciences, University of Wisconsin-Madison, Madison, Wisconsin 53706*

CLEMENT E. FURLONG (23), *Division of Medical Genetics, Center for Inherited Diseases, University of Washington, Seattle, Washington 98155*

HOWARD E. GANTHER (9, 35), *Department of Nutritional Sciences, University of Wisconsin-Madison, Madison, Wisconsin 53706*

LEONARD J. GARBER (61), *Institute for Photobiology, Brandeis University, Waltham, Massachusetts 02254*

JOHN GIOVANELLI (71, 76), *Laboratory of General and Comparative Biochemistry, Section on Alkaloid Biosynthesis, National Institute of Mental Health, Bethesda, Maryland 20212*

LOWELL A. GOLDSMITH (90), *Dermatology Unit, University of Rochester Medical Center, Rochester, New York 14642*

RICHARD K. GORDON (34), *Division of Biochemistry, Walter Reed Army Institute of Research, Washington, D.C. 20307*

ALICE A. GREENE (23), *Department of Pediatrics, University of California-San Diego, La Jolla, California 92093*

OWEN W. GRIFFITH (29, 38, 49, 50, 51, 53, 63, 68), *Department of Biochemistry, Cornell University Medical College, New York, New York 10021*

ANDRZEJ GURANOWSKI (73), *Katedra Biochemii, Akademia Rolnicza, ul. Wołyńska, PL-60-637 Poznań, Poland*

DAVID I. HALL (75), *Department of Botany, Ohio University, Athens, Ohio 45701*

PATRICK HAYDEN (41), *W. Alton Jones Cell Science Center, Lake Placid, New York 12946*

MICHAEL A. HAYWARD (51), *Department of Biochemistry, Cornell University Medical College, New York, New York 10021*

LAWRENCE L. HIRSCHBERGER (26), *Division of Nutritional Sciences, Cornell University, Ithaca, New York 14853*

H. ROBERT HORTON (89), *Department of Biochemistry, North Carolina State University, Raleigh, North Carolina 27695*

YU HOSOKAWA (67), *Division of Maternal and Child Nutrition, National Institute of Nutrition, Toyama, Shinjuku-ku, Tokyo 162, Japan*

JANETTE HOUK (21), *Environmental Protection Agency, Washington, D.C. 20460*

KAZUHIRO IMAI (12), *Branch Hospital Pharmacy, University of Tokyo, Bunkyo-ku, Tokyo 112, Japan*

MAKOTO ISHIMOTO (40, 87), *Department of Chemical Microbiology, Faculty of Pharmaceutical Sciences, Hokkaido University, Sapporo, Hokkaido 060, Japan*

KIMIKAZU IWAMI (74), *Department of Agricultural Chemistry, Kyoto Prefectural University, Shimogamo Nakaragicho, Sakyoku, Kyoto 606, Japan*

WILLIAM B. JAKOBY (36), *Laboratory of Biochemistry and Metabolism, National Institute of Diabetes and Digestive and Kidney Diseases, National Institutes of Health, Bethesda, Maryland 20892*

HIERONIM JAKUBOWSKI (73), *Katedra Biochemii, Akademia Rolnicza, ul. Wołyńska 35, PL-60-637 Poznań, Poland*

PETER C. JOCELYN (11, 46), *Department of Biochemistry, University of Edinburgh Medical School, Edinburgh EH8 9XD, Scotland*

CARL R. JOHNSON (52), *Department of Chemistry, Wayne State University, Detroit, Michigan 48202*

HAROLD KADIN (47), *Analytical Research and Development, The Squibb Institute for Medical Research, New Brunswick, New Jersey 08903*

HIROSHI KANZAKI (85), *Department of Agricultural Chemistry, Kyoto University, Kyoto 606, Japan*

H. KONDO (40, 87), *Department of Chemical Microbiology, Faculty of Pharmaceutical Sciences, Hokkaido University, Sapporo, Hokkaido 060, Japan*

YASUO KONISHI (18), *Monsanto Company AA4I, Chesterfield, Missouri 63198*

EDWARD M. KOSOWER (13, 48), *School of Chemistry, Sackler Faculty of Exact Sciences, Tel-Aviv University, Ramat-Aviv, Tel Aviv 69978, Israel*

NECHAMA S. KOSOWER (13, 48), *Department of Human Genetics, Sackler School of Medicine, Tel-Aviv University, Ramat-Aviv, Tel Aviv 69978, Israel*

JAN P. KRAUS (66), *Department of Human*

Genetics, Yale University School of Medicine, New Haven, Connecticut 06510

R. J. KRAUS (9, 35), *Department of Nutritional Sciences, University of Wisconsin-Madison, Madison, Wisconsin 53706*

K. KURIYAMA (28), *Department of Pharmacology, Kyoto Prefectural University of Medicine, Kawaramachi-Hirokaji, Kamimyo-Ku, Kyoto 602, Japan*

JOHN LANE (42), *Department of Biochemistry and Molecular Biology, The University of Chicago, Chicago, Illinois 60637*

GERALD LARSEN (41), *Metabolism and Radiation Research Laboratory, U.S. Department of Agriculture, North Dakota State University Station, Fargo, North Dakota 58105*

FRANZ-JOSEF LEINWEBER (4, 27), *Department of Drug Metabolism, Hoffman-La Roche, Inc., Nutley, New Jersey 07110*

DANIEL J. LOGAN (25), *Clayton Foundation Biochemical Institute, The University of Texas at Austin, Austin, Texas 78712*

MARJORIE F. LOU (20), *Alcon Laboratories, Inc., Fort Worth, Texas 76101*

JOHN H. MANGUM (65), *Department of Chemistry, Brigham Young University, Provo, Utah 84602*

ALTON MEISTER (57), *Department of Biochemistry, Cornell University Medical College, New York, New York 10021*

GEORGE A. MIURA (34), *Division of Biochemistry, Walter Reed Army Institute of Research, Washington, D.C. 20307*

KENNETH J. MONTY (4, 27), *Department of Biochemistry, University of Tennessee, Knoxville, Tennessee 37996*

TORU NAGASAWA (82, 85), *Department of Agricultural Chemistry, Kyoto University, Kyoto 606, Japan*

GERALD L. NEWTON (14, 15), *Department of Chemistry, University of California-San Diego, La Jolla, California 92093*

HACHIRO OZAKI (81), *Central Research Laboratories, Ajinomoto Co., Inc., Kawasaki 210, Japan*

LAWRENCE L. POULSEN (20), *Clayton Foundation Biochemical Institute, The*

University of Texas at Austin, Austin, Texas 78712

SENGODA G. RAMASWAMY (36), *Laboratory of Biochemistry and Metabolism, National Institute of Diabetes and Digestive and Kidney Diseases, National Institutes of Health, Bethesda, Maryland 20892*

BETH REDFERN (17), *Medical Research Laboratories, Veterans Administration Medical Center, Minneapolis, Minnesota 55417*

DONALD J. REED (16), *Department of Biochemistry, Oregon State University, Corvallis, Oregon 97331*

FRANCO RENOSTO (59), *Department of Biochemistry and Biophysics, University of California, Davis, California 95616*

RUSSELL B. RICHERSON (69), *Abbott Laboratories, North Chicago, Illinois 60064*

ALEXANDER B. ROY (37, 62), *Protein Chemistry Group, John Curtin School of Medical Research, Australian National University, Canberra ACT 2601, Australia*

TEKCHAND SAIDHA (58, 61), *Institute for Photobiology, Brandeis University, Waltham, Massachusetts 02254*

VALENTINE H. SCHAEFFER (41), *Department of Biochemistry and Biophysics, Center for Drugs and Biologics, Food and Drug Administration, Bethesda, Maryland 20892*

HAROLD A. SCHERAGA (18), *Baker Laboratory of Chemistry, Cornell University, Ithaca, New York 14853*

JEROME A. SCHIFF (58, 61), *Institute for Photobiology, Brandeis University, Waltham, Massachusetts 02254*

AHLERT SCHMIDT (77), *Botanisches Institut, Universität München, D-8000 Munich 19, Federal Republic of Germany*

JERRY A. SCHNEIDER (23), *Department of Pediatrics, University of California-San Diego, La Jolla, California 92093*

IRWIN H. SEGEL (59), *Department of Biochemistry and Biophysics, University of California, Davis, California 95616*

PETER A. SEUBERT (59), *Center for the Neurobiology of Learning and Memory, Uni-*

versity of California, Irvine, California 92717

ISAMU SHIIO (81), *Central Research Laboratories, Ajinomoto Co., Inc., Yokohama 244, Japan*

RAJEEVA SINGH (21), *Department of Chemistry, Harvard University, Cambridge, Massachusetts 02138*

WILLIAM E. SKIBA (65), *Department of Biochemistry, University of Miami School of Medicine, Miami, Florida 33101*

MARK X. SLIWKOWSKI (19), *Triton Biosciences, Inc., Alameda, California 94501*

IVAN K. SMITH (75), *Department of Botany, Ohio University, Athens, Ohio 45701*

MARGARET SMITH (23), *Department of Pediatrics, University of California-San Diego, La Jolla, California 92093*

KENJI SODA (24, 43, 44, 45, 54, 70, 78, 79, 86, 88), *Institute for Chemical Research, Kyoto University, Uji, Kyoto-Fu 611, Japan*

BO SÖRBO (1, 31), *Department of Clinical Chemistry, Linköping University, Regional Hospital, Linköping, Sweden*

JAMES L. STEVENS (41), *W. Alton Jones Cell Science Center, Lake Placid, New York 12946*

MARTHA H. STIPANUK (26), *Division of Nutritional Sciences, Cornell University, Ithaca, New York 14853*

DARRYL M. SULLIVAN (2), *Hazleton Laboratories America, Inc., Madison, Wisconsin 53705*

HAROLD E. SWAISGOOD (19, 89), *Department of Food Science and Biochemistry, North Carolina State University, Raleigh, North Carolina 27695*

HIDEHIKO TANAKA (43, 44), *Institute for Chemical Research, Kyoto University, Uji, Kyoto-Fu 611, Japan*

Y. TANAKA (28), *Department of Pharmacology, Kyoto Prefectural University of Medicine, Kawaramachi-Hirokoji, Kyoto 602, Japan*

THEODORE W. THANNHAUSER (18), *Baker Laboratory of Chemistry, Cornell University, Ithaca, New York 14853*

SEIZEN TOYAMA (88), *Department of Agricultural Chemistry, University of the Ryukyus, Nishihara, Okinawa 903-01, Japan*

TOSHIMASA TOYO'OKA (12), *Department of Foods, National Institute of Hygienic Sciences, Setagaya-ku, Tokyo 158, Japan*

D. C. TURNELL (22), *Department of Biochemistry, Coventry and Warwickshire Hospital, Coventry CV1 4FH, England*

TOSHIHIKO UBUKA (5), *Department of Biochemistry, Okayama University Medical School, Okayama 700, Japan*

LAWRENCE C. UHTEG (60), *Department of Biochemistry and Molecular Biology, The University of Chicago, Chicago, Illinois 60637*

JACK R. UREN (84), *Battelle Columbus Laboratories, Columbus, Ohio 43201*

RANDOLPH T. WEDDING (8), *Department of Biochemistry, University of California, Riverside, California 92521*

CATHERINE L. WEINSTEIN (38, 49, 68), *Department of Biochemistry, Cornell University Medical College, New York, New York 10021*

MARILYN S. WELLS (65), *Department of Medicine, Veterans Administration Medical Center, Miami, Florida 33101, and University of Miami School of Medicine, Miami, Florida 33101*

AIKO WESTLEY (42), *Department of Biochemistry and Molecular Biology, The University of Chicago, Chicago, Illinois 60637*

JOHN WESTLEY (6, 10, 42, 60), *Department of Biochemistry and Molecular Biology, The University of Chicago, Chicago, Illinois 60637*

MARK P. WESTRICK (52), *Department of Chemistry, Wayne State University, Detroit, Michigan 48202*

GEORGE M. WHITESIDES (21), *Department of Chemistry, Harvard University, Cambridge, Massachusetts 02138*

ELISABETH M. WOLF-HEUSS (33), *Analyti-*

cal Department, Degussa Pharma Group, D-6000 Frankfurt 1, Federal Republic of Germany

JOHN L. WOOD (7), Department of Biochemistry, University of Tennessee-Memphis, Memphis, Tennessee 38111

HIDEAKI YAMADA (82, 85), Department of Agricultural Chemistry, Kyoto University, Kyoto 606, Japan

SHUZO YAMAGATA (80, 83), Department of Biology, Faculty of General Education, Gifu University, Gifu 501-11, Japan

KENJI YAMAGUCHI (67), Division of Maternal and Child Nutrition, National Institute of Nutrition, Toyama, Shinjuku-ku, Tokyo 162, Japan

HIROSHI YANAGAWA (91), Mitsubishi-Kasei Institute of Life Sciences, Machida, Tokyo 194, Japan

SHANG FA YANG (72), Department of Vegetable Crops, University of California, Davis, California 95616

KYODEN YASUMOTO (74), Research Institute for Food Science, Kyoto University, Gokasho, Uji, Kyoto 611, Japan

KAZUO YONAHA (88), Department of Agricultural Chemistry, University of the Ryukyus, Nishihara, Okinawa 903-01, Japan

DANIEL M. ZIEGLER (20, 25, 69), Clayton Foundation Biochemical Institute, The University of Texas at Austin, Austin, Texas 78712

Preface

Because of the large number of diverse and critical roles played by sulfur in several of its oxidation states, the subject of testing and preparing derivatives of sulfur or of measuring, purifying, and characterizing enzymes associated with sulfur metabolism is represented in most of the individual volumes of *Methods in Enzymology*. For example, many of the methyl transferases which utilize S-adenosylmethionine have found a home in volumes that deal with nucleic acids, lipids, amines, or detoxication as the major topic; the polyamines are treated in a separate volume. Enzymes and procedures involving sulfur-containing metabolites such as coenzyme A are presented in most of the volumes dealing with metabolic enzymes, and aspects of biotin biochemistry have appeared in books in this series which range in emphasis from vitamins to affinity labeling. Since no single book could cover all of these topics in a reasonable manner, such dispersal is entirely appropriate. In fact, most of the examples cited above are not presented in this volume. It is because of the broad distribution of topics that we include, as an appendix, a brief index to other articles on some of the subject areas covered here.

Progress and interest in the work on sulfur and sulfur amino acids, on seleno and tellurium analogs, and on the separation and quantitation of sulfur-containing compounds have led to the suggestion that a single source book of methods would be useful. We have responded by providing descriptions of a large number of analytical methods for inorganic and organic sulfur compounds in their several oxidation states. This volume also includes general and specific methods of preparation for a number of organic sulfur compounds and presents procedures for the purification and assay of a group of enzymes that participate in inorganic sulfur metabolism as well as in the pathways of formation and utilization of cysteine, methionine, and related compounds.

We express our appreciation to the individual authors who have contributed to the volume and, in the process, have made their often unique expertise available to other investigators interested in the biochemistry of sulfur.

WILLIAM B. JAKOBY
OWEN W. GRIFFITH

METHODS IN ENZYMOLOGY

EDITED BY

Sidney P. Colowick and Nathan O. Kaplan

VANDERBILT UNIVERSITY
SCHOOL OF MEDICINE
NASHVILLE, TENNESSEE

DEPARTMENT OF CHEMISTRY
UNIVERSITY OF CALIFORNIA
AT SAN DIEGO
LA JOLLA, CALIFORNIA

I. Preparation and Assay of Enzymes
II. Preparation and Assay of Enzymes
III. Preparation and Assay of Substrates
IV. Special Techniques for the Enzymologist
V. Preparation and Assay of Enzymes
VI. Preparation and Assay of Enzymes (*Continued*)
    Preparation and Assay of Substrates
    Special Techniques
VII. Cumulative Subject Index

METHODS IN ENZYMOLOGY

EDITORS-IN-CHIEF

Sidney P. Colowick and Nathan O. Kaplan

VOLUME VIII. Complex Carbohydrates
Edited by ELIZABETH F. NEUFELD AND VICTOR GINSBURG

VOLUME IX. Carbohydrate Metabolism
Edited by WILLIS A. WOOD

VOLUME X. Oxidation and Phosphorylation
Edited by RONALD W. ESTABROOK AND MAYNARD E. PULLMAN

VOLUME XI. Enzyme Structure
Edited by C. H. W. HIRS

VOLUME XII. Nucleic Acids (Parts A and B)
Edited by LAWRENCE GROSSMAN AND KIVIE MOLDAVE

VOLUME XIII. Citric Acid Cycle
Edited by J. M. LOWENSTEIN

VOLUME XIV. Lipids
Edited by J. M. LOWENSTEIN

VOLUME XV. Steroids and Terpenoids
Edited by RAYMOND B. CLAYTON

VOLUME XVI. Fast Reactions
Edited by KENNETH KUSTIN

VOLUME XVII. Metabolism of Amino Acids and Amines (Parts A and B)
Edited by HERBERT TABOR AND CELIA WHITE TABOR

VOLUME XVIII. Vitamins and Coenzymes (Parts A, B, and C)
Edited by DONALD B. MCCORMICK AND LEMUEL D. WRIGHT

VOLUME XIX. Proteolytic Enzymes
Edited by GERTRUDE E. PERLMANN AND LASZLO LORAND

VOLUME XX. Nucleic Acids and Protein Synthesis (Part C)
Edited by KIVIE MOLDAVE AND LAWRENCE GROSSMAN

VOLUME XXI. Nucleic Acids (Part D)
Edited by LAWRENCE GROSSMAN AND KIVIE MOLDAVE

VOLUME XXII. Enzyme Purification and Related Techniques
Edited by WILLIAM B. JAKOBY

VOLUME XXIII. Photosynthesis (Part A)
Edited by ANTHONY SAN PIETRO

VOLUME XXIV. Photosynthesis and Nitrogen Fixation (Part B)
Edited by ANTHONY SAN PIETRO

VOLUME XXV. Enzyme Structure (Part B)
Edited by C. H. W. HIRS AND SERGE N. TIMASHEFF

VOLUME XXVI. Enzyme Structure (Part C)
Edited by C. H. W. HIRS AND SERGE N. TIMASHEFF

VOLUME XXVII. Enzyme Structure (Part D)
Edited by C. H. W. HIRS AND SERGE N. TIMASHEFF

VOLUME XXVIII. Complex Carbohydrates (Part B)
Edited by VICTOR GINSBURG

VOLUME XXIX. Nucleic Acids and Protein Synthesis (Part E)
Edited by LAWRENCE GROSSMAN AND KIVIE MOLDAVE

VOLUME XXX. Nucleic Acids and Protein Synthesis (Part F)
Edited by KIVIE MOLDAVE AND LAWRENCE GROSSMAN

VOLUME XXXI. Biomembranes (Part A)
Edited by SIDNEY FLEISCHER AND LESTER PACKER

VOLUME XXXII. Biomembranes (Part B)
Edited by SIDNEY FLEISCHER AND LESTER PACKER

VOLUME XXXIII. Cumulative Subject Index Volumes I–XXX
Edited by MARTHA G. DENNIS AND EDWARD A. DENNIS

VOLUME XXXIV. Affinity Techniques (Enzyme Purification: Part B)
Edited by WILLIAM B. JAKOBY AND MEIR WILCHEK

VOLUME XXXV. Lipids (Part B)
Edited by JOHN M. LOWENSTEIN

VOLUME XXXVI. Hormone Action (Part A: Steroid Hormones)
Edited by BERT W. O'MALLEY AND JOEL G. HARDMAN

VOLUME XXXVII. Hormone Action (Part B: Peptide Hormones)
Edited by BERT W. O'MALLEY AND JOEL G. HARDMAN

VOLUME XXXVIII. Hormone Action (Part C: Cyclic Nucleotides)
Edited by JOEL G. HARDMAN AND BERT W. O'MALLEY

VOLUME XXXIX. Hormone Action (Part D: Isolated Cells, Tissues, and Organ Systems)
Edited by JOEL G. HARDMAN AND BERT W. O'MALLEY

VOLUME XL. Hormone Action (Part E: Nuclear Structure and Function)
Edited by BERT W. O'MALLEY AND JOEL G. HARDMAN

VOLUME XLI. Carbohydrate Metabolism (Part B)
Edited by W. A. WOOD

VOLUME XLII. Carbohydrate Metabolism (Part C)
Edited by W. A. WOOD

VOLUME XLIII. Antibiotics
Edited by JOHN H. HASH

VOLUME XLIV. Immobilized Enzymes
Edited by KLAUS MOSBACH

VOLUME XLV. Proteolytic Enzymes (Part B)
Edited by LASZLO LORAND

VOLUME XLVI. Affinity Labeling
Edited by WILLIAM B. JAKOBY AND MEIR WILCHEK

VOLUME XLVII. Enzyme Structure (Part E)
Edited by C. H. W. HIRS AND SERGE N. TIMASHEFF

VOLUME XLVIII. Enzyme Structure (Part F)
Edited by C. H. W. HIRS AND SERGE N. TIMASHEFF

VOLUME XLIX. Enzyme Structure (Part G)
Edited by C. H. W. HIRS AND SERGE N. TIMASHEFF

VOLUME L. Complex Carbohydrates (Part C)
Edited by VICTOR GINSBURG

VOLUME LI. Purine and Pyrimidine Nucleotide Metabolism
Edited by PATRICIA A. HOFFEE AND MARY ELLEN JONES

VOLUME LII. Biomembranes (Part C: Biological Oxidations)
Edited by SIDNEY FLEISCHER AND LESTER PACKER

VOLUME LIII. Biomembranes (Part D: Biological Oxidations)
Edited by SIDNEY FLEISCHER AND LESTER PACKER

VOLUME LIV. Biomembranes (Part E: Biological Oxidations)
Edited by SIDNEY FLEISCHER AND LESTER PACKER

VOLUME LV. Biomembranes (Part F: Bioenergetics)
Edited by SIDNEY FLEISCHER AND LESTER PACKER

VOLUME LVI. Biomembranes (Part G: Bioenergetics)
Edited by SIDNEY FLEISCHER AND LESTER PACKER

VOLUME LVII. Bioluminescence and Chemiluminescence
Edited by MARLENE A. DELUCA

VOLUME LVIII. Cell Culture
Edited by WILLIAM B. JAKOBY AND IRA PASTAN

VOLUME LIX. Nucleic Acids and Protein Synthesis (Part G)
Edited by KIVIE MOLDAVE AND LAWRENCE GROSSMAN

VOLUME LX. Nucleic Acids and Protein Synthesis (Part H)
Edited by KIVIE MOLDAVE AND LAWRENCE GROSSMAN

VOLUME 61. Enzyme Structure (Part H)
Edited by C. H. W. HIRS AND SERGE N. TIMASHEFF

VOLUME 62. Vitamins and Coenzymes (Part D)
Edited by DONALD B. MCCORMICK AND LEMUEL D. WRIGHT

VOLUME 63. Enzyme Kinetics and Mechanism (Part A: Initial Rate and Inhibitor Methods)
Edited by DANIEL L. PURICH

VOLUME 64. Enzyme Kinetics and Mechanism (Part B: Isotopic Probes and Complex Enzyme Systems)
Edited by DANIEL L. PURICH

VOLUME 65. Nucleic Acids (Part I)
Edited by LAWRENCE GROSSMAN AND KIVIE MOLDAVE

VOLUME 66. Vitamins and Coenzymes (Part E)
Edited by DONALD B. MCCORMICK AND LEMUEL D. WRIGHT

VOLUME 67. Vitamins and Coenzymes (Part F)
Edited by DONALD B. MCCORMICK AND LEMUEL D. WRIGHT

VOLUME 68. Recombinant DNA
Edited by RAY WU

VOLUME 69. Photosynthesis and Nitrogen Fixation (Part C)
Edited by ANTHONY SAN PIETRO

VOLUME 70. Immunochemical Techniques (Part A)
Edited by HELEN VAN VUNAKIS AND JOHN J. LANGONE

VOLUME 71. Lipids (Part C)
Edited by JOHN M. LOWENSTEIN

VOLUME 72. Lipids (Part D)
Edited by JOHN M. LOWENSTEIN

VOLUME 73. Immunochemical Techniques (Part B)
Edited by JOHN J. LANGONE AND HELEN VAN VUNAKIS

VOLUME 74. Immunochemical Techniques (Part C)
Edited by JOHN J. LANGONE AND HELEN VAN VUNAKIS

VOLUME 75. Cumulative Subject Index Volumes XXXI, XXXII, XXXIV–LX
Edited by EDWARD A. DENNIS AND MARTHA G. DENNIS

VOLUME 76. Hemoglobins
Edited by ERALDO ANTONINI, LUIGI ROSSI-BERNARDI, AND EMILIA CHIANCONE

VOLUME 77. Detoxication and Drug Metabolism
Edited by WILLIAM B. JAKOBY

VOLUME 78. Interferons (Part A)
Edited by SIDNEY PESTKA

VOLUME 79. Interferons (Part B)
Edited by SIDNEY PESTKA

VOLUME 80. Proteolytic Enzymes (Part C)
Edited by LASZLO LORAND

VOLUME 81. Biomembranes (Part H: Visual Pigments and Purple Membranes, I)
Edited by LESTER PACKER

VOLUME 82. Structural and Contractile Proteins (Part A: Extracellular Matrix)
Edited by LEON W. CUNNINGHAM AND DIXIE W. FREDERIKSEN

VOLUME 83. Complex Carbohydrates (Part D)
Edited by VICTOR GINSBURG

VOLUME 84. Immunochemical Techniques (Part D: Selected Immunoassays)
Edited by JOHN J. LANGONE AND HELEN VAN VUNAKIS

VOLUME 85. Structural and Contractile Proteins (Part B: The Contractile Apparatus and the Cytoskeleton)
Edited by DIXIE W. FREDERIKSEN AND LEON W. CUNNINGHAM

VOLUME 86. Prostaglandins and Arachidonate Metabolites
Edited by WILLIAM E. M. LANDS AND WILLIAM L. SMITH

VOLUME 87. Enzyme Kinetics and Mechanism (Part C: Intermediates, Stereochemistry, and Rate Studies)
Edited by DANIEL L. PURICH

VOLUME 88. Biomembranes (Part I: Visual Pigments and Purple Membranes, II)
Edited by LESTER PACKER

VOLUME 89. Carbohydrate Metabolism (Part D)
Edited by WILLIS A. WOOD

VOLUME 90. Carbohydrate Metabolism (Part E)
Edited by WILLIS A. WOOD

VOLUME 91. Enzyme Structure (Part I)
Edited by C. H. W. HIRS AND SERGE N. TIMASHEFF

VOLUME 92. Immunochemical Techniques (Part E: Monoclonal Antibodies and General Immunoassay Methods)
Edited by JOHN J. LANGONE AND HELEN VAN VUNAKIS

VOLUME 93. Immunochemical Techniques (Part F: Conventional Antibodies, Fc Receptors, and Cytotoxicity)
Edited by JOHN J. LANGONE AND HELEN VAN VUNAKIS

VOLUME 94. Polyamines
Edited by HERBERT TABOR AND CELIA WHITE TABOR

VOLUME 95. Cumulative Subject Index Volumes 61–74, 76–80
Edited by EDWARD A. DENNIS AND MARTHA G. DENNIS

VOLUME 96. Biomembranes [Part J: Membrane Biogenesis: Assembly and Targeting (General Methods; Eukaryotes)]
Edited by SIDNEY FLEISCHER AND BECCA FLEISCHER

VOLUME 97. Biomembranes [Part K: Membrane Biogenesis: Assembly and Targeting (Prokaryotes, Mitochondria, and Chloroplasts)]
Edited by SIDNEY FLEISCHER AND BECCA FLEISCHER

VOLUME 98. Biomembranes [Part L: Membrane Biogenesis (Processing and Recycling)]
Edited by SIDNEY FLEISCHER AND BECCA FLEISCHER

VOLUME 99. Hormone Action (Part F: Protein Kinases)
Edited by JACKIE D. CORBIN AND JOEL G. HARDMAN

VOLUME 100. Recombinant DNA (Part B)
Edited by RAY WU, LAWRENCE GROSSMAN, AND KIVIE MOLDAVE

VOLUME 101. Recombinant DNA (Part C)
Edited by RAY WU, LAWRENCE GROSSMAN, AND KIVIE MOLDAVE

VOLUME 102. Hormone Action (Part G: Calmodulin and Calcium-Binding Proteins)
Edited by ANTHONY R. MEANS AND BERT W. O'MALLEY

VOLUME 103. Hormone Action (Part H: Neuroendocrine Peptides)
Edited by P. MICHAEL CONN

VOLUME 104. Enzyme Purification and Related Techniques (Part C)
Edited by WILLIAM B. JAKOBY

VOLUME 105. Oxygen Radicals in Biological Systems
Edited by LESTER PACKER

VOLUME 106. Posttranslational Modifications (Part A)
Edited by FINN WOLD AND KIVIE MOLDAVE

VOLUME 107. Posttranslational Modifications (Part B)
Edited by FINN WOLD AND KIVIE MOLDAVE

VOLUME 108. Immunochemical Techniques (Part G: Separation and Characterization of Lymphoid Cells)
Edited by GIOVANNI DI SABATO, JOHN J. LANGONE, AND HELEN VAN VUNAKIS

VOLUME 109. Hormone Action (Part I: Peptide Hormones)
Edited by LUTZ BIRNBAUMER AND BERT W. O'MALLEY

VOLUME 110. Steroids and Isoprenoids (Part A)
Edited by JOHN H. LAW AND HANS C. RILLING

VOLUME 111. Steroids and Isoprenoids (Part B)
Edited by JOHN H. LAW AND HANS C. RILLING

VOLUME 112. Drug and Enzyme Targeting (Part A)
Edited by KENNETH J. WIDDER AND RALPH GREEN

VOLUME 113. Glutamate, Glutamine, Glutathione, and Related Compounds
Edited by ALTON MEISTER

VOLUME 114. Diffraction Methods for Biological Macromolecules (Part A)
Edited by HAROLD W. WYCKOFF, C. H. W. HIRS, AND SERGE N. TIMASHEFF

VOLUME 115. Diffraction Methods for Biological Macromolecules (Part B)
Edited by HAROLD W. WYCKOFF, C. H. W. HIRS, AND SERGE N. TIMASHEFF

VOLUME 116. Immunochemical Techniques (Part H: Effectors and Mediators of Lymphoid Cell Functions)
Edited by GIOVANNI DI SABATO, JOHN J. LANGONE, AND HELEN VAN VUNAKIS

VOLUME 117. Enzyme Structure (Part J)
Edited by C. H. W. HIRS AND SERGE N. TIMASHEFF

VOLUME 118. Plant Molecular Biology
Edited by ARTHUR WEISSBACH AND HERBERT WEISSBACH

VOLUME 119. Interferons (Part C)
Edited by SIDNEY PESTKA

VOLUME 120. Cumulative Subject Index Volumes 81–94, 96–101

VOLUME 121. Immunochemical Techniques (Part I: Hybridoma Technology and Monoclonal Antibodies)
Edited by JOHN J. LANGONE AND HELEN VAN VUNAKIS

VOLUME 122. Vitamins and Coenzymes (Part G)
Edited by FRANK CHYTIL AND DONALD B. MCCORMICK

VOLUME 123. Vitamins and Coenzymes (Part H)
Edited by FRANK CHYTIL AND DONALD B. MCCORMICK

VOLUME 124. Hormone Action (Part J: Neuroendocrine Peptides)
Edited by P. MICHAEL CONN

VOLUME 125. Biomembranes (Part M: Transport in Bacteria, Mitochondria, and Chloroplasts: General Approaches and Transport Systems)
Edited by SIDNEY FLEISCHER AND BECCA FLEISCHER

VOLUME 126. Biomembranes (Part N: Transport in Bacteria, Mitochondria, and Chloroplasts: Protonmotive Force)
Edited by SIDNEY FLEISCHER AND BECCA FLEISCHER

VOLUME 127. Biomembranes (Part O: Protons and Water: Structure and Translocation)
Edited by LESTER PACKER

VOLUME 128. Plasma Lipoproteins (Part A: Preparation, Structure, and Molecular Biology)
Edited by JERE P. SEGREST AND JOHN J. ALBERS

VOLUME 129. Plasma Lipoproteins (Part B: Characterization, Cell Biology, and Metabolism)
Edited by JOHN J. ALBERS AND JERE P. SEGREST

VOLUME 130. Enzyme Structure (Part K)
Edited by C. H. W. HIRS AND SERGE N. TIMASHEFF

VOLUME 131. Enzyme Structure (Part L)
Edited by C. H. W. HIRS AND SERGE N. TIMASHEFF

VOLUME 132. Immunochemical Techniques (Part J: Phagocytosis and Cell-Mediated Cytotoxicity)
Edited by GIOVANNI DI SABATO AND JOHANNES EVERSE

VOLUME 133. Bioluminescence and Chemiluminescence (Part B)
Edited by MARLENE DELUCA AND WILLIAM D. MCELROY

VOLUME 134. Structural and Contractile Proteins (Part C: The Contractile Apparatus and the Cytoskeleton)
Edited by RICHARD B. VALLEE

VOLUME 135. Immobilized Enzymes and Cells (Part B)
Edited by KLAUS MOSBACH

VOLUME 136. Immobilized Enzymes and Cells (Part C) (in preparation)
Edited by KLAUS MOSBACH

VOLUME 137. Immobilized Enzymes and Cells (Part D) (in preparation)
Edited by KLAUS MOSBACH

VOLUME 138. Complex Carbohydrates (Part E)
Edited by VICTOR GINSBURG

VOLUME 139. Cellular Regulators (Part A: Calcium- and Calmodulin-Binding Proteins)
Edited by ANTHONY R. MEANS AND P. MICHAEL CONN

VOLUME 140. Cumulative Subject Index Volumes 102–119, 121–134 (in preparation)

VOLUME 141. Cellular Regulators (Part B: Calcium and Lipids)
Edited by P. MICHAEL CONN AND ANTHONY R. MEANS

VOLUME 142. Metabolism of Aromatic Amino Acids and Amines
Edited by SEYMOUR KAUFMAN

VOLUME 143. Sulfur and Sulfur Amino Acids
Edited by WILLIAM B. JAKOBY AND OWEN GRIFFITH

VOLUME 144. Structural and Contractile Proteins (Part D: Extracellular Matrix) (in preparation)
Edited by LEON W. CUNNINGHAM

VOLUME 145. Structural and Contractile Proteins (Part E: Extracellular Matrix) (in preparation)
Edited by LEON W. CUNNINGHAM

VOLUME 146. Peptide Growth Factors (Part A) (in preparation)
Edited by DAVID BARNES AND DAVID A. SIRBASKU

VOLUME 147. Peptide Growth Factors (Part B) (in preparation)
Edited by DAVID BARNES AND DAVID A. SIRBASKU

VOLUME 148. Plant Cell Membranes (in preparation)
Edited by LESTER PACKER AND ROLAND DOUCE

VOLUME 149. Drug and Enzyme Targeting (Part B) (in preparation)
Edited by RALPH GREEN AND KENNETH J. WIDDER

VOLUME 150. Immunochemical Techniques (Part K: *In Vitro* Models of B and T Cell Functions and Lymphoid Cell Receptors) (in preparation)
Edited by GIOVANNI DI SABATO

VOLUME 151. Molecular Genetics of Mammalian Cells (in preparation)
Edited by MICHAEL M. GOTTESMAN

VOLUME 152. Guide to Molecular Cloning Techniques (in preparation)
Edited by SHELBY L. BERGER AND ALAN R. KIMMEL

Section I

Separation and Analysis

A. Inorganic Sulfur and Selenium Compounds
Articles 1 through 10

B. Thiols and Disulfides
Articles 11 through 21

C. Other Organic Sulfur and Selenium Compounds
Articles 22 through 37

[1] Sulfate: Turbidimetric and Nephelometric Methods

By Bo Sörbo

Inorganic sulfate is the major end product of sulfur metabolism in higher animals and is excreted mainly via the kidney. Its determination in urine thus allows the assessment of the dietary intake of sulfur amino acids in animals and human beings.[1] Inorganic sulfate is also used for different bioconjugation reactions[2] of glycosaminoglycans, glycolipids, cerebrosides, steroids, and drugs and may for analytical purposes be released from these conjugates by acid or enzymatic hydrolysis.

A variety of methods for the estimation of sulfate in biological materials have been proposed in the literature. Methods based on the precipitation of sulfate with benzidine[3] have been abandoned due to the carcinogenicity of benzidine. The colorimetric methods based on the release of chloranilate from barium chloranilate,[4,5] described earlier in this series, are seriously affected by slight changes of ionic concentration,[6] which hamper their application to many biological samples. They may be useful, however, when sulfate has to be determined in the presence of sulfated glycosaminoglycans, which interfere in the methods described below. Although elegant methods based on ion chromatography with conductimetric detection[7] and high-performance liquid chromatography with ultraviolet detection[8] have been described, they require special equipment and are fairly time consuming.

Methods currently favored are based on the precipitation of sulfate with barium ions, but the old gravimetric procedures, which are insensitive and tedious, have been superseded by turbidimetric or nephelometric methods as described below. Indirect methods, where the excess of barium ions remaining after precipitation of barium sulfate is determined by

---

[1] Z. I. Sabry, S. B. Shaderevian, J. W. Conan, and J. A. Campbell, *Nature* (*London*) **206,** 931 (1965).
[2] R. H. De Meio, in "Metabolic Pathways" (D. M. Greenberg, ed.), 3rd ed., Vol. 7, p. 287. Academic Press, New York, 1975.
[3] J. Lange and H. Tarver, this series, Vol. 3, p. 995.
[4] A. G. Lloyd, this series, Vol. 8, p. 664.
[5] J. C. Dittmer and M. A. Wells, this series, Vol. 14, p. 509.
[6] I. T. Sutherland, *Clin. Chim. Acta* **14,** 559 (1966).
[7] D. E. C. Cole and C. R. Scriver, *J. Chromatogr.* **225,** 359 (1981).
[8] B. J. Koopman, G. Jansen, B. G. Wolthers, J. R. Beukhof, J. G. Go, and G. K. Van der Henn, *J. Chromatogr.* **335,** 259 (1985).

flame photometry,[9] atomic absorption spectrophotometry,[10] or by using $^{133}BaCl_2$ for precipitation of sulfate,[11] are more time-consuming than turbidimetric or nephelometric methods and offer no obvious advantages.

Turbidimetric or nephelometric procedures require, however, that certain precautions be taken. They should be performed under acid conditions in order to avoid interference from phosphate and preformed barium sulfate, and a stabilizing agent should be present in the assay in order to obtain adequate precision. A number of compounds have been used as stabilizing agents: gelatin,[12–14] Tween 80,[15] glycerol,[16] dextran,[17] agarose,[18] and polyethylene glycol (PEG).[19] We prefer PEG as a stabilizing agent since the commercial product is not contaminated with sulfate (in contrast to gelatin) and its action is reproducible from batch to batch. Although PEG preparations of varying polymer size are commercially available and are all effective stabilizers, we found PEG 6000 to give the most linear standard curves in our assay.[19] This method has also been adopted for the determination of inorganic sulfate in plasma without deproteinization, with a centrifugal analyzer,[20] and for automated laser nephelometry[21] of serum sulfate after deproteinization with trichloroacetic acid.

Two procedures are described here for determination of inorganic sulfate with a Ba–PEG reagent containing a small amount of preformed $BaSO_4$. The turbidimetric procedure is convenient for biological samples containing very little protein but a relatively high concentration of sulfate as in samples of urine. The nephelometric procedure, on the other hand, involves a deproteinization step and is applicable to samples of lower concentrations of sulfate such as serum or plasma.

[9] R. D. Strickland and C. M. Maloney, *Am. J. Clin. Pathol.* **24,** 1100 (1954).
[10] D. Michalk and F. Manz, *Clin. Chim. Acta* **107,** 43 (1980).
[11] D. E. Cole, F. Mohyaddin, and C. R. Scriver, *Anal. Biochem.* **100,** 339 (1979).
[12] F. Berglund and B. Sörbo, *Scand. J. Clin. Lab. Invest.* **12,** 147 (1960).
[13] A. G. Lloyd, this series, Vol. 8, p. 673.
[14] N. Tudball and J. H. Thomas, this series, Vol. 17B, p. 363.
[15] L. Garrido, *Analyst* **89,** 61 (1984).
[16] B. W. Grunbaum and N. Pace, *Microchem. J.* **9,** 184 (1965).
[17] R. S. N. Ma and J. C. M. Chan, *Clin. Biochem. (Ottawa)* **6,** 82 (1978).
[18] S. G. Jackson and E. L. McCandless, *Anal. Biochem.* **90,** 802 (1978).
[19] P. Lundquist, J. Mårtensson, B. Sörbo, and S. Öhman, *Clin. Chem. (Winston-Salem, N.C.)* **26,** 1178 (1980).
[20] P. J. Pascoe, M. J. Peake, and R. W. Walmsley, *Clin. Chem. (Winston-Salem, N.C.)* **30,** 275 (1984).
[21] A. Franceschin, L. Dell'Anna, and U. Lippi, *Clin. Chem. (Winston-Salem, N.C.)* **30,** 1420 (1984).

Reagents

Ba–PEG reagent: A solution containing 9.77 g of $BaCl_2 \cdot 2H_2O$ and 150 g polyethylene glycol 6000 in deionized water (final volume 1000 ml) is prepared. To 100 ml of this solution in a 250-ml beaker is added dropwise 0.2 ml of 50 mM Na_2SO_4 with efficient magnetic stirring. This reagent is stable for 1 week.
HCl, 0.5 M
Trichloroacetic acid (TCA), 0.5 M
Na_2SO_4, 50 mM

Turbidimetric Procedure[19]

The sample, containing up to 2.5 μmol of sulfate, is diluted to 3 ml with deionized water and 1 ml of 0.5 M HCl is added, followed by 1 ml of the Ba–PEG reagent. After vortex-mixing, the absorbance of the sample at 600 nm is determined in a 1-cm cuvette 5–30 min later. (If the determination of absorbance must be delayed for more than 5 min after the addition of the Ba–PEG reagent, the sample should be vortex-mixed again 5 min before the determination.) The absorbance value obtained is corrected for that of a reagent blank prepared by substituting an equal volume of water for the sample. The amount of sulfate present in the sample is then obtained from a standard curve prepared with known amounts of sulfate (0.01–0.05 ml of 50 mM Na_2SO_4).

Nephelometric Procedure[22]

The sample is deproteinized by mixing with an equal volume of 0.5 M TCA. After at least 5 min at room temperature, the precipitate is removed by centrifugation. The supernatant fluid is filtered through a 0.22-μm Millex filter (Millipore Corporation, Bedford, MA) in order to obtain an absolutely clear filtrate. An aliquot containing up to 500 nmol sulfate is adjusted to a volume of 1.25 ml with 0.25 M TCA. To the sample is added 0.4 ml of Ba–PEG reagent, and the $BaSO_4$ formed is determined by nephelometry at 500–600 nm, 5–60 min after the addition of the reagent as described for the turbidimetric method. (An ordinary fluorimeter may be used for this purpose with the primary and secondary monochromator set at the same wavelength and any polarizing filter disconnected.) A blank is prepared by replacing the sample with deionized water. The amount of sulfate in the sample is obtained from a standard curve prepared with

[22] B. Sörbo and S. Öhman, unpublished observations (1984).

known amounts of sulfate. The nephelometer cuvette should be washed with 0.2 M Na$_2$EDTA after the measurements.

Comments

The turbidimetric method described here may also be used after deproteinization of the sample with TCA. It may also be used for determination of total sulfur after wet oxidation of the sample with a mixture of nitric acid and perchloric acid[19] and for measurements of sulfate esters after acid hydrolysis. It should be noted that alkyl sulfates, e.g., certain steroid sulfates, are more resistant to hydrolysis than acyl sulfates and require 24 hr at 100° and an acid concentration of 1 M for complete hydrolysis.[19]

As the suspension of BaSO$_4$ obtained in the assay is largely polydisperse, the linearity of the standard curve may depend on the wavelength of the light used for measurements. The wavelength used in the turbidimetric procedure, 600 nm, was chosen to minimize interference from colored compounds present in biological samples, but both sensitivity[19] and linearity[23] are improved if a shorter wavelength is used.

It must be kept in mind that sulfated glycosaminoglycans[12,22] strongly inhibit the precipitation of Ba$_2$SO$_4$. If plasma is taken for analysis, care should be taken that heparin is not used as an anticoagulant; EDTA, citrate, or oxalate are all effective.

Acknowledgments

This work was supported by Grant No. 03X-05644 from the Swedish Medical Research Council.

---

[23] P. J. Pascoe, M. J. Peake, and R. W. Walmsley, *Clin. Chem. (Winston-Salem, N.C.)* **30**, 1422 (1984).

[2] Sulfate Determination: Ion Chromatography

By DARRYL M. SULLIVAN

The determination of sulfate in complex biological mixtures is severely restricted by the relative insensitivity and nonspecificity of classical chemical methods.[1,2] In biological fluids, procedures that depend on selective barium sulfate precipitation suffer from incomplete reaction sequences if the sulfate concentration is much less than 15 mg/liter.[3,4] Sensitive precipitations of sulfate with organic reagents such as benzidine,[5,6] or precipitations of excess barium with chloranilate[7] or rhodizonate,[8] are not sufficiently specific to distinguish sulfate from physiological concentrations of phosphate and chloride ions.[1,9] Fortunately, ion chromatography (IC) provides a new technology that is well suited to the assay of biological samples.[10–14]

Principle

Ion chromatography is a form of ion-exchange chromatography that incorporates a separator column, a background ion-suppressor column, and various eluents in a novel procedure using conductance as a mode of detection (Fig. 1). A liquid sample is introduced at the top of a separator column and an ionic eluent is pumped through the ion chromatograph causing the ionic species of the sample to move down the separator column at rates determined by their affinity for the resin. The differential rate of migration separates the ions into discrete bands. In anion chroma-

---

[1] W. J. Williams, "Handbook of Anion Determination," p. 529. Butterworth, Toronto, 1979.
[2] R. Belcher, S. L. Bogdanski, I. H. B. Rix, and A. Townshend, *Mikrochim. Acta* **2**, 81 (1977).
[3] D. E. C. Cole, F. Mohyuddin, and C. R. Scriver, *Anal. Biochem.* **100**, 339 (1979).
[4] G. Tallgren, *Acta Med. Scand., Suppl.* **640**, 1 (1980).
[5] T. V. Letonoff and J. G. Reinhold, *J. Biol. Chem.* **114**, 147 (1936).
[6] C. R. Kleeman, E. Taborsky, and F. H. Epstein, *Proc. Soc. Exp. Biol. Med.* **91**, 480 (1956).
[7] B. Spencer, *Biochem. J.* **75**, 435 (1960).
[8] A. Swaroop, *Clin. Chim. Acta* **46**, 333 (1973).
[9] A. Waheed and R. L. Van Etten, *Anal. Biochem.* **89**, 550 (1978).
[10] C. Anderson, *Clin. Chem. (Winston-Salem, N.C.)* **22**, 1424 (1976).
[11] D. E. C. Cole and C. R. Scriver, *J. Chromatogr.* **225**, 359 (1981).
[12] P. Jong and M. Burggraaf, *Clin. Chim. Acta* **132**, 63 (1983).
[13] D. E. C. J. Shafai and C. R. Scriver, *Clin. Chim. Acta* **120**, 153 (1982).
[14] D. E. C. Cole and D. A. Landry, *J. Chromatogr.* **337**, 267 (1985).

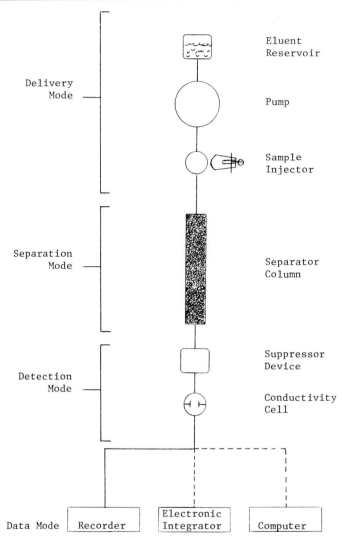

FIG. 1. Schematic of an ion chromatographic system for the determination of sulfate in physiological fluids.

tography the separation column generally consists of conventional cross-linked polystyrene resins to which basic (e.g., NH_3^+ or NR_3^+) groups have been attached.[15] In practice, it is assumed that exchanging ions are in chemical equilibrium with the column resin.

A second column, the suppressor, functions as a cation exchanger to

[15] F. Helfferich, "Ion Exchange." McGraw-Hill, New York, 1962.

chemically modify the ionic eluent of the separator column and convert it to a low conductive form. Thus, upon entering the suppressor column, the eluting base is removed by allowing the cations that are present to pass through a fiber membrane to a sulfuric acid stream which exists in the IC system. Through this mechanism, the analyte ions are simultaneously modified to highly conductive acids, which are ultimately detected by measuring the conductivity of the eluent stream. Quantitation is accomplished by comparing the computer-integrated peak areas of each analyte with those produced by standards of known concentration.

Materials and Apparatus

Sulfate determinations can be carried out using a Dionex Model 2020I IC (Sunnyvale, CA) or equivalent HPLC system with a conductivity detector. The separator column should be a Dionex Model HPIC-AS4 or equivalent anion-exchange column. The life expectancy of the separator column can be extended by using a precolumn that contains the same resin as the separator column (Dionex Model HPIC-AG4 or equivalent). The mobile phase eluent is 3.0 mM $NaHCO_3$ and 1.20 mM Na_2CO_3 prepared in ultrapure water; it is pumped through the system at 1–2 ml/min. The suppressor column can consist of any standard cation-exchange column or a Dionex Mode AFS anion fiber suppressor. Either of these column types should be regenerated with 25 mM H_2SO_4 as required.

A conductivity detector setting of 10 MS is usually adequate to measure sulfate concentrations in physiological fluids. This electrical output can be recorded using a strip chart recorder or any automated intergrator or laboratory computer system.

Analytical Procedure

All physiological fluids should be centrifuged before making sulfate determinations that use IC. After centrifugation, serum samples are generally diluted 10-fold and urine samples 100-fold with the sodium bicarbonate eluent. Other physiological fluids (i.e., saliva, sweat, cerebrospinal fluid) may require different dilutions. After dilution, the samples are filtered through a 0.45-μm filter.

Sulfate standard solutions are typically prepared in the mobile phase eluent at concentrations from 10 to 100 μg/ml. A 50- to 200-μl portion of standard or sample solution is then injected into the IC system. The sulfate peak elutes in 10–12 min.

Data Evaluation

In order to measure the performance of the chromatographic system for determining sulfate in physiological fluids, three parameters need to be examined. These parameters, column theoretical plate number (N), system capacity factor (K'), and column selectivity (α), allow determination of the overall system resolution (R_s).

Theoretical Plate Number. The N parameter is a measure of the efficiency of the ion-exchange column being used. The formula for calculating N is[16,17]

$$N = 16(t_r/t_w)^2$$

where t_r is the retention time of the sulfate peak and t_w the width of the sulfate peak at the baseline between tangents drawn to the peak. The larger the value for N, the more efficient the analytical column. The value of N can be improved by using different columns, increasing the system flow rate, or changing the ionic strength or pH of the mobile phase eluent.

System Capacity Factor. The K' parameter is a measure of analyte retention and is a valuable indicator in detecting system problems. The K' value should remain consistent in day-to-day operation of the IC system. The capacity factor is calculated using[16,17]

$$K' = (t_r - t_0)/t_0$$

Retention time (t_r) and dead time (t_0) can be measured from the chromatogram with a ruler or by using the time values recorded by an integrator or computer. The t_r is measured from the moment of injection to the center of the sulfate peak. The t_0 is measured from the moment of injection to the point at which the first baseline disturbance occurs.

Column Selectivity.[18] The α measures the separation of peaks adjacent to the sulfate peak and is the ratio of the capacity factors of the sulfate peak and any adjacent peak:

$$\alpha = \frac{K' \text{ of } SO_4^{2-}}{K' \text{ of adjacent peak}}$$

The α value is one of the most important parameters in that separation of two adjacent peaks is the basic goal of chromatographic separations.

System Resolution. The R_s measures the chromatographic system performance and is calculated using the three system parameters previously defined[18]:

$$R_s = 0.25(N) \frac{K'}{1 + K'} (\alpha - 1)$$

[16] J. H. Knox and M. Scleem, *J. Chromatogr. Sci.* **7**, 614 (1969).
[17] L. R. Snyder and J. J. Kirkeland, "Introduction to Modern Liquid Chromatography," 2nd ed. Wiley (Interscience), New York, 1979.
[18] L. R. Snyder, in "High-Performance Liquid Chromatography: Advances and Perspectives" (C. Horvath, ed.), Vol. 1. Academic Press, New York, 1980.

In an optimum chromatographic system for measuring sulfate, the R_s value should be 1.5 or greater. If the resolution is inadequate, the K' or N values can be modified by using different columns, changing the system flow rates, or by adjusting the ionic strength or pH of the mobile phase eluent.

[3] Sulfite Determination: Sulfite Oxidase

By HANS-OTTO BEUTLER

Sulfite and sulfite adducts (e.g., aldehyde-bisulfite addition compounds) are distributed in very low concentrations in plants and animals. Their physiological importance is unknown. In metabolism, sulfite is immediately oxidized to sulfate by sulfite oxidase (EC 1.8.3.1)[1] or by sulfite dehydrogenase (EC 1.8.2.1).[2] Sulfite can also be reduced to sulfide by sulfite reductase (NADPH) (EC 1.8.1.2).[3]

At higher concentrations, sulfite can inhibit plant growth as well as physiological reactions in animals. Because many microorganisms are sensitive to sulfite, it is applied in the processing of food as a preservative. As an enzyme inhibitor, sulfite is used in the processing of wine in order to inactivate phenol oxidases and thereby avoid enzymatic browning. The binding of sulfite to acetaldehyde and other carbonyl compounds formed during fermentation accounts in part for its protective action.

The usual methods for sulfite determination are based on its oxidation to sulfate followed by gravimetry or by titration of the sulfuric acid formed.[4] Sulfite has also been determined by iodometric titration after distillation.[5] More recently, a photometric method using pararosaniline and formaldehyde has been developed.[6] A reliable enzymatic method is presented here that allows specific measurement of sulfite in complex mixtures of food.[7-9]

[1] D. L. Kessler, J. L. Johnson, H. J. Cohen, and K. V. Rajagopalan, *Biochim. Biophys. Acta* **334**, 86 (1974).
[2] A. M. Charles and I. Suzuki, *Biochim. Biophys. Acta* **128**, 522 (1966).
[3] L. M. Siegel, M. J. Murphy, and H. Kamin, *J. Biol. Chem.* **248**, 251 (1973).
[4] J. F. Reith and J. J. L. Willems, *Z. Lebensm.-Unters. -Forsch.* **108**, 270 (1958).
[5] W. Weinmann and L. Walther, *Z. Lebensm.-Unters. -Forsch.* **87**, 49 (1944).
[6] M. Luise, *Vini Ital.* **11**, 60 (1969); *Chem. Abstr.* **71**, 11719h (1969).
[7] H. O. Beutler and I. Schütte, *Dtsch. Lebensm.-Rundsch.* **79**, 323 (1983).
[8] H. O. Beutler, in "Methods of Enzymatic Analysis" (H. U. Bergmeyer, J. Bergmeyer, and M. Grassl, eds.), 3rd Engl. ed., Vol. 7, p. 585. VCH Verlagsges., Weinheim, 1985.
[9] H. O. Beutler, *Food Chem.* **15**, 157 (1984).

Assay Principle

The procedure is based on the spectrophotometric determination of sulfite using a highly purified sulfite oxidase to catalyze the formation of sulfate and hydrogen peroxide [reaction (1)] coupled to the reduction of H_2O_2 in presence of NADH and NADH peroxidase (NADH : hydrogen-peroxide oxidoreductase, EC 1.11.1.1) [reaction (2)]. In the coupled reaction, the oxidation of NADH is related to the amount of sulfite present according to reactions (1) and (2).

$$SO_3^{2-} + O_2 + H_2O \rightarrow SO_4^{2-} + H_2O_2 \qquad (1)$$
$$H_2O_2 + NADH + H^+ \rightarrow 2\,H_2O + NADH^+ \qquad (2)$$

Reagents

All biochemicals were obtained from Boehringer-Mannheim GmbH, FRG. Other chemicals were from Merck, FRG.

Required Solutions

0.6 mM triethanolamine hydrochloride, pH 8.0 (solution A)

1% $NaHCO_3$, 7 mM β-NADH (solution B)

NADH peroxidase from *Streptococcus faecalis* with 45 units/ml, prepared by diluting the stock suspension (45 units/mg protein) with 3.0 M ammonium sulfate solution, pH 7.0 (suspension C) (1 unit of activity oxidizes 1 μmol of NADH/min)

Sulfite oxidase from chicken liver with 5 units/ml, prepared by diluting the stock suspension (1 unit/mg protein) with 3.0 M ammonium sulfate solution, pH 6.0 (suspension D) (1 unit of activity consumes 1 μmol of O_2/min under the test conditions)

0.94 mM sulfite standard solution, prepared by dissolving 120 mg Na_2SO_3 in 1 liter of water (solution E); this solution should be used for the assay within half an hour of preparation

General Method

Samples of sulfite-containing materials, e.g., white wine, were assayed directly. Red wines were adjusted to pH 7.5–8.0 with NaOH, 2 mol/liter, before using. Beverages and other liquid foodstuffs should be treated with L-ascorbate oxidase (EC 1.10.3.3) (10 units/ml sample) before the assay in order to remove interfering ascorbate and with polyvinylpolypyrrolidone (5 g/100 ml sample) in order to remove vegetable dyes. Solid foodstuffs are ground and extracted with water in a closed vessel at 60°. After centrifugation, the clear sample solution is suitable for assay. In some cases, it is necessary to remove heavy metal ions by adding 1 mg

EDTA/1 ml sample solution. In the presence of high concentrations of heavy metals (>1 mg/1 ml sample), as much as 2 mg of EDTA should be added.

Assay

The reaction mixture contains 1 ml of buffer (solution A), 0.1 ml of NADH solution (solution B), 10 μl of NADH peroxidase (suspension C), 1.9 ml of distilled water, and 0.1 ml of standard (solution E) or sample. After incubation for 5 to 20 min the absorbance of the mixture is measured at 339 nm.

Sulfite oxidase, 50 μl (suspension D), is added and the change in absorbance (ΔA) due to the resulting oxidation of NADH is measured after standing for 20–30 min at room temperature (corresponding to the end of the reaction). Blank measurement, in which water substitutes for the sample or for a standard solution, is carried out in the same way. Usually, the reaction is accompanied by an optical drift due to contaminating NADH oxidase in the sulfite oxidase; it can be ignored if the absorbances of the sample and the blank are read simultaneously, because the drift of the sample is normally the same as that of the blank.

Calculation

The sulfite oxidase-dependent absorbance change (ΔA) of sample and blank are calculated. After subtraction of the blank ΔA from the sample ΔA the net ΔA for sulfite is obtained.

The concentration of SO_3^{2-} in mg SO_3^{2-}/liter is equal to c as shown in Eq. (3), where 3.16, 80.05, 6.3, 1, and 0.1 are the assay volume, M_r of SO_3^{2-}, millimolar extinction coefficient of NADH, light path, and sample volume, respectively.

$$c = \frac{3.16 \times 80.05}{6.3 \times 1 \times 0.1} \times \Delta A = 401.5 \times \Delta A \tag{3}$$

To express millimolar concentration, this value has to be multiplied by 0.0125 and is thus $5.016 \times \Delta A$ mmol SO_3^{2-}/liter sample solution.

The values observed for 30 samples of white wine were between 94 and 232 SO_3^{2-}/liter, and for 24 samples of red wine, between 75 and 177 mg SO_3^{2-}/liter.

Discussion of the Method

To obtain a quantitative SO_3^{2-} conversion, three conditions must be fulfilled: (1) reaction in an alkaline medium to slow the chemical oxidation of sulfite to sulfate; (2) high affinity of the sulfite oxidase for sulfite as indicated by the low Michaelis–Menten constant ($K_m = 2.4 \times 10^{-5}$ mol/liter at pH 8.5; Tris buffer)[10] (large amounts of enzyme insure quantitative conversion of SO_3^{2-} to SO_4^{2-}); and (3) rapid removal of H_2O_2 produced according to Eq. (1) by NADH peroxidase which possesses a high affinity toward H_2O_2 ($K_m = 2.8 \times 10^{-5}$ mol/liter at pH 6.0; Tris buffer).[11] All conditions are fulfilled by the procedure described.

Optimal Reaction Conditions. The optimal pH of the coupled reaction system is 8.0. Triethanolamine is a particularly useful buffer and is best used at 0.1 to 0.25 mol/liter. The minimum amount of sulfite oxidase necessary to complete the reaction after 40 min is 0.02 units/ml. Complete conversion of sulfite within 15 min is obtained with 0.04 units/ml of sulfite oxidase and 0.05 units of NADH peroxidase/ml of assay solution. The most favorable ratio of NADH peroxidase to sulfite oxidase is 1.8.

Statistical Results. With freshly prepared and pure solutions of sodium sulfite the method is linear between 15 and 210 μmol sulfite/liter assay solution. The relative standard deviation of the method was determined with *CV* (coefficient of variation) equal to 0.67% in the medium concentration range (145 μmol SO_3^{2-}/liter assay solution; $n = 15$) and with *CV* equal to 1.5% in the upper concentration range (210 μmol SO_3^{2-}/liter assay solution; $n = 8$). The accuracy has been tested by comparison with a chemical method (titration with iodine solutions). A linear regression was calculated with r equal to 0.995 (correlation coefficient). The sensitivity of the sulfite determination is 0.8 μmol SO_3^{2-}/liter.

Specificity and Interferences. In addition to sulfite, sulfite-carbonyl addition compounds are also converted to SO_4^{2-} and therefore consume NADH. This behavior can be explained by the hydrolysis of such compounds in the alkaline assay milieu. Sulfides, thiosulfates, and organic sulfo compounds do not react under the assay conditions. Only isothiocyanates and their glycosides react like sulfite.

Some interferences in the assay could be observed in the presence of some organic acids: oxaloacetate (>25 μmol/liter assay solution) and pyruvate (>6 mmol/liter assay solution). Sulfite oxidase is inhibited by these substances, so the reaction time must be increased. Tannic compounds (e.g., pyrogallol), which occur in some vegetable juices, inhibit NADH peroxidase. The strongest interference is produced by ascorbic acid; greater than 4 μM ascorbate led to complete decomposition of NADH. Ascorbate in the sample should therefore be removed by addition of 10 units of L-ascorbate oxidase (EC 1.10.3.3) for each milliliter of sample.

[10] D. L. Kessler and K. V. Rajagopalan, *J. Biol. Chem.* **247**, 6566 (1972).
[11] Boehringer-Mannheim GmbH, German Patent DE 2,633,728 (1979).

[4] Sulfite Determination: Fuchsin Method

By FRANZ-JOSEF LEINWEBER and KENNETH J. MONTY

The method of Grant for the colorimetric determination of sulfur dioxide in extracts of animal and plant tissues[1] has been adapted to the study of microbial enzymes producing sulfite from substrates such as 3'-phosphoadenosine 5'-phosphosulfate (PAPS), thiosulfate and cysteine sulfinate (CSA), both *in vivo*[2-4] and *in vitro*.[5,6] Grant's method was also applied to the enzymatic determination of CSA.[7,8] The method relies on the formation of a red chromophore with basic fuchsin (pararosaniline).

$$\text{Basic fuchsin} \xrightarrow{H_2SO_3} \text{Schiff's reagent} \xrightarrow{RCHO} \text{colored aldehyde addition product}$$
$$\text{(colorless)}$$

Fuchsin Reagent

Grant's fuchsin reagent is prepared by adding 1.6 ml of a 3% ethanolic solution of basic fuchsin (Fisher) to 93.6 ml distilled water containing 4.4 ml concentrated H_2SO_4. After thorough mixing, 0.4 ml 40% formaldehyde is added. The solution is decolorized by addition of 200 mg activated carbon. After repeated shaking over a period of 15 min, the solution is clarified by filtration. Fresh reagent is prepared daily.

Standard Procedure

To a 2.5 ml sample, add 0.5 ml of 1% alcoholic KOH, mix thoroughly, and add 1.0 ml of a saturated aqueous solution of $HgCl_2$. After mixing and centrifuging, 1.0 ml of the clear supernatant fluid is mixed with 4.0 ml fuchsin reagent. After 15 min at room temperature, the absorbance of each sample is measured at 580 nm. The color is compared with that produced from standardized samples of sodium bisulfite ranging from 80 to 480 nmol.

[1] W. M. Grant, *Anal. Chem.* **19,** 345 (1962).
[2] F.-J. Leinweber and K. J. Monty, *Biochem. Biophys. Res. Commun.* **6,** 355 (1961).
[3] F.-J. Leinweber and K. J. Monty, *J. Biol. Chem.* **238,** 3775 (1963).
[4] F.-J. Leinweber and K. J. Monty, *J. Biol. Chem.* **240,** 782 (1965).
[5] F.-J. Leinweber and K. J. Monty, *Biochim. Biophys. Acta* **63,** 171 (1962).
[6] J. Dreyfuss and K. J. Monty, *J. Biol. Chem.* **238,** 3781 (1963).
[7] F.-J. Leinweber and K. J. Monty, *Anal. Biochem.* **4,** 252 (1962).
[8] See this volume [31].

Expanded Sensitivity

The sensitivity of the assay may be increased to 5 nmol by the following modification. To a 0.25 ml sample, add 50 μl 1% aqueous KOH followed by 50 μl of a saturated solution of $HgCl_2$ in absolute ethanol. Note the reversal of solvent for the two reagents between the two methods. Then combine 0.2 ml of the clear supernatant liquid obtained from that mixture with 1.0 ml fuchsin reagent, and measure the resulting color as described above.

Color Stability

Maximum color intensity is reached at approximately 15 min after addition of fuchsin reagent. Fading occurs slowly: after 60 min at room temperature approximately 85% of the color measured at 15 min remains.

Specificity

The selectivity of the method depends on the removal of potential interfering compounds by precipitation with $HgCl_2$.[1] Both versions of the method have been used successfully without interference from PAPS (0.28 mM), CSA (2 mM), cysteic acid, hypotaurine, thiosulfate (all at 16 mM), or bacterial protein (14 mg/ml).

Preparing a Sulfite Standard

A standardized solution of sodium bisulfite may be prepared titrimetrically, using sodium thiosulfate as a primary standard. An iodine solution is titrated with a standard solution of thiosulfate. A parallel iodine sample is reacted with an aliquot of the bisulfite solution, and the remaining iodine is titrated with thiosulfate. The decreased volume of thiosulfate used in the second titration indicates the number of equivalents of bisulfite introduced. Equations (1) and (2) explain the standardization:

$$3\ I_2 + 6\ S_2O_3^{2-} \rightarrow 3\ S_4O_6^{2-} + 6\ I^- \quad (1)$$
$$I_2 + SO_3^{2-} + H_2O \rightarrow 2\ HI + SO_4^{2-} \quad (2)$$

Reagents

0.1 M $Na_2S_2O_3$ (primary standard; may be purchased as a standard solution or standardized against a true primary standard such as KIO_3)

0.1 M iodine (6.35 g crystalline iodine plus 20 g KI dissolved in distilled water, final volume 500 ml)

0.1 M $NaHSO_3$ (1.0548 g in 100 ml water)

Starch reagent (make a paste with 1 g soluble starch in 5 ml H_2O, pour into 200 ml boiling water, boil 1 min and cool; preserve with a few drops of chloroform)

Titrate a 50 ml aliquot of iodine solution with thiosulfate until the color turns pale yellow. Add 3 ml starch reagent. Continue the titration until the blue color just disappears, and record the volume of thiosulfate used as V_a. To a second 50 ml aliquot of iodine solution, add 25 ml of the bisulfite solution to be standardized. Titrate as before with thiosulfate, and record the volume of thiosulfate as V_b. Calculate the molarity of the bisulfite solution as follows:

$$\text{Molarity}_{\text{bisulfite}} = \text{molarity}_{\text{thiosulfate}} \times (V_a - V_b)/50$$

The bisulfite solution is then diluted appropriately (e.g., 1:1000) for use as a standard in calibrating the fuchsin reaction.

[5] Sulfite and Thiosulfate: Using S-(2-Amino-2-carboxyethylsulfonyl)-L-cysteine

By TOSHIHIKO UBUKA

Sulfite and thiosulfate are important metabolites of cysteine sulfur in animals.[1-3] The method described here utilizes S-(2-amino-2-carboxyethylsulfonyl)-L-cysteine (ACESC) as a reagent for converting inorganic sulfite and thiosulfate into organic cysteine derivatives.[4,5] ACESC was first prepared by Toennies and Lavine by oxidizing L-cystine in nonaqueous media, and its structure was proposed to be cystine disulfoxide.[6]

---

[1] A. Meister, "Biochemistry of the Amino Acids," 2nd ed., Vol. 2, p. 757. Academic Press, New York, 1965.

[2] A. B. Roy and P. A. Trudinger, "The Biochemistry of Inorganic Compounds of Sulphur," p. 289. Cambridge Univ. Press, London and New York, 1970.

[3] L. M. Siegel, *in* "Metabolic Pathways" (D. M. Greenberg, ed.), 3rd ed., Vol. 7, p. 270. Academic Press, New York, 1975.

[4] T. Ubuka, M. Kinuta, R. Akagi, S. Kiguchi, and M. Azumi, *Anal. Biochem.* **126**, 273 (1982).

[5] T. Ubuka, N. Masuoka, H. Mikami, and M. Taniguchi, *Anal. Biochem.* **140**, 449 (1984).

[6] G. Toennies and T. F. Lavine, *J. Biol. Chem.* **113**, 571 (1936).

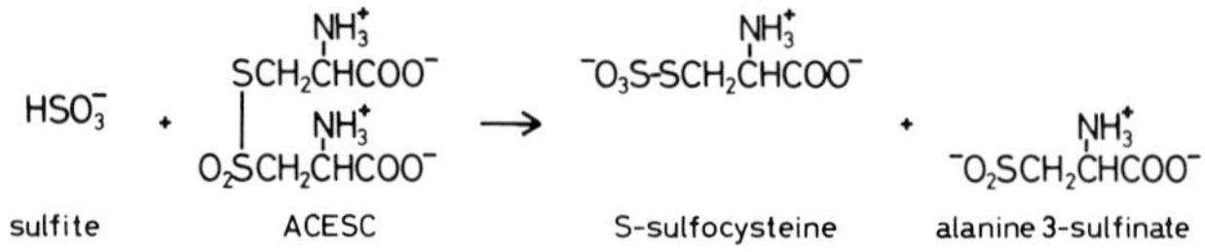

$$HSO_3^- \; + \; \begin{matrix} NH_3^+ \\ SCH_2CHCOO^- \\ | \\ NH_3^+ \\ O_2SCH_2CHCOO^- \end{matrix} \; \longrightarrow \; \begin{matrix} NH_3^+ \\ {}^-O_3S\text{-}SCH_2CHCOO^- \end{matrix} \; + \; \begin{matrix} NH_3^+ \\ {}^-O_2SCH_2CHCOO^- \end{matrix}$$

sulfite   ACESC   S-sulfocysteine   alanine 3-sulfinate

SCHEME 1. Reaction of sulfite and ACESC.

Subsequent studies have shown that it is not a disulfoxide but a thiosulfonate.[4,5,7–9] The reactivity of ACESC with thiol compounds such as cysteine[10] and glutathione[11] has been reported.

Principle

ACESC reacts readily and quantitatively with sulfite and thiosulfate in acidic solutions as shown in Schemes 1 and 2. The reaction products, S-sulfo-L-cysteine[12] and L-alanine sulfodisulfane,[13] respectively, are separated by ion-exchange chromatography on small Dowex 50W or Dowex 1 columns and are determined using the acid ninhydrin reagent 2 of Gaitonde.[14]

Reagents

S-(2-Amino-2-carboxyethylsulfonyl)-L-cysteine.[6]   L-Cystine (50 mmol; 12 g) is suspended in 250 ml of acetonitrile and mixed with a magnetic stirrer while cooling in an ice-water bath. Then, 0.1 mol of $HClO_4$ (14.36 g of 70% $HClO_4$), 300 ml of acetonitrile, and 21.4 ml of acetic anhydride are successively added. The mixture is stirred until the L-cystine dissolves completely. After the temperature is allowed to rise to about 15°, 0.12 mol (16.57 g) of perbenzoic acid,[15] dissolved in 150 ml of chloroform, is added with stirring and slight cooling. The total volume of the reaction mixture is then made up to 1 liter with acetonitrile. After standing at 20–25°C for 1 hr, the solution is extracted once with 250 ml and once with 100 ml of 1 N HCl. The combined extracts are extracted 6 times with 50-ml portions of chloroform. The aqueous layer is filtered and care-

[7] B. J. Sweetman, *Nature (London)* **183**, 744 (1959).
[8] T. Ubuka, S. Yuasa, M. Kinuta, and R. Akagi, *Physiol. Chem. Phys.* **12**, 45 (1980).
[9] T. Ubuka, S. Yuasa, M. Kinuta, and R. Akagi, *Physiol. Chem. Phys.* **11**, 353 (1979).
[10] T. Lavine, *J. Biol. Chem.* **113**, 583 (1936).
[11] B. Eriksson and S. A. Eriksson, *Acta Chem. Scand.* **21**, 1304 (1967).
[12] 2-Amino-2-carboxyethyl sulfosulfane or L-alanine sulfosulfane or S-sulfo-L-cysteine.
[13] 2-Amino-2-carboxyethyl sulfodisulfane or L-alanine sulfodisulfane.
[14] M. K. Gaitonde, *Biochem. J.* **104**, 627 (1967).
[15] Prepared according to G. Braun, *Org. Synth., Collect.* **1**, 431.

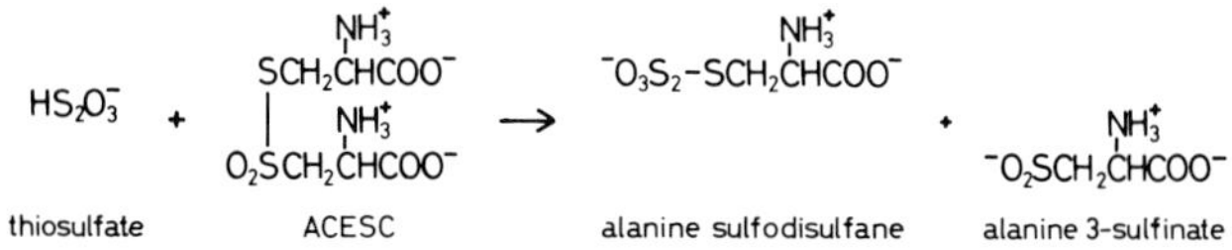

$$HS_2O_3^- + \begin{array}{c}\overset{+}{N}H_3\\SCH_2CHCOO^-\\|\\O_2SCH_2\overset{|}{C}HCOO^-\\\overset{|}{N}H_3^+\end{array} \rightarrow \begin{array}{c}\overset{+}{N}H_3\\{}^-O_3S_2-SCH_2\overset{|}{C}HCOO^-\end{array} + \begin{array}{c}\overset{+}{N}H_3\\{}^-O_2SCH_2\overset{|}{C}HCOO^-\end{array}$$

thiosulfate   ACESC   alanine sulfodisulfane   alanine 3-sulfinate

SCHEME 2. Reaction of thiosulfate and ACESC.

fully brought to pH 2.5 by adding 8 N and 1 N ammonia using a magnetic stirrer. After standing for about 10 min, the precipitate formed is collected on a sintered glass filter, washed with water, ethanol, and ether, and dried in a desiccator over silica gel. Yield is about 90% of the theoretical.

ACESC thus prepared should dissolve in water when a slight excess of sulfite is added. If the solution does not become clear, the product contains cystine and is not usable. ACESC can be stored in a desiccator at room temperature for several years.

Acid ninhydrin reagent 2 of Gaitonde.[14] Ninhydrin, 5 g, is stirred in 120 ml of glacial acetic acid until the ninhydrin dissolves. Then, 80 ml of 36% HCl (w/v) is added. The reagent is stable at room temperature for at least 1 month.

S-Sulfo-L-cysteine and L-alanine sulfodisulfane. S-Sulfo-L-cysteine and L-alanine sulfodisulfane are prepared as described[4,5] and used as standards. As shown in Fig. 1, color values of these compounds in the acidic ninhydrin reaction after dithiothreitol treatment are almost equal. Therefore, S-sulfo-L-cysteine may be used as the standard for the determination of both sulfite and thiosulfate.

Dowex 50W column. Small columns of Dowex 50W (X8, 200–400 mesh, 0.7 × 6.0 cm, H$^+$ form) are prepared using Econo-columns (0.7 × 10 cm; Bio-Rad Laboratories, Richmond, CA).

Dowex 1 column. Columns of Dowex 1 (X8, 200–400 mesh, 0.7 × 17 cm, Cl$^-$ form) are prepared using Econo-columns (0.7 × 20 cm), and the final washing with water is continued until the washing shows neutral pH.

Procedure

Determination of Sulfite as S-Sulfo-L-cysteine. To 1 ml of a sample solution containing sulfite (up to 10 μmol) are added 0.2 ml of glacial acetic acid and 2.5 ml of 99% ethanol in order to remove proteins. After standing for 10 min, the mixture is centrifuged at 1200 g for 10 min. The supernatant liquid is transferred to a test tube with a Teflon-lined screw cap (16 × 100 mm) containing 100 μmol of solid ACESC. Water, 4 ml, is added, and the tube is mechanically shaken longitudinally at about 170 strokes per minute for 30 min. The mixture is centrifuged at 1200 g for 10 min, and the resulting supernatant fluid is evaporated to dryness with a flash evaporator at 40°. The residue is dissolved in 1 ml of water, and 0.5 ml of the solution is applied to a column of Dowex 50W (0.7 × 6.0 cm). The column is washed 6 times with 0.5-ml portions of water, and the initial 3 ml of the effluent is pooled.

An aliquot of the pooled effluent in a graduated test tube (16 × 100 mm) is made up to 0.5 ml with water, and 0.5 ml of glacial acetic acid and 0.5 ml of acid ninhydrin reagent 2 are added. The mixture is heated in a boiling water bath for 10 min. After chilling in an ice-water bath, the volume is adjusted to 5.0 ml with 95% ethanol, and the absorbance at 560 nm is quickly determined. The sulfite content is calculated from a standard curve obtained with authentic S-sulfo-L-cysteine. The standard curve is linear up to 0.2 μmol per tube of the acidic ninhydrin reaction. Usually a standard containing 0.1 μmol of S-sulfo-L-cysteine is used in parallel with the sample. The recovery of sulfite as S-sulfo-L-cysteine in the above procedure was 96.8 ± 0.3%.[4]

Determination of Thiosulfate as L-Alanine Sulfodisulfane. Thiosulfate can be determined in the same manner as sulfite except that the acidic ninhydrin reaction is performed after dithiothreitol treatment because L-alanine sulfodisulfane does not produce color in the acidic ninhydrin reaction as shown in Fig. 1.

A sample solution containing up to 0.15 μmol L-alanine sulfodisulfane is brought to 0.4 ml with water. After addition of 0.1 ml of 50 mM dithiothreitol, the pH is adjusted to about 9 with sodium hydroxide solution, and the mixture is incubated for 30 min at room temperature. Then the acidic ninhydrin reaction is performed as above. Recovery of thiosulfate as L-alanine sulfodisulfane was 97.5 ± 0.2%.[5]

Determination of Sulfite and Thiosulfate. When a sample solution contains both sulfite and thiosulfate, S-sulfo-L-cysteine and L-alanine sulfodisulfane are produced by the reaction with ACESC. These products must be separated.[5]

The dried residue after the reaction with ACESC is dissolved in 1 ml of water, and 0.5 ml of the solution is charged onto a column of Dowex 1 chloride (X8, 200–400 mesh, 0.7 × 17 cm). Elution is with 1 M acetic acid containing 0.5 M NaCl. The initial 8 ml of effluent are discarded. S-Sulfo-L-cysteine is eluted in the next 12 ml (fraction A). Elution is continued with 1 M acetic acid containing 1.5 M NaCl: 12 ml of the effluent is pooled (fraction B). Fraction B contains L-alanine sulfodisulfane. S-Sulfo-L-cysteine in fraction A and L-alanine sulfodisulfane in fraction B are determined with the acidic ninhydrin reaction as described above. If a precipi-

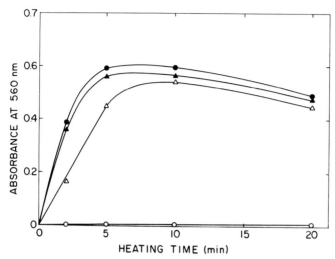

FIG. 1. Acidic ninhydrin reaction of L-alanine sulfodisulfane (●, ○) and S-sulfo-L-cysteine (▲, △). The compound (0.1 μmol) was heated with acid ninhydrin reagent 2 in a boiling water bath for the time indicated after dithiothreitol treatment (solid symbols) or without the treatment (open symbols). From Ubuka et al.[5]

tate of NaCl is produced upon addition of ethanol after this reaction, the precipitate must be removed by centrifugation before determining absorbance at 560 nm.

Recoveries of sulfite and thiosulfate as S-sulfo-L-cysteine and L-alanine sulfodisulfane, respectively, in the above procedure were over 95%.[5] When a large amount of L-alanine sulfodisulfane is applied to Dowex 1 column, some of the compound may be eluted in fraction A. This can be checked by paper electrophoresis.[5]

Comments

The method described utilizes the conversion of inorganic sulfite and thiosulfate to organic cysteine derivatives. It is, therefore, useful for quantitation of these compounds as well as for tracer experiments with sulfur compounds. The sensitivity of the present method is in the range of 0.1 μmol/ml of protein-free extracts. When both sulfite and thiosulfate are present in the sample, the sensitivity is about 1.0 μmol/ml.

Acidic sulfur-containing amino acids such as L-cysteic acid, taurine, and L-alanine 3-sulfinic acid[16] do not interfere. Since the presence of L-

[16] L-Cysteinesulfinic acid.

alanine sulfodisulfane in the acidic ninhydrin reaction of S-sulfo-L-cysteine reduced the color yield slightly, the two compounds must be separated before reaction.

The reaction of ACESC and sulfite or thiosulfate can be performed in either acetic or formic acid solutions. Acetic acid is preferable because purer commercial grades are available.

The present method was applied to the determination of sulfite which is formed by the transamination of L-alanine 3-sulfinic acid.[4] The results have shown that the sulfinopyruvic acid formed decomposed nonenzymatically and that stoichiometric amounts of sulfite and pyruvate were formed. The method has also been applied to thiosulfate sulfurtransferase, and thiosulfate, a substrate, and sulfite, a product, have been separately determined.[5]

[6] Thiocyanate and Thiosulfate

By JOHN WESTLEY

Several methods have been used for the analytical determination of inorganic thiocyanate.[1-3] These include procedures based on the formation of colored complexes, either with ferric ions or with cupric ions and pyridine, or on oxidation with bromine followed by determination of the product cyanogen bromide. An ion-selective electrode for thiocyanate analysis is also marketed, but its use is often limited by the presence of interfering ions. Similarly, inorganic thiosulfate has been determined by a variety of methods, including both the traditional iodometric titration and exhaustive cyanolysis followed by determination of the resulting thiocyanate.[4,5] Thiosulfate may also be readily determined polarographically.[6]

In each case, the choice of method ordinarily rests on the nature of the sample to be analyzed, but the following procedures, both based on the formation of the red ferric thiocyanate complex ion, have been found most generally useful and convenient. Thiocyanate itself requires only

---

[1] J. L. Wood, in "Chemistry and Biochemistry of Thiocyanic Acid and Its Derivatives" (A. A. Newman, ed.), p. 156. Academic Press, New York, 1975.
[2] F. D. Snell and C. T. Snell, "Colorimetric Methods of Analysis," 3rd ed., Vol. 2, p. 783. Van Nostrand-Reinhold, Princeton, New Jersey, 1949.
[3] J. L. Wood, this volume [7].
[4] A. Gutman, *Z. Anal. Chem.* **47**, 294 (1908).
[5] B. Sörbo, *Biochim. Biophys. Acta* **23**, 412 (1957).
[6] See this volume [10].

treatment with a soluble ferric salt in acid solution. Thiosulfate is subjected to quantitative cyanolysis in the presence of cupric ions prior to the ferric treatment.

Thiocyanate

Principle. Thiocyanate forms a highly colored complex ion with Fe^{3+} in acid solutions, as has been known since early in the nineteenth century.

$$Fe^{3+} + 6\ SCN^- \rightleftharpoons Fe(SCN)_6^{3-}$$

The absorption maximum is at 460 nm, and millimolar concentrations of SCN^- produce A_{460} values greater than 1.0 in cells of 1 cm path length. Absorbance is proportional to SCN^- content to very high values in the following standard procedure. Because the complex is formed reversibly, however, the absorption is not linear with simple dilution of completed analyses. For the same reason, samples containing very high concentrations of salts give smaller color yields, so that an empirical standard under conditions of the analysis is required.

Complex biological mixtures may also contain interfering colors or other materials that form colored complexes with Fe^{3+}, even in strong nitric acid solutions. Interference by the blue–violet ferric thiosulfate complex is avoided by the addition of formaldehyde.[7] For use with difficult mixtures, one of the alternative thiocyanate methods may be preferable. When proteins are removed from samples before analysis, e.g., by TCA precipitation, significant losses of SCN^- may be incurred. Such losses are usually negligible when proteins are left in the sample until after color development, during which they may be precipitated by the nitric acid. The precipitated proteins can then be removed from the completed analyses by centrifugation, ordinarily without loss of SCN^-.

Ferric thiocyanate colors are subject to photocatalyzed fading, and it is usual to avoid exposure to direct sunlight or strong fluorescent lighting. However, the fading problem, which is greatly exacerbated by the presence of certain materials (sulfinates, for example) in the sample, can be overcome by the addition of thiosulfate before the ferric treatment.[8] Colors developed when this precaution is observed are stable for hours.

Reagents

NH_4SCN, 0.10 M standard solution

Formaldehyde, 15%, prepared by addition of 1.5 volumes of water to 1 volume of formalin (38% commercial formaldehyde reagent)

---

[7] B. Sörbo, *Acta Chem. Scand.* **7**, 1129 (1953).
[8] R. Mintel and J. Westley, *J. Biol. Chem.* **241**, 3381 (1966).

Na$_2$S$_2$O$_3$, 0.25 M

Sörbo's ferric nitrate reagent[7]: 100 g of Fe(NO$_3$)$_3$ · 9H$_2$O and 200 ml of 65% HNO$_3$ per liter

Procedure. To an aliquot of sample or standard (10 nmol to 1 μmol) in 1 ml of water are added successively 0.5 ml of 15% formaldehyde, 0.5 ml of 0.25 M Na$_2$S$_2$O$_3$, and 1 ml of ferric nitrate reagent at room temperature. After mixing, absorbance at 460 nm is measured against a reagent blank. For samples that are themselves highly colored, a further blank value may need to be deducted. Samples that become turbid on addition of ferric nitrate reagent should be centrifuged before the photometry.

Thiosulfate

Principle. Thiosulfate can be completely cyanolyzed to thiocyanate and sulfite by prolonged treatment with alkaline cyanide[3,4] at 100°.

$$S_2O_3^{2-} + CN^- \rightarrow SCN^- + SO_3^{2-}$$

However, such conditions also result in the formation of SCN$^-$ with sulfur from a variety of other materials, including the cystinyl residues of GSSG and proteins.[9] The selectivity for thiosulfate sulfur can be greatly improved by use of a specific catalyst for the sulfur transfer from thiosulfate to cyanide. The enzyme rhodanese (thiosulfate : cyanide sulfurtransferase; EC 2.8.1.1, thiosulfate sulfurtransferase) is satisfactory, but the greater convenience afforded by Sörbo's use of cupric ions for this purpose makes his procedure the method of choice. Thiocyanate formed quantitatively from thiosulfate in the presence of Cu^{2+} and CN$^-$ is then determined colorimetrically as the ferric thiocyanate complex ion.

Reagents

Na$_2$S$_2$O$_3$, 0.10 M standard solution

KCN, 0.25 M

CuSO$_4$, 0.20 M

Sörbo's ferric nitrate reagent: 100 g of Fe(NO$_3$)$_3$ · 9H$_2$O and 200 ml of 65% HNO$_3$ per liter

Procedure. To an aliquot of sample or standard (10 nmol to 1 μmol) in 1.8 ml of water is added 0.10 ml of 0.25 M KCN at room temperature. After thorough mixing, 0.10 ml of 0.20 M CuSO$_4$ is added, and the solution is again thoroughly mixed. Addition of 1 ml of ferric nitrate reagent to this mixture causes the formation of a white precipitate, which gradually dissolves as the red color of Fe(SCN)$_3^{6-}$ develops. When all samples have become clear (the reagent blank is the last to clarify), absorbances at 460

[9] J. F. Schneider and J. Westley, *J. Biol. Chem.* **244**, 5735 (1969).

nm are determined. Only inorganic sulfide, which can be removed as the insoluble cadmium salt, and materials with very reactive sulfane sulfur atoms (e.g., per- and polysulfides and polythionates), which react rapidly with cyanide to produce SCN^- in uncatalyzed reactions, cause substantial positive interference. However, the contribution of such materials to the analytical values can usually be evaluated separately by omitting the cupric ions. Organic thiosulfonate anions do not yield thiocyanate in this procedure.

Acknowledgments

This work was supported by research grants from the National Science Foundation (DMB-8512836) and the National Institutes of Health (GM-30971).

[7] Sulfane Sulfur

By JOHN L. WOOD

The term, sulfane, designates sulfur atoms that are bonded covalently in chains *only* to other sulfur atoms. Examples are the outer sulfur of thiosulfate ($^-SO_3S^-$) and thiosulfonate ions (RSO_3S^-); the internal chain sulfurs of organic and inorganic polysulfides (RSS_nSR), where R represents an anion or organic group); persulfides (RSS^-); polythionates ($^-O_3SS_nSO_3^-$); and elemental sulfur (S_8).

Sulfane atoms are somewhat unstable, readily oxidizing in air, reducing with thiols, and decomposing slowly in dilute acid to release free sulfur. Most sulfane-containing compounds are stable in alkali, but polythionates decompose in strong alkali to form various products including thiosulfate, sulfite, sulfate, and elemental sulfur.

In strong alkali, some disulfides, such as cystine, form persulfides and other products.[1,2] Reaction with cystine in peptide combination is found above pH 8.[3,4] Also, sulfide ion reacts with cystine and cystinyl peptides to form persulfide groups. The persulfide groups on proteins are somewhat stabilized by the influence of neighboring groups.[5,6] Persulfide

---

[1] G. S. Rao and G. Gorin, *J. Am. Chem. Soc.* **24**, 749 (1957).
[2] J. P. Danehy, *Chem. Org. Sulfur Compd.* **2**, 337 (1966).
[3] D. S. Tarbell and D. P. Harnish, *Chem. Rev.* **49**, 11 (1951).
[4] N. Catsimpoolas and J. L. Wood, *J. Biol. Chem.* **239**, 4132 (1964).
[5] D. Cavallini, G. Federici, and E. Barboni, *Eur. J. Biochem.* **14**, 169 (1970).
[6] J. F. Schneider and J. Westley, *J. Biol. Chem.* **244**, 5735 (1969).

groups are generated in the catalytic cycles of the enzymes rhodanese[7] (EC 2.8.1.1, thiosulfate sulfurtransferase), mercaptopyruvate sulfurtransferase[8] (EC 2.8.1.2), and γ-cystathionase[9] (cystathionine γ-lyase, EC 4.4.1.1) acting on cystine. Other enzyme persulfides are found in a number of iron–sulfur proteins.[10] Persulfides readily react with nucleophilic reagents such as cyanide and sulfite ions,[11] decompose to free sulfur and thiols, and undergo alkylation.[12] The persulfide groups transfer to suitable acceptors, e.g., sulfite. The transfer is enhanced by the enzymes rhodanese and mercaptopyruvate sulfurtransferase.[13,14] Persulfides are formed as intermediates in the cyanolysis of polysulfides.[14]

Evidence for persulfide groups can be obtained from a broad band of absorbance at 335–350 nm. Unfortunately, the absorption coefficient is only about 310 at 335 nm. In practicality, sulfane sulfurs are identified by susceptibility to cyanolysis.

Cold Cyanolysis

Principle

Cyanolysis involves nucleophilic attack on a sulfur–sulfur bond by cyanide.[11] The uncatalyzed reaction indicates sulfane only if thiocyanate or thiosulfate is produced. (See also this volume [6].) The latter can be converted to thiocyanate which is easily determined by formation of the ferric thiocyanate complex.

Persulfides, polysulfides, aryl thiosulfonates, and higher polythionates quantitatively react with cyanide in alkaline solution at temperatures between 10° and ambient room temperature ("cold cyanolysis"). The rate of cyanolysis varies with the nature of the substituent groups of the compound[15] and the pH.[4] Some sulfanes are so unreactive as to require "hot cyanolysis" for analytical purposes. Cold cyanolysis of thiosulfate can be done if cupric ion is added as a catalyst.[16]

---

[7] A. Finazzi-Agrò, G. Federici, C. Giovagnoli, C. Cannella, and D. Cavallini, *Eur. J. Biochem.* **28,** 89 (1972).
[8] A. Meister, P. E. Fraser, and S. V. Tice, *J. Biol. Chem.* **206,** 561 (1954).
[9] D. Cavallini, C. DeMarco, B. Mondovi, and B. Gori, *Enzymologia* **22,** 161 (1960).
[10] J. L. Wood, *in* "Structure and Function Relationships in Biochemical Systems" (F. Bossa, E. Chiancone, A. Finazzi-Agro, and R. Strom, eds.), p. 327. Plenum, New York, 1982.
[11] O. Foss, *Acta Chem. Scand.* **4,** 404 (1950).
[12] M. Flavin, *J. Biol. Chem.* **237,** 768 (1962).
[13] B. Sörbo, *Biochim. Biophys. Acta* **7,** 32 (1953).
[14] R. Abdolrasulnia and J. L. Wood, *Bioorg. Chem.* **9,** 253 (1980).
[15] R. Mintel and J. Westley, *J. Biol. Chem.* **241,** 3381 (1966).
[16] B. Sörbo, *Biochim. Biophys. Acta* **23,** 412 (1957).

Cyanolysis is best carried out at pH 8.5–10. Depending on the compound, completion of the reaction requires from 5 to 40 min.[17] The reactivity of sulfane sulfur toward cyanolysis is approximately as follows: persulfide > polysulfide > thiosulfonate > polythionate thiosulfate > elemental sulfur.

Reagents

Potassium cyanide, 0.5 M
Ammonium hydroxide, 1 M
Hydrochloric acid, 0.2 M
Formaldehyde, 38%
Goldstein's reagent[18]: Dissolve 50 g of ferric nitrate nonahydrate in 500 ml of water, add 525 ml of nitric acid, specific gravity 1.4, and dilute to 2000 ml.
Standard potassium thiocyanate, 1 μmol/ml: Dilute with oxygen-free water to 0.1 μmol/ml before use.

Procedure

To the sample in 0.5 ml of water or 0.2 M hydrochloric acid are added 0.4 ml of 1 M ammonium hydroxide, 3.6 ml of water, and 0.5 ml of 0.5 M potassium cyanide. After 45 min at room temperature, 0.1 ml of 38% formaldehyde and 1 ml of Goldstein's reagent are added. The absorbance is determined at 460 nm. A blank is prepared excluding the sample. A standard curve is prepared with 0.25 and 0.75 μmol of potassium thiocyanate.

Comment

Ammonium hydroxide may be omitted if the sample is alkaline initially. Formaldehyde tends to stabilize the ferric thiocyanate complex by reacting with excess cyanide. Formaldehyde may be omitted in many instances. Light causes fading of the ferric thiocyanate color. The intensity increases with the concentration of ferric ion and with addition of organic solvents such as acetone. Very dilute potassium thiocyanate standard solutions over long periods of time undergo oxidation from absorbed air.

Hot Cyanolysis

Hot cyanolysis provides for analysis of alkyl thiosulfonates, thiosulfate, and trithionate, which react too slowly for cold cyanolysis.[19]

---

[17] J. C. Fletcher and A. Robson, *Biochem. J.* **87**, 553 (1963).
[18] F. Goldstein, *J. Biol. Chem.* **187**, 523 (1950).
[19] A. Gutman, *Z. Anal. Chem.* **47**, 294 (1908).

Tetra- and higher polythionates can be determined at 40° by cyanolysis according to the method of Koh[20] but most determinations are carried out at the temperature of a boiling water bath.

Reagents

Potassium cyanide, 0.1 M
Ammonium hydroxide, 1 M
Ferric nitrate reagent[21]: Dissolve 100 g of ferric nitrate nonahydrate in water, add 200 ml of nitric acid, specific gravity 1.4, and dilute to 1000 ml.

Procedure

To 0.5 ml of sample are added 0.2 ml of 1 M ammonium hydroxide and 0.7 ml of 0.1 M potassium cyanide. The solution is heated for 30 min in a boiling water bath. To the cooled solution is added 0.5 ml of ferric nitrate reagent and water to a final volume of 5 ml. The optical density is determined at 460 nm against an unheated blank.

Comment

Carbonic acid derivatives such as dixanthyldithiocarbamyl and diaryl disulfides react with cyanide to form thiocyanate and monosulfides.[11]

$$R_2NCS_2^{2-} + CN^- \rightarrow R_2N\underset{\underset{S}{\|}}{C}\underset{\underset{S}{\|}}{C}NR_2 + SCN^-$$

Thiosulfate and polythionates can be determined by cold cyanolysis if the reaction with cyanide is catalyzed by cupric ion. Apparently, the procedure has not been reported for the alkyl thiocyanates.

Cold Cyanolysis in Presence of Cupric Ion

Principle

Cyanolysis of polythionates, other than trithionate, occurs at room temperature or below.[12,19] Thiosulfate and $n-1$ thiocyanate ions are the products. The reactivity increases with the chain length.[2]

$$^-O_3SS_4SO_3^- + 3\ CN^- + OH^- \rightarrow 3\ SCN^- + {}^-SSO_3^- + HSO_4^- + H_2O$$

Catalysis by cupric ion is necessary for the quantitative conversion of thiosulfate to thiocyanate.

$$^-SSO_3^- + CN^- \xrightarrow{Cu^{2+}} SO_3^{2-} + SCN^-$$

[20] T. Koh and I. Iwasaki, *Bull. Chem. Soc. Jpn.* **39**, 703 (1966).
[21] E. L. Cosby and J. B. Sumner, *Arch. Biochem.* **7**, 457 (1945).

Reagents

Potassium cyanide, 0.1 M
Cupric chloride, 0.1 M
Ferric nitrate reagent: see above.

Procedure[16]

The sample, containing up to 1.5 μEq of sulfane sulfur, is adjusted to the blue color of thymophthalein (about pH 10). To 4.2 ml is added 0.5 ml of 0.1 M potassium cyanide and, after mixing, 0.3 ml of 0.1 M cupric chloride with vigorous mixing to form a finely dispersed precipitate. Ferric nitrate reagent, 0.5 ml, is added, and the suspension is mixed until the precipitate dissolves. The absorbance is determined at 460 nm against a blank prepared by adding the reagents in the order ferric nitrate reagent, potassium cyanide, and cupric chloride. One μEq of sulfane sulfur yields a corrected absorbance of about 0.58.

Comment

The absorbance of the blank should be less than 0.01. If the mixture is not vigorously mixed after addition of cupric chloride, the precipitate formed is not dissolved completely by the ferric nitrate reagent. Sulfide and thiols must be removed by preliminary treatment with cadmium chloride. The concentration of cyanide (10 mM) is optimum for this determination. A number of ions interfere with the determination, e.g., phosphate, 0.01 M, 25%; zinc ion, 60%; sulfite, 0.01 M, 30%.[16]

The use of this method to determine thiosulfate in urine has been reported.[22,23]

[22] V. E. Shih, M. M. Carney, and R. Mandell, *Clin. Chim. Acta* **95**, 143 (1979).
[23] B. Sörbo and S. Öhman, *Scand. J. Clin. Lab. Invest.* **38**, 521 (1978).

[8] Sulfide Determination: Ion-Specific Electrode

By RANDOLPH T. WEDDING

Principle. The decrease in sulfide ion from a solution is followed by monitoring the negative potential produced by a sulfide ion-specific electrode. This electromotive force is a logarithmic function of the concentration of S^{2-}.

Equipment. A sulfide ion-selective electrode (Orion Model 94-16A) is used with a double-junction calomel reference electrode (Orion Model 90-

02-00). The electrodes can be attached to a meter designed for use with selective ion electrodes or directly to a recording potentiometer giving full-scale deflection in the range 10–1000 mV. A bias or offset voltage must be supplied to permit reading a wide range of sulfide ion concentrations. A battery and potentiometer can provide a suitable range of bias voltage, and some recorders have adequate built-in offsets. The reaction cell consists of a water-jacketed Lucite block drilled to fit both electrodes and contains 1 ml of reaction medium. The temperature is controlled by circulating water from a temperature-regulated bath.

As an illustrative example, the sulfide electrode may be used for assay of O-acetyl-L-serine sulfhydrylase[1–3] or cysteine synthase[4] [EC 4.2.99.8, O-acetylserine (thiol)-lyase] as follows.

Reagents

Tris–HCl buffer, 0.1 M, pH 7.6
O-Acetyl-L-serine, 50 mM
Sodium sulfide, 10 mM, in the Tris buffer at pH 7.6

Procedure. An assay of 1.0 ml volume is made with 0.5 ml of Tris buffer, 0.2 ml of O-acetyl-L-serine, and 0.25 ml of sodium sulfide solution plus enzyme and water as needed to make 1.0 ml. The reading is brought on scale using the bias voltage, and the reaction is initiated by adding the required volume of enzyme. The change in EMF is recorded using a chart speed suitable to the enzyme activity used.

Data Reduction. Extraction of the rate of the enzyme reaction from the recordings of sulfide electrode potential is complex. While it can be accomplished by hand, it is best programmed for a computer. The steps in the process[1,4] are as follows:

1. Construct a standard curve of EMF versus the standard calomel electrode against the negative log of [S^{2-}]. Under the conditions described here this gives a linear relationship with a negative slope between sulfide concentrations of 10 mM and 1 μM. A line fitted to the linear portion of this standard curve gives the equation

EMF vs. Calomel electrode = $718.9 - 41.1(\log [S^{2-}])$

2. Using this equation, a point by point conversion of EMF to [S^{2-}] will accurately describe the time course of the reaction,[5] and the point at

---

[1] P. F. Cook and R. T. Wedding, *J. Biol. Chem.* **251**, 2023 (1976).
[2] B. L. Bertagnolli and R. T. Wedding, *Plant Physiol.* **60**, 115 (1977).
[3] See also this volume [81] and [82].
[4] P. F. Cook and R. T. Wedding, *Arch. Biochem. Biophys.* **178**, 293 (1977).
[5] T. T. Ngo and P. D. Shargool, *Anal. Biochem.* **54**, 247 (1973).

which the reaction ceases to be linear with time is readily determined. The slope of this secondary plot of $[S^{2-}]$ versus time over the period of linearity, determined by any of the usual methods, provides the rate of the reaction. The line can be fitted to a linear regression, and the slope used directly in calculating the rate.

3. Because of the volatility of the sulfide substrate, accurate measurements of the rate of sulfide disappearance due to enzyme activity require a correction for loss of sulfide from the assay. The rate of sulfide loss (starting at the same concentration as in the assay) is determined in the same way, and the rate of volatilization is subtracted from the observed enzyme rate.

Alternate Assay. While the assay with the sulfide electrode is accurate and (with appropriate equipment) rapid, a simpler method may be used if the enzyme cysteine desulfhydrase (EC 4.4.1.1, cystathionine γ-lyase) is available. This involves the following reactions:

$$O\text{-Acetyl-L-serine} + S^{2-} \rightarrow \text{L-cysteine} + \text{acetate} \tag{1}$$
$$\text{L-Cysteine} + H_2O \rightarrow \text{pyruvate} + NH_3 + H_2S \tag{2}$$
$$\text{Pyruvate} + NADH \rightarrow \text{lactate} + NAD^+ \tag{3}$$

The cysteine produced by O-acetylserine sulfhydrylase in reaction (1) is converted to pyruvate by cysteine desulfhydrase in (2), and the pyruvate is reduced by lactate dehydrogenase in (3), oxidizing NADH in the process, which can be followed at 340 nm in a spectrophotometer. Assay conditions like those described above may be used with the addition of appropriate quantities of the two coupling enzymes and 0.3 mM NADH to the assay.

With this assay the disappearance of NADH is followed, the rate determined as usual, and expressed in terms of cysteine produced in reaction (1). Both methods give comparable rates.

Assay of enzymes that produce sulfide is also feasible using the methods described. If the concentration of S^{2-} is initially zero and the assay is concluded within a relatively short period, no correction for sulfide volatility is required.

[9] Hydrogen Selenide and Methylselenol

By HOWARD E. GANTHER and R. J. KRAUS

Hydrogen selenide and related forms of selenium are difficult to handle.[1,2] The absence of direct methods for identification of H_2Se led us to develop a method[3] based on the use of Sanger's reagent (1-fluoro-2,4-dinitrobenzene, FDNB) to trap H_2Se in a stable form upon volatilization from acid medium in a stream of nitrogen [Eq. (1)]. The method works equally well for volatile selenols such as methylselenol [Eq. (2)]. The selenide derivatives are extracted into benzene and identified by TLC, HPLC, or mass spectrometry. The method also can be applied to the analogous sulfur compounds, H_2S and CH_3SH.

$$H_2Se + 2\ FDNB \rightarrow DNP—Se—DNP + 2\ HF \quad (1)$$
$$CH_3SeH + FDNB \rightarrow CH_3Se—DNP + HF \quad (2)$$

Materials and Methods

The preparation of compounds used as standards has been described elsewhere.[3] $H_2{}^{75}SeO_3$ is purified to remove selenate, using ascorbic acid reduction.[3] Nitrogen is purified by means of an Oxyclear cartridge (Pierce Chemical Co., Rockford, IL). Thin-layer chromatography is done on glass plates precoated with silica gel (Brinkmann). The TLC solvent systems and R_f values are described in Table I. For visualization of the DNP compounds, the plates are illuminated with an ultraviolet lamp. To increase the sensitivity, and to make a permanent copy, the plate is covered with print paper[4] and placed over a source of 366-nm light.[5] The photocopy technique can detect as little as a few nanomoles of DNP derivatives. If a region of the plate is to be scraped to recover or to analyze the selenium compounds, this should be done prior to making the photocopy.

Almost any HPLC apparatus can be used to separate the DNP derivatives on commercially available columns using isocratic elution and ultra-

---

[1] A. T. Diplock, C. P. J. Caygill, E. H. Jeffery, and C. Thomas, *Biochem. J.* **134**, 283 (1973).
[2] H. E. Ganther, *Adv. Nutr. Res.* **2**, 107 (1979).
[3] H. E. Ganther and R. J. Kraus, *Anal. Biochem.* **138**, 396 (1984).
[4] Graphic Arts Proof Paper 503-1, DuPont.
[5] TLC Photocopier, Brinkmann Instruments.

TABLE I
R_f VALUES AND ELUTION TIMES OF DNP DERIVATIVES

Derivative	R_f (TLC)		Elution time (min) (HPLC)	
	Solvent A	Solvent B	System 1	System 2[b]
2,4-Dinitrophenol	0.02	0.04	—[a]	9
1-Fluoro-2,4-dinitrobenzene (FDNB)	0.52	0.52	5.2	13
Di-DNP-monoselenide [(DNP)$_2$Se]	0.47	0.72	16.0	41
Di-DNP-diselenide [(DNP—Se—)$_2$]	0.61	0.79	6.4	90
CH$_3$Se—DNP	0.52	0.60	4.0	—[c]
Di-DNP-monosulfide [(DNP)$_2$S]	0.41	0.70	24.0	34
Di-DNP-disulfide [(DNP—S—)$_2$]	0.59	0.79	8.0	84

[a] Was not eluted.
[b] V. Tsiagbe and N. J. Benevenga (unpublished data).
[c] CH$_3$S—DNP had an elution time of 19 min.

violet absorption to monitor the separation. Table I gives the retention times for two systems. System 1 consists of a 5-μm Nucleosil NH$_2$-bonded column (4 × 200 mm) made by Macherey-Nagel (Rainin Instrument Co., Woburn, MA) eluted with n-heptane : chloroform (80:20, v/v) at 1.25 ml/min. System 2 is a reversed-phase column (4 × 250 mm, Alltech) eluted with acetonitrile : H$_2$O (50:50, v/v), containing 0.18 ml trifluoroacetic acid per ml, at a flow rate of 1 ml/min. The DNP derivatives of selenium and sulfur compounds have molar extinction coefficients[6] on the order of 10^4. They are susceptible to degradation by light, and the appropriate precautions[7] should be taken when handling them to minimize decomposition. Fractions can be collected and analyzed directly for Se by graphite furnace atomic absorption spectrometry using Zeeman background correction[8] or analyzed fluorometrically after wet ashing.[9]

Conclusive identification of the DNP—Se derivative is easily accomplished using mass spectrometry. To purify the compounds for this purpose, apply the benzene extract (containing about 10 μg of the compound) to a silica gel plate and chromatograph in the benzene–pyridine system (Table I). Allow the plate to dry, then scrape the plate at the desired R_f, and elute the compound with a 1:1 mixture of benzene–pyridine. Con-

[6] H. E. Ganther, R. J. Kraus, and S. J. Foster, this series, Vol. 107, p. 582.
[7] I. Smith, ed., "Chromatographic and Electrophoretic Techniques," Vol. 1, p. 143. Wiley (Interscience), New York, 1960.
[8] R. J. Kraus, S. J. Foster, and H. E. Ganther, *Biochemistry* **22**, 5853 (1983).
[9] S.-H. Oh, H. E. Ganther, and W. G. Hoekstra, *Biochemistry* **13**, 1825 (1974).

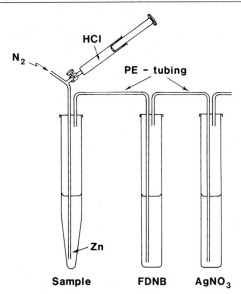

FIG. 1. Selenium volatilization apparatus. The sample vial is a 16 × 119 mm polystyrene centrifuge tube (Sarstedt Inc., Princeton, NJ, Catalog No. 55.461). This vial is fitted with a polyethylene cap (Sarstedt Catalog No. 65.816) having two holes drilled for inlet and outlet tubing (PE 240, O.D. 2.4 mm, Clay-Adams, Parsippany, NJ). The inlet tubing (160 mm) is fitted with a plastic Y-tube connector just above the cap; one arm of the Y-tube is connected to the N_2 gas supply, and the other arm is connected to a syringe for injecting the HCl. A hemostat or pinch clamp is attached below the syringe. The tip of the inlet tubing is placed near the bottom of the vial. The outlet tubing (300 mm) leads to trap 1, consisting of a 15 × 110 mm glass counting vial fitted with a polyethylene cap having two holes drilled for inlet and outlet tubing. A second glass vial is connected to trap 1 with a piece of polyethylene tubing (300 mm).

centrate the sample by evaporation at room temperature and transfer to a glass capillary using a 10-μl syringe (Hamilton). Mass spectra are obtained by direct probe analysis; 1–2 μg of compound is sufficient.

Selenium Volatilization Procedure

Volatile selenium is generated in a disposable apparatus (Fig. 1) using a procedure similar to that described by Diplock *et al.*[1] The selenium is trapped in scintillation counting vials so that ^{75}Se-labeled samples can be assayed directly in a well-type gamma counter.

Reagents

Dimethylformamide (DMF) (Mallinkrodt, St. Louis, MO). Redistill before use.

2,4-Dinitrofluorobenzene (trapping solution 1). Dissolve 150 μl FDNB (Sigma Chemical Co.) in 10 ml of redistilled DMF; protect from light. Pipette 1 ml of this stock solution, 0.4 ml distilled water, and 0.6 ml DMF into trapping vial 1 and add 14 mg of $NaHCO_3$.

0.1 M $AgNO_3$ (trapping solution 2). Dissolve 1.7 g of $AgNO_3$ in 100 ml of distilled water. Protect from light. Pipette 3 ml into trap 2.

Procedure

Place 350 mg of zinc dust, sample plus water (total volume 2 ml), and 2 drops of *n*-octanol in the sample vial. Connect vial in series to trap 1 (FDNB) and trap 2 (3 ml 0.1 M $AgNO_3$), and bubble N_2 gas through the entire system at a rate of 200 ml/min for 10 min. Without interrupting the gas flow, inject 3 ml of 12 N HCl as rapidly as possible into the sample vial, and continue gassing with N_2 for another 10 min. Remove the trapping vials and assay for ^{75}Se using a well-type scintillation counter. Add 1 ml of 1 N HCl to trap 1, and extract the contents 3 times with 4 ml of benzene; concentrate the pooled extracts by rotoevaporation. The benzene extracts are assayed for ^{75}Se and subjected to TLC, HPLC, or mass spectrometry to identify the selenium derivative.

Comments

The particular advantage of this method is that it permits the unequivocal identification of hydrogen selenide or other volatile selenols as stable, benzene-soluble derivatives. Other methods based on trapping with heavy metals[1,10] or arsenite,[11] or on reduction of 5,5′-dithiobis(2-nitrobenzoic acid)[10] are relatively nonspecific and are subject to possible interferences.

The technical problems involved in handling H_2Se have been discussed elsewhere.[1,2] We have used the procedure of Diplock *et al.*[1] for volatilization of H_2Se by reduction with Zn–HCl, as opposed to procedures based on borohydride reduction. There is some carryover of HCl fumes from the volatilization vial, but these are neutralized by the high concentration of bicarbonate in trap 1, so that the pH is maintained sufficiently high to trap the volatile selenol in the dissociated form. The reaction with FDNB is rapid and complete within 10 min,[3] and almost no H_2Se is recovered in the back-up trap containing $AgNO_3$. With large amounts (1 μmol or more) of H_2Se, a transitory orange color is formed rapidly in trap 1, followed by a return to the yellow color. This color change is probably

[10] N. Esaki, T. Nakamuro, H. Tanaka, and K. Soda, *J. Biol. Chem.* **257,** 4386 (1982).
[11] H. S. Hsieh and H. E. Ganther, *Biochemistry* **14,** 1632 (1975).

caused by the accumulation of the 2,4-DNP—Se⁻ intermediate, which then reacts with a second molecule of FDNP to form the di-DNP-monoselenide. The use of aqueous dimethylformamide rather than aqueous ethanol is preferred, since the reaction with FDNB is more rapid in this solvent.[3] Completeness of the reaction can be monitored by measuring how much Se is volatilized from trap 1 on acidification at the end of the trapping period.[3]

The percentage of ^{75}Se in trap 1 that is extractable into benzene ranges from 97 to 100%, over a sample range of 20–10,000 nmol Se.[3] Identification of the benzene-soluble selenium derivative is conveniently done by TLC, in which the major product (di-DNP-monoselenide) and a minor product (di-DNP-diselenide) have high mobility. If strictly anaerobic conditions are not maintained during volatilization, elemental Se may be formed by oxidation of H_2Se. A portion of this will be extracted into benzene but remains at the origin upon TLC.

The molecular ions of di-DNP-monoselenide and CH_3Se—DNP are prominent in both spectra.[3] For di-DNP-monoselenide the molecular ion ($m/e = 414$) is the base peak. Fragments containing the selenium isotope pattern are seen at m/e 247 (loss of DNP) and m/e 231 (DNP—Se minus oxygen). For CH_3Se—DNP the molecular ion at m/e 262 was the predominant peak, with selenium fragments at m/e 247 (loss of CH_3) and m/e 169 (loss of 2 NO_2). A peak corresponding to loss of CH_3Se was not observed, indicating a high degree of stability for the CH_3Se—DNP bond.

The formation of aniline derivatives by reduction of a nitro group on the DNP moiety is a possibility, especially if abundant quantities of H_2Se are formed, but has not been observed.[3] Reduction of nitro groups by sulfides is known to occur,[12] and reduction of aromatic nitro compounds by sodium telluride recently has been reported.[13] In applying the present procedure to the study of sulfur present in feathers (see below), V. Tsiagbe and N. J. Benevenga have detected the formation of aniline derivatives.

Any form of selenium that is reducible to hydrogen selenide can be identified using this procedure. Selenate (SeO_4^{2-}) is not reduced,[1,3] but other inorganic forms (selenite, Se^0, and selenide) as well as selenium in certain organoselenium compounds such as selenotrisulfides (RS—Se—SR) or methylselenenyl sulfides (RS—$SeCH_3$) will be released in volatile forms that are easily identified. Selenoproteins containing selenocysteine or selenomethionine residues should not release volatile sele-

[12] J. March, "Advanced Organic Chemistry," 2nd ed., p. 1125. McGraw-Hill, New York.
[13] H. Suzuki, H. Manabe, and M. Inouye, *Chem. Lett.* p. 1671 (1985).

nium,[3] but elemental selenium bound to proteins by hydrophobic interactions, as well as selenium that has reacted with protein sulfhydryl groups to form selenotrisulfide (—S—Se—S—) linkages, would be volatilizable. Determining the volatile selenium released from a putative selenoprotein would be useful in identifying true selenoproteins, such as nonheme iron proteins containing selenide, and help avoid pitfalls related to "selenoproteins" that may be artifacts. The decomposition of selenium in a selenoenzyme such as glutathione peroxidase also can be monitored. Glutathione peroxidase contains selenocysteine but is known to be very unstable in certain oxidation states[14] and releases Se in low molecular form. Selenium associated with other classes of compounds such as lipids, pigments, carbohydrates, or nucleic acids also could be studied using this procedure.

The interpretation of results obtained using the volatilization procedure must include an understanding that the ability of various selenium compounds to survive the volatilization procedure will vary. Selenium present in nucleic acids in forms such as selenopyrimidine or selenopurine derivatives, for example, might be released as inorganic selenium; this information would have to be correlated with other structural information and tests with model compounds before drawing conclusions about the form of selenium in the sample. It is also important to consider sample matrix effects, such as the presence of protein,[11] that interfere with the release of the volatile selenide. The matrix effect can be evaluated by spiking the sample with $H_2^{75}SeO_3$ (purified to remove contaminating selenate). Unlabeled selenium can be added as a carrier to samples in which ^{75}Se was used as a tracer, in order to improve the recovery of ^{75}Se.[3]

Other volatile forms of selenium, such as dimethyl selenide, are not trapped by FDNB; dimethyl diselenide is partly lost during the passage of N_2 prior to reduction and does not react with FDNB, but 75% was recovered as CH_3Se—DNP.[3] Identification of the common volatile forms of selenium can be accomplished by using sequential trapping solutions of 8 N HNO_3 (which traps dimethyl selenide but not H_2Se)[1] and FDNB (to trap H_2Se and CH_3SeH). We also have used benzyl chloride in xylene solution to trap dimethyl selenide, in order to permit assay of the volatile selenium by liquid scintillation counting.[15]

The procedure described here was developed for selenium, but also has applications to sulfur and tellurium compounds. We have applied the procedure to assay forms of selenium present in purified proteins such as

---

[14] H. E. Ganther and R. Kraus, this series, Vol. 107, p. 593.
[15] S. J. Foster, R. J. Kraus, and H. E. Ganther, *Arch. Biochem. Biophys.* **251**, 77 (1986).

glutathione peroxidase (see above) as well as forms of selenium isolated from urine. The procedure also has been used by others in this institution to characterize the forms of sulfur present in feathers of chickens fed various levels of methionine (V. Tsiagbe and N. J. Benevenga, unpublished data).

Acknowledgments

Contribution from the College of Agricultural and Life Sciences, University of Wisconsin—Madison. Research supported by the National Institutes of Health (AM 14184).

[10] Polarography of Sulfur Compounds

By HOWARD ADLER and JOHN WESTLEY

Many sulfur compounds, both inorganic and organic, exhibit electrochemical behavior that is useful for analytical purposes. Simple polarography with a dropping mercury electrode can provide a versatile method for the simultaneous identification and quantitative determination of several such materials in the same test solution. Nearly all of the common inorganic ions of sulfur, with the exception of the highly oxidized forms such as sulfate and dithionate, are polarographically active, and various organic forms, including thiols, disulfides, and persulfides, yield well-defined polarographic signals. So also do many nonsulfur compounds, such as cyanide and borohydride, that are useful reagents in sulfur biochemistry. The following account summarizes the electrochemical basis for polarographic determinations and lists some references to more complete treatments. It then goes on to describe typical procedures and the behavior of common sulfur compounds in practical polarographic systems.

Polarography

Components

A polarograph consists of three essential components:

1. A variable, direct current source of electrical potential with a span of at least 3 V. At its simplest, this can be an ordinary 6-V lantern battery attached to a 10-turn potentiometer with the null position at midscale.

Motorized advancement of the potentiometer provides a smooth continuous change of potential difference. Modern instruments sold commercially provide convenient all-electronic control of the applied potential.

2. A vessel or "cell" containing an electrolyte solution to which the test sample is added, with accommodations for the insertion of at least two electrodes. One is a polarograph capillary that delivers a continuous succession of tiny mercury drops, each of which serves as a polarizable microelectrode during the time it remains attached to the capillary column of mercury. The other is a relatively large, nonpolarizable electrode; for small current flows, an ordinary saturated calomel reference electrode with a relatively low resistance junction, such as a porous ceramic plug, is satisfactory.[1] The two electrodes are connected to the opposite poles of the potentiometer output in such a way that the potential of the dropping mercury electrode can be varied between $+0.3$ and -2.7 V relative to the saturated calomel electrode. Since oxygen is electroactive, it interferes with many determinations. Therefore, it is convenient also to have an arrangement that permits purging of the sample by passage of a stream of nitrogen or other inert gas through the sample solution in the cell.

3. A sensitive detector of current flow in the system. A strip chart recorder can be used for this component, with different resistors inserted to give full-scale deflections corresponding to 0.5–20 μA.

All polarographs sold commercially have these basic components and usually much more, primarily to extend the convenience and the range of use.

Principles and Applications

One analytically useful phenomenon that occurs in polarographic systems is oxidation–reduction behavior. The constantly renewed surface of the growing mercury drop can function as a cathode, serving as a source of electrons for the reduction of oxidized materials in the sample. When the solution contains no species that are reducible at the mercury surface, only a very small "residual current" flows, despite the substantial electrolyte concentration, because of the rapid polarization of the mercury drop. However, when a reducible compound is present, a greater current flows whenever the applied voltage is more negative than the redox potential of that compound.

---

[1] Systems that must accommodate large current flows employ three electrodes: The dropping mercury, an opposite electrode designed for the purpose, and a separate reference electrode. In simple two-electrode systems, current flows in excess of about 20 μA are often significantly distorted by the resistance of the liquid junction in common calomel electrodes.

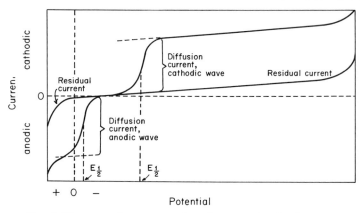

FIG. 1. Diagrammatic representation of a conventional polarogram.

The result is that a conventional residual polarogram, which is a plot of current flow as a function of applied potential in the absence of polarographically active substances, is essentially a straight line of low slope over most of the analytical range (Fig. 1). A polarogram of a solution containing a reducible compound, however, contains a sigmoid step or "wave." The potential at mid-step, the "half-wave potential" ($E_{1/2}$), is characteristic of the compound being reduced. For reversible systems, $E_{1/2}$ is simply the redox potential of that compound. The polarogram of a mixed solution of several reducible compounds having different redox behaviors features several successive waves, each relating to the presence of a particular component.

In ordinary electrolyte solutions, a voltage limitation is reached near −2.0 V, as the alkali metal ions start to be reduced. The range can be extended to beyond −2.5 V by the use of tetraalkylammonium salts as electrolytes. In acid solutions polarography reaches an earlier voltage limit caused by the reduction of H^+.

As might be expected, the reverse electrode process can also occur. The reduced forms of redox compounds can be oxidized at the mercury surface, which in this case acts as a anode, accepting electrons from each compound at applied potentials more positive than its redox potential. This process results in an electron flow through the system in the direction opposite that in the reductive process. Compounds such as lipoate, which participate in fully reversible redox processes with the mercury electrode, give "cathodic waves" for the oxidized forms and "anodic waves" for the reduced forms, with the same $E_{1/2}$ value. Compounds with redox behavior that is not fully reversible at the mercury surface may yield different $E_{1/2}$ values for their cathodic and anodic waves, but polarography is not for this reason any less useful for their identification and analysis.

If the solution in the polarograph cell is not stirred, the magnitude of each current that flows is limited by the rate of diffusion of the relevant compound from the bulk solution to the surface of the mercury drop. Since the diffusion rate is directly proportional to concentration, the polarographic "diffusion current" (the wave height) is also proportional to concentration. In practical analysis, diffusion currents measured from polarograms of test solutions are compared with those obtained with appropriate standard solutions to obtain quantitative estimates of concentrations.

A second analytically useful electrolytic phenomenon occurring in polarographic systems involves compounds that form stable complexes or insoluble precipitates with the ions of mercury. As the dropping mercury electrode is made positive relative to the saturated calomel electrode, there is an increasing tendency for Hg^+ and Hg^{2+} ions to enter the solution. The positive voltage limitation in polarography at about +0.3 V is caused by this phenomenon. Understandably, the ionization of mercury from the electrode can be facilitated by materials that combine with these ions. The applied potential at which the (anodic) current increases because of the ionization is thus a function of the affinity of the active solution component for mercury ions. Just as in the polarography of redox compounds, the magnitude of the anodic wave observed is proportional to the concentration of the active component, and multiple waves are obtained with mixtures of active components. Inorganic thiosulfate, sulfite, sulfide, and cyanide are examples of compounds that display characteristic anodic diffusion currents caused by complex formation. Although the half-wave potentials of these currents are not related to redox potentials and may be somewhat concentration dependent, they are no less useful for analytical purposes.

The applications of conventional polarographic analysis take advantage of all of the aforementioned properties of the system. The results with basic polarographic equipment are the qualitative identification (from the observed $E_{1/2}$ values) of a great variety of oxidizable and reducible compounds as well as mercury ion-reactive species, and a quantitative estimate of each concentration (from the magnitudes of the diffusion currents), with a precision of the order of 1% in the concentration range 10–5000 μM. Sample volumes of less than 1.0 ml can easily be used in a simple apparatus with cells made by drilling Lucite blocks, and purging through a fine polyethylene tube. Special cells that can handle substantially smaller volumes have also been used.

Both the precision and the sensitivity of the analyses can be improved by various electronic means that permit sampling the polarographic current at specified times and circumstances in the lifetime of the mercury drop. Excellent detailed accounts of both the basic theory and the refined practice of polarography are available,[2-4] as are also extensive collections of electrochemical data.[5]

Typical Procedure

In work with biochemical samples, it is useful to employ as solvent a supporting electrolyte that is a buffer with nonvolatile components. For example, a mixture of equal volumes of 0.1 M NaH_2PO_4 and 0.1 M $Na_2B_4O_7$ is a convenient, well-buffered electrolyte solution at pH 8.5.

A measured volume (typically 1.0 ml) of such a buffer is placed in a suitable polarograph cell equipped with dropping mercury and calomel electrodes and deaerated by passage of a stream of nitrogen through it for 5 min. The nitrogen is then diverted to sweeping the gas phase directly above the solution, and the mercury is started (4 ± 1 sec/drop). The potential applied to the mercury relative to the calomel electrode is then varied across the desired range within the span +0.2 to −2.0 V (with this supporting electrolyte) during the course of 2–5 min in conventional polarography. The current versus voltage curve recorded is the residual polarogram.

At this point, a sample for analysis, typically 0.01–1.0 μmol in 1–10 μl, is added directly to the solution in the cell. After a brief purging to remove the small amount of oxygen introduced with the sample, the current versus voltage curve is recorded as before. The difference between the residual and sample polarograms is then available for interpretation, directly if electronic recording is used, otherwise by superposition and manual measurement. Since the sample is virtually unaltered during the polarographic procedure, further additions can be made and successive polarograms recorded with the same sample.

When a new sample is desired, the cell is emptied, with care to control the disposal of mercury, the cell and electrodes are rinsed with a stream of water from a wash bottle, and another volume of supporting electrolyte is introduced for deaeration. It is often unnecessary to repeat recording residual polarograms with each fresh volume of buffer, as they are ordinarily quite reproducible.

[2] L. Meites, "Polarographic Techniques," 2nd ed. Wiley (Interscience), New York, 1966.
[3] P. Zuman and C. L. Perrin, "Organic Polarography." Wiley (Interscience), New York, 1969.
[4] A. J. Bard and L. R. Faulkner, "Electrochemical Methods." Wiley, New York, 1980.
[5] L. Meites and P. Zuman, "Electrochemical Data." Wiley (Interscience), New York, 1974.

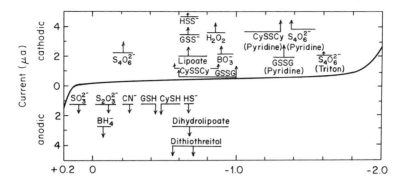

FIG. 2. Half-wave potentials of a variety of sulfur compounds and reagents used in sulfur biochemistry. The supporting electrolyte was a buffer at pH 8.5 containing phosphate and tetraborate, both at 0.05 M (residual current curve shown). Polarograms were obtained for the indicated materials at 0.5 mM.

Once the polarographic waves corresponding to particular components have been established, it is entirely feasible to make kinetic observations in this system by maintaining the voltage constant at the potential of full current flow for the component of interest (say, 0.1 V beyond $E_{1/2}$) and recording the change in current (reflecting concentration) with time. Obviously, in systems having several polarographically active components that change with time, it is necessary to buttress such observations with voltage scans at appropriate intervals.

Sulfur Compounds

The positions of half-wave potentials for a number of common sulfur compounds and reagents useful in sulfur biochemistry are indicated in Fig. 2. All of these materials have been much studied and, in general, their $E_{1/2}$ values are dependent on the pH and composition of the supporting electrolyte. The line through the center of the figure is an experimental residual current curve for the phosphate/borate buffer at pH 8.5, used for the present determinations. The $E_{1/2}$ values for cathodic waves in this particular system are displayed above the residual curve; anodic waves are indicated below the curve. Some compounds give double waves, of which the first (least negative $E_{1/2}$) is usually the best defined. The indicated current scale (10 μA full scale) is ordinarily suitable for determinations at concentrations of 0.2 to 1 mM.

It will be noted that many sulfur compounds exhibit half-wave potentials between -0.6 and -0.7 V. Some additional discrimination of these materials can often be gained by adding pyridine to the sample in the cell (10 μl/ml), a procedure that moves the $E_{1/2}$ values of many disulfides [e.g., cystine and glutathione disulfide (GSSG)], but not the corresponding persulfides, to much more negative potentials.

Interface Effects

Application of the methods of Stricks and Kolthoff[6] to the effect of pyridine on the half-wave potentials obtained for most net-anionic disulfides indicates that it is caused by electrocapillary effects at the mercury–buffer interface. The annoying fact that many polarographically active materials yield polarograms containing extrema, rather than the expected sigmoid waves, is also caused by adsorption and related phenomena at the interface. Fortunately, such "maxima," which hinder analytical work, can usually be suppressed by incorporation of a surface-active material into the supporting electrolyte. For example, Triton X-100 (10 μl of a 0.2% aqueous solution added to a 1.0-ml analytical sample) is often effective. Ionic detergents and other substances as diverse as thymol and proteins have also been used as maximum supressors.

Acknowledgments

This work was supported by research grants from the National Science Foundation (DMB-8512836) and the National Institutes of Health (GM-30971).

[6] W. Stricks and I. M. Kolthoff, *J. Am. Chem. Soc.* **74**, 4646 (1952).

[11] Spectrophotometric Assay of Thiols

By PETER C. JOCELYN

This chapter summarizes the chief methods in use for thiols in general. These methods exploit either of the two principal properties of the SH group, namely its capacity for oxidation or substitution. Reviews on this subject are found in Refs. 1–6.

[1] E. S. Guzman-Barrón, *Adv. Enzymol.* **11**, 201 (1951).
[2] F. P. Chinard and L. Hellerman, *Methods Biochem. Anal.* **1**, 1 (1954).
[3] R. Benesch and R. E. Benesch, *Methods Biochem. Anal.* **10**, 43 (1962).
[4] P. C. Jocelyn, "Biochemistry of the SH Group." Academic Press, London, 1974.
[5] J. F. Riordan and B. L. Vallee, this series, Vol. 25, p. 449.
[6] A. F. Habeeb, this series, Vol. 25, p. 457.

Preliminaries

Protein Precipitation. Assay of nonprotein thiols is normally preceded by removal of proteins. Much used reagents for their precipitation are (all at 2–3%) (1) trichloroacetic acid; (2) metaphosphoric acid; (3) sulfosalicylic acid; and (4) perchloric acid. The following considerations affect the choice to be made: reagents (1)–(3) are deliquescent solids; reagents (1) and (2) are not stable for long in solution; (2) and (3) are established antioxidants; (1) is extractable in ether; and (4) gives a potassium salt of low solubility, a property that may be an advantage in lowering the perchlorate concentration in extracts but may also cause problems with assays. Reagent (4) is also said to induce massive oxidation of thiols in erythrocyte extracts,[7] but no oxidative losses were reported from homogenates or standards even after heating at 100°.[8] The author has found perchloric acid a convenient precipitant for nonprotein thiol assays.

Choice of Buffer. Cacodylate buffer rapidly depletes thiols and must not be used for their assay.[9] Imidazole buffers may be unsuitable since they allow coupled autoxidation of thiol with NAD(P)H.[10] Addition of 0.1 mM EDTA is recommended in all cases.

Methods Based on Thiol Oxidation

Oxidizing agents may be of analytical value if they do not oxidize thiols further than disulfides (i.e., to sulfinic or sulfonic acids) and also give absorbance changes of sufficient magnitude. Aromatic disulfides fulfill these requirements best, and, as they are the reagents of first choice, they will be described in some detail. However, there may be circumstances which preclude their use, and previously described methods using other oxidants will also be reviewed.

Aromatic Disulfides

General Features. Aromatic disulfides dominate the field because they are easily reduced by thiol–disulfide exchange, the equilibrium overwhelmingly favoring the formation of aromatic thiol.

$$ArSSAr + 2\ RS^- \rightarrow RSSR + 2\ ArS^-$$

Resonance involving the ArS$^-$ group and the aromatic ring (resulting in thione formation in the case of the dithiopyridines) produces big shifts in

[7] S. Srivastava and E. Beutler, *Anal. Biochem.* **25**, 70 (1968).
[8] G. S. Tarnowski, R. K. Barclay, I. M. Mountain, M. Nakamura, H. G. Satterwhite, and E. M. Solney, *Arch. Biochem. Biophys.* **110**, 210 (1965).
[9] K. B. Jacobson, J. B. Murphy, and B. Das Sarma, *FEBS Lett.* **22**, 80 (1972).
[10] D. P. Baccanari, *Arch. Biochem. Biophys.* **191**, 351 (1978).

absorption maxima toward longer wavelengths, thus forming the basis for colorimetry.

Mixed disulfides are stable intermediate products with protein thiols, but with nonprotein thiols they are themselves rapidly reduced. In either case, the overall stoichiometry is one aromatic thiol generated for each biological thiol originally present. The design of particular reagents needs to meet the requirements of water solubility and appropriate values for λ_{max} and ε_m.

The first substance meeting these criteria, introduced by Ellman in 1955[11] and still by far the most popular, is 5,5'-dithiobis(2-nitrobenzoic acid) (DTNB) (**I**). Also in use are the heterocyclic disulfides 2,2'-dithiodipyridine (2,2'-DTP) (**II**) and 4,4'-dithiodipyridine (4,4'-DTP) (**III**).[12] (The last two are inappropriately named by their manufacturer as Aldrithiols.)

Reagents

DTNB, 10 mM: 398 mg solid is dissolved in 1 ml 0.5 M phosphate buffer at pH 7.2 and diluted to 100 ml with the same buffer. Kept in a dark bottle at 0° this is stable indefinitely, but it rapidly develops a yellow color if exposed to light at room temperature.[13] DTNB can be dissolved in an aqueous medium without buffer by carefully and slowly titrating a suspension in water with alkali[13] or with Tris base. However, if the pH is locally allowed to rise momentarily above 9, the disulfide is cleaved hydrolytically with formation of yellow thiol. Alternatively it can be stored indefinitely at 0° in

---

[11] G. L. Ellman, *Arch. Biochem. Biophys.* **82**, 70 (1959).
[12] D. R. Grassetti and J. F. Murray, *Arch. Biochem. Biophys.* **119**, 41 (1967).
[13] K. Brocklehurst and G. Little, *Biochem. J.* **133**, 67 (1973).

ethanol or dimethyl sulfoxide[13a] (nonvolatile) but care must be taken to provide for neutralization in the assay buffer.

2,2'-DTP and 4,4'-DTP: 40 mg of the solids are added to 50 ml water, stirred for several hours, and filtered to give solutions of approximately 1.5 and 3 mM, respectively. These are stable at 0°.[13] They can also be added at higher concentrations as solutions in water-miscible organic solvents.

Procedures. A convenient procedure for DTNB-based assays of non-protein thiols is to add 0.5 ml of the reagent dissolved in neutralizing buffer (e.g., 0.5 M potassium phosphate at pH 7.2) to the acid thiol extract (up to 0.5 ml containing 0–100 nmol thiol) at room temperature in a 1-ml cuvette. This will bring the mixture close to the buffer pH. The solution is stirred, and the absorbance measured at 412 nm after at least 2 min. (With 2.5% perchloric acid extracts, precipitation of perchlorate does not begin until somewhat later than this and subsequent brief centrifuging may be necessary.)

The absorbance does not change appreciably over a short time, 2–10 min, but two factors operate to alter the value over prolonged periods. There is a slow decline in absorbance due to autoxidation of the aromatic thiol and, opposing this, a slow increase due to hydrolysis. The former is delayed by the inclusion of EDTA in the buffer, and the latter by working near pH 7, at which the hydrolytic rate is only 10% of that at pH 8.[14] Blank absorbance values with the reagents alone must be determined and subtracted. They are small with freshly prepared reagents but increase with age.

For the product from reduction of DTNB, namely 5-mercapto-2-nitrobenzoate (MNB), ε_m is usually given as 13,600–13,800 M^{-1} cm^{-1} at 412 nm, but a careful study of the repurified substance has yielded a value of 14,150 M^{-1} cm^{-1} that is unchanged between at least pH 7.0–8.6.[14]

The procedure for dithiodipyridines is similar. For the thiol formed from 2,2'-DTP, ε_m is 7,060 M^{-1} cm^{-1} at 343 nm and for that from 4,4'-DTP it is 19,800 M^{-1} cm^{-1} at 324 nm.[11–13]

The amount of aromatic thiol produced, calculated from ε_m is, of course, equivalent to the thiol being assayed. With simple biological thiols the reaction goes to completion when reagent disulfide and thiol are initially present in the ratio of 1:2[15] showing that the intermediate mixed disulfide is easily reduced by thiol. This mixed disulfide has, however,

---

[13a] M. Marshall and P. P. Cohen, *J. Biol. Chem.* **255**, 7291 (1980).
[14] P. W. Riddles, R. L. Blakeley, and B. Zerner, *Anal. Biochem.* **94**, 75 (1979).
[15] Ming Man and R. G. Bryant, *Anal. Biochem.* **57**, 429 (1974).

been isolated in low yield using glutathione (GSH) as the thiol and with DTNB or 2,2'-DTP in excess.[16,17]

Factors Governing Choice of Aromatic Disulfide Reagent. The pK of the SH group of MNB is 4.4.[13] Since absorbance depends on the MNB anion concentration, the value for ε_m, therefore, declines with decreasing pH below 6. The thiol to be assayed itself reacts as RS$^-$, and so the rate also declines with decreasing pH, especially below pH 8–9.5 where the pK values for most biological thiols are to be found. However, as DTNB hydrolyzes spontaneously above pH 9, the rate varying with the composition of the medium,[18] the permissible pH range of the assay medium is normally 6.5–8.5. Sensitive thiols such as cysteine may undergo oxidation if maintained for long at these values.

DTNB is the more useful with turbid solutions since light scattering declines with increasing wavelength. Although color can be fully discharged by adding acid, thus allowing a correction to be made for turbidity, this is not recommended since the reagent tends to precipitate, thereby contributing to the turbidity. A better method is to decolorize by adding an excess of N-ethylmaleimide. More elegant still is a dual-wavelength spectrophotometer recording of the difference between absorption at 412 and at 520 nm.[19]

A possible future reagent is the selenium analog of DTNB, which is resistant to alkaline hydrolysis, even in dilute alkali, and is reduced by thiols to give an absorption peak of ε_m 10,000 M^{-1} cm^{-1} at 432 nm.[20]

For 2,2'-DTP and 4,4'-DTP, the pK of their corresponding thiols is 9.97 and 8.7, respectively,[13] but the absorbing species are thiones rather than thiol anions. For this reason, ε_m does not fall with decreasing pH, and rate alone limits the choice over the range 2.5–8.5. This feature of the dithiodipyridines allows them to be used for selective determination of thiols that have low pK values or are capable of thione forms. Thus, ergothioneine can be assayed at pH 1 in the presence of other thiols which are, of course, totally inactive.[21] As noted, 4,4'-DTP has a much higher ε_m than 2,2'-DTP and is nearly 1.5 times more sensitive than DTNB. Despite these differences, 2,2'-DTP may be the preferred reagent because of the greater electrophilicity of its monocation (pK for RH$^+$ is 2.15 as against

---

[16] S. Erikkson, P. Askelof, K. Axelsson, I. Carlberg, C. Guttenberg, and B. Mannervik, *Acta Chem. Scand., Ser. B* **B28,** 922 (1970).
[17] J. L. Aull and H. H. Daron, *Biochim. Biophys. Acta* **614,** 31 (1980).
[18] K. Hiramatsu, *Biochim. Biophys. Acta* **490,** 209 (1977).
[19] E. J. Harris and H. Baum, *Biochem. J.* **186,** 725 (1980).
[20] N. P. Luthra, R. B. Dunlap, and J. D. Odom, *Anal. Biochem.* **117,** 94 (1981).
[21] J. Carlsson, M. P. J. Kierstan, and K. Brocklehurst, *Biochem. J.* **139,** 221 (1974).

4.8 for 4,4'-DTP).[22] Another consideration is that with 4,4'-DTP the assay wavelength (343 nm) is near to the maximum absorbance of reduced nicotinamide nucleotides which accordingly may interfere.

The rate of reaction using either DTNB or DTP (either isomer) varies with the ionic strength and composition of the medium in ways which depend on the ionic environment around the SH group being assayed.[9]

DTNB reacts rapidly also with inorganic sulfide with a 1 : 1 stoichiometry.[23] Other reagents that cleave disulfide groups also assay as if they were thiols, e.g., $NaBH_4$, sulfite, cyanide, and even thiosulfate[15] (see this volume [6] and [7]). There is usually no reduction by NADPH or ascorbate. However, if cupric ions are present as well as ascorbate, DTNB is reduced, apparently by metal-stabilized monohydroascorbate radicals.[24]

Special Procedures for Protein Thiols. DTNB and the dithiodipyridines react with accessible SH groups in proteins. The reaction differs from that with simple thiols in that the mixed disulfide, protein—SS—Ar is the end product, but as before the aromatic thiol produced is stoichiometric with the SH content.

Although local environments may allow variations of a millionfold in rates of reaction with DTNB,[25] SH groups in pure proteins are usually described as either fast reacting, slow reacting, or unreactive in their response to this reagent.

The protein (or cell fraction) in buffer at pH 7.2 to 8.0 is treated with a 20- to 30-fold excess of DTNB over the amount expected to react, and the absorbance increase is recorded continuously. For very fast reacting groups, stopped-flow instrumentation may be necessary. A biphasic or polyphasic semilog plot of absorbance against time is obtained. The slow rate that may continue for perhaps 1 hr can be used to correct the fast early rate to allow partition and calculation of numbers of fast and slow reacting groups. Details have been presented in this series.[6] SH groups of GSH and bovine albumin can be distinguished by assaying the former with 1 mM DTNB at pH 7, and both together by working with 10 mM DTNB at pH 8.[26] A similar system has also been used as a model for kinetic analysis.[27]

Suitable denaturing agents for inclusion when total SH groups are

---

[22] J. P. Malthouse and K. Brocklehurst, *Biochem. J.* **185,** 217 (1980).
[23] R. P. Hausinger and J. B. Howard, *J. Biol. Chem.* **258,** 13486 (1983).
[24] W. L. Baker and K. L. Smiley, *Biochem. J.* **218,** 995 (1984).
[25] G. H. Snyder, M. J. Cennerazzo, A. J. Karalis, and D. Field, *Biochemistry* **20,** 6509 (1981).
[26] P. C. Jocelyn, *Biochem. J.* **85,** 480 (1962).
[27] A. F. Boyne and G. L. Ellman, *Anal. Biochem.* **46,** 639 (1972).

required are 1% sodium dodecyl sulfate (SDS) (final concentration), 8 M urea and 4 to 6 M guanidinium chloride. These reagents do not affect the absorbance due to the released MNB. Reaction under such conditions is usually complete in a few minutes. Simultaneous denaturation, precipitation, and thiol assay by addition of methanol containing DTNB[6] is also being found useful.[28]

The ability of ArSSAr to react with inorganic sulfide implies that there may be difficulties when iron–sulfur centers are present. Normally such groups will be inaccessible to the reagent but damage to the protein may release some sulfide and thereby give spurious values for SH groups.[23,29]

A List of Protein SH Assay Conditions. Conditions used since 1973 for a number of different proteins with aromatic disulfides are shown in Table I together with the number of SH groups determined.

Mixed Disulfides

PROTEIN—SS—AR. Titration of proteins with DTNB or dithiodipyridine labels the SH groups by formation of protein—SS—Ar. Separation from the reagent and addition of a simple thiol then leads to release of 1 ArSH per protein SH. This can be used as an alternative or confirmatory method. Insoluble proteins, which can be removed from the medium by centrifugation and washing, present an obvious example to which this technique can be readily applied, e.g., for purified membranes such as those from erythrocytes or platelets.[30] In other cases, gel filtration would be required.[31] Appropriate releasing conditions (see Table I) include the use of dithiothreitol (0.1–2 mM at pH 7.5–8.0). Protein—SS—Ar are also useful because the sulfur–sulfur bond can be cleaved asymmetrically, with sulfite, for example, forming protein—S—SO$_3^-$ with stoichiometric release of ArS$^-$.[32]

AR—SS—R. An alternative method for assaying SH groups, and one that achieves S-substitution, involves the use of synthesized nonprotein mixed disulfides of the form ArSSR. These react with protein SH groups to yield the corresponding S-substituted product while releasing an equivalent of aromatic thiol. The method can be extended to substances of the form ArSR although, in this case, some protein—SS—Ar may also form

---

[28] M. F. Roberts, R. L. Switzer, and K. R. Schubert, *J. Biol. Chem.* **250**, 5364 (1975).
[29] R. N. Thorneley and R. R. Eady, *Biochem. J.* **133**, 405 (1973).
[30] Y. Ando and M. Steiner, *Biochim. Biophys. Acta* **311**, 26 (1973).
[31] G. Voordouw, S. M. Van Der Vies, C. Veeger, and K. J. Stevenson, *Eur. J. Biochem.* **118**, 541 (1981).
[32] G. R. Stark and L. V. Crawford, *Nature (London) New Biol.* **237**, 146 (1972).

as it does with 2-nitro-5-thiocyanobenzoate. Table II illustrates reagents which have been used in this way.

$$PrS' + ArS(S)R \rightarrow PrS(S)R + ArS'$$

Other Methods Using Aromatic Disulfides. Analogs of DTNB containing either positive or negative groups have been synthesized and, from their rate of reaction, conclusions have been drawn as to the environment around the SH group.[33] Another way of studying this is to add to the protein DTNB and, in turn, a different catalytic disulfide. The latter may be reduced to thiol by protein SH groups inaccessible to DTNB itself and then be reoxidized back by the reagent. Rates differ with the ionic charge on the specific catalytic disulfide.[34]

DTNB and dithiopyridines may themselves differ in their rate of reaction with proteins for the same reason. Thus, DTNB (charged) reacts with hemoglobin SH groups (it has to be assayed at 450 nm in this case because of heme absorption). The data show that a positive charge near the SH group increases the rate of reaction and a negative one decreases it; there is no such effect therefore with 2,2'-DTP (neutral).[35]

Heterocyclic disulfides much more active than 2,2'-DTP—at least with the SH groups of lactate dehydrogenase—have been described recently. One of them, 2,2'-dithiodiazole, on reduction generates absorption at 310 nm with ε_m of 14,800 M^{-1} cm^{-1}.[36]

Other Organic Oxidants

Although various other organic oxidants lack the specificity of the aromatic disulfides, in many biological assays thiols are the only significant reducing substances present and reagent specificity is not required. Hence some of the oxidants used analytically in the last 20 years that remain commercially available are described.

Iodosobenzoic Acid. Iodosobenzoic acid (**IV**) is currently used for oxidizing SH groups, especially in enzymes[37,38] where it has been consid-

---

[33] G. Legler, *Biochim. Biophys. Acta* **405**, 136 (1975).
[34] J. M. Wilson, D. Wu, R. Motiu-De Grood, and D. J. Hupe, *J. Am. Chem. Soc.* **102**, 359 (1980).
[35] B. E. Halloway, B. E. Hedlund, and E. S. Benson, *Arch. Biochem. Biophys.* **203**, 332 (1980).
[36] V. Schmidt, G. Pleiderer, and F. Bartkowiak, *Anal. Biochem.* **138**, 217 (1984).
[37] C. S. Andreo, R. A. Ravizzini, and R. H. Vallejos, *Biochim. Biophys. Acta* **547**, 370 (1979).
[38] M. Ariko and A. E. Shamoo, *Biochim. Biophys. Acta* **734**, 83 (1983).

TABLE I
ASSAY OF PROTEIN SH GROUPS BY AROMATIC DISULFIDES

Protein	Fast SH	Total SH	Assay[a]	PrSSar[b]	Ref.[c]
Alanyl-tRNA synthetase (subunit) (alanine–tRNA ligase)	2	6	B 0.05–0.46, pH 7.6, S		pp
Aldehyde dehydrogenase			P 0.012		ss
5-Aminolevulate dehydratase (porphobilinogen synthase)	2	4	B 0.1–0.5, pH 6.8		a
ATPase, complex 5			B or P		b
Na$^+$,K$^+$-ATPase (dogfish)	6	34	B 10, pH 7.4, S		hh
ATPase, F$_1$ (*E. coli*)		19	B 0.05, pH 8, U		qq
ATPase (sarcoplasm)	5	20	B 1, pH 8, S		aa
Chloroplast CF$_1$	8	14	P, pH 8, U		ee
CoA transferase			B 0.4, pH 8.1	DT-2	c
Complement C2		1	B 0.1, pH 8.2, Gu		rr
Dihydrofolate reductase	1	2	B, pH 7.2, Gu	DT	d
Elongation factor G	1	4	B 0.08, pH 8, Gu S		e
Erythrocyte membrane			B 0.1, pH 8, S		f
Fibronectin (200 kDa)	1.5		B or P 0.2, pH 7.4, U S		g, h
Fructose-bisphosphate aldolase		32	B, pH 3.5, S		y
Fumarylacetoacetase (subunit)	2	3	B 0.1, pH 7.3, U		jj
Glutamylcysteine synthetase (glutamate–cysteine ligase)		3–4	B 0.067, pH 8, U S		x
Glutathione transferase	2	4	B 0.05, pH 7		ll
Glycerol-3-phosphate dehydrogenase	2	8	B 0.25, pH 7.4		gg
Histone H3		1	B, pH 6.6		kk
Lipase	1	2	B		i

Malate dehydrogenase (subunit)		3	B 0.067, pH 8, S Gu	z
Malic enzyme	7+		B 0.08, pH 7	bb
Microsomes (brain)			B 1.25	j
Myosin		40	B 0.4, pH 6.5, U	u
Ornithine transcarbamylase (subunit) (ornithine carbamoyltransferase)	1		B 10, pH 8	k
5-Oxoprolinase	6	27	B	l
Oxyhemoglobin			P 1–6, pH 7–8	m
Pantoate dehydrogenase	2	8	B 1, pH 7, S	ff
3-Phosphoglycerate kinase	2	7	B 0.1–0.2, pH 7.5, U	n
Platelet membranes			B or P 0.25, pH 8.1	o
PRPP-synthetase	3	4	B 2–5, pH 7.5	p
Pyruvated diphosphate kinase		18	B 0.5, pH 7.5	q
Pyruvate kinase		4	B 0.5, pH 8, U	ii
Rhodanese (thiosulfate sulfurtransferase)	1	4	B, pH 8	oo
Rhodopsin (bovine rod)	2	6	B	mm
Serine hydroxymethyltransferase	2	8	B 0.5, 90°, S	r
Succinyl-CoA synthetase (dimer)	2	4	B 1, pH 8	nn
Thiosulfate reductase (thiosulfate–thiol sulfurtransferase)		1	B, pH 7.4–8.6 S	dd
Thymidylate synthase	2	3	B 1, pH 7.3, S	t
Transhydrogenase (*Azobacter*) [NAD(P)+ transhydrogenase]	1	3	B, Gu	s
Trytophanyl-RNA synthetase (tryptophan–tRNA ligase)		1	B 0.2, pH 7.8, S	w
Urease		15	B 0.22, pH 7.3	cc
Xanthine oxidase	1		B 0.1	v

Additional columns: DT (Malate dehydrogenase), DE (Myosin), DT (PRPP-synthetase), Na$_2$S$_2$O$_3$/KCN (Rhodanese), DT (Thymidylate synthase), DT (Trytophanyl-RNA synthetase).

[a] Concentrations are mM. B, DTNB; P, dithiodipyridine; Gu, guanidine hydrochloride; S, sodium dodecyl sulfate; U, urea.

[b] Protein—SS—aromatic mixed disulfide isolated. Cleavage agent is given as follows: DT, dithiothreitol; DE, dithio-erythritol.

(*continued*)

c References to Table I:
a. J. S. Seehra, M. G. Gore, A. G. Chaudhry, and P. M. Jordan, *Eur. J. Biochem.* **114**, 263 (1981).
b. C. Godinot, D. C. Gautheron, Y. Galante, and Y. Hatefi, *J. Biol. Chem.* **256**, 6776 (1981).
c. H. White, F. Solomon, and W. P. Jencks, *J. Biol. Chem.* **251**, 1700 (1976).
d. M. N. Williams and C. D. Bennett, *J. Biol. Chem.* **252**, 6871 (1977).
e. R. L. Marsh, G. Chinali, and A. P. Parmeggiai, *J. Biol. Chem.* **250**, 8344 (1985).
f. C. W. Haest, D. Kamp, and B. Deuticke, *Biochim. Biophys. Acta* **643**, 319 (1981).
g. D. F. Mosher and R. B. Johnson, *J. Biol. Chem.* **258**, 6595 (1983).
h. D. E. Smith, D. F. Mosher, R. B. Johnson, and L. T. Furcht, *J. Biol. Chem.* **257**, 5831 (1982).
i. F. Benkouka, A. A. Guidoni, J. D. De Carlo, J. J. Bonicel, P. A. Desnuelle, and M. Rovery, *Eur. J. Biochem.* **128**, 331 (1982).
j. E. C. Dinovo, B. Gruber, and E. P. Noble, *Biochem. Biophys. Res. Commun.* **68**, 975 (1976).
k. M. Marshall and P. P. Cohen, *J. Biol. Chem.* **255**, 7291 (1980).
l. J. M. Williamson and A. Meister, *J. Biol. Chem.* **257**, 9161 (1982).
m. N. Makino and Y. Sugita, *J. Biol. Chem.* **257**, 163 (1982).
n. I. Cserpan and M. Vas, *Eur. J. Biochem.* **131**, 157 (1983).
o. Y. Ando and M. Steiner, *Biochem. Biophys. Acta* **311**, 26 (1973).
p. M. F. Roberts, R. L. Switzer, and K. R. Schubert, *J. Biol. Chem.* **250**, 5364 (1975).
q. H. Yoshida and H. G. Wood, *J. Biol. Chem.* **253**, 7650 (1978).
r. F. Gavilanes, D. Petersen, and L. Schirch, *J. Biol. Chem.* **257**, 11431 (1982).
s. G. Voordouw, S. M. Van Der Vies, C. Veeger, and K. J. Stevenson, *Eur. J. Biochem.* **118**, 541 (1981).
t. P. C. Plese and R. B. Dunlap, *J. Biol. Chem.* **252**, 6139 (1977).
u. H. Wiedner, R. Wetzel, and F. Eckstein, *J. Biol. Chem.* **253**, 2763 (1978).

v. R. Hille and V. Massey, *J. Biol. Chem.* **257**, 8898 (1982).
w. G. V. Kuehl, M. L. Lee, and K. H. Muench, *J. Biol. Chem.* **251**, 3254 (1976).
x. G. F. Seelig and A. Meister, *J. Biol. Chem.* **259**, 3534 (1984).
y. M. K. Offerman, M. J. McKay, M. W. Marsh, and J. S. Bond, *J. Biol. Chem.* **259**, 8886 (1984).
z. R. Scheibe, *Biochim. Biophys. Acta* **788**, 241 (1984).
aa. M. Ariki and A. E. Shamoo, *Biochim. Biophys Acta* **734**, 83 (1983).
bb. V. A. Reddy, *Biochim. Biophys. Acta* **743**, 268 (1983).
cc. P. W. Riddles, R. K. Andrews, R. L. Blakeley, and B. Zerner, *Biochim. Biophys. Acta* **743**, 115 (1983).
dd. T. R. Chauncey and J. Westley, *Biochim. Biophys. Acta* **744**, 304 (1983).
ee. C. S. Andreo, R. A. Ravizzini, and R. H. Vallejos, *Biochim. Biophys Acta* **547**, 370 (1979).
ff. T. Myohanen and P. Mantsala, *Biochim. Biophys. Acta* **614**, 266 (1980).
gg. R. E. Smith and R. Macquarrie, *Biochim. Biophys. Acta* **567**, 269 (1979).
hh. M. Esmann, *Biochim. Biophys. Acta* **688**, 251 (1982).
ii. D. P. Bloxham, S. J. Coghlin, and R. D. Sharma, *Biochim. Biophys. Acta* **525**, 61 (1978).
jj. M. P. Nagainis, W. Pu, B. Cheng, K. E. Taylor, and D. E. Schmidt, *Biochim. Biophys. Acta* **657**, 203 (1981).
kk. J. Palau and J. R. Daban, *Biochim. Biophys. Acta* **536**, 323 (1978).
ll. T. Carne, E. Tipping, and B. Ketterer, *Biochem. J.* **177**, 433 (1979).
mm. W. J. DeGrip, G. L. DeLar, F. J. Daemen, and S. L. Bonting, *Biochim. Biophys. Acta* **325**, 315 (1973).
nn. A. R. Prasad, J. Ybarra, and S. Nishimura, *Biochem. J.* **215**, 513 (1983).
oo. B. Pensa, M. Costa, L. Pecci, C. Cannella, and D. Cavallini, *Biochim. Biophys. Acta* **484**, 368 (1977).
pp. Z. Chen, J. P. Kim, C. Lai, and A. H. Mehler, *Arch. Biochem. Biophys.* **233**, 611 (1984).
qq. H. Stan-Lotter and P. D. Bragg, *Arch. Biochem. Biophys.* **229**, 320 (1984).
rr. M. A. Kerr and C. Parkes, *Biochem. J.* **219**, 391 (1984).
ss. T. M. Kitson, *Arch. Biochem. Biophys.* **234**, 487 (1984).

TABLE II
Some ArS—R Reagents

NBS: 3-carboxy-4-nitrophenyl thio-
NS: 4-nitrophenyl thio-
2PS: 2-pyridyl thio-

ArS—	—R	Ref.[a]
NS—	—S (17β)-3-hydroxy-1,3,5(10)-estratriene	a
NBS—	—SCoA	b
NBS—	—SCH$_2$C$_6$H$_5$	c
NBS—	—SCH$_2$CH$_2$OH	j
NBS—	—SG	d
NBS—	—CN	e
NBS—	—SO$_3$	f
2PS—	—SG (also polymer bound; commercial)	g
2PS—	—S—CH$_3$	k
2PS—	—S—CH$_2$CH$_2$COOH (and succinimido ester)	l
2PS—	—HgC$_6$H$_4$(2-OH)(5-NO$_2$)	h
2PS—	—S—(CH$_2$)$_2$NH—(4-NO$_2$-benzofurazan)	i

[a] References to Table II:
a. M. Ikeda, *Biochim. Biophys. Acta* **718**, 66 (1982).
b. G. Williamson and P. C. Engel, *Biochem. J.* **211**, 559 (1983).
c. J. Henkin, *J. Biol. Chem.* **252**, 4293 (1971).
d. S. A. Ericksson and B. Mannervik, *Biochim. Biophys. Acta* **212**, 518 (1970).
e. M. Silvergerg, J. T. Dunn, L. Garen, and A. P. Kaplan, *J. Biol. Chem.* **255**, 7281 (1980).
f. T. W. Thannhauser, Y. Konishi, and H. A. Scheraga, *Anal. Biochem.* **138**, 181 (1984).
g. J. L. Aull and H. H. Daron, *Biochim. Biophys. Acta* **614**, 31 (1980).
h. B. S. Baines and K. Brocklehurst, *Biochem. J.* **179**, 701 (1979).
i. T. Stuchbury, M. Shipton, R. Norris, J. P. Malthouse, K. Brocklehurst, J. A. Herbert, and H. Suschitzky, *Biochem. J.* **151**, 417 (1975).
j. R. J. Ackerman and J. F. Robyt, *Anal. Biochem.* **50**, 656 (1972).
k. T. Kimura, R. Matsueda, Y. Nakagawa, and E. T. Kaiser, *Anal. Biochem.* **122**, 274 (1982).
l. J. Carlsson, H. Dreven, and R. Axén, *Biochem. J.* **173**, 723 (1978).

ered as a reagent for vicinal thiols.[39] It is possible that with isolated SH groups, a sulfenic acid (RSOH) is first formed, but this has not been established.[40] In a spectrophotometric assay, the thiol is added to a known amount of the reagent. The reagent is reduced to iodobenzoate with consequent loss of absorption at 285 nm (ε_m, approximately 2,100 M^{-1} cm^{-1}).[41]

Diphenylpicrylphenylhydrazine. The reagent (V) (0.3 ml of a solution made by dissolving 10 mg of diphenylpicrylphenylhydrazine in 1 ml acetone and diluting to 50 ml with 4% Triton X-100 at pH 6) is added to the thiol solution. It absorbs at 522 nm (ε_m, 11,800 M^{-1} cm^{-1}), and this absorbance is lost after reaction with thiol. The fall in absorbance, ε_m = 12,000 M^{-1} cm^{-1}, is independent of the structure of the thiol as determined with a number of different thiols. Interfering substances could first be allowed for by a prior incubation with mercuribenzoate. Ascorbate, quinones, and adrenaline also react.[42]

[39] E. M. Valle, N. Carrillo, and R. H. Vallejos, *Biochim. Biophys. Acta* **681**, 412 (1982).
[40] D. J. Parker and W. S. Allison, *J. Biol. Chem.* **244**, 181 (1969).
[41] J. Leslie and F. Varrichio, *Can. J. Biochem.* **46**, 625 (1968).
[42] H. M. Klouwen, *Arch. Biochem. Biophys.* **99**, 116 (1962).

Benzofuroxan. Benzofuroxan (**VI**) oxidizes thiols to disulfides and is reduced to a dioxime (**VII**), thereby generating absorption at 430 nm (ε_m, 5,200 M^{-1} cm^{-1}). The pH range for the assay is 8–10. The method has been used for the estimation of cysteine and mercaptoethanol.[43]

4,4'-Bisdimethylaminodiphenylcarbinol. The reagent (**VIII**) dissociates in acid solution to give an intense blue color which is discharged by thiols and has been used, therefore, as a microassay. The technique has been to add the thiol (in up to 22.5 μl) to 10.0 μl of the reagent (made up in acetone to give 65 mg/100 ml). The mixture is then diluted with 40 mM sodium acetate at pH 5.1 containing 4 M guanidine chloride to 1 ml, and the absorbance determined at 612 nm. However, the assay response is not linear with concentration at low (below 2 μM) or high (above 20–30 μM) levels of thiols. A calibration curve is therefore essential.

Some idea of the sensitivity is apparent from the ε_m at 612 nm of the reagent itself, which is 70,800 M^{-1} cm^{-1}.[44] It has also been used as a spray to visualize thiols on thin-layer plates.[45]

Quinones. Dichlorophenolindophenol (DCPIP) (**IX**) is reduced by thiols at pH 7 with loss of absorbance at 600 nm. Allowance can be made for ascorbate since, unlike thiols, the former remains reactive at pH 2. DCPIP is a nonspecific but sensitive reagent (at 600 nm, ε_m is 19,100 M^{-1} cm^{-1}). The reaction is a mixed one because the reagent adds to thiols as well as oxidizes them. Fast reacting compounds, e.g., thiosalicylate, are mainly oxidized (2:1 stoichiometry) whereas slow ones, e.g., GSH and cysteamine mainly add (1:1 stoichiometry). Cysteine is intermediate[46]; however, in the presence of copper ions (almost stoichiometric with the thiol) and at a pH below 6.5, oxidation may become the only significant reaction.[47] Clearly, standard curves should be prepared for individual thiols.

Trinitrobenzenesulfonic Acid. Trinitrobenzenesulfonic acid, a well-known reagent for amino groups, also reacts quantitatively with thiols at pH 7. The value of ε_m for the cysteine derivative is 14,000 M^{-1} cm^{-1} at 345 nm.[48]

---

[43] M. Shipton and K. Brocklehurst, *Biochem. J.* **167,** 799 (1977).
[44] M. S. Rohrbach, B. A. Humphries, F. Yost, W. G. Rhodes, S. Boatmen, R. G. Hiskey, and J. H. Harrison, *Anal. Biochem.* **52,** 127 (1973).
[45] R. J. Burt, B. Ridge, and H. N. Rydon, *J. Chromatogr.* **118,** 240 (1976).
[46] R. E. Basford and F. M. Huennekens, *J. Am. Chem. Soc.* **77,** 3873 (1955).
[47] D. S. Coffey and L. Hellerman, *Biochemistry* **3,** 394 (1964).
[48] A. Kotaki, M. Harada, and K. Yagi, *J. Biochem. (Tokyo)* **55,** 553 (1964).

Inorganic Oxidants

Nitroprusside. The reagent reacts with thiols to give a product absorbing at 520 nm. Although the nature of the reaction has not been established, it was popular as a colorimetric assay before the advent of DTNB. The most recent modification[50] of the original method[51] is presented. To a protein-free thiol extract in sulfosalicylic acid (3 ml) is added 50% K_2CO_3 (2 ml) containing 0.086% sodium cyanide. A 1% nitroprusside solution, 2 ml, is added, and the absorbance, which is linear with thiol concentration, is read at 525 nm.

There are great differences in color formation between individual thiols. Approximate values for ε_m (M^{-1} cm^{-1}) are as follows: thioglycolate, 23,000; homocysteine, 20,000; mercaptoethanol, 14,000; cysteine, 13,500; coenzyme A, 8,000; GSH, 4,000 (extracted from[52]). Dithiothreitol is reported to give only 4% of the color found for cysteine, allowing the latter to be assayed in the presence of the former.[53] The absorbance of some of these thiols is also considerably increased in the presence of urea or guanidine.

A disadvantage of nitroprusside is that the color is unstable and fades quickly. Moreover, nitroprusside also reacts to give the same color with acetone, acetoacetate, and creatinine.

Ferricyanide. The reduction of ferricyanide by thiol can be followed spectrophotometrically at 410 nm and pH 7 from the decrease in absorbance.[41] The ε_m value for ferricyanide is, however, only about 1,000 M^{-1} cm^{-1}. A direct assay of the ferrocyanide formed is also possible because, at 237 nm, this has an ε_m of 9,200 M^{-1} cm^{-1} whereas ferricyanide has an ε_m of only 800 M^{-1} cm^{-1}.[54] Alternatively, there is an indirect method in which the ferrocyanide formed can be converted to Prussian blue with ferric sulfate.[55]

The ferricyanide reaction is inhibited by EDTA.[41] The reagent lacks any claim to specificity (see also Singer and Kearney[56]).

---

[49] D. J. Birkett, N. C. Price, G. K. Radda, and A. G. Salmon, *FEBS Lett.* **6,** 346 (1970).
[49a] K. Nitta, S. C. Bratcher, and M. J. Kronman, *Biochem. J.* **177,** 385 (1979).
[50] E. Mortensen, *Scand. J. Clin. Lab. Invest.* **16,** 87 (1964).
[51] R. R. Grunerts and P. H. Phillips, *Arch. Biochem. Biophys.* **30,** 217 (1951).
[52] R. Benesch, R. E. Benesch, and W. I. Rogers, in "Glutathione Symposium" (S. Colowick, ed.), p. 31. Academic Press, New York, 1954.
[53] W. W. Clelland, *Biochemistry* **3,** 480 (1964).
[54] D. K. Kidby, *Anal. Biochem.* **28,** 230 (1969).
[55] H. L. Mason, *J. Biol. Chem.* **86,** 623 (1930).
[56] T. P. Singer and E. B. Kearney, *Methods Biochem. Anal.* **4,** 321 (1966).

Cupric Copper. Cupric copper has been used in conjunction with a cuprous copper reagent, 2,9-dimethyl-1,10-phenanthroline (neocupreine). The reaction mixture contains 50 μM cupric chloride and 25 μM neocupreine plus the thiol at pH 7. Absorbance develops at 455 nm from zero (ε_m ~8,600 M^{-1} cm^{-1}). The color is said to be independent of pH over the range of pH 5 to 9 and to be stable for several minutes.[57]

Permanganate. Though chemically nonspecific, permanganate has found application for oxidizing certain SH groups to sulfonic acid because, at low concentrations, it may behave in a site-specific way with enzymes utilizing phosphate as substrate. Permanganate does appear sterically similar to the phosphate ion.[28,58] Decrease in permanganate can be followed at 530 nm.[58a]

Iodine. The iodine–iodate titrimetric method[61] has fallen into disuse because of lack of specificity and the overoxidation that it causes.

Methods Based on Thiol Substitution

Substitution products exploited for spectrophotometric assay include mercurials, nitrous acids, maleimides, halides, and others.

Mercurials

The remarkable affinity of sulfhydryl (mercaptan) for mercury ions has been exploited in the design of monovalent organomercury compounds that can be used with high specificity for the colorimetric assay of thiols. Most frequently used are *p*-chloromercuribenzoic acid (pCMB), *p*-hydroxymercuribenzoic acid (pHMB), and the more easily soluble *p*-chloromercuribenzenesulfonic acid (pCMBS).

The main disadvantage of mercurials is low solubility, but they can be dissolved first in the minimum of alkali, conditions under which pCMB is converted to pHMB,[62] and then diluted with buffer to achieve the desired pH. pHMB is also available commercially as the sodium salt, which is soluble directly in neutral medium.

Reaction with thiol produces changes in the pK of the mercurial acid group, and, since the absorption characteristics of the neutral and ionized

---

[57] E. Frieden, *Biochim. Biophys. Acta* **27**, 414 (1958).
[58] W. F. Benisek, *J. Biol. Chem.* **246**, 3151 (1971).
[58a] S. R. Dickman, *Anal. Chem.* **24**, 1064 (1952).
[59] S. L. Gonias and S. V. Pizzo, *J. Biol. Chem.* **258**, 5764 (1983).
[60] S. Akerfeld and G. Lovgren, *Anal. Biochem.* **8**, 223 (1964).
[61] G. E. Woodward and E. G. Fry, *J. Biol. Chem.* **97**, 465 (1932).
[62] P. D. Boyer, *J. Am. Chem. Soc.* **76**, 4331 (1954).

forms differ, the resulting spectral shifts form the basis for spectrophotometric assay at an appropriate wavelength and pH.

In Boyer's original technique,[62] aliquots of the thiol (e.g., 0.1 ml) are added to 3 ml of solution containing pCMB. The absorbance at 250 or 255 nm increases progressively until mercuration of the thiol is complete. An obvious alternative is to add repeatedly small and equal amounts of pCMB to two cuvettes, one with (A) and one without (B) the thiol, noting the absorbance increases after each addition. When the value of the absorbance difference, $A - B$, becomes constant, the endpoint has been reached. At pH 7 and 255 nm, ε_m for pCMB is 4,400 M^{-1} cm^{-1} and for the thiol adduct is 12,000 M^{-1} cm^{-1}; the method is, therefore, reasonably sensitive.

pCMBS (and phenylmercurinitrate) are less useful in this context because the wavelength must be lowered to 235 nm and 220 nm, respectively, to achieve the same incremental increase in adsorbance. At these low wavelengths, protein absorption interferes.

Another approach is to precipitate the protein, e.g., with metaphosphoric acid, after adding an excess of pCMB and then spectrophotometrically to measure the loss of mercurial in the acid supernatant fluid. A suitable concentration of pCMB (without loss on acidification due to low solubility) is 50 μM.[62a] Alternatively, the protein-bound mercurial is assayed directly by flame photometry after acid digestion.[63]

The rate of the reaction of mercurial with thiol varies considerably depending on the anionic composition of the medium[62] and, as the precise value of the increase in ε_m may also vary, it is best to standardize the mercurial and then infer the thiol concentration from the amount that has to be added to reach the endpoint. Standardization is achieved either by iodine titration or by using the ε_m value for CMB at pH 7 of 16,900 M^{-1} cm^{-1}.[62]

Boyer's method is suitable for low molecular weight thiols, but its main use has been with protein SH groups, especially as an alternative to DTNB. Among more recent applications, it has been used to investigate SH group reactivity in lysine monoxygenase[63a] and in carboxyhemoglobin.[64]

Sometimes the unreacted mercurial may have to be removed rapidly, as when studying the rate of dissociation of subunits due to SH block-

---

[62a] T. E. Isles and P. C. Jocelyn, *Biochem. J.* **88**, 84 (1963).
[63] J. B. Carlsen, *Anal. Biochem.* **64**, 53 (1975).
[63a] T. Yamuchi, S. Yamamoto, and O. Hayaishi, *J. Biol. Chem.* **250**, 7127 (1975).
[64] Q. H. Gibson, *J. Biol. Chem.* **248**, 1281 (1973).

age.[65] This can be done within 20 sec by extraction of the mercurial with dithizone (diphenylthiocarbazone) in carbon tetrachloride.[66]

Although pCMB remains the most used mercurial, others have been devised which absorb in the visible region. Chief among them are mercury derivatives of nitrophenol.[67] For these compounds the pK of the phenolic hydroxyl group changes on mercaptide formation. Thus, 2-chloromercuri-4-nitrophenol when uncharged, i.e., in acid medium, has ε_m values of 9,300 M^{-1} cm^{-1} at 327 nm and 0 at 410 nm, whereas when charged, i.e., in alkali, it has ε_m values of 2,150 M^{-1} cm^{-1} at 327 nm and 17,400 M^{-1} cm^{-1} at 410 nm. (These values change somewhat in the presence of denaturing agents.) At pH 7, addition of a simple thiol lowers the pK to give an increase in ε_m of 6,600 M^{-1} cm^{-1} at 410 nm.[67a]

Phenolic mercurials have been used in determining differential reactivity and accessibility of SH groups in proteins.[68,69] An indirect assay using them has also been described in a study of the rate of reaction of 14 different reagents with the SH groups of aspartate transcarbamylase. The method involves assaying unreacted mercurial, after incubation with the protein, with MNB. The mercurial reacts with this thiol (from DTNB reduction), abolishing its absorption at 412 nm.[70]

Another development is the synthesis of a derivative with 2-thiopyridine [2-(2'-pyridylmercapto)mercuri-4-nitrophenol], which reacts with reactive SH groups. However the two SH groups compete so that the reaction does not go to completion.[71]

$$\text{PyrS—HgAr} + \text{RS}^- \rightarrow \text{RS—HgAr} + \text{PyrS}^-$$

One great advantage of mercurials is that their adducts with thiols are reversible. This is of especial use with enzymes when the addition of an excess of a simple thiol such as cysteine or GSH, regenerates the protein SH group and, often, lost enzyme activity. It has also been reported that mercurials such as pCMB can protect SH groups from later attack by NEM, an irreversible inhibitor.[72]

Although mercurials are more specific than other known thiol agents, evidence of inactivation of transcortin at high mercurial concentration,

---

[65] S. Subramani and H. K. Schachman, *J. Biol. Chem.* **256**, 1255 (1981).
[66] I. Fridovitch and P. Handler, *Anal. Chem.* **29**, 1219 (1957).
[67] C. H. McMurray and D. R. Trentham, *Biochem. J.* **115**, 913 (1969).
[67a] R. A. Stinson, *Biochemistry* **13**, 4523 (1974).
[68] D. R. Evans, C. H. McMurray, and W. N. Lipscomb, *Proc. Natl. Acad. Sci. U.S.A.* **69**, 3638 (1972).
[69] D. M. Miller, M. E. Newcomer, and F. A. Quiocho, *J. Biol. Chem.* **254**, 7521 (1979).
[70] D. R. Evans and W. N. Lipscomb, *J. Biol. Chem.* **254**, 10679 (1979).
[71] B. S. Baines and K. Brocklehurst, *Biochem. J.* **179**, 701 (1979).
[72] R. Haguenauer-Tsapis and A. Kepes, *Biochim. Biophys. Acta* **551**, 157 (1979).

not due to reaction with SH, has been demonstrated,[73] emphasizing the need for caution.

Nitrous Acid

Thiols react with nitrous acid, initially to form the derivative RS—NO. This absorbs at 320 nm and has been used as an assay for GSH.[74] Because of the reactivity of the reagent it lacks specificity; other groups are attacked with consequential changes in absorption spectra.

The adduct itself is quite stable, and, after removal of excess nitrous acid, that bound to the thiol group can be later removed and assayed. This forms the basis for a highly sensitive method[75] which has been used by several groups recently.[76-78] A modification suitable for both low molecular thiols and for protein SH groups will therefore be described in detail.[78]

Principle. After removing excess nitrous acid with ammonium sulfamate, that bound as RS—NO is displaced by Hg^{2+} and assayed by using it for the diazotization of sulfanilamide. The product is coupled with naphthylethylenediamine, yielding a highly colored product.

Reagents

Solution A: 1 volume 0.12% sodium nitrite plus 4 volumes 1 M sulfuric acid, prepared freshly

Solution B: 1% ammonium sulfamate

Solution C: 2 volumes 6.9% sulfanilamide in 0.4 M HCl plus 1 volume 1% mercuric chloride in 0.4 M HCl, prepared freshly

Solution D: 0.2% N-1-naphthylethylenediamine hydrochloride in 0.4 N HCl.

Technique. To 0.2 ml thiol-containing sample in 0.25 M HCl add 50 μl of solution A. After 5 min at 20°, add to this mixture, with vigorous shaking, 20 μl solution B. After a further 1 to 2 min, follow with 100 μl solution C and 150 μl solution D. Leave for at least 5 min; as much as 30 min may be needed with protein SH groups. The color that develops is stable, and its intensity is measured at 550 nm. Specificity for thiols is tested by omitting $HgCl_2$ from solution C when color does not develop.

[73] F. Le Gaillard, H. Azam, G. Favre, and M. Dautrevaux, *Biochim. Biophys. Acta* **749,** 289 (1983).
[74] D. E. Kizer and B. A. Howell, *Proc. Soc. Exp. Biol. Med.* **112,** 967 (1963).
[75] B. Saville, *Analyst* **83,** 670 (1958).
[76] D. Di Monte, D. Ross, G. Bellomo, L. Eklow, and S. Orrenius, *Arch. Biochem. Biophys.* **235,** 334 (1984).
[77] T. Higashi, N. Tateishi, A. Naruse, and Y. Sakamoto, *J. Biochem. (Tokyo)* **82,** 117 (1977).
[78] P. Todd and M. Gronow, *Anal. Biochem.* **28,** 369 (1969).

This method has two virtues: it functions in an acid medium and, therefore, removes much of the danger of thiol autoxidation, and it is considerably more sensitive than the DTNB method with ε_m values of the order of 50,000 M^{-1} cm^{-1} (i.e., 12 nmol thiol gives an absorbance of 1.0 in 0.52 ml).

Limitations of the method are not immediately apparent but they may involve the binding of nitrous acid to other groups from which it may slowly be released by hydrolysis. It is less likely that, however released, a fraction of it may be lost to side reactions with the thiol rather than participate in diazotization of the sulfanilamide.

Maleimides

The general property of thiols that allows addition to activated double bonds has been exploited most of all with N-ethylmaleimide (NEM) (R = Et). This water-soluble substance reacts readily with thiol anions to form

$$\begin{array}{c} \text{HC—C} \\ \| \\ \text{HC—C} \end{array} \begin{array}{c} \text{O} \\ \diagdown \\ \diagup \\ \text{O} \end{array} \text{N—R}$$

an adduct with loss of its absorbance peak at 305 nm. This is the basis for a method of measuring the concentration of an added thiol since the stoichiometry is 1:1 and NEM has an ε_m value of 620 M^{-1} cm^{-1} at this wavelength. The technique has been described in detail repeatedly.[5]

Disadvantages are that the value for ε_m is very small compared with most other colorimetric methods and, hence, scale amplification is a necessity if sensitivity is required. On the other hand, this may be an advantage when the thiol concentration to be measured approaches 1 mM.

The reagent, and its adducts, is not stable for long in aqueous medium at pH values above 7 due to hydrolysis of the imide to give the corresponding maleamic acid,[79] but it can be stored indefinitely as a concentrated solution in isopropanol. As it reacts with thiol anions rather than thiols it cannot be used at a pH more than 1–2 units below the pK of the thiol to be assayed.

NEM–thiol adducts give a red color when alkaline. To the thiol in 0.5 ml is added 0.5 ml of 0.2 M NEM. After 10 min, 1.5 ml 2 M sodium carbonate is added. The resulting absorbance at 520 nm is a function of

---

[79] H. Gehring and P. Christen, *Anal. Biochem.* **107**, 358 (1980).

the specific thiol. It is most intense ($\varepsilon_m \sim 10{,}000\ M^{-1}\ cm^{-1}$) with sulfide[80] whereas the cysteine adduct has an extinction less than 10% of this value.

NEM is not entirely specific for thiols. At a 10- to 60-fold molar excess it may react slowly (e.g., 1 hr) at pH 7.0–7.4 with α-amino groups in peptides and proteins (e.g., the N-terminals in hemoglobin)[80a] and at pH 8, in smaller excess, with lysine and histidine residues in proteins.[81] The moral is to keep the excess and the pH as low as is consistent with thiol derivatization. NEM also reacts rapidly not only with sulfide, but also with sulfite and thiosulfate, forming stable nonabsorbing adducts.[80]

The main present use of NEM is as a protein thiol blocking agent, e.g., for ferrochetalase[80a] and fibronectin,[82] where the advantage over other such reagents is that the binding can be followed directly spectrophotometrically.[83] The reaction can also be stopped when required by adding an excess of dithiothreitol,[82] mercaptoethanol,[72,83] or other thiol. Since the formation of the NEM adduct is irreversible, this does not interfere with the extent of SH blockage which has occurred but rapidly removes all remaining free NEM.[84]

Substitution of the R group has led to the introduction of many different maleimides into proteins. (For a review, see Ref. 85, which also describes other reagents introducing chromophoric groups.) In some of them, R is a simple alkyl or aryl group,[86] and these are useful for determining whether protein SH groups are in or near hydrophobic environments.[87] Others are bifunctional reagents,[88] and there are some with chromophoric substituents used to visualize the adduct after chromatographic separation, e.g., N-(4-hydroxy-1-naphthyl)isomaleimide[89] and N-(dimethylamino-3,5-dinitrophenyl)maleimide[90], which were developed especially for this purpose. However, none of the substituted maleimides has been applied to the colorimetric, as distinct from fluorometric, assay of thiols.

[80] R. J. Ellis, *Biochim. Biophys. Acta* **85**, 335 (1964).
[80a] D. G. Smyth, O. O. Blumenfeld, and W. Konigsberg, *Biochem. J.* **91**, 589 (1964).
[81] C. F. Brewer and J. P. Riehm, *Anal. Biochem.* **18**, 248 (1967).
[81a] H. A. Dailey, *J. Biol. Chem.* **259**, 2711 (1984).
[82] P. J. McKeown-Longo and D. F. Mosher, *J. Biol. Chem.* **259**, 12210 (1984).
[83] F. Iborra, B. Labouesse, and J. Labouesse, *J. Biol. Chem.* **250**, 6659 (1975).
[84] B. M. Schoot, S. E. Emst-De Vries, P. M. Van Haard, J. J. Depont, and S. C. Bonting, *Biochim. Biophys. Acta* **602**, 144 (1980).
[85] K. Brocklehurst, *Int. J. Biochem.* **10**, 259 (1979).
[86] R. E. Dubler and B. M. Anderson, *Biochim. Biophys. Acta* **659**, 70 (1981).
[87] R. Barr and F. L. Crane, *Biochim. Biophys. Acta* **681**, 139 (1982).
[88] J. V. Moroney and R. E. McCarty, *J. Biol. Chem.* **254**, 8951 (1979).
[89] C. A. Price and C. W. Campbell, *Biochem. J.* **65**, 512 (1957).
[90] A. Witter and H. Tuppy, *Biochim. Biophys. Acta* **45**, 429 (1960).

A number of acrylic acid derivatives, benzoylacrylate is one, have been used to derivatize SH groups in enzymes, but they have not been exploited for colorimetric assay.[91] The well-known uncoupling agents based on carbonyl cyanide phenylhydrazones form adducts with thiols absorbing at or near 264 nm: ε_m values vary with the thiol in the range of 2,000–11,000 M^{-1} cm^{-1} [92]

Halides

Fluoropyruvate reacts with thiols to give S-pyruvyl derivatives absorbing at 280–310 nm,[93] the precise wavelength maximum being dependent on the structure of the thiol. The method has been used to assay cysteamine in biological extracts using an ε_m value of 6,800 M^{-1} cm^{-1} at 300 nm: comparable values are obtained with cysteine, homocysteine, and penicillamine; GSH, however, has a much smaller ε_m value.[94]

7-Chloro-4-nitrobenzo-2-oxa-1,3-diazole, a close relative of benzofuroxan (**VI**), gives an absorbance increase with thiols at 400–550 nm, but the wavelength maximum and ε_m depend on the specific thiol. For N-acetylcysteine, the ε_m value is 13,000 M^{-1} cm^{-1} at 425 nm. With this value, the reagent has been used to assay protein SH groups.[49] If there is an amino group adjacent to the SH group, it is also substituted by S → N migration. Thus, cysteine gives N,S-disubstitution whereas GSH yields only S-substitution. S-Substitution and N-substitution can be distinguished by the characteristics of the absorption spectrum. The reaction is not straightforward; thiols are bound by addition as well as by elimination of chloride.[49a]

2-Nitro-5-azidophenylsulfenyl chloride has a photoreactive group. It is used as a 40 mM solution in acetic acid. The thiol, in 90% acetic acid, is added to an equal volume of the reagent with residual reagent then extracted with ethyl acetate. The absorbance of the aqueous phase is determined at 318 nm. The GSH derivative has an ε_m value of 11,500 M^{-1} cm^{-1} and is stable up to pH 8.[95]

N-(Iodoacetylaminoethyl)-5-naphthylamine 1-sulfonate easily forms thiol adducts which, after removal of excess reagent, have ε_m values of 6,100 M^{-1} cm^{-1} at 366 nm. This reagent is, however, chiefly of value as a fluorescent label.[96]

[91] B. M. Anderson, M. L. Tanchoco, and A. D. Pozzo, *Biochim. Biophys. Acta* **703**, 204 (1982).
[92] L. Drobnica and E. Sturdik, *Biochim. Biophys. Acta* **585**, 462 (1979).
[93] Y. Avi-Dor and J. Mager, *J. Biol. Chem.* **222**, 249 (1956).
[94] K. A. Herrington, K. Pointer, O. M. Friedman, and A. Meister, *Cancer Res.* **27**, 130 (1967).
[95] K. Muramoto and J. Ramachandran, *Biochemistry* **20**, 3376 (1981).
[96] R. F. Steiner, L. Greer, R. Bhat, and J. Oton, *Biochim. Biophys. Acta* **611**, 269 (1980).

2-Bromacetamido-4-nitrophenol, added in 20% methanol to thiols in 1 mM HCl, labels their SH groups with the nitrophenol chromophore. In proteins, spectral shifts may result from changes in the phenolic pK induced by the local environment. After total hydrolysis of proteins or peptides treated with, and then separated from, the reagent, cysteine derivatives can be assayed at 445 nm where the ε_m value is 13,500 M^{-1} cm^{-1}.[97]

Miscellaneous

Platinum salts have considerable preference for thiols, with *cis*-dichlorodiammineplatinum binding specifically to the SH group of plasma albumin.[59] The bound metal can later be assayed with stannous chloride at 403 nm.

Palladium ions react with divalent sulfur to yield products absorbing at 350 nm. Disulfides can be distinguished from thiols because they react more slowly.[60]

[97] S. D. Lewis and J. A. Shafer, *Biochemistry* **13**, 690 (1974).

[12] Fluorometric Assay of Thiols with Fluorobenzoxadiazoles

By KAZUHIRO IMAI and TOSHIMASA TOYO'OKA

Two derivatives of 7-fluoro-2,1,3-benzoxadiazole have proved useful for the assay of thiols, both in macromolecules and in compounds of smaller molecular weight, by reason of their formation of fluorescent products. Methods are presented for the analysis of such products by use of ammonium 7-fluoro-2,1,3-benzoxadiazole 4-sulfonate (abbreviated SBD-F) and 4-(aminosulfonyl)-7-fluoro-2,1,3-benzoxadiazole (ABD-F).

Determination of Thiols with SBD-F

Ammonium 7-fluoro-2,1,3-benzoxadiazole 4-sulfonate (SBD-F) is a thiol-specific, fluorogenic reagent[1] derived from the corresponding chloro analog, ammonium 7-chloro-2,1,3-benzoxadiazole 4-sulfonate (SBD-Cl).[2]

[1] K. Imai, T. Toyo'oka, and Y. Watanabe, *Anal. Biochem.* **128**, 471 (1983). The reagent is now commercially available from Dojindo Laboratories, 2861 Kengunmachi, Kumamoto-shi 862, Japan, and from Fluka A G, Buchs, Switzerland.

[2] J. L. Andrews, P. Ghosh, B. Ternai, and M. W. Whitehouse, *Arch. Biochem. Biophys.* **214**, 386 (1982).

Replacement of Cl by F increases the reactivity toward thiols about 50-fold. SBD-F is freely soluble in water and yields fluorescent thiol adducts which have several properties advantageous to their research use. The adducts are, for example, stable for more than 1 week at pH 9.5 and 0°, exhibit fluorescence at a long wavelength (515 nm), and show spectral shifts correlating to environmental hydrophobicity.[3-6]

$$\text{SBD-F} \xrightarrow[\text{Sodium Borate}]{\text{RSH}} \text{SBD-SR} \quad (1)$$

Low molecular weight thiols such as mercaptoethanol, captopril (see [47]), coenzyme A, cysteamine, cysteine, and glutathione and high molecular weight compounds such as bovine serum albumin (BSA) react with SBD-F to give derivatives detectable at levels below 1 nmol/ml.[3] The reagent is also applicable to the search for the site of S—S linkages of cystine residues in proteins.[6] Some thiols (cysteine, glutathione, and captopril) have also been determined by reacting with SBD-F followed by separation and quantitation using high-performance liquid chromatography.[4,5]

Procedure for the Determination of Thiols and Disulfides in Solution

Reduced Thiols[3]. To 1.0 ml of 10 μM thiol in 0.1 M potassium borate at pH 9.5, containing 2 mM EDTA (disodium salt), is added an equal volume of 1.0 mM SBD-F in the 0.1 M borate buffer. The solution is thoroughly mixed and heated at 60° for 1 hr. After cooling on ice, the fluorescence intensity is measured at ambient temperature with emission at 515 nm (excitation, 380 nm). The reagent blank without thiol is treated in the same manner as above. For the determination of cysteine or homocysteine, the chilled solution is adjusted to pH 2 with 2 M HCl, and the fluorescence intensity is measured at the appropriate wavelengths (cysteine: emission 505 nm, excitation 375 nm; homocysteine: emission 515 nm, excitation 380 nm).

Under these conditions, the detection limits (i.e., net fluorescence

---

[3] T. Toyo'oka and K. Imai, *Analyst* **109**, 1003 (1984).
[4] T. Toyo'oka and K. Imai, *J. Chromatogr.* **282**, 495 (1983).
[5] T. Toyo'oka, K. Imai, and Y. Kawahara, *J. Pharm. Biomed. Anal.* **2**, 473 (1984).
[6] T. Sueyoshi, T. Miyata, S. Iwanaga, T. Toyo'oka, and K. Imai, *J. Biochem. (Tokyo)* **97**, 1811 (1985).

TABLE I
DETECTION LIMITS FOR SOME THIOLS WITH SBD-F[a]

Thiol	Detection limit ($S/B = 2$, pmol/ml)[b]	
Mercaptoethanol	43	
α-Mercaptopropionylglycine	72	
Glutathione	100	
Coenzyme A	120	
Captopril	150	
Cysteamine	160	
Homocysteine	330	120[c]
N-Acetylcysteine	390	
Cysteine	2600	520[d]
Cystine	ND	
Alanine	ND	
Proline	ND	
Tyrosine	ND	
Serine	ND	
HPEA	ND	
MHPEA	ND	
BSA	9.2 μg (135 pmol/ml)[e]	

[a] Thiols and 500 μM SBD-F were reacted at 60° for 1 hr in 0.1 M borate (pH 9.5) containing 1 mM EDTA (disodium salt). From Ref. 3.
[b] At 515 nm (excitation 380 nm); S, sample fluorescence intensity; B, blank fluorescence intensity; ND, Not detected.
[c] Adjusted to pH 2 with 2 M HCl.
[d] Adjusted to pH 2 with 2 M HCl; at 505 nm (excitation 375 nm).
[e] At 495 nm (excitation 380 nm).

intensity ≥ twice the reagent blank fluorescence intensity) for most thiols are in the range of 40–2600 pmol/ml. However, detection limits for cysteine and homocysteine at pH 2 are somewhat lower (Table I).

Oxidized Thiols.[6] The oxidized thiol in 700 μl of 2.5 M potassium borate at pH 9.5, containing 4 mM EDTA (disodium salt), 50 μl of SBD-F solution (0.4 mg/ml of water), and 10 μl of tributylphosphine solution (0.2 g/ml of dimethylacetoamide or dimethylformamide), are vigorously mixed and heated at 60° for 1 hr. After cooling on ice, the reaction medium is treated as described above for reduced thiols.

Comments on the Reaction of Thiols with SBD-F. The rate of reaction of thiols with SBD-F increases gradually with increasing pH. At the pH employed, 9.5, it is important to include 1 mM EDTA to prevent metal-catalyzed oxidation of the thiols. Of the several buffers examined (ammonium, borate, carbonate, glycine, or Veronal), borate gives the highest yield of SBD-thiol. A concentration of borate buffer higher than 50 mM is necessary for completion of the reaction. The fluorescence intensities of SBD-thiols are dependent on pH, but high relative fluorescence intensities are observed from pH 2 to 12 with all the SBD-thiols except SBD-cysteine or SBD-homocysteine. For the latter, the highest fluorescence intensity is obtained at about pH 2. Considering these results, an acidic medium (about pH 2) is preferable for fluorescence measurement of thiol-containing amino acids. With an increase in temperature of 10° the reaction rate of thiols with SBD-F approximately doubles[3]; a reaction temperature of 60° permits complete derivatization in a reasonably short period.

Disulfides do not react with SBD-F and must be reduced to thiols. Because tributylphosphine does not interfere with the fluorogenic reaction and may be added directly to the reaction medium, its use is recommended.

Determination of SBD-Thiols in Plasma by HPLC (Reduced and Oxidized)

SBD-thiols may be fractionated by gradient elution HPLC on an analytical column of μBondapak C_{18} (300 × 3.9 mm, i.d., 8–10 μm) connected through a guard column of Bondapak C_{18}-Corasil (20 × 3.9 mm, i.d., 37–50 μm). The required phosphate buffer (pH 6.0) and acetate buffer (pH 4.0) used as mobile phases are prepared with 0.1 M sodium dihydrogen phosphate and 0.1 M disodium hydrogen phosphate, and 0.1 M acetic acid and 0.1 M sodium acetate, respectively. All buffer solutions are filtered through a Type HA filter (0.45 μm, Millipore), mixed with methanol, and degassed just prior to use. The flow rate of the eluent is 1 ml/min at room temperature. The void volume of the column is measured with SBD-cysteine as a marker, eluted with methanol–water (1:1). A Hitachi 650-10S fluorescence spectrophotometer, equipped with an 18-μl flow cell and 150-W xenon lamp, is used with excitation and emission set at 385 ± 10 nm and 515 ± 10 nm, respectively. Any apparatus capable of gradient elution should be suitable for the chromatography. As shown in Fig. 1, an appropriate separation of SBD-thiols was achieved using two gradient elution programs when an aliquot of a reaction medium obtained as described above was injected.

To determine plasma thiols, blood is drawn with a heparinized syringe and immediately centrifuged at 1850 g and 0° for 5 min to separate the plasma. To 0.5 ml of plasma is added 0.5 ml of 10% trichloroacetic acid, containing 1 mM EDTA (disodium salt). The mixture is immediately vortex-mixed for about 10 sec and centrifuged at 1850 g and 0° for 5 min. To

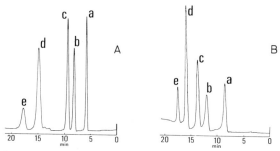

FIG. 1. HPLC separation of thiols derivatized with SBD-F. (A) Gradient elution: methanol–0.1 M sodium acetate (pH 4.0) (2:98) to methanol–0.1 M sodium phosphate (pH 6.0) (5:95) over 20 min (0–100%); (B) gradient elution: methanol–0.1 M sodium phosphate (pH 6.0) (1:99) to methanol–50 mM potassium biphthalate (pH 4.0) (5:95) over 20 min (0–100%). Fluorescence detection: at 515 nm (excitation 385 nm) at a flow rate of 1 ml/min. Key to compounds: (a) SBD-cysteine, 30 pmol; (b) SBD-homocysteine, 7.0 pmol; (c) SBD-cysteamine, 3.0 pmol; (d) SBD-glutathione, 14 pmol; (e) SBD-N-acetylcysteine, 15 pmol. From Ref. 4.

0.2-ml portions of the supernatant liquid are added 0.4 ml of 2.5 M potassium borate at pH 10.5 (prepared from 2.5 M boric acid and 2.5 M potassium hydroxide), containing 4 mM EDTA (disodium salt), and 0.2 ml of SBD-F (0.4 mg/ml 2.5 M potassium borate at pH 9.5). The solution is divided into two equal portions, to one of which is added 10 μl of tributylphosphine (0.1 ml/ml dimethylformamide) for reduction of disulfides. Each solution is vigorously mixed and allowed to react at 60° for 1 hr. A 20-μl aliquot of the reaction mixture is subjected to HPLC.

Two peaks, derived from SBD-cysteine and SBD-glutathione, were identified (Fig. 2A). Another peak co-elutes with cysteinylglycine[7] but is not yet clearly identified. The same compounds were observed in plasma treated with tributylphosphine added for the reduction of oxidized thiols (Fig. 2B).

Determination of Thiols with ABD-F

4-(Aminosulfonyl)-7-fluoro-2,1,3-benzoxadiazole[8] (ABD-F) reacts with low molecular weight thiols more rapidly than does SBD-F (e.g., the k value in min^{-1} for homocysteine is 1.16×10^{-1} with ABD-F versus 3.55×10^{-3} with SBD-F at pH 7.0 and 60°). With ABD-F a 90% complete reaction can be obtained in 5 min at 50°. ABD-F is soluble in water to ~10

---

[7] R. Tkachuk, *Can. J. Biochem.* **48**, 1029 (1970).
[8] T. Toyo'oka and K. Imai, *Anal. Chem.* **56**, 2461 (1984). The reagent is commercially available from Dojindo Laboratories, 2861 Kengunmachi, Kumamoto-shi 862, Japan.

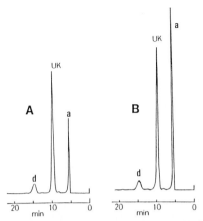

FIG. 2. Chromatograms obtained from plasma derivatized with SBD-F. (A) Plasma sample; (B) plasma sample treated with tributylphosphine. Gradient elution: methanol–0.1 M sodium acetate (pH 4.0) (2:98) to methanol–0.1 M sodium phosphate (pH 6.0) (5:95) over 20 min (0–100%). Fluorescence detection: at 515 nm (excitation 385 nm) at a flow rate of 1 ml/min. Key to compounds: (a) SBD-cysteine; (d) SBD-glutathione; (UK) unknown. From Ref. 4.

mM, and its adducts with thiols have similar spectral properties except that, to date, spectral shifts have not been correlated to environmental hydrophobicity.

$$\text{ABD-F} \xrightarrow{\text{RSH}} \text{ABD-SR} \tag{2}$$

Procedure for the Determination of Thiols in Solution[8]

Thiol (5 μM) and ABD-F (500 μM) are reacted in 2.0 ml of 0.1 M potassium borate at pH 8.0, containing 1 mM EDTA (disodium salt), at 50° for 5 min. After reaction, 0.6 ml of 0.1 M HCl is added, and fluorescence emission is measured (final pH, 2) at 510 nm (excitation, 380 nm). Under these conditions, i.e., 100-fold molar excess of ABD-F over thiol, the fluorescence intensity of cysteamine was the greatest whereas that of captopril was the smallest (Table II). The differences in the fluorescence intensities for all the thiols examined were, however, less than a factor of 3. In contrast, amino acids (alanine or proline) including the disulfide cystine did not react with ABD-F. The excitation and emission maxima of

TABLE II
Relative Fluorescence Intensities and Maximal Wavelengths for Various Thiols using ABD-F[a]

Thiol	λ_{max} (nm) Excitation	λ_{max} (nm) Emission	RFI[b]
Cysteamine	377	504	127
Mercaptoethanol	389	519	122
α-Mercaptopropionylglycine	377	505	119
Coenzyme A	389	513	118
Glutathione	381	512	108
Homocysteine	382	511	100
N-Acetylcysteine	375	508	65
Cysteine	374	500	62, 65[c]
Captopril	381	511	58
Cystine	—	—	ND
Alanine	—	—	ND
Proline	—	—	ND

[a] Thiol (5 μM) and ABD-F (500 μM) were reacted in 0.1 M borate (pH 8.0) containing 1 mM EDTA (disodium salt) at 50° for 5 min. After the reaction, 0.6 ml of 0.1 M HCl was added to 2.0 ml solution and measured (final pH, 2). Fluorescence intensity in homocysteine was arbitrarily taken as 100. From Ref. 8.
[b] Recorded fluorescence intensity at 510 nm (excitation 380 nm). ND, not determined.
[c] At 500 nm (excitation 375 nm).

the thiols were in the range of 374–389 nm and 500–519 nm, respectively (Table II).

HPLC Separation and Detection of ABD-Thiols[8]

ABD-thiols are readily separated by HPLC. A typical procedure, using an apparatus similar to that described above, follows: To a 5-ml glass tube are added 1 ml of ABD-F (1 mM) in 0.1 M potassium borate at pH 8.0 and 1 ml of mixed thiols (a few micromolar each) in the same borate buffer, containing 2 mM EDTA (disodium salt). The reaction tube is vortex-mixed, capped, and heated at 50° for 5 min. After the reaction tube is cooled on ice, 0.6 ml of 0.1 M HCl is added (final pH, 2), and a 10-μl aliquot of the acidic solution is injected into the column for HPLC. The eluting solvent is acetonitrile–potassium biphthalate (0.05 M) at pH 4.0 (8:92). The flow rate is 1 ml/min with the eluate monitored at 510 nm (excitation, 380 nm).

Retention times of ABD derivatives decreased with increased concentration of acetonitrile and at higher pH. Figure 3 shows the separation of

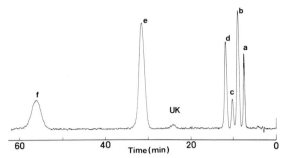

FIG. 3. Chromatogram of thiols derivatized with ABD-F. Column: μBondapak C_{18} (300 × 3.9 mm, i.d., 8–10 μm); eluent: CH_3CN–0.05 M potassium biphthalate (pH 4.0) (8:92); fluorescence detection: at 510 nm (excitation 380 nm) at a flow rate of 1.0 ml/min. Key to compounds: (a) ABD-cysteine, 8.3 pmol; (b) ABD-glutathione, 8.0 pmol; (c) ABD-N-acetylcysteine, 9.9 pmol; (d) ABD-homocysteine, 7.4 pmol; (e) ABD-cysteamine, 9.5 pmol; (f) ABD-F, about 5 nmol; (UK) unknown. From Ref. 8.

ABD-cysteine, ABD-glutathione, ABD-N-acetylcysteine, ABD-homocysteine, and ABD-cysteamine by a simple binary mixture [acetonitrile–potassium biphthalate (0.05 M) at pH 4.0 (8:92)]. Peak (f) is excess ABD-F, whereas the small unknown peak eluting at about 24 min is tentatively attributed to the hydrolysis product of ABD-F. Under these conditions, the detection limits (signal to noise ratio of 3) for ABD-cysteine, ABD-glutathione, ABD-N-acetylcysteine, ABD-homocysteine, and ABD-cysteamine are 0.6, 0.4, 1.9, 0.5, and 0.5 pmol, respectively.

Reaction of Enzymes and Proteins with ABD-F

ABD-F reacts with some, but often not with all, half-cystine residues of proteins. It is, therefore, useful for both the identification and differentiation of such residues. As described below, and shown previously for lysophospholipase L_2,[9,10] half-cystine residues buried within the protein structure can generally be reacted with ABD-F in the presence of SDS. The reaction of ABD-F with egg albumin is illustrative of the use of this reagent for protein thiols.

Preparation of ABD-Egg Albumins [ABD1- and ABD4(SDS)-Egg Albumin].[11] To 10 ml of 9.2 mM ABD-F in 0.1 M potassium borate at pH 8.0, containing 2 mM EDTA (disodium salt) and with or without 0.5% sodium dodecyl sulfate (SDS), is added 5 mg of egg albumin. The mixture is

[9] T. Toyo'oka, K. Karasawa, T. Kobayashi, I. Kudo, K. Inoue, and K. Imai, *Anal. Biochem.* (submitted).
[10] T. Kobayashi, I. Kudo, K. Karasawa, H. Mizushima, K. Inoue, and S. Nojima, *J. Biochem. (Tokyo)* **98**, 1017 (1985).
[11] T. Toyo'oka and K. Imai, *Anal. Chem.* **57**, 1931 (1985).

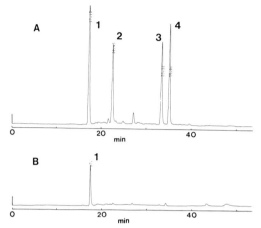

FIG. 4. Chromatograms of ABD-labeled peptides obtained from α-chymotryptic digestion of ABD-egg albumin. ABD-egg albumin (500 μg) dissolved in 500 μl of 0.1 M borate buffer at pH 8.0 was digested with α-chymotrypsin (500 μg) at 37° for 24 hr. (A) ABD4(SDS)-egg albumin; (B) ABD1-egg albumin. Column: μBondapak C_{18} (300 × 3.9 mm, i.d., 8–10 μm); gradient elution: CH_3CN–H_2O (5 : 95) containing 0.1% trifluoroacetic acid to CH_3CN–H_2O (80 : 20) containing 0.1% trifluoroacetic acid over 60 min (0–100%). Fluorescence detection: at 510 nm (excitation 380 nm) at a flow rate of 1.0 ml/min. Peak 1, ABD—Cys-Ile-Lys (No. 367–369); peak 2, Gly-Arg-ABD—Cys-Val-Ser-Pro (No. 380–385); peak 3, ABD—Cys-Pro-Ile-Ala-Ile-Met (No. 30–35); peak 4, ABD—Cys-Phe-Asp-Val-Phe (No. 11–15).

allowed to stand at 60° (or 40° without SDS) for 1 hr. To remove excess ABD-F and SDS, the solution is dialyzed at 3 ± 1° for 3 days against 6 changes of 6 liters of deionized, distilled water. With hydrophobic proteins such as lysophospholipase L_2,[10] gel-permeation chromatography (Sephadex G-15) may be more effective than dialysis. After dialysis, ABD1-egg albumin (sample treated without SDS), and ABD4(SDS)-egg albumin (sample treated with 0.5% SDS) are lyophilized.

The number of cysteine residues modified in ABD4(SDS)- and ABD1-egg albumin were determined spectrophotometrically as 4.2 and 1.1, respectively, using S-ABD-homocysteine as standard.

α-*Chymotryptic Digestion of ABD-Egg Albumins and Isolation of ABD-Labeled Peptides.*[11] Five hundred micrograms of ABD4(SDS)-egg albumin or ABD1-egg albumin dissolved in 500 μl of 0.1 M potassium borate at pH 8.0, containing 500 μg of α-chymotrypsin (Worthington Biochemicals)[12] is digested at 37° for 24 hr. After digestion, 5 μl of the solution is subjected to HPLC. The HPLC conditions and the resolution of labeled peptides are given in Fig. 4.

---

[12] Other proteolytic enzymes may be used as shown by A. D. Nisbet, R. H. Saundry, A. J. G. Moir, L. A. Fothergill, and J. E. Fothergill, *Eur. J. Biochem.* **115**, 335 (1981).

[13] Thiol Labeling with Bromobimanes

By NECHAMA S. KOSOWER and EDWARD M. KOSOWER

Some years ago, we discovered and developed convenient syntheses for a new class of heterocyclic molecules, the bimanes or 1,5-diazabicyclo[3.3.0]octadienediones.[1-5] In the course of preparing the bromo derivatives, we found fortuitously that proteins were fluorescently labeled by bromobimanes. We soon established that bromobimanes reacted preferentially with thiols and demonstrated the usefulness of such labeling for both small and large molecules in biological systems.[6-11] Many applications of the agents have been reported.[12-23] Of special importance is the

[1] E. M. Kosower, B. Pazhenchevsky, and E. Hershkowitz, *J. Am. Chem. Soc.* **100**, 6516 (1978).
[2] E. M. Kosower, J. Bernstein, I. Goldberg, B. Pazhenchevsky, and E. Goldstein, *J. Am. Chem. Soc.* **101**, 1620 (1979).
[3] E. M. Kosower and B. Pazhenchevsky, *J. Am. Chem. Soc.* **102**, 4983 (1980).
[4] E. M. Kosower, B. Pazhenchevsky, H. Dodiuk, H. Kanety, and D. Faust, *J. Org. Chem.* **46**, 1666 (1981).
[5] E. M. Kosower, B. Pazhenchevsky, H. Dodiuk, M. Ben-Shoshan, and H. Kanety, *J. Org. Chem.* **46**, 1673 (1981).
[6] N. S. Kosower, E. M. Kosower, G. L. Newton, and H. M. Ranney, *Proc. Natl. Acad. Sci. U.S.A.* **76**, 3382 (1979).
[7] N. S. Kosower, G. L. Newton, E. M. Kosower, and H. M. Ranney, *Biochim. Biophys. Acta* **622**, 201 (1980).
[8] N. S. Kosower, E. M. Kosower, J. Zipser, Z. Faltin, and R. Shomrat, *Biochim. Biophys. Acta* **640**, 748 (1981).
[9] H. Gainer and N. S. Kosower, *Histochemistry* **68**, 309 (1980).
[10] R. C. Fahey, G. L. Newton, R. Dorian, and E. M. Kosower, *Anal. Biochem.* **107**, 1 (1980).
[11] R. C. Fahey, G. L. Newton, R. Dorian, and E. M. Kosower, *Anal. Biochem.* **111**, 357 (1981).
[12] P. C. Chinn, V. Pigiet, and R. C. Fahey, *Fed. Proc., Fed. Am. Soc. Exp. Biol.* **44**, 1620 (1985).
[13] H. Borochov-Neori and M. Montal, *Biochemistry* **22**, 197 (1983).
[14] F. Vogel and L. Lumper, *Biochem. J.* **215**, 159 (1983).
[15] R. Cosstick, L. W. McLaughlin, and F. Eckstein, *Nucleic Acids Res.* **12**, 1791 (1984).
[16] C. Pande and A. Wishnia, *Biochem. Biophys. Res. Commun.* **127**, 49 (1985).
[17] T. T. F. Huang, N. S. Kosower, and R. Yanagimachi, *Biol. Reprod.* **31**, 797 (1984).
[18] D. K. Smith and J. Palek, *Blood* **62**, 1190 (1983).
[19] N. K. Burton, G. W. Aherne, and V. Marks, *J. Chromatogr. (Biomed. Appl.)* **309**, 409 (1984).
[20] A. I. Minchinton, *Int. J. Radiat. Oncol. Biol. Phys.* **10**, 1503 (1984).
[21] R. C. Fahey and G. L. Newton, *in* "Functions of Glutathione: Biochemical, Physiological, Toxicological, and Clinical Aspects" (A. Larsson *et al.*, eds.), Nobel Conf., p. 251. Raven, New York, 1983.

development by Fahey *et al.* of HPLC methods for the quantitative determination of individual thiols at the picomole level in nonprotein fractions of biological systems, methods which have seen considerable use (this volume [14] and [15]).[10-12, 21-24a]

Fluorescent labeling agents[25] ideally add minimal size fluorescent moieties ("labels") to target molecules, usually proteins, but sometimes nucleic acids or polysaccharides. The smaller the label, the less likely it is to change the biological, physical, and chemical properties of the labeled molecule as compared with the unlabeled molecule. This principle is apart from the changes expected because the groups (such as thiols) that are labeled are themselves central to a particular activity. Fluorescent labels are preferred over radioactive labels (in cases for which either may be used) because of the simplicity of detection and the avoidance of the necessity for disposal of radioactive materials.

Bromobimanes are derived from a basic structure (two fused five-membered rings) with the requisite minimal size. This chapter describes the use of three bromobimanes for fluorescent labeling of biological systems. The three bromobimanes,[26] mBBr (**1**), bBBr (**2**), and qBBr (**3**), are commercially available.[27] The short form[3] names are given under the formulas.

Chemical and Photophysical Background

1
mBBr
syn-(BrCH$_2$,CH$_3$)(CH$_3$,CH$_3$)B

2
bBBr
syn-(BrCH$_2$,CH$_3$)B

3
qBBr
syn-(BrCH$_2$,CH$_3$)[(CH$_3$)$_3$N$^+$CH$_2$,CH$_3$]B,Br$^-$

---

[22] G. L. Newton and B. Javor, *J. Bacteriol.* **161**, 438 (1985).

[23] R. C. Fahey, G. L. Newton, B. Arrick, T. Overdank-Bogart, and S. B. Aley, *Science* **224**, 70 (1984).

[24] G. L. Newton, R. Dorian, and R. C. Fahey, *Anal. Biochem.* **114**, 383 (1981).

[24a] R. C. Fahey and G. L. Newton, this volume [14] and [15].

[25] The class description for such agents is somewhat confusing. In general, fluorescent labeling agents are compounds which give rise to fluorescent derivatives of target molecules. The latter thus carry fluorescent labels. The agents themselves may or may not be fluorescent.

[26] Formal names are as follows: mBBr, 4-bromomethyl-3,6,7-trimethyl-1,5-diazabicyclo[3.3.0]octa-3,6-diene-2,8-dione; bBBr, 4,6-bis(bromomethyl)-3,7-dimethyl-1,5-diazabicyclo[3.3.0]octa-3,6-diene-2,8-dione; qBBr, 4-bromomethyl-3,7-dimethyl-6-trimethylammoniomethyl-1,5-diazabicyclo[3.3.0]octa-3,6-diene-2,8-dione.

[27] Calbiochem-Behring, La Jolla, CA 92093: mBBr (Thiolyte MB), bBBr (Thiolyte BB), qBBr (Thiolyte QB).

TABLE I
ABSORBANCE AND FLUORESCENCE OF BROMOBIMANES AND
LABELED MATERIALS

Bromobimanes and sB-labeled[a] materials	Absorbance λ_m (ε_m)		Fluorescence λ_m (ϕ_F)[c]
	CH_3CN	Buffer[b]	
mBBr	381 (6000)	396 (5300)[d]	(0)[e]
bBBr	392 (6000)	—	(0)
qBBr	—	378 (5700)[f]	(0)
mB-globin[g]	—	385 (~4200)	484 (~0.18–0.23)
mB-SR(MPA)[h]	—	392	484 (~0.26–0.28)
mB-SG[i]	—	390 (~5300)	482 (~0.3)
bB-globin[g]	—	385 (~4000)	477 (~0.25–0.33)
qB-globin[g]	—	370 (~4000)	480 (~0.08–0.13)
qB-SR (MPA)[h]	—	378	475 (~0.08)
qB-SG[i]	—	375 (~5300)	482 (~0.07–0.09)

[a] sB-labeled is a general term for any of the *syn*-bimane products, mBBr, bBBr, and qBBr yield the mB-, bB-, and qB-labeled materials, respectively.
[b] Buffer containing 135 mM NaCl–10 mM phosphate, pH 7.4.
[c] Quantum yield of fluorescence was based on quinine sulfate fluorescence as reference (ϕ_F).
[d] An mBBr solution in acetonitrile was diluted with the buffer described in note *b*.
[e] The quantum yields of fluorescence of the bromobimanes are less than 0.001. However, measurement must be done with proper care since photolysis may give rise to some fluorescence.
[f] Dissolved in buffer immediately before measurements.
[g] Labeled globin was prepared from hemoglobin reacted with bromobimane. It was stored as an air-dried powder and dissolved in H_2O for measurements.
[h] Metaphosphoric acid extract (MPA) of labeled erythrocytes, which contain NPSH, mainly GSH (see this volume [14] and [15]).
[i] Labeled GSH prepared by reacting GSH in buffer.

Two isomeric series of bimanes are known, those with the carbonyl oxygens on the same side (*syn*) and those with the carbonyl oxygens on opposite sides (*anti*). The *syn* series is generally fluorescent, the *anti* series phosphorescent. The use of the latter in biological systems remains to be studied.

The neutral agents are moderately soluble in medium polarity organic solvents (acetonitrile, dichloromethane), and slightly soluble in water. The quaternary salt, qBBr, is soluble in water, but less soluble in organic solvents. Bromobimanes are yellow. The absorption maxima for the agents and those of some derivatives are given along with emission maxima in Table I.

The bromo derivatives are essentially nonfluorescent and are relatively stable when stored in the dark. Chromatography on thin-layer silica yields a yellow nonfluorescent spot which develops a blue fluorescence after several minutes' exposure to 360 nm light. The change is a convenient characteristic for identification and also suggests that there is some sensitivity to light, with the nature of the photochemical processes still undetermined.

The reactions of bromobimanes with the tripeptide thiol, glutathione (GSH), are second order and dependent on pH, the active nucleophile being the thiolate anion, GS^-.[28] The reaction of bromobimane with a thiolate [Eqs. (1) and (2)] converts the nonfluorescent agent into a water-soluble fluorescent derivative, which also has an absorption spectrum different from that of the agent.

$$mBBr + RS^- \rightarrow mBSR + Br^- \quad (1)$$
$$bBBr + 2\ RS^- \rightarrow RSBSR \quad (2)$$

Bromobimanes are less reactive toward other nucleophiles (amines, carboxylates), especially in neutral aqueous solution in which the amines are largely in the ammonium ion form. Nevertheless, two points must be considered. (1) Buffer anions are much less active as nucleophiles than thiolate anions but are present in much greater concentrations. Therefore, nonnucleophilic buffers should be used, and dilute buffers are to be preferred. (2) In a labeling reaction involving multifunctional molecules like proteins, the first step involves displacement of the bromine of bBBr by a thiol, but the second bromine may react with a less reactive nucleophilic group if its local concentration is high, an expression of the neighboring group effect.

Biological Properties of Bromobimanes and Derivatives

The bromobimanes mBBr (**1**) and bBBr (**2**) are neutral and easily penetrate intact, live cells; positively charged qBBr (**3**) does not penetrate the live cell. Alkaline conditions are required for many other fluorescent labeling reactions, such as those with isothiocyanate or sulfonyl chloride linking groups. In contrast, bromobimanes can be used at physiological pH, exhibiting fast reactions with small thiols[28] [complete labeling of intracellular nonprotein thiols (NPSH) within seconds to minutes] and

---

[28] A. E. Radkowsky and E. M. Kosower, *J. Am. Chem. Soc.* **108**, 4527 (1986).

somewhat slower reactions with protein thiols (minutes to hours). One may thus not only identify reactive thiol species, but also learn how accessible the reactive groups are *under physiological conditions*.

Monobromobimane, mBBr, is the most frequently used in the series, not only as a labeling agent for thiols, but also for studies on the consequences of the alkylation of reactive thiols in biological systems. In analyses, mBBr may be added to intact cells before processing, obviating oxidative loss of thiols during fraction preparation. In addition to total thiol determination, the mBBr procedure makes possible quantitation of thiols in individual proteins and NPSH.[24a]

The bisbromobimane, bBBr, is used for cross-linking of reactive groups within the same chain or between different polypeptides in isolated proteins or in the intact cell (see below).[6,7,29-31] The charged bromobimane, qBBr, labels extracellular thiols as well as whole dead cells or lysed cells; the labeled material can be taken up by some live cells via endocytosis or phagocytosis.

The bimane-labeled materials are stable to air, to light, to chemical and biochemical procedures, and are resistant to irradiation (e.g., do not fade or fade very slowly during observation with fluorescence microscopy). Labeled materials exhibit good quantum yields of fluorescence, and the small fluorescent moiety (being only two pentacyclic rings) minimally perturbs macromolecular conformation.

Labeling Procedures

Bromobimane Solutions

Stock solutions of mBBr and bBBr in acetonitrile are easily prepared with concentrations of 50–150 mM. The acetonitrile solutions are stable at room temperature, provided they are kept in the dark (exposure to light causes some photolysis, with some conversion to a fluorescent bimane). Small volumes of the stock solutions are used in labeling experiments. mBBr can be diluted in aqueous media prior to use (see below), but bBBr, which is less soluble in aqueous media, is usually not diluted. qBBr is dissolved directly in buffer immediately before use. Some buffers may contain reactive nucleophiles (see remarks under Chemical and Photophysical Background) and should be checked for the extent of reaction with bromobimanes before use.

---

[29] N. S. Kosower and E. M. Kosower, *Abstr., Int. Biophys. Meet., 7th*, p. 34 (1981).
[30] G. Zimmer, L. Mainka, and B. M. Heil, *FEBS Lett.* **150**, 207 (1982).
[31] D. Mornet, K. Ue, and M. F. Morales, *Proc. Natl. Acad. Sci. U.S.A.* **82**, 1658 (1985).

Reactions in Solution

A solution of small thiol or protein in buffer is added with rapid mixing to a small volume of bromobimane stock solution. The mBBr can be first diluted with buffer before mixing. The reaction vessel should be protected from light to avoid photolysis. The reaction can be followed by changes in light absorption or fluorescence emission using aliquots which are clear, colorless solutions. The reaction can be stopped by the addition of acid. In the case of mBBr and bBBr, the reaction can also be terminated by extraction of the unreacted reagents from the aqueous reaction mixture with CH_2Cl_2. The fluorescence of the products from certain proteins (hemoglobin, chlorophyll-associated proteins) is measured after removal of the chromophore. Aliquots of the reaction mixture are mixed with 0.15 N HCl–acetone or 2.5% oxalic acid–acetone; the colorless, precipitated protein is washed and then redissolved for fluorescence measurements.

To expose thiol groups which are conformationally inaccessible in macromolecules dissolved in the usual aqueous buffers, solvents such as dimethyl sulfoxide (DMSO) may be added to the media used for the reaction with mBBr.[16]

Reactions in Cells and Tissues

A cell suspension in buffer is added with rapid mixing to an appropriate bromobimane solution, the amounts and reaction times varying according to the cell type and number and purpose of labeling. As an example, most of the bromobimane-reactive thiols in human red cells are labeled by adding 1 ml of a 10% cell suspension to 15–25 μl of 100 mM mBBr in acetonitrile and incubating the suspension for 30–45 min at 37°.[6,8,32] For labeling of live spermatozoa, a suspension of 5–10 × 10^6 cells/ml is incubated with 0.05–0.1 mM mBBr for 5–10 min, conditions which do not interfere with sperm motility or fertilizing ability.[17,33] Final concentrations of 50–500 μM mBBr are used for labeling of reactive thiols in various other cell suspensions, cell monolayers, and subcellular fractions. The reaction can be terminated by washing the samples, by the addition of acid, or by extracting residual free mBBr or bBBr with CH_2Cl_2.

The labeling of protein thiols in tissues is conveniently carried out by treating tissue sections with mBBr.[9] Suitable samples include sections of frozen tissues dried and fixed on slides as well as paraffin sections of fixed

---

[32] N. S. Kosower, Y. Zipser, and Z. Faltin, *Biochim. Biophys. Acta* **691**, 345 (1982).

[33] J. M. Cummins, A. D. Fleming, N. Crozet, T. J. Kuehl, N. S. Kosower, and R. Yanagimachi, *J. Exp. Zool.* **237**, 375 (1986).

tissues, after treatment with xylene to remove the paraffin and washing with ethanol. Sections on slides are washed with buffer, covered with 0.1 mM mBBr solution in buffer and left in the dark at room temperature for 5–30 min. Sections are then washed and covered with glycerol and coverslips for microscopic observations. Once the labeling reaction is over, the slides can be kept in the light without any precautions. A similar procedure is used for fixed subcellular fractions on slides, e.g., chromosome preparations.[34]

In addition to thiol labeling in tissues, mBBr can be used for the histochemical detection of disulfides. Tissue sections are first covered with 10–50 mM N-ethylmaleimide solution and incubated for a few minutes in order to block free SH groups, washed with buffer, and treated with 10–50 mM dithiothreitol (DTT) for 5–10 min. Slides are then washed thoroughly with buffer and labeled with mBBr as above. To label both thiols and disulfides, slides are treated only with DTT and washed prior to mBBr treatment.[9]

Analytical Procedures

Spectroscopic Data for Bromobimanes and Bimane-Labeled Products

Spectroscopic data for bromobimanes and bimane-labeled products are given in Table I. The values vary somewhat from material to material and should be determined for accurate analyses of specific derivatives.

Fluorescence Microscopy

Cells, cell fractions, and tissue sections can be examined with an epiillumination-type fluorescence microscope, equipped with appropriate filters for excitation between 360 and 400 nm, and emission at wavelengths longer than 440–450 nm. Photographs are made with sensitive films such as Kodak Tri-X or Ektachrome. In most cases, thorough observations and photography are possible, without an appreciable decrease in fluorescence intensity.

Absorption and Fluorescence Measurements of Cells and Cell Fractions

The total reactive thiol content can be measured in whole samples or in any cellular fraction, provided clear, colorless solutions are obtained [e.g., whole cells dissolved in sodium dodecyl sulfate (SDS); chlorophyll-

---

[34] A. T. Sumner, *J. Cell Sci.* **70**, 177 (1984).

free chloroplast stromal proteins and coupling factor (CF_1); nonprotein fractions]. For quantitative determination, the fluorescence intensity is compared with that of a known labeled small thiol or protein such as labeled globin.

Individual thiol-containing protein bands are readily observed by fluorescence following separation by gel electrophoresis. Gels, fixed by standard methods, are viewed using a long-wavelength ultraviolet lamp. Gels are photographed by placing them on a long-wavelength ultraviolet transilluminating box, using a Kodak Wratten gelatin filter No. 8; fixed gels may be stored in the fixative for days before photography. After fluorescence photography, the gels are stained with dyes such as Coomassie blue. A densitometric profile of fluorescent and stained protein bands may be obtained, using either the gels or films of the gels. The thiol content of the proteins in the bands is quantitated by comparison with the profile from a labeled known protein run on the same gel.

The thiol content of protein fractions separated by column chromatography can be measured by the mBBr procedure. Thiol-containing peptides can be identified in peptide maps following enzymatic degradation of labeled protein.

Other Applications of Thiol Labeling

1. Nonthiol molecules can be converted to thiol-containing derivatives, then labeled with mBBr and used as probes and fluorescent substrates. Examples are (a) (Fab)₂ was reduced by DTT, then reacted with mBBr.[35] The labeled fragment has been used in immunocytochemical experiments.[35] (b) Pepstatin was converted to an SH derivative (by coupling to cysteamine), treated with mBBr, and the mB-derivative used as a fluorescent probe for the subcellular location of cathepsin D with selectivity for the active conformation of the enzyme.[36] (c) fMet-tRNA, containing thiouridine, was reacted with mBBr and the mB-derivative shown to be useful for studying the tRNA–ribosome interaction.[16]

2. The dynamics of microtubule assembly and disassembly can be studied. Labeling of assembled microtubules was carried out with qBBr.[37] The derivatized proteins appeared to undergo association–dissociation in the same way as the unlabeled materials.[37]

3. The bimane moiety may serve as a component of a donor–acceptor system in energy transfer studies, as in the case of rhodopsin.[13]

[35] I. T. W. Matthews, *J. Immunol. Methods* **51**, 307 (1982).
[36] I. T. W. Matthews, R. S. Decker, and C. G. Knight, *Biochem. J.* **199**, 611 (1981).
[37] P. Wadsworth and R. D. Sloboda, *Biochemistry* **21**, 21 (1982).

4. As a substrate for glutathione (GSH) transferase, mBBr was shown to react at least 3 times faster with GSH in the presence of GSH transferase.[38]

5. For immobilization of motile cells, mB-labeling of spermatozoa had no effect on sperm motility; however, excitation of the labeled sperm by fluorescent microscopy resulted in almost instantaneous immobilization of the spermatozoa. Immobilization depended on the midpiece being irradiated, suggesting a mitochondrial site of action.[33]

6. As for diversion of electrons generated in chloroplast photosystem II (PSII), the mBBr is converted to a radical anion, mBBr$^{\overline{\cdot}}$, by electron transfer from the plastoquinone anion of PSII. Although most of the radical anion, mBBr$^{\overline{\cdot}}$, is converted to syn-(CH$_3$,CH$_3$)B via protonation and further reduction, a significant proportion yields a reactive free radical via dissociation of bromide ion. The free radical reacts with a PSII protein to yield a sB-protein.[39]

7. In the use for formation of a sulfide from two thiols via reaction with syn-(1-bromoethyl,methyl)bimane, glutathione sulfide (GSG) has been produced from glutathione.[40]

Acknowledgments

The authors are grateful for the support of their research on thiols and thiol agents by the National Institutes of Health, the United States–Israel Binational Science Foundation, the Israel Academy of Science, the Chief Scientist's Office, Israel Ministry of Health, the European Research Office, the Petroleum Research Fund, the J. S. Guggenheim Foundation, and the National Science Foundation.

---

[38] P. B. Hulbert and S. I. Yakubu, *J. Pharm. Pharmacol.* **35**, 384 (1982).

[39] A. Melis, N. S. Kosower, M. A. Crawford, E. Kirowa-Eisner, M. Schwarz, and E. M. Kosower, *Photochem. Photobiol.* **43**, 583 (1986).

[40] A. E. Radkowsky, E. M. Kosower, D. Eisenberg, and I. Goldberg, *J. Am. Chem. Soc.* **108**, 4532 (1986).

[14] Determination of Low-Molecular-Weight Thiols Using Monobromobimane Fluorescent Labeling and High-Performance Liquid Chromatography

By ROBERT C. FAHEY and GERALD L. NEWTON

The bromobimanes constitute a class of compounds introduced by Kosower and co-workers[1,2] (see also this volume [13]) as useful reagents for labeling of the thiol group. Monobromobimane (mBBr) has proven to be especially useful for the analysis of low molecular weight thiols. The compound itself is only weakly fluorescent but selectively reacts with thiols to yield highly fluorescent and stable thioethers (mBSR) which can be easily detected at the picomole level [Eq. (1)]. The bimane derivatives

$$\text{mBBr} + \text{RSH} \rightleftharpoons \text{mBSR} + \text{HBr} \quad (1)$$

can be separated by electrophoresis,[3] paper chromatography,[3] and ion-exchange chromatography,[4] but high-performance liquid chromatography (HPLC) has proved to be the most useful separation method.[5] This chapter describes methods for the preparation and HPLC analysis of monobromobimane derivatives of low molecular weight thiols in extracts of biological samples and discusses typical problems encountered in the development and application of these methods.

Materials

Materials and sources were as follows: HPLC grade sodium perchlorate and 50% sodium hydroxide—Fisher Scientific; methanol and acetoni-

---

[1] E. M. Kosower, B. Pazhenchevsky, and E. Hershkowitz, *J. Am. Chem. Soc.* **100**, 6516 (1978).
[2] N. S. Kosower, E. M. Kosower, G. L. Newton, and H. M. Ranney, *Proc. Natl. Acad. Sci. U.S.A.* **76**, 3382 (1979).
[3] R. C. Fahey, G. L. Newton, R. Dorian, and E. M. Kosower, *Anal. Biochem.* **107**, 1 (1980).
[4] R. C. Fahey, G. L. Newton, R. Dorian, and E. M. Kosower, *Anal. Biochem.* **111**, 357 (1981).
[5] G. L. Newton, R. Dorian, and R. C. Fahey, *Anal. Biochem.* **114**, 383 (1981).

trile—Burdick and Jackson; 5-sulfosalicylic acid—MCB; tetrabutylammonium phosphate—Kodak; Puriss grade methanesulfonic acid—Fluka; glacial acetic acid, sodium sulfide, sodium sulfite, sodium thiosulfate, potassium chloride, dibasic potassium phosphate, sodium acetate, and boric acid—Mallincrodt; diethylenetriaminepentaacetic acid (DTPA), 5,5′-dithiobis(2-nitrobenzoic acid) (DTNB), N-ethylmaleimide (NEM), 2-pyridyl disulfide, reduced glutathione, L-homocysteine, ergothioneine, pantetheine, 2-mercaptoethanesulfonic acid, N-acetylcysteine, penicillamine, thioglycolic acid, and N-ethylmorpholine—Sigma; methyl disulfide—Aldrich; monobromobimane (mBBr), N-2-hydroxyethylpiperazine-N'-3-propanesulfonic acid (HEPPS), cysteine hydrochloride, high purity dithiothreitol (DTT), and 2-mercaptoethanol—Calbiochem; L-cystinylbisglycine—Vega; WR-1065—Dr. Leonard Kedda, National Cancer Institute. A sodium methanesulfonate stock solution was prepared from methanesulfonic acid by addition of one equivalent of sodium hydroxide and dilution to 4 M. γ-Glutamylcysteine was prepared by the method of Strumeyer and Bloch[6] and 4′-phosphopantetheine by the method of Wang, Shuster, and Kaplan.[7]

Preparation of Samples

Several different types of samples are typically needed in the course of a given application, and these include the thiol standard sample, the reagent blank, the unknown sample, and the unknown control. The thiol standard solutions were prepared from authentic stock solutions of the individual thiols of interest. Thiol stock solutions were prepared at high concentration (10–20 mM) and low pH (10 mM methanesulfonic acid) to minimize losses due to oxidation, except for sulfide which was prepared in 50 mM HEPPS at pH 8.0. Varying levels of purity and degree of hydration often made it difficult to prepare stocks of precisely known concentration and it was, therefore, convenient to check the final concentration by titration of the thiol with DTNB.[8] Standard solutions of methanethiol, 2-mercaptopyridine, and cysteinylglycine were prepared from the corresponding disulfides by reaction with one equivalent of DTT in 50 mM HEPPS, pH 8.0, containing 5 mM DTPA for 10 min at 50° under argon.

The derivatization reaction was carried out at high concentration to maximize the reaction rate with mBBr and minimize competing side reactions; a 1–2 mM stoichiometric excess of mBBr was used to ensure rapid

---

[6] D. Strumeyer and K. Bloch, *Biochem. Prep.* **9**, 52 (1962).
[7] T. P. Wang, L. Shuster, and N. O. Kaplan, *J. Biol. Chem.* **206**, 299 (1954).
[8] G. L. Ellman, *Arch. Biochem. Biophys.* **82**, 70 (1959).

and complete reaction. In a typical reaction, the thiol (1 mM final concentration) was added to a solution containing 50 mM HEPPS, 5 mM DTPA, and 2–3 mM mBBr at pH 8.0 and the reaction allowed to proceed for 10 min under dim lighting at room temperature. Methanesulfonic acid was added to 25 mM prior to storage. Acetic acid has also been used to lower the pH for storage, but acetate reacts with mBBr to yield a fluorescent product which may interfere in some analyses. Under these conditions the half-life for reaction of glutathione (GSH) and other cysteine derivatives was about 20 sec whereas sulfite ($t_{1/2}$ ~2 min), penicillamine ($t_{1/2}$ ~1 min), and CoA ($t_{1/2}$ ~1.5 min) reacted more slowly but were >95% reacted at 10 min. Since heavy metal ion levels were low and the thiol concentration was high, losses due to oxidation were negligible and exclusion of air during preparation of standard samples was not required. When 10 μM standard solutions of the mBBr derivatives of cysteine, GSH, and WR-1065 in 1% acetic acid were stored as single aliquots in Eppendorf centrifuge tubes, no significant loss was observed at 4, −20, or −70° for 20 months. However, standard solutions repeatedly frozen at −20° and thawed showed significant losses over short periods. The bimane derivatives are photosensitive and should be protected from light during preparation and storage.

Preparation of cell extracts for analysis is more complicated, and the optimal procedure varies with cell type. Loss of thiol due to oxidation is an important problem which is minimized by using heavy metal chelators and limiting the exposure of the thiol to oxygen while at high pH. Trace levels of peroxide in buffers can also oxidize thiols so that high purity, slowly autoxidized buffers, such as Tris and HEPPS, have been found preferable to easily autoxidized buffers, e.g., N-ethylmorpholine, when low thiol levels are to be analyzed. Loss of thiol or thiol derivative due to enzymatic degradation represents another major problem; thus, attempts to extract kidney, which is rich in γ-glutamyltranspeptidase (γ-glutamyltransferase) and dipeptidase activities, without enzyme inactivation results in extensive conversion of GSH to cysteine. Enzymes must, therefore, be inactivated during the extraction process, and this can be accomplished by extracting either in acid or in organic solvent. The reagents selected for such extraction and for subsequent derivitization of cell extract thiols must not react significantly with mBBr. Although mBBr reacts preferentially with thiols, it also reacts slowly with amines, phosphate, carboxylates, and other nucleophiles when these are present at high millimolar or molar concentrations to yield micromolar quantities of fluorescent products which can interfere with the analysis.

As a guide to selection of appropriate reagents we include in Table I the retention times for reagent-derived products obtained with various components. Each analysis protocol should include as a reagent blank a

TABLE I
RETENTION TIMES OF THIOL-mBBr DERIVATIVES AND REAGENT-DERIVED PRODUCTS

Compound, abbreviation	Retention time (min)		
	1	2	3
Thiols			
N-Acetylcysteine, NAC	24	19.2	10.3
Coenzyme A, CoA	43[a]	25.9	13.2[a]
Coenzyme M, CoM	13	21.4	8.8
Cysteamine, CyA	35[a]	2.9	10.6
Cysteine, Cys	8	5.8	6.4
Cysteinylglycine, CG	17	5.1	8.7
Dephosphocoenzyme A, dpCoA	41[a]	23.7	14.5[a]
Dithiothreitol, DTT	48	20.5	19.8
Ergothioneine, Ergo	28	7.2	10.5
Glutathione, GSH	16	14.3	8.4
α-Glutamylcysteine, α-GC	17	10.1	8.7
γ-Glutamylcysteine, γ-GC	12	15.1	7.7
Homocysteine, hCys	18	8.5	8.8
Homoglutathione, hGSH	22	12.7	9.8
2-Mercaptoethanol, 2-ME	38	16.0	14.8
2-Mercaptopyridine, 2-PySH	57	26.7	20.6
Methanethiol, MeSH	48.5	20.9	20.0
2-Nitro-5-mercaptobenzoic acid	40	25.3	19.9
Pantetheine, Pant	44	18.8	19.3
Penicillamine	29	12.5	10.2
4'-Phosphopantetheine, 4'-p-Pant	33[a]	21.5	13.1
Sulfide, H_2S	49	20.9	20.0
Sulfite, SO_3^{2-}	4	18.5	4.6
Thioglycolic acid, TG	32	19.1	12.4
Thiosulfate, SSO_3^{2-}	7	21.1	8.1
WR-1065	None	2.2	12.0
Reagents[b]			
Water–mBBr, unbuffered	<u>21</u>, 41	9.0, 11.4, 16.2, <u>21.1, 21.9</u>	<u>9.4, 14.5, 20.3</u>
N-Ethylmorpholine, pH 8	None	None	None
HEPPS, pH 8	38[a]	3.0	10.8
Potassium chloride, pH 8[c]	<u>43</u>	None	<u>1.9</u>
Potassium phosphate, pH 8	4	<u>16.8, 17.7</u>	3.8
Sodium acetate, pH 8[c]	<u>37</u>	<u>18.1</u>	<u>19.5</u>, 12.3
Sodium borate, pH 9	<u>56, 57, 58</u>	<u>27.3, 28.3</u>	None
Sodium methanesulfonate, pH 8[c]	None	None	None
Sulfosalicylic acid, pH 8[c]	<u>3</u>	<u>22</u>	<u>1.8, 13.2</u>
Tris–methanesulfonate pH 8	<u>12</u>	<u>3.1</u>	<u>6.2</u>

[a] Broad or unsymmetrical peak.
[b] Peaks listed for water–mBBr also occur for other aqueous reagents. All reagents tested at 1 M. Major peaks are underlined, others are less intense.
[c] Buffered with 10 mM HEPPS.

sample treated identically to the unknown sample but with the cells or tissue omitted; this sample can be used to identify peaks arising from the reagents. A second blank, the unknown control, is prepared by reacting the thiols present in the cell extract with NEM or DTNB prior to derivatization with mBBr. Since most thiols react with NEM and DTNB to form derivatives that are unreactive with mBBr, this sample serves as a check on the assignment of thiols in unknown samples and allows the identification of fluorescent nonthiol components contributed by cells. 2-Pyridyl disulfide can be used in place of DTNB but produces greater background due to the more intense fluorescence of the 2-mercaptopyridine bimane derivative.[9]

Extraction of animal tissues using methanesulfonic acid to denature enzymes is illustrated by the following protocol: ~200 mg of fresh or frozen tissue was homogenized for 1 min in 1 ml ice-cold 200 mM methanesulfonic acid using a Tekmar or Brinkman polytron homogenizer. An equal volume of 4 M sodium methanesulfonate was added and the sample centrifuged 5 min in an Eppendorf microcentrifuge. The clear supernatant was diluted 1:3 into 200 mM HEPPS–methanesulfonate (pH 8.0) containing 5 mM DTPA and 3 mM mBBr, and reaction was allowed to proceed 10 min under dim light. After addition of methanesulfonic acid to 100 mM, the sample was centrifuged again and stored at $-70°$ until analyzed. For preparation of unknown control samples, 5 mM NEM (or 2 mM DTNB) was substituted for mBBr and the reaction allowed to proceed for 5 min after which mBBr was added to 2 mM and the reaction continued for 10 min before acidification. A procedure for extraction of animal tissue with sulfosalicylic acid, labeling with mBBr, and HPLC analysis for glutathione has been described by Anderson.[10]

Use of acetonitrile as a protein denaturant is illustrated in the following protocol for analysis of mung bean: Dried mung beans were ground to a fine powder in a mortar and pestle. The powder (200 mg) was placed in a septum-capped 3-ml vial and the vial flushed with argon. The extraction buffer (50% aqueous acetonitrile containing 50 mM HEPPS at pH 8.0, 5 mM DTPA, and 2 mM mBBr) was preheated to 60° and flushed with argon. A syringe was used to transfer 1 ml of extraction buffer to the vial containing the mung bean powder, and the vial was then intermittently vortexed while heating at 60° for 10 min. After cooling the vial, methanesulfonic acid was added to 50 mM. The sample was transferred to a 1.5-ml Eppendorf centrifuge tube and centrifuged 5 min in an Eppendorf micro-

---

[9] R. C. Fahey and G. L. Newton, *in* "Functions of Glutathione: Biochemical, Physiological, Toxicological, and Clinical Aspects" (A. Larsson *et al.*, eds.), Nobel Conf., p. 251. Raven, New York, 1983.

[10] M. E. Anderson, this series, Vol. 113, p. 548 (1985).

centrifuge. Samples were stored at $-20°$ and diluted at least 1:1 in 10 mM methanesulfonic acid to lower the acetonitrile content prior to HPLC analysis. In the control sample, 5 mM NEM (or 2 mM DTNB) was used in place of mBBr and reaction allowed to proceed 10 min, after which mBBr was added to 2 mM and the reaction continued an additional 10 min, all at 60°. The elevated temperature was required because organic solvents decrease the rate of reaction of mBBr with most thiols.

Under these conditions the $t_{1/2}$ for GSH is <15 sec but the $t_{1/2}$ for CoA is ~1.5 min. Even at this elevated temperature it was considered desirable to test the use of a higher pH to ensure complete derivatization. This was complicated by the fact that amine buffers react more extensively with mBBr at higher pH. However, substitution of 50 mM sodium borate at pH 9.0 for HEPPS in the described protocol resulted in minimal reagent peaks (Fig. 2A) and gave quantitative results which were only slightly higher than achieved using HEPPS at pH 8.0.

Acidified cell extracts have generally been found to be stable for weeks when stored at $-70°$ and showed no change when allowed to stand at room temperature in an autoinjector for up to 24 hr. An exception is the mBBr derivative of CoA which is converted to the dephospho-CoA derivative with a half-life of about 10 hr in 1% acetic acid at room temperature.

Chromatography

All analyses were carried out on a Varian Model 5060 high-performance liquid chromatograph equipped with a Waters WISP Model 710B autoinjector, an LDC Fluoromonitor III (Model 1311) fitted with a standard flow cell and 360 nm excitation/410–700 nm emission filters, and a Nelson Model 444 data system. Three HPLC protocols have proved useful for different types of sample analysis. These are illustrated with various standard mixtures in Fig. 1, and retention times for standard thiol derivatives are tabulated in Table I.

Separation of the greatest number of thiol derivatives was achieved with Method 1, which utilized a 4.6 × 250 mm Altex Ultrasphere ODS 5 μm analytical column equipped with a Brownley MPLC guard column containing an OD-GU 5 μm C_{18} cartridge. Solvent A was 0.25% aqueous acetic acid titrated to pH 3.5 with concentrated sodium hydroxide; solvent B was methanol. Solvents were filtered through 0.2 μm nylon filters (Rainin Scientific) in a Millipore solvent filter apparatus. The elution protocol (24°, 1.2 ml/min) employed linear gradients as follows: 0 min, 15% B; 5 min, 15% B; 15 min, 23% B; 45 min, 42% B; 65 min, 75% B; 67 min, 100% B; 70 min, 15% B; 85 min, 15% B; reinject.

Method 2 was developed to provide quantitative analytical data for coenzyme A and dephospho-CoA, which elute as broad peaks with

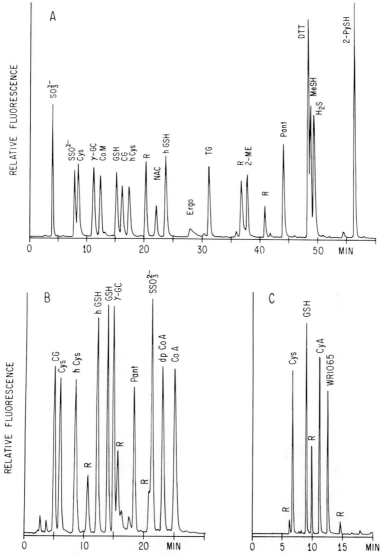

FIG. 1. HPLC chromatograms of thiol standard samples: (A) method 1; (B) method 2; (C) method 3. Peaks present in corresponding reagent blank samples are designated R. Other abbreviations are defined in Table I.

method 1, while retaining the capacity to simultaneously analyze for glutathione and cysteine. A 4.0 × 250 mm Lichrosorp RP-8 10 μm analytical column (E. Merck) was used with a Brownley MPLC guard column fitted with an RP-GU 10 μm C_8 cartridge. Solvent A was prepared by dilution of

100 ml methanol, 2.5 ml acetic acid, and 3.4 g tetrabutylammonium phosphate to 1 liter with water and adjusting the pH to 3.4 with sodium hydroxide or acetic acid. Solvent B was prepared by addition of 2.5 ml acetic acid and 3.4 g tetrabutylammonium phosphate to 100 ml water and diluting to 1 liter with methanol. The elution protocol (24°, 1.0 ml/min) with linear gradients was as follows: 0 min, 10% B; 15 min, 50% B; 30 min, 75% B; 40 min, 100% B; 42 min, 10% B; 55 min, 10%, B; reinject. Several different 5 μm C_8 columns were also tested but gave broader peaks for CoA than the Lichrosorp RP-8 column.

Method 3 utilizes sodium perchlorate ion pairing to enhance the analysis of cysteamine and WR-1065 derivatives in samples also containing GSH and cysteine. The method was designed for rapid, routine analysis of radioprotective drugs in animal tissue samples and has not been optimized for a wide range of thiol derivatives. The column and guard column used are as described for Method 1. Solvent A was prepared by diluting 2.5 ml acetic acid, 50 ml acetonitrile, and 7.03 g sodium perchlorate to 1 liter with water and adjusting the pH to 3.4 with concentrated sodium hydroxide. Solvent B was prepared by addition of 2.5 ml acetic acid and 7.03 g sodium perchlorate to 200 ml water and diluting with acetonitrile to 1 liter. The elution protocol (linear gradients, 24°, 1.5 ml/min) was as follows: 0 min, 2% B; 5 min, 13% B; 14 min, 13% B; 16 min, 100% B; 20 min, 2% B; 35 min, 2% B; reinject.

With the methods described above it is possible to detect 0.1 pmol of the mBBr derivative of GSH with a signal-to-noise ratio of 5:1. The response factor is sensitive to solvent and increases as the organic modifier is increased. Most other thiol derivatives have response factors comparable to that for the GSH derivative, but the factor for derivatives of aromatic thiols such as ergothioneine and 5-mercapto-2-nitrobenzoic acid can be an order of magnitude or more lower. Response factors vary with the equipment used and with the age of the fluorometer lamp and must be determined from standard samples analyzed concurrently with each series of unknown samples. The response factor (integrated area/pmol of standard) has been found to be constant with an increase in concentration for samples containing 0.1 to 200 pmol of the GSH derivative.

Analysis of Mung Bean

Figure 2 illustrates the application of HPLC Method 1 to the analysis of an extract of mung bean prepared in 50% aqueous acetonitrile containing sodium borate at pH 9. The dominant peak in the reagent blank (Fig. 2A) derives from mBBr and water (Table I). Additional peaks appear in the unknown control (Fig. 2B), the dominant one occurring at ~38 min.

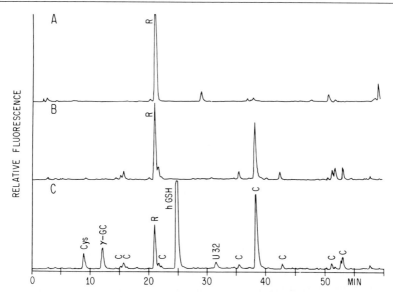

FIG. 2. HPLC chromatograms obtained for mung bean extracted in aqueous acetonitrile containing borate buffer at pH 9.0 (see Preparation of Samples). (A) reagent blank; (B) unknown control; (C) unknown. Peaks designated R and C represent peaks present in the reagent blank and unknown control samples, respectively. U32 designates an unidentified thiol derivative eluting at 32 min. Other abbreviations are given in Table I.

This is where *syn*-(methyl,methyl)bimane, the compound obtained by replacing Br in mBBr by H, elutes. In addition to undergoing nucleophilic substitution reactions, mBBr can serve as an electron acceptor, ultimately being reduced to *syn*-(methyl,methyl)bimane.[11] Constituents in cell extracts, especially those from photosynthetic organisms, appear to include electron donors which can participate in this reaction and lead to a peak for *syn*-(methyl,methyl)bimane in the control sample which is blocked by NEM and DTNB to varying degrees depending on the sample. Interpretation of the unknown control is complicated to some extent by the fact that a few thiol compounds do not react fully with NEM and DTNB. All of the thiols listed in Table I are blocked by NEM and DTNB with the exception of ergothioneine (which does not react with either), 2-mercaptopyridine (which reacts incompletely with DTNB), and thiosulfate (which is unreactive with NEM and only partially reactive with DTNB).

[11] A. Melis, N. S. Kosower, N. A. Crawford, E. Kirowa-Eisner, M. Schwartz, and E. M. Kosower, *Photochem. Photobiol.* **43**, 583 (1986).

The reagent blank and unknown control sample are used to identify peaks in the unknown sample (Fig. 2C) which do not correspond to thiol derivatives. The nonthiol peaks in the unknown sample are not always of the same intensity as those in the reagent blank and unknown control samples, presumably because the reaction of thiol in the unknown sample with mBBr depletes the reagent and changes the level of formation of other reagent-derived peaks. Four significant peaks represent thiols in Fig. 2C, three of which can be ascribed to cysteine, γ-glutamylcysteine, and homoglutathione. Analysis of the extract using Method 2 (not shown) verified these assignments and also revealed the presence of a low level of CoA which cannot be analyzed by Method 1 because the peak for CoA is too broad. The fourth peak, eluting at 32 min, does not correspond to any of our standards and is designated as an unknown (U32). It has been shown that mung bean contains homoglutathione rather than glutathione.[12,13]

Identification of the thiols in mung bean was relatively straightforward. In other instances we have found it necessary to collect individual peaks as they elute using Method 1 and to reinject them using Methods 2 and 3 in order to obtain unequivocal assignments. Thiols that cannot be assigned based on known standards can be purified by methods described elsewhere in this volume [15] for structure elucidation using other techniques.

Quantitative Determinations

For analytical applications it is important to establish that the method selected is capable of quantitatively measuring the thiols of interest in the specific system under study. This is usually done by recovery experiments in which known amounts of the thiols to be measured are added at levels comparable to the endogenous levels during the extraction step. Such recovery studies have shown that the methanesulfonic acid extraction method in combination with HPLC Method 3 results in recoveries of the radioprotective drug WR-1065 of at least 80%.[14,15] Application of the aqueous acetonitrile extraction (borate buffer at pH 9.0) in combination with HPLC Methods 1 and 2 to analysis of photosynthetic bacteria (*Chromatium vinosum*) was tested for recovery of cysteine, GSH, thiosulfate,

---

[12] C. A. Price, *Nature (London)* **180,** 148 (1957).
[13] P. R. Carnegie, *Biochem. J.* **89,** 459 (1963).
[14] J. F. Utley, N. Seaver, G. L. Newton, and R. C. Fahey, *Int. J. Radiat. Oncol. Biol. Phys.* **10,** 1525 (1984).
[15] P. M. Calabro-Jones, R. C. Fahey, G. D. Smoluk, and J. F. Ward, *Int. J. Radiat. Biol.* **47,** 23 (1985).

and CoA and found to give 80% or better recovery.[16] Recovery of GSH, CoA, and thiosulfate (0.05 μmol/g) during extraction of mung bean in acetonitrile (borate at pH 9) was shown to be 85% or better. Such results cannot be safely generalized to other thiols or other cell systems, and it is essential that recovery experiments be conducted with the specific system of interest if quantitative data are required. The analytical results can sometimes be improved through the use of an internal standard added to the initial extract and used to compensate for mechanical losses during sample processing. Use of a thiol as an internal standard to compensate for losses due to oxidation or other chemical reactions would likely prove unsatisfactory owing to differing rates of reaction for different thiols.

Aside from enzymatic degradation, thiol oxidation constitutes the most serious difficulty to be overcome in designing extraction methods that will yield quantitative results. DTPA has been described as a more effective than EDTA in reducing iron-mediated formation of peroxide from oxygen,[17] and we have generally found less oxidation of thiols during labeling with mBBr in the presence of DTPA than with EDTA. However, even in the presence of DTPA we have found with some cell systems that oxidative loss competes with the bimane labeling reaction,[16] despite the fact that the half-life for reaction of GSH with mBBr under typical labeling conditions is only about 20 sec. In such cases it is necessary to exclude oxygen from the sample during the labeling process. When acid extraction methods are used, the acid must be neutralized to permit the labeling reaction to occur at high pH. Large amounts of buffer are often needed for this purpose, and it is important that the buffer not contain traces of peroxide which can oxidize the thiol. Such neutralizations should not be conducted with strong base since production of local regions of very high pH can enhance oxidative loss if oxygen is not being excluded.

Following extraction of cells, intracellular thiols can potentially come in contact with cellular disulfides or thioesters with which they might react. Such thiol–disulfide exchange or transacylation reactions can change the thiol composition of the sample. However, these reactions, as well as the mBBr labeling reaction, occur via the thiolate anion (RS^-) but are significantly slower than the labeling reaction. Thus, only in the case of a highly reactive disulfide or thioester are such reactions likely to compete with formation of mBSR.

The methods described here can also be used to determine the thiols

[16] R. C. Fahey, R. M. Buschbacher, and G. L. Newton, *J. Mol. Evol.* (1987).
[17] G. Cohen and P. M. Sinet, in "Chemical and Biochemical Aspects of Superoxide and Superoxide Dismutase" (J. V. Bannister and H. A. O. Hill, eds.), p. 27. Am. Elsevier, New York, 1980.

that are present in cells in disulfide and thioester forms. For this purpose cells should be extracted in the presence of NEM to block free thiols[18] and the disulfide or thioester converted to the corresponding thiol for labeling with mBBr. Although it is often assumed that only disulfides are cleaved by DTT, thioesters can also be cleaved by transacylation reactions with DTT under conditions similar to those which lead to disulfide reduction. Methods for selective cleavage of disulfides and thioesters, with accompanying formation of their mBBr derivatives using hydroxylamine in combination with DTT and NEM, have been described.[19]

Acknowledgments

We thank the National Institutes of Health (Grants CA-32333 and CA-39582) and the National Aeronautics and Space Administration (Grant NAGW-342) for support of this research.

[18] F. Tietze, *Anal. Biochem.* **27**, 502 (1969).
[19] S. S. Fenton and R. C. Fahey, *Anal. Biochem.* **154**, 34 (1986).

[15] Purification of Thiols from Biological Samples

By GERALD L. NEWTON and ROBERT C. FAHEY

The dominant low molecular weight thiol in most eukaryotes is glutathione, but other thiols, including those of unknown identity, also occur in eukaryotes and many prokaryotes.[1] In order to identify novel thiols it is usually necessary to obtain a pure sample. Here, a two-step purification procedure is described that allows a low molecular weight thiol component in a biological extract to be isolated as the monobromobimane derivative (see this volume [13] and [14]) in highly purified form. The thiols present in a deproteinized extract were first isolated on a thiol–agarose gel by a thiol–disulfide exchange reaction. The thiols were then eluted with dithiothreitol and derivatized with monobromobimane. The derivative was purified to homogeneity by preparative HPLC. A procedure for electrolytic reduction of the bimane derivative was developed which allows regeneration of the thiol form of the purified product. Application of the method is illustrated for isolation of a major thiol component found in

[1] R. C. Fahey and G. L. Newton, *in* "Functions of Glutathione: Biochemical, Physiological, Toxicological, and Clinical Aspects" (A. Larsson *et al.*, eds.), Nobel Conf., p. 251. Raven Press, New York, 1983.

Halobacterium halobium, whose structure was shown to correspond to γ-glutamylcysteine.[2]

Materials

Reagents and sources were as follows: glacial acetic acid, pyridine, mercury, and ammonium hydroxide—Mallinckrodt; diethylenetriaminepentaacetic acid (DTPA), 5,5'-dithiobis(2-nitrobenzoic acid) (DTNB), Trizma base (Tris), and 2,2'-dithiodipyridine—Sigma; high-purity dithiothreitol (DTT) and monobromobimane (mBBr)—Calbiochem-Behring; acetonitrile and methanol—Burdick and Jackson; ethyl acetate—Pierce; Puriss grade methanesulfonic acid—Fluka. The 2-thiopyridyl mixed disulfide of thiopropyl-Sepharose 6B was obtained from Sigma or prepared from Sepharose 6B by the method of Axén, Drevin, and Carlsson[3] but using a 6-fold higher level of epichlorohydrin.

Purification

Halobacterium halobium R-1 cells (0.6 g wet weight) were dispersed with a Tekmar homogenizer into 5 ml of 50% aqueous acetonitrile containing 5 mM DTPA and 25 mM methanesulfonic acid while heating at 60° for 10 min. After cooling on ice the sample was centrifuged 10 min at 14,000 g in a Sorvall RC-5 centrifuge. The clear supernatant liquid contained 0.9 μmol of thiol by DTNB titration[4] and 0.77 μmol of an unidentified thiol (later shown to be γ-glutamylcysteine) by derivatization with mBBr and HPLC analysis (peak eluting at 10 min, Fig. 1A). The supernatant was diluted 2-fold into 200 mM Tris at pH 8.0, containing 5 mM DTPA and 1.0 mM DTT (added to reduce disulfides and protect thiols toward oxidation).

A 6 × 20 mm column was packed with 0.75 ml of 2-thiopyridyl-activated thiopropyl-Sepharose 6B (25 μmol disulfide) and equilibrated with 25% acetonitrile in water containing 100 mM Tris at pH 8.0 and 5 mM DTPA. The thiol-containing supernatant liquid was applied to the column at a flow rate of 0.5 ml/min after which the column was washed with 25 volumes of the equilibration buffer. Binding of thiol to the column was monitored based on the absorbance of 2-mercaptopyridine ($\varepsilon_{343} \cong 7000$)[5] in the effluent which indicated that all of the applied thiol had reacted with the gel. The column was further washed with 25 column volumes of 50

---

[2] G. L. Newton and B. Javor, *J. Bacteriol.* **161**, 438 (1985).
[3] R. Axén, H. Drevin, and J. Carlsson, *Acta Chem. Scand.* **29**, 471 (1975).
[4] G. L. Ellman, *Arch. Biochem. Biophys.* **82**, 70 (1959).
[5] D. R. Grassetti and J. F. Murray, *Arch. Biochem. Biophys.* **119**, 41 (1967).

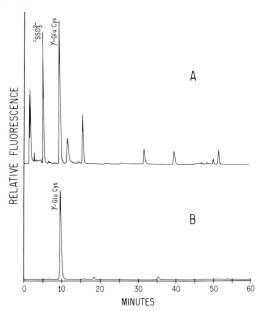

FIG. 1. HPLC analysis (method 1, this volume [14]) of mBBr-labeled samples: (A) crude acid–acetonitrile extract of *H. halobium* R-1; (B) purified mBBr derivative from *H. halobium* R-1.

mM Tris at pH 8.0 containing 1 mM DTPA and then eluted with 3 mM DTT in the same buffer. The thiol content of the eluate was determined by titration with DTNB and 1.5 equivalents of mBBr added as a 200 mM solution in acetonitrile. Reaction was allowed to proceed 10 min in the dark and was stopped by addition of acetic acid to 1%. Excess mBBr and the mBBr derivatives of 2-thiopyridine and DTT were removed by extracting 6 times with an equal volume of ethyl acetate. The recovery of the unknown derivative as determined by HPLC analysis using Method 1 (see this volume [14]) was 96% at this stage. The product was protected from light and lyophilized.

Preparative HPLC purification was accomplished with an Altex Ultrasphere ODS 5 μm 10 × 250 mm column operated at room temperature and a flow rate of 3 ml/min. Buffer components, selected for their volatility, were as follows: buffer A, 0.25% aqueous acetic acid titrated to pH 3.5 with triply distilled pyridine; buffer B, methanol. The elution program with linear gradients was the following: 0 min, 10% B; 20 min, 25% B; 25 min, 100% B; 27 min, 100% B; 28 min, 10% B; 40 min, reinjection. The derivatized and lyophilized thiopropyl-Sepharose eluate (see above) was taken up in 0.5% aqueous acetic acid, 100–200 μl aliquots were injected on the preparative column, and the peak eluting at 20 min was manually collected, pooled, and lyophilized. HPLC analysis of the isolated product (Fig. 1B) showed that it was highly purified relative to the crude material and was obtained in an overall yield of greater than 90%.

Electrolytic Reduction

A mercury pool electrolytic reduction cell, similar to that used by Kadin and Poet,[6] was constructed from a 30-mm section of 10-mm glass tubing fitted at one end with a rubber septum cap through which a platinum wire serving as cathode was inserted. Approximately 400 μl of mercury and a 1 × 3 mm magnetic stirring bar were placed in the tube, and 1 ml of the thiol-mBBr derivative in 100 mM aqueous acetic acid added on top. A U-shaped 1/8 in. O.D. Teflon tube filled with 2% agar in saturated sodium chloride served as an agar bridge to connect the electrolysis cell with a 10-ml beaker containing saturated sodium chloride and the platinum wire anode. The electrodes were connected to a Hewlett-Packard Model 6299A power supply and the electrolysis performed at constant voltage. Voltages given below refer to applied voltage.

The progress of the reduction was monitored by HPLC analysis of the remaining thiol-mBBr derivative, and the thiol production was determined from the increase in the derivative found on analysis after derivatization of the thiol formed with mBBr. At low voltage (~2 V) reduction of an 0.2 mM solution of the GSH-mBBr derivative was slow ($t_{1/2}$ ~25 min) and HPLC analysis showed that formation of *syn*-(methyl,methyl)bimane occurred in consort with disappearance of the mBBr derivative and appearance of the free thiol. The yield of thiol was greater than 95% when tested with the derivative of glutathione (see Meliss et al.[7] for a discussion of the electrochemistry of bimanes). At higher voltages (3–10 V) the reaction was faster but the *syn*-(methyl,methyl)bimane itself was reduced to nonfluorescent products (Fig. 2), this process becoming so rapid at 10 V that no *syn*-(methyl,methyl)bimane was detected during monitoring of the reaction progress by HPLC analysis. The yield also decreased with increasing voltage, the recovery of GSH from the corresponding derivative being 90% after 15 min at 3 V and 75% after 10 min at 10 V. Electrolytic reduction of the purified thiol derivative from *H. halobium* at 10 V for 10 min resulted in a 75% recovery of the thiol.

After lyophilization, the product was oxidized to the corresponding disulfide in 1 ml of 0.1 M ammonium hydroxide under a stream of dry

[6] H. Kadin and R. B. Poet, *J. Pharm. Sci.* **72,** 1134 (1982); see also this volume [47].
[7] A. Melis, N. S. Kosower, N. A. Crawford, E. Kirowa-Eisner, M. Schwarz, and E. M. Kosower, *Photochem. Photobiol.* **43,** 583 (1986).

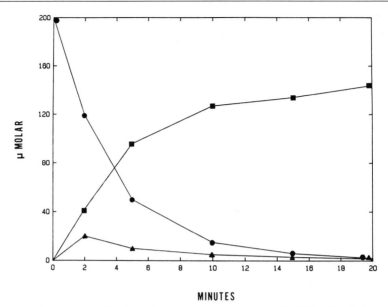

FIG. 2. Progress curve for electrolytic reduction of 1 ml 0.2 mM GSH-mBBr derivative at 5 V: (●) GSH-mBBr derivative assayed by HPLC; (■) GSH assayed by rederivatization with mBBr and HPLC; (▲) *syn*-(methyl,methyl)bimane assayed by HPLC.

oxygen, the progress being monitored by titration with DTNB and the ammonium hydroxide being replenished as needed. This material was used for amino acid analysis, N-terminal amino acid determination, and other structural studies without removal of bimane coproducts and was shown to correspond to γ-glutamylcysteine.[2] Removal of *syn*-(methyl, methyl)bimane following reduction is possible by extraction with ethyl acetate, but its electrolytic reduction products have not been identified and their removal may require an additional preparative HPLC step or other isolation procedure.

General Comments

The monobromobimane derivative of a thiol has unique characteristics that make it especially attractive as the form to be purified in the isolation of an unknown thiol compound. Blocking of the thiol with the bimane group stabilizes it toward oxidation and provides a fluorescent label that can be easily followed and quantitated. The bimane label also greatly increases the affinity of simple ionic thiols for reversed-phase HPLC column packings and enhances the degree of purification that can be obtained by preparative HPLC procedures. The fluorescent label allows the purified derivative to be readily characterized with respect to its behavior on electrophoresis and thin-layer chromatography, and facilitates structural studies of the derivative. Finally, the ability to electrolytically remove the bimane label to regenerate the thiol provides access to the unmodified form of the unknown compound.

Monobromobimane labeling has proved useful in the labeling of protein thiol groups and in the study of the peptides derived from the labeled protein.[8] The present methods should extend this utility by providing a technique for the purification of cysteine-containing peptides generated in protein structural studies.

Acknowledgments

Support of this research by the National Institutes of Health (Grant CA-39582) and the National Aeronautics and Space Administration (Grant NAGW-342) is gratefully acknowledged.

[8] N. S. Kosower, G. L. Newton, E. M. Kosower, and H. M. Ranney, *Biochim. Biophys. Acta* **622,** 201 (1980).

[16] High-Performance Liquid Chromatography of Thiols and Disulfides: Dinitrophenol Derivatives

By MARC W. FARISS and DONALD J. REED

High-performance liquid chromatography (HPLC) provides a rapid, sensitive, and reproducible means of separating and quantifying simultaneously a variety of sulfur-containing amino acids and related derivatives. The HPLC method described in this chapter was modified from the procedure of Reed *et al.*[1] and is based on the initial formation of S-carboxymethyl derivatives of free thiols followed by the conversion of free amino groups to 2,4-dinitrophenyl (DNP) derivatives. Following derivatization, nanomole levels of individual sulfur-containing amino acids are measured using UV detection at 365 nm after separation by reverse-phase ion-exchange HPLC. Because of the versatility of this HPLC method, biological specimen preparation as well as derivatization and HPLC analysis procedures are discussed.

[1] D. J. Reed, J. R. Babson, P. W. Beatty, A. E. Brodie, W. W. Ellis, and D. W. Potter, *Anal. Biochem.* **106,** 55 (1980).

Sample Preparation

Principle. The accurate measurement of sulfur-containing compounds in biological specimens relies, in part, on sample preparation, i.e., the rapid termination of metabolic processes in fluids, cells, and tissues and the prevention of thiol oxidation and thiol–disulfide interchange during the assay procedure.[2] To satisfy these criteria, biological specimens are treated promptly with perchloric acid (PCA; 1 M or 10%) containing a metal chelator such as ethylenediaminetetraacetic acid (EDTA) (2 mM).[3] Because EDTA inteferes with the HPLC measurement of glutathione disulfide (GSSG), we are currently using 1 mM bathophenanthrolinedisulfonic acid (BPDS), a water-soluble phenanthroline derivative with excellent iron- and copper-chelating properties.

Reagents

Perchloric acid (PCA), 70% (double distilled, G. F. Smith Chemicals).

Perchloric acid (PCA), 10% v/v in metal-free water (Milli-Q Reagent Water System, Millipore Corp.) containing 1 mM bathophenanthrolinedisulfonic acid (BPDS) (G. F. Smith Chemicals).

γ-Glutamyl glutamate (γ-Glu-Glu, Vega Biochemicals) 15 mM, in 0.3% PCA, stored at $-10°$, stable at room temperature for several days, used as an internal standard.

γ-Glutamyl glutamate, 10 mM, in metal-free water, stored at $-10°$; used as an internal standard (protein–glutathione mixed disulfide assay).

Dibutyl phthalate ($d = 1.046$).

Bathophenanthrolinedisulfonic acid (BPDS), 15 mM, in metal-free water.

N-Morpholinopropanesulfonic acid (MOPS) buffer at pH 8.0 containing MOPS, 50 mM, and 1,4-dithiothreitol (DTT), 25 mM.

Digitonin buffer at pH 7.4 containing mannitol (250 mM), MOPS (17 mM), EDTA (2.5 mM), and digitonin (0.8 mg/ml, Sigma Chemical Co., St. Louis, MO).

Standard Procedure

Tissue specimens to be analyzed are frozen in liquid nitrogen and pulverized to powder. The frozen, powdered tissue (50–100 mg) is weighed in a microcentrifuge tube, and 1 ml of 10% PCA with 1 mM

[2] P. C. Jocelyn, "Biochemistry of SH Group," p. 94. Academic Press, New York, 1972.
[3] T. P. M. Akerboom and H. Sies, this series, Vol. 77, p. 373.

BPDS is added. The acid extract is sonicated, frozen and thawed, and centrifuged at 15,000 g for 3 min. A portion of the resulting PCA extract (0.5 ml) is added to a tube containing 50 μl γ-Glu-Glu (15 mM); it is derivatized and analyzed by HPLC.

Cells, subcellular fractions in suspension, cell medium, plasma, or serum (0.5 ml) are treated by adding 50 μl γ-Glu-Glu (15 mM), 50 μl BPDS (15 mM), and 100 μl 70% PCA. Following mixing with a vortex mixer, freezing, thawing, and centrifugation, a portion of the 10% PCA extract (0.5 ml) is derivatized and analyzed by HPLC.

Derivatization

Principle. The derivatization procedure includes the reaction of iodoacetic acid with thiols to form S-carboxymethyl derivatives followed by chromophore derivatization (DNP) of primary amines with Sanger's reagent, 1-fluoro-2,4-dinitrobenzene.

Reagents

Iodoacetic acid, 100 mM, dissolved in a 0.2 mM m-cresol purple solution, stored at 4°, stable for at least 7 days.

Iodoacetic acid, 1 M, dissolved in a 0.2 mM m-cresol purple solution (used in the protein–glutathione mixed disulfide assay).

m-Cresol purple, 0.2 mM, dissolved in metal-free water, stored in the dark in a container for light-sensitive compounds.

KOH (stock), 10 M, dissolved in metal-free water.

KHCO$_3$ (stock), 3 M, dissolved in metal-free water.

KOH (2M)–KHCO$_3$ (2.4 M), working solution.

1-Fluoro-2,4-dinitrobenzene, 1% v/v in ethanol, stored at 4° in the dark in a container for light-sensitive compounds.

Lysine, 1 M.

Procedure

To 0.5 ml of acid extract (10% PCA with 1 mM BPDS) containing the internal standard, γ-Glu-Glu, 50 μl of 100 mM iodoacetic acid in a 0.2 mM m-cresol purple solution is added (use 1 mM iodoacetic acid in the protein–glutathione mixed disulfide assay). The acidic solution (pink in color) is brought to pH 8–9 (purple in color) by the addition of 0.48 ml of KOH (2 M)–KHCO$_3$ (2.4 M) and allowed to incubate in the dark at room temperature for 10 min. Next, 1 ml of 1% fluorodinitrobenzene is added, and the reaction mixture is capped, mixed with a vortex mixer, and stored at 4° overnight. Lysine (1 M, 50 μl) may be added to the derivatization solution to eliminate unreacted fluorodinitrobenzene. The final solution is stored in the dark at 4° until HPLC analysis.

Comments

The incubation periods required for complete derivatization are as follows [with only glutathione (GSH), GSSG, cysteinyl-glutathione disulfide (CySSG), and γ-Glu-Glu being tested]: (1) S-carboxymethylation (iodoacetate and KOH–KHCO$_3$) occurs within 5 min, (2) N-dinitrophenyl derivatization (fluorodinitrobenzene) occurs within 1 hr or within 4 hr at 4°, and (3) elimination of excess fluorodinitrobenzene (lysine) occurs within 4 hr at 4°. The S-carboxymethyl and N-DNP derivatives are stable at 4° in the dark for at least 2 weeks. Accurate and reproducible measurements of derivatized samples stored for up to 6 weeks, however, can be obtained using the internal standard method (γ-Glu-Glu) during HPLC analysis.

The modifications of the HPLC method of Reed *et al.*[1] that are reported here were adapted to further decrease sample preparation- and derivatization-induced thiol oxidation. Attempts to protect against thiol oxidation include the following: (1) the use of metal-free water and PCA; (2) the presence of BPDS, a metal chelator, during sample preparation and derivatization; and (3) rapid S-carboxymethyl derivatization resulting from the instantaneous pH change (KOH–KHCO$_3$) in the presence of iodoacetate. Using the assay conditions described above, less than 0.7% of the GSH is converted to GSSG during sample preparation and derivatization as determined by HPLC analysis. GSSG levels are routinely corrected for this oxidative artifact (0.7% GSH).

The presence of *m*-cresol purple in the iodoacetate solution not only functions as a pH indicator during derivatization but also acts as a preservative of iodoacetate while in solution. Iodoacetate dissolved in water diminishes over several hours, but iodoacetate dissolved in a *m*-cresol purple solution maintains alkylating activity for at least 7 days at 4°, thus eliminating the need for hourly preparation.

HPLC Analysis

Principle. DNP derivatives of acidic amino acids (including thiol-containing compounds) are separated on a 3-aminopropyl column by reversed-phase ion-exchange HPLC. In the mobile phase, methanol is used to elute rapidly the excess 2,4-dinitrophenol and the DNP derivatives of basic and neutral amino acids. Acetic acid is present in the mobile phase to maintain the bonded-phase amino groups in the protonated form. By increasing the sodium acetate concentration of the mobile phase, selective elution of acidic DNP derivatives is accomplished. The eluted DNP derivatives are measured by detection at 365 nm.

Reagents and Equipment

> Sodium acetate solution (stock), 2.6 M, prepared by adding 1 kg sodium acetate (HPLC grade) and 448 ml of metal-free water to 1.39 liters of glacial acetic acid (HPLC grade), is stirred with low heat until sodium acetate is in solution, then is cooled and filtered with a 4.0–5.5 μm polypropylene filter.
> Mobile phase A, 80% methanol, prepared by adding 800 ml of metal-free water to 3.2 liters of methanol (HPLC grade).
> Mobile phase B, 0.5 M sodium acetate in 64% methanol, prepared by adding 800 ml of the stock sodium acetate solution to 3.2 liters of 80% methanol.
> HPLC column: 3-Aminopropyl-Spherisorb, 20 cm × 4.6 mm, 5-μm particles (Custom L. C. Inc., or see Reed et al.[1] for preparation).
> HPLC system: Gradient capability with a UV detector (365 nm), a recording integrator, and an autosampler (optional).

Procedure

DNP derivatives are separated and measured using an HPLC system equipped with a 3-aminopropyl column and a UV detector (365 nm). Following a 100-μl injection of the derivatization solution (after centrifugation), the mobile phase is maintained at 80% A, 20% B for 5 min followed by a 10-min linear gradient to 1% A, 99% B at a flow rate of 1.5 ml/min. The mobile phase is held at 99% B until the final compound (usually GSSG) has eluted (5–10 min). Table I contains a list of compounds that have been separated and measured using this technique, and Figure 1 shows the separation of a select mixture of DNP derivatives.

Comments

An internal standard method is routinely used to determine the concentration of amino acids and related compounds in the derivatization sample following HPLC analysis. The method, with γ-Glu-Glu as the internal standard, accurately corrects for alterations in derivatization efficiency (*N*-DNP), derivatization stability (*N*-DNP), chromatographic conditions, and injection volumes (autosampler). Thus, the accuracy of the method with γ-Glu-Glu is greater than 95% as determined for GSH, GSSG, and CySSG.

TABLE I
AMINO ACIDS AND RELATED COMPOUNDS MEASURED BY HPLC ANALYSIS (DNP)[a]

Order of elution	Compound	Retention time (min)
Sulfur-containing compounds		
(12)	Cysteine (CyS)	12.2
(4)	Cystine (CySS)	8.4
(16)	Glutathione (GSH)	16.7
(17)	Glutathione disulfide (GSSG)	18.9
(15)	Cysteinyl-glutathione disulfide (CySSG)	15.5
(3)	Cystathionine	
(11)	Cysteic acid	
(7)	Cysteinylglycine	
(9)	γ-Glutamylcysteine	
(6)	Homocysteine	
(1)	Homocystine	
(5)	Homocysteic acid	
(13)	Homocysteinyl-glutathione disulfide	
(8)	Penicillamine	
Other compounds		
(10)	Aspartate (AsP)	10.7
(2)	Glutamate (Glu)	6.4
(14)	γ-Glutamyl glutamate (ISTD)	14.6

[a] The retention times for these compounds, if not listed here, have been previously reported by Reed et al.[1] using similar HPLC conditions. Retention times vary depending on the mobile-phase gradient and the length and age of the HPLC column.

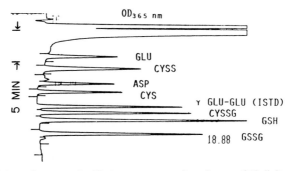

FIG. 1. High-performance liquid chromatogram of a mixture of N-dinitrophenyl-S-carboxymethyl derivatives of amino acids and related compounds under conditions described in the text. The abbreviations are given in Table I.

The sensitivity of the HPLC method is related, in part, to the number of amines available for chromophore (DNP) derivatization. Consequently, disulfides (CySSG and GSSG) are measured at concentrations as low as 0.5 nmol/ml or 25 pmol/injection whereas the limit of detection for thiols (GSH) is 1.0 nmol/ml or 50 pmol/injection.

The elimination of excess fluorodinitrobenzene in the derivatization mixture (by addition of lysine) is an attempt to extend the life of the HPLC column by preventing the reaction of the reagent with the amino-bonded phase. The addition of lysine is considered optional because data suggesting column preservation by its addition are insufficient at this time. We can only speculate that the presence of lysine in the derivatization mixture will extend column life past the 1000 analyses per column observed without lysine. The addition of lysine, however, does not adversely affect derivatization or HPLC analysis.

Applications

Measurement of Hepatocyte Intracellular Thiols and Disulfides

Because thiols and disulfides function in many intracellular compartments, defining a cellular role for these compounds requires the simultaneous measurement of intracellular, subcellular, and extracellular components. To facilitate these measurements, we have recently developed a rapid separation technique for hepatocytes in suspension.[4] This technique offers the advantage of analyzing only viable cells for changes in cellular thiol and disulfide content and thus eliminates the measurement of cellular alterations that occur after cell death (nonviable cells).[4] Separation of viable hepatocytes is accomplished in a microcentrifuge tube by layering a sample of hepatocyte suspension over a dibutyl phthalate oil layer, which is layered over PCA, and then centrifuging. Only viable hepatocytes are forced through the oil layer and into the 10% PCA, which then causes release of intracellular contents. The acid-soluble, sulfur-containing compounds (see Table I) are then measured in the 10% PCA layer, the protein–thiol mixed disulfides are measured in the resulting PCA precipitate, and the medium above the oil layer is analyzed for extracellular thiol and disulfides (Fig. 2). Following alteration of the cell suspension by digitonin, a process by which membranes become selectively permeable, the mitochondrial thiols and disulfides are measured in the PCA layer following centrifugation through the dibutyl phthalate oil layer (Fig. 2).

---

[4] M. W. Fariss, M. K. Brown, J. A. Schmitz, and D. J. Reed, *Toxicol. Appl. Pharmacol.* **79**, 283 (1985).

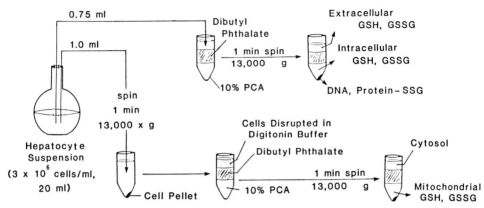

FIG. 2. Sample preparation and analysis of intracellular, extracellular, and mitochondrial thiols and disulfides by the dibutyl phthalate separation method.[4]

Intracellular. To use the dibutyl phthalate separation method[4] (Fig. 2), 0.75 ml of hepatocyte suspension is layered over dibutyl phthalate, layered over PCA with 1 mM BPDS (0.55 ml), and then centrifuged for 1 min. Intracellular thiol and disulfide levels (nonprotein) are measured in the supernatant liquid by transferring 0.5 ml of the acid to a tube containing 50 μl γ-Glu-Glu (15 mM, ISTD) followed by derivatization and analysis by HPLC.

Extracellular. A novel *in vitro* method for determining the selective efflux of glutathione (GSH) and glutathione disulfide (GSSG) from cells has recently been described.[5,6] This method takes advantage of the thiol–disulfide interchange reaction[7] between cystine (0.2 mM), which is added to the medium, and extracellular GSH, resulting in the stoichiometric formation of cysteinyl-glutathione disulfide (CySSG). Because the formation of the mixed disulfide does not affect extracellular GSSG, the cellular efflux of GSH and GSSG can be distinguished by analyzing cell medium for CySSG and GSSG concentrations, respectively, by the HPLC method reported here.

Protein–Glutathione Mixed Disulfides. Following treatment of the biological specimen with 10% PCA, the resulting protein precipitate is washed with 0.5 ml cold ethanol[8] to denature the protein and eliminate

---

[5] M. W. Fariss and D. J. Reed, in "Isolation, Characterization and Use of Hepatocytes" (R. A. Harriss and N. W. Cornell, eds.), p. 349. Elsevier Biomedical, New York, 1983.
[6] M. W. Fariss, K. Olafsdottir, and D. J. Reed, *Biochem. Biophys. Res. Commun.* **121,** 102 (1984).
[7] P. C. Jocelyn, "Biochemistry of SH Group," p. 121. Academic Press, New York, 1972.
[8] J. C. Livesey and D. J. Reed, *Int. J. Radiat. Oncol. Biol. Phys.* **10,** 1507 (1984).

residual nonprotein thiols. After mixing with a vortex mixer and centrifugation, the ethanol is discarded and 50 μl γ-Glu-Glu (10 mM in water) and 0.5 ml of a MOPS–DDT buffer are added to the protein pellet. The mixture is sonicated and incubated with shaking for 60 min at 37°, permitting the release of GSH (thiols) from protein–glutathione (thiol) mixed disulfides.[9] At the end of the incubation period, 50 μl BPDS (15 mM) and 0.1 ml of 70% PCA are added, and the sample is processed in a vortex mixer, frozen, thawed, and centrifuged. The 10% PCA extract (0.5 ml) is derivatized and analyzed by HPLC. Aliquots of a solution of GSH-derivatized bovine serum albumin (0.25, 0.1, 0.05, 0.025 ml) are used as controls for the measurement of protein–glutathione mixed disulfides (protein–SSG).[10] The above procedure, adapted from the work of Meredith[10] and Livesey and Reed,[8] may also be applicable to the measurement of nonglutathione thiol–protein mixed disulfides.

Mitochondrial. The rapid isolation of mitochondria from parenchymal liver cells (Fig. 2) was adapted from the method of Meredith and Reed.[11] Hepatocytes (3×10^6 cells) are suspended in 0.6 ml of digitonin buffer and incubated at for 1 min at room temperature, thus solubilizing cholesterol-containing membranes (plasma membrane) while leaving mitochondrial membranes intact. After the incubation period, the digitonin–cell suspension is layered over dibutyl phthalate (0.45 ml) which is layered over 10% PCA with 1 mM BPDS (0.55 ml) and centrifuged for 1 min (Fig. 2). The dense mitochondria are forced through the dibutyl phthalate layer and into the 10% PCA layer thereby releasing mitochondrial contents. A portion of the PCA layer (0.5 ml) is transferred to a tube containing 50 μl γ-Glu-Glu (15 mM) and is derivatized and analyzed by HPLC.

Acknowledgments

This work was supported by Grants ES01978 and ES07060 from the National Institute of Environmental Health Sciences and by a grant-in-aid, an A. D. Williams grant, and a School of Medicine grant from the Medical College of Virginia, Virginia Commonwealth University.

[9] See this volume [20], for a different means of measuring protein–thiol disulfides.
[10] M. Meredith, *Anal. Biochem.* **131**, 504 (1983).
[11] M. J. Meredith and D. J. Reed, *J. Biol Chem.* **257**, 3747 (1982).

[17] High-Performance Liquid Chromatography of Hepatic Thiols with Electrochemical Detection

By EUGENE G. DEMASTER and BETH REDFERN

Introduction

The application of HPLC with electrochemical detection to the separation and quantitation of biologically relevant thiols was first described by Rabenstein and Saetre.[1] This procedure was subsequently adapted for the measurement of reduced glutathione (GSH), cysteine, and homocysteine in liver homogenates.[2,3] Graphite, glassy carbon, and mercury have all been used as the working electrode for the measurement of thiols; however, mercury is the preferred electrode material for this application. The operating potential of the mercury-based electrode is significantly lower than the more commonly used carbon-based electrodes,[4] and, therefore, certain electrochemically active substances such as uric acid and ascorbate that can interfere with thiol analysis when using carbon-based electrodes are not detected with mercury-based electrodes.[4] The attractive features of this method for the analysis of hepatic thiols include (1) ease of sample preparation, (2) high detector sensitivity for thiols, and (3) detector insensitivity toward electrochemically inactive substances present in biological samples.

Principle. Hepatic GSH, cysteine, and homocysteine are separated by HPLC using a cation-exchange column. The half-reaction which occurs at the surface of the mercury-based working electrode of the amperometric detector is

$$2\ RSH + Hg \rightarrow Hg(SR)_2 + 2\ H^+ + 2\ e^-$$

A known, fixed quantity of mercaptoethylglycine is added to each sample as an internal standard.

Reagents

Perchloric acid, 0.3 M, in tetrasodium EDTA, 5 mM
Dipotassium hydrogen phosphate, 0.15 M, containing tripotassium citrate, 10 mM, and tetrasodium EDTA, 5 mM

[1] D. L. Rabenstein and R. Saetre, *Anal. Chem.* **49**, 1036 (1977).
[2] I. Mefford and R. N. Adams, *Life Sci.* **23**, 1167 (1978).
[3] E. G. DeMaster, F. N. Shirota, B. Redfern, D. J. W. Goon, and H. T. Nagasawa, *J. Chromatogr.* **308**, 83 (1984).
[4] L. A. Allison and R. E. Shoup, *Anal. Chem.* **55**, 8 (1983).

Mercaptoethylglycine (MEG),[5] 1.0 mM, in tetrasodium EDTA, 5 mM
N-Acetylcysteine, 0.2 mM
Solutions of GSH, cysteine, and homocysteine prepared in the perchloric acid–EDTA mixture above as external standards
Mobile phase: disodium hydrogen phosphate, 10 mM, containing citric acid, 10 mM, adjusted to pH 2.1 with metaphosphoric acid (ionic strength $\cong 0.038\ M$)

Assay Procedure

HPLC–Electrochemical Detection System. The chromatography system can be assembled from commercially available modular components. The system used by us consisted of a Rheodyne 20-μl loop sample injector (Model 7125, Berkeley, CA), a Milton Roy minipump (Model 396, Riviera Beach, FL), a Bioanalytical Systems electrochemical detector (Model LC-4, West Lafayette, IN) equipped with a thin-layer transducer (Model TL-6A) composed of a Au/Hg working electrode and a glassy carbon auxillary electrode, and a Hewlett Packard integrator (Model 3390A, Palo Alto, CA).

Separation of thiols is accomplished using a glass column (0.2 × 50 cm) and precolumn (0.2 × 5 cm), each dry packed with Zipax SCX (Dupont, Wilmington, DE). The precolumn is changed after every 500 biological samples. The mobile phase is filtered through a 0.45-μm Millipore filter (Bedford, MA) and degassed under reduced pressure for 10 min before use. An oxygen-free eluent is maintained by keeping the mobile phase reservoir (2-liter round bottom flask) between 50 and 55° using a heating mantle and by continuously flushing the head space of the reservoir with a slow flow of nitrogen gas. The column flow rate is set at 0.7 ml/min. The mobile phase effluent from the detector must be directed to waste and not recycled back to the reservoir. Contamination of the reservoir with mercury from the working electrode causes a splitting of the thiol peaks.

The potential of the working electrode is set at 0.0 V versus the Ag/AgCl reference electrode. Before each injection, a 10-sec cleansing voltage (-0.2 V) is applied to the working electrode. Omission of this cleansing pulse results in a gradual decline in detector response.

Sample Preparation and Chromatography. Rat liver thiols are separated and quantitated by HPLC–electrochemical detection as follows: After perfusing the liver *in situ* with isotonic saline, an approximate 0.5-g section of the lower median lobe of the liver is excised and weighed, and

---

[5] MEG was synthesized as described by H. T. Nagasawa, J. A. Elberling, and E. G. DeMaster, *J. Med. Chem.* **23**, 140 (1980).

then immediately homogenized in a volume (~2.5 ml) of cold perchloric acid–EDTA equivalent to 5 times the weight of the liver section. Homogenization is carried out using a Potter–Elvehjem tissue grinder with a serrated Teflon pestle. The homogenate is centrifuged at 12,000 g for 15 min yielding a clear supernatant fluid. An aliquot of 1 ml of the protein-free solution is mixed with an equal volume of cold phosphate–citrate–EDTA solution. After the mixture is allowed to stand for a minimum of 30 min at 0°, the sample is centrifuged for 15 min at 12,000 g to remove the potassium perchlorate precipitate. For analysis, 0.1 to 0.4 ml of the perchlorate-free solution (depending on the anticipated levels of thiols), 0.1 ml of MEG (internal standard) and 0.1 ml of N-acetylcysteine are added to 1.4 to 1.7 ml of the mobile phase for a final volume of 2.0 ml. After bubbling with a stream of nitrogen gas to remove dissolved oxygen, an aliquot of 20 μl of the final solution is injected onto the column.

The retention times for GSH, cysteine, homocysteine, and MEG are 1.7, 2.6, 3.8, and 6.8 min, respectively. N-Acetylcysteine, which elutes from the column before GSH, is added to mask a negative peak which interferes with the electronic integration of the GSH peak area.[1]

Standard Curves, Recoveries, and Sample Storage. Standard curves for each thiol injected, including MEG, are linear over a range of 0–1 nmol as are the peak area ratios of the thiol standards prepared over a concentration range of 0 to 2 mM and carried through the sample preparation procedure with MEG as the internal standard when plotted against thiol concentrations. The recovery for thiols in standard solutions processed through the sample preparation procedure is 100%. When standard thiols are added to a rat liver homogenate prepared with perchlorate–EDTA, typical recoveries for GSH, cysteine and homocysteine are 96, 94, and 77%, respectively.[3]

Although it is generally prudent to analyze biological samples as soon as possible after preparation, the samples can be stored (70°) as their perchlorate-free supernatant solutions for at least 11 days without significant losses.[3] Solutions of standard thiols in perchloric acid–EDTA and their perchlorate-free supernatant solutions are stable at −20 or 4° for a minimum of 2 weeks.

Calculations. Liver thiol levels are calculated using peak area ratios as follows:

$$\text{Concentration of unknown} = \frac{\text{Peak areas of unknown/MEG} \times 6.0 \text{ (dilution factor)} \times \text{concentration of external standard}}{\text{Peak areas of external standard/MEG}}$$

TABLE I
HEPATIC GSH, CYSTEINE, AND HOMOCYSTEINE LEVELS IN PRETREATED RATS[a]

Treatment	Electrochemical detection–HPLC				Spectrophotometry, total thiols[b,c]
	GSH	Cysteine	Homocysteine	Total thiols[d]	
Saline	4.07 ± 2.01	0.05 ± 0.032	0.087 ± 0.006	4.20 ± 0.71	4.67 ± 0.23
Pargyline (1.0 mmol/kg)	0.53 ± 0.14	0.023 ± 0.007	0.099 ± 0.008	0.66 ± 0.14	0.83 ± 0.14
L-Cysteine (1.25 mmol/kg)	5.04 ± 0.33	0.187 ± 0.025	0.075 ± 0.010	5.30 ± 0.31	5.61 ± 0.29

[a] The drugs were administered, i.p., to male, overnight fasted rats (n = 5 or 6) 1 hr before sacrifice. Thiol levels are from DeMaster et al.,[3] and are given in μmoles per gram wet weight tissue.
[b] Using Ellman's reagent according to H. S. Buttar, A. Y. K. Chow, and R. H. Downie, Clin. Exp. Pharmacol. Physiol. **4**, 1 (1977).
[c] Total thiols for the two methods were compared using the paired Student's t test; the differences were found to be nonsignificant.
[d] Equals sum of the individual thiols. Data are not corrected for recovery.

The combined external standard containing GSH, cysteine, and homocysteine in perchloric acid–EDTA should be carried through the sample preparation procedure.

Comments

The total hepatic thiol levels determined in this manner are in good agreement with those measured spectrophotometrically using Ellman's reagent (Table I). This relation holds for livers of control animals as well as for livers from animals depleted of GSH using pargyline, a hepatotoxic substance,[6] and for livers from animals administered L-cysteine, a biochemical precursor of GSH.

Electrochemical detection in conjunction with HPLC is particularly useful for studies where (1) the levels of thiols other than GSH are of interest and (2) sulfhydryl substances, e.g., L-cysteine or its prodrug forms, are administered for therapeutic intervention. In the latter case the level of total hepatic thiols may be significantly altered. However, for studies which do not involve exogenous sulfhydryl substances and in which only GSH levels are of interest (and also do not involve exogenous sulfhydryl compounds), a reasonable estimate of the actual GSH levels may be achieved spectrophotometrically using Ellman's reagent, since GSH constitutes more than 95% of the total hepatic thiols in normals and about 80% of the total thiols in GSH-depleted livers of rats (Table I).

This technique should also be readily applicable to the analysis of thiols in (1) brain[2] and other tissues, (2) body fluids,[7-9] (3) isolated cell preparations, and (4) model *in vitro* systems.[10] With dual Au/Hg electrodes placed in series, thiols and their disulfides can be quantified in biological samples.[11]

---

[6] E. G. DeMaster, H. W. Sumner, E. Kaplan, F. N. Shirota, and H. T. Nagasawa, *Toxicol. Appl. Pharmacol.* **65,** 390 (1982).
[7] D. L. Rabenstein and R. Saetre, *Clin. Chem.* **24,** 1140 (1978).
[8] R. Saetre and D. L. Rabenstein, *Anal. Biochem.* **90,** 684 (1978).
[9] F. Kreuzig and J. Frank, *J. Chromatogr.* **218,** 615 (1981).
[10] S. M. Lunte and P. T. Kissinger, *J. Chromatogr.* **317,** 579 (1984).
[11] S. M. Lunte and P. T. Kissinger, *J. Liq. Chromatogr.* **8,** 691 (1985).

[18] Analysis for Disulfide Bonds in Peptides and Proteins

By THEODORE W. THANNHAUSER, YASUO KONISHI, and HAROLD A. SCHERAGA

Introduction

Disulfide bonds make major contributions to the stability of the native conformations of proteins. The determination of the number of disulfide bonds per molecule is therefore crucial to structural studies of proteins. Many methods for the determination of disulfide bond concentration have been proposed.[1-17] Unfortunately, none of these methods has proved suitable as a routine assay procedure because they are either insensitive,[3-11] nonquantitative,[10-13] or cumbersome.[14-17] Recently, a new method involving the use the reagent 2-nitro-5-thiosulfobenzoate (NTSB) has been introduced.[18] This method has significant advantages when compared to those cited above, the two most important being that it is both sensitive and quantitative.[18] Furthermore, the reaction of NTSB with thiols and disulfides can be carried out in the presence of both dissolved oxygen and the reducing agent, sodium sulfite, which eliminates the inaccuracy and inconvenience associated with the necessity to work under an oxygen-free atmosphere as well as the need to remove the reducing agent.

The NTSB assay is actually composed of two sequential reactions.

---

[1] R. Cecil and J. R. McPhee, *Biochem. J.* **60**, 496 (1955).
[2] J. R. McPhee, *Biochem. J.* **64**, 22 (1956).
[3] R. Cecil and J. R. McPhee, *Adv. Protein Chem.* **14**, 255 (1959).
[4] J. L. Bailey and R. D. Cole, *J. Biol. Chem.* **234**, 1733 (1959).
[5] S. J. Leach, *Aust. J. Chem.* **13**, 520 (1960).
[6] S. J. Leach, *Aust. J. Chem.* **13**, 547 (1960).
[7] J. L. Bailey, "Techniques in Protein Chemistry," p. 108. Am. Elsevier, New York, 1962.
[8] R. Cecil, *in* "The Proteins" (H. Neurath, ed.), 2nd ed., Vol. 1, p. 379. Academic Press, New York, 1963.
[9] K. S. Iyer and W. A. Klee, *J. Biol. Chem.* **248**, 707 (1973).
[10] W. L. Zahler and W. W. Cleland, *J. Biol. Chem.* **243**, 716 (1968).
[11] W. L. Anderson and D. B. Wetlaufer, *Anal. Biochem.* **67**, 493 (1975).
[12] D. Cavalini, M. T. Graziani, and S. Dupré, *Nature (London)* **212**, 294 (1966).
[13] A. F. S. A. Habeeb, *Anal. Biochem.* **56**, 60 (1973).
[14] S. Moore, *J. Biol. Chem.* **238**, 235 (1963).
[15] A. M. Crestfield, S. Moore, and W. H. Stein, *J. Biol. Chem.* **238**, 622 (1963).
[16] D. D. Gilbert, *Anal. Chem.* **41**, 1567 (1969).
[17] P. D. J. Weitzman, *Anal. Biochem.* **76**, 170 (1976).
[18] T. W. Thannhauser, Y. Konishi, and H. A. Scheraga, *Anal. Biochem.* **138**, 181 (1984).

The first reaction is the cleavage of a disulfide bond with sodium sulfite[4,19]:

$$RSSR' + SO_3^{2-} \rightleftharpoons RSSO_3^- + R'S^- \qquad (1)$$

The kinetics and equilibrium of this reaction have been studied in detail.[4,6,20] It has been shown that, above pH 9, reaction (1) is a reversible bimolecular reaction. Since it is reversible, an excess of sodium sulfite is required to ensure the complete and rapid cleavage of the disulfide bond. Furthermore, the reactivity of bisulfite toward most disulfides is much less than that of sulfite. Therefore, to ensure that all the sodium sulfite is in its most reactive form, the pH of the reaction must be much greater than the pK_2 of sulfurous acid (literature value 6.8[4]).

The second reaction involves the nucleophilic attack of the thiolate produced in reaction (1) on NTSB to yield 1 mol each of a thiosulfonate and 2-nitro-5-thiobenzoate (NTB).[18,21,22]

$$\text{NTSB} + R'S^- \rightleftharpoons \text{NTB} + R'SSO_3^- \qquad (2)$$

Since 1 mol of disulfide reacts quantitatively to yield 1 mol of NTB, the concentration of disulfide bonds can be calculated from the absorbance measured at 412 nm and the extinction coefficient of NTB (13,600 M^{-1} cm^{-1}).[23] Damodaran[24] has reported that the NTB produced in reaction (2) is readily converted to a nonchromophoric derivative in the presence of room light. The reaction that is responsible for the loss of NTB is not

---

[19] It should be noted that this reaction is represented in this manner for simplicity of presentation. It is not meant to imply that the attack of the sulfite is directed toward any one of the constituent thiols of an asymmetric disulfide.

[20] W. Stricks and I. M. Kolthoff, *J. Am. Chem. Soc.* **73**, 4569 (1951).

[21] The mechanism of this reaction has not been studied extensively but the fact that the rate of reaction (2) shows a dependence on the sulfite concentration[18] suggests that it involves an intermediate asymmetric disulfide (R'—SS—NTB), which subsequently reacts with sulfite to give a thiosulfonate and NTB.

[22] It should be noted that NTSB will also react with any free thiols present in the original protein as well as with those generated by the sulfitolysis. To calculate the number of disulfide bonds per molecule for proteins or peptides that contain both thiols and disulfides, it is first necessary to determine the number of thiols by a suitable method.[23] The number of disulfides can then be calculated by difference.

[23] G. L. Ellman, *Arch. Biochem. Biophys.* **82**, 70 (1959).

[24] S. Damodaran, *Anal. Biochem.* **145**, 200 (1985).

known; however, evidence indicates that the reaction is photoactivated.[24] Therefore, it is recommended that the NTSB–protein reaction be carried out in the dark before measuring the absorbance at 412 nm.

Preparation of NTSB

NTSB is not commercially available. However, it can be prepared easily from a common laboratory reagent, 5,5-dithiobis-2-nitrobenzoic acid, i.e., Ellman's reagent[23]; see reaction 3 of Thannhauser *et al.*[18] The details of this synthesis are as follows: 100 mg of Ellman's reagent (0.253 mmol) is dissolved in 10 ml of 1 M Na_2SO_3. The pH is adjusted to 7.5. The bright red solution is brought to 38°, and oxygen is bubbled through it with a gas dispersion tube. The reaction is judged to be complete when the solution turns to a pale yellow (~45 min). This stock solution can be stored for up to 1 year at −20°. The NTSB assay solution is prepared from the stock solution by diluting it 1:100 with a freshly prepared solution that is 2 M in guanidine thiocyanate,[25] 50 mM in glycine, 100 mM in sodium sulfite, and 3 mM in EDTA. The pH should be adjusted to 9.5. The assay solution is used directly to measure disulfide bond concentrations and is stable for up to 2 weeks when stored at room temperature.

Recommended Disulfide Assay Procedure

The recommended assay procedure is as follows: 10–200 μl of a peptide or protein solution is pipetted into 3 ml of the NTSB assay solution. The reaction mixture is incubated in the dark for 5 min for peptides or 25 min for proteins. The absorbance at 412 nm is then recorded against a blank of 3 ml of NTSB assay solution and the appropriate amount of water.

The absorbance at 412 nm was found to increase linearly with disulfide bond concentration.[18] The extinction coefficient of NTB was determined by a least-squares fit of a plot of absorbance against disulfide bond concentration.[18] It was found to be 13,900 M^{-1} cm^{-1}, which is consistent with the range of values reported in the literature.[23,26] This manual procedure has proved to be applicable to a variety of proteins and peptides and can be used routinely to measure disulfide bond concentrations with as little as 8 nmol of disulfide.[18]

[25] It is recommended that the guanidine thiocyanate be omitted if the disulfide bonds of the protein or peptide of interest are sufficiently exposed to allow the reaction to proceed in the absence of a denaturant.
[26] P. W. Riddles, R. L. Blakeley, and B. Zerner, *Anal. Biochem.* **94**, 75 (1979).

Applications

This assay procedure has several important applications, the most obvious being the determination of the disulfide bond concentration of protein or peptide solutions. If the concentration of the protein is known by an independent method, then the number of disulfide bonds per protein molecule can easily be calculated. Conversely, if the number of disulfide bonds per molecule is already known, then this assay can be used as a convenient way of determining protein concentration.

The speed and simplicity of this procedure have made it easy to automate. A particularly useful example of an automated adaptation of this procedure has been the development of a thiol- and disulfide-specific detection system for the continuous monitoring of HPLC column effluents.[27] This automated system extends the range of sensitivity of the NTSB assay down to 5 pmol, and it makes the clear identification of the disulfide- or thiol-containing peptides in HPLC chromatograms easy. The introduction of this automated system has led to the development of a two-dimensional reversed-phase HPLC technique for determining disulfide bond pairings in proteins.[27]

Equations (1) and (2) indicate that, in addition to the quantitative production of NTB, the thiols and disulfides of proteins are converted quantitatively to S-sulfocysteine. S-Sulfoproteins have many desirable properties,[28] and a slight modification of the NTSB assay procedure can be used to prepare large quantities of such materials.[29] Since reactions (1) and (2) are specific to thiols and disulfide bonds,[4,29] the only additional purification necessary after this modification is the desalting of the protein. The absence of any side reactions during sulfonation makes this an ideal procedure to prepare proteins for enzymatic digestion and subsequent peptide mapping.[27] Furthermore, the sulfonation of thiols is a reversible reaction. This has led to renewed interest in sulfonation as a preparative technique because of its obvious industrial biotechnological applications.[30] The original thiols can be regenerated by incubation in the presence of an excess of another thiol such as dithiothreitol.[31] The disulfide bonds may then be regenerated by oxidation in air.[32] Alternatively,

[27] T. W. Thannhauser, C. A. McWherter, and H. A. Scheraga, *Anal. Biochem.* **149**, 322 (1985).
[28] J. F. Pechère, G. H. Dixon, R. H. Maybury, and H. Neurath, *J. Biol. Chem.* **233**, 1364 (1958).
[29] T. W. Thannhauser and H. A. Scheraga, *Biochemistry* **24**, 7681 (1985).
[30] K. J. Hayenga, V. B. Lawlis, and B. R. Snedecor, European Patent Office, 0,114,507(A1) (1984).
[31] J. M. Swan, *Nature (London)* **180**, 643 (1957).
[32] U. T. Rüegg, this series, Vol. 47, p. 123.

the disulfide bonds can be regenerated directly from the sulfonated protein by incubation with a mixture of reduced and oxidized glutathione under appropriate conditions.[29]

Since thiosulfonates are more stable than free thiols, it is convenient to store proteins in their sulfonated form and regenerate the thiols when needed. The relative stability of sulfoproteins makes them preferable to the reduced form as the starting material for oxidative refolding studies.[29] Since they are stable, they can be characterized easily, enabling one to start the refolding study from well-defined starting material. If there is some unexplained chemical heterogeneity observed during the refolding process (as in the case of ribonuclease A[33]), then the stability of the protein can be studied in the absence of the reductant. In this way, it was demonstrated recently that the chemical heterogeneity observed during the refolding of ribonuclease A[33] is due to the specific deamidation of Asn-67,[29] and not due to reactions with free radical or peroxide contaminants in the reducing agent, as was also suggested.[33]

Conclusions

The determination of disulfide bond concentration by sulfitolysis followed by reaction of the resulting thiols with NTSB is a simple and reliable procedure. It enables one to measure concentrations manually on the nanomole level routinely. The simplicity of this method also has made it easy to automate. The automated method extends the range of detection to the picomole level, and it can be used to monitor HPLC column effluents continuously for thiol- or disulfide-containing peptides. A modification of this procedure has been used to produce preparative quantities of S-sulfoproteins and has possible industrial biotechnological applications.

[33] T. E. Creighton, *J. Mol. Biol.* **129**, 411 (1979).

[19] Assay of Enzyme-Catalyzed Oxygen-Dependent Disulfide Bond Formation

By MARK X. SLIWKOWSKI and HAROLD E. SWAISGOOD

Enzyme-catalyzed oxidation of thiols to disulfides with oxygen as the electron acceptor is mechanistically different from metal ion-catalyzed oxidation.[1] Thus, sulfhydryl oxidase-catalyzed thiol oxidation results in

[1] H. P. Misra, *J. Biol. Chem.* **249**, 2151 (1974).

the production of stoichiometric amounts of H_2O_2.[2] Two methods are presented in this chapter for assay of sulfhydryl oxidase activity and determination of H_2O_2 in the presence of thiols. One measures the reduction in the stoichiometry of O_2 consumption in the presence of catalase, while the other measures H_2O_2 from the horseradish peroxidase-catalyzed oxidation of a dye. The utility of these procedures has been demonstrated in both general assays[2] and in mechanistic studies.[3] Another method for assaying sulfhydryl oxidase activity is presented elsewhere in this volume.[4]

Method I: Oxygen Electrode–Catalase-Coupled System

To confirm the generation of H_2O_2 as a stoichiometric product of enzyme-catalyzed thiol oxidation, the Clark electrode can be used in conjunction with catalase [reaction (2)]. If glutathione (GSH) is used as substrate, then glutathione reductase and NADPH are added, ensuring that O_2 is the only limiting substrate throughout the course of the reaction [reaction (3)]. Under these conditions, the reactions occurring in the oxygraph cell are summarized by reaction (4).

$$2\ GSH + O_2 \rightarrow GSSG + H_2O_2 \quad (1)$$
$$H_2O_2 \rightarrow H_2O + \tfrac{1}{2} O_2 \quad (2)$$
$$GSSG + NADPH + H^+ \rightarrow NADP^+ + 2\ GSH \quad (3)$$
$$\tfrac{1}{2} O_2 + NADPH + H^+ \rightarrow NADP^+ + H_2O \quad (4)$$

The phenazinemethosulfate–NADH method of Robinson and Cooper[5] is used to calibrate the Clark electrode. By performing this calibration procedure, the concentration of dissolved oxygen, rather than its activity, can be measured directly on solutions containing a variety of salts, detergents and sugars.

Assay Procedure

The rate of O_2 consumption was measured with a Yellow Springs Instruments (Yellow Springs, OH) oxygen monitor equipped with a Clark electrode. The instrument was modified as described by Fridovich[6] by increasing the available bucking potential, which allows for greater sensi-

---

[2] M. X. Sliwkowski, M. B. Sliwkowski, H. R. Horton, and H. E. Swaisgood, *Biochem. J.* **209,** 731 (1983).

[3] M. X. Sliwkowski, H. E. Swaisgood, D. A. Clare, and H. R. Horton, *Biochem. J.* **220,** 51 (1984).

[4] H. E. Swaisgood and H. R. Horton, this volume [89].

[5] J. Robinson and J. M. Cooper, *Anal. Biochem.* **33,** 390 (1970).

[6] I. Fridovich, *J. Biol. Chem.* **245,** 4053 (1970).

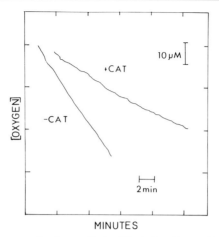

FIG. 1. Oxygen consumption by purified sulfhydryl oxidase with and without catalase (800 units).

tivity at high O_2 concentrations. Mixtures of air, O_2, and N_2, were reproducibly mixed by use of a flow meter. Equilibration is accomplished by vigorously bubbling assay solutions with gas mixtures (via a Pasteur pipette) while stirring in the thermally equilibrated oxygraph cell for at least 8 min. Prior to positioning the oxygen probe, enzymes and cofactors are added in a minimal volume. Incubations included 800 units of catalase, 0.45 units of glutathione reductase, and NADPH at a final concentration of 0.35 mM. Sulfhydryl oxidase (or other effectors) is added at this time and a stable baseline is established. Reactions are initiated by the addition of a small volume of concentrated substrate solution (e.g., 50 µl of 48 mM GSH). An example of this type of experiment, using purified bovine milk sulfhydryl oxidase from a Holstein cow[2] is shown in Fig. 1. As expected, the rate of oxygen consumption in the presence of catalase is one-half the rate for sulfhydryl oxidase alone.

In separate experiments, the addition of nitro blue tetrazolium (0.28 mM or 2.6 mM) to the assay mixture did not significantly inhibit the rate of oxygen consumption. Thus, sulfhydryl oxidase-catalyzed thiol oxidation proceeds by a direct two-electron transfer process and does not appear to produce superoxide anion. By contrast, control experiments with Cu^{2+} bis-o-phenanthroline complexes,[7] which act as a nonenzymatic catalyst for thiol oxidation, displayed marked inhibition of O_2 consumption by nitro blue tetrazolium.

[7] K. Kobashi, *Biochim. Biophys. Acta* **158**, 239 (1968).

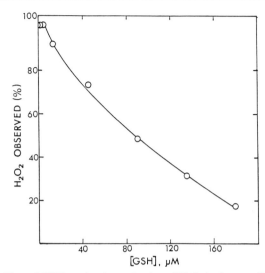

FIG. 2. The effect of GSH on the determination of H_2O_2 by horseradish peroxidase and 2,2'-azinodi(3-ethylbenzthiazolinesulfonic acid). Similar results were observed when o-dianisidine was used as a substrate for horseradish peroxidase. All incubations contained H_2O_2 at a concentration of 200 μM.

Method II: Horseradish Peroxidase System

Randall[8] was one of the first to address the problem of interference by thiols in peroxide quantitation. In this early work, it was found that thiols could be oxidized by peroxidase in the presence of peroxide and that many of the resulting oxidized dyes could then be equally well reduced by thiols. Iodometric determination of peroxide also could not be used since it was found that thiols reacted with the iodine that was liberated from the iodide by hydrogen peroxide.

The effect of GSH on the quantitation of 200 μM H_2O_2, using horseradish peroxidase and 2,2'-azinodi(3-ethylbenzthiazolinesulfonic acid) (ABTS) is shown in Fig. 2. Even 10 μM thiol will interfere with the peroxide–peroxidase reaction. Pretreatment of thiols with N-ethylmaleimide prior to incubation with peroxidase and ABTS totally alleviates thiol interference. Standard curves generated in the presence of 120 μM GSH yield no color development even at the highest concentrations of H_2O_2.

[8] L. O. Randall, *J. Biol. Chem.* **164**, 521 (1946).

TABLE I
Purification of Sulfhydryl Oxidase by Covalent Affinity Chromatography[a]

	Type of assay					
	Depletion of SH[b]		Formation of H_2O_2[c]		O_2 Consumption	
Step	Total units	Specific activity (units/mg)	Total units	Specific activity (units/mg)	Total units	Specific activity (units/mg)
Whey	40	0.005	18.2	0.002	21	0.003
Solubilized whey	17	0.031	8.6	0.015	6.6	0.012
Covalent chromatography	13	24	4.0	7.27	6	10.9

[a] Assay mixtures contained 0.8 mM GSH in 0.05 M sodium phosphate at pH 7.0. All assays were at 35°. Adapted from data presented by Sliwkowski *et al.*[2]
[b] Depletion of SH groups was followed using the DTNB reaction.[4]
[c] H_2O_2 formation was measured with horseradish peroxidase using 2,2'-azinodi(3-ethylbenzthiazolinesulfonic acid) after prior reaction with N-ethylmaleimide as described.

Assay Procedures

Standard thiol oxidation mixtures typically contain 0.5–3.0 mM GSH or other thiol in 50 mM sodium phosphate at pH 7.0. A 0.5-ml portion of the assay solution is treated with 0.2 ml of 20 mM N-ethylmaleimide in 200 mM sodium acetate, pH 5.5, for 10 min to block thiols that have not reacted. Peroxide is measured after addition of 0.5 ml of the resulting solution to 2.5 ml of the acetate buffer containing 3 μmol of ABTS and 75 μg of horseradish peroxidase. The absorbance of the resulting solution is read at 436 nm.

Comparison of Sulfhydryl Oxidase Assays

The formation of H_2O_2, in conjunction with O_2 and sulfhydryl group depletion, was substantiated throughout the purification of bovine milk sulfhydryl oxidase. Data presented in Table I confirm the stoichiometry of the enzymatic reaction shown in Eq. (1) and illustrate typical results for assay of sulfhydryl oxidase activity by these two methods, involving measurement of O_2 consumption or H_2O_2 formation, *vis-à-vis* assays of thiol group depletion as described elsewhere in this volume.[4]

[20] Cellular Protein-Mixed Disulfides

By Marjorie F. Lou, Lawrence L. Poulsen, and Daniel M. Ziegler

Potential regulation of metabolic reactions by reversible oxidation–reduction of accessible enzyme thiols has renewed interest in the redox state of cellular proteins as a function of different metabolic states.[1] While the oxidation of thiols can lead to the formation of protein-mixed disulfides as well as intramolecular disulfides, the latter are difficult to detect in crude tissue extracts. Changes in the concentration of protein-mixed disulfides are, at least in principle, more readily accessible to analysis, and a number of methods for their estimation in animal tissues have been described.[2-4] However, the estimation of endogenous protein-mixed disulfides in crude preparations presents formidable technical problems that, although generally recognized,[2] are difficult to overcome. The major problems are the prevention of artifacts during sample preparation and the analysis of low molecular weight components linked to proteins by disulfide bonds.

The concentration of thiols relative to disulfides is quite large in whole cells,[5] and oxidation of even a small fraction of thiols during sample preparation will give erroneously high values for tissue protein-mixed disulfides. It is, therefore, essential that any technique claiming reasonable accuracy must employ methods that prevent oxidation of cellular thiols during cell breakage and analysis. Whereas disrupting tissue in strong acid retards oxidation of thiols by oxidants that react with the thiolate ion, it does not prevent their oxidation by peroxides or other oxidants that react rapidly with thiols in acid solutions.[6] The addition of *N*-ethylmaleimide (NEM), successfully used to block oxidation of glutathione (GSH) to glutathione disulfide (GSSG) in tissue extracts above pH 6.0,[2] reacts rapidly only with the thiolate ion and is largely ineffective in blocking oxidation of thiols in acid solution. Since oxygenated tissues undoubtedly contain endogenous peroxides in addition to those formed upon denaturation of oxygenated hemoproteins, the use of agents to block

---

[1] D. M. Ziegler, *Annu. Rev. Biochem.* **54**, 305 (1985).
[2] T. P. M. Akerboom and H. Sies, this series, Vol. 77, p. 373.
[3] H. G. Modig, M. Edgren, and L. Revesz, *J. Radiol. Biol.* **22**, 257 (1971).
[4] A. F. S. A. Habeeb, *Anal. Biochem.* **56**, 60 (1973).
[5] A. Meister and M. E. Anderson, *Annu. Rev. Biochem.* **52**, 711 (1983).
[6] J. W. Finley, E. L. Wheeler, and S. C. Witt, *J. Agric. Food Chem.* **29**, 404 (1981).

oxidation of cellular thiols to disulfides in acid solution appears essential for accurate estimation of cellular protein-mixed disulfides.

The identification and analysis of tissue components bound to proteins by disulfide bonds is also critical. The presumed natural occurrence of a large pool of protein-GSH and protein-cysteine mixed disulfides in a variety of animal tissues was based on the observation that cellular proteins precipitated with acid liberate acid-soluble thiol upon reduction with borohydride. However, the more detailed studies of Modig et al.[3] suggest that unprecipitated protein and protein fragments account for most of the thiol present in the acid-soluble fraction, and the presumed high concentration of protein-mixed disulfide appear to be experimental artifacts. On the other hand, very small amounts of protein-mixed disulfides are apparently present in normal tissues. Current methods for their determination are described in the following sections.

Preparation of Samples

The concentration of cellular protein-mixed disulfides is presumed to change as a function of the thiol:disulfide redox potential maintained by opposing reductive and oxidative reactions.[1] Since the latter reactions are ultimately dependent on oxygen it is essential that metabolic reactions be quenched as quickly as possible in adequately oxygenated tissues. Because of these constraints, estimation of protein-mixed disulfides is usually restricted to isolated cells in culture, perfused organs, or tissue taken by freeze-clamp technique from anesthetized animals.[2] However, in the intact animal, variable amounts of blood relative to tissue in some organs (heart, kidney, lung) can interfere with measurements and more reproducible values are obtained with organs perfused with oxygenated buffer.

The formation of disulfide during cell breakage is blocked by homogenizing the tissue in buffer at pH 4.5 containing 4,4'-bis(dimethylamino)-benzhydrol (DAB-OH). At this pH, DAB-OH dissociates in aqueous solution to the resonance-stabilized carbonium-immonium ion [Eq. (1)]. The carbonium ion (BDC^+) reacts with sulfhydryls yielding the corresponding sulfides [Eq. (2)], effectively blocking the oxidation of free thiols by peroxides or other oxidants produced during cell rupture.[7]

[7] M. S. Rohbach, B. A. Humphries, E. J. Yost, Jr., W. G. Rhodes, S. Boatman, R. G. Hiskey, and J. H. Harrison, *Anal. Biochem.* **52,** 127 (1973). BDC^+ is one of the few reagents tested that can quantitatively alkylate thiols in less than 1 sec at 0–5° in acidic aqueous solutions. Thiols apparently react much faster with BDC^+ than with peroxides since addition of up to 50 μM H_2O_2 or *tert*-butyl hydroperoxide to the homogenization media did not affect concentration of either GSSG or protein–GSH mixed disulfides in rat liver (L. L. Poulsen, unpublished studies in this laboratory).

$$\left(\begin{array}{c} H_3C \\ H_3C \end{array}\!\!N\!\!-\!\!\langle\ \rangle\!\!-\right)_2\!\!CHOH \rightleftharpoons \left(\begin{array}{c} H_3C \\ H_3C \end{array}\!\!N\!\!-\!\!\langle\ \rangle\!\!-\right)_2\!\!\overset{\bullet}{C}H^+ + OH^- \quad (1)$$

$$BDC^{\ddagger}$$

$$BDC^{\ddagger} + RSH \longrightarrow \left(\begin{array}{c} H_3C \\ H_3C \end{array}\!\!N\!\!-\!\!\langle\ \rangle\!\!-\right)_2\!\!CHSR + H^+ \quad (2)$$

Reagents

10 mM 4,4'-(Dimethylamino)benzhydrol in 0.5 M sodium acetate at pH 4.5. This solution, prepared daily, is purged with argon for about 10 min at 0–5° and transferred to an oxygen-free glove box kept under positive nitrogen pressure.

3 M Trichloroacetic acid (TCA)

6 M Potassium hydroxide

Procedure. Tissue collected by freeze-clamp technique is immediately immersed in liquid nitrogen and transferred to the glove box.[8] After purging the box for about 20 min to remove oxygen introduced during tissue transfer, about 0.5 g (no less than 0.4 or more than 0.6 g) of the frozen tissue is weighed to ±1 mg on a top-loading electric balance and transferred to a 10-ml glass-Teflon homogenizer containing 5 ml of the BDC‡ solution in acetate buffer. The sample is removed from the glove box and immediately homogenized at 1,500 rpm for 1 min or less. Then, 0.6 ml 3 M TCA is added and the contents thoroughly mixed by further homogenization. A 4-ml aliquot of the uniform homogenate is transferred to a 30-ml Corex tube and precipitated protein collected by centrifugation at 10,000 rpm for 10 min.

The protein pellet is washed 3 times by resuspension and resedimentation from 20 ml of 0.3 M TCA and then from 24 ml methanol–diethyl ether (1:5 v/v). The final protein pellet is dried under reduced pressure.[9]

[8] Any of a variety of inexpensive commercially available units that can be kept under positive nitrogen pressures are suitable. While the use of a glove box is not absolutely essential, we have not found a more convenient alternative for minimizing exposure of tissue to air in the time required to fragment and weigh the frozen tissue. The concentration of disulfide increases at a slow, but perceptible, rate in frozen tissue exposed to oxygen. However, in an atmosphere of nitrogen, the concentration of disulfide does not change, and tissue can also be stored in liquid nitrogen for days without affecting the concentration of disulfide.

[9] Small quantities of protein are presumably lost in this step but this is the minimum number of washes required to remove the last traces of radiolabeled GSSG added during homogenization. The final wash with organic solvents removes unreacted BDC‡ and soluble DAB–thiol conjugates which can interfere with the estimation of acid-soluble thiols after reduction.

Liberation of Protein-Bound Thiols

Low molecular weight thiols bound to proteins by disulfide bonds can be liberated either by oxidation or reduction, and the procedures for both methods are described in the following sections.

Oxidative Method

Principle. At 0° disulfides are oxidatively cleaved by performic acid [Eq. (3)] to sulfonic acids without appreciable hydrolysis of protein pep-

$$\text{Protein—SSR} \rightarrow \text{Protein—SO}_3\text{H} + \text{RSO}_3\text{H} \qquad (3)$$

tide bonds.[10] The acid-soluble sulfonic acids are readily separated from proteins and their nature as well as concentration determined on an amino acid analyzer system.

Reagents

Performic acid prepared just prior to use by mixing (at 0–5°) 19 ml 99% formic acid with 1 ml 30% H_2O_2
Ninhydrin solution prepared and stored as described by Moore[11]
0.7 M Sulfosalicyclic acid

Procedure. The dried protein powder is suspended in 5 ml of cold performic acid and incubated at 0° for 2.5 hr. At the end of this period, the mixture is quantitatively transferred to a large round-bottomed flask, diluted with 400 ml of distilled water (prechilled to 0°) and concentrated to about 40 ml by freeze-drying. The material is quantitatively transferred to a centrifuge tube, insoluble protein removed by centrifugation, and the clear supernatant fluid, which contains acid-soluble sulfonic acids, is taken to dryness by lyophilization. The dried residue is suspended in 0.3–0.5 ml 15% sulfosalicylic acid, and, after removing precipitated protein, an aliquot of the supernatant liquid (100–150 μl) is analyzed with an amino acid analyzer.

The analyzer, modified to accommodate an anion-exchange system similar to that described by Tabor and Tabor,[12] consists of a single-column chromatographic system used according to the procedure described by Hamilton.[13] The unit is fitted with a jacketed 15 × 0.9 cm column (LDC/Milton Roy Co.) and packed with Bio-Rad Aminex A-28 anion-exchange resin. The column, maintained at 55°, is equilibrated with 0.3 M sodium acetate at pH 5.1. After injecting the sample, the column is

---

[10] C. H. W. Hirs, *J. Biol. Chem.* **219**, 611 (1956).
[11] S. Moore, *J. Biol. Chem.* **243**, 6281 (1968).
[12] C. W. Tabor and H. Tabor, *Anal. Biochem.* **78**, 543 (1977).
[13] P. B. Hamilton, *Anal. Chem.* **35**, 2055 (1963).

eluted with the same buffer at 1 ml/min for 40 min before switching to 0.5 M sodium acetate at the same pH. The eluate, mixed with ninhydrin (0.5 ml/min), is passed through a Teflon reaction coil heated in a boiling water bath (transit time 15 min) before measuring absorbance at 570 nm with a Dupont 410 photometer.

The concentrations of cysteic acid and glutathionesulfonic acid, the two principal sulfonic acids present in the extract, are calculated from standard curves obtained with the authentic compounds.[14] In this system, cysteic acid and glutathionesulfonic acid elute at 26 and 63 min, respectively, and both are well resolved from contaminating ninhydrin-positive peptides liberated by performic acid oxidation. This method has been used to measure the concentration of protein-mixed disulfides in normal and oxidatively stressed lens tissue from rats and monkeys.[14] The procedure is quite reproducible and gives values for protein-bound GSH and cysteine that are considerably less than that obtained by other methods.

Reductive Method

Principle. Thiols linked to proteins by disulfide bonds can be released from proteins by reduction [Eq. (4)]. The acid-soluble thiols are readily separated by precipitating the protein with trichloroacetic or perchloric acid.

$$\text{Protein—SSR} \rightarrow \text{Protein—SH} + \text{RSH} \tag{4}$$

While most procedures[2-4] employ either sodium borohydride or dithiothreitol (DTT) to reduce disulfides, neither reductant alone or in combination has proven satisfactory. The large amount of DTT required for complete reduction frequently interferes with subsequent assays for acid-soluble thiols, and reduction of disulfides in crude tissue preparations by borohydride is rarely quantitative.[15] We have found that reduction by a system consisting of NADPH and glutathione reductase in the presence of a small amount of DTT gives the most consistent results.

Reagents

A solution containing 25 mM sodium pyrophosphate at pH 8.4, 5 mM EDTA, 1 mM DTT, 0.2 mM NADP$^+$, 2.5 mM glucose 6-phosphate, 2 μg/ml glucose-6-phosphate dehydrogenase (*Leuconostoc mesen-*

---

[14] M. F. Lou, R. McKeller, and O. Chyan, *Exp. Eye Res.* **42,** 607 (1986).

[15] The reasons for lack of quantitative reduction are not immediately apparent; however, it is not related to lack of dispersion of denatured proteins. The yield of GSH from liver proteins by borohydride reduction was not improved by dispersing the pellet in urea or detergents, and was consistently less than that obtained with the DTT–glutathione reductase-dependent system.

teroides; Sigma Type XXIII), and 4 μg/ml glutathione reductase (yeast; Sigma Type III) is prepared daily from stock solutions and stored on ice until needed.

Procedure. The dried protein powder is resuspended in 1.5 ml of the NADPH–glutathione reductase solution and incubated at 37° for 10 min with occasional mixing. After incubation any protein in solution is precipitated by adding 0.15 ml 3 M perchloric acid. Specific acid-soluble thiols liberated by reduction are then measured by procedures described in detail in other chapters of this volume. GSH in the extract liberated from protein of perfused rat liver, measured by the sensitive enzymatic recycling technique,[16] gives values that vary from 12 to 17 nmol/g (wet weight) liver.

[16] F. Tietze, *Anal. Biochem.* **27**, 502 (1969).

[21] Measurement of Thiol–Disulfide Interchange Reactions and Thiol pK_a Values

By JANETTE HOUK, RAJEEVA SINGH, and GEORGE M. WHITESIDES

Thiol–disulfide interchange (SH/S$_2$ interchange) reactions involving proteins are important in a number of biochemical processes including formation and cleavage of structural cystines,[1] control of enzyme activities by reversible redox reactions of enzyme thiols and disulfides,[2,3] and redox processes requiring thiols.[4] The reaction is mechanistically simple: it involves initial ionization of thiol to thiolate anion, followed by nucleophilic attack of thiolate anion on the sulfur–sulfur bond of the disulfide [Eq. (1)]:

$$\text{RSH} \rightleftharpoons \text{RS}^- + \text{H}^+ \quad (1a)$$
$$\text{RS}^- + \text{ESSE} \rightleftharpoons \text{RSSE} + \text{ES}^- \quad (1b)$$
$$\text{ES}^- + \text{H}^+ \rightleftharpoons \text{ESH} \quad (1c)$$

Equation (1) makes it evident that three types of parameters must be determined to characterize fully a SH/S$_2$ interchange reaction: (1) the rates at which the displacement steps occur; (2) the values of pK_a of the

[1] T. Y. Liu, *in* "The Proteins" (H. Neurath and R. L. Hill, eds.), 3rd ed., Vol. 3, p. 239. Academic Press, New York, 1977.
[2] D. M. Ziegler, *Annu. Rev. Biochem.* **54**, 305 (1985).
[3] H. F. Gilbert, this series, Vol. 107, p. 330.
[4] P. C. Jocelyn, "Biochemistry of SH Group." Academic Press, New York, 1972.

participating thiols; and (3) the positions of the equilibria between the thiol (thiolate) and disulfide species. Here we discuss the characteristics of each of these parameters and describe methods of determining them. We place particular emphasis on methods that are useful for biologically relevant thiols and cysteine groups in proteins.

Rates of Thiol–Disulfide Interchange Reactions

Ellman's reagent [5,5'-dithiobis(2-nitrobenzoic acid)][5,6] (see this volume [11]) has been widely used in studies[7–9] of rates of SH/S_2 interchange reactions because of several reasons. (1) It is water soluble; (2) forward rates are seldom complicated by back reactions; (3) the reduction of Ellman's reagent is easily followed spectrophotometrically; and (4) Ellman's reagent has been widely used for the determination of sulfhydryl groups, and information concerning rates of its reduction by thiols is useful in these applications.

SH/S_2 interchange between most aliphatic thiols and Ellman's reagent proceeds under thermodynamic control at pH 7 to complete reduction of the Ellman's reagent. Thus, Ellman's reagent *cannot* normally be used to explore the influence of the structure of the reducing thiol on SH/S_2 interchange equilibria. Reduction of typical cystine moieties by aliphatic thiols is thermodynamically much less favored than is reduction of Ellman's reagent. Thus, Ellman's reagent, though convenient to study, is a poor model for protein cystine groups.

Glutathione disulfide [GSSG, (γ-glutamylcysteinylglycine)$_2$] is a better cystine-containing peptide for use as a model in the study of rates of SH/S$_2$ interchange.[10] The rate of release of glutathione (GSH) on reduction of GSSG by thiols can be determined enzymatically. GSH can be converted to *S*-lactoylglutathione (GS-lac) by reaction with methylglyoxal in the presence of glyoxalase I (lactoylglutathione lyase), and the concentration of GS-lac monitored spectrophotometrically at 240 nm.[10,11] Equations (2)–(4) list reactions occurring in the assay for reduction of GSSG by a monothiol (RSH). The conditions of reaction can be adjusted such that

---

[5] G. L. Ellman, *Arch. Biochem. Biophys.* **82**, 70 (1959).
[6] P. W. Riddles, R. L. Blakeley, and B. Zerner, this series, Vol. 91, p. 49.
[7] G. M. Whitesides, J. E. Lilburn, and R. P. Szajewski, *J. Org. Chem.* **42**, 332 (1977).
[8] G. M. Whitesides, J. Houk, and M. A. K. Patterson, *J. Org. Chem.* **48**, 112 (1983).
[9] J. M. Wilson, D. Wu, R. M. DeGrood, and D. J. Hupe, *J. Am. Chem. Soc.* **102**, 359 (1980).
[10] R. P. Szajewski and G. M. Whitesides, *J. Am. Chem. Soc.* **102**, 2011 (1980).
[11] D. L. Vander Jagt, E. Daub, J. A. Krohn, and L.-P. B. Han, *Biochemistry* **14**, 3669 (1975).

the rate of conversion of GSH to GS-lac is fast relative to the rate of formation of GSH.

$$RS^- + GSSG \underset{k_{-1}}{\overset{k_1}{\rightleftharpoons}} RSSG + GS^- \qquad (2)$$

$$RS^- + RSSG \underset{k_{-2}}{\overset{k_2}{\rightleftharpoons}} RSSR + GS^- \qquad (3)$$

$$GSH + CH_3COCHO \rightleftharpoons$$
$$GS-CH(OH)COCH_3 \xrightarrow[\text{glyoxalase I}]{k_3} GS-COCH(OH)CH_3(GS\text{-lac}) \qquad (4)$$

Using suitable concentrations of methylglyoxal and glyoxalase I, and assuming steady-state concentration for GSH, the initial rate of formation of GS-lac is given by Eqs. (5) and (6).

$$\frac{d(GS\text{-lac})}{dt} = k_1(RS^-)(GSSG) = k_1^{obsd}\,[(RS^-) + (RSH)](GSSG) \qquad (5)$$

where parentheses denote molar concentrations and

$$k_1 = k_1^{obsd}\,(1 + 10^{\,pK_a^{RSH}-pH}) \qquad (6)$$

Mass balance and integration of Eq. (5) gives Eq. (7),

$$k_1^{obsd}\,t = \frac{1}{(S)_0 - (GSSG)_0}\ln\left[\frac{(GSSG)_0}{(S)_0} \times \frac{(S)_0 - (GS\text{-lac})/n}{(GSSG)_0 - (GS\text{-lac})/n}\right] \qquad (7)$$

In Eq. (7), for monothiols, $(S)_0 = [(RS^-) + (RSH)]$, $n = 1$; for dithiols, $(S)_0 = [(^-SRS^-) + (HSRS^-) + (HSRSH)]$, $n = 2$. The following assay conditions were found suitable: 5 mM thiol, 0.77 mM methylglyoxal, 0.35 mM GSSG, 2.4 units/ml glyoxalase I.[10] This procedure is not completely general—aromatic thiols absorb at 240 nm, the wavelength used to monitor GS-lac; aminothiols (cysteine, N,N-diethylcysteamine) react rapidly with methylglyoxal and form species that absorb at 240 nm; 2,3-dimercaptopropanol inhibits glyoxalase I.[10]

Typical rate constants (k_1, M^{-1} min^{-1}) for reduction of GSSG by thiols are as follows: 2-mercaptoethanol ($k_1^{obsd} = 8.7$, $k_1 = 3.4 \times 10^3$); dithiothreitol (DTT) ($k_1^{obsd} = 14.1$, $k_1 = 2.2 \times 10^3$) (in 66 mM phosphate at pH 7.0, 30°, under argon).[10] Typical rate constants (k, M^{-1} min^{-1}) for reduction of Ellman's reagent by thiols under similar conditions are the following: 2-mercaptoethanol ($k_1^{obsd} = 3.7 \times 10^4$, $k_1 = 1.2 \times 10^7$, $k_2^{obsd} = 1.7 \times 10^3$, $k_2 = 5.2 \times 10^5$); dithiothreitol ($k_1^{obsd} = 1.5 \times 10^5$, $k_1 = 1.7 \times 10^7$, k_2^{obsd} and k_2 could not be determined).[7] Since the fraction of thiol present in the reactive thiolate form in solution depends upon the thiol pK_a and solution pH, k^{obsd} provides a more direct measure of reactivity of the thiol toward disulfide than does k.

Hupe et al. have described a spectrophotometric method for reaction of disulfides with buried thiol groups in bovine serum albumin.[9]

Brønsted Plots for Rates of Reduction of Disulfides with Thiols

Rates of SH/S$_2$ interchange follow a Brønsted correlation. A Brønsted plot (log k_1 versus pK_a of thiols) for reduction of Ellman's reagent with thiols gave $\beta_{nuc} = 0.36$.[7] This plot included data for both aryl and alkyl thiols. A similar study by Hupe et al. showed separate correlation lines for alkyl and aryl thiols, with $\beta_{nuc}^{alkyl} = 0.49$ and $\beta_{nuc}^{aryl} = 0.48$.[12] The Brønsted coefficients for reduction of GSSG by thiols were $\beta_{nuc} = 0.50$,[10] ($\beta_c + \beta_{lg}$) $= -1.0$[10,13] [nuc = nucleophilic; c = central; lg = leaving group (see below)]. Other Brønsted correlations have been determined with 4,4'-dipyridyl disulfide ($\beta_{nuc} = 0.34$)[14] and with 2,2'-dipyridyl disulfide ($\beta_{nuc} = 0.23$).[15] Mechanistically, reductions of Ellman's reagent and glutathione disulfide by thiols appear to be closely related reactions. Both show similar values of Brønsted coefficients. Neither shows any curvature in the Brønsted plot (a kinetic feature suggesting a change in the mechanism).

Assuming that the rate of SH/S$_2$ interchange [Eq. (8)] is described by an equation of the form of Eq. (9), experimental evaluation of coefficients gives Eq. (10). Equation (10) was found to be a useful kinetic model for SH/S$_2$ interchange.[10]

$$R^{nuc}S^- + R^cSSR^{lg} \rightarrow R^{nuc}SSR^c + {}^-SR^{lg} \quad (8)$$
$$\log k = C + \beta_{nuc}pK_a^{nuc} + \beta_c pK_a^c + \beta_{lg}pK_a^{lg} \quad (9)$$
$$\log k = 7.0 + (0.50)pK_a^{nuc} - (0.27)pK_a^c - (0.73)pK_a^{lg} \quad (10)$$

Determination of Thiol pK_a Values

Small Molecules

Values of pK_a for structurally simple, low molecular mass thiols are easily determined by conventional acid/base titration. This method can be used satisfactorily to determine values of pK_a for many thiols and dithiols.[7,8] The procedure used for low molecular mass thiols follows.

Solution pH values were measured with a Radiometer PH M82 standard pH meter equipped with a REA 160 titrigraph module, a REA 260 derivation unit, and a 25-ml thermostated titration vessel. All manipulations of thiol-containing solutions were carried out under argon. The pH meter was standardized against pH 7.00 and pH 10.00 standard buffer solutions at 25°. The water used to prepare all solutions was deionized and

---

[12] J. M. Wilson, R. J. Bayer, and D. J. Hupe, J. Am. Chem. Soc. **99**, 7922 (1977).
[13] T. E. Creighton, J. Mol. Biol. **96**, 767 (1975).
[14] C. E. Grimshaw, R. L. Whistler, and W. W. Cleland, J. Am. Chem. Soc. **101**, 1521 (1979).
[15] M. Shipton and K. Brocklehurst, Biochem. J. **171**, 385 (1978).

doubly distilled, once from glass. Degassed distilled water (10 ml) was allowed to equilibrate under argon in the thermostated titration vessel.

Mercaptoethanol (70 μl), 1,3-dithiopropan-2-ol (50 μl), or dithiothreitol (77.2 mg) was added to the titration vessel yielding a 0.1 N solution of the thiol. Thiols were titrated against 0.151 M carbonate-free potassium hydroxide.[16] All titrations were in triplicate. Complete titration curves (pH versus volume of titrant added) were obtained by using the stepped curve mode on the autotitrimeter. The KOH solution (about 7 ml) required to neutralize the thiol was added in 20 equal aliquots, and the pH of the solution was measured 1 min after each addition. The titrant was added to the thiol solution slowly enough to maintain the desired temperature throughout the titration.

For monothiols, e.g., mercaptoethanol, the pH of the solution at half-equivalence point of the titration was taken to be the pK_a (for mercaptoethanol, $pK_a = 9.5$).[7] For dithiols, e.g., 1,3-dithiopropan-2-ol or dithiothreitol, assuming C to be the initial concentration of the dithiol and B the amount of base added, the following expressions can be derived for K_1 and K_2 (the first and second acid dissociation constants of the dithiol),

$$K_1 = Z/(K_2Y - X)$$
$$K_2 = (Z + K_1X)/K_1Y$$

where $X = (B + [H^+] - C)[H^+]$; $Y = 2C - B - [H^+]$; and $Z = [H^+]^2(B + [H^+])$.

Two points on the titration curve equidistant from the midpoint of the titration curve were chosen and K_1 and K_2 were calculated as follows,

$$K_1 = (Y_1Z_2 - Y_2Z_1)/(X_1Y_2 - X_2Y_1)$$
$$K_2 = (X_1Z_2 - X_2Z_1)/(Y_1Z_2 - Y_2Z_1)$$

Using a computer program, several values of K_1 and K_2 were calculated for several pairs of points on the titration curve.[7,8,16,17] Values of pK_{a1} and pK_{a2} were calculated from the average values of K_1 and K_2. Values of pK_a obtained by this procedure (1,3-dithiopropan-2-ol, 9.2, 10.7; DTT, 9.3, 10.3) agreed with literature values.[7,18]

---

[16] A. Albert and E. P. Serjeant, "Ionization Constants of Acids and Bases," p. 52. Methuen, London, 1962; H. T. S. Britton, "Hydrogen Ions," 4th ed., pp. 217ff. Chapman & Hall, London, 1954.

[17] Z. Shaked, R. P. Szajewski, and G. M. Whitesides, *Biochemistry* **19**, 4156 (1980).

[18] J. J. Christensen, L. D. Hansen, and R. M. Izatt, "Handbook of Proton Ionization Heats." Wiley, New York, 1976; J. P. Danehy and C. J. Noel, *J. Am. Chem. Soc.* **82**, 2511 (1960); E. L. Loechler and T. C. Hollocher, *ibid.* **102**, 7312 (1980); see also H. Fukada and K. Takahashi, *J. Biochem. (Tokyo)* **87**, 1105 (1980).

Values of pK_a for thiols containing groups with proton affinities similar to that of SH can be determined spectrophotometrically by measuring the UV absorption of thiolate anion as a function of pH.[19] This method was first described by Benesch and Benesch[19] for the determination of values of pK_a of sulfur and nitrogen functions in aminothiols. Ionization constants of cysteine thiols in small peptides can be determined using this method, providing the peptide does not contain tyrosines (a group that resembles cysteine in that it shows a pH-dependent absorption at 245 nm).[20]

Proteins

The determination of values of pK_a for thiol groups in proteins is less straightforward. Several investigators have developed indirect methods for studying the ionization behavior of protein SH groups. Shafer has estimated the manner in which ionization of SH groups in low molecular mass thiols and bovine serum albumin affects the behavior of other ionizable groups in the molecule.[21] Parente *et al.* have described a kinetic method for estimating values of thiol pK_a.[22] They studied the reaction kinetics of Ellman's reagent and 2,2'-dithiopyridine with low molecular mass thiols, model peptides, and monomeric bovine seminal ribonuclease (Cys-31 and Cys-32) as a function of pH. Plots of the apparent second-order rate constant for the SH group versus pH resemble titration curves. The pH at the inflection point in these curves corresponds to the thiol pK_a value. They obtain good results with model thiols and peptides. The titration curve of the monomeric enzyme exhibits, however, a discontinuity that is attributed to a pH-dependent change in protein tertiary structure.

A kinetic method based on Eq. (9) permits estimation of the pK_a values of thiol moieties in proteins. The method has been tested with papain (Cys-25), adenylate kinase, DNase, and lysozyme.[17] The rates of reduction of the disulfide groups of these several proteins and protein derivatives by low molecular mass thiols appear to follow normal Brønsted relationship(s). This observation provides a method for determining the value of the pK_a of a protein cysteine thiol group indirectly by measuring the rates of SH/S_2 interchange involving this thiol. For example, DNase and lysozyme contain two and four cystines, respectively. Reduc-

---

[19] G. H. Snyder, M. J. Cennerazzo, A. J. Karalis, and D. Field, *Biochemistry* **20**, 6509 (1981); G. H. Snyder, *J. Biol. Chem.* **259**, 7468 (1984); P. H. Connett and K. E. Wetterhahn, *J. Am. Chem. Soc.* **107**, 4282 (1985); R. E. Benesch and R. Benesch, *ibid.* **77**, 5877 (1955).

[20] G. H. Snyder, M. K. Reddy, M. J. Cennerazzo, and D. Field, *Biochim. Biophys. Acta* **749**, 219 (1983).

[21] S. D. Lewis, D. C. Misra, and J. A. Shafer, *Biochemistry* **19**, 6129 (1980).

[22] A. Parente, B. Merrifield, G. Geraci, and G. D'Alessio, *Biochemistry* **24**, 1098 (1985).

tion of the disulfide groups leads to complete loss of activity, and rates of reduction can be followed by observing the loss in activity. Papain has an active site cysteine moiety. The S-thioalkyl derivative of papain (papain-SSR) has no enzymatic activity; reduction of papain-SSR leads to restoration of activity as papain-SH is regenerated. Adenylate kinase (AdK) contains two cysteine SH groups. The bisthiomethyl derivative of the kinase [AdK(SSCH$_3$)$_2$] has 70% of the activity of the native enzyme. Reduction of the active site cysteine (SSCH$_3$) can be followed by restoration of enzymatic activity.

Activation of Papain and Adenylate Kinase. Commercial papain and adenylate kinase have little or no activity as a result of oxidation of essential thiol groups (in major part to disulfides). Both enzymes were activated before modification with CH$_3$SSO$_2$CH$_3$, by incubation with reducing solution as follows: for 3 mg/ml papain (<0.1 unit/mg), a solution containing 50 mM L-cysteine, 60 mM 2-mercaptoethanol, and 10 mM EDTA was used in a 1-hr incubation at 30°; for 1 mg/ml adenylate kinase (340 units/mg), 80 mM DTT was used in a 1-hr incubation at 30°. The enzyme was separated from excess reducing agent by placing in a stirred ultrafiltration cell (Amicon Diaflo, PM10 membrane) and passing 2 liters (papain) or 0.6 liter (AdK) of degassed 50 mM phosphate at pH 7.0 (0.1 M in KCl) through the cell at 4°. The resulting enzymes typically had specific activities of 3.8 units/mg (for hydrolysis of benzoyl-DL-arginine p-nitroanilide) for papain and 400 units/mg for AdK.

Methyl Methanethiolsulfonate. Methyl disulfide (14.1 g, 160 mmol) was dissolved in 60 ml of glacial acetic acid in a 250-ml three-necked flask fitted with a reflux condenser and a 125-ml dropping funnel. The flask was cooled to 0°, and H$_2$O$_2$ (34 g of 30% solution; 10.2 g = 0.30 mol of H$_2$O$_2$) was added slowly while maintaining the temperature below 5°. *Caution:* Mixtures of H$_2$O$_2$ and organic solvents are potentially hazardous and should be manipulated behind a shield. The solution was stirred for 30 min at room temperature, and the flask was slowly warmed to 50° for about 1 hr. After destroying the excess of peroxide by additional heating at 50° for 1 hr and testing for peroxide with starch–iodide paper, the glacial acetic acid was removed under reduced pressure. The residual oil was treated with 50 ml of saturated NaHCO$_3$ solution to neutralize residual acid. The oil was separated and diluted with chloroform. After drying over anhydrous MgSO$_4$, the chloroform was removed and yellow oil was distilled. A colorless liquid (7.2 g, 57 mmol) was obtained, bp 60–70° (0.3 torr) [lit. 54° (0.04 torr)].[23] The yield was 36%. The ^{1}H-NMR spectrum (CDCl$_3$) showed peaks at δ 2.65 (s, 3 H) and 3.22 (s, 3 H).

---

[23] H. E. Wijers, H. Boelens, A. van der Gen, and L. Brandsma, *Recl. Trav. Chim. Pays-Bas* **88**, 519 (1969).

Caution: Alkyl methanethiolsulfonates should be synthesized and handled with care in the hood! They all have a very unpleasant odor and some cause dizziness and headache.

Preparation of Papain(SSR) and AdK(SSR)$_2$. The following papain derivatives of the general structure papain(SSR) were prepared: papain(SSCH$_3$), papain(SSCH$_2$CH$_2$CH$_3$), papain(SSCH$_2$CH$_2$OH), papain(SSCH$_2$CF$_3$), and papain(SSCH$_2$CF$_2$CF$_3$). The preparation of papain(SSCH$_3$) is presented in detail; the others were prepared by analogous procedures. Treatment of all the papain(SSR) and AdK(SSR)$_2$ derivatives with a small excess of typical thiol agents restored more than 98% of the native enzyme. A degassed solution (100 ml; 50 mM phosphate at pH 7.0, 100 μM in EDTA and 0.1 M in KCl) containing 0.3 g (13 μmol) of completely activated papain was treated with 35.2 mg (280 μmol, a 20× excess) of CH$_3$SSO$_2$CH$_3$ under argon. The decrease in activity was monitored: after 1–2 hr at 30° no residual enzymatic activity (<0.1%) was observed. Excess CH$_3$SSO$_2$CH$_3$ was separated by placing the reaction mixture in an ultrafiltration cell (Amicon Diaflow, PM10 membrane), separating the protein from the rest of the solution, and passing 2 liters of 0.1 M phosphate at pH 7.0 (100 μM in EDTA, 0.1 M in KCl) through the cell at 4°.

An analogous procedure was used for adenylate kinase, starting with treatment of 0.1 g (4.8 μmol, 400 units/mg) of protein in 100 ml of degassed solution with 12 mg (20× molar excess) of CH$_3$SSO$_2$CH$_3$. This mixture reached a constant activity corresponding to 70% (±3%) of the activity of the native enzyme after a 30-min incubation at 30°. Excess CH$_3$SSO$_2$CH$_3$ was removed as described for papain(SSCH$_3$).

Rates of Reduction of DNase, Lysozyme, Papain-SSR, and AdK(SSCH$_3$)$_2$. Reduction of enzymes by several low molecular mass thiols were measured by the loss or recovery of the native enzymatic activity. These rates follow a Brønsted relationship with slopes (β_{nuc}) ranging from 0.36 (DNase) to 0.65 [AdK(SSCH$_3$)$_2$].

A representative procedure is that for DNase. DNase (2000 Kunitz units, ~1 mg of electrophoretically purified protein) was transferred to a small polypropylene vial which had been rinsed with 0.1 M phosphate buffer at pH 7.0 and flushed with argon. Additional degassed buffer (10 ml) containing 5 mM EDTA was added, and the solution was equilibrated under argon in a 30 ± 0.5° constant-temperature bath. The enzyme solution was assayed. Sufficient DTT (100 μl) was added at $t = 0$ to make the solution 21 mM in DTT, and an initial aliquot (20 μl) was removed and used to check the concentration of thiol groups in solution by using Ellman's reagent (0.5 mM, 0.1 mM in EDTA). Aliquots (10 μl) were removed every 1–5 min and added to cuvettes containing 1 ml of assay solution. These solutions were analyzed immediately for residual enzymatic activity. Manipulations of all solutions containing thiols were conducted under a static argon atmosphere, maintained by inserting a hypodermic needle connected to an argon line through the top of the polypropylene reaction vial. This procedure provided adequate protection against autoxidation for all but the most slowly reacting solutions. At the end of each run, the thiol concentration was checked again by using Ellman's assay.

Those reactions (lysozyme with 2-mercaptoethanol, N-acetyl-L-cysteine, and others), that required 2–5 days to proceed to completion were carried out in an S-40325-50 Sargent-Welch glovebox under argon, and higher than usual concentrations of thiols were used (0.05–2.5 M). Dimethyl sulfoxide (1–3% of the solution) was used to ensure complete solubility of the less-soluble thiols at these high concentrations. The activity of lysozyme is not influenced by these dimethyl sulfoxide concentrations.

Satisfactory Brønsted correlations observed for the rates of SH/S_2 interchange reactions between low molecular mass thiols and protein disulfide bonds suggest that for these thiols and proteins, steric and electronic effects characteristic of protein tertiary structures are less important than the electronic effects responsible for the Brønsted reactivity–basicity correlations. The rates of SH/S_2 interchange reactions involving glutathione and protein disulfides are, however, slower than would be anticipated based on correlations with lower molecular mass species.

Use of Eq. (10) and rates of reduction of papain($SSCH_3$) by DTT at various values of pH (4–9) give the pK_a of Cys-25 of papain as 4.1 (pH 6) and 8.4 (pH 9). The pK_a of Cys-25 of papain is abnormally low at pH 6 due to the interaction of Cys-25 thiolate with imidazolium ion of His-159, which has a pK_a of about 7.5.[17] The pK_a of Cys-25 of AdK is 7.5 (at pH 7). Assuming that the values of pK_a of the two cysteine thiol groups generated on reduction of the cystine moieties of DNase and lysozyme are the same, we estimate these values of pK_a to be 8.8 and 11.0, respectively (at pH 7).

This kinetic method of determination of values of thiol pK_a has the useful feature that it can be carried out as a function of pH. It has two disadvantages: First, that it is necessary to convert protein thiols into disulfides when the thiol of interest is not naturally present as a disulfide; this derivatization might, in principle, change the conformation of the active site. Second, that the particular protein thiol (disulfide) involved in reduction may not be obvious in proteins having more than one. The values of pK_a obtained by this method should be regarded as semiquantitative due to the combined uncertainties in measurements of rates of SH/ S_2 interchange and in the analysis based on Eq. (10). Since these uncertainties are quite different from those encountered in other methods for evaluating thiol pK_a, this method provides a useful, independent method for characterization of protein thiol and disulfide groups.

Determination of Values of Thiol Equilibrium Constants

Equilibration of a symmetrical disulfide, ESSE, to a thiol, ESH, with concomitant oxidation of a reducing thiol, RSH, to a disulfide, RSSR, occurs in two steps via an unsymmetrical disulfide [Eqs. (11)–(13)]. Both thiol and thiolate anion may be present in appreciable concentration in

$$(RSH + RS^-) + ESSE \xrightleftharpoons{K_1} RSSE + (ESH + ES^-) \quad (11)$$

$$(RSH + RS^-) + RSSE \xrightleftharpoons{K_2} RSSR + (ESH + ES^-) \quad (12)$$

$$2(RSH + RS^-) + ESSE \xrightleftharpoons{K^{obsd}} RSSR + 2(ESH + ES^-) \quad (13)$$

$$K^{obsd} = K_1 K_2 = [RSSR][ESH + ES^-]^2/[ESSE][RSH + RS^-]^2 \quad (14)$$

solution, and measured equilibrium constants will contain terms for the concentration of both species [Eq. (14)]. In these equations, and subsequently, an equilibrium constant referring to an interchange reaction involving a mixture of thiol and thiolate species will be denoted by the superscript "obsd."

A large contribution to SH/S_2 interchange equilibrium constants can, in certain circumstances, be attributed to the relative values of pK_a of the reducing thiol and the thiol derived from the disulfide. This contribution is dependent on the pH of the solution and is proportional to the difference between the values of thiol pK_a. It is largest when both thiols are present entirely as thiolate anion (i.e., when the equilibrium considered is thiolate–disulfide interchange). It is this difference in pK_a that provides the driving force for the ability of Ellman's reagent to oxidize aliphatic thiols: values of pK_a for aryl thiols are typically 5 to 6, while values of pK_a for alkyl thiols are 9 to 10.

Equilibrium constants for SH/S_2 interchange of monothiols with similar values of pK_a are often close to unity. Many dithiols capable of forming cyclic disulfides are, however, more strongly reducing than are the corresponding monothiols. This effect is attributable in major part to a much higher forward rate for the second step of Eq. (15) than for that of Eq. (16); its origin is, in large part, entropic. This principle is, of course, the structural basis for the well-established value of dithiothreitol (DTT) as a reducing agent.[24] There is a clear correlation between the reducing

---

[24] W. W. Cleland, *Biochemistry* **3**, 480 (1964).

$$\text{HSRSH} + \text{ESSE} \underset{k_{-1}}{\overset{k_1}{\rightleftharpoons}} \text{HSRSSE} + \text{ESH} \underset{k_{-2}}{\overset{k_2}{\rightleftharpoons}} \overset{\frown}{\text{SRS}} + 2\,\text{HSE} \quad (15)$$

$$2\,\text{RSH} + \text{ESSE} \underset{k'_{-1}}{\overset{k'_1}{\rightleftharpoons}} \text{RSH} + \text{RSSE} + \text{ESH} \underset{k'_{-2}}{\overset{k'_2}{\rightleftharpoons}} \text{RSSR} + 2\,\text{HSE} \quad (16)$$

potential of an α,ω-dithiol and the size of the ring formed on its oxidation. The maximum equilibrium advantage is observed for dithiols capable of forming six-membered rings, although five- and seven-membered rings can have good stability as well.[10]

The characterization of dilute solutions of disulfides and thiols under equilibrating conditions can present substantial difficulties. The SS and SH groups are not themselves intrinsically readily detected, and differentiation between different disulfides is especially problematic. Rigorous determination of values of equilibrium constants requires a knowledge of the values of pK_a of the thiols and of the concentration of all thiol and disulfide species present in solution at equilibrium. Snyder et al.[20] have measured values of K^{obsd} for the reaction between N,N'-diacetylcystine (AA) and a cysteine-containing peptide (B). Mixtures of AA + B and BB + A were equilibrated in sealed vials. Reactions were quenched with acid (acidification converts thiolate anion to thiol and slows thiol–disulfide interchange by orders of magnitude). The components of the mixture were separated by gel filtration chromatography and their concentrations determined by UV absorption.

Separation and identification of all thiol and disulfide species is the most straightforward way to determine the concentrations of solution components at equilibrium. Such methods are, however, time consuming, and it is not always clear that the quantities of each species found after separation are truly those present at equilibrium. Rabenstein and Theriault have used an NMR method to determine equilibrium constants for captopril, penicillamine[25] (RSH), and glutathione disulfide (GSSG).[25,26] The high-field ^{1}H-NMR spectra of RSH, RSSR, and RSSG are sufficiently different to allow the relative concentrations of each species to be determined by integration. This method allows analysis of the thiol–disulfide mixture *at* equilibrium, but is not general in that the NMR spectra of many thiols are not substantially different from the spectra of the corresponding disulfides.

Equilibrium constants for reduction of lipoamide disulfide by thiols[10] can also be determined using an enzymatic procedure first described by Cleland.[24] A representative equilibration with an α,ω-dithiol (HSRSH) is presented in Eqs. (17)–(20). The lipoamide dehydrogenase-mediated reaction between lipoamide and NAD$^+$ was used as a shuttle to obtain reduc-

---

[25] D. L. Rabenstein and Y. Theriault, *Can. J. Chem.* **63**, 33 (1985).
[26] D. L. Rabenstein and Y. Theriault, *Can. J. Chem.* **62**, 1672 (1984).

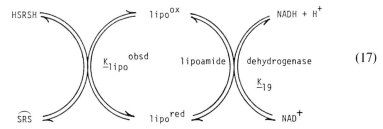

$$K_{lipo}^{obsd} = [SRS][lipo^{red}]/[HSRSH][lipo^{ox}] \quad (18)$$
$$K_{19} = [lipo^{ox}][NADH]/[lipo^{red}][NAD^+] \quad (19)$$
$$K_{NAD}^{obsd} = [SRS][NADH]/[HSRSH][NAD^+] = K_{lipo}^{obsd} K_{19} \quad (20)$$

tion potentials of dithiols relative to NADH. The concentration of NADH at equilibrium was measured, and straightforward calculations based on known initial concentrations of α,ω-dithiol, lipoamide disulfide, and NAD$^+$ yielded the concentrations required to calculate K_{lipo}^{obsd}. This experimental procedure measures directly only the concentration of NADH and, by inference, that of reduced lipoamide; it does not identify the structures or concentrations of other thiols and disulfides present. The procedure does, however, provide a straightforward method of linking the reducing ability of a wide range of thiols to the biochemically relevant reducing agent NADH. Equilibrium constants for several thiols determined using this procedure are as follows (pK_{a1}, pK_{a2}, K_{lipo}^{obsd}): 1,2-dithio-3-propanol, 8.6, 10.5, 5.8 M^{-1}; 1,3-dithio-2-propanol, 9.0, 10.3, 0.51; dithiothreitol, 9.2, 10.1, 15; mercaptoethanol, 9.6, —, 2.0 × 10^{-3} M^{-1}; glutathione, 8.7, —, 1.8 × 10^{-3} M^{-1}.[10] These values are in good agreement with literature values.[10]

Acknowledgment

This research was supported by the National Institutes of Health, Grant GM 34411.

[22] Cysteine and Cystine: High-Performance Liquid Chromatography of o-Phthalaldehyde Derivatives

By JOHN D. H. COOPER and D. C. TURNELL

Introduction

The selectivity and sensitivity of separation techniques are increased by using either pre- or postcolumn derivatization. Unlike many other amino acid derivatizing reagents, o-phthalaldehyde reacts rapidly with amino acids at normal ambient temperatures under aqueous conditions to yield intensely fluorescent derivatives. The relatively low polarity of these derivatives is ideally suited to separation by reversed-phase HPLC[1-3] with a sensitivity of 1 pmol for each amino acid fluorophore.[1]

The active derivatization reagent is prepared by treating o-phthalaldehyde with an excess of a thiol compound, e.g., 2-mercaptoethanol, to form a thiol adduct.[4] Primary amines react with the adduct to form 1-alkylthio-2-alkyl-substituted isoindoles.[5] The rapid fluorescence decay of some amino acid isoindoles[1] is not a technical limitation since the derivatives are stable once injected onto the reversed-phase column.[6]

Due to the excess thiol present in the prepared o-phthalaldehyde reagent, the derivative of a sulfhydryl amino acid is the reduced amino acid isoindole. Thus, cystine or cysteine will yield the same isoindole derivative. The fluorescence intensity of cysteine isoindole can be as little as 2% of that of other amino acid isoindoles.[4,7] Methods used to increase the fluorescence yield of the cysteine isoindole have included the oxidation of cysteine to cysteic acid with performic acid[8] and the reaction of the sulfhydryl of cysteine with 4-vinylpyridine,[9] ethyleneimine,[10] or iodoacetate.[11] Cysteine derivatives formed with these reagents subsequently yield isoindoles that have approximately the same fluorescence intensity as

---

[1] D. C. Turnell and J. D. H. Cooper, *Clin. Chem. (Winston-Salem, N.C.)* **28**, 527 (1982).
[2] M. Griffin, S. J. Price, and T. Palmer, *Clin. Chim. Acta* **125**, 89 (1982).
[3] D. L. Hogan, K. L. Kraemer, and J. I. Isenberg, *Anal. Biochem.* **127**, 17 (1982).
[4] M. Roth, *Anal. Chem.* **43**, 7 (1971).
[5] S. S. Simmons, Jr. and D. F. Johnson, *J. Org. Chem.* **43**, 2886 (1978).
[6] J. D. H. Cooper, G. Ogden, J. McIntosh, and D. C. Turnell, *Anal. Biochem.* **142**, 98 (1984).
[7] K. S. Lee and D. G. Drescher, *Int. J. Biochem.* **9**, 457 (1978).
[8] S. Moore, *J. Biol. Chem.* **238**, 235 (1963).
[9] J. S. Wall, *J. Agric. Food Chem.* **19**, 619 (1971).
[10] M. A. Raferty and R. D. Cole, *Biochem. Biophys. Res. Commun.* **10**, 467 (1963).
[11] J. D. H. Cooper and D. C. Turnell, *J. Chromatogr.* **227**, 158 (1982).

other amino acid isoindoles. However, the alkylation of cysteine with iodoacetate to form S-carboxymethylcysteine has the advantages of being a rapid reaction at ambient temperatures and compatible with the o-phthalaldehyde reaction. Other alkylating agents such as iodoacetamide can be used to alter the retention of the isoindole on the HPLC column if required.

Before biological samples can be injected onto the HPLC column, high molecular weight compounds that would otherwise progressively reduce the performance of the analytical column and directly interfere with the separation and quantitation, must be removed. Techniques for removing proteins prior to cysteine analysis have included precipitation with perchloric acid[11] and acetonitrile.[1] 2-Mercaptoethanol is added to the protein precipitant to reduce cystine to cysteine before reacting with iodoacetate. Alternatively, an automated method that involves dialysis has been used to deproteinize samples for amino acid analysis using precolumn derivatization with o-phthalaldehyde.[12]

Recommended Protocol for Sample Preparation

Reagents. All chemicals should be of analytical grade; the products from the Sigma Chemical Company (St. Louis, MO) have been used. High-purity water should be used for all reagent preparation.

Borate buffer (400 mM, pH 9.5): Dissolve 24.7 g of boric acid in approximately 800 ml of water, adjust the pH to 9.5 with 4 M sodium hydroxide solution, and bring the volume to 1 liter with water. Store at 4°.

Iodoacetate reagent: Dissolve 5 g of iodoacetic acid in approximately 80 ml of the borate buffer. Adjust to pH 9.5 with 4 M sodium hydroxide solution, and bring the volume to 100 ml with water. Store at 4°.

o-Phthalaldehyde/2-mercaptoethanol reagent: Dissolve 100 mg of o-phthalaldehyde in 10 ml methanol, bring the volume to 100 ml with borate buffer, and add 400 μl of 2-mercaptoethanol. This working solution has been found to be stable for at least 2 months when stored at ambient temperature in a dark glass bottle.

Standard cystine solution (500 μM): Dissolve 120 mg of L-cystine in 1 liter of water and store at $-20°$.

Internal standard solution (1.0 mM): Dissolve 183 mg of DL-homocysteic acid in 1 liter of water.

---

[12] D. C. Turnell and J. D. H. Cooper, *J. Autom. Chem.* (in press).

Acetonitrile precipitation reagent: Add 200 µl of 2-mercaptoethanol to 100 ml of acetonitrile. Prepare each day.

HPLC solvents: Solvent A, 8 volumes of acetonitrile and 92 volumes of 75 mM sodium phosphate at pH 6.6. Solvent B, 60 volumes of methanol and 40 volumes of water. Filter solvents through a 0.45-µm pore size filter and degas before use.

Procedures

Sample Preparation. To 20 µl of standard cystine solution or sample add 20 µl of internal standard solution and 200 µl of acetonitrile reagent. After vortex-mixing, centrifuge at 12,000 g for 2 min. To 20 µl of the supernatant fluid add 100 µl of the iodoacetate reagent and mix. Add 100 µl of the phthalaldehyde/mercaptoethanol reagent, and exactly 1 min later inject 20 µl of the derivatized sample onto the HPLC column using a filled loop technique.

HPLC Procedures. Biological samples will usually contain many free amino acids and amines that will derivatize with the phthalaldehyde reagent and be detected on HPLC. For this reason, a gradient HPLC system is required to minimize analysis times. Amino acid isoindoles fluoresce at 455 nm in response to excitation at 230 or 340 nm.[4,13] Greater quantum yields are obtained by exciting at 230 nm.

Numerous types of reversed-phase HPLC columns are available. Columns, 4.6 mm (i.d.) × 150 mm long, packed with 5 µm diameter silica bonded with octadecylsilane (ODS) are commonly used for full amino acid analysis. However, if only quantitation of S-carboxymethylcysteine isoindole is required, a 4.6 × 100 mm column is adequate. S-carboxymethylcysteine isoindole can be resolved from other isoindoles on 5 µm Ultrasphere ODS (Beckman Instruments Inc., Fullerton, CA) or 5 µm Rainin ODS (Rainin Instrument Co., Inc., Woburn, MA). Using this derivatization method and running solvent A isocratically at a flow rate of 2.5 ml/min, elution of the amino acid isoindoles derivatives found in biological fluids occurs in the following order: phosphoserine, aspartate, glutamate, S-carboxymethylcysteine, and aminoadipic acid. The internal standard, homocysteic acid, elutes between glutamate and S-carboxymethylcysteine. Elution times of S-carboxymethylcysteine are of the order of 4 min. After this peak emerges, a rapid gradient can be imposed, using solvent B, to elute all the other amino acid derivatives. The purging and equilibration of the column can be accomplished within 10 min.

---

[13] D. Hill, L. Burnworth, W. Skea, and R. Pfeifer, *J. Liq. Chromatogr.* **5**, 2369 (1982).

[23] Cystine: Binding Protein Assay

By Margaret Smith, Clement E. Furlong, Alice A. Greene, and Jerry A. Schneider

The cystine-binding protein is utilized for cystine transport through the osmotic shock-sensitive transport system of *Escherichia coli*.[1] This protein is located in the periplasmic space and can be isolated following osmotic shock of this bacterium.[1] It reversibly binds cystine with a dissociation constant of 10 nM.[1] When purified, the cystine-binding protein may be used as described below to measure cystine levels in physiological fluids.[2]

Purification of the Cystine-Binding Protein

Choice of Bacterial Strain. The cystine-binding protein has been purified from several strains of *E. coli*, any one of which is satisfactory for the following purification. The strains include *E. coli* W (ATCC 9637), its derivative D_2W,[1] and mutant strains derived from D_2W, particularly strain HA12-GA7.[3]

Growth of Cells. Prepare the following medium (per 1000 ml at pH 6.5): K_2HPO_4, 13.9 g; KH_2PO_4, 4.1 g; NH_4Cl, 5.35 g.[4] Autoclave. Add sterile glycerol (from 50% stock) to a final concentration of 2%, $MgCl_2$ to a final concentration of 2 mM, and 1 ml of trace element solution[5,6] per liter.

Grow the desired volume of cells to late exponential or early stationary phase in a well-aerated fermenter. Cell density is followed by light scattering at 600 nm (A_{600}).[5] The pH is maintained at 6.5 with concentrated ammonium hydroxide. At A_{600} of 3, an additional 0.5 ml of trace elements,

---

[1] E. A. Berger and L. A. Heppel, *J. Biol. Chem.* **247**, 7684 (1972).
[2] R. G. Oshima, R. C. Willis, C. E. Furlong, and J. A. Schneider, *J. Biol. Chem.* **249**, 6033 (1974).
[3] G. D. Schellenberg and C. E. Furlong, *J. Biol. Chem.* **252**, 9055 (1977).
[4] R. C. Willis, R. G. Morris, C. Cirakoglu, G. D. Schellenberg, N. H. Gerber, and C. E. Furlong, *Arch. Biochem. Biophys.* **161**, 64 (1974).
[5] Trace elements are prepared by heating a 100 ml solution containing 0.5 g $CaCl_2$, 0.018 g $ZnSO_4 \cdot 7H_2O$, 0.016 g $CuSO_4 \cdot 5H_2O$, 0.018 g $CoCl_2 \cdot 6H_2O$, and 2 g disodium EDTA (ethylenediaminetetraacetic acid) until the components dissolve. After cooling, 2 g $FeCl_2 \cdot 6H_2O$ and 100 μl 1 M HCl are added. This solution is sterilized by filtration and stored frozen in aliquots. For A_{600} above 1.0 the sample was diluted for measurement.
[6] A. D. Kellmers, C. W. Hancher, E. F. Phares, and G. D. Novelli, this series, Vol. 20, p. 3.

an additional 1% glycerol, and 2 mmol/liter $MgSO_4$ are added. At A_{600} of 18, 1 mmol/liter $NaSO_4$ and at A_{600} of 27, 2 mmol/liter $MgSO_4$ are added. This procedure should provide a yield of approximately 30 to more than 50 g wet cell paste/liter of medium. Antifoam (Dow Corning FG 10) is used as necessary to control foaming.

Osmotic Shock Procedure. In late exponential or early stationary phase, sufficient volumes of 1 M NaCl and 1 M Tris–HCl at pH 7.3 are added directly to the culture to achieve a final concentration of 30 mM each. The cells are harvested by centrifugation and suspended in 7 to 10 volumes (w/v) of 33 mM Tris–HCl at pH 7.3. An equal volume of a solution of 33 mM Tris–HCl at pH 7.3, 40% sucrose, and 2 mM EDTA is slowly added with stirring. Following the harvest of the plasmolyzed cells by centrifugation, they are rapidly suspended in 20 volumes cold deionized water, and $MgCl_2$ is added to a final concentration of 1 mM within 1 min so as to prevent lysis of the cells. The cells are removed by centrifugation, and the supernatant solution (osmotic shock fluid), which contains the released periplasmic proteins, is fractionated as described below.

Alternative Periplasmic Protein Extraction Procedure. A recently developed chloroform extraction procedure[7] has been found to extract efficiently the glutamine, glutamate/aspartate, and phosphate-binding proteins from *E. coli* on a preparative scale and should work equally well for the cystine-binding protein.

Purification of the Cystine-Binding Protein. The osmotic shock fluid is adjusted to pH 4.5 with 7.5% acetic acid. The precipitate is removed by centrifugation or filtration, and the supernatant fraction is loaded onto a CM-cellulose column equilibrated with 10 mM sodium acetate, pH 4.5, at a protein load of 7–10 mg/10 ml resin bed volume. The column is washed with 3–5 volumes of 10 mM sodium acetate at pH 4.5, and protein is then eluted with 4 column volumes of 0.3 M sodium chloride in 10 mM sodium acetate at pH 4.5. Column fractions are assayed for cystine binding on membrane filters by mixing 10 μl of radiolabeled 10 μM cystine (approximately 50 μCi/μmol) with 100 μl of column fraction and 6 μl 1 M sodium acetate at pH 5.0. Filtration and scintillation counting are as described for the quantitative assay. An equilibrium dialysis method may also be used for assay.[4,8]

The fractions containing cystine-binding activity are pooled, adjusted to pH 7.8, and dialyzed against 10 mM Tris–HCl, pH 7.8. The dialyzed fraction is loaded onto a DEAE-cellulose column (height-to-diameter ra-

---

[7] G. F.-L. Ames, C. Prody, and S. Kustu, *J. Bacteriol.* **160**, 1181 (1985).
[8] C. E. Furlong, R. G. Morris, M. Kandrach, and B. P. Rosen, *Anal. Biochem.* **47**, 514 (1972).

TABLE I
PURIFICATION OF CYSTINE-BINDING PROTEIN[a]

Step	Volume (ml)	Protein (mg)	Units	Specific activity (units/mg)
Crude shock fluid	1390	676	232	0.34
CM-Cellulose	146	59	264	4.5
DEAE-Cellulose	10.2	6.2	188	30

[a] Based on starting with 100 g wet cell paste.

tio of approximately 30:1) at a protein load of 7–10 mg protein/ml resin bed volume. The column is eluted with a linear gradient (total of 40 resin bed volumes) of 0–150 mM NaCl in 10 mM Tris–HCl, pH 7.8. Fractions containing cystine-binding activity are pooled, divided into small portions and frozen. The specific activity of the purified protein should be approximately 30 nmol cystine bound (at 10 μM free cystine)/mg protein.[2] Two peaks of binding activity are often observed.[1] Both may be used.

A summary of the purification procedure is presented in Table I.

Use of Cystine-Binding Protein for Quantitative Cystine Assay in Biological Samples

Principle. Nanomolar concentrations of the disulfide, cystine, are assayed based on the competitive binding between [^{14}C]cystine and the nonradioactive cystine present in a sulfosalicylic acid extract of the biological material. The cystine-binding protein does not bind reduced glutathione or other thiols which are present in much larger amounts than cystine in most biological samples, eliminating the need for purification procedures prior to assay. The specificity of the cystine binding has been reported in detail.[2]

Reagents

12% Sulfosalicylic acid (SSA)
1.66 N NaOH
N-Ethylmaleimide, 5 mM prepared fresh in 0.01 M potassium phosphate buffer, pH 7.3
Sodium acetate buffers, 1.0, 0.1, and 0.01 M acetate, pH 5.0
[^{14}C]Cystine, 0.4 μM in 0.1 M sodium acetate buffer, pH 5.0 (about 350 mCi/mmol)

Cystine-binding protein[9] diluted in 0.1 M sodium acetate buffer at pH 5.0 which contains 0.4 mg/ml bovine serum albumin so that 25 μl will bind about 10 pmol of cystine.

Cystine standards from 0.1 to 1.0 μM. These are conveniently made from a 500 μM stock solution prepared by adding 30 mg of L-cystine to 250 ml of 0.01 N HCl; heat and stir the solution until all cystine is dissolved. Dilute in 0.1 M sodium acetate buffer to the required concentrations. Standards are stable if frozen.

Procedure. Tissue is disrupted by homogenization or sonication in 3 volumes of ice-cold N-ethylmaleimide solution. An aliquot of 300 μl is removed and 100 μl of 12% SSA is added. Proteins and nucleic acids are precipitated on ice for 15 min and removed by centrifugation. The supernatant is removed, and the SSA is neutralized with 50 μl of 1.66 N NaOH. The sample is buffered by the addition of 100 μl of 1.0 M sodium acetate, pH 5, and brought to a final volume of 1.0 ml. Samples are stable when kept frozen.

For assay, 50 μl of sample or an appropriate dilution in 0.1 M sodium acetate is added to a 1-ml plastic centrifuge tube (Fisher). Then 25 μl of 0.4 μM [^{14}C]cystine (10 pmol), 25 μl of cystine-binding protein, and 10 μl of 1.0 M sodium acetate are added, for a total volume of 110 μl. These last three components can be premixed and stored as an assay cocktail which is stable at 4°. Tubes with cystine standards, and as a blank, a tube containing 50 μl of 0.1 M sodium acetate buffer (which measures the maximum [^{14}C]cystine counts bound), are also included. After mixing, the samples are incubated at room temperature for approximately 30 min; then 100 μl is filtered through 24 mm diameter nitrocellulose filters (0.45 μm pore size) which have been previously hydrated in 10 mM sodium acetate buffer. The filters are supported by the bottom half of a 25 mm Swinex filter unit placed on a side-arm vacuum flask maintained at 127 torr. The filter is then rapidly rinsed with 0.6 ml of 0.01 M sodium acetate buffer and dried under a heat lamp. The [^{14}C]cystine–protein complex bound to the filter is measured by liquid scintillation counting.

Both initial binding of cystine to the protein and binding of the protein complex to nitrocellulose are pH sensitive with a sharp optimum around pH 5.0.[2] Levels of extraneous protein above that equivalent to 30 μg of bovine serum albumin per assay decrease cystine binding. The protein complex is increasingly lost from the filter when wash volumes greater than 0.6 ml are used or when the wash remains on the filter for longer than 5 sec. It is essential that all buffers be free of bacteria.

---

[9] Purified cystine-binding protein is available from Analytical Applications, P.O. Box 1476, Bellevue, WA 98009.

Calculation of Results. Since the cystine-binding protein has been diluted so that it is already saturated by the 10 pmol of [^{14}C]cystine added to the assay, nonradioactive cystine present in sample and standard solutions will compete directly for binding sites on the protein. If f is the fraction of [^{14}C]cystine bound in the sample as compared to the maximum amount bound, c the concentration of radioactive cystine added, and x the concentration of cystine present in the unknown samples then

$$f = c/(c + x)$$

Using the data from standards a plot of the reciprocal $1/f$ versus x is made and should give a straight line. For 10 pmol as the saturating concentration for the binding protein, a line with slope of about 5 is usually obtained (with a y intercept of 1).

Acknowledgments

Supported by the National Institutes of Health, Grants AM 18434 and GM 15253.

[24] Selenocysteine

By Nobuyoshi Esaki and Kenji Soda

Cysteine can be determined easily and accurately by the method of Gaitonde.[1] When the reaction is carried out as described in the literature, selenocysteine reacts with Gaitonde's reagent and develops an intense, purple color which quickly disappears. We describe here a modified Gaitonde's method for the determination of selenocysteine.

Reagents

50 mM NaBH$_4$ dissolved in 0.1 N NaOH
1% (w/v) Ninhydrin dissolved in acetic acid
15 M Phosphoric acid
0.1 M Lead acetate

Procedure

To 50 μl of selenocysteine solution, 50 μl of NaBH$_4$ solution is added, and the mixture is incubated at room temperature for 15 min to ensure

[1] M. K. Gaitonde, *Biochem. J.* **104,** 627 (1967).

complete reduction of selenocystine. The ninhydrin solution and phosphoric acid are mixed in a ratio of 3:1 (v/v), and 1 ml of the mixture is added to the reduced selenocysteine sample. The solution is heated at 70° for 10 min, and cooled quickly to room temperature (about 25°). From 10 μM (A_{570} = 0.024) to 500 μM (A_{570} = 1.2) of selenocysteine can be determined under these conditions.

Note on Procedure

The ninhydrin chromophore of selenocysteine is relatively stable at 25°; absorbance at 570 nm decreases by less than 1% per hour under the conditions used. However, when the sample is vigorously aerated, the red color disappears within 30 min. None of the following compounds interfere with determination of selenocysteine: 5 mM serine, serine O-sulfate, O-acetylserine, O-acetylhomoserine, methionine, or dithiothreitol.

[25] Cysteamine and Cystamine

By Michael W. Duffel, Daniel J. Logan, and Daniel M. Ziegler

Cysteamine (2-aminoethanethiol) is formed in mammalian tissues by the enzymatic hydrolysis of pantetheine,[1] and the role of cysteamine in the biosynthesis of hypotaurine has been reviewed.[2] In addition, cysteamine has been used therapeutically to treat cystinosis[3] and sickle cell anemia,[4] and has been used as a protective agent against hepatotoxins[5] and ionizing radiation.[6] However, despite its clinical applications, only a few procedures have been described for the determination of cysteamine in body fluids of animals treated with this aminothiol,[7–9] and none are

[1] G. D. Novelli, F. J. Schmetz, and N. O. Kaplan, *J. Biol. Chem.* **206**, 533 (1954).
[2] D. Cavallini, R. Scandurra, S. Dupre, G. Federici, L. Santoro, G. Ricci, and D. Barra, in "Taurine" (R. Huxtable and A. Barbeau, eds.), p. 59. Raven Press, New York, 1976.
[3] J. G. Thoene, G. R. Oshima, J. C. Crawhall, D. L. Olson, and J. A. Schneider, *J. Clin. Invest.* **58**, 180 (1976).
[4] W. Hassan, Y. Beuzard, and J. Rossa, *Proc. Natl. Acad. Sci. U.S.A.* **73**, 3288 (1976).
[5] F. C. DeFarreyra, O. M. DeFenos, A. S. Bernacchi, C. R. Decastro, and J. A. Castro, *Toxicol. Appl. Pharmcol.* **48**, 221 (1979).
[6] Z. M. Bacq, "Chemical Protection Against Ionizing Radiation." Thomas, Springfield, Illinois, 1965.
[7] M. Hsiung, Y. Y. Yeo, K. Itiaba, and J. C. Crawhall, *Biochem. Med.* **19**, 305 (1978).
[8] A. J. Jonas and J. A. Schneider, *Anal. Biochem.* **114**, 429 (1981).
[9] S. Ida, Y. Tanaka, S. Ohkuma, and K. Kuriyama, *Anal. Biochem.* **136**, 352 (1984).

readily adaptable for measuring both cysteamine and its disulfide, cystamine, in a single sample. This chapter describes two different methods for the determination of these aminothiols. The first is a rapid, specific enzymatic procedure that is readily adapted for the simultaneous estimation of both cysteamine and cystamine in plasma from treated animals. The second is a more sensitive, but also more laborious, colorimetric method for the estimation of endogenous (protein-bound + free cysteamine) in animal tissues.

Enzymatic Method

Principle. Cysteamine dioxygenase (EC 1.13.11.19) catalyzes the oxidation of cysteamine to hypotaurine [Eq. (1)]. The reaction is essentially

$$HN_2CH_2CH_2SH + O_2 \rightarrow H_2NCH_2CH_2SO_2^- + H^+ \quad (1)$$

irreversible, and in a closed reaction vessel the concentration of cysteamine is directly proportional to oxygen uptake. The liver cytosolic enzyme (see this volume [69]) is quite specific for cysteamine, and this high specificity coupled to the low K_m of cysteamine, 1 μM or less, permits direct estimation in body fluids without prior sample preparation. Since cystamine is not a substrate it does not interfere with the estimation of cysteamine. However, on reduction by excess glutathione (GSH) [Eq. (2)], cystamine and cysteamine mixed disulfides yield cysteamine. There-

$$(H_2NCH_2CH_2S-)_2 + 2\ GSH \rightarrow 2\ H_2NCH_2CH_2SH + GSSG \quad (2)$$

fore, by measuring oxygen uptake before and after reduction by GSH the concentration of cysteamine and of cystamine are readily estimated in the same sample.

Reagents

1.0 M Phosphate buffer, pH 7.8
20 mM Ethylenediaminetetraacetic acid (EDTA)
0.5 M GSH
5 mM Hydroxylamine hydrochloride
Cysteamine dioxygenase[10]

---

[10] Only the enzyme isolated from hog liver by the procedure described elsewhere in this volume [69] has been used by us. However, cysteamine dioxygenase, purified from liver of other species, should be suitable. Relatively crude preparations of the liver enzyme can be used, but the best results are obtained with fractions purified at least 50-fold from the cytosol, since cysteamine-independent oxygen uptake, common with less pure fractions, interferes with the measurements. To avoid adding large amounts of liquids, the enzyme solution should contain no less than 1 unit/ml. One unit is defined as the concentration of enzyme that catalyzes the oxidation of 1 μmol cysteamine/min.

Procedure. The following description is based on equipment in routine use in this laboratory, but the procedure can be adapted to any of a number of commercially available oxygen monitors capable of measuring micromolar changes in oxygen concentration in a 1- to 2-ml reaction vessel.

The reaction is carried out in a 2-ml, thermostatted (37°) vessel (Gilson Medical Electronics, Middleton, WI) fitted with a Clark-type oxygen probe (YSI 4004, Yellow Springs Instrument Co., Yellow Springs, OH). The signal from the electrode is recorded with a strip chart recorder capable of calibrated zero suppression and scale expansion.

The following reagents, warmed to 37°, are transferred to the reaction vessel in the order listed: 0.2 ml phosphate buffer, 0.1 ml hydroxylamine hydrochloride, 0.1 ml EDTA, an aliquot of a standard or unknown solution containing 10–100 nmol cysteamine and/or cystamine, and water to a final volume of 2.0 ml. The reaction vessel is closed, and the oxygen concentration is recorded. After temperature equilibration (3–4 min) there should be no observed change in oxygen concentration, and about 1 min after reaching constant oxygen, 10 μl of the dioxygenase is added through the capillary access port. The decrease in oxygen concentration is directly proportional to cysteamine concentration [Eq. (1)], and, depending on the activity of the dioxygenase, the reaction is complete in 1–2 min as indicated by no further changes in oxygen uptake. Then, 5 μl of 0.5 M GSH is added through the capillary access port, and oxygen concentration is recorded until the rate again returns to zero or near zero. The concentration of cysteamine and of cystamine are readily calculated from total oxygen uptake before and after the addition of GSH.

As shown for a typical analysis carried out with the highly purified dioxygenase (Fig. 1), the rate of oxygen uptake after GSH addition is limited by the reduction of cystamine [Eq. (2)]. On the other hand, crude preparations of the liver dioxygenase are frequently contaminated with thioltransferase[11] which catalyzes the reduction of cystamine by GSH; the use of the partially purified preparations usually decreases the time required for this phase of the assay. However, if the ratio of cystamine (or cysteamine mixed disulfides) relative to cysteamine is 5:1 or greater, the time required for reaction may still be excessive. In this event, an aliquot of the unknown is preferably preincubated with GSH for about 10 min before adding the dioxygenase, and the concentration of disulfide forms of cysteamine is calculated from the difference in oxygen uptake measured on samples analyzed before and after reduction.

Cysteamine or cystamine added to thiol-free media (from 5 to 200

---

[11] B. Mannervik, K. Axelson, and K. Larson, this series, Vol. 77, p. 281.

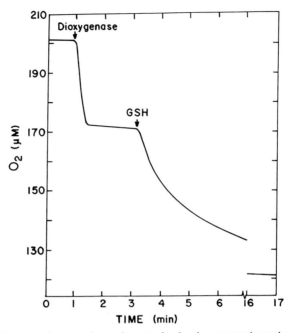

FIG. 1. Estimation of cysteamine and cystamine by the enzymatic method. Cysteamine dioxygenase and GSH were added, at times shown, to the oxygraph cell containing equal concentrations (30 μM) of cysteamine and cystamine.

μmol/ml) can be measured quantitatively, and the method can also be used to measure these compounds in blood plasma. However, 2–3 nmol/ml cysteamine or cystamine, added to rat plasma, react irreversibly with some plasma component(s) and cannot be recovered (as cysteamine) even after prolonged incubation with GSH. Higher concentrations added to plasma are, however, measured essentially quantitatively (96–97%).

Whereas the enzymatic method is simple and convenient, it cannot be used to measure the small amounts of endogenous cysteamine present in tissues. A more sensitive colorimetric procedure, developed for this purpose, is necessary.

Colorimetric Method

Principle. Cysteamine readily forms colored diadducts with 2,4-dinitrofluorobenzene (DNFB) in an alkaline solution as shown in Eq. (3). The water-insoluble N,S-bis(2,4-dinitrophenyl)cysteamine[12] is extracted into

---

[12] This compound, synthesized essentially by the procedure described in Tabor and Rosenthal[13] for the synthesis of the analogous diadduct of putrescine, has no detectable solubil-

Cysteamine + 2 DNFB → N,S-bis(2,4-dinitrophenyl)cysteamine + 2 F$^-$ (3)

dichloroethane, and the concentration is determined by HPLC. This procedure is essentially a modification of the method described by Tabor and Rosenthal[13] for the estimation of diamines.

Reagents

3.0 M Trichloroacetic acid (TCA)
35 mM Sodium dodecyl sulfate (SDS) in 0.05 M Tris at pH 8.0 containing 5 mM dithiothreitol (DTT) (DTT is added just prior to use)
2,4-Dinitrofluorobenzene (DNFB)
1.0 M Sodium aspartate (pH 9.0)
1.0 M Sodium bicarbonate
6.0 M Sodium hydroxide
1,2-Dichloroethane
Carbon tetrachloride

Procedure. Tissues are removed as quickly as possible, with 0.5 g (wet weight) homogenized in 2 ml of the SDS–Tris–DTT solution. The pH of the homogenate should be between 7.8 and 8.0 for quantitative reduction of cysteamine–protein mixed disulfides, which is essentially complete after a few minutes incubation at room temperature. SDS serves two functions: it makes the protein mixed disulfides more accessible to DTT and also inactivates tissue cysteamine dioxygenase which is capable of oxidizing cysteamine as rapidly as it is liberated from protein.

To ensure complete reduction the homogenate is incubated at room temperature (without further mixing) for 10 min. Then, 0.25 ml TCA is added to precipitate most of the protein. After sedimenting insoluble material by centrifugation, an aliquot (1.5 ml) of the supernatant fraction is transferred to a test tube containing 5 μl of DNFB. The pH is quickly raised to 10–10.5 by adding 0.2 ml bicarbonate and 0.05 ml NaOH. The solution is thoroughly mixed and incubated at room temperature with occasional mixing for 10 min. Next, 50 μl aspartate is added to react with any remaining DNFB, and, after a few minutes of incubation, nonpolar DNFB adducts are extracted into 0.5 ml dichloroethane by thorough mixing on a vortex test tube mixer. After separating the phases by brief centrifugation, about 0.2–0.3 ml of the organic phase is withdrawn and transferred to a clean test tube on ice so as to minimize changes in concentration due to evaporation of solvent.

---

ity in water, benzene, hexane, carbon tetrachloride, or diethyl ether. It is slightly soluble in tetrahydrofuran, acetonitrile, or methanol. It is somewhat more soluble in ethyl acetate, acetone, and dichloroethane, but, even in the latter solvent, solubility does not exceed 1 mM at room temperature.

[13] C. W. Tabor and S. M. Rosenthal, this series, Vol. 6, p. 615.

The concentration of bis(dinitrophenyl)cysteamine in the dichloroethane extract is determined by HPLC with a system consisting of a Kratos Spectroflow 400 equipped with a Pharmacia HR 10/10 column packed with polyanion SI-17 anion-exchange resin (Code 17-0526-01). The absorbance of the effluent is monitored at 360 nm with a Kratos Spectroflow 773. The mobile phase consists of 80:20 (v/v) dichloroethane–CCl_4 at a flow rate of 1 ml/min. In this solvent mixture the cysteamine diadduct has an absorption peak at 358 nm; absorptivity determined with the pure compound is 20.5 cm^{-1}.

The cysteamine diadduct elutes at 10.8 ml and is fully resolved from other nonpolar dinitrophenyl derivatives present in the dichloroethane extract.[14] The concentration of cysteamine can be calculated from the peak area by triangulation, but, in practice, it is more convenient to inject a constant volume (20 μl) of the extract and calculate concentration from peak height. Peak height is directly proportional to concentration upon injecting 20 μl of 0.2–100 μM bis(dinitrophenyl)cysteamine in trichloroethane.[15]

Without further modification, this method can detect as little as 0.5 nmol cysteamine/g tissue, and the recovery of cysteamine (or cystamine) added to the homogenate, between 1 and 20 nmol/g tissue, is usually better than 90%. Low recovery is usually due to incomplete reduction by DTT or reoxidation of cysteamine. Delay in formation of the dinitrophenyl derivative after preparation of the acid extract must be avoided.

The concentration of total cysteamine in rat liver and kidney, estimated by this method, is between 18 and 20 nmol/g tissue, virtually all of which appears protein bound. Less than 0.2 nmol cysteamine/g is present in extracts prepared in the absence of DTT, but whether this represents the status in the intact tissue or is due to autoxidation of this aminothiol upon homogenization is not known. Cystamine is an excellent protein thiol oxidant,[16] and any formed during tissue disruption would react almost immediately with protein thiols.

[14] The resolution increases with increasing CCl_4 in the mobile phase but retention time also increases. The 80:20 mixture appears optimal for full resolution with the shortest retention time but some variation in resolution with different batches of the polyanion resin have been observed. However, by changing the ratio of CCl_4 to dichloroethane, baseline resolution can always be obtained with this resin. Our system is programmed to flush the column for 10 min with dichloroethane and then for 10 min with the mobile phase after each run. Analysis of each sample takes about 35 min.

[15] In our system, 1 μmol of the diadduct injected in 20 μl has a peak absorbance of 0.025 cm^{-1}. However, this will vary with column size and other differences in configuration of the unit, and the value should be checked periodically with standards of known concentration.

[16] J. M. Wilson, D. Wu, R. Molice-DeGrood, and D. Hupe, *J. Am. Chem. Soc.* **102,** 359 (1980).

[26] Cysteinesulfinic Acid, Hypotaurine, and Taurine: Reversed-Phase High-Performance Liquid Chromatography

By MARTHA H. STIPANUK, LAWRENCE L. HIRSCHBERGER, and JAMES DE LA ROSA

Although a number of methods for the determination of taurine and other metabolites of cysteine have been reported,[1-9] none of these methods is appropriate for the simultaneous determination of cysteinesulfinic acid, hypotaurine, and taurine in biological samples. In particular, separation of these metabolites from alanine, glutamate, cysteic acid, and β- or γ-aminobutyrate has been a problem with other reversed-phase chromatographic methods; separation from O-phosphorylethanolamine and glycerophosphorylethanolamine has been a problem with analysis of taurine by ion-exchange methods. Additionally, preliminary purification of biological samples prior to chromatographic analysis of taurine often involves procedures that remove much of the cysteine sulfinate and/or hypotaurine.

Because cysteine sulfinate and hypotaurine are formed in the intermediate metabolism of cysteine to taurine via the cysteine sulfinate pathway, it is frequently of interest to measure all three compounds. The method described here allows quantitative analysis of cysteine sulfinate, taurine, and hypotaurine as well as other amino acids in biological samples. It is a modification of a method described previously.[10] We have used it for analysis of these compounds in animal tissues and plasma and for measurement of their production by tissue preparations incubated with cysteine *in vitro*.

[1] D. C. Turnell and J. D. H. Cooper, *Clin. Chem.* (*Winston-Salem, N.C.*) **28**, 527 (1982).
[2] M. Elsami and J. D. Stuart, *J. Liq. Chromatogr.* **7**, 1117 (1984).
[3] D. Hill, L. Burnworth, W. Skea, and R. Pfeifer, *J. Liq. Chromatogr.* **5**, 2369 (1983).
[4] G. H. T. Wheler and J. T. Russell, *J. Liq. Chromatogr.* **4**, 1281 (1981).
[5] J. D. Stuart, T. D. Wilson, D. W. Hill, F. H. Walters, and S. Y. Feng, *J. Liq. Chromatogr.* **2**, 809 (1979).
[6] W. G. Martin, B. L. Robeson, and L. S. Cadle, *Fed. Proc., Fed. Am. Soc. Exp. Biol.* **38**, 288 (1979).
[7] S. Gurusiddaaiah and R. W. Brosemer, *J. Chromatogr.* **223**, 179 (1981).
[8] J. Garvin, *Arch. Biochem. Biophys.* **91**, 219 (1960).
[9] T. L. Perry, D. Stedman, and S. Hansen, *J. Chromatogr.* **38**, 460 (1968).
[10] L. L. Hirschberger, J. De La Rosa, and M. H. Stipanuk, *J. Chromatogr.* **343**, 303 (1985).

Apparatus

The chromatographic system consists of two Waters Model 510 pumps, a Waters Model 680 gradient controller, a WISP Model 710B automatic sample injector, a column temperature control block, a Nova-Pak C_{18} column (15 cm × 3.9 mm, 4-μm spherical particles) and a guard column dry-packed with BondaPak C_{18}/Corasil 37–50 μm bulk packing (all from Millipore, Waters Chromatography Division, Milford, MA). Fluorescence is measured with a Spectro-Glo fluorometer (Gilson Medical Electronics, Middleton, WI) equipped with a 5-μl flow cell and filters for excitation and emission peaks at 360 and 455 nm, respectively. The output signal from the fluorometer is set at 10 mV full scale, and the photomultiplier sensitivity switch is set at X1 and the range at 10 or 20. The fluorometer is connected to an HP 3392A integrator (Hewlett-Packard, Palo Alto, CA).

We obtain superior separation on the Nova-Pak C_{18} column compared to other C_{18} columns tried for this purpose. The column is stored in acetonitrile : water (1 : 1, v/v) when not in use; methanol should not be used because storage of the Nova-Pak in it results in a substantial loss of resolving power.

Sample Preparation

Fresh or frozen tissues are homogenized in 5% (w/v) 5-sulfosalicylic acid to prepare 20% (w/v) homogenates. Blood is collected into tubes that contain heparin and is centrifuged at 2000 g for 10 min to obtain plasma; care is taken not to disturb the buffy coat when plasma is removed. The plasma protein is precipitated by addition of 1 volume of 35% (w/v) 5-sulfosalicylic acid to 9 volumes of plasma. The samples are centrifuged at 5000 g for 10 min to remove the precipitated protein, and the pH of the supernatants is adjusted to pH 6–7 with 10 N KOH. The supernatant fractions are chromatographed immediately or frozen for later analysis. Sulfosalicylic acid extracts of incubation mixtures are prepared similarly; perchloric acid extracts are neutralized with 10 N KOH and centrifuged to remove potassium perchlorate. Samples may be diluted with water prior to HPLC analysis if appropriate. The recovery of cysteine sulfinate, hypotaurine, and taurine from samples prepared as described ranges from 90 to 100%. (Heating samples in order to precipitate protein is not recommended as it results in substantial losses of cysteine sulfinate, apparently due to conversion of cysteine sulfinate to hypotaurine.)

Although not necessary for most samples, deproteinized tissue supernatant fluids may be passed through Sep-Pak C_{18} clean-up cartridges (Millipore, Waters Chromatography Division) prior to HPLC analysis. Before use, these cartridges are prepared by sequentially washing them with 20 ml of methanol, 20 ml of 0.1% (v/v) trifluoroacetic acid (TFA) in water and 10 ml of 1% TFA in methanol:water (20:80, v/v). A 1-ml aliquot of the sample is mixed with 2 ml of 0.1% TFA in methanol:water (30:70, v/v) and pushed through the cartridge with a syringe. This is followed with 2 ml of 0.1% TFA in methanol:water (30:70, v/v) to completely elute the amino acids. The entire 5 ml of the eluate is collected and adjusted to a final volume of 10 ml. Recovery of cysteine sulfinate, hypotaurine, and taurine from samples passed through the Sep-Pak cartridges ranges from about 90 to 100%.

Standard Solutions

Stock standard solutions of cysteinesulfinate, hypotaurine, taurine, and other compounds (Sigma Chemical Co., St. Louis, MO) are prepared in water, dispensed in small volumes suitable for daily use, and stored at $-20°$. Standards may also be prepared in 5-sulfosalicylic acid and neutralized in a manner that corresponds to that by which samples are prepared.

Chromatographic Procedure

The o-phthaldialdehyde/2-mercaptoethanol derivatizing reagent[11] is prepared each day by dissolving 120 mg of o-phthaldialdehyde in 0.5 ml of 95% ethanol and mixing this with 50 ml of 100 mM sodium borate (pH 10.4) and 0.1 ml of 2-mercaptoethanol. [A lower concentration of o-phthaldialdehyde, about 0.07% (w/v) compared to the 0.24% (w/v) described here, may be used if samples do not contain high concentrations of thiols.] This derivatizing reagent should be stored in an opaque vial at room temperature where it is stable for at least 24 hr. Samples and standards are derivatized by mixing equal volumes of the sample or standard solution and the derivatizing reagent; the WISP Model 710B automatic sample injector is used to perform the derivatization.

Two solutions are used to create the gradient for the mobile phase. Both contain a final concentration of 100 mM potassium phosphate buffer and 3% (v/v) tetrahydrofuran and are adjusted to pH 7.0. One of the solutions (Buffer B) also contains 40% (v/v) acetonitrile. HPLC-grade reagents are used for preparation of the buffers. Before use, solutions are filtered through a 0.5-μm filter (Millipore, Bedford, MA) and degassed under reduced pressure.

[11] M. Roth, *Anal. Chem.* **43**, 880 (1971).

Under conditions of no flow and no pressure in the HPLC system, 10 μl of the derivatizing reagent and then 10 μl of the sample or standard solution are sequentially injected into the tubing between the injector and the guard column (200 cm × 0.23 mm including that in the temperature-control block).

We begin our gradient program with a 2.25-min delay period (no flow) to allow time for injection and reaction of the derivatizing reagent and sample or standard. After the sample or standard is injected, the flow rate is increased from 0 to 0.1 ml/min and held there for 0.75 min to facilitate thorough mixing of the derivatizing reagent and sample or standard. The flow rate is then increased to 0.95 ml/min, the integrator is started, and chromatography is carried out at the flow rate of 0.95 ml/min and a column temperature of 44°. The mobile phase is run isocratically for 4 min at 0% Buffer B (no acetonitrile). The amount of Buffer B is then increased to 32% (12.8% acetonitrile) over 18 min using curve profile 3 of the Waters Model 680 gradient controller. The percentage of Buffer B is then linearly increased from 32 to 100% (40% acetonitrile) over 2 min and held at 100% for 6 min. At 33 min, the percentage of Buffer B is returned to 0% by a linear decrease over 2 min, and the column is allowed to equilibrate for at least 10 min. The system is then returned to conditions of no flow and no pressure over 0.5 min and held there for 2 min before the next sample is analyzed. Total run time is 47.5 min per sample or about 50 min if the time for injection and derivatization is included.

If manual injection of samples is used, samples may be derivatized by mixing equal volumes of sample and derivatizing reagent and injecting 20 μl of the final solution onto the column within 1 min of mixing. If this method is used, the flow rate is maintained at 0.95 ml/min, and the delay period between runs that allows the system to return to conditions of no flow and no pressure is omitted. The initial isocratic phase of 0% Buffer B may also be decreased from 4.0 to 1.5 min. The total run time with the manual injection method is about 39.5 min per sample.

We routinely use standard solutions interspersed with sample solutions and sometimes use internal standards as an additional control. Standards are necessary for each batch of derivatizing reagent.

Linearity, Sensitivity, and Reproducibility

The fluorescence intensity is proportionate to the amino acid content of standard solutions between 0 and 1000 pmol per 10-μl aliquot for cysteine sulfinate, hypotaurine, and taurine. The fluorescence yield for the three compounds is similar. As little as 10 pmol of any one of the three compounds can be measured. The separation is very reproducible; the

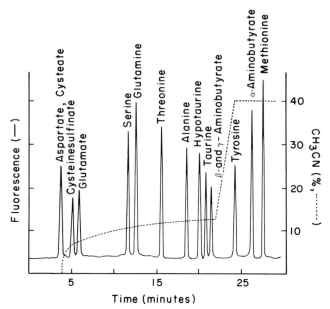

FIG. 1. Representative chromatogram for the analysis of a mixture of amino acids. The percentage of acetonitrile in the mobile phase is indicated by the dotted line. The origin of the chromatogram (time 0) corresponds to the point at which the flow rate was increased to 0.95 ml/min.

coefficients of variation for the retention times of cysteine sulfinate, hypotaurine, and taurine on our system are all about 0.25%. The coefficients of variation we have obtained for determinations of cysteine sulfinate, hypotaurine, and taurine in biological samples are in the range of 1–2%. Very high concentrations of thiols, i.e., cysteine or cysteamine, in samples may decrease fluorescence yield. The use of 0.24% (w/v) o-phthaldialdehyde in the derivatizing reagent minimizes this effect, but internal standards should be used with such samples as an additional precaution.

Separation of Cysteine Sulfinate, Hypotaurine, and Taurine from Other Compounds

As shown in Fig. 1, this method separates alanine, hypotaurine, taurine, β-aminobutyrate, and γ-aminobutyrate. β-Aminobutyrate elutes slightly (about 0.2 min) earlier than γ-aminobutyrate when standards are run separately, but they produce a single peak when both are present in the same mixture of standard amino acids. Both cysteine and cystine give very low fluorescence yields with o-phthaldialdehyde/2-mercaptoethanol;

they are below the limits of detection even when as much as 10 nmol of either compound is derivatized and injected. The elution positions of cysteine and cystine under standard conditions are well before cysteine sulfinate. Neither O-phosphorylethanolamine nor glycerophosphorylethanolamine coelute with any of the metabolites of cysteine in this system.

Changes in either tetrahydrofuran concentration or column temperature have substantial effects on this separation. Tetrahydrofuran is essential for adequate separation of hypotaurine, alanine, and taurine. Heating the column to about 44° is essential for samples that contain β- or γ-aminobutyrate. For samples that do not contain β- or γ-aminobutyrate, improved resolution of taurine and hypotaurine may be obtained by reducing the column temperature to 38° or ambient; when lower column temperatures are used, the mobile phase for the initial isocratic phase should contain 2–4% (v/v) acetonitrile. The initial isocratic phase is necessary for resolution of cysteine sulfinate. If only taurine and hypotaurine are of interest, the sample run time may be decreased by using a higher initial concentration of acetonitrile (i.e. 6%, v/v, at ambient column temperature) and using a similar gradient without an isocratic phase at the beginning of the run.

[27] Cysteinesulfinic Acid: Fuchsin Method

By Franz-Josef Leinweber and Kenneth J. Monty

Cysteinesulfinic acid (CSA), or β-sulfinyl-α-aminopropionic acid, is accepted as a structural analog of aspartic acid by the L-aspartate aminotransferase (glutamic-oxaloacetic transaminase; EC 2.6.1.1) from both animal and bacterial sources.[1–3] Evidence for that conclusion is found in the competitive relationship between CSA and aspartate as alternative substrates for the enzyme.[3] When accepting CSA as a substrate. The aminotransferase catalyzes the formation of β-sulfinyl pyruvate, which spontaneously and rapidly decomposes to release inorganic sulfite [Eq. (1)]. That series of events explains the utilization of CSA as a sole source of sulfur in bacteria.[2,3] When augmented with a colorimetric determina-

[1] F.-J. Leinweber and K. J. Monty, *Anal. Biochem.* **4**, 252 (1962).
[2] F.-J. Leinweber and K. J. Monty, *Biochem. Biophys. Res. Commun.* **6**, 355 (1961).
[3] F.-J. Leinweber and K. J. Monty, *Biochim. Biophys. Acta* **63**, 171 (1962).

$$\begin{array}{c} \text{COOH} \\ | \\ H_2N\text{CH} \\ | \\ \text{CH}_2 \\ | \\ \text{SO}_2\text{H} \end{array} \xrightarrow{\alpha\text{-ketoglutarate} \rightarrow \text{glutamate}} \left[\begin{array}{c} \text{COOH} \\ | \\ \text{C}=\text{O} \\ | \\ \text{CH}_2 \\ | \\ \text{SO}_2\text{H} \end{array} \right] \xrightarrow{H_2O} \begin{array}{c} \text{COOH} \\ | \\ \text{C}=\text{O} \\ | \\ \text{CH}_3 \\ + \\ H_2SO_3 \end{array} \quad (1)$$

tion of the sulfite released, the reaction of CSA with α-ketoglutarate in the presence of aspartate aminotransferase provides a convenient, reliable, and precise method for the determination of CSA. The original method was applied to as little as 50 nmol of CSA,[1] but that sensitivity can be improved.

Commercially available aspartate aminotransferase of pig heart is a convenient source of suitable enzyme. The enzyme should be dissolved in 40 mM sodium phosphate, pH 6.8, at a final protein concentration of approximately 2 mg/ml. Dialysis against the same buffer is recommended to remove potential interference by salts which contaminate the enzyme. Such preparations of the pig heart enzyme should retain 85% of their activity after storage for 8 days at 3°.

The aminotransferase can be assayed by incubating the enzyme in 3 ml of a solution containing 300 μmol L-monosodium glutamic acid, 7.5 μmol oxaloacetic acid, and 250 μmol sodium phosphate at pH 8.0. The absorbance of the solution at 280 nm is followed as a function of time at 37°. For the purposes of this method, a unit of enzyme is defined as that amount catalyzing an initial rate of change of 0.185 absorbance units per minute at 280 nm (1 cm light path), a change which corresponds to 1 μmol of oxaloacetate.

To demonstrate the adequacy of the method, appropriate amounts of a neutral solution of CSA are combined with 15 μmol α-ketoglutarate and 210 μmol of the phosphate buffer in a total volume of 2.4 ml. The reaction is initiated by the addition of 25 units of enzyme in a volume of 100 μl. After incubation at 37° for 15 min, the reaction is terminated by addition, in quick succession, of 0.5 ml 1% alcoholic KOH and 1 ml saturated aqueous $HgCl_2$. After mixing thoroughly, the solutions are subjected to brief centrifugation to obtain a clear supernatant layer. One milliliter of that supernatant fluid is added to 4 ml fuchsin reagent.[4] Precisely 15 min later, the absorbance (A) of the mixture is determined at 580 nm.

A control for inorganic sulfite indigenous to the samples is performed by adding the KOH and $HgCl_2$ prior to addition of the aminotransferase. Such controls are not incubated at 37° but are centrifuged directly and combined with fuchsin reagent.

---

[4] See this volume [4].

The amount of CSA present in the 2.5 ml incubation mixture is calculated as follows:

nmol of CSA = $(A_{\text{incubated sample}} - A_{\text{control}})517$

The proportionality constant, shown above as 517, was determined with a solution of sulfite standardized by iodimetric titration[4] and is valid for a Beckman Model DU spectrophotometer with a slit width of 0.03 mm and a light path of 1 cm. This value should be determined empirically for alternative instrumentation.

Seven samples of CSA have been measured in parallel, in amounts ranging from 7.66 to 76.6 μg. The results displayed an average of 99.4% of the theoretical amount of sulfite, with a standard deviation of 2.7% for the seven determinations.

Specificity

Specificity of the method depends on attributes of both the enzyme and the fuchsin reaction. The high degree of specificity that can be expected from the latter is discussed elsewhere in this volume,[4] and that reaction has proved to be highly satisfactory in studies of sulfur metabolism in microorganisms. The possibility of release of inorganic sulfite from two compounds similar to CSA had been of concern. Hypotaurine (2-aminoethanesulfinic acid) produced no color in the enzyme-dependent assay or in the sulfite control described above. Homocysteinesulfinic acid, a structural analog of glutamate, was not available for test; it would be expected to be deaminated enzymatically to γ-sulfinyl-α-ketobutyric acid, but it is not likely that this product would hydrolyze to release sulfite under the assay conditions.

Interfering Compounds

Sources of potential interference in the method also reflect properties of the enzyme or of the fuchsin reaction. Acceptable substrates for transamination by the pig heart enzyme include glutamic, aspartic, cysteic, and homocysteic acids, and their corresponding keto acids, α-ketoglutaric, oxaloacetic, β-sulfonylpyruvic (β-sulfonyl-α-ketopropionic), and γ-sulfonyl-α-ketobutyric acids, respectively. Each of these compounds at sufficiently high concentrations can retard the rate of release of sulfite from CSA by competing for the substrate binding site on the enzyme. Indeed, any dicarboxylic acid with a 2- or 3-carbon chain separating the carboxyl groups is likely to associate reversibly with the substrate binding site, and may be a competitive inhibitor for the release of sulfite from CSA. In analyzing samples of unknown composition for the presence of

CSA, possible interference by such divalent organic anions may be detected and overcome by increasing the amount of aminotransferase used until a maximum amount of sulfite is formed. The spontaneous decomposition of β-sulfinyl pyruvate guarantees the ultimate stoichiometric conversion of CSA to sulfite if oxaloacetate is not limiting and an aminotransferase is present.

Inorganic sulfite is susceptible to oxidation at pH 8.0, although the presence of α-ketoglutarate significantly retards the rate of this oxidation. Inorganic sulfite has been carried through the analytical procedure with a recovery of better than 99%. However, increasing the incubation at 37° to 30 min decreased recovery to 95%. These observations argue that the preferred way of ensuring more complete reaction of CSA is to increase the amount of enzyme present rather than to increase the duration of the incubation.

A more detailed discussion of possible interferences with the fuchsin reaction appears elsewhere in this volume.[4] In dealing with biological specimens, remember that disulfides react rapidly with inorganic sulfite in the neutral to basic pH range. When the presence of significant quantities of cystine, glutathione disulfide, or other disulfide compounds is suspected, care should be taken to remove those compounds before attempting this assay. A convenient method for the removal of thiols and disulfides from CSA using alumina has been described by Fromageot *et al.*[5]

The ultimate proof that an unknown sample is not interfering with the assay is to show stoichiometric recovery of sulfite from a known amount of CSA added to the sample.

Expanding Sensitivity

The sensitivity of this method can be improved significantly by minimizing the volumes employed in the enzyme incubation and in the measurement of sulfite.[4] The use of alternative methods for the determination of sulfite should also be considered.[6] Baba *et al.*[7] enhanced the sensitivity very considerably by replacing the sulfite determination with a series of enzyme-catalyzed reactions employing aspartate aminotransferase, lactic dehydrogenase, alcohol dehydrogenase, and malate dehydrogenase; a linear response in the range of 5 to 50 pmol has been reported.

---

[5] C. Fromageot, M. Jutisz, and E. Lederer, *Biochim. Biophys. Acta* **2**, 487 (1948).
[6] See this volume [3] and [5].
[7] A. Baba, S. Yamagami, H. Mizuo, and H. Iwata, *Anal. Biochem.* **101**, 288 (1980).

[28] Cysteinesulfinic Acid and Cysteic Acid: High-Performance Liquid Chromatography

By K. Kuriyama and Y. Tanaka

L-Cysteinesulfinic acid is a key intermediate in the biosynthesis of taurine from cysteine. Cysteic acid is another important intermediate in taurine biosynthesis from cysteinesulfinic acid or sulfate. For the assay of cysteine sulfinate, the ninhydrin reaction has usually been employed after separating this compound by column chromatography.[1] Reaction with ninhydrin, however, is not sufficiently sensitive to measure the low concentrations of cysteine sulfinate present in biological samples. Although a sensitive enzymatic assay method has been reported,[2] its disadvantages include the complexity of assay procedures and unavailability of this assay system for the simultaneous determination of cysteic acid. The following assay procedures, using HPLC, are simple and specific for cysteinesulfinic acid and cysteic acid, and are readily applicable to the determination of these amino acids in biological samples.[3-6]

Assay Method

Preparation of Samples

Biological samples are homogenized in 10 volumes of ice-cold 10% trichloroacetic acid (TCA), and the homogenates are centrifuged at 8000 g for 30 min at 4°. For the assay of amino acids other than the subject of this article, this supernatant liquid is used directly. For cysteine sulfinate and cysteic acid, the supernatant liquid is extracted 4 times with 3 volumes of water-saturated ether to remove TCA. Aliquots of the extracted solution are lyophilized and dissolved in distilled water. Each extract, 100 μl, is applied to a column (6 × 120 mm), packed with cation-exchange resin (Dowex 50W-X8, 200–400 mesh, H$^+$ form) and is eluted with water. The initial 1 ml of eluate is discarded; the following 1 ml of eluate, containing

---

[1] T. P. Singer and E. B. Kearney, *Arch. Biochem. Biophys.* **61**, 397 (1956).
[2] A. Baba, S. Yamagami, H. Mizuo, and H. Iwata, *Anal. Biochem.* **101**, 288 (1980).
[3] S. Ida and K. Kuriyama, *Anal. Biochem.* **130**, 95 (1983).
[4] K. Kuriyama, S. Ida, and S. Ohkuma, *J. Neurochem.* **42**, 1600 (1984).
[5] S. Ida, S. Ohkuma, M. Kimori, K. Kuriyama, N. Morimoto, and Y. Ibata, *Brain Res.* **344**, 62 (1985).
[6] S. Ida, Y. Tanaka, S. Ohkuma, and K. Kuriyama, *Anal. Biochem.* **136**, 352 (1984).

cysteine sulfinate, cysteic acid, and taurine is collected, lyophilized, and then dissolved in water. Aliquots of this solution are subjected directly to HPLC. Samples are stored at $-70°$ until analysis.

Liquid Chromatography System

The HPLC system used is Model LC3A (Shimadzu Seisakusho Ltd., Kyoto, Japan). Two different column and mobile phases used are as follows:

1. Determination of other amino acids: Column (15 cm × 4 mm i.d.) packed with ISC-07/S1504 (cation-exchange resin); temperature, 55°; eluent, 0.067 N sodium citrate (pH 3.25 and 4.25) : 0.2 N sodium citrate (pH 9.00); time program of the stepwise gradient elution system (pH 3.25, 4.25, and 9.00) employed, 15, 35, and 35 min, respectively.

2. Determination of cysteine sulfinate and cysteic acid: Column (25 cm × 4 mm i.d.) packed with ISA-07/S2504 (anion-exchange resin); temperature, 55°; eluent, 0.05 M KH_2PO_4; detection, effluents from each HPLC column are mixed with o-phthalaldehyde, and fluorescence of each eluent is measured spectrofluorometrically (excitation wavelength 348 nm, emission wavelength 450 nm). To quantify each amino acid, it is advisable to use an on-line computing integrator for the measurement of each peak area. The peak for each amino acid is identified by comparison with the retention time obtained from each authentic compound.

Profile of Amino Acid Separation in Biological Samples

The separation of amino acids in a standard mixture containing 1 nmol of various amino acids and in biological samples using the cation-exchange resin is essentially identical. Amino acids used and the order of appearance close to the void volume are as follows: (1) cysteinesulfinic acid, cysteic acid, O-phosphoserine; (2) O-phosphoethanolamine; (3) taurine; (4) aspartic acid; (5) threonine; (6) serine; (7) glutamic acid; (8) proline; (9) glycine; (10) alanine; (11) cystine; (12) valine; (13) methionine; (14) isoleucine; (15) leucine; (16) tyrosine; (17) phenylalanine; (18) γ-aminobutyric acid; (19) histidine; (20) lysine; (21) arginine. Taurine is separated from other amino acids although O-phosphoethanolamine, one of the o-phthalaldehyde-reactive compounds having a similar retention time in HPLC, is eluted close to the peak of taurine. Cysteine sulfinate, cysteic acid, and O-phosphoserine, however, are coeluted in the same fraction when a cation-exchange column (ISC-07/S1504) is used. This constitutes a problem when applied to biological samples, since the elution peak of this fraction is the nearest to the void volume. Resolution results by the use of a strong-base anion-exchange column (ISA-07/ S2504). This chromatographic system clearly separates cysteinesulfinic acid (retention time 22.5 ± 0.3 min, mean ± SEM), cysteic acid (retention time 34.2 ± 0.5 min), and O-phosphoserine (retention time 40.2 ± 0.6 min) from other compounds. In contrast, the retention time for taurine in this system is 7.2 ± 0.1 min. This assay is linear between 20 pmol and 5 nmol for cysteine sulfinate and 10 pmol and 5 nmol for cysteic acid. In the case of biological samples, pretreatment with a cation-exchange column (Dowex 50W-X8, see Preparation of Samples) is essential. The recovery of both cysteine derivatives from the column-chromatographic procedures is approximately 100%.

The contents of cysteine sulfinate and cysteic acid in whole brain, kidney, and liver of male Wistar rats (weighing 180–200 g), measured by the above HPLC procedure, are as follows: cysteine sulfinate, whole brain, 21.8 ± 1.4, kidney, 1.18 ± 0.10, liver, 1.67 ± 0.15 nmol/g wet weight, $N = 5$; cysteic acid, whole brain, 10.1 ± 1.8, kidney, 18.6 ± 1.5, liver, 2.03 ± 0.19 nmol/g wet weight, $N = 5$.

[29] Resolution of Cysteine and Methionine Enantiomers

By OWEN W. GRIFFITH and ERNEST B. CAMPBELL

Introduction[1]

Studies of the metabolism of cysteine and methionine generally address either implicitly or explicitly the enantiomeric specificity of the reactions involved. Whereas mammalian enzymes other than D-amino-acid oxidase (EC 1.4.3.3) are relatively or absolutely specific for L-enantiomers, plant and microbial systems contain a variety of enzymes and transport systems active toward D-methionine and D-cyst(e)ine.[2-6] Studies of the substrate specificity of enzymes as well as the evaluation of the enantiomeric purity of reagent amino acids (particularly isotopically labeled amino acids) require analytical methods for the resolution of

---

[1] These studies were supported in part by National Institutes of Health Grants AM26912 and GM32907. O.W.G. is an Irma T. Hirschl Career-Scientist.
[2] A. Schmidt and I. Erdle, *Z. Naturforsch., C: Biosci.* **38C**, 428 (1983).
[3] A. Schmidt, *Z. Pflanzenphysiol.* **107**, 301 (1982).
[4] J. Poland and P. D. Ayling, *Mol. Gen. Genet.* **194**, 219 (1984).
[5] F. Schlenk, C. H. Hannum, and A. J. Ferro, *Arch. Biochem. Biophys.* **187**, 191 (1978).
[6] H. Ohkishi, D. Nishikawa, H. Kumagai, and H. Yamada, *Agric. Biol. Chem.* **45**, 2397 (1981).

methionine and cysteine enantiomers. Recently developed methods for the separation of amino acid enantiomers by HPLC in the presence of chiral mobile[7-9] or stationary[10-12] phases provide convenient procedures for such analytical resolutions. Preparative resolution of unlabeled DL-methionine, DL-cyst(e)ine, and other naturally occurring sulfur amino acids is rarely needed since the pure enantiomers are either available commercially or are easily synthesized from chiral precursors. Preparative resolution of isotopically labeled sulfur amino acids remains of considerable interest since only racemates of the materials of interest are available in many cases. The procedures described here illustrate the analytical and preparative resolution of some of the sulfur amino acids.

Analytical Resolution of Sulfur Amino Acids

The enantiomers of methionine and other S-alkylhomocysteines, S-carboxamidomethylcysteine, S-benzylcysteine, S-benzylhomocysteine, and several of their corresponding α-methyl and α-ethyl derivatives are resolved by HPLC on conventional C_{18} reversed-phase silica columns using Cu^{2+}–L-proline as a chiral mobile phase. The method is a direct extension of procedures developed by Gil-Av et al.[8]

Materials. α-Ethyl-DL-methionine,[13] S-benzyl-α-methyl-DL-cysteine,[14] and α-methyl-DL-cysteine[14] were synthesized as described. α-Methyl-DL-methionine (from Vega-Fox) and other amino acids (from Sigma) are commercially available. The o-phthalaldehyde reagent is prepared as follows: o-Phthalaldehyde, 800 mg, is dissolved completely in 10 ml of absolute ethanol, and that solution is added to 1 liter of 400 mM sodium borate buffer at pH 9.7 containing 1 g of Brij detergent, 2.5 g of disodium EDTA, and 9.7 ml of 2-mercaptoethanol.[8] The reagent should be stored under nitrogen or prepared fresh every 2–3 days.

S-Carboxamidomethyl derivatives of cysteine, α-methylcysteine, and homocysteine are prepared by dissolving the amino acid (0.1 μmol) in 1 ml of 40 mM sodium borate buffer at pH 9.7 and adding to that mixture 10 μl of freshly prepared 50 mM iodoacetamide. The reaction is complete after standing for 10 min at 25°. If the amino acid solution contains disulfides, they can be reduced and derivatized by adding 30 μl of 10 mM

---

[7] H. Nakazawa and H. Yoneda, *J. Chromatogr.* **160**, 89 (1978).
[8] E. Gil-Av, A. Tishbee, and P. E. Hare, *J. Am. Chem. Soc.* **102**, 5115 (1980).
[9] S. Weinstein, M. H. Engel, and P. E. Hare, *Anal. Biochem.* **121**, 370 (1982).
[10] W. H. Pirkle, M. H. Hyun, and B. Bank, *J. Chromatogr.* **316**, 585 (1984).
[11] W. H. Pirkle, D. W. House, and J. M. Finn, *J. Chromatogr.* **192**, 143 (1980).
[12] G. Gubitz, W. Jellenz, and W. Santi, *J. Chromatogr.* **203**, 377 (1981).
[13] O. W. Griffith and A. Meister, *J. Biol. Chem.* **253**, 2333 (1978).
[14] O. W. Griffith, *J. Biol. Chem.* **258**, 1591 (1983).

dithiothreitol to the amino acid–borate mixture 15 min prior to the addition of 20 μl of iodoacetamide.

Apparatus. Waters HPLC equipment (Model 510 pumps, Model 680 automatic gradient controller, and Model 420-AC fluorescence detector) was used with a Rainin model 7125 injection valve and a Phase Separations ODS2 column (4.5 × 250 mm containing fully end-capped C_{18} 5-μm spherical silica). The *o*-phthalaldehyde reagent was introduced into the column effluent using an Eldex A-30 pump and a T-connection. To allow time for *o*-phthalaldehyde to react, the resulting mixture was passed through a 9-foot coil of 0.02 in. i.d. tubing prior to entering the fluorometer. For analysis of α-alkylamino acids, which react slowly with *o*-phthalaldehyde, the 9-foot coil was replaced by a 35-foot coil of the same size of tubing maintained at 55° by immersion in a water bath. Data were recorded and analyzed with an Apple IIe computer using data acquisition hardware and software from Interactive Microware (State College, PA). Any similar equipment capable of isocratic elution and able to detect and quantitate fluorescent species in an eluent stream should be suitable.

Chromatographic Procedure. For the resolution of most sulfur amino acids the column is equilibrated and eluted with 17 mM L-proline and 8 mM cupric acetate (pH 5.5 without adjustment) (buffer A) at a flow rate of 0.5 ml/min.[8] *o*-Phthalaldehyde reagent is mixed with the column effluent at a flow rate of 0.5 ml/min. For buthionine and *S*-benzyl derivatives of amino acids the equilibrating and eluting buffer is 17 mM L-proline and 8 mM cupric acetate in 40% methanol (buffer C). Buffer B is a 50:50 mixture of buffers A and C. Samples (20 μl) are dissolved either in water or in the carboxamidomethyl derivitization mixture described above; samples of 2 nmol are appropriate for methionine and *S*-carboxamidomethylcysteine; larger samples (5–10 nmol) are required for similar fluorescent yield from α-alkylamino acids. Samples containing much smaller amounts of amino acid can be accommodated by adjustment of the sensitivity of the fluorescence detector.

Results. The resolution of the enantiomers of *S*-carboxamidomethyl-DL-cysteine and DL-methionine is shown in Fig. 1. Cysteinesulfonic acid, the enantiomers of which are not resolved, is included as a void volume marker. Table I lists the elution times of the enantiomers of several additional sulfur amino acids; for all of the *S*-substituted cysteine and homocysteine derivatives examined, the D-enantiomer eluted prior to the L-enantiomer. Since resolved samples of the α-alkylamino acids were not available, the order in which their enantiomers elute is unknown. Several commerical samples of D- and L-methionine and D- and L-cysteine were examined for enantiomeric purity using sample loads of 100 nmol; all samples examined contained <0.1% enantiomeric contaminant. It may be

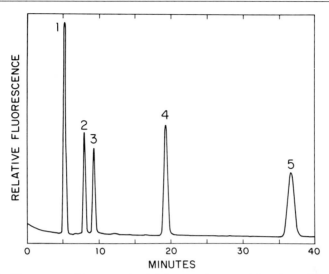

FIG. 1. Chromatographic separation of the enantiomers of S-carboxamidomethyl-DL-cysteine and DL-methionine. The amino acids were chromatographed as described using buffer A. The peaks correspond to 4 nmol of L-cysteic acid (peak 1), 2 nmol of S-carboxamidomethyl-D-cysteine (peak 2), 2 nmol of S-carboxamidomethyl-L-cysteine (peak 3), 4 nmol of D-methionine (peak 4), and 4 nmol of L-methionine (peak 5).

noted that in some cases the fluorescent peak representing a large amount of S-carboxamidomethyl-D-cysteine may interfere with the accurate quantitation of a small amount of the L-enantiomer, which elutes later. In such cases, changing the elution buffer to Cu^{2+}–D-proline reverses the order of enantiomer elution and facilitates quantitation.

Preparative Resolution of DL-[^{14}C]Cysteine

Detailed experimental procedures for the enzymatic resolution of DL-cystine and DL-methionine using hog kidney acylase I or D-amino-acid oxidase have been presented by Greenstein and Winitz.[15] The procedures described for DL-methionine are convenient and are easily adapted to the small-scale resolution of isotopically labeled methionine (O. Griffith, unpublished). Enzymatic resolution of cystine on a small scale is less convenient. Since most preparations of DL-cystine also contain *meso*-cystine, enzymatic resolution of cystine generally yields three products that on a small scale are not easily separated. In addition, the amount of pure D- or

---

[15] J. P. Greenstein and M. Winitz, "Chemistry of the Amino Acids," Vol. 3, pp. 1879 and 2125. Wiley, New York, 1961.

TABLE I
RESOLUTION OF SULFUR AMINO ACID ENANTIOMERS

Compound	Solvent system[a]	Elution time of enantiomers[b] (min)	
		Peak 1	Peak 2
S-Carboxamidomethyl-DL-cysteine	A	2.76 (D)	4.07 (L)
S-Methyl-DL-cysteine	A	9.20 (D)	27.05 (L)
S-Benzyl-DL-cysteine	C	12.00 (D)	14.25 (L)
DL-Methionine	A	13.97 (D)	31.23 (L)
	B	2.71 (D)	3.93 (L)
DL-Ethionine	B	7.43 (D)	9.80 (L)
DL-Prothionine	B	24.06 (D)	31.73 (L)
DL-Buthionine	C	19.10 (D)	20.40 (L)
S-Carboxamidomethyl-DL-homocysteine	A	5.89 (D)	7.66 (L)
S-Benzyl-DL-homocysteine	C	17.43 (D)	19.25 (L)
S-Carboxamidomethyl-α-methyl-DL-cysteine	A	5.78	6.31
S-Methyl-α-methyl-DL-cysteine	A	20.54	29.35
S-Benzyl-α-methyl-DL-cysteine	C	13.84	17.27
α-Methyl-DL-methionine	A	30.72	57.68
	B	4.61	6.27
α-Ethyl-DL-methionine	A	98.18	106.72
	B	8.11	9.42

[a] The composition of solvent systems A, B, and C are given in the text.
[b] The configuration of the enantiomer, where known, eluting in peak 1 and peak 2 is shown in parentheses.

L-cyst(e)ine obtained is reduced by the amount of *meso*-cystine remaining. Enzymatic resolution of molecules containing only one cysteine moiety offers one solution to these difficulties. Although cysteine itself oxidizes too easily to be conveniently resolved, both S-benzyl-DL-cysteine[15] and S-(methanethio)-DL-cysteine[16] are useful derivatives in this respect. The S-benzyl derivative has the advantage of greater stability during the resolution procedure but has the considerable disadvantage of requiring more drastic conditions for the regeneration of free cysteine after the resolution is complete. The resolution of N-acetyl-S-(methanethio)-DL-cysteine is described below.[16]

Synthesis of N-Acetyl-S-(methanethio)-DL-[1-^{14}C]cysteine. DL-Cysteine · HCl · H$_2$O (833 mg, 5 mmol) and 1.21 g of Tris base (10 mmol) are dissolved in 30 ml of water containing 77 mg of dithiothreitol (0.5 mmol) and 500 μCi of DL-[1-^{14}C]cystine (1.62 mg, 74 mCi/mmol). The mixture is

[16] O. W. Griffith, *J. Biol. Chem.* **257**, 13704 (1982).

stirred (magnetic stirrer) for 30 min at 25° while a slow stream of argon or nitrogen is carefully allowed to bubble through the solution. There should be no insoluble cystine present after 30 min. Radiolabeled DL-cysteine can be substituted for cystine, but the amount of dithiothreitol should not be reduced. The solution is chilled to 0°, and a cold solution of 1.1 g of methyl methanethiosulfonate (S-methyl thiomethanesulfonate, 10 mmol) in 9 ml of ethanol is added in about 1 ml portions over a period of 5 min. The reaction should be carried out in an exhaust hood and care should be taken to avoid exposure to toxic methyl methanethiosulfonate. A white precipitate of S-(methanethio)-DL-cysteine forms as the mixture is allowed to stir for 30 min under argon.

Although this intermediate can be isolated in 60–70% yield by filtration,[16,17] it is preferable to solubilize it by adding 1 ml of concentrated HCl directly to the reaction mixture. The solution is then filtered to remove a small amount of unidentified insoluble material. The filter is washed with about 40 ml of water, and the combined filtrates, containing the desired mixed disulfide contaminated by about 5% cystine, are applied to a column (1.2 × 8 cm) of Dowex 50-X8 (H^+ form, 200–400 mesh). After the resin is washed well with water, S-(methanethio)-DL-cysteine is eluted with 1.2 M HCl. Fractions containing product are pooled and rotary evaporated to dryness under reduced pressure (the bath temperature should not exceed 40°). The product, free of cystine, is obtained in about 85% yield. It is essential that the product not be exposed to thiols, base, or other agents able to cleave disulfide bonds since any cleavage produces free thiols which will catalyze the dismutation of the mixed disulfide into cystine and methyl disulfide.

The dried S-(methanethio)-DL-[1-^{14}C]cysteine is suspended in 25 ml of glacial acetic acid containing 410 mg of anhydrous sodium acetate (5 mmol). Acetic anhydride (2 ml, 20 mmol) is added to the mixture, and it is stirred or swirled in a 100° oil bath for 15 min. The mixture is cooled and brought to dryness with a rotary evaporator at reduced pressure. The residue is dissolved in 50 ml of water, and the resulting solution is passed through a column (1.2 × 8 cm) of Dowex 50 (H^+). N-Acetyl-S-(methanethio)-DL-[1-^{14}C]cysteine is eluted with water. The yield is quantitative based on radioactivity.

Resolution of N-Acetyl-S-(methanethio)-dl-[1-^{14}C]cysteine. The Dowex 50 eluent described above (~100 ml) is adjusted to pH 7 by addition of a saturated solution of Tris base; the solution is stirred well during this procedure to avoid local concentrations of base. To the neutral solution is added 6 mg of acylase I (Sigma grade III, 6300 units/mg), and the solution is allowed to stand at 25° with occasional swirling. It is important

---

[17] D. J. Smith, E. T. Maggio, and G. L. Kenyon, *Biochemistry* **14**, 766 (1975).

to use a highly purified grade of acylase I in order to avoid the introduction of adventitious protein thiols that could catalyze dismutation of the mixed disulfide; in some cases we have also found it useful to either dialyze the protein or pass it through a column of Sephadex G-25 to remove contaminating low molecular weight thiols and amino acids. The selective hydrolysis of the L-enantiomer is complete within 90 min as judged by amino acid analysis. The solution is then applied to a column (1.2 × 24 cm) of Dowex 50 (H$^+$), and N-acetyl-S-(methanethio)-D-[1-^{14}C]cysteine is washed through the resin with 100 ml of water. The S-(methanethio)-L-[1-^{14}C]cysteine is eluted with 1.2 M HCl. If fractions of 6 ml are collected, the product elutes in fractions 90 to 110; a small amount of [^{14}C]cystine is eluted later as a separate peak. Appropriate fractions are pooled and evaporated to dryness at reduced pressure.

S-(Methanethio)-L-[1-^{14}C]cysteine is obtained in a yield of 40–45% based on radioactivity (80–90% of the theoretical yield). The mixed disulfide is converted to free cysteine by treating an aqueous solution at pH 8 with 1.1 equivalents of dithiothreitol; nitrogen or argon is bubbled through the solution to remove the methanethiol. Further bubbling with oxygen in the presence of a few milligrams of FeCl$_3$ catalyst can be carried out to convert the cysteine to cystine. The latter can be conveniently isolated from other components of the reaction mixture by chromatography on Dowex 50 using 1.2 M HCl as eluent (see above).

[30] Isethionic Acid

By JACK H. FELLMAN

$$HO-CH_2-CH_2-\overset{\overset{\displaystyle O}{\|}}{\underset{\underset{\displaystyle O}{\|}}{S}}-OH$$

Isethionic acid

Introduction

The occurrence of isethionate as the major anion in squid axoplasm[1] encouraged the view that this unusual sulfonate had an important role in the bioelectric behavior of the squid axion. While isethionate is found in

[1] B. A. Koechlin, *Proc. Natl. Acad. Sci. U.S.A.* **40**, 60 (1954).

mammalian tissue at only a small fraction of the 200 µmol/g concentration observed in squid axoplasm, additional interest was stimulated by the structural relationship and possible biological relationship between isethionate (2-hydroxyethyl sulfonate) and taurine (2-aminoethyl sulfonate). Indeed early studies using isotopically labeled taurine revealed the *in vivo* and *in vitro* conversion taurine to isethionate.[2-4]

More recent investigations[5,6] including studies with axenic mice and rigorous analytical procedures indicate that (1) the mammalian tissues studied cannot metabolize taurine to isethionate, (2) certain bacteria, most notably gut flora, can carry out this conversion, and (3) a number of acyl derivatives of taurine, among them bile acids and other taurine conjugates, are not separated from isethionate by the anion-exchange column chromatographic procedures used previously. These compounds have been mistakenly characterized as isethionate. The putative isethionate found in human and other mammalian tissue and attributed to taurine metabolism thus is either genuine isethionate formed in the gut by microbial metabolism of taurine derived from taurocholic acid or is comprised of acyl derivatives of taurine (since they are found to be hydrolyzable to taurine). It is clear that rigorous analytical procedures are required in investigation of isethionate metabolism.

Early analytical procedures involved separation of the isethionate from contaminating anions using ion-exchange resins followed by digestion with nitric acid. The sulfate formed was isolated and determined gravimetrically as the barium salt.[1] The conflicting results reported by various workers on the metabolic origin and fate of isethionate encouraged the development of gas–liquid chromatographic procedures.[5-9]

Gas–Liquid Chromatography

Several gas–liquid chromatographic methods have been described which use different derivatives of isethionate to permit analysis.[5-9] Three

[2] J. D. Welty, W. O. Read, and E. H. Shaw, Jr., *J. Biol. Chem.* **237**, 1160 (1962).
[3] W. O. Read and J. D. Welty, *J. Biol. Chem.* **237**, 1521 (1962).
[4] E. J. Peck, Jr. and J. Awapara, *Biochim. Biophys. Acta* **141**, 499 (1967).
[5] J. H. Fellman, E. S. Roth, and T. S. Fujita, in "Taurine and Neurological Disorders" (A. Barbeau and R. Huxtable, eds.), p. 19. Raven, New York, 1978.
[6] J. H. Fellman, E. S. Roth, N. A. Avedovech, and K. D. McCarthy, *Arch. Biochem. Biophys.* **204**, 560 (1980).
[7] O. Stokke and P. Helland, *J. Chromatogr.* **146**, 132 (1978).
[8] M. A. Remtulla, D. A. Applegarth, D. G. Clark, and I. H. Williams, *Life Sci.* **20**, 2029 (1977).
[9] S. W. Schaffer, A. Sevem, and J. Chovan, in "Taurine and Neurological Disorders" (A. Barbeau and R. Huxtable, eds.), p. 25. Raven, New York, 1978.

methods are discussed here which use methyl ether/methyl ester, silyl, or chloroethylsulfonylchloride derivatives.

Methyl Ether/Methyl Ester Derivatives[8]

Preparation of Samples. HEART AND BRAIN TISSUE. Samples, 2.5 g, of pooled brain or heart tissue are minced before homogenization in 10 ml of 50% methanol/water (v/v) using a Sorvall Omnimixer. The homogenate is centrifuged at 3,000 g for 5 min. The homogenizing chamber was rinsed 3 times with 5 ml of 50% methanol/water (v/v), and the rinse is added to the centrifuge pellet which is suspended in the rinsing solution with a vortex mixer. The resulting suspension is centrifuged for 5 min and the supernatant fluid removed.

To the combined supernatant liquids, an equal volume of Folch solvent (chloroform : methanol, 2 : 1, v/v) is added. The solutions are thoroughly mixed and centrifuged to separate the layers. The upper, aqueous layer is removed and evaporated to dryness in a rotary evaporator under reduced pressure. After evaporation, 2 ml of a purified cation-exchange resin (AG 50W-X8, 50–100 mesh, H^+ form, Bio-Rad Laboratories, Richmond, CA, prewashed in methanol and suspended in an equal volume of methanol) is added to the flask. The residue is triturated with the resin mixture, and the entire suspension is transferred to a 5 ml glass-stoppered conical centrifuge tube. After centrifugation, the methanol layer is removed and dried in a vacuum desiccator over sulfuric acid.

SQUID AXOPLASM. Axoplasm from the squid giant axon may be obtained in freeze-dried form. The sample, about 80 mg fresh weight, is dispersed in water with a glass homogenizer. The resultant turbid solution is brought to a volume of 10 ml with water. An aliquot of the solution is mixed with an equal volume of absolute methanol, and this mixture is treated with an equal volume of Folch solvent. The remainder of the procedure is the same as that described above for isolation of isethionate from heart and brain.

Methylation of Samples. Note: The derivatization should be carried out in an exhaust hood to avoid exposure to toxic and explosive diazomethane. Isethionate standards, or samples obtained from a tissue extract, are dissolved in 0.1 ml of a 1 mM solution of salicylic acid (Sigma Chemical Co.) in methanol. Salicylic acid serves as an internal standard. The vials are stoppered with Teflon caps lined with laminated discs and placed in an ice-bath for 5 min. An ethereal diazomethane solution, prepared from Diazald (see technical bulletin from the supplier, Aldrich Chemical Co., Milwaukee, WI), is introduced into the vials slowly, with mixing, until the yellow color persists. An additional three drops of diazomethane solution are added and the mixture allowed to stand for about 30 min in ice. The total volume in the vial is generally 200 μl or less; when the total volume is greater, excess solvent is carefully evaporated under a stream of dry nitrogen and the sample remethylated for an additional 30 min. Methylation reactions are all stopped by adding one drop of 50% acetic acid in water (v/v). The remethylation procedure does not affect the standard calibration curve for isethionate.

Gas–Liquid Chromotography: Flame Ionization Detection. A Bendix, Model 2500, gas–liquid chromatograph, equipped with a flame ionization detector, has been used with columns of 6 foot by 4 mm (i.d.) glass U-tubes. The stationary phases are 5% OV-1 and 5% OV-17 (Gas Chromatographic Specialities Ltd., Brockville, Ontario) on 80–100 mesh H. P. Chromosorb W. Analyses are performed isothermally at temperatures ranging from 100 to 150°. The optimal temperature for OV-1 is 115° and for OV-17, 135°. Nitrogen serves as the carrier gas at a flow rate of 40 ml/min. Using a column of 5% OV-1, a single peak with a retention time of 1.6 min is obtained for methylated isethionic acid. Using a column of 5% OV-17, two peaks at retention times of 3.5 and 4.0 min are observed. Mass spectral analysis revealed the first (minor) peak to be the methyl ether/methyl ester derivative of isethionate while the second (major) peak was the methyl ester of isethionate.

Silylation[9]

Preparation of Sample. Male Wistar rats (240–280 g) are decapitated and the hearts perfused within 45 sec with Krebs–Henseleit buffer supplemented with 5 mM glucose and 5×10^{-3} units of insulin/ml. After 20 min, the hearts are rapidly frozen with Wollenberger clamps cooled in liquid nitrogen. The tissue samples are lyophilized, and a known weight of ventricular tissue (0.1–0.15 g) is powdered and then extracted with 3 ml of 6% perchloric acid. Following centrifugation, the supernatant fluid is neutralized to pH 5.5–6.0 with 3 N K_2CO_3. The potassium perchlorate precipitate is removed by centrifugation.

A known volume of extract is applied to an AG 1-X8 column (0.5 × 2.5 cm) equilibrated with 0.1 M formic acid. The column is first washed with 10 ml of 0.1 M formic acid, followed by 1.0 ml of 1 M formic acid. The isethionic acid fraction is eluted from the column with 10 ml of 8 M formic acid.

Derivatization and Analysis. Following either lyophilization or evaporation of the 8 M formic acid eluent, the residue is dissolved in 0.1 ml of dry pyridine containing 500 mg of α-naphthol per liter, and 0.2 ml of Sylon H.T.[10] is added. The reaction is allowed to proceed for 45 min at 115 ° in

---

[10] Sylon H.T. is available from Supelco, Inc., Bellefonte, PA, and consists of a mixture (3 : 1) of hexamethyldisilazane and trimethylchlorosilane.

tightly sealed vials. The derivitized samples are analyzed with a Perkin-Elmer Model 3920 gas chromatograph equipped with flame ionization detectors. Either an OV-1 or an OV-17 (3%) on chromasorb W column may be used, with helium as the carrier gas. The column, detector, and injection port temperatures are maintained at 105, 305, and 350°, respectively. Helium gas flow is arranged at 60 ml/min, and the flow rates for hydrogen and oxygen are 35 and 800 ml/min, respectively. Isethionate is observed to emerge from the OV-1 column in approximately 7 min.

Chloroethylsulfonylchloride Derivative

Preparation of Isethionic Acid Standard Solutions. Sodium isethionate, 0.2 g (1.4 mmol), in 1 ml water is placed on the top of a 0.9 × 15 cm column of AG 50-X8 (200–400 mesh, H^+ form) ion-exchange resin and eluted with water. The isethionic acid emerges quantitatively in the void volume and is further diluted as required to give solutions of known concentrations.

Isolation of Isethionic Acid from Animal Tissue. Rat liver, heart, or other tissue is divided into two portions. Isethionic acid, 5 ml of a 0.14 μM solution, is added to one portion of 1 g; 5 ml of water is added to an equal portion. The two samples are homogenized separately, and 1.0-ml aliquots placed in extraction tubes. The protein is precipitated with 0.5 ml of 1 M perchloric acid. After centrifugation, 0.5 ml of 1 M KCl is added to precipitate excess perchlorate. The tubes are centrifuged, and 1.0 ml of each sample is charged onto a 0.9 × 15 cm column of AG 50 resin (200–400 mesh, H^+ form). The column is eluted with water, and the fraction emerging from 5 to 20 min (approximately 10.0 ml) is collected. The eluate is placed on a Bio Rex 5 (100–200 mesh, chloride form) 0.8 × 5 cm column. The resin is washed at a flow rate of 0.75 ml/min with 2 ml of water followed by 0.5 M HCl. Fractions emerging from 5 to 15 min (approximately 7.5 ml) are collected and diluted to 10 ml.

Isolation of Isethionic Acid from Squid Axon. Five squid axons (250 mg) in 0.25 ml of 0.5 M $HClO_4$ are homogenized. The homogenate is diluted to 1.5 ml with water and placed in an extraction tube. After centrifugation, 0.25 ml of 1 M KCl is added to precipitate excess perchlorate. The tube is again centrifuged, and isethionic acid is isolated from 1.0 ml of the supernatant as described above.

Derivatization of Isethionic Acid. Zinc chloride in hydrochloric acid was used to convert the alcohol group of isethionic acid to the alkyl chloride. Thus, 10 μl of a solution of zinc dissolved in concentrated hydrochloric acid (8 mg Zn/ml) is added to a tube containing a dry sample of isethionic acid and fitted with a Teflon-sealed screw cap. The tube is warmed for 1 hr at 80°. The HCl is blown off with a stream of nitrogen at 80°, and 10 μl of a solution of PCl$_5$ (0.006 g/ml) in freshly distilled POCl$_3$, is added. The tube is warmed for 1 hr at 80°, and the excess reagent removed with a stream of nitrogen. The residual material is dissolved in 50 μl of toluene and 2-μl aliquots are injected for assay by gas–liquid chromatography.

Gas Chromatography. Gas chromatography has been carried out isothermically with a Hewlett Packard 5710 using an electron capture detector. A 2 mm × 10 foot 3% OV-25 at 150° or a 2 mm × 10 foot 3% OV-225 glass column at 130° and 30 ml/min gas flow is used. A 5% methane and 95% argon gas mixture is used and an injector set at 200°. Chromatography on the 3% OV-25 column at 150° produced a peak emerging at 3.58 min for the derivatized isethionic acid. The 3% OV-225 column at 130° exhibited a peak emerging at 3.77 min. Identity is confirmed by comparison to an authentic sample of 2-chloroethane sulfonyl chloride by gas chromatography–mass spectrometry.

Employment of the chlorinated derivative extended the sensitivity of the gas chromatographic analysis manyfold due to the intrinsic sensitivity of the electron capture detector. With this method, sensitivity was greater than 5 ng/g tissue.

Paper Electrophoresis

High-voltage paper electrophoresis has been carried out in a formic/acetic acid (pH 1.9) buffer (240 ml formic acid, 960 ml acetic acid, made to 12 liters with distilled water) with a Savant Instruments high voltage electrophoresis apparatus. Electrophoresis proceeds at 4500 V for 1 hr. The dried electrophoretograms are scanned with a Packard 7220/21 radiochromatograph scanner. [^{35}S]Isethionate is used in parallel runs as a standard. Isethionate can be separated from *N*-acyl derivatives of taurine and from a large number of anionic substances by this method.

[31] 3-Mercaptopyruvate, 3-Mercaptolactate and Mercaptoacetate

By BO SÖRBO

3-Mercaptopyruvate, 3-mercaptolactate, and mercaptoacetate are compounds formed in the transamination pathway of cysteine.[1] In this pathway cysteine is first converted to 3-mercaptopyruvate via transamination. 3-Mercaptopyruvate may then be reduced by lactate dehydrogenase to 3-mercaptolactate or converted to mercaptoacetate by oxidative decarboxylation.[2] All of these compounds are found in the urine of higher animals as mixed disulfides with cysteine; the disulfides are presumably formed by a thiol–disulfide exchange with cystine in the extracellular space. It may be noted that the mixed disulfides of 3-mercaptopyruvate and mercaptoacetate with cysteine may also be formed through the action of L-amino-acid oxidase on cystine,[3] but these reactions are of questionable physiological importance.[1]

The determination of the transaminative metabolites of cysteine is not only of interest for studies of their metabolism but is also of clinical interest as human beings may be affected by an inborn error of metabolism, 3-mercaptolactate cysteine disulfiduria.[4,5] This condition is characterized by an increased excretion of the mixed disulfides between cysteine and 3-mercaptopyruvate, 3-mercaptolactate, and mercaptoacetate and is due to a deficiency of 3-mercaptopyruvate sulfurtransferase. This enzyme transfers the sulfur atom of 3-mercaptopyruvate to various sulfur acceptors.[6]

The determination of high urinary concentrations of the mixed disulfides between cysteine and its transaminative metabolites found in subjects with 3-mercaptolactate cysteine disulfiduria may be performed with a conventional amino acid analyzer.[7] A method for the assay of 3-mercaptopyruvate cysteine disulfide formed by the action of L-amino-acid oxi-

---

[1] A. J. L. Cooper, *Annu. Rev. Biochem.* **52**, 187 (1983).
[2] J. Mårtensson, L. Nilsson, and B. Sörbo, *FEBS Lett.* **176**, 334 (1984).
[3] T. Ubuka and K. Yao, *Biochem. Biophys. Res. Commun.* **55**, 1305 (1973).
[4] J. C. Crawhall, R. Parker, W. Sneddon, E. P. Young, M. G. Ampola, M. Efrom, and E. M. Bixby, *Science* **160**, 419 (1968).
[5] U. Hannestad, J. Mårtensson, R. Sjödahl, and B. Sörbo, *Biochem. Med.* **26**, 106 (1981).
[6] B. Sörbo, *in* "Metabolic Pathways" (D. M. Greenberg, ed.), Vol. 7, p. 433. Academic Press, New York, 1975.
[7] J. C. Crawhall, K. Bir, P. Purkiss, and J. B. Stanbury, *Biochem. Med.* **5**, 109 (1971).

dase has been reported,[8] where the disulfide is separated by ion-exchange chromatography, reduced with dithiothreitol, and the 3-mercaptopyruvate formed is determined with lactate dehydrogenase. 3-Mercaptolactate has also been determined after aminoethylation with bromoethylamine using an amino acid analyzer,[9] and mercaptoacetate has been determined by high-performance liquid chromatography after derivatization with N-(1-pyrene)maleimide.[10]

The simultaneous analysis of the transaminative metabolites of cysteine, however, is best performed by gas chromatography.[5,11] In this method the metabolites are first liberated from their mixed disulfides by reduction with an insoluble, thiol-containing polymer, and free thiols are then subjected to a cleanup procedure. They are thus adsorbed to an organomercurial resin and then eluted with cysteine. The latter (as well as other thiol compounds containing a positively charged group) is removed on the hydrogen form of a strong cation-exchange resin. The transaminative cysteine metabolites, which in addition to a thiol group also contain a carboxyl group, are then derivatized before gas chromatography by extractive alkylation with benzyl bromide. As originally described[11] the method did not include 3-mercaptopyruvate as this compound is fairly labile due to the presence of a keto group. In an improved version[5] of the method 3-mercaptopyruvate was stabilized by conversion to the ethoxime with ethylhydroxylamine. The procedure described here is a modification of this method in which 3-mercaptopyruvate is converted to the methoxime[2] with methylhydroxylamine, a reagent more easily accessible than ethylhydroxylamine.[12]

Reference Compounds

3-Mercaptopyruvate is easily prepared as its ammonium salt from 3-bromopyruvate.[13] An excellent description of the chemical properties of 3-mercaptopyruvate and especially its tendency to form an acyclic aldol dimer (which changes its properties as an enzyme substrate) is found in Ref. 14 which should be consulted by anyone contemplating experimental work with 3-mercaptopyruvate. The racemic form of 3-mercaptolactate

---

[8] T. Ubuka, Y. Ishimoto, and K. Kasahara, *Anal. Biochem.* **67,** 66 (1975).
[9] B. Pensa, M. Costa, and D. Cavallini, *Anal. Biochem.* **145,** 120 (1985).
[10] B. Kågedal and M. Källberg, *J. Chromatogr.* **229,** 409 (1982).
[11] U. Hannestad and B. Sörbo, *Clin. Chim. Acta* **95,** 189 (1979).
[12] However, the separation of the ethoxime derivative of 3-mercaptopyruvate from intefering compounds in urine is better than that of the methoxime.
[13] E. Kun, *Biochim. Biophys. Acta* **25,** 135 (1957).
[14] A. J. L. Cooper, M. T. Haber, and A. Meister, *J. Biol. Chem.* **257,** 816 (1982).

may be obtained from 3-mercaptopyruvate by borohydride reduction[15] or synthesized from 3-chlorolactic acid by a somewhat more laborious procedure.[16] The latter reference also describes the preparation of the two optical isomers of 3-mercaptolactic acid. Mercaptoacetic acid (thioglycolic acid) is commercially available from many firms and may be used without further purification.

Reagents

Disodium EDTA, 0.18 M
Methylhydroxylamine hydrochloride, 1.25 M in 2.4 M sodium acetate
Ammonia, 5 M
Thiopropyl-Sepharose 6B (Pharmacia Fine Chemicals, Uppsala, Sweden) is obtained as its mixed disulfide with 2-thiopyridine. It is converted to the free thiol form by treatment with dithiothreitol according to the manufacturer's instruction and should then contain about 35 μmol thiol groups per milliliter of swollen gel.
Acetic acid, 4 M
p-Acetoxymercurianiline-Sepharose 4B (PAMAS) is prepared according to Sluyterman and Wijdenes[17] and should contain about 10 μmol organomercurial groups per milliliter of swollen gel.
Cysteine hydrochloride, 0.01 M
HCl, 0.01 M.
AG 50W-X8 (100–200 mesh), a purified grade of Dowex 50, is obtained from Bio-Rad Laboratories, Richmond, CA, and is equilibrated with 0.01 M HCl before use.
Tetrabutylammonium hydrogen sulfate (TBAHS), 1 M, is prepared from a commercial product which has been purified by recrystallization from methyl isobutyl ketone.
NaOH, 2 M
Dichloromethane containing 0.04 M benzyl bromide (purified by distillation *in vacuo*) and 20 μM nonadecanoic acid as an internal standard
n-Hexane
H_2SO_4, 0.01 M
Ethanol, 95% (v/v)

Procedure. To a 5-ml aliquot of sample, e.g., urine, is added 0.2 ml of 0.18 M disodium EDTA followed by 0.2 ml of 1.25 M methylhydroxyl-

---

[15] M. Costa, B. Pensa, C. Iavarone, and D. Cavallini, *Prep. Biochem.* **12**, 417 (1983).
[16] D. B. Hope and M. Wälti, *J. Chem. Soc. C* p. 2475 (1970).
[17] L. A. A. E. Sluyterman and J. Wijdenes, *Biochim. Biophys. Acta* **200**, 593 (1970).

amine hydrochloride in 2.4 M sodium acetate. This mixture is then left to stand at room temperature for 2 hr. (If the sample does not contain significant amounts of mercaptopyruvate this step may be omitted.) The pH is then adjusted to 9.8–10.0 with 5 M ammonia, and the volume made up to 6.0 ml with water. After addition of 1 ml of thiopropyl-Sepharose the reaction mixture is shaken for 30 min on a rotary mechanical shaker. The pH is then adjusted to 3.5–4.0 with 4 M acetic acid, and the sample is centrifuged.

An aliquot of the supernatant, containing up to 2 μmol thiol as determined by Ellman's reagent,[18] is then applied to a 0.7 × 1.3 cm column of PAMAS, and the resin is washed with 2 ml water. The adsorbed thiol compounds are then eluted with 3 ml of 0.01 M cysteine hydrochloride, and the eluate transferred to a 0.5 × 2.5 cm column of AG 50W. The latter is washed with 1 ml of 0.01 M HCl, and the total eluate is transferred to a test tube which may be closed with a PTFE-lined screw cap.

To the total eluate are added 0.2 ml of 0.18 M disodium EDTA, 0.25 ml of 1 M TBAHS, 0.25 ml of 2 M NaOH, and 2 ml of dichloromethane containing 0.04 M benzyl bromide and 20 μM nonadecanoic acid as an internal standard. The tube is then shaken at 37° for 2 hr. After centrifugation the aqueous phase is removed by aspiration, and the dichloromethane is evaporated by blowing nitrogen over the sample at 45°. The residue is taken up in 2 ml of n-hexane and treated with 3 ml of 0.01 M H_2SO_4 for 5 min on a rotary shaker.[19] After transfer of the organic phase to a small tapered centrifuge tube, the solvent is removed by a stream of nitrogen at 45°.

The residue is dissolved in 25 μl ethanol, and a 2-μl aliquot of this solution is analyzed by gas chromatography. This step is performed with a gas chromatograph equipped with a flame ionization detector and a silanized glass column 180 × 0.3 cm (i.d.) containing Chromosorb W (HP), 100–200 mesh, coated with 3% OV-17. Nitrogen is used as a carrier gas at a flow rate of 35 ml/min. The sample is injected on-column and other conditions are as follows: Column temperature 250°, injector temperature 270°, and detector temperature 290°. Mercaptoacetate appears first on the chromatogram, followed by 3-mercaptolactate, 3-mercaptopyruvate, and the internal standard (Fig. 1). It may be noted that the flame ionization detector response is significantly better for mercaptoacetate than for 3-mercaptopyruvate or 3-mercaptolactate (Fig. 1). The relative peak heights versus the internal standard are then calculated for the compounds ana-

---

[18] A. F. S. A. Habeeb, this series, Vol. 25, p. 457.
[19] This step removes compounds which otherwise interfere in the final gas chromatographic step.

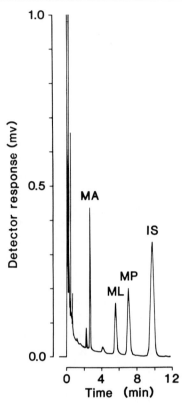

FIG. 1. Gas chromatographic separation of an equimolar standard mixture of 3-mercaptopyruvate (MP), 3-mercaptolactate (ML), and mercaptoacetate (MA). IS = Internal standard. Each peak represents about 5 nmol sulfur compound.

lyzed, and their concentrations in the original sample are obtained from standard curves.

Acknowledgment

This work was supported by Grant No. 03X-05644 from the Swedish Medical Research Council.

[32] Hypotaurine Aminotransferase Assay: [2-³H]Hypotaurine

By JACK H. FELLMAN

$$R-CH_2-\overset{O}{\underset{\|}{C}}-\overset{O}{\underset{\|}{C}}-O^-$$

$$+$$

$$^-O-\overset{O}{\underset{\|}{S}}-CH_2-\overset{+}{\underset{^3H_2}{C}}-NH_3$$

$$\xrightleftharpoons{}$$

$$R-CH_2-\overset{\overset{+}{N}H_3}{\underset{H}{C}}-C\overset{\overset{O}{\|}}{\underset{O^-}{}}$$

$$+$$

$$\left[O-\overset{O}{\underset{\|}{S}}-CH_2-C\overset{O}{\underset{^3H}{\|}} \right] + {}^3H_2O$$

$$\downarrow$$

$$\underset{\|}{S=O} + CH_3-\overset{O}{\underset{\|}{C}}\overset{^3}{H}$$

Hypotaurine aminotransferase catalyzes the α-ketoglutarate-dependent transamination of hypotaurine to form sulfinoacetaldehyde, an unstable intermediate which rapidly degenerates to acetaldehyde and SO_2. The overall process is irreversible. Although the rate of the enzyme reaction can be monitored by determining the rate of formation of glutamate, SO_2, or acetaldehyde, the rate is most easily and specifically quantitated by following ^3H_2O release when [2-³H]hypotaurine is the substrate. Use of [³H]hypotaurine is particularly convenient when assaying enzyme activity in crude tissue fractions including the heavy mitochondrial fraction with which the mammalian enzyme is associated. The assay of hypotaurine aminotransferase is presented here as is the synthesis of the substrate of the enzyme, [2-³H]hypotaurine.

Synthesis of [2-³H]Hypotaurine[1]

[2-³H]Taurine (1.0 mCi, 14 Ci/mmol) is added to 1.25 mg of taurine in 1 ml of 1 M HCl in a 10 × 100 mm Pyrex test tube fitted with a Teflon-sealed screw-top cap. The solution is evaporated to dryness and 0.3 ml of a solution containing 250 mg PCl_5 and 250 mg $AlCl_3$ dissolved in 5 ml of freshly distilled $POCl_3$ is added. The tube is capped and heated at 70° for 90 min in a controlled-temperature heating block. Excess $POCl_3$ and PCl_5 are evaporated under a gentle stream of dry nitrogen. The residue is dissolved in 2 ml of dry methanol, and 200–500 mg of powdered zinc is

---

[1] J. H. Fellman, *J. Labelled Compd. Radiopharm.* **18**, 765 (1980).

added. The mixture is shaken for a few minutes and filtered. The methanol is evaporated, and the residue is dissolved in 1 ml of water and placed on a column (0.9 × 7 cm) of AG 50-X8 (200–400 mesh, H^+ form) ion-exchange resin. The column is washed with 14 ml of water followed by 14 ml of 2 N NH_4OH. The eluate from the latter contains all of the hypotaurine. Both ^{14}C- and ^{35}S-labeled hypotaurine can be synthesized from the appropriately labeled taurine precursors by the above procedure. The yield is consistently greater than 97%.

The radiolabeled hypotaurine cochromatographs with authentic hypotaurine in several solvent systems on thin-layer chromatographic plates [cellulose; n-butanol/acetic acid/water (4 : 1 : 1) and silica gel; 2-propanol/water (5 : 1)] and on cation-exchange resin columns. The product is oxidized to taurine under suitable oxidizing conditions, e.g., treatment with dilute hydrogen peroxide.

Assay of Hypotaurine Aminotransferase[2,3]

Principle. Transamination of [2-^{3}H]hypotaurine forms 1 equivalent each of ^3H_2O and [1-^{3}H]sulfinoacetaldehyde; the latter decomposes spontaneously to [1-^{3}H]acetaldehyde and sulfite. After the reaction is terminated by addition of acid, ^3H_2O is selectively distilled from the reaction mixture and quantitated by liquid scintillation counting.

Reagents

Buffer: 250 mM imidazole–HCl at pH 8.2
Pyridoxal 5'-phosphate, 10 mM
α-Ketoglutarate, 100 mM, adjusted to pH 8.2 with NaOH
2-Mercaptoethanol, 100 mM
[2-^{3}H]Hypotaurine, 100 mM (approximately 10 μCi/ml)

Procedure. A working solution is prepared containing 5 ml of buffer, 20 μl of pyridoxal phosphate, 200 μl of 2-mercaptoethanol, and water to 10 ml. Assays are carried out in Thunberg tubes in a final volume of 1.0 ml; the reaction mixtures contain 0.50 ml of working solution, 5 μl of α-ketoglutarate, 50 μl of [^{3}H]hypotaurine, and a source of hypotaurine aminotransferase. After incubation for 30 min at 37° in a shaking water bath, the reaction is stopped by the addition of 0.2 ml of 20% sulfosalicylic acid. The Thunberg tubes are evacuated briefly with a water pump, and the ^3H_2O formed is distilled at 60° into the Thunberg tube cap which

---

[2] J. H. Fellman, E. S. Roth, N. A. Avedovech, and K. D. McCarthy, *Life Sci.* **27**, 1999 (1980).
[3] J. H. Fellman and E. S. Roth, *Adv. Exp. Med. Biol.* **139**, 99 (1980).

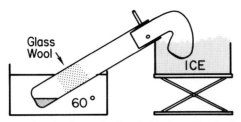

FIG. 1. Arrangement of a Thunberg tube for distillation of ^3H_2O after completion of the enzymatic reaction.

reclines on chipped ice (Fig. 1). It is important that the distillation is allowed to proceed to completion since there is a small difference between the boiling points of ^3H_2O and H_2O. Portions of the distillate (0.5 ml) are submitted to liquid scintillation counting, and the amount of product is calculated from the known specific activity of the hypotaurine.

Comments. Hypotaurine aminotransferase activity appears to be restricted to the "heavy" mitochondrial fraction in liver and brain preparations; the activity from hog brain mitochondria copurifies with γ-aminobutyrate aminotransferase and has been purified by the procedure of Bloch-Tardy.[4] Hypotaurine aminotransferase activity is optimal at a pH near 8.2 and is stimulated 20% by the inclusion of 10 μM pyridoxal phosphate in the assay solution. The activity is inhibited by reagents that bind to pyridoxal phosphate (e.g., aminooxyacetic acid and hydroxylamine) and by established inhibitors of γ-aminobutyrate aminotransferase (e.g., 4-amino-5-fluoropentanoic acid[5] and vinyl-GABA.[3] Both selenotaurine and selenohypotaurine[6] are inhibitors of hypotaurine aminotransferase.[3] Taurine is neither a substrate nor an inhibitor.[3]

---

[4] M. Bloch-Tardy, B. Rolland, and P. Gonnard, *Biochimie* **56**, 823 (1974).
[5] R. B. Silverman and M. H. Levy, *Biochem. Biophys. Res. Commun.* **95**, 250 (1980).
[6] A. Rinaldi, P. Floris, P. Cossu, and C. De Marco, *Bull. Mol. Biol. Med.* **3**, 234 (1978).

[33] Penicillamine

By ELISABETH M. WOLF-HEUSS

Introduction

D-Penicillamine (2-amino-3-mercapto-3-methylbutanoic acid) is a trifunctional thiol amino acid.[1] It is widely used for the treatment of a variety of diseases including rheumatoid arthritis,[2] Wilson's disease,[3] cystinuria,[4] heavy metal poisoning,[5] chronic active hepatitis,[6] and primary biliary cirrhosis.[7] The determination of D-penicillamine in biological fluids of patients being treated with this drug is complicated by the fact that D-penicillamine occurs in many different forms; free thiol, internal disulfide, the mixed disulfide with cysteine, the metabolite S-methyl-D-penicillamine, and D-penicillamine bound to plasma proteins may all be present in blood and urine samples.[8]

Assay methods employing high-performance liquid chromatography (HPLC), gas–liquid chromatography, amino acid analysis, colorimetry, and radioimmunoassay have been described.[9,10]

Liquid Chromatography

Determination by Electrochemical Detection

Principle. D-Penicillamine and other sulfhydryl-containing compounds are separated on a cation-exchange column or reversed-phase HPLC column and quantified by electrochemical detection. Saetre and Rabenstein used a mercury-based electrochemical detector,[11,12] which was

[1] W. M. Weigert, H. Offermanns, and P. Scherberich, *Angew. Chem.* **87**, 374 (1975).
[2] I. J. Jaffe, *Arthritis Rheum.* **8**, 1064 (1965).
[3] J. M. Walshe, *Am. J. Med.* **21**, 487 (1956).
[4] J. C. Crawhall, E. F. Scowen, and R. W. E. Watts, *Br. Med. J.* **1**, 588 (1963).
[5] L. T. Zimmer and D. E. Carter, *Life Sci.* **23**, 1025 (1978).
[6] R. B. Stern, S. P. Wilkinson, P. N. J. Howorth, and R. Williams, *Gut* **18**, 19 (1977).
[7] E. R. Dickson, C. R. Fleming, M. C. Geall, J. T. McCall, and A. H. Baggenstoss, *Gastroenterology* **72**, 1049 (1977).
[8] A. O. Muijsers, R. J. van de Stadt, A. M. A. Henrichs, and J. K. van der Korst, *Clin. Chim. Acta* **94**, 173 (1979).
[9] D. R. Lecavalier and J. C. Crawhall, *J. Rheumatol.* **8**, Suppl. 7, 20 (1981).
[10] J. C. Crawhall, *Clin. Invest. Med.* **7**, 31 (1984).
[11] R. Saetre and D. L. Rabenstein, *Anal. Chem.* **50**, 276 (1978).
[12] A. S. Russell, R. Saetre, P. Davis, and D. L. Rabenstein, *J. Rheumatol.* **6**, 15 (1979).

slightly modified by Bergstrom et al.,[13] and Carruthers et al.[14] used a commercially available gold–amalgam electrode.

In the technique of Saetre and Rabenstein, blood samples are centrifuged to separate the plasma, and plasma proteins are precipitated with metaphosphoric acid prior to analysis. In order to determine both reduced and oxidized D-penicillamine in one analytical run, an aliquot of the protein-free plasma filtrate is electrolyzed for 10 min in an electrolysis cell to reduce D-penicillamine disulfides to free penicillamine. Protein-bound D-penicillamine and S-methyl-D-penicillamine cannot be measured by this method. It is essential to deproteinate blood samples immediately upon collection so as to avoid the loss of free penicillamine by binding to human serum albumin.[15]

Dithiothreitol has been used to reduce possible disulfides of D-penicillamine and protein-bound penicillamine for the determination of total penicillamine in plasma.[16] Two independent assay methods were used for quantitating free penicillamine or penicillamine disulfide by direct injection of unreduced plasma filtrates using reversed-phase ion-pairing chromatography with electrochemical detection.

A rapid and sensitive HPLC method, described by Kreuzig and Frank,[17] will be outlined here. It is essential to stabilize the samples immediately after collection by addition of EDTA because of the instability of D-penicillamine in solution against oxidation particularly at neutral pH and in the presence of heavy metals. A low sample pH is necessary.

Procedure. PREPARATION OF PLASMA SAMPLES. To 1 ml of freshly drawn blood add 100 µl of a 10% EDTA solution. After shaking, the mixture is centrifuged. Mix 1 volume of the supernatant fluid with 2 volumes of a solution prepared by dissolving 4.39 g of diammonium hydrogen citrate in about 900 ml of water, adding 100 g of metaphosphoric acid, and adjusting the volume to 1 liter with water. The precipitated proteins are removed by centrifugation. If the analysis is not performed immediately, the samples must be frozen with liquid nitrogen.

PREPARATION OF URINE SAMPLES. To 1 ml of fresh urine, add 100 µl of a 10% EDTA solution. Mix 1 volume of this sample with 2 volumes of the aqueous solution of metaphosphoric acid in diammonium hydrogen citrate described above. If turbidity results, the solution is centrifuged.

CHROMATOGRAPHIC CONDITIONS. The HPLC separation is performed

---

[13] R. F. Bergstrom, D. R. Kay, and J. G. Wagner, *J. Chromatogr.* **222,** 445 (1981).
[14] G. Carruthers, M. Harth, D. Freemann, D. Weir, R. Rothwell, and M. Butler, *Clin. Invest. Med.* **7,** 35 (1984).
[15] R. F. Bergstrom, D. R. Kay, and J. G. Wagner, *Life Sci.* **27,** 189 (1980).
[16] M. A. Abounassif and T. M. Jefferies, *J. Pharm. Biomed. Anal.* **1,** 65 (1983).
[17] F. Kreuzig and J. Frank, *J. Chromatogr.* **218,** 615 (1981).

using a cation-exchange column (Nucleosil 5 SA 200/6/4, No. 715300 Machery-Nagel & Co, Düren, FRG). The column is operated at a flow rate of 1.5 ml/min corresponding to a back pressure of 150 bars. The electrochemical detection unit consists of a flow-through cell (EA 1096/2, Metrohm, Switzerland) equipped with a three-electrode system (working electrode, gold, EA 286/3; reference electrode, silver–silver chloride with 3 M potassium chloride solution, EA 442; auxiliary electrode, glassy carbon, EA 286/1, Metrohm, Switzerland) and an amperometric detector (EA 611, Metrohm). The polarization voltage is adjusted to +800 mV. Plasma samples are detected at a signal sensitivity of 12.5×10^{-8} A, urine samples at a signal sensitivity of 12.5×10^{-7} A. In the case of plasma samples 20 mM diammonium hydrogen citrate is taken as eluent, whereas for urine samples 10 mM diammonium hydrogen citrate is used. The pH of the eluent is adjusted to 2.2 by means of phosphoric acid, followed by filtration through a 0.5-μm Millipore filter and degassing prior to use. Standard solutions are prepared containing 1 and 5 μg of D-penicillamine, and 100 μg of EDTA per milliliter of eluent, for plasma and urine samples, respectively. The injection volume is 20 μl. After injection of five samples, a standard solution is injected.

The retention time of D-penicillamine in plasma samples is 6 min with a detection limit of 50 ng/ml. When assaying urine samples under these chromatographic conditions, the retention time of D-penicillamine is 7.4 min with a detection limit of 0.2 μg/ml. The coefficient of variation is 2.9% ($n = 10$).

Specificity. The assay method is specific for reduced D-penicillamine.

Determination by Derivatization

Principle. An HPLC method for the quantitative determination of D-penicillamine in urine has been described using 5,5'-dithiobis(2-nitrobenzoic acid) (Ellman's reagent) for postcolumn derivatization of free sulfhydryl groups.[18]

Analytical methods for assaying D-penicillamine in plasma by selective precolumn derivatization of the sulfhydryl group are described. Separations are performed by reversed-phase liquid chromatography and fluorescence detection. Fluorescence derivatization prevents oxidation of D-penicillamine during analysis and enhances the sensitivity of the assay. 5-Dimethylaminonaphthalene-1-sulfonylaziridine[19] and N-[p-(2-benzoxazolyl)phenyl]maleimide (BOPM) in 4- to 5-fold excess[20] have been used as derivatization reagents.

[18] D. Beales, R. Finch, and A. E. M. McLean, *J. Chromatogr.* **226,** 498 (1981).
[19] E. P. Lankmayr, K. W. Budna, and K. Müller, *J. Chromatogr.* **222,** 249 (1981).
[20] J. O. Miners, I. Fearnley, K. J. Smith, and D. J. Birkett, *J. Chromatogr. (Biomed. Appl.)* **275,** 89 (1983).

Reagents and Standards. D-Penicillamine stock solution is prepared by dissolving D-penicillamine standard substance in 0.1% EDTA to give a final concentration of 5 mM. This solution is further diluted with 0.1% EDTA and with plasma to prepare calibration standards in the concentration range of 1 to 100 μM.

Procedure. STEP 1. PROTEIN PRECIPITATION. Immediately after collection, blood samples are transferred to 1.5-ml Eppendorf microtubes and centrifuged at 6500 g for 30 sec to obtain the plasma. An aliquot of 1 ml plasma is treated in a second Eppendorf microtube with 0.15 ml of 25% trichloroacetic acid. The samples are vortexed and cooled in an ice bath for 10 min in order to complete the plasma protein precipitation process. The proteins are separated by centrifugation at 6500 g for 2 min.

STEP 2. DERIVATIZATION WITH BOPM. Plasma supernatant, 0.5 ml, is transferred into a 5-ml glass tube and neutralized with 0.2 ml of 1% aqueous sodium hydroxide solution. The pH is adjusted to 5.0 by addition of 0.25 ml of a 0.5 M sodium citrate solution (titrated to pH 5.0 with perchloric acid). The derivatization reagent, 1 ml (1 mM BOPM in ethanol), is added. The mixture is incubated at 37° overnight. The penicillamine–BOPM complex is stable for at least 24 hr.

CHROMATOGRAPHIC CONDITIONS. The separation is done on a 10 μm reversed-phase μBondapak C$_{18}$ column (30 cm × 3.9 mm i.d., Waters Assoc.). The mobile phase consists of methanol : 0.1 mM sodium acetate (48 : 52). A flow rate of 2.0 ml/min is maintained. Detection is with a Model 970 fluorescence detector (Spectra Physics). The excitation wavelength is set at 319 nm, and emission is measured using a 360 nm cut-off filter. The injection volume is 50 μl. Unknown concentrations of D-penicillamine in patient plasma are determined by comparison of the penicillamine peak heights with those of the calibration curve.

The retention time of the penicillamine–BOPM derivative is 5.5 min. Fluorescence detection enables the quantitation of plasma penicillamine concentrations in the range of 0.25 to 500 μM.

Specificity. The method is specific for reduced D-penicillamine. The sulfhydryl-specific reagent BOPM does not react with the major metabolites of penicillamine, e.g., penicillamine disulsulfide or penicillamine–cysteine disulfide. Disulfides may be determined as described above after Determination by Electrochemical Reduction.[11,17] Under the chromatographic conditions employed, BOPM complexes formed with cysteine and glutathione do not interfere.

Determination by Amino Acid Autoanalysis

Principle. Quantitation of D-penicillamine in human plasma and urine by an amino acid autoanalyzer method has been described.[8] D-Penicillamine and its disulfides are oxidized with performic acid to penicillaminic acid, which is separated from other ninhydrin-positive compounds as cysteic acid by anion-exchange chromatography. The possible metabolite, S-methyl-D-penicillamine,[21] is oxidized by performic acid to a sulfone and separated by cation-exchange chromatography.

Procedure. Freshly prepared performic acid, 2 ml (98% formic acid, 30% H_2O_2, 9:1 v/v), are added to 1 ml of serum or urine. The mixtures are incubated 1 hr at room temperature and 4 hr at 0°. HBr, 0.3 ml of a 48% solution, is added and centrifuged at 3000 g for 30 min. The supernatant liquid is brought to dryness and dissolved in 67 mM sodium citrate (pH 2.2) prior to analysis. Standards for calibration purposes are prepared by treating D-penicillamine and S-methyl-D-penicillamine standard substance in drug-free serum or urine.

CHROMATOGRAPHIC CONDITIONS: SEPARATION OF D-PENICILLAMINIC ACID. Separation is done with a Beckmann Multichrom M amino acid analyzer equipped with a 40 × 0.4 cm column of Aminex A-25 (Bio-Rad Laboratories). The column is regenerated with 0.2 M NaOH, briefly equilibrated with 0.8 M sodium citrate (pH 1.8), and equilibrated with 0.01 M acetic acid. Sample sizes of 50 to 150 µl are injected. Elution is performed at a column temperature of 65° with 10 mM acetic acid at a flow rate of 18 ml/hr. D-Penicillaminic acid is detected by postcolumn reaction with ninhydrin (10 ml/hr flow rate for ninhydrin).

SEPARATION OF S-METHYL-D-PENICILLAMINE SULFONE. A 40 × 0.4 cm cation-exchange column packed with Beckman resin M-902 is used. After regeneration with 0.2 M NaOH, the column is equilibrated with 67 mM sodium citrate (pH 2.2). Elution is performed with the same citrate buffer at a column temperature of 65° and at a flow rate of 25 ml/hr.

Specificity. The method enables the determination of all forms of D-penicillamine. Protein-bound D-penicillamine is released by performic acid oxidation. The detection limit is 2 µM D-penicillamine in serum and urine samples.

Assay by Colorimetry

Principle. A simple colorimetric procedure is based on derivatization of reduced D-penicillamine with 5,5'-dithiobis(2-nitrobenzoic acid) (Ellman's reagent).[22]

Procedure. Ten parts of blood sample and 1 part of Tris–EDTA (25 mM Tris, 100 mM EDTA, 120 mM NaCl, and 5 mM KCl, adjusted to pH

---

[21] D. Perrett, W. Sneddon, and A. D. Stephens, *Biochem. Pharmacol.* **25,** 259 (1976).
[22] J. Mann and P. D. Mitchell, *J. Pharm. Pharmacol.* **31,** 420 (1979).

7.4) are mixed in an ice bath and centrifuged at 800 g for 10 min. An aliquot of 1.1 ml of plasma is mixed with 0.1 ml 3 M HCl, frozen in a CO_2–methanol bath and lyophilized overnight. Ethanol, 1.5 ml, is added to the freeze-dried plasma sample. The suspension is sonicated for 5 to 10 sec and vortexed for 30 sec. After centrifugation at 800 g for 10 min, 1 ml of the supernatant fluid is added to 1 ml of 0.45 M Tris–HCl of pH 8.2 and mixed with 20 μl of Ellman's reagent (10 mM in 0.1 M potassium phosphate at pH 7.0). Diethyl ether, 4 ml, is added, the tube is shaken for 10 min and centrifuged at 800 g for 10 min. The ethanol–ether layer is discarded, and the absorbance of the aqueous layer is measured at 412 nm. The calibration curve is obtained by adding known amounts of D-penicillamine standard substance to drug-free plasma and treating the samples as described.

Specificity. The detection limit of D-penicillamine is 2 μg/ml. The calibration curve is linear up to 50 μg/ml. Ellman's reagent has a limited range of application because it fails to distinguish between the drug and other thiols present in plasma samples. Penicillamine disulfides are not determined by this method.

[34] S-Adenosylmethionine and Its Sulfur Metabolites

By RICHARD K. GORDON, GEORGE A. MIURA, TERESA ALONSO, and PETER K. CHIANG

Introduction

Biological reactions in which S-adenosylmethionine (AdoMet)[1] is the methyl donor are key processes in many diverse cellular functions, such as transmethylation, transsulfuration, and polyamine biosynthesis.[2] Some of the key metabolites of AdoMet are S-adenosylhomocysteine (AdoHcy), decarboxylated S-adenosylmethionine (dcAdoMet), and methylthioadenosine (MeSAdo). The use of specific inhibitors of S-

[1] The following abbreviations are used: AdoMet, S-adenosylmethionine; AdoHcy, S-adenosylhomocysteine; AdoCys, S-adenosylcysteine; Ade, adenine; dzAdo, 3-deazaadenosine; dzAdoHcy, S-3-deazaadenosylhomocysteine; dcAdoMet, decarboxylated S-adenosylmethionine [S-adenosyl-(5′)-3-methylthiopropylamine]; dcAdoHcy, decarboxylated S-adenosylhomocysteine [S-adenosyl-(5′)-3-thiopropylamine]; MeSAdo, 5′-deoxy-5′-methylthioadenosine; InoHcy, S-inosylhomocysteine.

[2] E. Usdin, R. T. Borchardt, and C. R. Creveling, "Biochemistry of S-Adenosylmethionine and Related Compounds." Macmillan, London, 1982.

adenosylhomocysteine hydrolase (EC 3.3.1.1, adenosylhomocysteinase) results in the accumulation of AdoHcy and subsequent inhibition of transmethylation reactions, and a variety of biological and biochemical perturbations have been attributed to an increase in the ratio of AdoHcy (or S-nucleosidylhomocysteine) to AdoMet.[3] Of particular interest are antiviral effects,[4] modulation of cellular differentiation,[5] and immunological effects.[6] Described here is a high-performance liquid chromatographic (HPLC) procedure that quantitatively separates most of the AdoMet metabolites, including S-inosylhomocysteine (InoHcy) and S-adenosylcysteine (AdoCys),[7] and also detects the formation of novel AdoMet analogs.

Procedure

Chemicals. AdoMet and AdoHcy were purchased from Boehringer-Mannheim (Indianapolis, IN). dcAdoMet, dcAdoHcy, and dzAdo were synthesized by Southern Research Institute (Birmingham, AL). InoHcy and AdoCys or other S-nucleosidylhomocysteine standards can be synthesized by using purified AdoHcy hydrolase[8]; S-neplanocinylmethionine or other S-nucleosidylmethionine standards can also be synthesized by using AdoMet synthetase (EC 2.5.1.6) or by the method of Zimmerman *et al.*[9] using red blood cells.

HPLC Instrumentation. Two HPLC systems are used; the first one consists of two Beckman model 110A pumps, a 420 controller, a Glenco 5480 UV monitor (254 nm filter), a Waters U6K injector, and a 3390A Hewlett-Packard integrator. When an autosampler is used, the system consists of a Perkin-Elmer Series 4 Chromatography system, a 3600 Data Station, an ISS-100 autosampler, and a Waters 481 spectrophotometer. The compounds are separated by cation-exchange chromatography with a Waters Associates Radial-Pak SCX cartridge, 10 μm, 8 × 100 mm in a Z-Module.

[3] P. K. Chiang, H. H. Richards, and G. L. Cantoni, *Mol. Pharmacol.* **13,** 939 (1977).
[4] J. Bader, N. R. Brown, P. K. Chiang, and G. L. Cantoni, *Virology* **89,** 494–505 (1978).
[5] P. K. Chiang, *Science* **211,** 1164 (1981).
[6] I. Garcia-Castro, J. M. Mato, G. Vasanthakumar, W. P. Wiesman, E. Schiffmann, and P. K. Chiang, *J. Biol. Chem.* **258,** 4345 (1983).
[7] G. A. Miura, J. R. Santangelo, R. K. Gordon, and P. K. Chiang, *Anal. Biochem.* **141,** 161 (1984).
[8] A. Guranowski, J. A. Montgomery, G. L. Cantoni, and P. K. Chiang, *Biochemistry* **20,** 110 (1981).
[9] T. P. Zimmerman, R. D. Deprose, G. Wolberg, and G. S. Duncan, *Biochem. Biophys. Res. Commun.* **91,** 997 (1979).

HPLC Chromatography of AdoMet and Related Metabolites

There are two HPLC procedures that can be used. The first (Method A) requires more time but separates more sulfur metabolites; the second (Method B) is shorter for routine use in the assay of AdoHcy and AdoMet.

Method A. The elution gradient is arranged with two buffers [buffer A: 1.0 mM NH$_4$COOH adjusted to pH 4 with formic acid; buffer B: 0.2 M NH$_4$COOH (pH 4) and 0.8 M (NH$_4$)$_2$SO$_4$]. All buffers are made up in deionized water processed to 18 mohm (Milli Q Reagent Water System), filtered (Nalgene, 0.45 μm), and degassed prior to use. The elution, in four steps at a rate of 5 ml/min, is programmed as follows: (1) 0–10 min, isocratic conditions of buffer A only (wash step); (2) 10–30 min, a linear gradient to 3% buffer B; (3) 30–50 min, a linear gradient to 20% buffer B; and (4) 50–60 min, a linear gradient to 80% buffer B. The column is regenerated with 80% buffer B for 10 min followed by equilibration of the column with the starting buffer A for 20 min. When radioactive precursors are used, fractions from the column are collected into scintillation vials and the radioactivity subsequently determined after adding scintillation fluid.

Method B. The elution can be modified for faster separation and the specific quantification of AdoMet and AdoHcy without sacrificing resolution. Moreover, less buffers are required. The modified protocol, at a flow rate of 2.5 ml/min, is programmed as follows: (1) 0–3 min, isocratic conditions of buffer A only (wash step); (2) 3–9 min, a linear gradient to 12% buffer B; (3) 9–24 min, a linear gradient to 35% buffer B; and (4) 24–35 min, a linear gradient to 80% buffer B. The column is regenerated with buffer A for 8 min prior to (auto)injection of the next sample.

Identification and Evaluation of Peaks. Each column is first conditioned by injecting 5% sulfosalicylic acid extracts of biological samples (Preparation of Sample Extracts) and chromatographed until the retention time of the standards is stabilized, typically twice. Peaks are identified by (1) retention times, (2) coelution with standards, or (3) by analysis of UV spectra with a Hewlett-Packard 1040A scanning diode array detector. The peak heights and areas of each compound are directly proportional to the quantities of authentic materials injected. The method is very sensitive and can detect 50 pmol of AdoHcy or AdoMet as a lower limit with UV absorbance.

Preparation of Sample Extracts

Tissue Extracts. Excised tissues from a rat, e.g., liver or adrenal glands, are frozen immediately in liquid nitrogen and stored at −70°. To

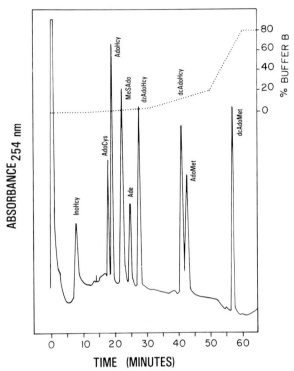

FIG. 1. Method A HPLC analysis of AdoMet and metabolites. Quantities injected were 2 nmol InoHcy, 0.5 nmol AdoCys, 2 nmol AdoHcy, 3 nmol MeSAdo, 1 nmol Ade, 4.8 nmol dzAdoHcy, 3 nmol dcAdoHcy, 3 nmol AdoMet, and 3 nmol dcAdoMet at 0.1 absorbance units full scale.

extract the metabolites, approximately 1 g of liver is homogenized with 4 ml of 5% ice-cold sulfosalicylic acid (w/v) in a blender, and then filtered; optionally, 40 to 60 mg of adrenal glands is homogenized in 1 ml of 5% sulfosalicylic acid with a glass homogenizer.

Tissue Cultures. Cultured cells, e.g., HeLa cells or HL-60 cells, are centrifuged after treatment, and the cell pellet is extracted with sulfosalicylic acid (1 ml, 5% w/v, per approximately 10^7 cells). Alternatively, the metabolites can be labeled radioactively by incubating cells for 2 hr or more with [^{35}S]methionine, which becomes AdoMet via AdoMet synthetase; ultimately, the ^{35}S is incorporated in AdoHcy and other metabolites. This procedure is especially suited for small samples of cells.

Sample Clarification. Prior to the injection of sample extracts, only filtration (Gelman, HPLC Acro-LC13) is required. The sulfosalicylic acid is effectively removed in the void volume during the wash step (100% buffer A).

Applications

Figure 1 depicts a standard HPLC separation of InoHcy, AdoCys, AdoHcy, MeSAdo (methylthioadenosine), Ade (adenine), dzAdoHcy (S-3-deazaadenosylhomocysteine), dcAdoHcy, AdoMet, and dcAdoMet using the protocol of Method A. If the AdoHcy hydrolase inhibitor dzAdo (3-deazaadenosine) is present, it coelutes with Ade and can interfere with the detection of Ade. However, dzAdo is a synthetic compound that is absent in normal biological samples. A novel AdoMet analog, S-neplanocinylmethionine, is detected when HL60 cells are incubated with [^{35}S]methionine and 10 μM neplanocin A for 2 hr; it elutes at 47 min, immediately after AdoMet (42 min).

The HPLC procedure described here is readily adaptable to automated operation. The use of such radioactive precursors as [^{35}S]methionine in tissue cultures allows the detection of AdoMet, AdoHcy, and other novel metabolites that are otherwise not detectable by UV absorbance because of small sample size. The procedure permits a complete analysis of AdoMet and its metabolites. The tissue/cell extraction with 5% sulfosalicylic acid is a one-step procedure that enables small samples to be quantitated, with no other handling except filtration prior to injection. Method A allows the complete separation to be finished in 60 min, with additional regeneration and wash steps totaling 30 min. Method B has a shorter elution time of 35 min and 8 min of regeneration, and it uses less than half the amount of buffers of Method A. Method B can easily be used for the enzymatic assay of AdoHcy hydrolase.

Acknowledgments

Teresa Alonso is a recipient of a grant from the Government of Canary Islands, Spain.

[35] Trimethylselenonium Ion

By HOWARD E. GANTHER, R. J. KRAUS, and S. J. FOSTER

The trimethylselenonium ion (TMSe) has been identified as a metabolite of inorganic and organic forms of selenium in rats and has been reported to occur in human urine.[1] TMSe can be a major metabolite, depending on the dose and the form of selenium administered, and is the

[1] H. E. Ganther, *in* "Trace Element Analytical Chemistry in Medicine and Biology" (P. Braetter and P. Schramel, eds.), Vol. 3, p. 3. de Gruyter, Berlin, 1984.

only urinary metabolite of Se that has been identified conclusively. Selenium in excess of requirements is excreted in urine, and formation of TMSe is regarded as a detoxication mechanism. With the increased use of supplemental selenium in animals and humans, methods for the quantification of this metabolite in urine are needed. We have developed two different methods for this purpose. One is based on precipitation of TMSe with ammonium Reineckate at neutral pH. The other method uses HPLC cation-exchange chromatography of desalted urine.[2]

General Procedures

TMSe is commercially available as the iodide salt[3] or can be synthesized by a number of methods.[4] [^{75}Se]TMSe is very useful in standardizing procedures or as an internal standard and is conveniently prepared from [^{75}Se]selenocysteine.[4] Trimethylsulfonium iodide, used as a carrier in Method 1, is commercially available.[5]

For quantitative analysis, TMSe is oxidized with a mixture of nitric/sulfuric/perchloric acid plus hydrogen peroxide and complexed with 2,3-diaminonaphthalene for fluorometric measurement.[6] TMSe is unusually stable to oxidation and may be underestimated using wet oxidation procedures that are otherwise adequate for selenium compounds.[7,8] Direct analysis of TMSe by graphite furnace atomic absorption is difficult because of the ease with which dimethyl selenide is volatilized during the steps prior to atomization.

For qualitative analysis, the Dragendorf reagent is useful for locating TMSe on plates following TLC or TLE[4]; sensitivity for TMSe is higher than for trimethylsulfonium ion or for selenonium compounds having free carboxyl groups. The R_f of TMSe is much affected by the type of absorbent (silica gel or cellulose) when basic solvent systems are used. TLE at pH 1.6 or 5.3 is useful in distinguishing TMSe, which has a permanent positive charge, from other selenium compounds.[4] Direct probe mass spectrometry of TMSe produces fragments corresponding to CH_3SeCH_3 and CH_3Se, but no molecular ion.[9]

---

[2] R. J. Kraus, S. J. Foster, and H. E. Ganther, *Anal. Biochem.* **147**, 432 (1985).
[3] Organometallics, Inc., East Hampstead, NH 03826.
[4] S. J. Foster and H. E. Ganther, *Anal. Biochem.* **137**, 205 (1984).
[5] Eastman Kodak Co., Rochester NY 14650.
[6] S. H. Oh, H. E. Ganther, and W. G. Hoekstra, *Biochemistry* **13**, 1825 (1974).
[7] O. E. Olson, I. S. Palmer, and E. Cary, *J. Assoc. Off. Anal. Chem.* **58**, 117 (1975).
[8] B. Welz and M. Melcher, *Anal. Chim. Acta* **165**, 131 (1984).
[9] J. L. Byard, *Arch. Biochem. Biophys.* **130**, 556 (1969).

Method 1 (Reineckate Precipitation)

Precipitation with ammonium Reineckate was used to isolate and identify TMSe in urine of animals given radioactive selenium.[9] We developed a modified method to identify TMSe in urine of human subjects ingesting dietary selenium from ordinary foods,[10] and for measuring [^{75}Se]TMSe excretion in rats.[11]

Principle. Urine is chromatographed on a strong cation exchanger (SP-Sephadex), and TMSe is eluted with ammonium formate. TMSe is precipitated with ammonium Reineckate at neutral pH in the presence of trimethylsulfonium ion as a carrier. TMSe is almost quantitatively precipitated whereas selenonium compounds having a carboxylate moiety remain in solution. The method is based on an earlier technique used to separate choline and betaine.[12]

Reagents

1 M Ammonium formate buffer (pH 4): To prepare a stock solution, add 425 ml of 23.6 N formic acid to approximately 800 ml of distilled water; titrate to pH 4.0 with 6 N ammonium hydroxide and bring to 1 liter. Prepare 0.25 M and 0.5 M buffer solutions by dilution.

Ammonium Reineckate (2%): Dissolve 20 g of Reinecke salt (Sigma Chemical Co.) in 100 ml of distilled water; protect from light.

Trimethylsulfonium iodide (1%): Dissolve 0.1 g of $(CH_3)_3S^+I^-$ (Eastman Kodak) in 10 ml of distilled water.

Procedure. The procedure described here has been used for assay of TMSe in rat urine.[11] The urine is collected in flasks containing toluene (0.5 ml). Duplicate aliquots of urine (up to 10 ml) are applied to SP-Sephadex ion-exchange columns (1.5 × 12 cm, H$^+$ form) previously equilibrated with 0.01 N HCl. Each column is washed with 50 ml 0.01 N HCl and 25 ml of water, followed by (1) 25 ml 0.25 M ammonium formate buffer, (2) 25 ml 0.5 M ammonium formate, and (3) 25 ml 1 M ammonium formate; 25 ml of eluate is collected after each application of the three buffers. TMSe should be present in the 0.5 and 1 M ammonium formate eluates; adjust the pH of these to 6–7 with NH$_4$OH, add 100 μl of 1% trimethylsulfonium iodide and 25 ml of 2% ammonium Reineckate, and allow the samples to

---

[10] H. E. Ganther, *in* "Selenium in Biology and Medicine" (G. F. Combs, Jr., J. A. Spallholtz, D. A. Levander, and J. E. Oldfield, eds.). Avi Publ. Co., Westport, Connecticut, 1986.

[11] S. J. Foster, R. J. Kraus, and H. E. Ganther, *Arch. Biochem. Biophys.* **251**, 77 (1986).

[12] D. Glick, *J. Biol. Chem.* **156**, 643 (1944).

stand at 4° overnight. The precipitate is filtered (Whatman #1 paper) and washed with 5 ml of cold 1% ammonium Reineckate. For ^{75}Se-labeled samples the precipitate is transferred to scintillation vials and assayed in a well-type scintillation counter. For unlabeled samples, the Reineckate anion should be removed prior to chemical analysis. Dissolve the precipitate in a small volume of 50% ethanol; apply the solution to a small column of QAE-Sephadex (acetate form) and elute the TMSe with water. Residual Reineckate ion can be removed by rechromatography of the TMSe on SP-Sephadex, as described above.

Comments. When [^{75}Se]TMSe was added to rat urine and isolated by this procedure, 92–100% of the TMSe was precipitated.[11] When trimethylsulfonium carrier was not added, results were often erratic, especially with urine containing small amounts of TMSe. Care must be taken not to overload the cation-exchange column with urine, since the high level of Na$^+$ and other substances in urine can interfere with binding of the TMSe. With human urine, a bed volume of SP-Sephadex equal to about 1 ml per ml of urine gave good recovery of [^{75}Se]TMSe; with 2- or 4-fold higher volumes of urine, recovery fell to 44 and 24%.[2] If TMSe is to be measured following the administration of selenium compounds that might be partially precipitated by Reineckate under the conditions of this method, the HPLC procedure (Method 2) is recommended.

Method 2 (Cation HPLC using Gradient Elution)

The Reineckate precipitation procedure (Method 1) does not separate TMSe from certain other selenonium compounds. The HPLC procedure was developed in order to achieve the higher resolution necessary for studying the metabolism of selenonium compounds.[2]

Principle. Urine is desalted prior to HPLC using low temperature ethanol precipitation. The supernatant fraction is applied to a strong cation HPLC column and eluted by means of an ammonium phosphate gradient. Fractions are collected and assayed for selenium.

Reagents

^{75}Se-labeled standard compounds can be synthesized from [^{75}Se]selenocystine and [^{75}Se]selenomethionine (both are available from Amersham Radiochemicals, Arlington Heights, IL) as described elsewhere.[4,13]

Chromatography. One liter of 0.5 M $(NH_4)_2HPO_4$ stock solution is prepared using 33.75 ml H_3PO_4 (85%) titrated to pH 4 with NH_4OH. This

---

[13] S. J. Foster, R. J. Kraus, and H. E. Ganther, *J. Labelled Compd. Radiopharm.* **22**, 301 (1985).

stock solution is diluted with water to give 3 mM $(NH_4)_2HPO_4$ (pH 4) (solvent A) and 0.33 M $(NH_4)_2HPO_4$ (pH 4) (solvent B). The solutions are filtered through a Metricel GN-6 filter, pore size 0.45 μm (Gelman, Ann Arbor, MI) and degassed prior to use on the HPLC column. The column is a 4 × 200 mm Nucleosil 5 μm/SA strong cation exchanger (Rainin Instrument Co., Woburn, MA), equilibrated with solvent A.

Procedure. Inorganic ions and other interfering substances present in urine are removed using a method adapted from that described by Labardie *et al.*[14] To the urine sample in a centrifuge tube is added 10 volumes of absolute ethanol, and the mixture is chilled in a solid CO_2/acetone bath for 15–20 min. The chilled solution is centrifuged at 7000 rpm for 15 min at −15°, and the ethanol is removed by decanting. The ethanol extract is dried by rotoevaporation. The pellet is dissolved in 0.3 ml 50% aqueous ethanol, and a 100-μl aliquot applied to the column. If large volumes of urine are used, the first ethanol extract can be dried, dissolved in 1 ml of water, and carried through the desalting procedure again. Recoveries of TMSe and various other selenium compounds added to urine range from 85 to 92%.[2]

After sample application, elution with solvent A is continued at a rate of 0.8 ml/min, and fractions of 0.8 ml are collected. After fraction 11 is obtained, a linear gradient composed of 40 ml each of solvents A and B is used to elute fractions 12–82. Fractions 83–120 are eluted with 0.5 M $(NH_4)_2PO_4$ (pH 4).

TMSe has a retention time of about 60 min, and is well separated from four other selenonium compounds; neutral selenoamino acids are not retained under these conditions.[2] The retention time of TMSe is somewhat variable, but it is eluted well after most other compounds have been eluted. Of the components tested, only the methyl ester of selenobetaine was found to be retained more strongly by the column; good separation of this compound from TMSe was obtained, whereas these compounds cannot be separated by Reineckate precipitation at neutral pH (Method 1). The resolution obtainable with the HPLC procedure is also demonstrated by the separation of *Se*-dimethylselenocysteineselenonium and *Se*-methylselenomethionineselenonium, which differ by only one methylene carbon. Recovery of [^{75}Se]TMSe off the column is about 90–95%, and recovery for the other compounds is routinely about 85%; when desalted urine from a rat injected with [^{75}Se]TMSe was applied, 85% of the urinary ^{75}Se was recovered as [^{75}Se]TMSe.[2]

Comments. The procedure as described above was developed for assay of TMSe in rat urine. We have learned that the analysis of human

---

[14] J. Labardie, W. A. Dunn, and N. N. Aronson, Jr., *Biochem. J.* **160,** 85 (1976).

urine using this procedure gave lower values for TMSe than expected. When we added [^{75}Se]TMSe to human urine to explore the cause of this problem, recovery of TMSe in the desalting procedure was satisfactory (85%), but during HPLC the [^{75}Se]TMSe was eluted as a very broad band, rather than a peak, and earlier than would be expected for the pure compound. The elution behavior is greatly improved when the desalted urine is put through a short clean-up column (Sep-Pak, Waters Associates, Inc., Milford, MA) prior to chromatography. The procedure is as follows: Human urine (10 ml) containing [^{75}Se]TMSe (24 ng Se) is desalted and taken up in 6 ml of water, then passed through a cartridge of Sep-Pak C_{18} (previously washed with methanol, followed by water) fitted onto a syringe. After washing the cartridge with 10 ml of water, the eluate plus washings are combined. Most of the urinary pigments are retained on the Sep-Pak. The light yellow aqueous solution is dried by rotary evaporation, dissolved in 0.8 ml of 20% ethanol, and 0.5 ml applied to the HPLC column. Following gradient elution, the fractions containing [^{75}Se]TMSe are pooled and assayed for total selenium. A correction is applied for recovery of the internal standard.

General Comments

The Reineckate method was used to measure urinary TMSe in an extensive study of the metabolism of TMSe and various methylated selenoamino acids in rats.[11] The method works well and recovery of TMSe is good, provided that trimethylsulfonium is added as a carrier. Method 2 has been used[2] to measure TMSe in urine of rats given [^{75}Se]dimethylselenocysteineselenonium or [^{75}Se]methylselenomethionineselenonium. TMSe was well separated from the starting compounds and from other urinary metabolites.

TMSe was conclusively identified in human urine[10] using Method 1 and 2 combined. Urine was acidified to pH 2 and applied to a SP-Sephadex column; the fraction eluted with 0.5–1 M ammonium formate (pH 4) was spiked with a small amount of [^{75}Se]TMSe as an internal standard and subjected to Reineckate precipitation at neutral pH, in the presence of trimethylsulfonium carrier. Nearly all of the [^{75}Se]TMSe was precipitated but more than half of the endogenous cationic Se remained in solution; acidification of the Reineckate supernatant liquid to pH 2 did not precipitate any additional Se. The precipitated Se was passed through cation and anion exchangers to remove Reineckate anion, and then subjected to HPLC (Method 2). All of the endogenous Se was eluted in the same position as the [^{75}Se]TMSe marker, showing that the Se precipitated from human urine by Method 1 was composed entirely of TMSe. TMSe was estimated to account for about 12% of the total Se in human urine.

100 μl of enzyme. Incubate for 30 min at 37° and stop the reaction by addition of 0.5 ml methylene blue solution, followed by 2 ml of chloroform. The two phases are agitated with a vortex mixer and separated by brief centrifugation at about 2000 g. The chloroform layer is transferred to a tube containing between 50 and 100 mg of anhydrous sodium sulfate which is used to dry and thereby clarify the solvent. Absorbance is determined at 651 nm. By this method, 1 μmol of 1-napthol sulfate would yield an A_{651} of 30.

Amine Sulfotransferase

  Reagents

   Bicine, 1 M, adjusted to pH 8.0 with KOH
   PAPS, 10 mM
   2-Naphthylamine, 20 mM, is prepared by treating the crystalline free amine in water with a few drops of 1 N HCl to dissolve the crystals, and then diluting. The concentration of the solution can be checked from the known molar extinction coefficient of 5,040 in water at 276 nm.

  Procedure. Into 13 × 100 mm tubes are added 100 μl buffer, 40 μl 2-naphthylamine–HCl, 40 μl PAPS, and up to 200 μl of enzyme. Incubation is for 60 min at 37° and is terminated by addition of 0.5 ml of methylene blue reagent followed by 2 ml of chloroform. Subsequent treatment is as outlined for the phenol sulfotransferases (above). 2-Naphthyl sulfamate carried through this procedure would yield an A_{651} of 6 per μmol.

Thin-Layer Chromatography[15]

  Principle. Substrates are incubated with enzyme and [^{35}S]PAPS (commercially available), and the radioactive products are subsequently separated by chromatography on small TLC strips (cellulose Eastman chromatogram without fluorescence indicator). Radioactivity in PAPS, its contaminants, and its breakdown products are found at a low R_f in the chromatographic systems used whereas the products of sulfuryl transfer have a high R_f. The results can be evaluated qualitatively by autoradiography on X-ray film or quantitatively by cutting out the appropriate section of the TLC strip and measuring radioactivity of that section.

  Procedure. A number of assay systems have proved useful for the several substrate catagories of the sulfotransferases, and these are listed below. After stopping the reaction by the means indicated, aliquots of 5 or 10 μl are applied to a 2 × 10 cm strip of a plastic-backed TLC plate. Drying is aided by hot air from a hair dryer, and efforts are made to retain liquid within a spot of about 3 mm diameter. After the solvent front moves about 8 cm (20 min), the strips are dried in air (20–30 min) and a section from the solvent front to 2 cm below it are cut out and placed in a scintillation vial containing 10 ml of Aquasol, or similar solvent, for measurement of radioactivity. These methods are useful when at least 200 pmol of product are formed.

The sulfated products from amines, phenols, and steroids are separated on Eastman 6064 cellulose strips developed with 1-propanol : ammonia : water (6 : 3 : 1); phenols are developed at 4° but all other classes of compounds are at room temperature. Sulfates of the aliphatic alcohols are separated on Eastman 13181 silica gel, developed with 2-propanol : chloroform : methanol : water (10 : 10 : 5 : 2) at room temperature.

PHENOLS.[6] Phenols are assayed in a total volume of 50 μl containing 0.10 M sodium phosphate at pH 6.5, 5 mM 2-mercaptoethanol, 1 mM 2-naphthol, and 0.5 mM [^{35}S]PAPS, with enzyme added last. After 30 min at 37°, the reaction is terminated by addition of 13 μl of 2 M acetic acid.

AMINES.[14] Amines are assayed in a total volume of 50 μl containing 0.2 M Bicine at pH 8.0, 2 mM 2-naphthylamine, and 1.0 mM [^{35}S]PAPS (0.5 μCi), with enzyme added last. After 30 min at 37° the reaction is terminated by addition of 20 μl of 2 M acetic acid. Under these conditions, 0.1–2 nmol of sulfamate is readily measured.

STEROIDS.[7,8,15] Steroids are assayed in a total volume of 50 μl containing 0.2 M sodium acetate at pH 5.5, 50 μM steroid, 60 μM [^{35}S]PAPS, and enzyme. After incubation for 30 min at 37° the tubes are placed in a boiling water bath for 2 min.

ALIPHATIC ALCOHOLS.[7] Aliphatic alcohols are assayed in a total volume of 100 μl containing 0.1 M Tris–HCl at pH 7.5, 80 μM [^{35}S]PAPS (170 μCi/μmol), and the appropriate alcohol, e.g. 5 mM 1-butanol. After addition of enzyme, and following incubation for 20 min at room temperature, the reaction is stopped with 50 μl of acetone.

Ecteola-Cellulose Chromatographic Assay[18]

Principle. Sulfate esters, the products of the reaction, should be separable from [^{35}S]PAPS by chromatography on Ecteola-cellulose. Phenol sulfotransferases from rat and human tissues have been assayed by this method.

Procedure. The reaction mixture, in a total volume of 1.0 ml, contains 10 mM sodium phosphate at pH 6.4, 0.3 μM [^{35}S]PAPS (0.5 μCi), 5–50

---

[18] R. T. Borchardt, A. Baranczyk-Kuzma, and C. L. Pinnick, *Anal. Biochem.* **130**, 334 (1983).

μM phenolic acceptor, and enzyme. After 5 min at 37°, the reaction is terminated by cooling to 4°. The pH is raised to 8.2 by addition of 1 ml of 20 mM ammonium bicarbonate, and the mixture is chromatographed on a column (1 × 6 cm) of Ecteola-cellulose that had been equilibrated with 20 mM ammonium bicarbonate. The ion-exchange material is prepared by washing 100 g of resin successively with 2 liters of 0.1 M HCl, 4 liters water, 2 liters 0.1 M NaOH, 4 liters water, 2 liters 0.5 M formic acid, 4 liters water, and 2 liters 20 mM NH$_4$HCO$_3$ and is suspended in 1 liter of 20 mM NH$_4$HCO$_3$.

After loading with the reaction mixture, each column is eluted sequentially in a stepwise manner with 20 ml of 20 mM, 50 mM, 100 mM, 200 mM, and 350 mM NH$_4$HCO$_3$. Fractions of 2 ml each are collected, with sulfate esters usually found in the eluate between 50 and 200 mM. PAPS elutes at 350 mM salt. The suggestion has been made that the course of the reaction may also be followed by the decrease in radioactivity of residual PAPS. By either measurement, linearity has been achieved as a function of protein concentration and time (up to 10 min of incubation).[18]

Methods using Radioactive Sulfuryl Group Acceptors

Some product sulfates are separable from their radioactive acceptor substrates by either ion-exchange, as illustrated with [^{14}C]aniline, or by a filter-binding technique, illustrated for steroid alcohols.

Amine Sulfotransferase Assay with [^{14}C]Aniline[14]

Reagents

[^{14}C]Aniline, 50 mM, specific activity of about 3 μCi/μmol
Bicine, 1 M at pH 8.0
PAPS, 5 mM
Magnesium acetate, 100 mM

Procedure. Enzyme is assayed in a total volume of 100 μl containing 20 μl bicine, 10 μl aniline, 10 μl PAPS, 5 μl magnesium acetate, and enzyme. After 30 min at 37°, 0.3 ml of water is added, and the reaction is stopped by placing in a boiling water bath for 1 min. With crude enzyme extracts, brief centrifugation is required at about 2,000 g. Aliquots of 200 μl are charged onto a small column of Dowex 50-X8 (0.5 ml bed volume in a Pasteur pipette). The column is washed with a total volume of 0.8 ml of water, and the entire effluent is collected in a scintillation vial in which it can be measured after addition of 20 ml of Hydrofluor or comparable material.

The amine is bound to the resin whereas the sulfate is not bound and is washed out with water. Although there is a significant blank value in the absence of PAPS, the method is useful when 0.5–12 nmol of sulfate is produced during the 30 min incubation.

Binding Assay for Sulfotransferases Acting on Hydroxysteroids[15]

Steroid sulfotransferase activity is measured in reaction mixtures of 50 μl total volume containing 0.25 M sodium acetate at pH 5.5, 50 mM radioactive sterol (about 10 μCi/μmol), 1.8 mM PAPS, and 4% acetone.

In preparing the assay mixture it is convenient to first evaporate the radioactive steroid, mixed with unlabeled steroid, to dryness and to dissolve the dry compound in acetone. When the resulting solution is added to the sodium acetate buffer and agitated with a vortex mixer, a dispersion of the sterol is formed that is stable for several hours. Aliquots are removed for assays, to which are added the other components; enzyme is added last to initiate the reaction. After 30 min at 37°, the reaction vessel is placed in a boiling water bath for 2 min. Upon cooling, 50 μl of 0.5 mM carrier steroid in absolute ethanol is added and mixed. A 20 μl-portion of the reaction mixture is applied to the center of a 3-cm square of silica gel-impregnated glass fiber sheet (Gelman Instrument Co., ITLC-SG). After the spot has dried, the square is placed on a Hoefer FH-12 filtration apparatus. A circular disk, with the radioactive material in the middle, is easily cut from the square filter by placing the cylindrical steel sleeve onto the filter apparatus with a twisting motion. The resultant filter disk is washed under slightly reduced pressure 3 times with portions of 3 ml 25% dioxane in hexane (v/v). Under these conditions, sterol sulfates remain bound to the filter whereas unesterified sterol is removed by solvent. The entire filter disk is placed in a scintillation vial and counted in 10 ml of Aquasol.

Evaluation

The ion-pair extraction method with methylene blue, where applicable, is the method of choice because of its simplicity and the rapidity with which it can be used. It requires much more enzyme, i.e., is less sensitive, than the TLC technique and requires standardization for each substrate because of the different extinction coefficients for each ion pair and, possibly, differences in the extent of partition of specific ion pairs. The method is sensitive to ionic strength, resulting in increased solubility of methylene blue in chloroform at higher concentrations of salt. Such conditions are encountered when eluting columns with a salt gradient but can be compensated for by using a control reaction mixture from which PAPS is omitted.

Of the several methods that have been presented, the most generally useful and highly sensitive is probably the TLC technique in which the sulfated products are separated from the sulfate donor for quantitation by radiometry. The process has been effective with a large variety of substrates without need for modification or change of solvent from that originally proposed. It permits processing of many samples at once, is limited by the time needed to transfer sample to the chromatography matrix, but has the added advantage of allowing preliminary screening by autoradiography.

The chromatography system employing Ecteola-cellulose is also broadly applicable but requires a separate column for each sample and prior standardization for the elution of each substrate. Methods employing radioactive sulfuryl group acceptors are of relatively limited use because of the unavailability of a wide range of radioactive substrates. When they are available, this approach allows for a sensitive assay, particularly when filter binding systems can be applied as in the case with steroids.

[37] Arylsulfatases: Colorimetric and Fluorometric Assays

By ALEXANDER B. ROY

$$ROSO_3^- + H_2O \rightarrow ROH + H^+ + SO_4^{2-}$$

Arylsulfatases (EC 3.1.6.1) catalyze the above reaction, where ROH is a phenol or at least has some phenolic properties. They form a diverse group of enzymes of rather broad specificity with pH optima occurring over a very wide range. It is impossible to give a completely general method for their assay which in crude preparations is complicated by several factors: (1) the ubiquitous occurrence of phosphate which inhibits many arylsulfatases; (2) the ability of some tissues to metabolize both the substrates and the derived phenols[1]; and (3) the fact that tissues generally contain several enzymes showing arylsulfatase activity. Most mammalian tissues contain two soluble lysosomal arylsulfatases, sulfatases A and B (with multiple forms of each of these), and an insoluble microsomal arylsulfatase, sulfatase C.[2] These three enzymes show overlapping specificities, and although the determination of sulfatase C is simple,[3] the separate

[1] K. S. Dodgson and B. Spencer, *Biochem. J.* **53**, 444 (1953).
[2] A. B. Roy, *Aust. J. Exp. Biol. Med. Sci.* **54**, 111 (1976).
[3] D. W. Milsom, F. A. Rose, and K. S. Dodgson, *Biochem. J.* **128**, 331 (1972).

determination of sulfatases A and B in mixtures is fraught with difficulty[4] and depends on either the separation of the enzymes[4] or the use of modifiers.[5,6] Species differences are important; for example, although the classical method[5] for the determination of sulfatases A and B in human urine can be used with human tissues, it cannot be used with rat tissues.[4]

The three mammalian arylsulfatases show other types of sulfatase activity which reflect their physiological roles. Sulfatase A is a cerebroside sulfatase which hydrolyzes the galactose 3-sulfate residues in sulfolipids,[7] sulfatase B hydrolyzes N-acetylgalactosamine 4-sulfate residues in oligosaccharides derived from glycosaminoglycans,[8] and sulfatase C is a steroid sulfatase.[9] These more specific activities should be used for the unambiguous determination of the enzymes for clinical purposes.

Principles

The assay of arylsulfatases is simple in principle, and any of the reaction products may be determined. Spectrophotometric methods using indicators[9a] for the determination of H^+ do not appear to have been adapted for the assay of these enzymes although they could be useful and complement methods using the pH-stat[10] or protonometry.[11] Spectrophotometric methods are available, although not entirely satisfactory, for SO_4^{2-} [12,13] but are most suited to the determination of ROH. The general method is simple and requires only that the reaction mixture, after incubation with the enzyme, be made alkaline so that the liberated phenol can be determined, as the phenoxide, either spectrophotometrically or fluorometrically. With crude enzyme preparations deproteinization with, for example, ethanol or trichloroacetic acid may be necessary but any acid treatment should be as mild and as brief as possible because all aryl sulfates are acid-labile. In some cases, the spectra of the phenol and its sulfate ester differ sufficiently at the pH of arylsulfatase activity for the formation of the former to be followed directly so that continuous spec-

[4] M. Worwood, K. S. Dodgson, G. E. R. Hook, and F. A. Rose, *Biochem. J.* **134**, 183 (1973).
[5] H. Baum, K. S. Dodgson, and B. Spencer, *Clin. Chim. Acta* **4**, 453 (1959).
[6] H. Christomanou and K. Sandhoff, *Clin. Chim. Acta* **79**, 527 (1977).
[7] A. Jerfy and A. B. Roy, *Biochim. Biophys. Acta* **293**, 178 (1973).
[8] A. A. Farooqui and A. B. Roy, *Biochim. Biophys. Acta* **452**, 431 (1976).
[9] J. O. Dolly, K. S. Dodgson, and F. A. Rose, *Biochem. J.* **128**, 337 (1972).
[9a] H. Gutfreund, *in* "An Introduction to the Study of Enzymes," p. 145. Blackwell, London (1965).
[10] A. B. Roy, *Biochim. Biophys. Acta* **526**, 489 (1978).
[11] A. B. Roy, *Anal. Biochem.* **116**, 123 (1981).
[12] A. Waheed and R. L. Van Etten, *Anal. Biochem.* **89**, 550 (1978).
[13] A. B. Roy and J. Turner, *Biochim. Biophys. Acta* **704**, 366 (1982).

trophotometric assays are possible. Unhydrolyzed substrate can be determined spectrophotometrically (this volume [62], pp. 362–363).

The lysosomal sulfatases A and B have pH optima at about 5, and acetate buffers, or formate buffers at lower pH, are generally suitable for their assay. These enzymes are inhibited by phosphate. The microsomal sulfatase C of mammalian tissues, and the arylsulfatases of mollusks and of microorganisms, have pH optima in the neutral or slightly alkaline region so that imidazole or Tris buffers may be used in their assay. Sulfatase C is much less sensitive than sulfatases A and B to inhibition by phosphate so that buffers containing the latter can conveniently be used[3] to minimize activity due to the lysosomal arylsulfatases which usually contaminate the microsomal enzyme.

Arylsulfatases may be quite sensitive to changes in the salt[14,15] or buffer[15-17] concentration of the reaction mixture so that careful control of the assay conditions is essential. Finally it should be noted that sulfatases A are hysteretic enzymes[18] which undergo a substrate-induced inactivation with a half-time of a few minutes. For their assay the reaction time must be short and the initial velocity obtained by suitable computation.[10]

Substrates

The structures of some useful substrates are shown in Fig. 1, and some pertinent data for these, and the phenols formed on their hydrolysis, are given in Table I. Only nitroquinol sulfate and methylnitrocatechol sulfate are not available commercially.

Nitrocatechol sulfate[19]

Nitrocatechol sulfate (mono- or dipotassium 2-hydroxy-5-nitrophenyl sulfate[20]) is the most common substrate for the arylsulfatases and is particularly useful for sulfatases A and B. The possibility of contamination by a nitropyrogallol bissulfate[21] should be noted. The hydroxyl group in nitrocatechol sulfate has a pK of 6.4 and, therefore, buffers significantly in the pH region of interest. Further, over the pH range of about 5 to 8 the compound changes from a mono- to a dianion. Hydrolysis gives 4-nitroca-

---

[14] K. Stinshoff, *Biochim. Biophys. Acta* **276**, 475 (1972).
[15] H. J. Jeffrey and A. B. Roy, *Anal. Biochem.* **77**, 478 (1977).
[16] K. S. Dodgson and C. H. Wynn, *Biochem. J.* **68**, 387 (1958).
[17] A. B. Roy, *Biochim. Biophys. Acta* **227**, 129 (1971).
[18] A. B. Roy, *Comp. Biochem. Physiol. B* **82B**, 333 (1985).
[19] The name *p*-nitrocatechol sulfate is obviously incorrect and should not be used.
[20] A. B. Roy, *Biochem. J.* **68**, 519 (1958).
[21] A. B. Roy and L. M. H. Kerr, *Nature (London)* **178**, 376 (1956).

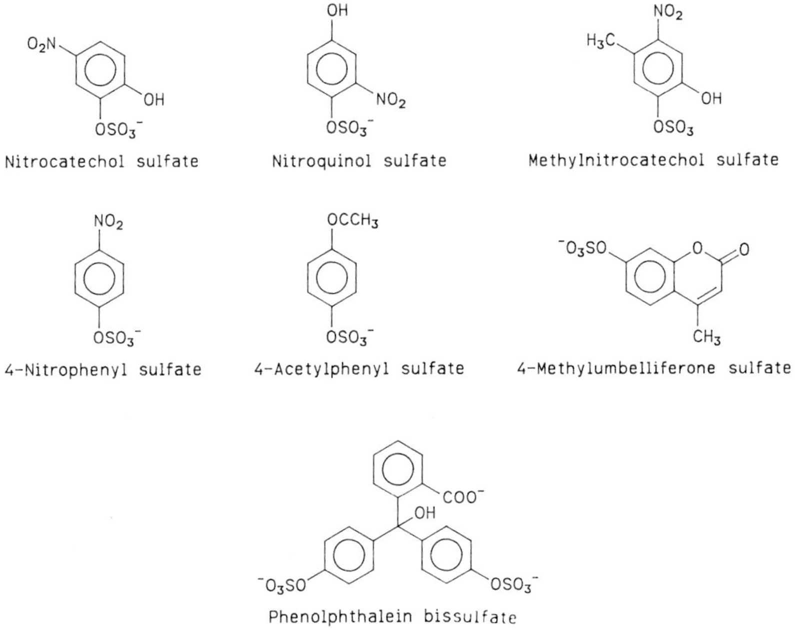

FIG. 1. Structures of substrates.

techol which absorbs strongly at 510 nm in 0.1 M NaOH. Blanks are generally negligible. With crude enzyme preparations, such as liver extracts, a rather rapid fading of the color may occur, but this can be overcome by including quinol in the final solution.[22]

Conditions for the assay of some arylsulfatases are given in Table II. Again, it must be stressed that the differences between sulfatases A and B are small, and that the conditions in Table II will not separately determine the two enzymes in a mixture.

Nitrocatechol sufate cannot be used for the continuous spectrophotometric assay of the arylsulfatases.

Nitroquinol Sulfate

The hydroxyl group in nitroquinol sulfate (potassium 4-hydroxy-2-nitrophenyl sulfate) is considerably less acidic than that in nitrocatechol sulfate (Table I), but the sulfate group is rather more acid-labile, although less so than that in 4-nitrophenyl sulfate.[15] Nitroquinol sulfate has been

[22] A. B. Roy, *Biochem. J.* **53**, 12 (1953).

TABLE I
SOME SUBSTRATES FOR ARYLSULFATASES AND SPECTRAL PROPERTIES
OF THE PHENOLS FORMED ON THEIR HYDROLYSIS

Sulfate ester			Phenol			
Trivial name	Phenolic pK	Preparation (reference)	Solvent	λ_{max} (nm)	ε_{max}	Assay[a]
Nitrocatechol sulfate	6.4	20	0.1 M NaOH	510	12400	d
Nitroquinol sulfate	8.2	15	0.1 M NaOH	535	4100	c, d
Methylnitrocatechol sulfate	8.2	25	1.7 M NaOH	494	11900	c
4-Nitrophenyl sulfate	—	3	0.1 M NaOH	400	17800	c, d
4-Acetylphenyl sulfate	—	3	0.1 M NaOH	323	20500	d
			0.166 M NaOH 66% ethanol	327	21700	d
Phenolphthalein bissulfate	—	31	Glycine, pH 10.4	540	41000	d
4-Methylumbelliferone sulfate	—	32	Carbonate, pH 10.3	362	18400	d

[a] c, continuous spectrophotometric assay, d, point assay.

TABLE II
CONDITIONS FOR THE ASSAY OF SOME ARYLSULFATASES

Substrate	Assay[a]	Enzyme		Buffer				S_{opt} (mM)	K_m (mM)	Ref.
		Source	Type	Type	Concentration (M)	pH	Additive			
Nitrocatechol sulfate	d	Ox liver	Sulfatase A	Acetate	0.125	4.9		3.0	0.8	34
			Sulfatase B	Acetate	0.5	5.4		10	1.9	35
		Human liver	Sulfatase A	Acetate	0.5	5.0		6.0	—	36
			Sulfatase B	Acetate	0.5	6.1		10	1.8	16
		Kangaroo liver	Sulfatase A	Imidazole	0.5	6.9		10	2.7	17
			Sulfatase B	Acetate	0.5	5.8		10	1.4	17
		Patella vulgata		Acetate	0.125	5.5		5	0.7	37
		Helix pomatia		Tris	0.25	7.5		15	0.73	24
		Aspergillus oryzae		Acetate	0.5	5.9		2.5	0.35	23
		Alcaligenes metalcaligenes		Phosphate	0.1	8.0		15	0.22	38
Nitroquinol sulfate	d	A. oryzae		Acetate	0.5	6.0		2.0	0.13	23
	c	Ox liver	Sulfatase A	Acetate	0.5	5.5		17	7.4	15
			Sulfatase B	Acetate	0.5	5.5		2.5	—	15

Substrate	Assay[a]	Source	Enzyme	Buffer	[Conc]	pH	Additive			Ref
Methylnitrocatechol sulfate	c	*Haliotis iris*		Acetate	0.5	5.6	±3 mM CuCl$_2$	—	0.037	25
4-Nitrophenyl sulfate	d	Ox liver	Sulfatase B	Acetate	0.2	5.1	0.2 M NaCl	1.2	—	27
			Sulfatase C	Tris	0.125	8.0		10	2.0	39
		H. pomatia		Tris	0.25	7.5		15	5.1	24
		A. oryzae		Acetate	0.5	6.2		15	0.17	23
		A. metalcaligenes		Phosphate	0.1	8.75		15	0.48	38
	c	Ox liver	Sulfatase B	Acetate	0.13	5.6	0.13 M NaCl	2.0	10	27
		Aerobacter aerogenes		Tris	0.25	7.1		8.0	3.3	29
4-Acetylphenyl sulfate	d	Rat liver	Sulfatase C	Phosphate	0.1	8.0		40	9.0	3
				Tris	0.1	8.0		12	2.5	3
		A. metalcaligenes		Phosphate	0.1	8.75		3.0	0.90	38
Phenolphthalein bissulfate	d	*Mycobacterium piscium*		Phosphate		6.3		3.0	—	31
4-Methylumbelliferone sulfate	d	Human brain	Sulfatase A	Acetate	0.2	5.7		>15	12.5	40
			Sulfatase B	Acetate	0.2	5.4		>20	8.3	40
		P. vulgata		Acetate	0.12	5.4		>15	2.5	32

[a] c, Continuous spectrophotometric assay; d, point assay.

used[16,23,24] in discontinuous assays similar to, but with no advantage over, those with nitrocatechol sulfate. Indeed, there are disadvantages because both ε_{max} and the stability of nitroquinol in alkali are considerably less than those of 4-nitrocatechol.

On the other hand, at pH values below about 7 the spectra of nitroquinol and nitroquinol sulfate differ considerably ($\Delta\varepsilon_{400}$ about 2900, almost constant between pH 4 and 6.8[15]) so that the latter is a useful substrate for the continuous spectrophotometric assay of the arylsulfatases[15] (Table II). It should be stressed, however, that the absorbance of nitroquinol sulfate at 400 nm is by no means negligible so that the method is not without its disadvantages.

Methylnitrocatechol Sulfate

Only recently prepared, methylnitrocatechol sulfate (potassium 2-hydroxy-5-methyl-4-nitrophenyl sulfate) has been used[25] in two types of continuous spectrophotometric assay. In the first, the liberated 5-methyl-4-nitrocatechol is measured directly by the change in absorbance at 374 nm ($\Delta\varepsilon_{374}$ about 3900 at pH 5.6, varying with pH[25]). In the second, 3 mM $CuCl_2$ is included in the reaction mixture so that the 5-methyl-4-nitrocatechol forms its 1:1 Cu^{2+} complex which is detected by the absorbance increase at 432 nm. The sensitivity of the second method is about double that of the first.[25] Conditions for the assay of the arylsulfatase of the abalone, *Haliotis iris,* are given in Table II.

The method involving complex formation should, in principle, be widely applicable because nitrocatechols in general form complexes with Cu^{2+} and other transition[25] and rare earth[7] ions. On the other hand, some arylsulfatases are inhibited by Cu^{2+}.[26]

4-Nitrophenyl Sulfate

4-Nitrophenyl sulfate is another commonly used substrate. It is, however, relatively labile, and care is needed in its storage and purification. It serves for discontinuous assays in which the reaction mixture, after incubation with the enzyme, is made 0.1 M in NaOH to produce the intensely yellow 4-nitrophenoxide ion. This substrate has been particularly useful for the assay of sulfatase B, for which a high concentration of Cl^- must be present.[27] It is a poor substrate for sulfatase A, having a K_m of more than

---

[23] D. Robinson, J. N. Smith, B. Spencer, and R. T. Williams, *Biochem. J.* **51**, 202 (1952).
[24] K. S. Dodgson and G. M. Powell, *Biochem. J.* **73**, 672 (1959).
[25] A. G. Clark, D. A. Jowett, and J. N. Smith, *Anal. Biochem.* **118**, 231 (1981).
[26] W. Bleszynski and A. Lesnicki, *Enzymologia* **33**, 373 (1967).
[27] E. C. Webb and P. M. F. Morrow, *Biochem. J.* **73**, 7 (1959).

0.2 M.[10] Conditions for this, and some other arylsulfatases, are given in Table II. Because of the lability of 4-nitrophenyl sulfate, which is accentuated in alkali by proteins and some amino acids,[28] careful controls are required and blanks are not always negligible. Further, 4-nitrophenol can be reduced to 4-aminophenol by crude extracts of animal tissues, so giving poor recoveries.[1]

4-Nitrophenyl sulfate can also be used in continuous spectrophotometric assays of the arylsulfatases at pH values around neutrality, but careful control of pH is required in these methods because the pK of 4-nitrophenol, 7.15, is such that the apparent extinction coefficient, ε', of the phenoxide varies greatly with pH in this region. The relationship is

$$\varepsilon'_{400} = 17800 \times \frac{10^{pH-7.15}}{1 + 10^{pH-7.15}}$$

from which it follows that a change of pH of 0.1 causes a change in ε'_{400} of about 1070 (6%) near the pK. The common use of Tris buffers which have high values of dpH/dT (about -0.028 per °C) accentuates this problem.

The absorption spectrum of 4-nitrophenyl sulfate is almost invariant with pH (λ_{max}, 278 nm; ε_{278}, 9000) but that of 4-nitrophenol is not, therefore, the optimum wavelength for use in continuous assays varies with pH. At pH 5.6, 330 nm ($\Delta\varepsilon_{330}$ about 6200) has been recommended,[27] while at pH 7.4, 420 nm ($\Delta\varepsilon_{420}$ about 6750) has been used.[29] At wavelengths less than about 340 nm the absorbance of 4-nitrophenyl sulfate itself is troublesome (ε_{340}, 1000).

4-Acetylphenyl Sulfate

4-Acetylphenyl sulfate is the substrate of choice for the assay of sulfatase C^3 because, first, it is a poor substrate for sulfatases A and B and, second, the acetyl group is not metabolized by animal tissues.[1] Conditions for this and other arylsulfatases are given in Table II. When necessary, reaction mixtures are deproteinized with ethanol,[3] and care is needed to ensure that a slight turbidity does not contribute to the measured absorbance at 327 nm. The absorbance of the substrate at this wavelength is significant and careful controls are necessary.[3]

Phenolphthalein Bissulfate

The presence of two sulfate groups in phenolphthalein bissulfate means that it is not an ideal substrate for arylsulfatases because the ki-

---

[28] K. S. Dodgson and B. Spencer, *Biochem. J.* **65**, 665 (1957).
[29] L. R. Fowler and D. H. Rammler, *Biochemistry* **3**, 230 (1964).

netic difficulties could be formidable and with at least some enzymes, for example the arylsulfatase of *Helix pomatia,* the progress curves are nonlinear with a pronounced lag phase. Also, at the pH optima of most arylsulfatases further complications can be introduced by the slow lactonization of the form of the substrate shown in Fig. 1. It has proved useful in screening microorganisms for arylsulfatase activity,[30] and for more detailed studies of one such enzyme[31] (Table II), but it is not hydrolyzed by all microbial arylsulfatases.[29]

As with the similar, but more useful, method for the determination of β-glucuronidase with phenolphthalein monoglucuronide as substrate, the reaction mixture should be brought to pH 10.4 with a concentrated glycine buffer to give a stable color with the liberated phenolphthalein.

4-Methylumbelliferone Sulfate

Fluorimetric methods for the determination of arylsulfatase are few, and the only useful substrate is 4-methylumbelliferone sulfate. Details of procedures yielding products giving satisfactorily low blanks have been provided.[32,33] The potassium salt has a rather low solubility (<10 mM at 4°, about 20 mM at 37°) and as the values of K_m may be of this order the salt of 2-amino-2-methylpropan-1-ol (solubility >20 mM at 4°) is preferable.[33]

The reaction mixture is brought to pH 10.3 by adding a concentrated carbonate buffer when the liberated 4-methylumbelliferone can be determined either spectrophotometrically at 362 nm or fluorometrically (excitation, 359 nm; emission, 451 nm). The availability of these two methods means that a wide range of enzyme concentrations can be studied with the single substrate. For purely enzymological purposes this substrate seems to have little to offer but it is undoubtedly important in clinical chemistry where extreme sensitivity is required.[6,33]

A General Method

As is obvious from Table II,[34–40] no one set of conditions can be satisfactory for the assay of all arylsulfatases and the classical review of

---

[30] M. Barber, B. W. L. Brooksbank, and S. W. A. Kuper, *J. Pathol. Bacteriol.* **63,** 57 (1951).
[31] J. E. M. Whitehead, A. R. Morrison, and L. Young, *Biochem. J.* **51,** 585 (1952).
[32] W. R. Sherman and E. F. Stanfield, *Biochem. J.* **102,** 905 (1967).
[33] H. Rinderknecht, M. C. Geokas, C. Carmack, and B. J. Haverback, *Clin. Chim. Acta* **29,** 481 (1970).
[34] A. B. Roy, *Biochem. J.* **55,** 653 (1953).
[35] F. Allen and A. B. Roy, *Biochim. Biophys. Acta* **168,** 243 (1968).
[36] K. S. Dodgson, B. Spencer, and C. H. Wynn, *Biochem. J.* **62,** 500 (1956).

Dodgson and Spencer[41] remains an important source of information to be read before embarking on such assays. The following procedure should prove suitable for a preliminary search for such an enzyme, although it should be noted that arylsulfatase activity is generally low in tissues and microorganisms. Nitrocatechol sulfate is the substrate of choice because it is hydrolyzed by most, if not all, arylsulfatases and usually has a low K_m. Of the other commercially available compounds, 4-nitrophenyl, 4-acetylphenyl, and 4-methylumbelliferone sulfates can have unsuitably high values of K_m, and phenolphthalein bissulfate is not a general substrate. The enzyme should be dialyzed or otherwise freed from phosphate. Two sets of reaction mixtures containing the enzyme in 25 mM and 2.5 mM nitrocatechol sulfate at pH 5 (acetate), pH 6 (acetate or imidazole), pH 7, and pH 8 (Tris) should be incubated at 25 or 37°, one set for 1 hr, the other for 24 hr. Both sets are required because substrate inhibition is common in the arylsulfatases. Account must be taken of the high buffer capacity of nitrocatechol sulfate between pH 5.5 and 6.5 and suitable controls set up. Unless the protein concentration is very high, when deproteinization with trichloroacetic acid may be needed, the reactions can be stopped by making the mixtures approximately 0.1 M in NaOH, and the amount of liberated 4-nitrocatechol may be obtained from the absorbance at 510 nm.

---

[37] A. B. Roy, *Biochem. J.* **62**, 41 (1956).
[38] K. S. Dodgson, T. H. Melville, B. Spencer, and K. Williams, *Biochem. J.* **58**, 182 (1954).
[39] A. B. Roy, *Biochem. J.* **64**, 651 (1956).
[40] B. C. Harinath and E. Robins, *J. Neurochem.* **18**, 245 (1971).
[41] K. S. Dodgson and B. Spencer, *Meth. Biochem. Anal.* **4**, 211 (1957).

Section II

Preparative Methods

A. Preparation of Specific Metabolites
Articles 38 through 45

B. General Preparative Techniques
Articles 46 through 54

C. Nutritional Methods
Articles 55 through 57

[38] β-Sulfopyruvate

By OWEN W. GRIFFITH and CATHERINE L. WEINSTEIN

$$\text{Li}^+ \; {}^-\text{O}_3\text{S}-\text{CH}_2-\overset{\overset{\displaystyle O}{\|}}{\text{C}}-\text{COO}^- \; \text{Li}^+$$

β-Sulfopyruvic acid (2-carboxy-2-oxoethanesulfonic acid) is the α-keto acid corresponding to cysteinesulfonic acid (cysteic acid) and is formed from that amino acid either by transamination[1] or by amino-acid oxidase activity.[2] It is of biochemical interest in part because cysteinesulfonate is consumed in the diet and transaminated *in vivo;* the further metabolism of β-sulfopyruvate has not been elucidated. β-Sulfopyruvate is also of interest as an isoelectronic and reasonably isosteric analog of oxaloacetate and β-sulfinylpyruvate. The last compound is formed by transamination of cysteinesulfinate, a major cysteine catabolite in many species. Whereas β-sulfinylpyruvate is unstable and spontaneously decomposes to sulfite and pyruvate, β-sulfopyruvate is relatively stable and is thus of possible use in studies of the metabolism of oxidized cysteine derivatives. β-Sulfopyruvate has been synthesized by the action of L-amino-acid oxidase on cysteine sulfonate and was isolated as the barium salt.[2] The chemical synthesis described here is more convenient and yields the biologically acceptable lithium salt.

Synthesis of Lithium β-Sulfopyruvate

Sodium sulfite (5.67 g, 45 mmol) and 25 ml of water are placed in a 100-ml, 3-neck, round-bottomed flask fitted with a thermometer, reflux condenser, dropping funnel, and magnetic stirring bar. The stirred solution is blanketed with Ar or N_2 gas and is warmed on an oil bath to 80°. Bromopyruvic acid (3.33 g, 20 mmol), dissolved in 20 ml of water, is added from the dropping funnel over a period of 30 min. The resulting light brown solution is stirred at 80° for an additional hour and is then allowed to cool to room temperature. Enzymatic assay of the crude reaction mixture (see below) indicates that sulfopyruvate is formed in a yield greater than 90%.

The entire reaction mixture is applied to a column (2.5 × 45 cm) of Dowex 1-X8 (200–400 mesh, chloride form) and is washed into the resin with 100–200 ml of water. The products are eluted with a linear gradient formed between 750 ml of water and 750 ml of 2 M HCl; the gradient is

---
[1] S. Darling, *Congr. Int. Biochim., C. R., 2nd, 1952,* p. 304 (1953).
[2] A. Meister, P. E. Fraser, and S. V. Tice, *J. Biol. Chem.* **206,** 561 (1954).

followed by 200 ml of 2 M HCl. Fractions of 20 ml are collected, and those with A_{280} values greater than 0.5 are assayed for β-sulfopyruvate. Fractions not containing β-sulfopyruvate are assayed for sulfate and sulfite by adding a drop of 0.5 M BaCl$_2$ to the column fraction and noting the appearance of a white precipitate; sulfite is distinguished from sulfate on the basis of its ability to decolorize malachite green.[3]

β-Sulfopyruvate elutes as a sharp, well-formed peak between fractions 70 and 85; it is preceded by sulfite (fractions 8–14) and sulfate (fractions 45–65). The appropriate fractions are pooled and concentrated with a rotary evaporator at reduced pressure to yield a tan oil. Water (25 to 50 ml) is added and then evaporated several times to assure that the HCl has been adequately removed. The resulting oil is dissolved in about 100 ml of water, and that solution is adjusted to pH 6–7 by addition of dilute LiOH. Care is taken not to exceed pH 7. The neutralized solution is evaporated to dryness. If it does not spontaneously crystallize, it is stored in a vacuum desiccator over P$_2$O$_5$ until it solidifies. The product, obtained as hard, white flakes, typically weighs about 3.3 g.

A number of preparations of dilithium β-sulfopyruvate have been submitted for elemental analysis and have given results consistent with compositions containing 1, 1.5, or 2 mol of water. Solutions prepared by weight (taking into account the water of hydration) contain 90–98% of the expected β-sulfopyruvate when assayed with malate dehydrogenase as described below. Since the state of hydration may change with storage, it is preferable to assay each solution of β-sulfopyruvate when it is prepared; solutions are stable for at least several weeks when stored at −20°. The synthetic β-sulfopyruvate yields cysteinesulfonate when treated with glutamate and aspartate aminotransferase and is free of chloride (AgNO$_3$ test) and pyruvate (assayed with lactate dehydrogenase). The product melts with decomposition at about 280°, but the melting point is not sufficiently sharp to be of use in characterization.

Assay of β-Sulfopyruvate

Principle. β-Sulfopyruvate, an analog of oxaloacetate, is reduced by NADH and malate dehydrogenase. NADH oxidation is followed spectrophotometrically and related to the amount of β-sulfopyruvate present.

Reagents

Potassium phosphate at pH 7.5, 100 mM
NADH, 10 mM, prepared in the phosphate buffer

---
[3] F. Feigle, "Spot Tests in Inorganic Analysis" (R. E. Oesper, trans.), 5th Engl. ed., p. 311. Elsevier, Amsterdam, 1958.

Malate dehydrogenase (porcine heart, mitochondrial), 10 units/μl (1 unit of activity converts 1 μmol · min of oxaloacetate to malate in the presence of NADH at 25°)

Procedure. Phosphate buffer (0.95 ml), NADH solution (25 μl), and an aliquot of the solution to be assayed (1–25 μl, 0–0.15 μmol of β-sulfopyruvate) are placed in a cuvette of 1.0 cm light path. The volume is brought to 1.0 ml with water. After the A_{340} is recorded, malate dehydrogenase (1 μl) is added. Absorbance at 340 nm is monitored for several minutes until the rate of change becomes very small or zero. The change in A_{340}, corrected for the small change observed in the same time period in the absence of β-sulfopyruvate, is used to calculate the amount of NADH oxidized (ε = 6.2 × 10^3); the resulting value is assumed to equal the amount of β-sulfopyruvate present.

Acknowledgments

Studies from the authors' laboratory were supported in part by the National Institutes of Health, Grant AM26912. O.W.G. is a recipient of an Irma T. Hirschl Career-Scientist award. We thank Ernest B. Campbell and Michael A. Hayward for excellent technical assistance.

[39] Alanine Sulfodisulfane

By Jack H. Fellman

Alanine sulfodisulfane (or cysteine thiosulfonate)

Synthesis

From Cystine Disulfoxide and Thiosulfate.[1] Sodium thiosulfate (10 mmol) dissolved in 10 ml of water is added dropwise over a 5-min period to a magnetically stirred suspension of cystine disulfoxide [S-(2-amino-2-carboxyethylsulfonyl)-L-cysteine][2] (10 mmol) in 20 ml of 0.15 M formic acid at room temperature. After an additional 10-min period of stirring, the reaction mixture is placed onto a column (1.5 × 40 cm) of Dowex 50W-X8 (H^+ form, 200–400 mesh) and is eluted with water. The strongly anionic alanine sulfodisulfane is not retained and is eluted in an early fraction (20–65 ml). This fraction was subjected to rotary evaporation under reduced pressure, and the residue dissolved in 10 ml of water and adjusted to pH 6 with dilute sodium hydroxide. The solution is treated with 7 volumes of ethanol and, after standing overnight, the crystalline precipitate is filtered and washed with absolute ethanol and diethyl ether. It should be stored in a vacuum desiccator. The yield is in excess of 90%.

From Cystine and Thiosulfate.[3,4] Cystine (1 mmol) is dissolved in 5 ml 15 M NH_4OH. Sodium thiosulfate (5 mmol) dissolved in 5 ml of water is added, and the solution is heated on a steam bath for 4 hr. The solution is cooled, taken to dryness with a rotary evaporator, and the residue triturated with 10 ml of water. The filtrate obtained after removal of the unreacted cystine is chromatographed using the 1-butanol, acetic acid, water system described below. Two ninhydrin-reacting spots were observed, one of which cochromatographed with sulfocysteine ($R_f = 0.18$, minor impurity) and the other of which was alanine sulfodisulfane. The product is isolated as described above in a yield of approximately 80%.

Analytical Properties

Paper Chromatography. Using Whatman 3MM paper developed with 1-butanol:acetic acid:water (12:3:5) alanine sulfodisulfane exhibits an R_f of 0.24; developed with phenol:water (8:2), the R_f is 0.28.[4]

---

[1] T. Ubaka, N. Masuoka, H. Mikami, and M. Taniguchi, *Anal. Biochem.* **140**, 449 (1984).
[2] G. Toennies and T. F. Lavine, *J. Biol. Chem.* **113**, 571 (1936).
[3] T. W. Szczepkowski, *Nature (London)* **182**, 934 (1958).
[4] J. H. Fellman and N. Avedovech, *Arch. Biochem. Biophys.* **218**, 303 (1982).

Anion-Exchange Chromatography. The identification and quantitative determination of alanine sulfodisulfane in tissue extracts can be achieved with a dual column procedure. Liver homogenates and other tissue extracts are deproteinized with trichloroacetic acid (final concentration = 2.75%). The supernatant fluid obtained after centrifugation is applied to the top of a column (0.8 × 16 cm) of AG 50-X8 (H^+ form, 200–400 mesh) and eluted with water. The highly acid sulfonic amino acids are not retained. The eluate is applied to the top of a column (0.8 × 5 cm) of Bio-Rex 5 (Cl^- form, 100–200 mesh)[4] or a column (8 × 16 cm) of Dowex 1 (Cl^- form, 200–400 mesh)[1] and eluted with a linear gradient prepared from 1 volume of 1 M acetic acid in the mixing chamber of an equal volume of 1 M acetic acid containing 2 M sodium chloride in the reservoir. The flow rate is regulated at 40 ml/hr. The amino acid fractions are determined with Gaitonde's acidic ninhydrin reagent 2 after treatment with dithiothreitol.[5] Alanine sulfodisulfane can thus be separated from related amino acids (taurine, cysteic acid, sulfocysteine, alanine sulfodisulfane, emerging from the column in that order).

Paper Electrophoresis. Paper electrophoresis successfully separates alanine sulfodisulfane from related amino acids. Using a Savant Instruments high voltage electrophoresis apparatus at 4500 V for 120 min, a formic acid–acetic acid buffer at pH 1.9 (240 ml of formic acid and 960 ml of acetic acid made to a volume of 12 liters with distilled water) is employed.[4] Alanine sulfodisulfane migrates 9.5 cm under these conditions. Separation of related strongly anionic amino acids is also achieved in pyridine–acetic acid–water (0.5:10.0:79.5) pH 3.1 at 85 V/cm for 45 min.[1] Alanine sulfodisulfane migrates 13.5 cm under these conditions.

Chemical Properties. Alanine sulfodisulfane can be cleaved with reducing agents such as dithiothreitol to yield cysteine and thiosulfate.[1] Zinc and hydrochloric acid reduce the amino acid to cysteine, hydrogen sulfide, and sulfite.[4] Oxidation of alanine sulfodisulfane with hydrogen peroxide yields cysteic acid.[4]

Biochemistry. Rat liver homogenates produce substantial amounts of alanine sulfodisulfane when incubated with cystine. The metabolic pathway requires oxygen. Based on various lines of evidence, including the inhibition of the pathway by propargylglycine [a cystathionine γ-lyase (EC 4.4.1.1) inhibitor], a biosynthetic pathway has been proposed[4] (Fig. 1).

This hypothesis is predicated, as shown, on the initiation of the sequence by the cytosolic enzyme, cystathionine γ-lyase. The metabolism of cystine by cystathionine γ-lyase (EC 4.4.1.1) has been demonstrated *in*

---

[5] M. K. Gaitonde, *Biochem J.* **104,** 627 (1967).

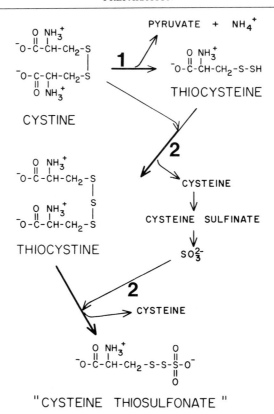

FIG. 1. Formation of "cysteine thiosulfonate." Reaction 1 is catalyzed by cystathionine γ-lyase; reactions labeled 2 are not enzymatic.

vitro and occurs by β-elimination leading to thiocysteine, ammonia, and pyruvate.[6,7] Substantial evidence for thiocysteine and thiocystine formation in this process has been presented.[7,8] It is proposed that a nonenzymatic nucleophilic attack on cystine by the thiocysteine, to form thiocystine and cysteine, continues the pathway. Evidence for such a reaction has also been reported.[8] Cysteine generated in this sequence is converted into cysteine sulfinate by the O_2-dependent enzyme, cysteine dioxygenase (EC 1.13.11.20); cysteine sulfinate is subsequently desulfinated directly[9] or is transaminated and thus desulfinated.

[6] D. Cavallini, C. DeMarco, B. Mondovi, and B. G. Mori, *Enzymologia* **22**, 161 (1961).
[7] D. Cavallini, B. Mondovi, C. DeMarco, and A. Scioscia-Santoro, *Enzymologia* **24**, 253 (1962).
[8] T. W. Szczepkowski and J. L. Wood, *Biochim. Biophys. Acta* **139**, 469 (1967).
[9] P. K. Rathod and J. H. Fellman, *Arch. Biochem. Biophys.* **238**, 435 (1985).

It was observed that rats on normal chow diet did not excrete detectable amounts of alanine sulfodisulfane. Rats on diets supplemented with 1% cystine were observed to excrete trace but detectable amounts of alanine sulfodisulfane; rats fed 5, 7, or 10% cystine-supplemented chow excreted much greater amounts of alanine sulfodisulfane. In the latter groups, the amounts excreted by individual animals were roughly equal as judged by the area and intensity of the ninhydrin spot obtained as outlined above.

[40] Thetin

By H. Kondo and Makoto Ishimoto

Dimethyl-β-propiothetin chloride, $(CH_3)_2S^+(Cl^-)\ CH_2CH_2COOH$, was first isolated in 1947 by Challenger and Simpson from a red alga, *Polysiphonia fastigiata*. The compound evolves dimethyl sulfide on exposure of the alga to air.[1]

Chemical Synthesis of Propiothetin Bromide

The procedure is based on that reported by Challenger and Simpson.[1]

$$BrCH_2CH_2COOH + (CH_3)_2S \rightarrow (CH_3)_2S^+(Br^-)\ CH_2CH_2COOH$$

β-Bromopropionic acid (15.3 g, 0.1 mol) is mixed with dimethyl sulfide (6.2 g, 0.1 mol) in a 50-ml round-bottomed flask. The mixture is refluxed at 55–60° for 6 hr and allowed to solidify on cooling. The solid cake is triturated and washed with dry ether to remove unchanged reactants. The white solid is crystallized from absolute alcohol and dried over phosphorus pentoxide under reduced pressure. The product melts at 112–113°. This compound was identified as dimethyl-β-propiothetin bromide by element analysis and mass spectroscopy.

Comments. Challenger and Simpson obtained a melting point of 112–114°. They mentioned that β-chloropropionic acid and dimethyl sulfide did not react under the conditions described here and that dimethyl-β-propiothetin chloride could be obtained from the bromide and silver chloride.[1] Chemical synthesis of acetothetin bromide from bromoacetic acid and dimethyl sulfide, or its chloride analog, has been noted.[2,3]

---

[1] F. Challenger and M. I. Simpson, *J. Chem. Soc., Part 3*, p. 1591 (1948).
[2] C. Brown and E. A. Letts, *Ber. Dtsch, Chem. Ges.* **7**, 695 (1874).
[3] V. du Vigneaud, A. M. Moyer, and J. P. Chendler, *J. Biol. Chem.* **174**, 477 (1948).

[41] Cysteine S-Conjugates

By Patrick Hayden, Valentine H. Schaeffer, Gerald Larsen, and James L. Stevens

Introduction

As one of the amino acids of the tripeptide glutathione, cysteine provides the nucleophilic sulfur used in the formation of S-glutathione conjugates.[1] The corresponding S-cysteine conjugates are produced from the glutathione conjugates following the cleavage of the glutamate and glycine.[2] Detoxication is achieved by N-acetylation of the xenobiotic S-cysteine conjugate and excretion of the resulting mercapturate.[3] In addition, S-cysteine conjugates are metabolized to other species by a number of pathways, some of which may lead to toxicity.[4]

Since S-cysteine conjugates of most xenobiotics are not readily available, general methods are described here for their synthesis and purification with special emphasis on radiolabeled compounds. An extensive list of physical data on synthetic S-cysteine conjugates will not be presented, although characterization of standards for each conjugate should be carried out prior to performing the radiolabeled synthesis.

General Considerations

Each reaction described relies on the attack of the nucleophilic cysteine sulfur on an electrophilic center. Since the cysteine molecule has three potential nucleophiles—sulfur, nitrogen, and oxygen—the selectivity of the nucleophilic addition reaction must be considered. Of the three, sulfur is the most nucleophilic followed closely by nitrogen, with the oxygen of the carboxylate last. The simplest solution to solving the specificity problem is to begin with a radiolabeled electrophile and add an excess of cold cysteine. Under such conditions, the reaction can ap-

---

[1] D. J. Reed, *in* "Bioactivation of Foreign Compounds" (M. W. Anders, ed.), p. 71. Academic Press, Orlando, Florida, 1985.

[2] W. B. Jakoby, J. L. Stevens, M. W. Duffel, and R. A. Weisiger, *Rev. Biochem. Toxicol.* **6**, 97 (1984).

[3] M. W. Duffel and W. B. Jakoby, *Mol. Pharmacol.* **21**, 444 (1982).

[4] Published data [S. P. James and D. Needham, *Xenobiotica* **3**, 207 (1973); J. E. Bakke *et al.*, *Science* **210**, 433 (1980)], as well as our unpublished work show that S-cysteine conjugates are subject to sulfur oxidation, oxidation to α-keto acids and to decarboxylation. See footnote 9 for references to toxic reactions of S-cysteine conjugates.

proach quantitative radiochemical yield with a single substitution at sulfur. Unfortunately, radiolabel in cysteine is more readily available than in most xenobiotics of interest. Therefore, adding an excess of cysteine is not practical because the radiochemical yield will be poor. In order to avoid substitution at nitrogen, and to maximize yield from radiolabeled cysteine, the electrophile to cysteine stoichiometry could be limited to near unity. This would improve radiochemical yield but also increase the probability of unwanted substitution at nitrogen. Separation of the unwanted N-conjugates from the S-conjugates is more difficult than separation of S-conjugate from excess cysteine and cystine.

Another consideration is whether to use radiolabeled cysteine or cystine. Cysteine is generally shipped as an aqueous solution in 10 mM dithiothreitol (DTT); DTT must be removed since it will participate in the reaction and form an undesirable thioether. Although DTT can be extracted with ethyl acetate from the aqueous solution at pH 1, we prefer to use radiolabeled cystine. Since, the actual amount of labeled cystine is negligible compared to the carrier cysteine added to the reaction mixture, disulfide interchange will rapidly reduce the labeled cystine, forming radiolabeled cysteine, and obviate the need for an extraction step.

Regardless of the method used, trial syntheses with unlabeled reactants are advised. When scaling down reactions for radiolabeled syntheses, the stoichiometry may differ from large-scale preparations. The consumption of base equivalents or adsorption of reactants to surfaces can decrease yields; therefore, it may be necessary to add greater than theoretical amounts of reactants.

Finally, the choice of the reaction medium is limited since cysteine is freely soluble only in polar solvents. Since many xenobiotic S-cysteine conjugates are derived from hydrophobic xenobiotics, solubilizing both cysteine and the xenobiotic of interest in polar solvents is difficult. Methanol is a reasonable compromise since nonpolar molecules will be soluble, and cysteine free base can be dissolved in dry methanol as a 1% solution with the addition of solid sodium.

Thin-Layer Chromatography. The reaction progress can be followed easily by thin-layer chromatography on silica gel plates in either System A (2-propanol:acetic acid:water, 7:1:3) or System B (1-butanol:acetic acid:water, 4:1:5, upper phase). R_f values in System A and B, respectively, are as follows: cysteine, 0.38 and 0.12; cystine, 0.15 and 0.1. The R_f values for cysteine conjugates will be lower in System B, but this solvent is most useful for separations on columns of silica gel. The propanol system is faster and, therefore, desirable for following the progress of a reaction. Cysteine, cystine, and all the conjugates can be detected by spraying the TLC plate with 0.1% ninhydrin in acetone and heating gently to visualize the conjugate. Iodine vapor, ultraviolet absorbance, or detection of divalent sulfur compounds with the system of Knight and Young[5] may also be useful. More than one method of detection is desirable since ninhydrin will not react with the N-linked derivatives and other possible contaminants.

Synthesis using Methanol and Sodium Methoxide as Base

This procedure, adapted from a method for the synthesis of mercapturic acids,[6] can be used for the synthesis of most cysteine or glutathione conjugates when an appropriate substrate for nucleophilic addition is available. The ease of preparing and handling dry, degassed methanol makes this procedure preferable to those using liquid ammonia.

S-Benzyl-L-cysteine. To a 50-ml round-bottomed flask, add [^{35}S]cystine (1 mCi, generally shipped in 0.1 N HCl) and 10 mg cysteine (83 μmol). Add an additional 5 ml of 0.1 N HCl to dissolve all the cysteine and bring the solution to dryness under reduced pressure; dry the contents of the flask further over phosphorus pentoxide. Prepare a 0.06% solution of sodium methoxide by adding sodium metal to dry methanol which has been flushed with nitrogen. Transfer 10 ml of the sodium methoxide solution (430 μmol of sodium methoxide) to the flask. Add a stirring bar, and flush the flask with nitrogen after capping with a serum stopper. Stir the solution until cysteine · HCl is dissolved; add additional sodium methoxide to attain solution if necessary. Add 10 μl of benzyl bromide (84 μmol) through the stopper with a microliter syringe. Stir the reaction for 2 hr before removing the cap and reducing the contents to dryness under reduced pressure. Dissolve the residue in water and adjust the pH to about 9–10. Although cysteine conjugates are more stable under less basic conditions, the pH cannot be reduced to the range of 5–7 since isoelectric precipitation occurs.[7] The product can be purified using the HPLC procedure described below and stored as an aqueous solution at $-70°$. Radiochemical yield of 80% with a purity of >98%, determined by TLC and HPLC, can be achieved (*S*-benzyl-L-cysteine, $R_f = 0.74$, System A; $R_f = 0.39$, System B). The unlabeled compound is available commercially.

The methanol procedure is very useful when a labeled electrophile such as [^{14}C]ethyl iodide is the labeled reactant. In this case, increase the cysteine to a 3-fold molar excess, with respect to the ethyl iodide, maintaining constant the ratio of sodium methoxide to cysteine. Separate the radiolabeled product from excess cysteine using silica gel or reversed-

---

[5] R. H. Knight and L. Young, *J. Org. Chem.* **29**, 2203 (1958).
[6] P. J. van Bladderen, W. Buys, D. D. Breimer, and A. van der Gen, *Eur. J. Med. Chem.* **15**, 495 (1980).
[7] Isoelectric precipitation is an effective method for crystallizing *S*-cysteine conjugates.

phase chromatography. Radiochemical yields of 95 and purity of 99% can be achieved.

S-2-Chloro-1,1,2-trifluoroethyl-L-cysteine. Cysteine conjugates of toxic gases[8] such as 2-chloro-1,1,2 trifluoroethylene can also be prepared in methanol. However, careful control of base addition is necessary to avoid unwanted side reactions; base-catalyzed polymerization of the vinyl monomers can occur.

The following procedure was found to provide the best yields: Radiolabeled cystine (in 0.1 N HCl) is added to a 25 ml round bottomed flask and evaporated to dryness. An aqueous solution of unlabelled cysteine·HCl is prepared in a separate flask at a concentration such that the desired amount of cysteine for the reaction can be transferred in a volume of 0.5 ml. The pH of the solution is adjusted to 9.0 with sodium hydroxide. This should be 3 : 1 stoichometry of base with respect to cysteine. Aqueous base gave better yields in this reaction, presumably because the presence of water facilitates protonation of the addition intermediately to form product rather than further polymerization.

Cysteine solution, 0.5 ml, is transferred to the flask containing the radiolabeled cystine. The mixture is swirled to achieve solution and reduction of the labeled cystine, after which 10 ml of methanol are added. The flask is capped with a serum stopper and flushed with nitrogen by use of 19 gauge needles inserted through the stopper.

Finally, the gas is bubbled through the reaction mixture, again using 19-gauge needles inserted through the stopper to introduce gas and vent the flask. The solubility and reactivity of the gas will influence the length of time necessary to achieve good yields. This should be determined first in a trial preparation before attempting the radiolabeled synthesis.

For 2-chloro-1,1,2-trifluoroethylene, 5 min of bubbling is sufficient. The reaction is continued for a total of 30 min, after which the mixture is evaporated to dryness, dissolved in 0.5 ml of water, and purified by HPLC using one of the two systems described in the purification section. The product, *S*-2-chloro-1,1,2-trifluoroethyl-L-cysteine was achieved in 50–80% yields (based on ^{35}S-cystine) by this method with 99% radiochemical purity. For TLC, R_f = 0.72 in System A, R_f = 0.39 in System B; m.p. = 150–152° (uncorrected).

Synthesis using Liquid Ammonia and Sodium Amide

S-1,2-Dichlorovinyl-L-cysteine is a commonly used model for studying cysteine conjugate-induced renal toxicity.[9,10] The synthetic method is

[8] D. R. Dohn, A. J. Quebbeman, R. F. Borch, and M. W. Anders, *Biochemistry* **24**, 5137 (1985).
[9] A. E. Elfarra, I. Jakobson, and M. W. Anders, *Biochem. Pharmacol.* **33**, 3729 (1984).
[10] J. L. Stevens, P. Hayden, and G. Taylor, *J. Biol. Chem.* **261**, 13076 (1986).

based on the procedure of McKinney et al.[11] Most of the reactions discussed here can be performed in liquid ammonia with sodium amide as base, but, due to the ease of handling, the methanol procedure is preferred in most instances.

S-1,2-Dichlorovinyl-L-cysteine. Dissolve 5 mg of cysteine (41 μmol) and the [^{14}C]cystine in 5 ml of 0.1 N HCl in a 25-ml round-bottomed flask, and dry it as described for the methanol procedure. Place a small glass stirring bar in the flask and insert a sleeve-type serum stopper in the neck. Do not use a Teflon stirring bar since sodium amide will react with Teflon. Place the flask in a Dry Ice–ethanol bath with a stirring plate underneath and stir continuously while condensing 10–15 ml of liquid ammonia into the flask through a 19-gauge needle, using a second needle as a vent. When all of the cysteine has dissolved, remove the flask from the Dry Ice bath and add 5 mg (220 μmol) of sodium after removing the cap. Removing the flask from the Dry Ice bath assures a slight positive pressure from the escaping ammonia, thus preventing water from condensing into the flask when sodium is added. Return the flask to the Dry Ice bath and continue to stir the ammonia solution. As the sodium reacts with the ammonia to form sodium amide, a characteristic blue color develops. When the blue color dissipates, add 4 μl of trichloroethylene (45 μmol) through the serum stopper with a microliter syringe.[12] Stir the reaction mixture for 30 min, leaving the vent in place, and remove the flask from the Dry Ice bath. Remove the cap and allow the ammonia to evaporate, aided by a gentle stream of nitrogen directed into the flask.

The stoichiometry of 5:1:1 for sodium:cysteine:trichloroethylene, respectively, was determined as optimum from preliminary trials and differs from the 3:1:1 stoichiometry for larger scale reactions starting with cysteine·HCl. The method results in radiochemical yields of 70% with radiochemical purity of 99% as determined by HPLC. Data for *S*-1,2-dichlorovinyl-L-cysteine are as follows: $R_f = 0.72$, System A; $R_f = 0.38$, System B; m.p. = 155–157° (uncorrected).

Other Synthetic Routes

Although the simplest methods have been described, alternate routes to *S*-cysteine conjugates may be useful under specific circumstances.

Beginning with *N*-acetyl-L-cysteine can be effective when a starting material which is soluble in a variety of organic solvents is desired. *N*-

---

[11] L. L. McKinney, J. C. Picken, R. B. Weakely, A. D. Eldridge, R. E. Campbell, J. C. Cowan, and H. E. Beister, *J. Am. Chem. Soc.* **81**, 909 (1959).

[12] Add the trichloroethylene soon after the blue dissipates, since cysteine will oxidize to cystine over time.

Acetyl derivatives can be prepared from radiolabeled cysteine by acetylation with acetyl chloride or acetic anhydride. N-Acetylcystine is reduced to N-acetylcysteine by cysteine or glutathione in water, and the free acid of N-acetylcysteine is extracted into ethyl acetate under mild acidic conditions. N-Acetylcysteine conjugates, i.e., the mercapturates, can be prepared by one of the procedures described above, or in another organic solvent with the appropriate base. Using N-acetyl derivatives has the disadvantage that a blocking step is required; the conjugate must also be sufficiently stable to survive removal of the N-acetyl blocking group.

When the appropriate electrophile is not available, but the mercaptan is, e.g., benzyl mercaptan instead of benzyl bromide, it is sometimes possible to prepare the S-cysteine conjugate from N-carbobenzyloxy-O-tosylserine benzyl ester (NTSB). NTSB is prepared from commercially available N-carbobenzyloxy-serine by acid-catalyzed esterification in benzene with p-toluenesulfonic acid as catalyst. After tosylation with tosyl chloride, the O-tosyl can be displaced by the mercaptan. Two disadvantages are obvious: the attack by the mercaptan may encounter steric hindrance, and synthesis requires two steps for the preparation of starting material as well as a third deprotecting step. Details of the esterification and tosylation procedures are available.[13]

A similar approach can be taken starting with β-chloro-L-alanine. Under basic conditions, the mercaptan can displace chlorine from the β-carbon to yield the cysteine conjugate. The synthesis of radiolabeled β-chloro-L-alanine has been described.[14]

Purification of Labeled Cysteine Conjugates

Following synthesis, the product can be purified by reversed phase HPLC, silica gel chromatography, or chromatography on LH-Sephadex. The HPLC system is also useful for assessing purity, particularly when supplemented by TLC. Purity can be assessed by TLC and autoradiography or by radiochemical detection and HPLC.

Reversed-Phase HPLC Separation. A Waters μBondapack C-18 column, 3.9 × 300 mm (Waters Associates, Milford, MA) operating at a 1.0 ml/min flow rate, was used. Two solvent systems have given successful separations.

SYSTEM A. Potassium phosphate, 0.1 M, pH 3.0/methanol. Sample injections were made at 0% methanol and hydrophobic compounds were eluted by increasing the methanol concentration. The following gradient

---

[13] J. P. Greenstein and M. Winitz, "Chemistry of the Amino Acids." Wiley, New York, 1961.
[14] C. Walsh, A. Schonbrunn, and R. H. Abeles, *J. Biol. Chem.* **246,** 6855 (1971).

program was used: 0–5 min, 0% methanol; 5–15 min, 0–40% methanol; >15 min, 40% methanol. Concentrations greater than 60% methanol can cause precipitation of phosphate on the column.

SYSTEM B. 0.1% Trifluoroacetic acid/acetonitrile water. The aqueous trifluoroacetic acid solution must be made fresh daily. Gradient elution with acetonitrile can be used in a manner similar to System A, with the further advantage that concentrations of up to 100% acetonitrile are possible. Less hydrophobic compounds that are not retained with System A are retained well with System B and, as TFA can be removed under reduced pressure, further separation of purified compounds from buffer is not a problem.

These solvents can also be used to assay conjugates in biological systems. The S-1,2-dichlorovinyl conjugates of glutathione, cysteine, N-acetylcysteine, and cysteinylglycine, for example, have been purified from incubation mixtures by this means.

As exact retention times will vary depending on the specific column or HPLC system used, all columns should be calibrated with standards of the appropriate conjugates.

Silica Gel Chromatography. Best results are obtained with small particle size (10 μm) silica. Low pressure glass column systems or preparative HPLC columns are most useful. Elute the column with System B described above (Thin-Layer Chromatography). Preliminary chromatography on TLC will serve as a guide for the separation.

LH-20 Sephadex Chromatography. A column of LH-20 Sephadex can be useful in the absence of silica gel chromatography or HPLC.[15] Pour the column in water and equilibrate with water. Apply the conjugate in water, and elute by continued washing with water.

Desalting on Amberlite XAD-2. If it is necessary to desalt the conjugate prior to additional purification, Amberlite XAD-2 resin may be useful. This procedure has been employed for the cysteine conjugate of propachlor.[15]

Safety. The dangers inherent with radioactive chemistry will not be covered here but should not be overlooked. However, two safety comments are important. First, the liquid ammonia reactions need to be monitored so that pressure does not develop in the flask. Second, volatile sulfur compounds can be generated during the reactions; therefore, care needs to be exercised when reducing solvent under vacuum.

---

[15] J. C. Pekas, G. L. Larsen, and V. J. Feil, *J. Toxicol. Environ. Health* **5,** 653 (1979).

[42] Chromogenic and Fluorigenic Substrates for Sulfurtransferases

By MARY R. BURROUS, JOHN LANE, AIKO WESTLEY, and JOHN WESTLEY

The enzymes rhodanese (thiosulfate:cyanide sulfurtransferase; EC 2.8.1.1, thiosulfate sulfurtransferase) and thiosulfate reductase catalyze reactions in which a sulfane sulfur atom is transferred from inorganic thiosulfate anion or other suitable donor substrate to a nucleophilic acceptor substrate, which is typically CN^- or a thiolate or sulfinate anion. Bovine liver rhodanese has been much studied, and its kinetic[1] and chemical[2] mechanisms as well as its primary[3] and detailed X-ray crystallographic[4] structures have all been worked out. There has been much less investigation of thiosulfate reductase, but it has been purified to apparent homogeneity from yeast,[5] and both its formal mechanism and some of its molecular parameters have been determined.[6,7] The phylogenetic,[8] tissue,[9] and subcellular[10] distributions of the sulfurtransferases have also been examined, and the whole body of data has been reviewed at intervals in various contexts.[11-16]

In view of the substantial volume of work done with these enzymes, it is striking that nearly all of it has been carried out with discontinuous

[1] M. Volini and J. Westley, *J. Biol. Chem.* **241**, 5168 (1966).
[2] P. Schlesinger and J. Westley, *J. Biol. Chem.* **249**, 780 (1974).
[3] J. Russell, L. Weng, P. S. Keim, and R. L. Heinrikson, *J. Biol. Chem.* **253**, 8102 (1978).
[4] J. H. Ploegman, G. Drent, K. H. Kalk, and W. G. J. Hol, *J. Mol. Biol.* **123**, 557 (1978).
[5] T. R. Chauncey and J. Westley, *Biochim. Biophys. Acta* **744**, 304 (1983); also see this volume [60].
[6] L. C. Uhteg and J. Westley, *Arch. Biochem. Biophys.* **195**, 211 (1974).
[7] T. R. Chauncey and J. Westley, *J. Biol. Chem.* **258**, 15037 (1983).
[8] H. Schiewelbein, R. Baumeister, and R. Vogel, *Naturwissenschaften* **56**, 416 (1969).
[9] A. Koj, M. Michalik, and H. Kasperczyk, *Bull. Acad. Pol. Sci., Ser. Sci. Biol.* **25**, 1 (1977).
[10] A. Koj, J. Frendo, and L. Woitczak, *FEBS Lett.* **57**, 42 (1975).
[11] J. Westley, *Adv. Enzymol.* **39**, 327 (1973).
[12] B. Sörbo, in "Metabolic Pathways" (D. M. Greenberg, ed.), 3rd ed., Vol. 7, p. 433. Academic Press, New York, 1975.
[13] J. Westley, in "Bioorganic Chemistry" (E. E. Van Tamelen, ed.), Vol. 1, p. 371. Academic Press, New York, 1977.
[14] J. Westley, in "Enzymatic Basis of Detoxication" (W. B. Jakoby, ed.), Vol. 2, p. 245. Academic Press, New York, 1980.
[15] J. Westley, in "Cyanide in Biology" (B. Vennesland, E. E. Conn, C. J. Knowles, J. Westley, and F. Wissing, eds.), p. 61. Academic Press, London, 1981.
[16] J. Westley, H. Adler, L. Westley, and C. Nishida, *Fundam. Appl. Toxicol.* **3**, 377 (1983).

assay methods, mostly by one or another variant of Sörbo's procedure based on the formation of the red $Fe(SCN)_6^{3-}$ complex ion in strong acid solution.[17] An exception is Volini's work[1] that established the kinetic mechanism of rhodanese action by a procedure based on the absorbance of lipoate at 335 nm. This method has not been adopted extensively, probably because of the inconvenience of handling dihydrolipoate solutions strictly anaerobically. A continuous method that relies on the difference in ultraviolet absorbance between inorganic thiosulfate and sulfite has seen less use,[18,19] as has a procedure based on the absorbance difference between benzene thiosulfonate and benzene sulfinate[20]; in both cases, the wavelength at which the maximum effect occurs is so low (<250 nm) that many interferences intrude. Two procedures based on visible absorbance changes in multiple coupled redox indicator systems have also been described but not extensively used.[21,22] A pyridine nucleotide-based procedure for thiosulfate reductase, which utilizes glutathione reductase as coupling agent, contributed strongly to the analysis of the formal mechanism of that sulfurtransferase,[7] but the method involves an annoying inherent lag period and also cannot be adapted for rhodanese.

The foregoing considerations, together with Sörbo's original observation of an absorbance difference between aromatic thiosulfonate anions and the corresponding sulfinates,[20] prompted an effort to develop chromogenic and fluorigenic substrates for the sulfurtransferases. The former are based on the azo dye series in which the common indicator dye methyl orange is the sulfonate. The latter are related in the same way to the common fluorescent materials that contain dansyl groups. Both developments provide practical, continuous assay procedures for both rhodanese and thiosulfate reductase.

Chromogenic Substrates: 4-(Dimethylamino)-4′-azobenzene (DAB) Sulfinate and Thiosulfonate Anion

$$DAB\text{-}SO_2^- + SSO_3^{2-} \rightleftharpoons DAB\text{-}S(O_2)S^- + SO_3^{2-} \quad (1)$$
$$DAB\text{-}S(O_2)S^- + CN^- \rightarrow DAB\text{-}SO_2^- + SCN^- \quad (2)$$
$$DAB\text{-}S(O_2)S^- + GS^- \rightleftharpoons DAB\text{-}SO_2^- + GSS^- \quad (3)$$

[17] B. Sörbo, *Acta Chem. Scand.* **7,** 1129 (1953).
[18] B. Davidson and J. Westley, *J. Biol. Chem.* **240,** 4463 (1965).
[19] S. F. Chow, P. M. Horowitz, J. Westley, and R. Jarabak, *J. Biol. Chem.* **260,** 2763 (1985).
[20] B. Sörbo, *Acta Chem. Scand.* **16,** 243 (1962).
[21] A. J. Smith and J. Lascelles, *J. Gen. Microbiol.* **42,** 357 (1966).
[22] C. Cannella, R. Berni, and G. Ricci, *Anal. Biochem.* **142,** 159 (1984).

Syntheses. DAB sulfinic acid is made by reduction of commercial DAB sulfonyl chloride (Aldrich) with excess inorganic sulfite,[23] following the general method of Kulka.[24] Five grams of solid DAB-S(O$_2$)Cl is added

$$\text{DAB-S(O}_2\text{)Cl} + \text{SO}_3^{2-} + \text{H}_2\text{O} \rightarrow \text{DAB-SO}_2\text{H} + \text{SO}_4^{2-} + \text{HCl}$$

to 100 ml of well-stirred aqueous 2.5 M Na$_2$SO$_3$ at 65°. The reaction mixture is stirred at this temperature until thin-layer chromatographic analysis (see below) shows the reaction to be essentially complete (~2.5 hr). The mixture is acidified (pH < 1) with HCl, chilled, and centrifuged. The sedimented solids are extracted with 50 ml of hot ethanol, and the product is allowed to crystallize from the extract overnight at −20°. Recrystallization is from ethanol by the same procedure. The twice recrystallized sulfinic acid, which contains only traces of the sulfonate (i.e., methyl orange) is stored under dry nitrogen. The yield is typically greater than 90%. At its visible absorption maximum (460 nm) in aqueous buffer at pH 5, DAB sulfinate has a molar absorptivity of $1.88 \times 10^4 \, M^{-1} \, \text{cm}^{-1}$.

This product as isolated is suitable for use as a sulfurtransferase acceptor substrate [Eq. (1)]. For critical applications, it can be cleared of its slight contamination with the sulfonate by isocratic column chromatography on silica with 5:2 ethyl acetate, 65% ethanol (v/v) in water as the solvent. In this system, as in the related thin-layer system described below, the sulfinate has a lower R_f value than the sulfonate.

DAB thiosulfonate anion is made by treatment of DAB sulfonyl chloride (Aldrich) with 1 analytical equivalent of inorganic sulfide,[23] following modifications of the general method originated by Traeger and Linde.[20,25,26] Five grams of solid DAB-S(O$_2$)Cl is added to 50 ml of well-

$$\text{DAB-S(O}_2\text{)Cl} + \text{S}^{2-} \rightarrow \text{DAB-S(O}_2\text{)S}^- + \text{Cl}^-$$

stirred aqueous 0.31 M Na$_2$S at 65°. The reaction mixture is stirred at this temperature until thin-layer chromatographic analysis (see below) shows the reaction to be essentially complete (~2 hr). The mixture is then flushed with N$_2$ and dried under reduced pressure. The solid residue is extracted twice with 25-ml portions of hot ethanol, and the product is allowed to crystallize from the pooled extracts overnight at −20°. Recrystallization is from ethanol by the same procedure. The recrystallized sodium DAB thiosulfonate, which contains traces of the sulfinate, is stored under dry nitrogen. The yield is typically 60%. At its visible absorption

---

[23] M. R. Burrous and J. Westley, *Anal. Biochem.* **149**, 66 (1985).
[24] M. Kulka, *J. Am. Chem. Soc.* **72**, 1215 (1950).
[25] J. Traeger and O. Linde, *Arch. Pharm. (Weinheim, Ger.)* **239**, 121 (1901).
[26] R. Mintel and J. Westley, *J. Biol. Chem.* **241**, 3381 (1966).

maximum (466 nm) in pH 5 aqueous buffer, DAB thiosulfonate anion has a molar absorptivity of $1.98 \times 10^4\ M^{-1}\ cm^{-1}$.

This product as isolated is suitable for use as a sulfurtransferase donor substrate [Eqs. (2) and (3)]. For critical applications, it can be cleared of its slight contamination with the sulfinate by isocratic column chromatography on silica with 5:2 ethyl acetate, 65% ethanol (v/v) in water as the solvent. In this system, the thiosulfonate has a higher R_f value than either the sulfinate or the sulfonate.

Characterization. The substrates synthesized by these methods are most conveniently characterized by thin-layer chromatography on 25 × 100 mm glass plates precoated with a hard layer of silica gel (Analtech).[27] Small aliquots (1 μl) of either aqueous or alcoholic solutions are applied, and, after air drying, plates are subjected to ascending chromatography at room temperature in 4:1:1 ethyl acetate, water, methanol. The chromatography requires 15–20 min, and the thiosulfonate, sulfonate, and sulfinate give well-separated, compact orange spots with R_f values of 0.61, 0.54, and 0.48, respectively. For greater viewing sensitivity, dried chromatograms may be exposed to HCl vapor, which turns the spots bright red.

Sulfurtransferase Assays.[23] DAB thiosulfonate and sulfinate anions differ maximally in visible absorbance at 500 nm, where the molar absorbance of the former is approximately 3000 greater than that of the latter. This difference is the basis for the three sulfurtransferase assay systems indicated in Eqs. (1)–(3). All of these assays are carried out in 0.10 M acetate buffer, pH 5, containing glycine at 0.1 M, conditions that help stabilize the enzymes. For use in routine assays, DAB substrates are made up in concentrated stock solutions of which microliter volumes are then added to assay buffer. Stock solutions may be stored overnight at 0–5° after brief flushing with N_2. With final DAB substrate concentrations in the range of 50 to 150 μM, the reactions can be followed readily by continuous recording at 500 nm with a spectrophotometer set at a sensitivity of 0.100 absorbance units full scale. All three assays give initial velocities proportional to enzyme concentration.

For assay of rhodanese by transfer of sulfane sulfur from $S_2O_3^{2-}$ to DAB sulfinate [Eq. (1)], 1 mM $Na_2S_2O_3$ in 150 μM DAB-SO_2^- is a suitable system for observing the time course of reaction catalyzed by rhodanese at 20–200 nM. This is a system in which absorbance increases with time. A colorless organic thiosulfonate anion, e.g., methane thiosulfonate, may be used instead of $S_2O_3^{2-}$, but the reaction is then subject to rather strong substrate inhibition whenever the ratio of methane thiosulfonate concentration to DAB sulfinate concentration exceeds 5.

[27] A. Westley and J. Westley, *Anal. Biochem.* **142**, 163 (1984).

For assay of rhodanese by cyanolysis of DAB thiosulfonate [Eq. (2)], 1 mM KCN in 50 μM DAB-S(O$_2$)S$^-$ provides a suitable system for observing the progress of the reaction catalyzed by rhodanese in this same concentration range. In this system, absorbance decreases with time.

For assay of thiosulfate reductase by the catalyzed thiolysis of DAB thiosulfonate [Eq. (3)], 1 mM GSH in 50 μM DAB-S(O$_2$)S$^-$ provides a suitable system for observing the progress of the reaction catalyzed by the enzyme at concentrations in the range of 0.1 to 1.0 μM. In this system, absorbance decreases with time.

Fluorigenic Substrates: 5-Dimethylamino-1-naphthalene (DAN) Sulfinate and Thiosulfonate Anion[28]

$$\text{DAN-SO}_2^- + \text{SSO}_3^{2-} \rightleftharpoons \text{DAN-S(O}_2)\text{S}^- + \text{SO}_3^{2-} \quad (4)$$
$$\text{DAN-S(O}_2)\text{S}^- + \text{CN}^- \rightarrow \text{DAN-SO}_2^- + \text{SCN}^- \quad (5)$$
$$\text{DAN-S(O}_2)\text{S}^- + \text{GS}^- \rightleftharpoons \text{DAN-SO}_2^- + \text{GSS}^- \quad (6)$$

The same procedures, outlined above for synthesis of chromogenic substrates, can be used to obtain fluorigenic substrates, with the sole alteration that an equivalent amount of 5-dimethylamino-1-naphthalenesulfonyl chloride (dansyl chloride) is substituted for the dabsyl chloride used as starting material. Moreover, the same thin-layer chromatographic procedures can be used to characterize the products, with the difference that the materials are located on developed plates by fluorescence under long-wavelength UV illumination rather than by color. The R_f values for the thiosulfonate, sulfonate, and sulfinate are 0.60, 0.47, and 0.41, respectively. Column chromatography on silica can also be used, as before, for additional purification of substrates in critical applications.

Moreover, all of the assay systems outlined for the chromogenic substrates can also be used, with the exception that fluorescence emission at 500–510 nm rather than absorbance is monitored. The excitation wavelength is 320–330 nm. Since DAN sulfinate yields a much larger fluorescence output than DAN thiosulfonate anion, the direction of change of the signal monitored in each of these assays is the opposite of that in the corresponding chromogenic assay. Thus, reaction (4) proceeds with a decrease in recorded fluorescence, while reactions (5) and (6) yield fluorescence that increases with time. All of the assays give initial velocities proportional to enzyme concentration.

Acknowledgments

This work was supported by research grants from the National Science Foundation (DMB-8512836) and the National Institutes of Health (GM-30971).

[28] J. Lane and J. Westley, unpublished observations.

[43] Selenocysteine

By Hidehiko Tanaka and Kenji Soda

$$\begin{array}{l} \text{Se—CH}_2\text{—CH(NH}_2\text{)—COOH} \\ | \\ \text{Se—CH}_2\text{—CH(NH}_2\text{)—COOH} \end{array}$$

Selenocysteine (2-amino-3-hydroselenopropionic acid) has been demonstrated as an integral moiety in polypeptide chains of several mammalian and bacterial enzymes. Selenocysteine is highly oxidizable; its synthesis is very difficult under aerobic conditions. However, selenocysteine can be prepared easily from its oxidized form, selenocystine, with sodium borohydride or with thiols such as dithiothreitol, 2-mercaptoethanol, and 2,3-dimercapto-1-propanol.[1]

Selenocystine was first prepared chemically by Fredga in 1936.[2] Methods for its synthesis were reviewed by Zdansky,[1] but most are tedious and time consuming; they involve multiple steps with a low overall yield. A facile procedure for synthesis of selenocystine with a high yield has been developed, which is applicable to synthesis of the labeled compounds on a small scale.[3] It is based on the reaction of 3-chloroalanine with disodium diselenide and represents a modification of the method of Klayman and Griffin for synthesis of dialkyl diselenides.[4]

Procedures

Note: Selenium compounds are toxic and extremely irritating to tissues. Procedures should be conducted in a hood and gloves should be worn to avoid contact with the reagents.

Disodium Diselenide. Elemental powdered selenium is reduced to disodium diselenide with sodium borohydride.[4]

Selenocystine. To 60 ml of 1 M disodium diselenide solution in a nitrogen-flushed, stoppered 300-ml three-necked flask is added dropwise 3-chloro-DL-alanine[5] (3.2 g, 26 mmol) dissolved in 50 ml of water (pH 9.0), over a period of 2 hr. After stirring at 37° for 16 hr, the pH is lowered to 2

---

[1] G. Zdansky, in "Organic Selenium Compounds: Their Chemistry and Biology" (D. L. Klayman and W. H. H. Gunther, eds.), p. 579. Wiley, New York, 1973.
[2] A. Fredga, *Sven. Kem. Tidskr.* **48**, 160 (1936).
[3] P. Chocat, N. Esaki, H. Tanaka, and K. Soda, *Anal. Biochem.* **148**, 485 (1985).
[4] D. L. Klayman and S. T. Griffin, *J. Am. Chem. Soc.* **95**, 197 (1973).
[5] 3-Chloroalanine is commercially available; it can also be prepared from L- or DL-serine by the method of C. Walsh, A. Schonbrumm, and R. T. Abeles, *J. Biol. Chem.* **246**, 6855 (1971).

with 6 N HCl, and 218 mg (4 mmol) of hydroxylamine hydrochoride are added to reduce contaminating elemental selenium to hydrogen selenide. The reaction mixture is flushed with nitrogen gas for 2 hr, and the exhaust gas was passed twice through saturated solution of lead acetate to trap hydrogen selenide. Elemental selenium precipitated in the solution was removed by filtration. The pH of the yellow filtrate was adjusted between 6.0 and 6.5 with 6 N NaOH. After standing overnight at 4°, yellow crystals of selenocystine were collected. Thin-layer chromatography in n-butanol:acetic acid:water (4:1:1) showed formation of only a trace amount of selenolanthionine, but 3-chloroalanine was not observed. The crude product (2.9 g) was dissolved in 50 ml of 1 N HCl, and insoluble materials were removed by filtration. After adjustment to pH 6.0–6.5 as described above, the solution was left overnight at 4° to crystallize selenocystine; selenolanthionine remains in solution under these conditions. DL-Selenocystine (2.71 g, 62% yield) was obtained as thin hexagonal plates, m.p. 184–185° (with decomposition).

L-Selenocystine has been prepared from 3-chloro-L-alanine in the same way. The optical purity of the selenocystine synthesized was above 99.6% when measured by reversed-phase high-performance liquid chromatography after reduction with dithiothreitol and conversion to Se-ethylselenocysteine with ethyl iodide.[3] The retention times of the enantiomers of Se-ethylselenocysteine were 10.3 (D-isomer) and 18.2 min (L-isomer) under the following conditions[6]: column, C_{18} octadecyl silica gel column (4.6 × 150 mm); temperature, 40°; mobile phase, 17 mM L-proline containing 8 mM cupric acetate; flow rate, 1.0 ml/min. The observed $[\alpha]_D^{25}$ of +173° (1% solution in 0.5 N HCl), is in good agreement with the reported value.[2] L-Selenocystine thus prepared is a starting material of various Se-substituted L-selenocysteines.[2]

Note on Procedures

It should be noted that selenolanthionine is formed as a by-product by the procedure. To suppress the formation of selenolanthionine, the amount of disodium diselenide should be raised to more than twice the total amount of 3-chloroalanine added. 3-Chloroalanine is known to undergo α,β-elimination at alkaline pH. Therefore, 3-chloroalanine should be added dropwise and slowly to the disodium diselenide solution. The present procedure is applicable on a smaller scale to preparation of the labeled compounds with ^{14}C and ^{75}Se in the alanyl or the diselenide products.

[6] E. Gil Av, A. Tishbee, and P. E. Hare, *J. Am. Chem. Soc.* **73**, 2659 (1980).

Selenohomocystine is similarly synthesized from 2-amino-4-chlorobutyric acid and disodium diselenide in ethanol, in a yield of about 50%.[7] 2-Amino-4-chlorobutyric acid is prepared from homoserine either via 2-benzamido-4-butyrolactone[8] or 3,6-bis(2-chloroethyl)-2,5-diketopiperazine,[9] followed by hydrolysis by refluxing in 2 N HCl for 2 hr. The product is obtained in both cases as a mixture of homoserine (and its lactone) and 2-amino-4-chlorobutyric acid and is used without further purification for the synthesis of selenohomocystine. L-Selenohomocystine is synthesized in the same manner from L-homoserine. The L-selenohomocystine obtained contains less than 1% of the D-enantiomer when L-2-amino-4-chlorobutyric acid is synthesized via 3,6-bis(2-hydroxyethyl)-2,5-diketopiperazine, but it contains 10% of the D-enantiomer when L-2-amino-4-chlorobutyric acid is prepared via 2-benzamido-4-butyrolactone.[7] Optically active selenocystine and selenohomocystine can also be prepared with microbial enzymes such as O-acetylhomoserine sulfhydrylase [(thiol)-lyase; EC 4.2.99.10].[10] In addition, Se-alkyl and Se-aryl derivatives of selenocysteine and selenohomocysteine have been synthesized enzymatically.[10]

Chemical and Physical Properties

Solubility in water at 25° and pH 7.0 is given as 0.63 mM and 0.235 mM for L-selenocystine and DL-selenocystine, respectively.[1] The stability of selenocystine in 6 N HCl at 110° is as follows: after hydrolysis for 6 hr, only 5% of the initial selenocystine remains.[11] The pK values for the ionizations of carboxyl, selenohydryl, and amino groups of DL-selenocysteine are given as well.[1] The ¹H-NMR spectrum of selenocystine in ²H₂O with sodium 3-(trimethylsilyl)[²H₄]propionate as an internal standard shows the following peaks: 3.52 ppm (4H, multiplet, β-CH$_2$) and 4.34 ppm (2H, quartet, α-CH).

Selenohomocystine has a melting point of 214° (decomposition). The compound is eluted as a single peak at 50.4 min between homocystine (48 min) and lysine (51.8 min), when analyzed with a Hitachi high-performance amino acid analyzer 835 equipped with a 4 × 150 mm column. The infrared spectrum of selenohomocystine (KBr) exhibits the following assignable peaks: 3500–2200 cm^{-1} (strong, NH), 1600 cm^{-1} (medium, NH),

---

[7] P. Chocat, N. Esaki, H. Tanaka, and K. Soda, *Agric. Biol. Chem.* **49**, 1143 (1985).
[8] M. Frankel and Y. Knobler, *J. Am. Chem. Soc.* **80**, 3147 (1968).
[9] H. R. Snyder, J. H. Andreen, G. N. Cannon, and C. F. Peters, *J. Am. Chem. Soc.* **64**, 2082 (1942).
[10] N. Esaki and K. Soda, this volume [54].
[11] R. E. Huber and R. S. Criddle, *Arch. Biochem. Biophys.* **122**, 164 (1967).

1580 cm^{-1} (strong, C=O), 1400 cm^{-1} (strong, C=O), and 530 cm^{-1} (medium, NH). The ^{1}H-NMR spectrum in ^{2}H$_2$O gives signals at 2.32 ppm (4H, multiplet, β-CH$_2$), 3.05 ppm (4H, triplet, γ-CH$_2$), and 3.9 ppm (2H, triplet, α-CH).

[44] Selenodjenkolic Acid

By HIDEHIKO TANAKA and KENJI SODA

$$\begin{array}{l} \text{Se—CH}_2\text{—CH(NH}_2\text{)—COOH} \\ | \\ \text{CH}_2 \\ | \\ \text{Se—CH}_2\text{—CH(NH}_2\text{)—COOH} \end{array}$$

L-Selenodjenkolic acid [3,3'-methylenediselenobis(2-aminopropionic acid)] is a peculiar amino acid in which two selenocysteine moieties are linked through a methylene group. The compound has not been demonstrated in nature. It is expected to undergo various enzymatic reactions, although its physiological functions have not been studied. It is chemically synthesized in a manner analogous to its naturally occurring sulfur counterpart, djenkolic acid.[1]

Procedure

L-Selenocystine[2] (117 mg, 0.5 mmol) suspended in 15 ml of water is treated with 38 mg (1 mmol) of sodium borohydride, and 270 mg (1 mmol) of methylene iodide is added to the solution under N$_2$, followed by stirring at room temperature for 2 hr. The resulting solution is applied to a Dowex 50-X8 column (H$^+$ form, 1 × 40 cm). After sequentially washing the column with 100 ml of water, 30 ml of 2 N HCl, and 100 ml of water, the product is eluted with 130 ml of 4 N NH$_4$OH. The eluate is evaporated to dryness at 4° under reduced pressure. The residue is dissolved in a minimum volume of water, and precipitated by addition of acetone. The product crystallizes from a small volume of water by addition of acetone (84 mg, 48% yield), m.p. 204–205° (decomposition).[3] The ^{14}C, ^{75}Se, ^{3}H or other labeled compounds can be synthesized in the same way.

[1] V. du Vigneaud and W. I. Patterson, *J. Biol. Chem.* **114**, 533 (1936).
[2] H. Tanaka and K. Soda, this volume [43].
[3] P. Chocat, N. Esaki, H. Tanaka, and K. Soda, *Anal. Biochem.* **148**, 485 (1985).

Properties

Selenodjenkolate is only slightly soluble in water, although its exact solubility has not been determined. When it is dissolved in 1 N HCl, and neutralized with 2 N NaOH, solutions of at least 10 mM can be prepared in 0.1 M potassium phosphate buffer (pH 7.2) or other similar buffers. Most organic selenium compounds are highly oxidizable, but selenodjenkolate is relatively stable. When 2 mM selenodjenkolate is incubated in 0.1 M Tris–HCl buffer (pH 8.0) at 37° for 2 hr, 86% of it remains, whereas only 48% of selenocystine survives treatment under the same conditions.

The ^{1}H-NMR spectrum in ^{2}H$_2$O with sodium 3-(trimethylsilyl)-[^{2}H$_4$]propionate as an internal standard exhibits the following peaks: 3.32 ppm (4H, doublet, β-CH$_2$), 3.83 ppm (2H, singlet, Se—CH$_2$—Se), and 4.06 ppm (2H, triplet, α-CH). It is closely similar to that of the sulfur analog, djenkolic acid, although the β-CH$_2$ and Se—CH$_2$—Se peaks are shifted to the lower field by about 5 Hz. The infrared spectrum (KBr) shows assignable peaks at 3500–2200 cm^{-1} (strong, NH), 1640 cm^{-1} (medium, NH), 1580 cm^{-1} (strong, C=O), 1500 cm^{-1} (medium, NH), 1400 cm^{-1} (strong, C=O), and 530 cm^{-1} (medium, NH). The compound is eluted as a single peak at 33 min between selenocystine (32.5 min) and methionine (34 min), when analyzed with a Hitachi high-performance amino acid analyzer 835 equipped with a 4 × 150 mm column. The specific rotation of selenodjenkolate, measured as a 0.25% solution in 6 N HCl at 25°, is −14° at 589 nm. Selenodjenkolate is completely decomposed by methionine γ-lyase (EC 4.4.1.11) of *Pseudomonas putida*, which catalyzes α,γ- and α,β-elimination reactions of the L-isomers of sulfur and selenium amino acids,[4] indicating that the compound synthesized is the L-isomer.

The degradation products of selenodjenkolate by methionine γ-lyase have been identified as 2 equivalents each of pyruvate, ammonia and elemental selenium, and 1 equivalent of formaldehyde. The formation of elemental selenium as an inherent product has not been shown but is inferred on the basis of the formation of hydrogen selenide after treatment with sodium borohydride.[3,5] Djenkolate is also degraded with the enzyme in a similar way. Kinetic constants are as follows: L-selenodjenkolate, K_m = 2.3 mM, V_{max} = 0.56 μmol/min/mg/; L-djenkolate, K_m = 2.2 mM, V_{max} = 0.83 μmol/min/mg.

---

[4] T. Nakayama, N. Esaki, K. Sugie, T. T. Berezof, H. Tanaka, and K. Soda, *Anal. Biochem.* **138**, 421 (1984).
[5] N. Esaki, T. Nakamura, H. Tanaka, and K. Soda, *J. Biol. Chem.* **257**, 4386 (1982).

[45] Te-Phenyltellurohomocysteine and Te-Phenyltellurocysteine

By Nobuyoshi Esaki and Kenji Soda

C_6H_5—Te—CH_2—CH_2—CH(NH_2)—COOH C_6H_5—Te—CH_2—CH(NH_2)—COOH
Te-Phenyltellurohomocysteine   Te-Phenyltellurocysteine

Tellurium is a homolog of sulfur and selenium, and it is classified as a metal. The chemistry of organotellurium compounds has not been established as well as for organosulfur and organoselenium compounds. Only a few telluroamino acids have been prepared.[1]

Reagents

β-Chloro-L-alanine hydrochloride
L-α-Amino-γ-bromobutyrate hydrobromide
Phenyltellurocyanate
Sodium borohydride
Ethanol

Procedure

Te-Phenyltellurohomocysteine and Te-phenyltellurocysteine are prepared as follows. Phenyltellurocyanate[2] and α-amino-γ-bromobutyrate hydrobromide[3] are prepared according to the reported methods. Phenyltellurocyanate (1.5 mmol) is dissolved in 20 ml of ethanol and reduced with 3 mmol of sodium borohydride. The red color of phenyltellurocyanate disappears during the reaction. A solution, 20 ml, of 2 mmol of L-α-amino-γ-bromobutyrate hydrobromide in ethanol, is added dropwise to the above solution under N_2. Te-Phenyltellurohomocysteine precipitates and is collected by filtration, washed with ethanol, and recrystallized from water (pH 5.0) with a yield of 39% based on phenyltellurocyanate.

Te-Phenyltellurocysteine is prepared in the same manner as described above except that L-α-amino-γ-bromobutyrate is replaced by β-chloro-L-alanine (commercially available from Vega). The yield is about 36% based on phenyltellurocyanate.

---

[1] F. F. Knapp, Jr., *J. Org. Chem.* **44**, 1007 (1979).
[2] F. Ogura, H. Yamaguchi, T. Otsubo, and K. Chikamatsu, *Synth. Commun.* **12**, 131 (1982).
[3] J. Baddiley and G. A. Jamieson, *J. Chem. Soc.* p. 4280 (1954).

[46] Chemical Reduction of Disulfides

By PETER C. JOCELYN

Disulfides are easily and specifically reduced by thiols, which are the most used reagents for this purpose. However, the excess of the thiol used as a reductant has to be removed before it is possible to assay the newly generated SH groups. Other methods can, therefore, compete with thiols in overall convenience of disulfide reduction. Direct chemical reduction is achieved with sodium borohydride or sodium amalgam whereas more specialized reagents include the phosphines and phosphorothioates. For cleavage generating half thiol only, there is sulfitolysis and cyanolysis. Less well controllable and defined is alkaline hydrolysis. These reagents will be considered in turn.

Reduction by Thiols

Thiol–disulfide exchange occurs in two stages:

$$RS^- + XSSX \rightarrow RSSX + XS^-$$
$$RSSX + RS^- \rightarrow RSSR + XS^-$$

For the completion of each of the reactions in the forward direction it is necessary either to use a large excess of thiol (RSH) (as in method b), or to use a thiol with a redox potential much more negative than that of the thiol to be generated by the reduction (as in method a).

Method a: Dithiothreitol and Dithioerythritol

Dithiothreitol (DTT) and dithioerythritol (DTE), first described by Cleland,[1] are usually the reagents of choice; more than 500 papers describe their application to specific problems. They are water-soluble solids with little odor and a redox potential (E_o') of -331 mV. Their pK values are for DTT, 8.3 and 9.5; for DTE, 9.0 and 9.9.[2] Since they function only when ionized, at least to their monoanions, it follows that (1) reducing power increases considerably over the range pH 7–9.5, (2) at lower pH, DTT is more effective than DTE, the reason for its greater use, and (3) the reaction can be stopped by acidifying e.g., with acetic acid to pH 3.[3]

[1] W. W. Cleland, *Biochemistry* **3**, 480 (1964).
[2] W. L. Zahler and W. W. Cleland, *J. Biol. Chem.* **243**, 716 (1968).
[3] S. S. Ristow and D. B. Wetlaufer, *Biochem. Biophys. Res. Commun.* **50**, 544 (1973).

TABLE I
DISULFIDE REDUCTIONS BY DTT OR DTE

Disulfide	Conditions	Ref.
Collagen III (rat skin)	DTT 10 mM, pH 8, 3 hr. +U	a
Thioredoxin	DTT 5 mM, pH 7.6, 0.5 hr, 23°	b
Lysozyme	DTT 65 mM, pH 8.6, 2 hr, 20°, +U	c
Gonadotropin(chorionic)	DTE 0.7 mM, pH 8.5, 0.5 hr, 20°	d
Insulin receptor	DTT 20 mM, pH 7.4, 0.5 hr, 23°, +SDS	e
Lactalbumin	DTE 20-fold excess, pH 8.5, 25°	f
C1-Esterase inhibitor	DTT 200 mM, pH 7.3, 0.5 hr, +Gu	g
Thiolproteinase inhibitor	DTT 10 mM, pH 8, 0.33 hr, 20°, +Gu	h
Enterokinase	DTE 50 mM, pH 9, 2 hr, 4°	i
Fibronectin	DTT 10 mM, pH 7.4, 1 hr, 20°	j
Thyroid-stimulating hormone	DTT 10 mM, pH 8.5, 4°	k
Cholinesterase	DTT 0.4 mM, pH 8.8, 33 hr, 8, 0°	l

[a] U, urea. D. T. Cheung, P. DiCesare, P. D. Benya, E. Libaw, and M. E. Nimmi, *J. Biol. Chem.* **258,** 7774 (1983).
[b] S. Adler and P. Modrich, *J. Biol. Chem.* **258,** 6956 (1983).
[c] S. S. Ristow and D. B. Wetlaufer, *Biochem. Biophys. Res. Commun.* **50,** 544 (1973).
[d] T. Mise and O. P. Bahl, *J. Biol. Chem.* **256,** 6587 (1981).
[e] SDS, sodium dodecyl sulfate. J. Massague and M. P. Czech, *J. Biol. Chem.* **257,** 6729 (1982).
[f] T. Segawa, K. Kuwajima, and S. Sugai, *Biochim. Biophys. Acta* **668,** 89 (1981).
[g] Gu, guanidine hydrochloride. T. Nilsson and B. Wiman, *Biochim. Biophys. Acta* **705,** 271 (1981).
[h] N. Watmatsu, E. Kohinami, K. Taklo, and H. Katunuma, *J. Biol. Chem.* **259,** 13832 (1984).
[i] A. Light and P. Fonseca, *J. Biol. Chem.* **259,** 13195 (1984).
[j] E. C. Williams, P. A. Janmey, R. B. Johnson, and D. F. Mosher, *J. Biol. Chem.* **258,** 5911 (1983).
[k] J. G. Pierce, L. C. Giudice, and J. R. Reeve, *J. Biol. Chem.* **251,** 6388 (1976).
[l] O. Lockridge, H. W. Eckerson, and B. N. Ladu, *J. Biol. Chem.* **254,** 8324 (1979).

Table I illustrates recently used conditions for the reduction. Table I also shows that it is common practice to use at least a 20-fold excess of reagent, with final concentrations of 0.5–20 mM, and to work at 20–30° and pH 8–9. In one example, reduction was complete in 2–6 hr at pH 8.5 but incomplete at pH 7 after 24 hr.[4]

Reduction of disulfides in intact proteins by DTT or DTE is often only

[4] J. G. Pierce. L. C. Giudice, and J. R. Reeve, *J. Biol. Chem.* **251,** 6388 (1976).

partial because of inaccessibility. Here the hydrophilicity of the reagents may be a hindrance to their penetration. Completion may require a denaturant such as urea, guanidine hydrochloride, or sodium dodecyl sulfate (SDS).

DTT and DTE easily chelate metal ions and thereby prevent deleterious effects on sensitive enzymes. However, this property also facilitates the rapid metal-catalyzed autoxidation of the dithiols. DTT at pH 7.4 and 0.3 mM is almost completely oxidized at 30° in the presence of 10 μM Fe^{2+}.[5] It is, therefore, desirable for disulfide reductions to be performed under nitrogen and in the presence of a chelating agent such as EDTA.

DTT and DTE have also been extensively used to reduce soluble nonprotein disulfides such as cystine[6] and mixed homocysteine–cysteine disulfides.[7] An advantage here is that a specific colorimetric assay for cysteine[8] can be applied without removing excess reagent. The disulfide need not be soluble. Thus, sulfur–sulfur bonds are reduced in cystine kidney stones,[9] membrane suspensions,[10] and collagen.[11] DTT also reduces diselenides.[12]

An advantage of DTT or DTE is that the reaction can be followed spectrophotometrically because the oxidized forms are ring structures (dithiolanes) which, unlike the dithiols, absorb with peaks at 310 and 283 nm.[13,14] However, since sensitivity is low (ε_m at 310 nm is 110 M^{-1} and at 283 nm, 275 M^{-1} cm^{-1}), this technique is only of value if scale expansion is available.

After the reduction excess DTT or DTE can be removed at low pH by dialysis, precipitation, or column separation in the case of proteins. For some ionic nonprotein disulfides (e.g., amino acid disulfides), extraction with ethyl acetate, in which the dithiols are soluble, may be convenient.[15]

Another method, which masks the reagent thiol, is utilizing the ability of DTT and DTE to form arsenite derivatives in which their SH groups are blocked. Here DTE is superior to DTT as it forms a more stable

---

[5] D. O. Lambeth, G. R. Ericson, M. A. Yorek, and P. D. Ray, *Biochim. Biophys. Acta* **719**, 501 (1982).
[6] J. C. Crawhall and S. Segal, *Biochem. J.* **105**, 891 (1967).
[7] J. A. Schneider, K. H. Bradley, and J. E. Seegmiller, *J. Lab. Clin. Med.* **71**, 122 (1968).
[8] M. K. Gaitonde, *Biochem. J.* **104**, 627 (1967).
[9] J. M. Walshe, *Lancet* **2**, 263 (1970).
[10] Y. Ando and M. Steiner, *Biochim. Biophys. Acta* **311**, 38 (1973).
[11] D. T. Cheung, P. DiCesare, P. D. Benya, E. Libaw, and M. E. Nimmi, *J. Biol. Chem.* **258**, 7774 (1983).
[12] W. H. H. Gunther, *J. Org. Chem.* **32**, 3931 (1967).
[13] K. S. Iyer and W. E. Klee, *J. Biol. Chem.* **248**, 707 (1973).
[14] T. E. Creighton, *J. Mol. Biol.* **96**, 767 (1975).
[15] J. Butler, S. P. Spielberg, and J. D. Schulman, *Anal. Biochem.* **75**, 674 (1976).

complex. Sodium arsenite, added in 25-fold excess over the amount of DTT or DTE originally present allows the reaction at pH 8.1 to be complete in 5 min at 20°. The new SH groups from the reduced disulfide can then be assayed directly, e.g., with DTNB. The method, originally developed for nonprotein disulfides[2,16] has been applied to the assay protein disulfide bonds[10]; however, arsenite binding by proteins with vicinal SH groups could lead to error.

Method b: 2-Mercaptoethanol

Mercaptoethanol, the simplest water-soluble organic thiol, is "the poor man's DTT," i.e., mole for mole it is 150–175 times less expensive. However, it is normally used in concentrations of 0.1–0.7 M, i.e., 10 times those required with DTT. This reflects its higher pK of 9.5[17] (versus 8.3 for the pK of DTT) and its probably less negative redox potential.[18]

Mercaptoethanol was the reductant of choice before 1963 and is still much used especially in conjunction with SDS–gel electrophoresis. Thus, the separation of proteins on the basis of molecular weight by this technique requires linear peptide chains and, hence, reduction of any disulfide bridges; in this case, estimation of their number may not be required. The preferred technique, using mercaptoethanol, has been well described.[19,20] In brief, the protein is heated at 100° with SDS and 0.15–0.7 M mercaptoethanol at pH 7 for 2–3 min before applying to the gel.

For structural studies of homogeneous proteins a greater variety of conditions are appropriate. Some recent examples are shown in Table II.

If sensitive SH groups are formed by reduction with mercaptoethanol, it may be necessary to block them at once. One way of doing this is to cyanoethylate by adding acrylonitrile in excess over the amount of mercaptoethanol used.[20]

There are some disadvantages in using mercaptoethanol. At the high concentrations required, it has considerable absorption in the UV[21] which may interfere with any spectrophotometry after reduction. It also affects protein estimations by the Lowry or Biuret methods. It can, of course, easily be removed from protein solutions by dialysis. Since it is volatile

---

[16] K. A. Walsh, R. M. McDonald, and R. A. Bradshaw, *Anal. Biochem.* **35,** 193 (1970).
[17] K. Weber, J. R. Pringle, and M. Osborn, this series, Vol. 26, p. 3.
[18] M. Calvin, in "Glutathione" (S. Colowick, A. Lazarow, E. Racker, D. R. Schwarz, E. R. Stadtman, and H. Waelsch, eds.), p. 20. Academic Press, New York, 1954.
[19] B. R. Olson, H. P. Hoffman, and D. J. Prockop, *Arch. Biochem. Biophys.* **175,** 341 (1976).
[20] T. S. Seibles and L. Weil, this series, Vol. 11, p. 204.
[21] L. Spero, *Biochim. Biophys. Acta* **671,** 193 (1981).

TABLE II
REDUCTION OF DISULFIDES BY MERCAPTOETHANOL

Protein	Treatment	Ref.
ATPase	ME 0.3 M, pH 7.5, 53°	a
Bence Jones	ME 0.1 M, pH 7.4, 4 hr, 22°	b
Bronchial mucus	ME 0.1 M, pH 8, 1 hr	c
Enterotoxins (A, B)	ME 0.7 M, pH 8.6, 3 hr (C requires ME 2.4 M)	d
Fibrinogen	ME 0.1 M, pH 7.5, 0.5 hr (ME, 4SS; DTT, 10SS)	e
Lysozyme	ME 0.3 M, pH 8, 18 hr, 25°, U 8 M	f
Papain	ME 0.3 M, U 8 M (U partial, Gu complete)	g
Protamines	ME 50-fold excess, pH 8	h
Ribonuclease	ME 1 μl/mg protein, pH 8.6, U 8 M	i

[a] M. Kawamura and K. Nagano, *Biochim Biophys. Acta* **774,** 188 (1984)
[b] K. R. Ely, R. L. Girling, M. Schiffer, D. E. Cunningham, and A. B. Edmundson, *Biochemistry* **12,** 4233 (1973).
[c] G. P. Roberts, *Arch. Biochem. Biophys.* **173,** 528 (1976).
[d] L. Spero, *Biochim. Biophys. Acta* **671,** 193 (1981).
[e] A. Z. Budzynski and M. Stahl, *Biochim. Biophys. Acta* **175,** 282 (1969).
[f] U, urea. A. S. Acharya and H. Taniuchi, *J. Biol. Chem.* **251,** 6934 (1976).
[g] Gu, guanidine hydrochloride. R. Arnon and E. Shapira, *J. Biol. Chem.* **244,** 1033 (1969).
[h] T. Tobita, H. Suzuki, K. Soma, and M. Nakano, *Biochim. Biophys. Acta* **748,** 461 (1983).
[i] C. B. Anfinsen and E. Haber, *J. Biol. Chem.* **236,** 1361 (1961).

(b.p.$_{760}$ = 157°, b.p.$_{13}$ = 55°), an alternative is by evaporation. For low concentrations and small volumes, 10 min at 100° may be sufficient,[22] but more certain treatment is drying under reduced pressure.[23,24] It may be inadvisable to leave a protein in contact in acid solution for long with large amounts of mercaptoethanol because of the possibility, difficult to disprove, that some esterification of carboxyl residues by the hydroxyl group might occur, resulting in subsequent overestimation of SH groups.

Method c: Other Thiols

A carboxyl-bound polymer of dihydrolipoate, immobilized in a solid matrix (Sephadex, Sepharose, cellulose, or polyacrylamide) has been described which quantitatively reduces simple disulfides, e.g., glutathione disulfide (GSSG) and cystine, and also some in proteins, and which can itself be later regenerated with borohydride. Yields of 100% are claimed

[22] K. K. Tan, *Anal. Biochem.* **86,** 327 (1978).
[23] H. P. Makkear, O. P. Sharma, and S. S. Negi, *Anal. Biochem.* **104,** 124 (1980).
[24] H. P. Makkear, O. P. Sharma, and S. S. Negi, *Analyst* **107,** 579 (1982).

as the result of simply shaking for a few minutes with the thiol at pH 7.5 under nitrogen and then removing the polymer by filtration.[25]

A slow reduction of simple thiols results by mixing them at pH 6–9 with thiophenol. The latter is later extracted, together with any diphenyl disulfide formed, with benzene.[26]

There is evidence[27] that, in a medium of low ionic strength, thiols with a nearby anionic group react more readily than neutral thiols with accessible disulfide groups adjacent to cationic residues. The simplest thiol in this category is thioglycolate, i.e., 2-mercaptoacetate. (The simplest cationic disulfide is 2-mercaptoethylamine, but this has not been used as a disulfide reductant probably because of its pronounced tendency to form mixed disulfides.)

Thioglycolic acid reduces protein disulfide groups in ribonuclease. In 8 M urea it was found to be as effective as mercaptoethanol although it has a somewhat higher pK value (10.4). The substance must be freshly distilled since it forms polythioglycolides on storage which can acylate amino groups and lead to spuriously high estimates of SH groups.[28]

Conditions for full reduction by thioglycolate include 400 mol/mol of enzyme in 8 M urea at pH 8.5 for 4.5 hr at 20°.[29] At high concentration and long exposure, e.g., 18 hr, it is able to reduce disulfide bonds in wool at pH 5, leading to solubilization.[30]

Another anionic thiol, glutathione (GSH) is largely ineffective with protein disulfide groups, presumably because of difficulties in penetration, but able easily to reduce simple disulfides.[31] Since it has a more negative redox potential, it is more effective than cysteine.[31a]

Direct Reduction

A great advantage of direct reduction is the absence of any adventitious thiol to remove afterward.

Borohydride

A fresh solution of sodium or potassium borohydride (0.75 M) is added to the protein at pH 9 to give a final concentration of 0.3 M together with a few drops of an antifoam agent (e.g., octanol), and the mixture incubated.

[25] M. Goreck and A. Patchornik, *Biochim. Biophys. Acta* **303,** 36 (1973).
[26] G. R. Limb and A. M. Dollar, *Anal. Biochem.* **29,** 100 (1969).
[27] M. K. Reddy, M. J. Cennerazzo, and D. Field, *Biochim. Biophys. Acta* **749,** 219 (1983).
[28] F. H. White, *J. Biol. Chem.* **235,** 383 (1960).
[29] M. Sela, F. H. White, and C. B. Anfinsen, *Science* **125,** 691 (1957).
[30] H. Lindley and R. W. Cranston, *Biochem. J.* **139,** 515 (1974).
[31] F. J. R. Hird, *Biochem. J.* **85,** 320 (1962).
[31a] P. C. Jocelyn, *Eur. J. Biochem.* **2,** 327 (1967).

After acidifying the excess borohydride decomposes; the last traces can be removed by adding a little acetone.

Under these conditions, all 17 disulfide bonds of bovine serum albumin are reduced without also adding a denaturant at 50° in 1 hr[32] or, with 8 M urea present, at 37° in 0.5 hr.[33] With less borohydride and a lower temperature there may be selective reduction of native proteins. Here are some examples:

    Trypsin and trypsinogen, 0°, 0.1 M borohydride, 2 hr[34]
    Trypsin inhibitor, 2°, 0.2 M borohydride, 0.5 hr[35]
    Taka-amylase A, 30°, 1.8 M borohydride, 2 hr[36]
    Tryptophanyl-RNA ligase, 23°, 0.2 M borohydride, pH 7.8[37]

Borohydride has also been used to reduce mixed protein–nonprotein disulfides in tissue extracts (see this volume [19 and 20]). After the reduction, protein can be precipitated with perchloric or trichloroacetic acid to leave the nonprotein thiol in solution.[38,39] Simple disulfides are reduced easily by borohydride[40] and are then ready for assay by the DTNB method.

"Nascent Hydrogen"

Zinc and acid were widely used at one time for reduction of a variety of simple disulfides.[41] More recently sodium mercury amalgam (containing 0.5% sodium) was reported to reduce cystine quantitatively at pH 7.5 in a few minutes in acid hydrolyzates of a protein (lipophilin).[42] Apparently, there is no problem with mercaptide formation and the liberated thiol can be assayed with DTNB.

Sodium hydride in anhydrous dimethyl sulfoxide can reduce some freeze-dried proteins, e.g., albumin.[42a]

---

[32] W. B. Brown, *Biochim. Biophys. Acta* **44**, 365 (1960).
[33] D. Cavallini, M. T. Graziani, and S. Dupré, *Nature (London)* **212**, 294 (1966).
[34] A. Light and N. K. Sinha, *J. Biol. Chem.* **242**, 1358 (1967).
[35] L. F. Kress and M. Laskowski, *J. Biol. Chem.* **242**, 4925 (1967).
[36] B. K. Seon, *J. Biochem. (Tokyo)* **61**, 606 (1967).
[37] G. V. Kuehl, M. L. Lee, and K. H. Muench, *J. Biol. Chem.* **251**, 3254 (1976).
[38] K. R. Harrap, R. C. Jackson, P. G. Riches, C. H. Smith, and B. T. Hill, *Biochim. Biophys. Acta* **310**, 104 (1973).
[39] H. G. Modig, *Biochem. Pharmacol.* **22**, 1623 (1973).
[40] G. L. Ellman and H. Lysko, *J. Lab. Clin. Med.* **70**, 518 (1967).
[41] G. E. Woodward and E. G. Fry, *J. Biol. Chem.* **97**, 465 (1932).
[42] S. A. Cockle, R. M. Epand, J. G. Stollery, and M. A. Moscarello, *J. Biol. Chem.* **255**, 9182 (1980).
[42a] L. H. Krull and M. Friedmann, *Biochem. Biophys. Res. Commun.* **29**, 373 (1967).

Mercurous Acetate

This reagent reduces and mercurates thiols at pH 4.5 to give the mercuric dithiol derivative.[43] However yields with cystine are not quantitative even after 2 days.

$$RSSR + 2\ Hg^+ \rightarrow RSHgSR + Hg^{2+}$$

Tributylphosphine

$$Bu_3P + RSSR + H_2O \rightarrow Bu_3PO + 2\ RSH$$

Tributylphosphine (Bu_3P) (commercially available) is a liquid with sufficient volatility to be highly toxic. Simple disulfides such as dipropyl or dibutyl disulfide in 10% methanol are reduced by shaking them with 0.01 volume of 0.1 M Bu_3P for 1 hr at 20°.[44] Excess reagent can be extracted either with benzene or with chloroform containing 1% sulfur.[45] Its advantage is that it is not easily autoxidized and that its affinity for disulfide groups is such that only a slight excess (5–20%) is required.

The accessible disulfide bonds of several soluble proteins (bovine albumin, vasopressin, insulin, ribonuclease), with or without urea or guanidine, are reduced under the following conditions: Bu_3P (0.05–0.25 M) in propanol is added to the protein in a 1 : 1 mixture of 0.5 M sodium bicarbonate and propanol (pH about 8–8.3). Reduction is complete in 30–40 min at 20°.[46]

In a more recent variant, toluene sultone is included with this mixture. The sultone immediately converts the thiol released to an S-2-sulfobenzyl derivative.[47]

RSH + [toluene sultone structure: benzene ring with –CH$_2$–O–SO$_2$– forming a ring] → [benzene ring with –CH$_2$SR and –SO$_3$H substituents]

Disulfide bonds in wool, which are resistant to thioglycolate, are reduced by immersion in a 10 mM solution of Bu_3P in 20% propanol for 3 days.[48]

Reduction with Substitution

A number of reagents add to the disulfide bond to give only one molecule of SH group, the other being released as an S-substituted derivative:

$$RSSR + X^- \rightarrow RS^- + RSX$$

---

[43] N. M. David, R. Spetling, and I. Z. Steinberg, *Biochim. Biophys. Acta* **359**, 101 (1974).
[44] R. E. Humphrey and J. L. Potter, *Anal. Biochem.* **37**, 164 (1965).
[45] A. Kirkpatrick and J. A. McClaren, *Anal. Biochem.* **56**, 137 (1973).
[46] U. T. Ruegg and J. Rudinger, this series, Vol. 47, p. 111.
[47] U. T. Ruegg, D. Jarvis, and J. Rudinger, *Biochem. J.* **179**, 127 (1979).
[48] H. Lindley and R. W. Cranston, *Biochem. J.* **139**, 515 (1974).

Sulfitolysis

$$\text{RSSR} + \text{SO}_3^{2-} \rightarrow \text{RS}^- + \text{RS—SO}_3^-$$

Sulfite cleaves disulfides to thiol and thiosulfate, RSSO_3^-. Kinetic analysis shows that the true reactant is SO_3^{2-}, not HSO_3^-, and that the reaction is consequently impeded if the disulfide is in a negatively charged environment.[49] Since these parameters both increase with increasing pH, a wide variation in the response of individual disulfides is to be expected. This is readily shown for simple disulfides. Thus, cystamine (dication) reacts 6–8-fold faster than cystine (neutral) and 50-fold faster than GSSG (dianion).

When sulfitolysis is performed on unsymmetrical, i.e., mixed, disulfides, there is also the question of the direction of cleavage. This was found to be random with GSS—cysteine at pH 7.2 and also with mixed disulfides containing a cysteine moiety linked to various proteins (i.e., Protein—SS—Cy), provided these were first denatured. If intact, there was a marked preference for attack on the nonprotein cysteine residue to give *S*-sulfocysteine.[50]

For proteins, sulfite has been used as a cleavage reagent over the concentration range 0.05–1 M and between pH 7 and 9.5. However, for assay of protein disulfide groups by subsequent thiol assay, its value is limited, partly because only half of the possible number of SH groups are released and partly because of interference by excess sulfite on the thiol assay itself, i.e., sulfite reacts with DTNB.

The second objection has now been overcome by an ingenious method[50a] described elsewhere in this volume.[50b] The first is minimized by trapping the SH groups with a mercurial which can then be assayed by polarography[51,52] or by using mercuribenzoate containing ^{203}Hg.[53]

With such techniques, it was found that only the two interchain disulfide bonds of insulin react with sulfite at pH 7 unless guanidine is also present. Estimates were also made of the disulfide groups in platelet membranes. Another recent application has been in studying the single disulfide present in xanthine oxidase.[54]

A different approach is to convert all (instead of half) the original

---

[49] R. Cecil and J. R. McPhee, *Adv. Protein Chem.* **14**, 255 (1959).
[50] N. J. Van Rensburg and O. A. Swanepoel, *Arch. Biochem. Biophys.* **121**, 729 (1967).
[50a] T. W. Thannhauser, Y. Konishi and H. A. Scheraga, *Anal. Biochem.* **138**, 181 (1984).
[50b] This vol., article by Konishi.
[51] R. Cecil and R. G. Wake, *Biochem. J.* **82**, 401 (1962).
[52] See also this volume [10].
[53] Y. Ando and M. Steiner, *Biochim. Biophys. Acta* **311**, 38 (1973).
[54] R. Hille and V. Massey, *J. Biol. Chem.* **257**, 8898 (1982).

disulfide to an *S*-sulfo derivative. By cyclically regenerating disulfide from the thiol formed, further attack by sulphite is made possible.

$$RSSR + SO_3^{2-} \rightarrow RSO_3 + RS^-$$
$$\underbrace{\phantom{RSSR + SO_3^{2-} \rightarrow RSO_3 + RS^-}}_{\text{oxidation}}$$

Full experimental details for autoxidation by 1 mM copper sulfate or dehydrogenation by iodosobenzoate or tetrathionate are available.[55,56]

Cyanolysis

$$RSSR + CN^- \rightarrow RS^- + RSCN$$

Simple aliphatic disulfides do not readily react with cyanide unless one of the products is simultaneously removed. However, if there is an adjacent amino group (as in cystine) the first formed thiocyanate cyclizes to a thiazoline in alkkaline solution thereby driving the reaction to completion.

$$\begin{array}{c} H \\ R-C-NH_2 \\ | \\ H_2C-SCN \end{array} \rightarrow \begin{array}{c} H \\ R-C-N \\ | \quad \| \\ H_2C \quad C-NH_2 \\ \diagdown \diagup \\ S \end{array}$$

This cyclization also occurs if the amino group of the cysteine residue is joined in peptide linkage, with its resultant hydrolysis. Thus the disulfide bonds of cystamine, cystine, and GSSG are all cleaved quantitatively by 5% KCN in 1 M ammonia in 30 min.[57] The free thiol groups are titrated by bringing the pH to 6 with acetic acid and assaying with *N*-ethylmaleimide.

To produce the protein thiocyanates for this reaction directly from protein SH groups, they can be cyanated with 2-nitro-5-thiocyanobenzoate.[58]

A better method is by cyanolysis of a mixed disulfide formed between protein SH groups and 3-carboxy-6-mercaptopyridine (i.e., Protein—SS—Pyr—COOH). In this instance, cyanide is specifically added to the sulfur atom of the cysteine residue. The technique is to add sodium cyanide (50 mM) at pH 6.9 whereupon cyanolysis is complete in 20 min.[59]

The spontaneous cyclization and hydrolysis of protein thiocyanates as described above constitutes an important method for specific cleavage adjacent to an SH group.[60]

---

[55] R. D. Cole, this series, Vol. 11, p. 206.
[56] S. Clark and L. C. Harrison, *J. Biol. Chem.* **257**, 12239 (1982).
[57] C. De Marco, M. T. Graziani, and R. Mosti, *Anal. Biochem.* **15**, 40 (1966).
[58] G. R. Jacobson, M. H. Schaffer, G. R. Stark, and T. C. Vanaman, *J. Biol. Chem.* **248**, 6583 (1973).
[59] M. Silverberg, J. T. Dunn, L. Garen, and A. P. Kaplan, *J. Biol. Chem.* **255**, 7281 (1980).
[60] Compare with this volume [6 and 7].

Phosphorothioate

$$RSSR + SPO_3^{3-} \rightarrow RSSPO_3^{2-} + RS^-$$

Introduced by Neumann,[61] phosphorothioate (80 mM) in Tris chloride at pH 8.5 cleaves disulfide bonds in ribonuclease when used in 10-fold excess. With incorporated ^{35}S or ^{32}P, it labels the released SH groups. Subsequent treatment with, for example, mercaptoethanol, frees them again for radioassay. The reagent has not found wider application.

Hydrolysis

$$RSSR + H_2O \rightarrow RSH + \text{other products}$$

Hydrolytic cleavage of a disulfide at high pH yields, by disproportionation, thiol plus oxidized products. Aromatic disulfides (e.g., DTNB) give a stoichiometry corresponding to the overall reaction:

$$2\ ArSSAr + 2\ H_2O \rightarrow 3\ ArSH + ArSO_2H$$

Aliphatic disulfides are resistant unless (1) there is sufficient heavy metal ion (Hg^{2+} or Ag^+) present to drive the reaction forward by complexing the thiol formed, in which case it proceeds as above,[62] or (2) there is a carbonyl α or β to the disulfide group. Thus, without metal ion present, cystamine, CoA, pantothine, and homocystine are only very slowly attacked by alkali, whereas dithiodiglycolate and cystine are hydrolyzed.

The most labile disulfides are cystinyl peptides, e.g., GSSG, and, most significantly, protein disulfide groups. GSSG gives 1.3 mol of GSH after 5 min in strong alkali, the procedure has been used as an analytical method.[63] Hydrolysis of oxytocin and insulin proceeds similarly.[64] Recent studies indicate that the mechanism is by β-elimination to give, initially, a perthiol (RSSH) and a dehydroalanine residue:

$$\rangle CH-CH_2-S-S-CH_2CH\langle \rightarrow CH-CH_2-S-SH + CH=CH$$

The former easily loses sulfur to become a thiol and the latter may also undergo secondary reactions.[65]

The complexity of alkaline hydrolysis makes it of little practical value as a method of disulfide cleavage.

---

[61] H. Neumann, I. Z. Steinberg, J. R. Brown, R. F. Goldberger, and M. Sela, *Eur. J. Biochem.* **3**, 171 (1967).
[62] R. Cecil and J. R. McPhee, *Adv. Protein. Chem.* **14**, 300 (1959).
[63] W. L. Anderson and D. B. Wetlaufer, *Anal. Biochem.* **67**, 493 (1975).
[64] T. M. Florence, *Biochem. J.* **189**, 507 (1980).
[65] A. J. Jones, E. Helmerhorst, and G. B. Stokes, *Biochem. J.* **211**, 499 (1983).

[47] Electrochemical Reduction of Disulfides

By HAROLD KADIN

Introduction[1]

Electrochemical reduction of disulfides has been reported[2-5] to give higher yields of thiols than alternative zinc–acid or borohydride reductions. Whereas zinc–acid reduction of disulfides was complete in pure aqueous solution, it was appreciably less than quantitative in blood or tissue samples. Presumably the acid pH of the zinc reduction favors thiol stability, but zinc ion and the alkaline pH of alternative borohydride reduction may promote partial reoxidation to disulfide during work-up. Cyanide[6] and tributylphosphine[7] are highly effective but very toxic reductants and should be generally avoided in routine laboratory operations. Other nucleophilic reductants, including dithiothreitol, mercaptoethanol, and sulfite, are less useful because they react with 5,5'-dithiobis(2-nitrobenzoic acid), Ellman's reagent, a widely utilized colorimetric reagent.[8] In the present discussion, the use of electrochemical reduction is exemplified by the low toxicity, high yield, selective analysis of urinary captopril [(S,S)-1-(3-mercapto-2-D-methyl-1-oxopropyl)-L-proline], an antihypertensive, angiotensin-converting enzyme inhibitor.

Disulfide metabolites in the acidified urine of captopril-dosed subjects were cleaved by electrochemical reduction prior to separation from endogenous sulfur amino acids by solvent partition and selective analysis using an automated version of Ellman's colorimetry.[1a] The method was used to measure the rate and extent of urinary excretion of "total" captopril which includes "free" captopril plus its symmetrical and predomi-

---

[1] Parts of the following article were reproduced, with permission of the copyright owner, the American Pharmaceutical Association, from a publication[1a] in the *Journal of Pharmaceutical Sciences*.
[1a] H. Kadin and R. B. Poet, *J. Pharm. Sci.* **71**, 1134 (1982).
[2] J. S. Dohan and G. E. Woodward, *J. Biol. Chem.* **129**, 393 (1939).
[3] T. Hata, *Bull. Res. Inst. Food Sci.* **3**, 63 (1950).
[4] S. K. Bhattacharya, J. S. Robson, and C. P. Stewart, *Biochem. J.* **60**, 696 (1955).
[5] R. Saetre, Ph.D. thesis, University of Alberta, Edmonton, Alberta, Canada (1978).
[6] H. Takahashi, Y. Nara, T. Yoshida, K. Tuzimura, and H. Meguro, *Agric. Biol. Chem.* **45**(1), 79 (1981).
[7] U. T. Rüegg and J. Rudinger, this series, Vol. 47, Part E, p. 114.
[8] G. L. Ellman and H. Lysko, *J. Lab. Clin. Med.* **70**, 518 (1967).

nantly[9] mixed captopril–cysteine disulfides. "Free" captopril was estimated by analysis without electrochemical reduction.

To minimize oxidation of "free" captopril to disulfides during long-term storage of urine samples, a solution of the very effective metal chelator, diethylenetriaminepentaacetic acid (DTPA) was mixed into freshly voided urine, and a solution of citric and oxalic acids was added soon after. The acidified urine was rapidly chilled and could then be stored at 3–5° for at least 30 days prior to analysis.[1a] The order of addition for DTPA, citric, and oxalic acids is important since DTPA is more effective at pH 5–6, the pH of human urine.

Method

Reagents

Captopril, captopril disulfide, and captopril–cysteine disulfide are prepared in the acid chelating mixture at concentrations of 25 to 100 μg per ml.

Acid chelating mixture: 0.4% DTPA, 1.05% citric acid monohydrate, and 0.63% oxalic acid dihydrate. To facilitate solution, DTPA was diluted from a stock solution of 0.8% DTPA in 0.08 M sodium hydroxide.

Saturated aqueous sodium chloride

Sodium carbonate, 0.5 M

Agar, Special Noble No. 0142-0, Difco

Purified mercury, Fisher M-140 is adequate

Electrochemical Reduction Cell

The cell illustrated in Fig. 1 is scaled-up from that previously reported,[10] and is used for most studies. The 5.05 × 1.85 cm (i.d.) main chamber contains a magnetically stirred mercury pool (about 3 ml) in electrical contact with a platinum wire extending to the exterior of the chamber. An inverted U side arm (3.5 mm i.d.) is attached about 2 cm from the bottom of the chamber. The latter contains a saturated sodium chloride–2% agar salt bridge constituted and maintained as described subsequently. The agar salt bridge isolates the negative, reductive mercury pool cathode from the positive, oxidative platinum anode. The anode is positioned at the tip of the bridge in a small container of saturated sodium chloride. The cells are connected in series with miniature alligator

---

[9] B. H. Migdalof, M. J. Antonaccio, D. N. McKinstry, S. M. Singhvi, S. J. Lan, P. Egli, and K. J. Kripalani, *Drug Metab. Rev.* **15**(4), 861 (1984).

[10] R. Saetre and D. L. Rabenstein, *Anal. Chem.* **50**, 276 (1978).

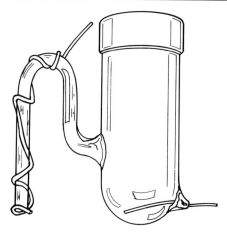

FIG. 1. Saetre–Rabenstein agar-bridged electrochemical reduction cell.[10]

clip jumper leads through a 0–5 milliammeter and a power supply (No. 6212A Hewlett-Packard, Palo Alto, CA) capable of supplying a constant current of 2.5 mA. The integrity of the agar bridge is maintained optimally by storage in saturated sodium chloride solution at both ends.

During electrolysis the mercury pool–urine interface is magnetically stirred (Micro V Stirrers, Cole Parmer, Chicago, IL). Speed is adjusted to minimize vortex formation and the formation of mercury droplets. Two cells are stirred over each stirrer and as many as 12 cells can be operated in series. To minimize chlorine generation at the anode, the saturated sodium chloride electrolyte is made alkaline, as recommended by Saetre.[5] Five to six drops of 0.5 M Na_2CO_3 are mixed in 9 ml of the anode electrolyte, and the reduction assembly is operated in a hood to sweep chlorine away. The alligator clips were placed well away from the anodic electrolyte so as to minimize corrosion by chlorine. To minimize chlorine retention in the gel, the bridge tip is removed from electrolyte soon after a run. Nearly quantitative reduction of disulfide is obtained without the widely recommended [2–5,11,12] blanketing with nitrogen of the cathode compartment. Almost complete sulfhydryl recovery is ensured by the DTPA–citric–oxalic thiol stabilization, the slight electrolytic hydrogen generation, and a minimum headspace beneath a pinhole-vented plastic cap (Polyethylene Containers, Nalgene No. 6250, Nalge Co., Rochester, NY).

[11] E. G. Sement, M. L. Girard, F. Rousselet, and M. Chemla, *Ann. Pharm. Fr.* **37**, 513 (1979).
[12] P. D. J. Weitzman, J. K. Hanson, and M. G. Parker, *FEBS Lett.* **43**, 101 (1974).

Preparing the Agar Salt Bridge

Distilled water, 90 ml in a 125 ml conical flask, containing a magnetic stirring bar are brought to a boil. The flask is removed from the heat source. Special Noble Agar, 2 g, is gradually added to the moderately stirred hot water. The agar gradually dissolves as evidenced by disappearance of turbidity. Three 10-g portions of sodium chloride are separately and gradually dissolved into the hot, moderately stirred (to prevent whipping and trapping of high electrical resistance air into the viscous liquid) agar solution. The mixture is warmed as necessary in a boiling water bath or a 100–110° oven. Pasteur pipets and empty, scaled-up Saetre–Rabenstein[10] cells are also warmed for about 15 min in the oven; paper wipes or insulated gloves facilitate handling and pipet bulb installation.

While depressing the bulb, the pipet tip is placed in the unstirred gel salt solution, and the bulb slowly released to draw gel into the pipet. While holding the hot cell upside down (with paper wipes or other insulated holder) the gel-filled, air-free tip of the pipet is placed in the open tip of the cell's side arm. Gel is allowed to gradually flow in by squeezing the rubber bulb so that the agar flows without interruption and fills the tilted side arm completely up to the cathode compartment. The pipet tip is slowly withdrawn while squeezing agar so that the gel fills flush or slightly above the side arm tip. If air is introduced, the process must be repeated with fresh gel to displace the previous air-interrupted gel. The cell is retained in the appropriate tilted position until the uninterrupted agar gel solidifies completely (2–3 min). Excess gel may be stored in a tightly stoppered flask for use after liquifaction at 100–110°. Cells are conveniently stored in large covered beakers with agar salt bridges moistened with saturated sodium chloride at both ends.

Electrochemical Reduction of Sample

Prior to analysis, the saturated sodium chloride and three subsequent water washes are aspirated from the mercury in each cell. The wet mercury and stirring bar are then poured into a 100-ml beaker containing a generous piece of filter paper on the bottom. The mercury is cleaned and dried by swirling it several times over the paper. The mercury and stirring bar are returned to the cell. The mercury should completely cover the platinum electrode at the bottom of the cell and allow the stirring bar to rotate freely. The cell is next clamp-mounted over a magnetic stirrer with its salt bridge exit tip immersed in a small container of the alkaline saturated sodium chloride. Washed, dried, and mounted cells are filled with 8 to 10 ml of urine sample, leaving minimal headspace beneath the pinhole-vented plastic caps. Moderate magnetic stirring is initiated.

The cells are securely connected in series to the milliammeter and power supply with the miniature alligator clip jumper leads as follows. The negative terminal of the power supply is attached to the platinum–mercury cathode of the first cell, then its positive salt bridge platinum anode was attached to the platinum–mercury cathode of the next cell in the series. This series connection between adjoining cells is continued until the last platinum salt bridge anode is connected to the positive terminal of the power source through the milliammeter. The power supply current is adjusted to 2.5 mA. The voltage across each cell is measured to locate those with abnormally high voltage drops due to a high resistance, deteriorated agar bridge. When a cell has a voltage drop greater than 5 V it is replaced. Electrolysis proceeds for 1 hr with adjustment of the 2.5 mA current, if required. Without delay exactly 5 ml of the reduced urine sample is solvent partitioned and analyzed as described.[1a]

Cells are prepared for the next electrolysis by first shutting off the power supply and stirrers. The urine and three successive, briefly stirred water washes are aspirated. Saturated sodium chloride is added to each cell. The electrolysis, as just described for urine samples, is performed for 30 min. After 30 min, the electrolysis leads at the power supply terminals are reversed for 10 sec. The power source is then turned off. Leads are disconnected. Cells are demounted for storage in large covered beakers with salt bridges moistened with saturated sodium chloride at both ends.

Comments

Electrochemical Reduction

The electrical characteristics of a typical cell included a resistance across the agar bridge of 1600–1700 ohms, a voltage drop across the cell of 3–4 V, and a cathode voltage, versus silver–silver chloride, of -0.9 to -1.0 V. These characteristics allow efficient disulfide reduction by minimizing current-wasting reductions of hydrogen ion, oxalic acid,[2] and DTPA–metal complexes. Excessive hydrogen ion reduction to hydrogen gas may increase the pH of the electrolysate by two or more pH units. This is undesirable since thiols may undergo air oxidation[4] to disulfides above pH 3.

Part of a pool of five normal human urines, stabilized with DTPA–citric–oxalic acids, was fortified with captopril disulfide to a final concentration of 50 μg/ml. Table I presents results of a time and current study of the electrochemical reduction of this fortified urine subsequently carried through the extraction analysis. The unfortified human urine pool (control), reduced electrochemically at 2.5 mA for 1 hr, yielded a colorimetric

TABLE I
EFFECT OF TIME AND CURRENT ON REDUCTION
OF CAPTOPRIL DISULFIDE[a] IN NORMAL HUMAN
URINE POOL

Time (min)	Recovery (%) at current (mA)		
	2	2.5	5
20	53.4	—	68.7
40	84.4	—	92.0
60	97.3	99.3, 98.1, 100.1	93.0

[a] 50 μg/ml.

response, through the extraction analysis, of 0.003 absorbance units, equivalent to a reagent blank.

Table I indicates a possible decrease of reduction at 5 mA. Subsequently, a current of 2.5 mA was used for 1 hr with the results indicated. Whereas a 1-hr reduction was necessary here, Saetre and Rabenstein[5,10] reported a complete 10-min reduction with cystine and with penicillamine and glutathione disulfides. The amino groups of the three disulfides may facilitate disulfide reduction by shifting electrons from the disulfide linkage.[13] The 1-hr electrochemical reduction was found to be valid for captopril–cysteine disulfide. However, a time–current study comparable to that in Table I was not done with the mixed disulfide.

In view of the fragility of the agar bridge in routine use, several substitutes have been studied, including a sturdy polyacrylamide gel,[14] sintered glass disks,[11] tightly rolled paper wads in each end of a glass U tube containing the saturated sodium chloride,[15] and 4-mm porous disks.[16] The porous glass (pore diameter 4 μm) acts as a semipermeable membrane providing ultra-low liquid leakage rates and minimum voltage drop through the tip.[16] A simple porous glass disk bridge (Fig. 2) has the same resistance as the agar bridge and yields captopril analyses very comparable to that with the agar bridge; it is considered a preferable substitute for the agar bridge. Incorporating the porous glass disk at its end, a tapered glass tube enters a hole in the previously needle-punctured plastic cap (from a 5-ml plastic container) at the top of the cell illustrated. The 4 × 2.9

[13] W. W. C. Chan, *Biochemistry* **7**, 4247 (1968).
[14] J. A. Rothfus, *Anal. Biochem.* **16**, 167 (1966).
[15] M. Szyper, Squibb Institute for Medical Research, New Brunswick, New Jersey.
[16] "Handling Instructions for Glassware Incorporating Vycor Tips," No. 11328-A-MD. Princeton Applied Research, Princeton, New Jersey.

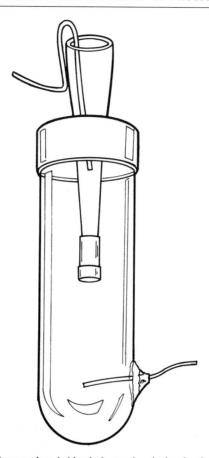

FIG. 2. Porous glass-bridged electrochemical reduction cell.[16]

mm (diameter × thickness) porous glass disk in heat shrunk polytetrafluoroethylene tube (disk and tubing, Catalog No. G0100, Princeton Applied Research, Princeton, NJ) previously had been slipped over the glass taper. The plastic tubing, well away from the porous glass disk, was then heat shrunk with a heat gun. The assembly was completed with a platinum electrode clipped to the wide top of the glass tube, which contains saturated aqueous sodium chloride. Both agar and porous glass bridged cells have been used routinely in our laboratories in an analysis for "total" urinary captopril, as has a subsequently developed HPLC with electrochemical detection.[17]

The reported electrochemical reduction utilizies an ammeter, a relatively sophisticated power supply, and an integrated reduction cell. Sim-

[17] P. Yeh, *East. Anal. Symp., 1980,* Abstracts, No. 60, p. 54 (1980).

pler versions using a battery, a variable resistor, beakers, and detached agar U bridges have been used.[12] We have also used[18] a simpler version for dye reduction with battery, variable resistor, three-necked flask, pencil lead (graphite) instead of platinum electrodes (obtained by burning off the wood casing), and the porous glass bridge. Potentials were measured with a pH meter in the millivolt mode.

Smaller cells[5] have been used for reduction of lesser volumes of blood and urine. Other reduction media such as 2% hydrochloric acid[3] or 3% sulfosalicylic acid[4] have been used for reduction of blood, urine, or tissue homogenates.

Acknowledgments

The author thanks Drs. Glen Brewer, Allen Cohen, Klaus Florey, Doris McKinstry, Bruce Migdalof, Sampat Singhvi, and Mr. Raymond Poet for support and suggestions; Peter Hazlett for construction of the electrochemical reduction cells; Jose Alcantara for designing Figs. 1 and 2; and Eva Johnson, Lois George and Rita Kalasin for technical assistance.

[18] M. Doremus and R. Robison, Squibb Institute for Medical Research, New Brunswick, New Jersey.

[48] Formation of Disulfides with Diamide

By NECHAMA S. KOSOWER and EDWARD M. KOSOWER

Some 20 years ago, we required a reagent capable of producing a rapid diminution of the tripeptide thiol, glutathione (GSH), within red blood cells. At that time, acetylphenylhydrazine was used to induce slow decreases in GSH concentration. Significantly, oxygen was required for its action, a clue which suggested that the diazene, acetylphenyldiazene, was the active agent. Several diazenecarbonyl derivatives were then found to be active for the conversion of thiols to disulfides. The most convenient and chemically simple agent was diazenedicarboxylic acid bis(N,N-dimethylamide), now known by the trivial name, diamide.[1]

$$(CH_3)_2NCON{=}NCON(CH_3)_2$$
Diamide

[1] N. S. Kosower, E. M. Kosower, B. Wertheim, and W. S. Correa, *Biochem. Biophys. Res. Commun.* **37,** 593 (1969).

Chemical Background and Reactions

The reaction of thiols with diazenecarbonyl derivatives occurs in two observable stages, with thiolate anions (RS^-) as the reactive species.[2] The reaction proceeds via addition and displacement steps. For GSH, the GS^- anion adds to the diazene double bond to form a sulfenylhydrazine [Eq. (1)], which, in a second step, reacts with a second GS^- anion at sulfur to yield a disulfide and a hydrazine [Eq. (2)]. The formation of mixed disul-

$$XCON=NCOX + GS^- \rightarrow XCON(SG)NHCOX \quad (1)$$
$$XCON(SG)NHCOX + GS^- \rightarrow GSSG + XCONHNHCOX \quad (2)$$

fides can occur and is easily explained with the two-step mechanism, since the RS^- anion in the displacement reaction (e.g., GS^-) can be replaced by another thiolate.

Acidic low molecular weight thiols are oxidized in preference to protein thiols. The latter are generally less acidic and more sterically hindered. GSH reacts with diamide rapidly, the rate constant being 300 M^{-1} sec^{-1} (aqueous buffer at pH 7.3). GSH is 3–10 times more reactive than other nonprotein thiols (NPSH) such as dithiothreitol, coenzyme A, cysteine, and lipoic acid.[3]

Proteins vary widely in their response to diamide. Some (human hemoglobin A, albumin) exhibit little or no reactivity while others (rat hemoglobin, thioredoxin, spectrin) react with diamide or the diamide–SR intermediate to yield disulfides or mixed disulfides.

Biological Background and Reactions: Thiol Status

The *thiol status* of a biological system may be described as the distribution of thiols, exogenous or endogenous, among their different chemical forms, including thiols (GSH, protein thiols), thiol esters (acyl-CoA), and disulfides (GSSG, CoASSCoA, protein–GS mixed disulfides, protein–protein disulfides).[4] A quantitative expression of thiol status can be given as follows: Thiol status (TS) = directly measurable forms of thiol [mM] per all forms of thiols and disulfides [mM]. (The total includes all SH and SS equivalents; some species may require special assays to reveal thiol or disulfide content as with thiol esters and certain disulfides.) The experimental basis for both quantities must be specified. For human red blood cells, TS = 13/35 (mBBr/hemolysis, denaturation, DTNB). Treat-

---

[2] E. M. Kosower and H. Kanety-Londner, *J. Am. Chem. Soc.* **98,** 3001 (1976).
[3] E. M. Kosower, W. Correa, B. J. Kinon, and N. S. Kosower, *Biochim. Biophys. Acta* **264,** 39 (1972).
[4] N. S. Kosower and E. M. Kosower, *Int. Rev. Cytol.* **54,** 109 (1978).

ment of cells with diamide perturbs the thiol status, decreasing the thiol content and increasing the amount of disulfides. The time course of TS perturbation can vary with the cell under treatment from variations in mixed disulfide formation and capacity to regenerate GSH.

GSH is the major nonprotein thiol in most cells, and is thus the species preferentially affected initially by the addition of stoichiometric amounts of diamide. With greater amounts of diamide, certain reactive intracellular and membrane protein thiols are oxidized. Diamide reacts with nonthiol substrates much more slowly then with thiols (e.g., NADH reacts 0.01 as fast as GSH).[3]

Properties of Diamide

Diamide is a yellow, nonhygroscopic solid, easily soluble in both water and organic solvents, and rather stable toward hydrolysis ($t_{1/2}$ ~3000 hr at pH 7.4, 25°). Stock solutions of 10 mM or more can be prepared in water or in any nonthiol buffer and frozen or kept at 0–4° for long periods. The UV–visible absorption maxima of diamide are as follows: (CH_2Cl_2) λ_{max} 290 nm (ε_{max} 1800), 435 nm (50), (H_2O, 135 mM NaCl–10 mM phosphate buffer at pH 7.4) λ_{max} 296 nm (ε_{max} 3000), 425nm (50). The reaction of diamide with other molecules can be followed spectrophotometrically between 300 and 325 nm. The diazene forms the hydrazine, which does not absorb down to 230 nm [Eqs. (1) and (2)]. The data yield rate constants for a second-order reaction. Diamide is commercially available from many sources.[5]

Diamide penetrates cell membranes within seconds and also reacts within the cell at a high rate (seconds to minutes) at physiological pH. Diamide is effective in the absence of oxygen, conditions under which metabolic activity is minimal. There is a simple stoichiometric relationship between the amount of agent added and the quantity of thiol reacted. Since the reaction of diamide with thiols has a low activation energy, thiol oxidation is fast at low temperatures. Reaction can be terminated through removal of diamide by washing of the cells or stopped instantaneously by the addition of acid. In most cases, diamide treatment does not cause any irreversible damage, and, after incubation of the cells with appropriate substrates at a suitable temperature, the original thiol status is recovered. Diamide treatment thus allows the study of cell functions altered by a perturbation in thiol status.

[5] Listed under diamide, azodicarboxylic bisdimethylamide, diazenedicarboxylic acid *bis* (*N,N*-dimethylamide).

Treatment Procedures

Reaction in Solutions. A solution of thiol (cysteine, glutathione, proteins, etc.) is mixed quickly with diamide solution. In certain cases, deoxygenation of the solutions is required. Diamide can be removed from aqueous reaction solutions by several extractions with CH_2Cl_2 (1 volume/ 3 volumes reaction solution). The amount of unreacted diamide can thus be determined even when there are interfering absorbing materials in the aqueous medium.

Reactions in Red Blood Cells. Diamide solution, 0.1–2.0 mM or more, in isotonic buffer is added to an equal volume of washed red blood cells (5–50% cells suspended in isotonic buffer) with rapid mixing with a vortex apparatus. The reaction is allowed to proceed for a few minutes at 1–4° or at any desired higher temperature. Dilute diamide solutions should be used for cell suspensions in order to ensure an efficient, rapid mixing of the agent with the total cell population. As diamide penetrates cell membranes very rapidly, small volumes of concentrated diamide solution might lead to extensive reaction within that part of the cell population first encountered by the added reagent.

About 0.6–0.7 mol of diamide is required to oxidize 1 mol of nonprotein thiol (NPSH), mostly GSH, in human erythrocytes [0.5 mol diamide is required per mol of GSH, Eqs. (1) and (2)]. One volume of 0.15 mM diamide, for example, would oxidize most of the cellular GSH when added to 1 volume of 10% human erythrocytes. Samples are removed from the reaction solution and analyzed by the desired method, after stopping the reaction with acid. To preserve the cells, reaction should be terminated by washing of the cells with buffer. Higher concentrations of diamide and 10 min or longer incubations at 1–4° or at 37° are used for oxidation of reactive membrane protein thiols to intrachain and interchain disulfides. (Human hemoglobin SH groups are not oxidized by diamide.) Larger amounts of diamide are required for the oxidation of GSH in erythrocytes containing diamide-reactive hemoglobin SH groups (rat Hb forms disulfides, Hb–SG, and Hb–membrane protein mixed disulfides[6]).

Regeneration of NPSH and reduction of protein disulfides is achieved in the normal human red cell by incubation of the diamide-treated cells with glucose at 37°. Cells either washed free of reagent or unwashed (containing diamide) can be used. With washed cells, almost complete regeneration of GSH occurs within 30 min. In the presence of diamide, the time for regeneration is lengthened by the time required to consume the remaining reagent. Regeneration does not occur in red cells lacking

[6] N. S. Kosower, E. M. Kosower, and R. L. Koppel, *Eur. J. Biochem.* **77**, 529 (1977).

NADPH (glucose-6-phosphate dehydrogenase deficiency)[7] nor in cells in which GSSG formation has been precluded by the production of mixed GS–protein disulfides and which lack the enzymatic capacity to reduce such disulfides (rat red cells).[6]

Analytical Procedures for Thiols in Diamide-Treated Red Cells. Either Ellman's reagent, DTNB,[8] or the fluorescent labeling agent, monobromobimane (mBBr),[9–11] can be used for analysis. For the DTNB method, it is usual to first prepare nonprotein, hemoglobin, and membrane fractions and then add DTNB to each fraction.[12] In the mBBr procedure, the reagent may be added to intact cells before processing, an approach which prevents loss of thiols during fractionation. The mBBr procedure makes possible quantitation of thiols in individual proteins and in the NPSH.[13–16] (see this volume [13] and [14]).

---

[7] N. S. Kosower, Y. Zipser, and Z. Faltin, *Biochim. Biophys. Acta* **691**, 345 (1982).

[8] G. L. Ellman, *Arch. Biochem. Biophys.* **82**, 70 (1959).

[9] E. M. Kosower, B. Pazhenchevsky, and E. Hershkowitz, *J. Am. Chem. Soc.* **100**, 6516 (1978).

[10] N. S. Kosower, E. M. Kosower, G. L. Newton, and H. M. Ranney, *Proc. Natl. Acad. Sci. U.S.A.* **76**, 3382 (1979).

[11] R. C. Fahey, G. L. Newton, R. Dorian, and E. M. Kosower, *Anal. Biochem.* **107**, 1 (1980).

[12] *Non-protein Acid Solution.* Residual diamide (unwashed cells) is extracted from the metaphosphoric acid solution with CH_2Cl_2 prior to raising the pH to 7.5. The reaction of diamide with the undissociated thiol present in acid solution is extremely slow; however, diamide is stable in the acid solution and should be removed to avoid a slow reaction with the anion, $^-SC_6H_3(COO^-)NO_2$, formed by the reaction of thiolate anion with the DTNB reagent.

Proteins of Diamide-Treated Cells. Electrophoresis is carried out in the absence and in the presence of dithiothreitol (DTT) or mercaptoethanol, in order to identify any high molecular weight fractions containing disulfide-crosslinked proteins. Two-dimensional electrophoresis (first dimension: −DTT, second dimension: +DTT) is used for separation of the proteins contributing to the high molecular weight material. For the detection of thiols in the gel protein fractions, the mBBr procedure is used. Details for the preparation of fractions and analyses of thiols are covered elsewhere in this volume.

[13] N. S. Kosower, E. M. Kosower, Y. Zipser, Z. Faltin, and R. Shomrat, *Biochim. Biophys. Acta* **640**, 748 (1981).

[14] N. S. Kosower and E. M. Kosower, in "Functions of Glutathione: Biochemical, Physiological, Toxicological, and Clinical Aspects" (A. Larson *et al.*, eds.), Nobel Conf., p. 307. Raven, New York, 1983.

[15] G. L. Newton, R. Dorian, and R. C. Fahey, *Anal. Biochem.* **114**, 383 (1981).

[16] R. C. Fahey and G. L. Newton, in "Functions of Glutathione: Biochemical, Physiological, Toxicological, and Clinical Aspects" (A. Larsson *et al.*, eds.), Nobel Conf., p. 251. Raven, New York, 1983.

Additional Substrates and Agents

Other Cells and Tissues. A variety of isolated cell suspensions can be treated by simple mixing with a diamide solution. The types of cells which have been studied include isolated fat cells,[17] neutrophiles,[18] lymphocytes,[18,19] bacteria,[20,21] sea urchin eggs,[20] *Chlamydomonas*,[22] Ehrlich ascites tumor cells,[23] mammalian spermatozoa,[24] and eggs.[25] Diamide can also be used on monolayers,[26] tissue slices,[27] neuromuscular preparations,[28] in the perfusion of liver,[29] and in intravenous injections.[30] It should be emphasized, however, that since diamide reacts so rapidly with reactive thiols and penetrates cells very fast, intramuscular or subcutaneous injection of the agent will not affect the total body thiol content directly.

The usual concentrations of diamide used for various cell suspensions (5 to 50 × 10^6 cells/ml) and tissue preparations are 0.05–0.5 mM. Exceptionally, cells are extremely sensitive to diamide, and lower concentrations are used (e.g., spermatozoa, for which concentrations of 1–10 μM are used).[24] The rapid regeneration of thiols, which occurs in many types of metabolically active cells, may necessitate higher concentrations of diamide than those calculated on the basis of the measured cellular thiol content.

Depending on the type of cell studied, diverse functional consequences are observed as the result of diamide oxidation of thiols to disulfides. These may be studied on the cellular level (e.g., bacterial growth,

---

[17] B. J. Goldstein and J. N. Livingston, *Biochim. Biophys. Acta* **513**, 99 (1978).
[18] J. M. Oliver, D. F. Albertini, and R. D. Berlin, *J. Cell Biol.* **71**, 921 (1976).
[19] N. S. Kosower, Z. Faltin, and E. M. Kosower, *J. Immunol. Methods* **41**, 215 (1981).
[20] N. S. Kosower and E. M. Kosower, *in* "Glutathione" (L. Flohe, H. C. Benoir, H. Sies, H. D. Waller, and A. Wendel, eds.), p. 276. Thieme, Stuttgart, 1974.
[21] P. Apontoweil and W. Berends, *Biochim. Biophys. Acta* **399**, 10 (1975).
[22] J. R. Warr and D. Quinn, *Exp. Cell Res.* **104**, 442 (1977).
[23] J. W. Harris, N. P. Allen, and S. S. Teng, *Exp. Cell Res.* **68**, 1 (1971).
[24] A. D. Fleming, N. S. Kosower, and R. Yanagimachi, *Gamete Res.*, **13**, 93–102 (1986).
[25] B. R. Zirkin, D. A. Soucek, T. S. K. Chang, and S. D. Perreault, *Gamete Res.* **11**, 349 (1985).
[26] M. D'Urso, C. Mareni, D. Toniolo, M. Piscopo, D. Schlessinger, and L. Luzzatto, *Somatic Cell Mol. Genet.* **9**, 429 (1983).
[27] J. Hewitt, D. Pillion, and F. H. Leibach, *Biochim. Biophys. Acta* **363**, 267 (1974).
[28] R. Werman, P. L. Carlen, M. Kushnir, and E. M. Kosower, *Nature (London) New Biol.* **233**, 120 (1971).
[29] T. P. M. Akerboom, M. Bilzer, and H. Sies, *J. Biol. Chem.* **259**, 5838 (1984).
[30] N. S. Kosower and E. M. Kosower, *in* "Glutathione" (L. Flohe, H. C. Benoir, H. Sies, H. D. Waller, and A. Wendel, eds.), p. 216. Thieme, Stuttgart, 1974.

radiation sensitivity, selection of mutants, capping of lymphocytes, acrosome reaction) or on biochemical and enzymatic reactions.

Other Diazene Agents. The rate of penetration of diazenecarbonyl agents into cells can be controlled by varying the chemical structure. A series of agents derived from piperazine has been prepared and used. These include DIP and DIP + 2.[31,32]

Acknowledgments

The authors are grateful for the support of their research on thiols and thiol agents by the National Institutes of Health, the United States–Israel Binational Science Foundation, the Israel Academy of Sciences, the Chief Scientist's Office, Israel Ministry of Health, the European Research Office, the Petroleum Research Fund, the J. S. Guggenheim Foundation, and the National Science Foundation.

[31] E. M. Kosower, N. S. Kosower, H. Kanety-Londner, and L. Levy, *Biochem. Biophys. Res. Commun.* **59**, 347 (1974).

[32] N. S. Kosower, E. M. Kosower, and L. Levy, *Biochem. Biophys. Res. Commun.* **65**, 901 (1975).

[49] Sulfinic Acids from Disulfides: L-[^{35}S]Cysteinesulfinic Acid

By OWEN W. GRIFFITH and CATHERINE L. WEINSTEIN

Introduction[1]

Sulfinic acids (RSO_2H) are moderately stable compounds of limited natural occurrence. Only cysteinesulfinate and hypotaurine have been detected in tissues or physiological fluids of normal animals. Two additional sulfinic acids, β-sulfinopyruvate and β-sulfinoacetaldehyde, are formed *in vivo* but spontaneously decompose with loss of SO_2.[2] Homocysteinesulfinate has been detected in the urine of patients with homocystinuria.[3] In addition to studies of their role as sulfur amino acid metabolites, sulfinic acids have received considerable attention as analog of carboxylic acids. Cysteinesulfinic acid, an analog of aspartic acid, is, for

[1] Studies from the authors' laboratory were supported by the National Institutes of Health, Grant AM26912. O.W.G. is an Irma T. Hirschl Career Scientist. We thank Ernest B. Campbell and Michael A. Hayward for excellent technical assistance.

[2] O. W. Griffith, this volume [63].

[3] S. Ohmori, H. Kodama, T. Ikegami, S. Mizuhara, T. Oura, G. Isshiki, and I. Uemura, *Physiol. Chem. Phys.* **4**, 286 (1972).

$$\text{2 mol Cystine} \xrightarrow{H_2O_2} \text{Thiosulfonate intermediate} \xrightarrow{NH_4OH} \text{Cysteinesulfinic acid (2 mol)} + \text{Cystine (1 mol)}$$

example, effectively transaminated by aspartate aminotransferase *in vitro* and *in vivo*[4] and is desulfinated by bacterial aspartate β-decarboxylase.[5] Cysteinesulfinate and homocysteinesulfinate are transported by the dicarboxylic amino acid carriers present in many biological membranes.[6-8] The ability of many enzymes and transport carriers to efficiently bind both carboxylic and sulfinic acids is somewhat surprising in that sulfinic acids are both larger than carboxylic acids and are nonplanar (the sulfur atom and its substituents are pyrimidal due to a nonbonding electron pair on sulfur). Sulfinic acids are stronger acids than carboxylic acids but are weaker than sulfonic acids. The —SO_2H pK_a values of cysteinesulfinic acid and homocysteinesulfinic acid are 1.50 and 1.66, respectively, whereas the α-COOH groups of those amino acids have pK_a values of 2.38 and 2.60, respectively.[8] For comparison, the pK_a of the β-COOH of aspartic acid is about 3.7.

The naturally occurring sulfinic acids and their analogs are generally made by reduction of the corresponding sulfonyl chloride,[9] reduction of a benzyl sulfone,[10,11] or by alkaline cleavage of a thio-sulfonate intermediate

---
[4] T. P. Singer and E. B. Kearney, *Biochim. Biophys. Acta* **14**, 570 (1954).
[5] K. Soda, A. Novogrodsky, and A. Meister, *Biochemistry* **3**, 1450 (1964).
[6] G. C. Gazzola, V. Dall'Asta, O. Bussolati, M. Makowske, and H. V. Christensen, *J. Biol. Chem.* **256**, 6054 (1981).
[7] H. N. Christensen and M. Makowske, *Life Sci.* **33**, 2255 (1983).
[8] F. Palmieri, I Stipani, and V. Iacobazzi, *Biochim. Biophys. Acta* **555**, 531 (1979).
[9] J. H. Fellman, this volume [32].
[10] O. W. Griffith, this volume [50].
[11] D. B. Hope, C. D. Morgan, and M. Wälti, *J. Chem. Soc.* p. 270 (1970).

derived from a symmetrical disulfide. The last method, which is presented here, is based on original studies by Emiliozzi and Pichat.[12] The chemistry of sulfinic acids including other synthetic methods has been reviewed.[13]

Synthesis of L-[^{35}S]Cysteinesulfinate

Radiolabeled Cystine and Other Amino Acid Disulfides. L-[^{35}S]Cystine and some other radiolabeled cystines are available commercially and can be used directly. If the specific activity supplied is higher than necessary, unlabeled cystine may be added to facilitate the manipulations and decrease loss of material due to radiolysis. If the material is in solution, the solvents should be removed using a stream of N_2 or by evaporation at reduced pressure. For some radiolabeled amino acids, the reduced derivative, e.g., L-[^{35}S]cysteine, is less expensive or is the only compound available, e.g., L-[^{35}S]homocysteine synthesized by sodium/liquid ammonia reduction of L-[^{35}S]methionine. In such cases it is necessary to carry out a preliminary oxidation to the disulfide. For amounts up to a few mmol, oxidation in aqueous solution with diamide [1,1'-azobis(N,N-dimethylformamide, Kosower's reagent[14]] is most convenient; for larger amounts of material, air should be bubbled through a solution of the thiol at pH 8 to 9 in the presence of a few milligrams of a ferric salt. With either procedure, small portions of the solution are tested with 5,5'-dithiobis(2-nitrobenzoic acid), Ellman's reagent,[15,16] to verify the complete disappearance of thiol. Following oxidation, the coproducts (e.g., reduced diamide, ferric salt, or oxidized dithiothreitol if the original thiol was shipped in a dithiothreitol containing solution) are removed by absorbing the amino acid disulfide to a small column of Dowex 50 (H^+) resin, washing noncationic contaminants through with water, and eluting the amino acid with 3 M ammonium hydroxide. Evaporation of that eluent generally gives a quantitative yield of the pure disulfide.

Formation of Thiosulfonate Intermediate. L-[^{35}S]Cystine (48 mg, 2.5 mCi) is placed in a 10-ml round-bottomed flask and suspended in 1 ml of 98% formic acid and 40 μl of concentrated HCl. The suspension is stirred with a magnetic stirring flea at about 20°, and 55 μl of 30% hydrogen peroxide is added. The suspension clears. After stirring for 2–3 hr, the solution is brought to dryness with a rotary evaporator at reduced pres-

---

[12] R. Emiliozzi and L. Pichat, *Bull. Soc. Chim. Fr.* **11–12**, 1887 (1959).
[13] K. K. Andersen, in "Comprehensive Organic Chemistry" (D. N. Jones, ed.), Vol. 3, p. 317. Pergamon, Oxford, 1979.
[14] N. S. Kosower and E. M. Kosower, this volume [13].
[15] P. C. Jocelyn, this volume [11].
[16] G. L. Ellman, *Arch. Biochem. Biophys.* **82**, 70 (1959).

sure (bath temperature: 25–30°), and the residue is dissolved in 0.5 ml of water. That solution is adjusted to pH 3.0–3.5 (<1 µl portions are tested with pH paper) by addition of concentrated NH_4OH (about 20 µl added in small portions with good mixing). The thiosulfonate intermediate begins to precipitate as the pH is lowered and eventually fills most of the volume. Precipitation is completed by chilling the solution overnight at 4°. Care must be taken in adjusting the pH since precipitates do not form at lower pH values, whereas the thiosulfonate is prematurely cleaved at higher pH. The precipitate is collected by centrifugation and washed twice with small portions of ice-cold water.

Cleavage of the Thiosulfonate Intermediate. The precipitate of thiosulfonate intermediate is suspended in 1 ml of water and treated at room temperature with 1 ml of concentrated NH_4OH. After 1 hr, the solution is evaporated to dryness at reduced pressure. A few milliliters of water are added and removed twice to assure complete removal of ammonia. The pH should be near neutrality and a precipitate of cystine, the coproduct, present. Cystine is removed by filtration or centrifugation and can be recycled; the filtrate contains cysteinesulfinate and is usually contaminated with small to moderate amounts of cysteinesulfonate.

Purification of L-[^{35}S]Cysteinesulfinate. The cysteinesulfinate-containing filtrate is applied to column (1.5 × 55 cm) of Dowex 1-X8 (200–400 mesh) that previously had been sequentially washed with 1 M ammonium bicarbonate and water. The load is washed into the resin with a few milliliters of water, and the products are eluted with a linear gradient formed between 400 ml of water and 400 ml of 1 M ammonium bicarbonate. Portions of the fractions are monitored for radioactivity; cysteinesulfinate elutes at about 0.5 M ammonium bicarbonate and is well separated from residual cystine, which elutes in the flow-through, and cysteinesulfonate, which elutes near the end of the gradient. Appropriate fractions are pooled and evaporated to dryness under reduced pressure. The large excess of ammonium bicarbonate can be removed by repeated addition and evaporation of water or, more conveniently, by acidification and passage of the resulting solution over Dowex 50 (H^+). Thus, the dried cysteinesulfinate pool is dissolved in a few milliliters of water, cautiously acidified to pH less than 3 with acetic acid (foaming), and the resulting solution is applied to a small column (0.5 × 5 cm) of Dowex 50. The resin is washed with about 20 ml of water which elutes cysteinesulfinate. Evaporation of that solution yields chromatographically pure L-[^{35}S]cysteinesulfinic acid which should be stored at $-20°$, either in solution or as the solid. The overall yield based on radioactivity is 30–50% (depending on the amount of cysteinesulfonate formed); as noted, much of the radioactivity is recovered as cystine and can be recycled.

Evaluation of Product. Solutions of cysteinesulfinate can be quantitated by either HPLC or conventional ion-exchange amino acid analysis. The purity of radiolabeled cysteinesulfinate is best determined by TLC and autoradiography. On silica gel plates developed in *tert*-butanol:formic acid:water, 75:10:15, cysteinesulfinic acid, cysteinesulfonic acid, and cystine migrate with R_f values of 0.2, 0.1, and 0.02, respectively.

Comments

Variants of the procedure described have been used to prepare large amounts of unlabeled cysteinesulfinate,[12] L-[1-^{14}C]- and L-[3-^{14}C]cysteinesulfinate and unlabeled α-methyl-DL-cysteinesulfinate.[17] Yields are generally in the range indicated here. Spears and Martin have adapted the procedure for use with very small amounts of material (1 mg cystine) by omitting the isolation of the thiosulfonate intermediate and simply treating the dried residue directly with NH_4OH. The product was purified by TLC on cellulose layers.[18] If this approach is taken, it is very important that all residual hydrogen peroxide be removed before the thiosulfonate is cleaved. If cysteinesulfinate is formed in the presence of peroxide, it is quickly and irreversibly oxidized to cysteinesulfonate.

[17] O. W. Griffith, *J. Biol. Chem.* **258**, 1591 (1983).
[18] R. M. Spears and D. L. Martin, *J. Labelled Compd. Radiopharm.* **18**, 1055 (1981).

[50] Amino Acid Sulfones: S-Benzyl-DL-α-methylcysteine Sulfone

By OWEN W. GRIFFITH

Introduction[1]

Sulfones are sulfur derivatives of the general formula RSO_2R' in which the sulfur atom is at an oxidation level between that of sulfoxides ($RSOR'$) and sulfonic acids (RSO_3H). The substituents on sulfur are arranged in a roughly tetrahedral pattern with the bonds to oxygen generally indicated

[1] Studies from the author's laboratory were supported by the National Institutes of Health, Grant AM26912, and the Irma T. Hirschl Foundation. I thank Ernest B. Campbell and Michael A. Hayward for excellent technical assistance.

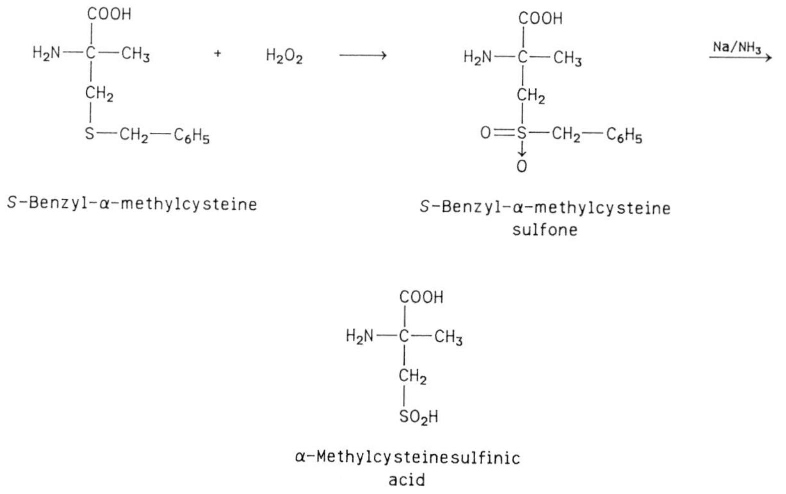

S-Benzyl-α-methylcysteine + H_2O_2 ⟶ S-Benzyl-α-methylcysteine sulfone $\xrightarrow{Na/NH_3}$

α-Methylcysteinesulfinic acid (structure: $H_2N-C(COOH)(CH_3)-CH_2-SO_2H$)

as mixed double–semipolar as shown above. It is noted, however, that independent of the bond representation, the two oxygens of sulfones are equivalent; sulfone sulfur is not a chiral center unless one of the oxygens is isotopically labeled. In general, sulfones exhibit great chemical stability and are reduced or oxidized only under vigorous reaction conditions. In contrast to sulfoxides in which the oxygen is strongly nucleophilic and capable of forming strong hydrogen bonds, sulfone oxygens form hydrogen bonds poorly and are essentially unreactive with conventional alkylating agents such as alkyl halides. In this respect sulfones differ also from the more or less isosteric sulfoximines [RS(O)NHR'] in which the nitrogen is strongly nucleophilic. The chemistry of sulfones has been reviewed in detail.[2]

Biochemical interest in sulfones is based on their stability to oxidation and on specific aspects of their chemistry and stereochemistry. Methionine sulfone, for example, has received attention as a stable derivative of methionine suitable for amino acid analysis[3] and has been tested as an isosteric but nonreactive analog of the convulsant methionine sulfoximine.[4] Sulfone analogs of N-(phosphonoacetyl)-L-aspartate are weak transition state inhibitors of aspartate transcarbamylase,[5] but penicillanic acid sulfone is a potent β-lactamase inhibitor in which sulfone sulfur acts

---

[2] T. Durst, in "Comprehensive Organic Chemistry" (D. N. Jones, ed.), Vol. 3, p. 171. Pergamon, Oxford, 1979.
[3] N. P. Neumann, S. Moore, and W. H. Stein, *Biochemistry* **1**, 68 (1962).
[4] R. A. Ronzio, W. B. Rowe, and A. Meister, *Biochemistry* **8**, 1066 (1969).
[5] G. K. Farrington, A. Kumar, and F. C. Wedler, *J. Med. Chem.* **28**, 1668 (1985).

as a leaving group.[6-8] Sulfone analogs of choline have been used to investigate the relative importance of hydrophobicity and volume in the binding of the quarternary ammonium group to acetylcholinesterase,[9] and the sulfone analog of amanitin, a peptide toxin, has been compared to the diastereomeric sulfoxide analogs in an effort to elucidate the importance of the orientation and bonding properties of the SO group.[10] In addition, several useful drugs (e.g., dapsone) are sulfones or are metabolized to and excreted as sulfones (e.g., sulindac).[11]

S-Benzyl sulfones can be converted to sulfinic acids (RSO_2H), compounds of interest both as naturally occurring metabolites and as isoelectronic and reasonably isosteric analogs of carboxylic acids.[12] This chapter describes the synthesis of S-benzyl-DL-α-methylcysteine sulfone from the corresponding sulfide; the procedure is generally applicable to the synthesis of amino acid and most other water-soluble sulfones. The cleavage of the S-benzyl group to yield α-methylcysteinesulfinic acid is also described. Both procedures were originally developed by D. B. Hope and colleagues.[12]

Synthesis of S-Benzyl-DL-α-methylcysteine Sulfone

Synthesis of S-Benzyl-DL-α-methylcysteine and Other S-Benzyl Amino Acids. S-Benzyl-α-methylcysteine is prepared by the method of Potts[13] in which 1-(benzylthio)-2-propanone is first synthesized from 1-chloro-2-propanone and benzyl mercaptan, and that intermediate is then reacted with ammonium bicarbonate and sodium cyanide to form a hydantoin that can be hydrolyzed to the desired amino acid. A detailed experimental protocol has been published.[14]

S-Benzyl derivatives of amino acids which are available in thiol or disulfide form, e.g., cysteine or homocystine, can be prepared directly by reaction with benzyl chloride. *Benzyl chloride is a potent lacrimator and should be used in an exhaust hood.* For thiol amino acids, the reaction

---

[6] D. G. Brenner and J. R. Knowles, *Biochemistry* **20**, 3680 (1981).
[7] D. G. Brenner and J. R. Knowles, *Biochemistry* **23**, 5833 (1984).
[8] R. Labia, V. Lelievre, and J. Peduzzi, *Biochim. Biophys. Acta* **611**, 351 (1980).
[9] S. G. Cohen, S. B. Chishti, J. L. Elkind, H. Reese, and J. B. Cohen, *J. Med. Chem.* **28**, 1309 (1985).
[10] T. Wieland, C. Götzendörfer, J. Dabrowski, W. N. Lipscomb, and G. Shoham, *Biochemistry* **22**, 1264 (1983).
[11] A. G. Gilman, L. S. Goodman, and A. Gilman, eds., "The Pharmacological Basis of Therapeutics," 6th ed. Macmillan, New York, 1980.
[12] D. B. Hope, C. D. Morgan, and M. Wälti, *J. Chem. Soc.* p. 270 (1970).
[13] K. T. Potts, *J. Chem. Soc.* p. 1632 (1955).
[14] O. W. Griffith, *J. Biol. Chem.* **258**, 1591 (1983).

should be carried out under nitrogen to avoid oxidation to the unreactive disulfide form. Typically, 10–50 mmol of amino acid are dissolved in 20–100 ml of water or, preferably, a water–ethanol mixture, and the pH is adjusted to 9–10 by addition of NaOH. Benzyl chloride (1.2 molar equivalents) is added, and the mixture is vigorously stirred until the reaction is complete. Aliquots can be tested for residual thiol with Ellman's reagent. The pH should be maintained in the indicated range by periodic addition of NaOH or by use of a carbonate/bicarbonate buffer system. After the reaction is complete, the product can often be immediately collected by filtration since the *S*-benzyl derivatives are less water soluble than the original thiols. If the product does not crystallize spontaneously, the reaction mixture is acidified and applied to a column of Dowex 50 (H^+ form). Anions and unreacted benzyl chloride are washed through with water or water–ethanol, and the amino acid is eluted with 3 M NH_4OH. Evaporation of that eluent at reduced pressure generally yields the pure *S*-benzyl derivative in nearly quantitative yield.

Disulfides must be reduced before benzylation. On a small scale it is possible to treat the disulfide with 1.2–2 equivalents of dithiothreitol at pH 7.5–9 and then to proceed with the protocol outlined above using enough benzyl chloride to react with all of the thiols present. If the amount of disulfide is greater than a few millimoles, it is more convenient to reduce the disulfide with sodium in liquid ammonia. *This procedure should be carried out in an exhaust hood to avoid exposure to ammonia and benzyl chloride.* The amino acid disulfide or its hydrochloride salt (0.1–0.25 mol) is placed in a 1-liter three-necked round-bottomed flask that is fitted with a mechanical stirrer (glass blade, not Teflon) and a Dry Ice condenser. The condenser is filled, and the flask is immersed in a similar Dry Ice/ethanol slush. Ammonia gas is condensed into the flask until about 500 ml accumulates. The slush bath is then removed and the flask is allowed to warm until the Dry Ice condenser is clearly containing the refluxing ammonia. Small pieces of sodium metal[15] are added to the stirred suspension until a blue color persists for a few minutes, and then solid ammonium chloride is added cautiously in small portions until the blue color is discharged. Benzyl chloride (1.2 equivalents) is added in 5-ml portions to the mixture, and stirring is continued for 30 min. The condenser is then removed and the ammonia is allowed to evaporate. Water is cautiously added to dissolve the salts and product, and the pH is ad-

---

[15] Sodium metal should be stored and manipulated under an inert nonprotic solvent such as xylene and can be cut free of the hydroxide crust and into smaller pieces with a stainless steel knife. Using forceps, small pieces are blotted dry with a tissue and dropped into the reaction mixture. Small amounts of sodium scrap are destroyed by placing them in a higher alcohol, e.g., *sec*-butanol.

justed to neutrality with HCl. The *S*-benzyl derivative is isolated by filtration or Dowex 50 chromatography as described above.

Oxidation of Sulfides to Sulfones. Ammonium molybdate (91 mg) is added to 2.7 ml of water, and 0.63 ml of 70% perchloric acid are cautiously added. The mixture is boiled for 5 min and filtered. *S*-Benzyl-DL-α-methylcysteine (4.51 g, 20 mmol) is added to the cooled filtrate in a 100-ml flask; an additional 5 ml of water is added to allow magnetic stirring of the thick slurry. The mixture is cooled to 0° in an ice bath, and 5.82 ml of 30% H_2O_2 is added in 0.5-ml portions at intervals of a few minutes. The temperature should remain near 0°. After addition is complete, the temperature is allowed to rise to ambient, and the mixture is stirred overnight. The product is collected by filtration and washed successively with 100-ml portions of ice water, ethanol, and ether. The product, *S*-benzyl-DL-α-methylcysteine sulfone, weighs about 4.5 g (18 mmol, 90% yield) and is pure by amino acid analysis. The procedure is general for oxidizing sulfides to sulfones and has been applied to *S*-alkylhomocysteines, e.g., methionine and buthionine.[16] In some cases, it is advantageous to destroy excess H_2O_2 by addition of mercaptoethanol prior to product isolation. Products with high water solubility can be crystallized from the reaction mixture by addition of ethanol or can be isolated by absorption to and elution from Dowex 50 (H^+) as described above.

Synthesis of DL-α-Methylcysteinesulfinic Acid

Cleavage of S-Benzyl Sulfones to Sulfinic Acids. *S*-Benzyl-DL-α-methylcysteine sulfone (3.58 g, 13.9 mmol) is placed in a 500-ml three-necked round-bottomed flask and dissolved in 200 ml of liquid ammonia (the apparatus and procedure are as described above). Small pieces of sodium (BB sized) are added until a blue color persists for 30 min, and then 1 g of ammonium chloride is added in small portions to discharge the blue color. The condenser is removed and the ammonia is allowed to evaporate. The residue is taken up in 100 ml of water, filtered to remove a small amount of insoluble material, and the filtrate is extracted with 50 ml of petroleum ether to remove toluene and dibenzyl. The aqueous portion is evaporated to dryness at reduced pressure and is dissolved in a small volume of water. That solution is applied to a column (1.0 × 25 cm) of Dowex 1-X8 (200–400 mesh, HCO_3^- form), and the resin is washed with 20 ml of water. The product is eluted with a linear gradient formed between 400 ml of water and 400 ml of 1.0 *M* ammonium bicarbonate (pH 8.2), and is identified by testing fractions with *o*-phthalaldehyde.[17] Sulfinic

[16] O. W. Griffith, *J. Biol. Chem.* **257,** 13704 (1982).
[17] J. R. Benson and P. E. Hare, *Proc. Natl. Acad. Sci. U.S.A.* **72,** 619 (1975).

acids elute at about 0.5 M ammonium bicarbonate and are well separated from any contaminating sulfonic acids, which elute later. Fractions containing product are pooled and evaporated to dryness at reduced pressure. Residual ammonium bicarbonate can be removed by adjusting a solution of the product to a pH less than 3 with acetic acid and passing the acidified solution through a small column of Dowex 50 (H^+) using water or 1 M acetic acid as eluent. Yields of sulfinic acids are typically 50–80% using this procedure.

[51] Sulfonic Acids: L-Homocysteinesulfonic Acid

By MICHAEL A. HAYWARD, ERNEST B. CAMPBELL, and OWEN W. GRIFFITH

Introduction[1]

Sulfonic acids (RSO_3H) are strongly acidic compounds containing organic sulfur in its most oxidized state. Whereas sulfenic acids (RSOH) and sulfinic acids (RSO_2H) are easily oxidized by addition of oxygen to sulfur, sulfonic acids are oxidized further, i.e., to sulfate, only by cleavage of the C—S bond, a process usually requiring more extreme reaction conditions. Sulfonic acids are thus among the most stable of organic sulfur derivatives. The chemical stability of cysteic acid (cysteinesulfonic acid) and the ease with which less oxidized forms of cyst(e)ine can be converted to that compound account for the frequent determination of protein cyst(e)ine as cysteic acid during ion-exchange amino acid analysis.[2,3] Cysteinesulfonate has also received attention in biochemical studies as an analog of aspartate and cysteinesulfinate.[4–6] Homocysteinesulfonate is of interest as a glutamate, homocysteinesulfinate, or aspartate

---

[1] Studies from the author's laboratory were supported in part by the National Institutes of Health, Grant AM26912. O.W.G. is recipient of an Irma T. Hirschl Career-Scientist Award.
[2] E. Schram, S. Moore, and E. J. Bigwood, *Biochem. J.* **57,** 33 (1954).
[3] S. Moore, *J. Biol. Chem.* **238,** 235 (1963).
[4] G. C. Gazzola, V. Dall'Asta, O. Bussolati, M. Makowske, and H. N. Christensen, *J. Biol. Chem.* **256,** 6054 (1981).
[5] A. Meister, P. E. Fraser, and S. V. Tice, *J. Biol. Chem.* **206,** 561 (1954).
[6] T. P. Singer and E. B. Kearney, *Arch. Biochem. Biophys.* **61,** 397 (1956).

analog in enzymological,[7–9] neurophysiological,[10–12] and amino acid transport[4,13] studies. It is notable that in comparison to thiols, disulfides, and sulfides, relatively few sulfonic acids occur in nature. In mammalian systems, taurine and its bile acid derivatives, e.g., taurocholic acid, account for most of the sulfonic acid present; cysteinesulfonic acid and β-sulfopyruvate occur in much smaller amounts. High concentrations (150 mM) of isethionic acid (2-hydroxyethanesulfonic acid) are found in squid giant axons[14] with much smaller amounts in mammalian tissues.

Sulfonic acids are easily prepared by oxidation of thiols, disulfides, sulfenic acids and sulfinic acids; the choice of starting material is governed by availability. A variety of oxidizing agents have been used: bromine,[15–17] hydrogen peroxide,[18,19] performic acid,[2,3] p-nitroperbenzoic acid,[20] and dimethyl sulfoxide.[21] The procedure given below is based on studies by Friedmann[15] and Greenstein and Winitz.[16]

Synthesis of L-Homocysteinesulfonic Acid

Note: Gloves should be worn to prevent contact with bromine.

Procedure. L-Homocystine (268 mg, 1.0 mmol) is dissolved in 8 ml of 3 M HCl contained in a 25-ml round-bottomed flask. The solution is chilled on an ice bath and stirred magnetically as bromine (0.8 g, 260 µl, 10 mmol) is added in several portions over a 30-min period. The portions initially added decolorize spontaneously as homocystine is oxidized, but the solution soon develops and maintains a reddish-brown bromine color. After stirring an additional 15 min, the solution is brought to dryness with a rotary evaporator under reduced pressure (water aspirator). The residue is dissolved in 10 ml of water, and the resulting solution is passed through

[7] A. Rosowsky, R. A. Forsch, J. H. Freisheim, R. G. Moran, and M. Wick, *J. Med. Chem.* **27**, 600 (1984).
[8] B. Jollès-Bergeret and M. Charton, *Biochimie* **53**, 553 (1971).
[9] B. Jollès-Bergeret, *Biochim. Biophys. Acta* **146**, 45 (1967).
[10] P. C. May and P. N. Gray, *Life Sci.* **37** 1483 (1985).
[11] D. W. G. Cox, M. H. Headley, and J. C. Watkins, *J. Neurochem.* **29**, 579 (1977).
[12] J. J. Hablitz, *Brain Res.* **247**, 149 (1982).
[13] F. Palmieri, I. Stipani, and V. Iacobazzi, *Biochim. Biophys. Acta* **555**, 531 (1979).
[14] G. G. J. Deffner and R. E. Hafter, *Biochim. Biophys. Acta* **42**, 200 (1960).
[15] E. Friedman, *Bietr. Chem. Physiol. Pathol.* **3**, 1 (1903).
[16] J. P. Greenstein and M. Winitz, "Chemistry of the Amino Acids," Vol. 3, p. 1908. Wiley, New York, 1961.
[17] H. T. Clarke, *Org. Synth., Collect. Vol.* **3**, 226 (1955).
[18] G. Toennies and M. A. Bennett, *J. Biol. Chem.* **121**, 497 (1935).
[19] S. H. Lipton, *J. Agric. Food Chem.* **26**, 1406 (1978).
[20] B. Jollès-Bergeret, *Bull. Soc. Chim. Biol.* **48**, 1265 (1966).
[21] O. G. Lowe, *J. Org. Chem.* **42**, 2524 (1977).

a small column (0.6 × 8 cm) of Dowex 50-X8 (H$^+$ form, 200–400 mesh). The product does not bind and is completely washed through with 10 ml of water. The combined eluents are evaporated to dryness, and the solid product is freed of most residual HCl and HBr by twice adding and then evaporating small portions of water. Residual bromine is removed from the light tan product by adding and evaporating small portions of chloroform 3 times. The final white product is dried in a vacuum desiccator over P$_2$O$_5$. The yield is 347 mg (95%); the m.p. (corrected) is 261–263° (dec.) (literature m.p., 271–172°[19]). The product is free of Cl$^-$ and Br$^-$ by test with AgNO$_3$/HNO$_3$ and is free of homocyst(e)ine by HPLC analysis.

Comments on the Synthesis. Comparable yields are obtained from homocysteine and homocysteine thiolactone. Although the D- and L-enantiomers of homocyst(e)ine are available commercially, for larger scale preparations it may be preferable to prepare them by reduction of the much less expensive D- and L-methionine using sodium in liquid ammonia.[22] With larger scale preparations, the temperature may increase markedly during the addition of bromine, e.g., to 60°.[17] Although such increases do not appear to affect the yield of cysteic acid from cystine,[17] it is noted that the oxidation proceeds smoothly at lower temperatures and that overoxidation of sensitive compounds is less likely if higher temperatures are avoided. Using the conditions recommended here ($0° < T < 10°$), methionine is oxidized to methionine sulfone, tyrosine is converted to a more hydrophobic product (possibly bromo- or dibromotyrosine), and tryptophan is totally destroyed. Serine and threonine are unaffected.

[22] O. W. Griffith, *J. Biol. Chem.* **257,** 13704 (1982).

[52] Sulfoxides

By CARL R. JOHNSON and MARK P. WESTRICK

The most common route to sulfoxides is the oxidation of the corresponding sulfide. This oxidation can be achieved by virtually every class of oxidants. The difficulty is to avoid overoxidation to the sulfone—a step that often occurs with almost the same facility as the first oxidation [Eq. (1)]. Readers of these volumes will be primarily interested in the preparation of sulfoxides from sulfides, the latter representing or structurally related to biological substances such as amino acids, coenzymes, and metabolic intermediates. These sulfides are generally highly polar, multiple functional group compounds that display fair to excellent water solubility. The conversion of a sulfide to the corresponding sulfoxide imparts additional polar character and usually increases the water solubility of the substance. The S=O group is highly polarized, and the oxygen terminus is an excellent hydrogen bonding acceptor. Depending on the metal, either oxygen or sulfur of the sulfoxide functionality can participate in coordination with metal ions. It may be useful to note that the sulfone functionality, on the other hand, is quite inert; it is not a good site for hydrogen bonding, and complexation with strong Lewis acids occurs only in the absence of other Lewis bases.

$$\underset{\text{Sulfide}}{\overset{R'}{\underset{R}{>}}S} \xrightarrow{[O]} \underset{\text{Sulfoxide}}{\overset{R'}{\underset{R}{>}}S\overset{O}{\underset{..}{\diagdown}}} \xrightarrow{[O]} \underset{\text{Sulfone}}{\overset{R'}{\underset{R}{>}}S\overset{O}{\underset{O}{\diagdown}}} \quad (1)$$

Prior to a discussion of preparative methods for sulfoxides, a brief survey of the chemistry of this class of substances will be presented.[1,2] The sulfoxide group is pyramidal; if the remaining lone electron pair on sulfur is consided as a fourth ligand, the shape of the sulfoxide becomes tetrahedral with the sulfur residing at the center of the tetrahedron. With appropriate substitution the sulfur becomes a stereogenic center and enantiomers and diastereomers can result. Sulfoxides undergo stereomutation by the application of heat (130–200°), light, or a number of chemical reagents, e.g., hydrochloric acid, dinitrogen tetroxide, and acetic anhydride. Sulfoxides are readily oxidized to sulfones. Hydrogen peroxide in acetic acid or the presence of metal catalysts (tungsten or molybdenum oxides) or peracids (*m*-chloroperbenzoic acid and potassium persulfate) are useful reagents for this transformation. Sulfoxides are readily reduced to sulfides by diverse reagents including iodide ion under acidic conditions, thiols, phosphines, sodium bisulfite, and a number of boron, silicon, and aluminum hydride reagents. The ability of sulfoxides to act as oxidizing reagents is used to advantage in the many variants of dimethyl sulfoxide (DMSO)-promoted oxidations of alcohols to aldehydes or ketones, most notably the Moffatt oxidation (DMSO, dicyclohexylcarbodiimide, pyridinium trifluoroacetate) and the Swern oxidation [Eq. (2)]. Many reactions of sulfoxides are initiated by electrophilic activation at the sulfoxide oxygen. In addition to the above examples, the Pummerer rearrangements are classics among such reactions. These rearrangements involve

---

[1] T. Durst, in "Comprehensive Organic Chemistry" (N. Jones, ed.), Vol. 3, p. 121. Pergamon, Oxford, 1979.
[2] S. Oae, in "Organic Chemistry of Sulfur" (S. Oae, ed.), p. 383. Plenum, New York, 1977.

reduction of sulfoxide to sulfide with concommitant oxidation of the adjacent carbon; typical is that shown in Eq. (3). The syn thermal elimination of sulfenic acids from sulfoxides bearing β-hydrogens [Eq. (4)] has seen use in the generation of alkenes and in the conversion of penicillin sulfoxides to penam and cepham derivatives. The sulfoxide functionality is capable of stabilizing adjacent carbanions. DMSO has a pK_a of approximately 35; the salt (MeSOCH$_2$Na) formed by reaction of DMSO with sodium hydride has found considerable application in synthetic chemistry as a base and nucleophile.

$$\text{CH}_3\text{-S(O)-CH}_3 + \text{R-CH}_2\text{OH} \xrightarrow{(\text{CF}_3\text{CO})_2\text{O}} \text{CH}_3\text{-S-CH}_3 + \text{R-CHO} \qquad (2)$$

$$\text{CH}_3\text{-S(O)-CH}_3 + (\text{CH}_3\text{CO})_2\text{O} \longrightarrow \text{CH}_3\text{-S-CH}_2\text{-OC(O)CH}_3 \qquad (3)$$

$$\text{(cis-stilbene sulfoxide)} \xrightarrow{\Delta} \text{PhC(CH}_3\text{)=CHPh} + [\text{PhSOH}] \qquad (4)$$

The physical and chemical properties resulting from the presence of another functionality in a sulfide substrate considerably influences the choice of oxidants for the conversion of sulfides to sulfoxides. Occasionally reagent choice is dictated by the stereochemistry desired in the sulfoxide product as various oxidants have been shown to exhibit different stereoselectivity.[3] There follows a brief discussion of the more common and useful oxidants for the conversion of sulfides to sulfoxides with some commentary with respect to their applicability to molecules of biological interest.

Sodium metaperiodate in water or water/methanol mixtures at 0° has the advantage of generally avoiding overoxidation and in that the method seldom fails.[4] The disadvantages are difficulty of isolation of highly water-soluble sulfoxides and/or separation of inorganic salts and high cost of the reagent.

Commercially available *m*-chloroperbenzoic acid (MCPBA) in dichloromethane at 0°,[3] or ethyl acetate at temperatures as low as −40°,[1] generally does not lead to overoxidation. The reagent is moderately expensive, the solvent systems are not always appropriate for highly polar

---

[3] C. R. Johnson and D. McCants, Jr., *J. Am. Chem. Soc.* **87**, 1109 (1965).
[4] N. J. Leonard and C. R. Johnson, *J. Org. Chem.* **27**, 282 (1962).

substances, and some difficulty can be experienced in separation of the *m*-chlorobenzoic acid by-product from the desired sulfoxide.

tert-Butyl hypochlorite in methanol at −40 to −70° avoids the overoxidation problem by the intervention of a sulfurane intermediate [Eq. (5)].

$$\text{R-S-R} + \text{t-Bu-OCl} \xrightarrow[-78°C]{\text{EtOH}} \underset{\underset{\text{Cl}}{|}}{\overset{\overset{\text{OEt}}{|}}{\text{R-S-R}}} \longrightarrow \overset{\overset{\text{O}}{\|}}{\text{R-S-R}} \quad (5)$$

The reagent is readily and rapidly prepared from household bleach, acetic acid, and *tert*-butyl alcohol. Lack of solubility of polar sulfides at the low temperatures that are necessary is a major disadvantage. Complications could arise from the presence of reactive groups including amide NH and electron-rich aryl moieties. The presence of water in the reaction mixture can lead to sulfone formation. Numerous other "positive halogen" reagents have been used in the oxidation of sulfides including *N*-bromo- and *N*-chlorosuccinimide, pyridine–bromine complex, the bromine complex with diazabicyclo[2.2.2]octane, iodobenzene dichloride, and bromine itself in cold, aqueous media.[1-3]

Singlet oxygen ($^1O_2$), generated photochemically in the presence of various sensitizers, oxidizes sulfides to sulfoxides. In the case of simple sulfides, the reaction apparently has the stoichiometry and mechanism illustrated in Eq. (6).[5] In the authors' laboratory singlet oxygen generated

$$\text{R-S-R} + {}^1O_2 \longrightarrow \underset{\text{R-S}^+\text{-R}}{\overset{O-O^-}{|}} \xrightarrow{\text{R-S-R}} 2\,\overset{\overset{O}{\|}}{\text{R-S-R}} \quad (6)$$

photochemically in methanol, using either methylene blue or rose bengal as sensitizer, is used in the large-scale conversion (>500 g) of methyl phenyl sulfide to the sulfoxide. The yield is better than 99% with no detectable formation of sulfone. For dialkyl sulfides some overoxidation can occur, and the reaction requires monitoring. Studies on the photooxidation of methionine, in free form or in protein, have appeared.[5] From a synthetic point of view, some difficulty can be experienced in the separation of the sensitizers from the sulfoxide product with highly polar molecules such as methionine sulfoxide. This problem can be circumvented by the use of polymer-bound rose bengal, but polymer-bound material, appropriate for hydrophilic solvents, is no longer commercially available.

Hydrogen peroxide has long been considered the classic reagent for the conversion of sulfides to sulfoxides. The procedures often result in a product contaminated by sulfone, but the extent of this difficulty varies. For substrates soluble in the medium, oxidation is best carried out in

---

[5] P. K. Sysak, C. S. Foote, and T.-Y. Ching, *Photochem. Photobiol.* **26**, 19 (1977).

acetone at 0°. A definitive advantage of the method is that the by-product of oxidation is water which minimizes difficulties in the separation of product.

A number of other reagents which have been used for the oxidation of sulfides to sulfoxides include iodosobenzene, chromic acid, nitric acid, ozone, and dinitrogen tetraoxide.[1-3] The last is used in a catalytic cycle for the air oxidation of dimethyl sulfide in industrial processes for dimethyl sulfoxide. Of considerable interest is the development of asymmetric induction methods of oxidation. To date, the most notable successes have been achieved by the Kagan group using reagents composed of Ti(O-i-Pr)$_4$/(+)- or (−)-diethyl tartrate/water/t-BuOOH (1:2:1:1) in dichloromethane at −20°; enantiomeric excesses ranged from 75 to 90% for alkyl aryl sulfoxides and from 50 to 71% for dialkyl sulfoxides.[6]

Nonoxidative methods for the preparation of sulfoxides include the elaboration of sulfinyl carbanions by addition of electrophiles and the reaction of sulfinyl chlorides with reactive carbon nucleophiles, e.g., silyl enol ethers. The most important nonoxidative process is the Andersen synthesis of optically pure sulfoxides by the reaction of Grignard reagents with optically pure menthyl sulfinates [Eq. (7)].[7]

$$\underset{\text{Ar}}{\overset{\overset{\overset{O}{\|}}{S}}{\diagdown}}\text{O-Menthyl} \quad \xrightarrow{\text{RMgX}} \quad \underset{\text{R}}{\overset{\overset{\overset{O}{\|}}{S}}{\diagdown}}\text{Ar} \quad (7)$$

Procedures

Sodium Metaperiodate Oxidations.[4] The sulfide (1 mmol) is added to a solution of sodium metaperiodate (225 mg, 1.05 mmol) in 25 ml of water cooled with an ice bath. The mixture is stirred at 0° for about 12 hr, and the precipitated sodium iodate is removed by filtration. For nonionic sulfoxides the filtrate is extracted with dichloromethane, and the extracts are dried over anhydrous magnesium sulfate prior to removal of the solvent by rotary evaporation. For highly water-insoluble sulfides, methanol can be added to the reaction mixture as a cosolvent. Nonvolatile, dichloromethane-insoluble sulfoxides can be isolated by lyophilization.

tert-Butyl Hypochlorite Oxidations.[3,8] Reagent: At room temperature in subdued light, 6 ml of acetic acid and 10 ml of *tert*-butyl alcohol are

---

[6] P. Pitchen, E. Dunach, M. N. Deshmukh, and H. B. Kagan, *J. Am. Chem. Soc.* **106**, 8188 (1984).

[7] K. K. Andersen, J. W. Foley, R. I. Perkins, W. Gaffield, and N. E. Papanikolaou, *J. Am. Chem. Soc.* **86**, 5637 (1964).

[8] C. R. Johnson and J. J. Rigau, *J. Am. Chem. Soc.* **91**, 5398 (1969); P. S. Skell and M. F. Epstein, *Abstr. Pap. Meet., 147th Am. Chem. Soc. 1964*, p. 26N (1964).

added to 150 ml of household hypochlorite bleach (aqueous NaOCl, 5.25%) in a separatory funnel. The contents are shaken and the *tert*-butyl hypochlorite is allowed to separate as an upper layer which is isolated (yield: 9 g of 99% purity). The product is dried over calcium chloride. It may be distilled at 77–78° prior to use but this is generally not necessary.

Oxidation: To a solution of the sulfide (1 mmol) in 10 ml of methanol cooled in a Dry Ice/acetone bath is added 1 equivalent of *tert*-butyl hypochlorite. The bath is removed and the mixture is allowed to come to room temperature slowly. Conversion of the intermediate to the sulfoxide is facilitated by the addition of solid sodium carbonate to the reaction mixture when it has reached a temperature of about $-40°$; this step should be omitted if the product has an acidic group. The salts are removed by filtration and the methanol is removed under reduced pressure to obtain the product sulfoxide.

Oxidation of Methionine to Methionine Sulfoxide with Hydrogen Peroxide.[5] *dl*-Methionine (1.5 g, 10 mmol) is dissolved in 50 ml of water at room temperature, and 30% hydrogen peroxide (2 ml, about 20 mmol) is added. After 3 hr, methanol (25 ml) and acetone (200 ml) are added to precipitate the methionine sulfoxide (a mixture of diastereomers) (yield: 1.5 g, 92%), m.p. 230–232° (browning prior to melting).

[53] Amino Acid Sulfoximines: α-Ethylmethionine Sulfoximine

By OWEN W. GRIFFITH

Introduction[1]

In 1950 Bentley *et al.*[2] and Reiner *et al.*[3] reported that the convulsive agent present in nitrogen trichloride-treated (i.e., "agenized") proteins is a derivative of methionine containing N, H, and O on sulfur. Both groups of investigators proposed and quickly confirmed the correct structure for methionine sulfoximine, the first reported example of this class of sulfur

---

[1] Studies from the author's laboratory were supported in part by the National Institutes of Health, Grant AM26912, and an Irma T. Hirschl Career-Scientist Award. Ernest B. Campbell provided excellent technical assistance.

[2] H. R. Bentley, E. E. McDermott, J. Pace, J. K. Whitehead, and T. Moran, *Nature (London)* **165**, 150 (1950).

[3] L. Reiner, F. Misani, T. W. Fair, P. Weiss, and M. G. Cordasco, *J. Am. Chem. Soc.* **72**, 2297 (1950).

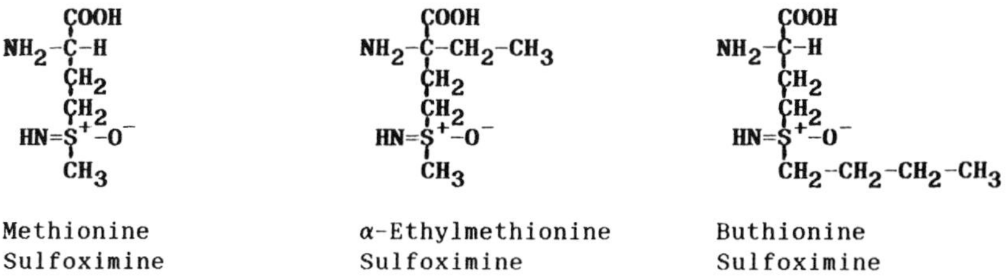

Methionine Sulfoximine	α-Ethylmethionine Sulfoximine	Buthionine Sulfoximine
$\text{NH}_2\text{-CH(COOH)-CH}_2\text{-CH}_2\text{-S}^+(\text{=NH})(\text{O}^-)\text{-CH}_3$	$\text{NH}_2\text{-C(COOH)(CH}_2\text{CH}_3)\text{-CH}_2\text{-CH}_2\text{-S}^+(\text{=NH})(\text{O}^-)\text{-CH}_3$	$\text{NH}_2\text{-CH(COOH)-CH}_2\text{-CH}_2\text{-S}^+(\text{=NH})(\text{O}^-)\text{-CH}_2\text{CH}_2\text{CH}_2\text{CH}_3$

FIG. 1. Structures of biochemically useful sulfoximines.

derivative. (Fig. 1).[4,5] Subsequent studies by Meister's group demonstrated that methionine sulfoximine is bound as a glutamate analog by both glutamine synthetase (glutamate–ammonia ligase, EC 6.3.1.2)[6,7] and γ-glutamylcysteine synthetase (glutamate–cysteine ligase, EC 6.3.2.2)[8,9] and that, in the case of glutamine synthetase, the sulfoximine nitrogen is enzymatically phosphorylated by ATP to form L-methionine S-sulfoximine phosphate.[10] Evidence supporting a similiar mechanism-based inactivation of γ-glutamylcysteine synthetase by sulfoximines has also been presented[9,11]; methionine sulfoximine is, therefore, among the earliest reported examples of a "suicide substrate."[12] Selective inhibition of either glutamine synthetase or γ-glutamylcysteine synthetase is possible *in vitro* and *in vivo* using analogs of methionine sulfoximine. Thus, α-ethylmethionine sulfoximine inhibits only glutamine synthetase[13] whereas prothionine sulfoximine,[14] buthionine sulfoximine,[11,15,16] and higher S-alkyl analogs of methionine sulfoximine[11] are specific inhibitors of γ-glutamylcysteine synthetase (Fig. 1). More recently, hexathionine sulfoximine and heptathionine sulfoximine[11] have been shown to inhibit the γ-glutamyl-histamine synthetase activity of *Aplysia* neurons (D. Weinreich and O. Griffith, unpublished). The syntheses of DL-buthionine (S,R)-sulfox-

---

[4] J. K. Whitehead and H. R. Bentley, *J. Chem. Soc.* p. 1572 (1952).
[5] F. Misani and L. Reiner, *Arch. Biochem. Biophys.* **27**, 234 (1950).
[6] R. A. Ronzio, W. B. Rowe, and A. Meister, *Proc. Natl. Acad. Sci. U.S.A.* **59**, 164 (1968).
[7] R. A. Ronzio, W. B. Rowe, and A. Meister, *Biochemistry* **8**, 1066 (1969).
[8] P. G. Richman, M. Orlowski, and A. Meister, *J. Biol. Chem.* **248**, 6684 (1973).
[9] R. Sekura and A. Meister, *J. Biol. Chem.* **252**, 2599 (1977).
[10] J. M. Manning, S. Moore, W. B. Rowe, and A. Meister, *Biochemistry* **8**, 2681 (1969).
[11] O. W. Griffith, *J. Biol. Chem.* **257**, 13704 (1982).
[12] A. Meister, in "Enzyme-Activated Irreversible Inhibitors" (N. Seiler, M. J. Jung, and J. Koch-Weser, eds.), p. 187. Elsevier, Amsterdam, 1979.
[13] O. W. Griffith and A. Meister, *J. Biol. Chem.* **253**, 2333 (1978).
[14] O. W. Griffith, M. E. Anderson, and A. Meister, *J. Biol. Chem.* **254**, 1205 (1979).
[15] O. W. Griffith and A. Meister, *J. Biol. Chem.* **254**, 7558 (1980).
[16] O. W. Griffith, this series. Vol. 77, p. 59.

imine[16] and L-buthionine (S,R)-sulfoximine[11] have been described in detail; the synthesis of DL-α-ethylmethionine (S,R)-sulfoximine is given below.

Although several procedures for the synthesis of sulfoximines (also known as sulfoximides) have been reported,[17] amino acid sulfoximines are nearly always synthesized by treatment of the corresponding sulfide or sulfoxide with hydrazoic acid. The procedure does not racemize α-amino acids but does yield an R,S mixture at the sulfoximine sulfur; the products are diastereomeric mixtures in which the two diastereomers obtained from an L-amino acid are formed in similiar, although not necessarily identical, amounts. Enantiomerically pure sulfoxides are racemized at sulfur on conversion to sulfoximines with hydrazoic acid. Although the reaction conditions employed (HN_3 in warm H_2SO_4) will cleave or rearrange most carbonyl and carboxyl groups (Schmidt reaction[18]), the adjacency of the α-amino group prevents destruction of the α-carboxyl of amino acids. Groups forming stable carbonium ions (*tert*-butyl, benzyl) are lost from sulfur during attempted sulfoximine formation with HN_3/H_2SO_4, and the method is thus not suitable for such compounds. The S-isopropyl group is borderline in this respect; S-(isopropyl)homocysteine sulfoxide is converted to the sulfoximine in fair yield, but S-(isopropyl)-homocysteine is converted mainly to homocysteine sulfonamide (O. Griffith and A. Meister, unpublished).

Conversion of enantiomeric sulfoxides to sulfoximines with little or no racemization at sulfur is possible using O-mesitylsulfonylhydroxylamine (MSH)[19]; the amino group of amino acids must be blocked (e.g., N-*t*-BOC) to prevent oxidation at that site. Although the MSH precursor, ethyl O-mesitylenesulfonyl acethydroxamate is now commercially available, MSH is an unstable reagent and somewhat difficult to work with; it has not yet been extensively used to prepare amino acid sulfoximines.

Synthesis of DL-α-Ethylmethionine (S,R)-Sulfoximine

Preparation of DL-α-Ethylmethionine

Note: The entire preparation should be carried out in an efficient fume hood to avoid exposure to methanethiol, cyanide, and hydrazoic acid. Glassware contaminated with methanethiol should be soaked in dilute Clorox before being removed from the hood.

---

[17] C. R. Johnson, *in* "Comprehensive Organic Chemistry" (D. N. Jones, ed.), Vol. 3, p. 223. Pergamon, Oxford, 1979.

[18] H. Wolff, *Org. React. (N.Y.)* **3**, 307 (1946).

[19] C. R. Johnson, R. A. Kirchhoff, and H. G. Corkins, *J. Org. Chem.* **39**, 2458 (1974).

Ethyl vinyl ketone (25 g, 0.3 mol) is placed in a 100-ml three-necked round-bottomed flask fitted with a N_2 inlet, thermometer, and reflux condenser (a Dry Ice condenser is preferred but a water-cooled condenser will suffice). The liquid is chilled to 0° by stirring magnetically in an ice/salt bath. Approximately 24 g (0.5 mol, ~27 ml) of methanethiol is condensed into a graduated tube held in a Dry Ice/acetone bath, and the cold thiol is added directly to the ketone under N_2. Piperidine (50 μl) is added as catalyst, and the temperature is held between 0 and 10° for at at least 60 min. The mixture then is allowed to warm to room temperature over about 30 min and is heated (oil bath) to 80° for 60 min to complete the reaction and drive off most of the excess thiol. The flask is then arranged for vacuum distillation, and the pressure is reduced cautiously to about 12 mm Hg (water aspirator). The oil bath is warmed to 115°, and the first 2–3 ml of distillate is discarded. Ethyl 2-(methylthio)ethyl ketone (38.5 g) distills as a clear, oily liquid (bp_{12} 90–91°, 96% yield).

Ethyl 2-(methylthio)ethyl ketone (31.5 g, 0.24 mol, 32.8 ml) is added to a fresh solution of ammonium carbonate (96 g, 1 mol) and sodium cyanide (16.2 g, 0.33 mol) in 700 ml of 50% ethanol contained in a 1-liter round-bottomed flask fitted with a wide-bore reflux condenser. The mixture is stirred magnetically and heated at 60°. The solution, which contains insoluble salts from the beginning, turns slowly brown. After 18 hr the mixture is cooled, and the apparatus is rearranged for distillation. The solution is then reheated, and about half of the solvent is removed by simple distillation. The cooling water to the condenser is temporarily turned off if the condenser begins to clog with ammonium carbonate; the distillate should be collected in an iced flask and disposed of as if containing cyanide. *It is possible that cyanide is evolved during the distillation and during the following procedure.* The cooled, concentrated reaction mixture is poured into a 2-liter beaker, mixed with 200 ml of water, and cautiously acidified to pH less than 3 (pH paper) by addition of concentrated HCl. The product, 5-(2-methylthioethyl)-5-ethylhydantoin, precipitates spontaneously when the mixture is chilled on ice *in the hood* for several hours. The hydantoin is collected by filtration and washed with cold water. After drying the product weighs 39 g (m.p. 136–137°, 81% yield).

The hydantoin (38 g, 0.19 mol) is hydrolyzed to DL-α-ethylmethionine by refluxing for 48 hr in 250 ml of 2 M NaOH (the flask will be etched by this procedure). The cooled reaction mixture is acidified to pH less than 3 with concentrated HCl (CO_2 evolution), and any precipitate of unhydrolyzed hydantoin and silicates is removed by filtration. The filtrate is evaporated to dryness at reduced pressure with a rotary evaporator, and the resulting residue is suspended in 100 ml of concentrated HCl. That mixture is filtered to remove insoluble NaCl, and the filtrate is again evaporated to yield a crude preparation of DL-α-ethylmethionine·HCl. The crude amino acid is dissolved in 100 ml of water and applied to a column (2.5 × 45 cm) of Dowex 50-X8 (200–400 mesh; H$^+$ form). The resin is washed with 1 liter of water, and the amino acid is eluted with 3 M NH$_4$OH. The amino acid-containing effluent is evaporated at reduced pressure with a rotary evaporator to yield 13.9 g of DL-α-ethylmethionine as a white powder (42% yield). It is recrystallized from aqueous ethanol [m.p. 285–286° (dec.)]. An additional 3.8 g of product is obtained by resuspending the recovered hydantoin in NaOH and repeating the hydrolysis procedure.

Preparation of DL-α-Ethylmethionine (S,R)-Sulfoximine

DL-α-Ethylmethionine (5.0 g, 0.03 mol) is placed in a 250-ml round-bottomed flask containing 60 ml of dry chloroform and 17 ml of concentrated H$_2$SO$_4$; the flask is fitted with a reflux condenser and a mechanical stirrer (if magnetic stirring is used, the operation must be closely watched since the viscous product will occasionally stop the stir bar). With vigorous stirring the mixture is warmed to 55° in an oil bath, and sodium azide (9.1 g, 0.14 mol) is added in small (~0.5 g) portions over a 6-hr period. Stirring is continued overnight at 55°. The mixture is then cooled to room temperature and mixed with 400 ml of ice water. The chloroform layer is separated and extracted with 100 ml of water, and the combined aqueous layers are applied to a column of Dowex 50 (H$^+$) as described above. The resin is washed with 1 liter water. (*Caution: the effluent contains toxic hydrazoic acid and should be disposed of in a fume hood drain with a large volume of water.*) The product is then eluted with 3 M NH$_4$OH. The appropriate fractions are evaporated to dryness at reduced pressure, and the crude product is crystallized from aqueous ethanol. DL-α-Ethylmethionine (S,R)-sulfoximine is obtained in 65–75% yield (m.p. 203–204°). DL-α-Ethylmethionine and DL-α-ethylmethionine (S,R)-sulfoximine elute immediately after methionine on a Durrum amino acid analyzer, using sodium citrate buffers and the standard "hydrolysis procedure" recommended by the manufacturer.[13]

Comments on the Synthetic Procedures

Several α-, β-, and γ-substituted methionine analogs can be synthesized from appropriate α,β-unsaturated aldehydes and ketones by the general procedure outlined above. The amount of piperidine catalyst should be reduced substantially with aldehydes (0.1–1 μl) and may have to be increased for efficient reaction of unsaturated ketones with higher molecular weight thiols.[14]

Within the limitations stated in the Introduction the procedure for forming sulfoximines directly from sulfides is general. Synthesis of sulfoximines from sulfoxides is carried out identically but requires less sodium azide (1–1.5 molar equivalents is usually sufficient). In the case of sensitive sulfoxides, e.g., S-isopropyl derivatives, it may be advantageous to reduce the reaction temperature by 10–20°. As noted, sulfoximines formed from amino acids are diastereomeric, and the synthetic mixtures are, in principle, separable by crystallization. Although this possibility has not been exploited synthetically, partial inadvertant resolution of the diastereomers could result in variable biologic activity being exhibited by different crops of inhibitors obtained from the same synthetic reaction mixture. Unless resolution of the diastereomers is intended, it is wise to avoid the generation of multiple small crops during the final crystallization. The diastereomers of L-methionine (SR)-sulfoximine have been resolved by fractional crystallization as the (+)-camphor 10-sulfonate salts[20] and by ion-exchange chromatography.[10]

[20] B. W. Christensen, A. Kjaer, S. Neidle, and D. Rogers, *Chem. Commun.* p. 169 (1969).

[54] Preparation of Sulfur and Selenium Amino Acids with Microbial Pyridoxal Phosphate Enzymes

By NOBUYOSHI ESAKI and KENJI SODA

A number of pyridoxal 5′-phosphate-dependent enzymes have been applied to the preparation of sulfur- and selenium-bearing amino acids. The reverse reaction of α,β-elimination [Eq. (1)], β-and γ-replacement reactions [(Eqs. (2) and (3), respectively)], and γ-addition reaction [Eq. (4)] have been utilized.

$$CH_3COCOOH + RXH + NH_3 \rightarrow RXCH_2CH(NH_2)COOH + H_2O \qquad (1)$$
$$RXCH_2CH(NH_2)COOH + R'X'H \rightarrow R'X'CH_2CH(NH_2)COOH + RXH \qquad (2)$$
$$RXCH_2CH_2CH(NH_2)COOH + R'X'H \rightarrow R'X'CH_2CH_2CH(NH_2)COOH + RXH \qquad (3)$$
$$CH_2=CH-CH(NH_2)COOH + RXH \rightarrow RXCH_2CH_2CH(NH_2)COOH \qquad (4)$$
$$X, X' = S \text{ or } Se$$

Yamada and co-workers have established enzymatic procedures for the synthesis of D- and L-cysteine with bacterial pyridoxal phosphate enzymes.[1] Here, we describe enzymatic methods for the syntheses of L-selenocystine, L-selenohomocystine, S-substituted L-homocysteines, S-

[1] K. Soda, H. Tanaka, and N. Esaki, *Bio Technology* **3**, 479 (1983).

substituted L-cysteines, and Se-substituted L-selenocysteines. Also included are simple methods for the specific labeling of sulfur amino acids with deuterium and tritium. The latter methods are based on the hydrogen exchange reactions catalyzed by L-methionine γ-lyase.

Preparation of S-Substituted L-Homocysteines with L-Methionine γ-Lyase

L-Methionine γ-lyase (EC 4.4.1.11) is a versatile pyridoxal phosphate enzyme as described elsewhere in this volume [79]. Various S-substituted L-homocysteines can be prepared by the enzyme-catalyzed γ-replacement reaction [Eq. (3)].

Procedure

S-(β-Hydroxyethyl)-L-homocysteine is prepared with a mixture (10 ml) containing 0.2 M potassium phosphate at pH 7.8, 4 mmol O-acetyl-L-homoserine, 8 mmol 2-mercaptoethanol, and 40 units L-methionine γ-lyase (20 units/mg).[2] After incubation at 37° for 6 hr, the pH is brought to 12 with 4 N KOH, and then to 2 with 12 N HCl. Remaining O-acetyl-L-homoserine is converted spontaneously to N-acetyl-L-homoserine under these conditions and is easily separated from the product with a Dowex 50-X8 (H$^+$) column (2 × 25 cm). The reaction mixture is applied to the column and washed thoroughly with deionized water. S-(β-Hydroxyethyl)-L-homocysteine elutes with 1 N NH$_4$OH. The product is dried by evaporation and crystallized from hot 95% methanol. The yield was 50% based on O-acetyl-L-homoserine.

Note on Procedure

Several alkanethiols (C$_2$–C$_7$), arylthioalcohols (e.g., benzenethiol), and their derivatives (e.g., cysteamine) can serve as the S-substituent donor in the γ-replacement reaction. Relative velocities of the syntheses of S-substituted L-homocysteines are summarized in Table I. Homologs higher than octanethiol are inert, and thiols with a charged group are generally poor S-substituent donors.

Preparation of S-Substituted L-Cysteines and Se-Substituted L-Selenocysteines with Tryptophan Synthase

Tryptophan synthase (EC 4.2.1.20) is a pyridoxal phosphate enzyme with multiple catalytic functions. The enzyme is applicable to the preparation of S-substituted L-cysteines[3] and Se-substituted L-selenocysteines.

[2] T. Nakayama, N. Esaki, K. Sugie, T. T. Berezov, H. Tanaka, and K. Soda, *Anal. Biochem.* **138,** 421 (1984), see also this volume [79].

[3] See also this volume [41] for the synthesis of S-alkyl and S-aryl cysteine derivatives.

TABLE I
RELATIVE VELOCITY OF SYNTHESIS OF
S-SUBSTITUTED HOMOCYSTEINES FROM
L-METHIONINE AND THIOLS BY L-METHIONINE
γ-LYASE

Thiol compound	Relative rate
Ethanethiol	100
1-Propanethiol	81
2-Propanethiol	38
1-Butanethiol	43
2-Methyl-1-propanethiol	75
1-Methyl-1-propanethiol	27
2-Methyl-2-propanethiol	26
1-Pentanethiol	43
3-Methyl-1-butanethiol	52
2-Methyl-1-butanethiol	12
2-Methyl-2-butanethiol	9
1-Hexanethiol	17
2-Methyl-2-pentanethiol	5
1-Heptanethiol	4
2-Methyl-2-hexanethiol	0
2-Methyl-2-heptanethiol	0
2-Methyl-2-octanethiol	0
1-Decanethiol	0
Cyclohexanethiol	10
Thiobenzyl alcohol	71
Benzenethiol	43
β-Naphthalenethiol	6
Thioglycolic acid	11
Thioglycolic acid ethyl ester	111
2-Mercaptoethanol	84
Thiolactic acid	7
Cysteamine	44
N-Acetylcysteamine	267

Procedure

S-Benzyl-L-cysteine can be prepared as follows: the reaction mixture contains 0.2 mmol L-serine, 20 nmol pyridoxal 5'-phosphate, 10 μmol potassium phosphate at pH 7.8, 0.48 mmol thiobenzyl alcohol, and 0.2 mg of the crystalline $\alpha_2\beta_2$ complex of tryptophan synthase[4] in a final volume of 1.0 ml. It is incubated at 30° for 12 hr under N_2. The product crystallizes in the reaction mixture at 4°. Crystals are collected and washed successively with cold water, cold ethanol, and ether. The yield was 89% based

[4] O. Adachi, L. D. Kohn, and E. W. Miles, *J. Biol. Chem.* **249**, 7756 (1974).

on L-serine. The synthesis of S-benzyl-L-cysteine proceeds as a function of time (0–8 hr) and enzyme concentration.

S-Ethyl-L-cysteine can be prepared essentially in the same manner except that 0.48 mmol of thiobenzyl alcohol is replaced by 0.58 mmol of ethanethiol. After the reaction is performed at 37° for 12 hr, the reaction mixture is chromatographed on Merck 60 F254 silica gel chromatoplates (2 mm thick) with 1-butanol–acetic acid–water (12:3:5, v/v/v) as solvent. The product eluted from the corresponding area is applied to a Dowex 50W-X8 (H$^+$) column (1.0 × 10 cm), and then eluted with 1 N NH$_4$OH. After evaporation, the product is crystallized from 80% hot ethanol. The yield was 83% based on L-serine.

Se-Benzyl-L-selenocysteine is prepared as follows: The reaction mixture contains 150 μmol L-serine, 300 μmol potassium phosphate at pH 7.8, 0.1 μmol pyridoxal 5′-phosphate, 1 mmol of selenobenzyl alcohol, and 6 mg of crystalline $\alpha_2\beta_2$ complex of tryptophan synthase[4] in a final volume of 5 ml in a sealed tube in which air is displaced by N$_2$, and the mixture is incubated at 30° for 16 hr. After deproteinization by addition of HCl, the supernatant solution is neutralized with NaOH, followed by chromatography on a preparative silica gel chromatoplate (Merck 60 F254, 2 mm thick) with 1-butanol–acetic acid–water (12:3:5, v/v/v) as a solvent. The product is eluted with 1 N HCl and applied to a Dowex 50-X8

TABLE II
RELATIVE REACTIVITIES OF VARIOUS THIOLS IN THE SYNTHESIS OF S-SUBSTITUTED CYSTEINES FROM L-SERINE

Thiol compound	Relative rate
Thiobenzyl alcohol	100
Methanethiol	19
Ethanethiol	21
1-Propanethiol	152
1-Butanethiol	113
2-Butanethiol	24
2-Methylpropane-1-thiol	40
2,2-Dimethylpropane-1-thiol	0
1-Pentanethiol	37
1-Hexanethiol	19
1-Heptanethiol	12
1-Octanethiol	3
1-Nonanethiol	1
1-Decanethiol	0
Benzenethiol	8

(H⁺) column (1 × 5 cm). After the column is washed thoroughly with water, the product is eluted with 1 N NH₄OH, followed by evaporation to dryness. The product is crystallized from water. The yield was 44% based on L-serine.

Se-Methyl-L-selenocysteine is prepared essentially as described above except that methaneselenol is substituted for selenobenzyl alcohol. The yield was 16% based on L-serine.

Notes on Procedure

Table II summarizes the reactivities of various thiols in the β-replacement reactions. 1-Propanethiol is most reactive, followed by 1-butanethiol and thiobenzyl alcohol. Bulky alkanethiols such as 2,2-dimethylpropane-1-thiol and 1-decanethiol are inert. Specific activities of the enzyme in β-replacement reaction with selenobenzyl alcohol and methaneselenol are 0.96 and 0.77, respectively, whereas those with thiobenzyl alcohol and methanethiol are 3.2 and 0.6, respectively.

Preparation of L-Selenocystine and L-Selenohomocystine with O-Acetylhomoserine Sulfhydrylase

O-Acetylhomoserine sulfhydrylase [(thiol)-lyase, EC 4.2.99.10][5] catalyzes only γ-and β-replacement reactions: no elimination reactions that lead to a decrease in the yield of product amino acids occur. Therefore, selenohomocystine and selenocystine can be prepared with the enzyme in high yields.

Procedure

Selenohomocystine is prepared as follows: Incubation is carried out at 37° for 3 hr in a reaction mixture containing 1 mmol O-acetyl-L-homoserine, 1.5 mmol potassium phosphate at pH 7.2, 1.5 μmol pyridoxal 5'-phosphate, 4.5 mmol Na₂Se₂, and 27 units of O-acetylhomoserine sulfhydrylase in a final volume of 15 ml. Na₂Se₂ is added in 3 portions over a period of 3 hr to avoid inhibition. Under these conditions, about 0.3 mmol of L-selenohomocystine is synthesized (60% yield based on O-acetyl-L-homoserine). After incubation for 3 hr, the solution is brought to pH 2 with 6 N HCl, and 100 mg of hydroxylamine hydrochloride is added to reduce unreacted Na₂Se₂ to H₂Se, which is removed by bubbling the solution with a stream of nitrogen for 2 hr. Elemental selenium that pre-

---

[5] S. Yamagata, K. Takeshima, and N. Naiki, *J. Biochem.* (*Tokyo*) **75**, 1221 (1974), see also this volume [80].

cipitates is removed by filtration. The yellow filtrate is applied to a Dowex 50-X8 column (15 × 350 cm; H$^+$ form), which is washed successively with water, 2 N HCl, water, and 2 N NH$_4$OH. The eluate with NH$_4$OH is evaporated to dryness under reduced pressure at 30°. The residue is dissolved in a minimal volume of water and crystallized by addition of acetone. The yield was 48% based on O-acetyl-L-homoserine.

L-Selenocystine is prepared as follows: The reaction mixture, containing 0.5 mmol L-serine O-sulfate, 500 mmol potassium phosphate at pH 7.2, 50 nmol of pyridoxal 5′-phosphate, 1 mmol of Na$_2$Se$_2$, and 40 units of O-acetylhomoserine sulfhydrylase[5] in a final volume of 5 ml is incubated at 37° for 3 hr. Na$_2$Se$_2$ is added in 4 portions over a period of 2 hr. After deproteinization, the product is isolated in the same manner as L-selenohomocystine. The yield was 64% based on L-serine O-sulfate.

Preparation of Deuterated and Tritiated L-Methionine and
S-Methyl-L-cysteine with L-Methionine γ-Lyase

Some of the reactions involved are presented along with the properties of the γ-lyase elsewhere in this volume [79].

Procedure

L-[α-^{2}H,β-^{2}H$_2$] Methionine is prepared as follows: The reaction mixture contains 2 mmol L-methionine, 2 mmol potassium phosphate at pH 8.6, and 0.2 mg L-methionine γ-lyase[2] in 5 ml of deuterium oxide. After 4 hr of incubation, the reaction is stopped by addition of 2 ml of 6 N HCl. After centrifugation, the supernatant solution is applied to a Dowex 50-X8 (H$^+$) column (1 × 10 cm), which is washed with water. Fractions containing methionine, eluted with 1 N NH$_4$OH, are pooled and concentrated to a small volume, followed by evaporation to dryness under reduced pressure. The residue is dissolved in a small volume of hot 80% ethanol and allowed to crystallize at 4°. Recrystallization is performed in the same manner. The yield was 81%. S-[α-^{2}H,β^{2}H$_2$] Methyl-L-cysteine can be prepared in the same manner with a yield of 84% when L-methionine is replaced by S-methyl-L-cysteine.

L-Methionine and S-methyl-L-cysteine can be labeled with tritium at the α and β positions by the action of L-methionine γ-lyase. A reaction mixture (0.2 ml) containing L-methionine or S-methyl-L-cysteine (40 μmol), pyridoxal 5′-phosphate (5 nmol), tritiated water (825 μCi), potassium phosphate at pH 8.0 (40 μmol), and enzyme (125 μg) is incubated at 30°. After incubation for 40 min, 50 μl 6 N HCl is added to stop the reaction. After deproteinization by centrifugation, the supernatant solution is applied to a column (0.5 × 6 cm) of Dowex 50-X8 (H⁺), which is washed with water and eluted with 1 N NH₄OH. The radioactive fractions are pooled and evaporated to dryness under reduced pressure. Specific radioactivities are as follows: L-methionine, 8,450; S-methyl-L-cysteine, 8,600 dpm/μmol.

[55] Construction of Assay Diets for Sulfur-Containing Amino Acids

By DAVID H. BAKER

Both methionine and cysteine are required for protein synthesis of simple-stomached mammalian and avian species. Thus, for maximal animal growth, diets must provide these two amino acids, or methionine alone, in a quantity sufficient to meet requirements for maintenance as well as growth. The physiological requirement for cysteine can be met by dietary cystine or by excess dietary methionine which provides metabolic cysteine. The molar efficiency of transsulfuration, i.e. methionine sulfur converted to cysteine sulfur, is 100%.[1,2] No cysteine sulfur is converted to methionine sulfur in higher organisms. From the standpoint of the diet, methionine alone is capable of providing all the needed body sulfur compounds with the exception of the two sulfur-containing vitamins, thiamin and biotin.

Biological Activity of Dietary Sulfur Compounds

D-Methionine is used almost as efficiently as L-methionine in most species, although carefully controlled bioassays suggest a bioefficacy value of D-methionine relative to L-methionine of less than 100%, particularly when the methionine isomers are added to crystalline amino acid diets.[3–5] Adult man obtains less bioactivity from D-methionine than other animals.[6,7] The mixed DL-isomer of methionine has generally been found to contain more bioefficacy than that predicted from its equal contributions of D- and L-methionine.

[1] G. Graber and D. H. Baker, *J. Anim. Sci.* **33**, 1005 (1971).
[2] K. M. Halpin and D. H. Baker, *J. Nutr.* **114**, 606 (1984).
[3] R. S. Katz and D. H. Baker, *Poult. Sci.* **54**, 1667 (1975).
[4] K. P. Boebel and D. H. Baker, *Fed. Proc., Fed. Am. Soc. Exp. Biol.* **42**, 543 (1983).
[5] R. G. Teeter, D. H. Baker, and J. E. Corbin, *J. Anim. Sci.* **46**, 1287 (1978).
[6] C. Kies, H. Fox, and S. Aprahamian, *J. Nutr.* **105**, 809 (1975).
[7] A. Y. Zezulka and D. H. Calloway, *J. Nutr.* **106**, 1286 (1976).

Foods and feeds are often supplemented with a biologically active methionine precursor compound, e.g., the methionine hydroxy analog DL-2-hydroxy-4-(methylthiobutyrate) and N-acetyl-L-methionine. The former has a molar bioefficacy of about 90% while the latter has a molar bioefficacy of 100%, relative to equimolar levels of DL- and L-methionine, respectively.[8-11] Interestingly, D-hydroxymethionine has a higher biological activity than L-hydroxymethionine.[12] N-Acetyl-D-methionine has no bioactivity because the deacetylase enzyme required to remove the acetyl moiety is specific for L-methionine. N-Acetylmethionine was developed as a methionine precursor for human food uses because it has more desirable organoleptic qualities than methionine or hydroxymethionine.

The D-isomer of cyst(e)ine has no biological efficacy when used as a diet supplement for animals.[13] This explains, in part, why dietary lanthionine (a mixture of DL- and *meso*-lanthionine in foods) has no more than 30% biopotency as a dietary cysteine precursor.[14] Lanthionine is found in considerable quantity in certain heat- and/or alkali-processed foodstuffs, e.g., soy flours and feather meals.[15,16] While it does deliver bioavailable cysteine to animals, conventional amino acid analytical methodology does not accurately quantify this cross-linked amino acid.

Another sulfur compound present in many food and feed products is glutathione. This cysteine-containing tripeptide delivers 100% of its cysteine content to the animal as bioavailable cysteine.[17-19] Conventional acid hydrolysis procedures, of course, cause release of the cysteine moiety such that the cysteine present in this compound is measured in the diet as cysteine, therefore being indistinguishable from the measured cysteine–cystine component of the diet.

Both methionine and cyst(e)ine in foods are subject to processing losses of bioactivity. Crystalline sources of both amino acids can react with free aldehyde groups of reducing sugars such as glucose or lactose, forming Maillard products that have no bioactivity.[20] Methionine,

[8] K. P. Boebel and D. H. Baker, *Poult. Sci.* **61**, 1167 (1982).
[9] J. M. Harter and D. H. Baker, *Proc. Soc. Exp. Biol. Med.* **156**, 201 (1977).
[10] J. T. Rotruck and R. W. Boggs, *J. Nutr.* **107**, 357 (1977).
[11] D. H. Baker, *J. Nutr.* **109**, 970 (1979).
[12] D. H. Baker and K. P. Boebel, *J. Nutr.* **110**, 959 (1980).
[13] D. H. Baker and J. M. Harter, *Poult. Sci.* **57**, 562 (1978).
[14] K. R. Robbins, D. H. Baker, and J. W. Finley, *J. Nutr.* **110**, 907 (1980).
[15] D. H. Baker, R. C. Blitenthal, K. P. Boebel, G. L. Czarnecki, L. L. Southern, and G. M. Willis, *Poult. Sci.* **60**, 1865 (1981).
[16] A. P. Degroot and P. Slump, *J. Nutr.* **98**, 45 (1969).
[17] H. M. Dyer and V. du Vigneaud, *J. Biol. Chem.* **115**, 543 (1936).
[18] J. M. Harter and D. H. Baker, *Proc. Soc. Exp. Biol. Med.* **156**, 20 (1977).
[19] K. P. Boebel and D. H. Baker, *Proc. Soc. Exp. Biol. Med.* **172**, 498 (1983).
[20] D. H. Baker, K. W. Bafundo, K. P. Boebel, G. L. Czarnecki, and K. M. Halpin, *J. Nutr.* **114**, 292 (1984).

whether free or in peptide linkage, can also undergo oxidative conversion to methionine sulfoxide or methionine sulfone. Cyst(e)ine, likewise, can undergo oxidation to cysteic acid. These oxidation products are of particular concern with certain dairy products because H_2O_2 is often used for sterilization purposes in milk handling.[21-26] Methionine sulfoxide has been estimated to possess 60% bioactivity relative to methionine, while methionine sulfone and cysteic acid have no sulfur amino acid bioactivity when given orally.[25]

Cysteic acid given orally is absorbed and converted largely to taurine. Taurine, also present in foods, has neither methionine nor cysteine bioactivity, although it does function importantly in bile acid conjugation and vision, and may have a role in neurotransmission as well.[27-30] Biosynthesis meets the physiological need for taurine in virtually all animal species. The cat, however, is (and the human infant may be) an exception. Felids experience retinal photoreceptor cell degeneration when taurine is absent from the diet.[29,31] Thus, preformed taurine is an obligatory dietary essential nutrient for cats, and neither excess methionine nor cyst(e)ine can eliminate its dietary requirement.

Homocyst(e)ine, although not present in foods, is an important sulfur-containing metabolite, knowledge of which is critical in formulating treatment regimens for homocystinuria.[32] It has also been implicated as a factor in atherosclerosis.[33] Many investigators studying the role of homocyst(e)ine in these and other disease states have used DL-homocysteine or DL-homocystine as a diet supplement instead of the more biologically active L-isomer of these compounds. L-Homocysteine has a methionine-sparing value of 60 to 70% for rats and chicks. The D-isomer, however, has a methionine-sparing value of only 7% relative to L-methionine. In terms of providing metabolic cysteine, L-homocysteine is 100% effective while D-homocysteine is 68% as effective as orally administered L-cystine.[34,35] These bioactivity values should be given careful consideration

[21] L. R. Njaa, *Br. J. Nutr.* **16,** 571 (1962).
[22] S. F. Yang, *Biochemistry* **9,** 5008 (1970).
[23] S. A. Miller, S. R. Tannenbaum, and A. W. Seitz, *J. Nutr.* **100,** 909 (1970).
[24] G. H. Anderson, G. S. K. Li, J. D. Jones, and F. Bender, *J. Nutr.* **105,** 317 (1975).
[25] G. H. Anderson, D. V. M. Ashley, and J. D. Jones, *J. Nutr.* **106,** 1108 (1976).
[26] J. L. Cuq, M. Provansal, F. Guilleux, and C. Cheftel, *J. Food Sci.* **38,** 11 (1973).
[27] C. E. Sasse and D. H. Baker, *J. Nutr.* **104,** 244 (1974).
[28] K. C. Hayes, *Nutr. Rev.* **34,** 161 (1976).
[29] K. Knopf, J. A. Sturman, M. Armstrong, and K. C. Hayes, *J. Nutr.* **108,** 773 (1978).
[30] A. N. Davison and L. K. Kaczmarek, *Nature (London)* **234,** 107 (1971).
[31] P. A. Anderson, D. H. Baker, J. E. Corbin, and L. C. Helper, *J. Anim. Sci.* **49,** 1227 (1979).
[32] Anonymous, *Nutr. Rev.* **42,** 180 (1984).
[33] K. S. McCully and R. B. Wilson, *Atherosclerosis* **22,** 215 (1975).
[34] J. M. Harter and D. H. Baker, *Proc. Soc. Exp. Biol. Med.* **157,** 139 (1978).

before deciding on which source and what quantity of homocyst(e)ine to use as a diet supplement in studies where oral homocyst(e)ine is part of the experimental design.

Inorganic sulfate deserves special consideration as a contributor to the useful sulfur pool of a diet. Sulfate contributes sulfur for synthesis of chondroitin and other sulfur-containing body compounds.[36-39] In so doing, it spares cysteine from having to serve as a precursor for these same compounds. Thus, work in several laboratories has confirmed that inorganic sulfate can spare roughly 15% of the physiological need for cyst(e)ine.[40] Moreover, only 200 mg/kg sulfate is required for such sparing to occur, whether in chicks or rats.[40] Preliminary indications suggest that humans use sulfate sulfur very efficiently as well.[40]

It is important to put sulfate-sparing of cyst(e)ine in proper perspective. Thus, only under conditions of a dietary deficiency of sulfur amino acids where cyst(e)ine is more limiting than methionine, and where diet and water sources of sulfate provide less than 200 mg/kg sulfate, would one expect to observe a growth response to oral administration of a sulfate salt. Most animal diets fortuitously provide more than 200 mg/kg dietary sulfate—from sulfate salts of trace-mineral additions, from phosphate supplements where sulfate is a contaminant, and from fish and meat products that may be present in the diet.[40] Thus, few diets by virtue of already containing more than 200 mg/kg sulfate would be expected to respond to additional sulfate supplementation. Nonetheless, it is important to recognize that were the dietary sulfate level extremely low, more dietary cyst(e)ine (or methionine as a source of cysteine) would be required to maximize growth of the animal in question.

Quantifying Methionine and Cyst(e)ine in Diets

It is generally accepted that the typical acid hydrolysis procedure for hydrolyzing intact proteins to free amino acids prior to chromatographic quantification of the amino acids therein results in considerable losses of both methionine and cyst(e)ine. Losses up to 20% of the methionine and 40% of the cyst(e)ine are not uncommon. A procedure that eliminates

---

[35] D. H. Baker and G. L. Czarnecki, *J. Nutr.* **115,** 1291 (1985).
[36] H. Bostrom, *J. Biol. Chem.* **196,** 477 (1952).
[37] R. J. Miraglia, W. G. Martin, D. G. Spaeth, and H. Patrick, *Proc. Soc. Exp. Biol. Med.* **123,** 725 (1966).
[38] W. G. Martin, N. L. Sass, L. Hill, S. M. Tarka, and R. C. Truex, *Proc. Soc. Exp. Biol. Med.* **141,** 632 (1972).
[39] A. L. Daniels and J. J. Rich, *J. Biol. Chem.* **36,** 27 (1918).
[40] D. H. Baker, "Sulfur in Nonruminant Nutrition." NFIA Publ. Co., West Des Moines, Iowa, 1977.

these losses involves preoxidation of the intact protein or diet sample with performic acid prior to the typical 24-hr acid hydrolysis procedure.[41] With this procedure, methionine is quantified as methionine sulfone and cyst(e)ine as cysteic acid. Lanthionine, if present, is converted to lanthionine sulfone which has a retention time similar to cysteic acid. Thus, to accurately quantify lanthionine, nonoxidized sample must be subjected to acid hydrolysis, followed by special chromatographic procedures to clearly separate the two lanthionine peaks from glutamic acid/proline.[42] Cyst(e)ine is then quantified by subtracting lanthionine from the cysteic acid value obtained via the performic acid preoxidation procedure.[15]

Assay Diets for Sulfur Amino Acid Studies

Rats, mice, dogs, and pigs have relatively low dietary requirements for sulfur amino acids compared with chicks and cats. In all of these species, cyst(e)ine has been found capable of supplying 50% of their total requirement (wt/wt) for growth. Because of the molecular weight difference between methionine and cysteine, a higher concentration of methionine, alone, is required to meet the requirement than would be the case if equal-weight portions of methionine and cyst(e)ine were used.[1,40] Thus, while transsulfuration is a biologically efficient process, a higher dietary concentration of methionine, alone, is needed to furnish the same sulfur content as a 1:1 mixture (wt/wt) of a methionine–cyst(e)ine combination.

The easiest way to control levels of the sulfur amino acids in a research diet is to use a purified crystalline amino acid diet. With such a diet there are fewer concerns about the plethora of bioavailability factors mentioned earlier. Moreover, because crystalline amino acid diets are "chemically defined," it is possible to set methionine and cystine levels precisely to accomplish deficiencies or excesses as the situation demands. Research during the past 20 years has resulted in amino acid diets for most species that give growth rates equal to those obtainable with intact protein or grain–soybean diets. The proper pattern of free amino acids in these diets is extremely important, however, in that voluntary food intake is particularly sensitive to imbalances and ratios of one amino acid to another.[43]

The carbohydrate source in purified amino acid diets should consist of nonreducing sugars such as cornstarch and/or sucrose. Regular commercial cornstarch works very well for most species, although canine species tend to have loose stools when fed this carbohydrate material. This is

[41] S. Moore, *J. Biol. Chem.* **238**, 235 (1963).
[42] M. Friedman, J. Finley, and L. S. Yeh, *in* "Protein Crosslinking—Nutritional and Medical Consequences," (M. Friedman, ed.), Vol. 86A. 1977.
[43] D. H. Baker, *Adv. Nutr. Res.* **1**, 299 (1977).

TABLE I
Amino Acid Diets for Various Species[a]

Ingredient	Dietary amount (g/100 g)			
	Rats/mice	Chicks	Dogs	Cats
Cornstarch[b]	45.79	59.24	29.08	24.29
Sucrose	25.90	—	29.00	24.28
Amino acid mixture	16.00	20.48	16.39	18.85
Corn oil	5.00	10.00	15.00	—
Poultry fat	—	—	—	25.00
Solka floc[c]	—	3.00	3.00	—
Mineral mixture	5.40	5.37	5.42	5.37
Vitamin mixture[d]	0.40	0.40	0.60	0.60
Ethoxyquin (antioxidant)	0.01	0.01	0.01	0.01
NaHCO$_3$[e]	1.50	1.50	1.50	1.50
Taurine	—	—	—	0.10

[a] The amino acids should be weighed and then premixed before being added to the feed mixer. All ingredients other than fat sources should be mixed dry for at least 5 min, after which the fat should be added and the entire diet mixed for another 5 min. The mixer should then be emptied and the diet forced through a 2.5 mm^2 screen to break up lumps. The diet should be placed back in the mixer for an additional 3 to 5 min of mixing.
[b] Pregelatinized cornstarch should be used for dogs.
[c] Purified wood cellulose.
[d] Some vitamin mixtures do not contain vitamin E (a viscous liquid in pure form) or choline chloride (a hygroscopic material); if not, these should be premixed with sucrose or cornstarch and added directly to the diet.
[e] This material is added to buffer the acidity introduced from use of the HCl forms of lysine, arginine, and histidine.

easily solved with dogs if pregelatinized cornstarch is used instead of regular cornstarch. Texture is also important in purified diet formulation. Thus, the proportions of fat and fiber can be varied to arrive at the most desirable texture for each individual species such that food consumption can be maximized and food wastage minimized. Complete vitamin and mineral mixes can be obtained commercially. Table I describes the composition of purified amino acid diets that have been shown to promote excellent growth in young rats,[44] mice,[44] chicks,[45] dogs,[46] and cats.[47] The

[44] D. A. Hirakawa, L. M. Olson, and D. H. Baker, *Nutr. Res.* **4,** 891 (1984).
[45] D. H. Baker, K. R. Robbins, and J. S. Buck, *Poult. Sci.* **58,** 749 (1979).
[46] D. A. Hirakawa and D. H. Baker, *Nutr. Res.* **5,** 631 (1985).
[47] P. A. Anderson, D. H. Baker, P. A. Sherry, and J. E. Corbin, *Am. J. Vet. Res.* **11,** 1646 (1980).

TABLE II
Composition of Amino Acid Mixtures[a]

Amino acid	Dietary amount (g/100 g)			
	Rats/mice	Chicks	Dogs	Cats
L-Arginine · HCl	1.00	1.15	1.15	1.00
L-Histidine · HCl · H$_2$O	0.60	0.45	0.45	0.40
L-Lysine · HCl	1.40	1.14	1.14	1.00
L-Tyrosine	0.40	0.45	0.45	0.50
L-Tryptophan	0.20	0.15	0.16	0.15
L-Phenylalanine	0.80	0.50	0.50	0.50
L-Methionine[b,c]	0.60	0.35	0.40	0.45
L-Cystine[c]	0.40	0.35	0.40	0.45
L-Threonine	0.80	0.65	0.65	0.80
L-Leucine	1.20	1.00	1.00	1.20
L-Isoleucine	0.80	0.60	0.60	0.60
L-Valine	0.80	0.69	0.69	0.60
Glycine	1.00	0.60	1.60	1.60
L-Proline	1.00	0.40	0.80	1.60
L-Glutamic acid	1.00	12.00	2.40	4.80
L-Alanine	1.00	—	1.60	0.80
L-Asparagine · H$_2$O	1.00	—	1.60	1.60
L-Serine	1.00	—	0.80	0.80
L-Glutamine	1.00	—	—	—
Total	16.00	20.48	16.39	18.85

[a] These dietary amino acid mixtures were designed for use with young actively growing animals. Adult animals would require much lower levels and perhaps different amino acid patterns as well.

[b] DL-Methionine is generally used in the chick amino acid diet.

[c] The minimal methionine (M) and cystine (C) levels required for maximal growth would be 0.25% M and 0.25% C for rats[4] and mice,[52] 0.30% M and 0.30% C for chicks,[2] 0.225% M and 0.225% C for dogs,[46] and 0.40% M and 0.40% C for cats.[53-55]

composition of the amino acid mixtures used in these diets is shown in Table II.

Because of great advances in the technology of producing free amino acids, cost of amino acid diets for small animal work is no longer a deterrent to their use.[48,49] There are several companies who have positioned themselves to produce and sell all of the 22 amino acids that might

---

[48] D. H. Baker, *Proc. Arkansas Nutr. Conf.*, p. 1 (1985).

[49] Anonymous, *Nutr. Rev.* **43**, 88 (1985).

be needed. With proper care given to purchasing arrangements, it should be possible to prepare the diets shown in Tables I and II at a reasonable cost. Another possibility, however, is to purchase a specific basal diet from a diet-preparation company, then prepare the individual assay diets in the investigator's own laboratory.

Levels of methionine and cystine in the diets shown in Table I and II are in excess of their minimal requirements for maximal growth of young animals (weanling rats, mice, cats, and dogs, and chickens from 1 to 4 weeks posthatching), as is also the case for the other amino acids—except for the feline diet which has most of the amino acids set at close to minimum levels for maximal growth. The requirement levels of methionine and cystine for each species are listed in Table II. The requirement values listed for the sulfur amino acids apply to these diets only and should not be extrapolated to other diets containing intact proteins with different levels of both total nitrogen and energy.

To work at different levels of sulfur amino acid adequacy requires knowledge of the exact requirement level for each species. The linear phase of the growth curve generally occurs between 30 and 70% of the sulfur amino acid requirement, or of the methionine or cystine requirement. Deviation from linearity, i.e., lesser slope, generally results between 0 and 30% and also between 70 and 100% of the requirement. Excess dietary cystine is well tolerated, but excess methionine can be very noxious, causing depressed growth, hemolytic anemia, and pancreatic damage among other disorders.[50] Thus, methionine is generally considered to be the most toxic of the (dietary) essential amino acids. A level twice the requirement is generally well tolerated, but at 3-fold or above, toxicity often results.

Methionine and cyst(e)ine are rapidly depleted from tissue pools during periods of feeding nitrogen- or methionine-free diets. Many investigators have shown that protein retention is enhanced and fractional protein synthesis rate elevated when methionine, alone, is added to a nitrogen-free diet. Apparently, depletion of the sulfur amino acids from tissue pools during periods of fasting or nitrogen-free diet feeding is greater than that occurring for the other essential amino acids, such that methionine supplementation elicits a small but measurable response whereas none of the other essential amino acids, by themselves, will do likewise.

Intact protein diets can likewise be used in sulfur studies, although most intact proteins that are deficient in sulfur amino acids are not deficient enough to allow a broad range of response. Casein is singularly deficient in cyst(e)ine (not methionine) for most animal species, including

[50] J. M. Harter and D. H. Baker, *J. Nutr.* **108,** 1061 (1978).

the rat. When dealing with avian species, however, casein must also be supplemented with arginine to achieve maximal growth. Feather meal and gelatin, also high-nitrogen intact proteins like casein, are more deficient in sulfur amino acids than is casein. Moreover, the deficient sulfur amino acid in these proteins is methionine rather than cyst(e)ine. Peanut meal, cottonseed meal, meat meal, and soybean meal are also limiting, with methionine generally more limiting than cyst(e)ine.[51]

For rat or mouse studies employing the intact proteins mentioned above, all except casein require amino acid supplementation other than just sulfur amino acids to correct inherent deficiencies so as to not limit the full range of the sulfur amino acid response. Feather meal must, therefore, be supplemented with lysine, tryptophan, and histidine; gelatin with all essential amino acids except arginine; oilseed meals (except soy) with lysine and threonine; and meat meal with lysine and tryptophan. The composition of some sulfur-deficient intact protein diets for rats is shown in Table III.

The most efficient way to mix purified diets containing graded increments of methionine, cystine, or both is to prepare a basal diet which contains all dietary ingredients except the amino acid under study. Thus, if 0, 50, 100, and 200% of the rat's methionine requirement (amino acid diet) were to be investigated, i.e., 0, 0.125, 0.25, and 0.50%, a basal diet would be prepared containing everything but methionine. It would thereby constitute 99.5% (cornstarch is generally used as the variable ingredient to accommodate additions or deletions). The four individual diets would then be prepared from the common basal diet, i.e., diet 1— 99.5% basal + 0.50% cornstarch, diet 2—99.5% basal + 0.125% methionine + 0.375% cornstarch, diet 3—99.5% basal + 0.25% methionine + 0.25% cornstarch, and diet 4—99.5% basal + 0.50% methionine. There can be little doubt that this method of preparing individual diets is both more efficient and more accurate than the alternative method of mixing each of the individual diets separately.

The rodent diets shown in Tables I and II[52-55] have a powdery consistency, as does the avian diet shown in Table I. These diets are designed to be fed from a feed cup or self feeder in an "as is" dry form. The dog and cat amino acid diets shown in Table I, on the other hand, have a pie dough consistency. They are readily consumed in this form and can easily be fed

---

[51] National Research Council, "United States–Canadian Tables of Feed Composition." Natl. Acad. Press, Washington, D.C., 1982.
[52] A. John and J. M. Bell, *J. Nutr.* **106,** 1361 (1976).
[53] R. G. Teeter, D. H. Baker, and J. E. Corbin, *J. Nutr.* **108,** 291 (1978).
[54] M. C. Schaeffer, Q. R. Rogers, and J. G. Morris, *J. Nutr.* **112,** 962 (1982).
[55] K. A. Smalley, Q. R. Rogers, and J. G. Morris, *Br. J. Nutr.* **49,** 411 (1983).

TABLE III
Sulfur Amino Acid-Deficient Intact Protein Diets for Rodents (g/100 g Diet)

Ingredient	Methionine-deficient diet		Cystine deficient diet[c]
	A[a]	B[b]	
Feather meal (87% CP)	20.00	—	—
Peanut meal (42% CP)	—	40.00	—
Casein (90% CP)	—	—	17.00
Cornstarch	46.50	32.42	46.22
Sucrose[d]	20.00	15.00	25.00
Solka floc	1.00	1.00	1.00
Corn oil	5.00	5.00	5.00
Mineral mix[e]	5.37	5.37	5.37
Vitamin mix[f]	0.40	0.40	0.40
L-Lysine · HCl	0.80	0.40	—
L-Tryptophan	0.12	—	—
L-Histidine · HCl · H$_2$O	0.30	—	—
L-Threonine	—	0.15	—
Ethoxyquin	0.01	0.01	0.01
NaHCO$_3$	0.50	0.25	—

[a] Severely deficient in methionine; and superadequate in cystine; contains 0.09% methionine, 0.84% cystine, and 0.34% lanthionine.
[b] Moderately deficient in methionine and adequate in cystine; contains 0.15% methionine and 0.25% cystine.
[c] Superadequate in methionine and deficient in cystine; contains 0.45% methionine and 0.05% cystine.
[d] Dextrose (i.e., glucose · H$_2$O) can be substituted for sucrose, but due to potential Maillard action, dextrose-containing diets should be kept refrigerated.
[e] Amount will vary depending on the particular mineral mix used.
[f] Amount will vary depending on the particular vitamin mix used.

on an "as is" basis in a twice-daily feeding schedule. Rats, mice, and chicks must be fed *ad libitum* if optimal weight gains are to be achieved, since these species are "nibblers" rather than "meal eaters."

Little has been said here about sulfur amino acid assay diets for adult animals. Certainly less is known about the efficacy of assay diets for adults than is the case for young, actively growing animals. A procedure that some investigators have used is to simply reduce the level of either protein and the supplemented amino acids (intact protein diets) or of all crystalline amino acids (free amino acid diet) by a constant proportion, e.g., to a dietary crude protein (Kjeldahl N × 6.25) level of, say, 5%. The key to such a procedure is that the diet arrived at must, when adequately supplemented with sulfur amino acids, facilitate a consumption level that will allow both weight maintenance and a nitrogen retention of at least zero. Sulfur requirements for maintenance of adult animals are impacted importantly by the continued growth of keratoid tissues such as skin, hair, and feathers. These tissues are high in cyst(e)ine and relatively low in methionine. Thus, cyst(e)ine may be capable of furnishing up to 70% of the total need for the sulfur amino acids of adult animals compared with the 50% value that is common in young animals whose sulfur requirement is dominated by soft tissue somatic growth.

[56] Production of Selenium Deficiency in the Rat

By RAYMOND F. BURK

Selenium is an essential nutrient for animals, and diets deficient in the element have been in use for a number of years. Most studies of selenium deficiency have been carried out in rats, and this chapter will focus on that species. It is common practice to feed mice and rats the same diet. Selenium-deficient diets for chickens,[1] pigs,[2] sheep,[3] fish,[4] guinea pigs,[5] monkeys,[6] and human beings[7] have been described but will not be presented here.

Assessing Severity of Selenium Deficiency

An integral part of producing selenium deficiency for research purposes is an assessment of its severity. Although clinical signs occur, biochemical measurements are more commonly used in characterizing the

---

[1] G. F. Combs, Jr., C. H. Liu, Z. H. Lu, and Q. Su, *J. Nutr.* **114,** 964 (1984).
[2] G. R. Ruth and J. F. Van Vleet, *Am. J. Vet. Res.* **35,** 237 (1974).
[3] P. D. Whanger, P. H. Weswig, J. A. Schmitz, and J. E. Oldfield, *J. Nutr.* **107,** 1288 (1977).
[4] D. M. Gatlin III, and R. P. Wilson, *J. Nutr.* **114,** 627 (1984).
[5] R. F. Burk, J. M. Lane, R. A. Lawrence, and P. E. Gregory, *J. Nutr.* **111,** 690 (1981).
[6] M. A. Beilstein and P. D. Whanger, *J. Nutr.* **113,** 2138 (1983).
[7] O. A. Levander, B. Sutherland, V. C. Morris, and J. C. King, *Am. J. Clin. Nutr.* **34,** 2662 (1981).

TABLE I
SUGGESTED CLASSIFICATION OF SELENIUM DEFICIENCY IN
THE RAT BY ITS SEVERITY

Degree of severity	Liver glutathione peroxidase activity[a] (% of control)	Effects on drug-metabolizing enzymes and plasma glutathione[b]
Mild	5–50%	Absent
Moderate	<5%	Absent
Severe	<5%	Present

[a] The 5% dividing line is based on assay of the selenoenzyme using H_2O_2 as substrate. The 5% figure was chosen because of the variety of assays used in different laboratories. In our experience (see Table III) it could be set at 1–2%.
[b] Described in Burk[9] and Reiter and Wendel.[10]

deficiency. Activity of the selenoenzyme glutathione peroxidase (EC 1.11.1.9) decreases as selenium deficiency develops.[8] Hepatic activity is very sensitive to selenium intake and is the most commonly used index of selenium deficiency; other tissues lose glutathione peroxidase activity more slowly than liver when rats are fed a selenium-deficient diet.

In recent years it has become apparent that selenium has biochemical functions other than glutathione peroxidase. Other selenoenzymes have not been identified, but effects on hepatic heme metabolism, hepatic drug-metabolizing activities, and plasma glutathione appear after hepatic glutathione peroxidase activity has fallen to very low levels.[9,10] This suggests that indices in addition to glutathione peroxidase activity could be useful in assessing severity of selenium deficiency states. Suggested guidelines for assessing selenium deficiency in the rat are shown in Table I.

Strategies for Producing Selenium Deficiency

Feeding a selenium-deficient diet is the key for the production of selenium deficiency. Since animals can conserve their selenium stores when deprived of the element, it is customary to begin depletion in weanlings weighing about 50 g. Male rats are generally preferred because they have a higher selenium requirement than females.

[8] D. G. Hafeman, R. A. Sunde, and W. G. Hoekstra, *J. Nutr.* **104**, 580 (1974).
[9] R. F. Burk, *Annu. Rev. Nutr.* **3**, 53 (1983).
[10] R. Reiter and A. Wendel, *Biochem. Pharmacol.* **33**, 1923 (1984).

To produce very severe selenium deficiency, investigators have fed low-selenium diets to female rats from weaning and then to their offspring. These "second generation" selenium-deficient rats grow very slowly, have sparse hair, and cannot reproduce.[11,12] Another approach to producing very severe deficiency is to use an amino acid diet.

Attempts have been made to induce selenium deficiency by administering substances which interfere with selenium utilization. Silver can exacerbate the effects of selenium deficiency,[13] but this approach has not been generally adopted because of effects unrelated to selenium.

The Selenium-Deficient Diet

The selenium-deficient diet currently used in our laboratory (Table II) is often modified to optimize specific experiments. Therefore, a discussion of the major components of the diet is provided.

Protein. Since the vast majority of selenium in feedstuffs is present in protein, this is the most critical ingredient of the selenium-deficient diet. Dried *Torula* yeast supplied by Rhinelander Paper Co. is the most reliable low-selenium protein source available. It contains 45–50% protein and is commonly used as 30–40% of the diet. The yeast protein is deficient in sulfur-containing amino acids, requiring the addition of methionine to obtain maximum growth rates. The selenium content of the yeast is specified to be less than 0.05 mg per kg. Assay of a recent batch received by us yielded a value of 0.015 mg selenium per kg.[14] Assay of the complete diet, mixed as in Table II, yielded a value of 0.008 mg selenium per kg, i.e., the *Torula* yeast accounted for over half the selenium in the diet.

Other protein sources have been used. Amino acid diets can produce more severe selenium deficiency than diets based on *Torula* yeast[12,15] but are very expensive. Casein from low-selenium areas can be used[7] but careful monitoring of selenium content is necessary.

Fat. Most investigators use corn oil or lard as the fat source with this diet. If a vitamin E-deficient diet is desired, vitamin E-stripped corn oil or lard (Eastman Organic Chemicals, Rochester, NY 14650) is used.

Minerals. The mineral mix in Table II is suitable for use with *Torula* yeast, which contains approximately 130 mg zinc per kg. If the protein

---

[11] K. E. M. McCoy and P. H. Weswig, *J. Nutr.* **98,** 383 (1969).
[12] H. D. Hurt, E. E. Cary, and W. J. Visek, *J. Nutr.* **101,** 761 (1971).
[13] R. P. Peterson and L. S. Jensen, *Poult. Sci.* **54,** 795 (1975).
[14] The analyses were performed by Dr. Ivan Palmer of South Dakota State University, Brookings, South Dakota.
[15] G. Siami, A. R. Schulert, and R. A. Neal, *J. Nutr.* **102,** 857 (1972).

TABLE II
COMPOSITION OF SELENIUM-DEFICIENT DIET
FOR RATS

Component	Percentage by weight
Torula yeast[a]	30
Sucrose	56.99
Corn oil	6.67
Mineral mix[b]	5
Vitamin mix[c]	1
DL-Methionine	0.3
Vitamin E[d]	0.04

[a] Lake States *Torula* Dried Yeast, U.S.P. XIX, Type B, supplied by Lake States Division, Rhinelander Paper Co., Rhinelander, WI 54501.

[b] Contains (in g per kg) $CaCO_3$, 543; KH_2PO_4, 225.2; KCl, 104.8; NaCl, 59.69; $MgCO_3$, 25; $MgSO_4$, 16; ferric ammonium citrate, 20.5; $MnSO_4 \cdot H_2O$, 3.44; NaF, 1; $CuSO_4$, 0.9; $CrCl_3 \cdot 6H_2O$, 0.1; KI, 0.08.

[c] Contains (per 100 g) sucrose, 88 g; choline chloride, 10 g; niacin, 1 g; calcium D-pantothenate, 200 mg; thiamin · HCl, 40 mg; riboflavin, 25 mg; pyridoxine · HCl, 20 mg; menadione sodium bisulfite, 20 mg; folic acid, 20 mg; biotin, 10 mg; vitamin B_{12}, 1 mg; retinyl palmitate, water dispersible dry form to provide 15,000 IU per kg diet; ergocalciferol to provide 240 IU per kg diet.

[d] DL-α-Tocopherol powder to provide 100 IU per kg diet.

comes from a different source, zinc should be added to the mineral mix. Other complete mineral mixes can be used, provided they do not contain selenium.

Vitamins. The vitamin mix in Table II is one we have used for many years[16] with some modifications. It omits vitamin E, which we add separately. Other vitamin mixes can be used. Many investigators find it convenient to purchase the AIN Vitamin Mixture 76[17] from a commercial supplier; the formulation can be prepared without vitamin E. It does not contain choline, so that this nutrient must be added separately.

[16] R. F. Burk, R. Whitney, H. Frank, and W. N. Pearson, *J. Nutr.* **95**, 420 (1968).
[17] AIN Committee on Standards for Nutritional Studies, *J. Nutr.* **110**, 1726 (1980).

Vitamin E. This diet is often used to produce vitamin E deficiency, either alone or combined with selenium deficiency. To do so, the fat source must be stripped of vitamin E and no exogenous vitamin E is provided.

Selenium. The control diet is the selenium-deficient diet with selenium added. The nutritional requirement of the rat for selenium is satisfied by 0.1 mg selenium per kg of diet, or less, for most commonly used forms of the element. Inorganic selenium compounds such as selenate or selenite are generally added to control diets to give a final concentration of 0.1–0.5 mg selenium per kg. We prefer to use selenate because of evidence that dietary selenite causes lipid peroxidation.[18,19] The selenium concentration we use is 0.5 mg per kg diet. Na_2SeO_4 is desiccated and 358.4 mg is added to 2 kg of sucrose in a jar. The jar is sealed and its contents are thoroughly mixed, preferably by the use of a rolling mill for several hours. This selenium–sucrose mixture is added as 0.67% of the diet (40 g per 6 kg batch) at the expense of sucrose. Organic forms of selenium such as selenomethionine are occasionally used but are subject to dietary influences which can confound interpretation of experimental results.[20]

Mixing. Investigators who mix their own diets can alter them to optimize experimental design and can directly control the quality of the diet. In addition, cost is less than when diets are purchased from commercial suppliers.

Torula yeast is a fine powder which makes a dust cover necessary for the mixer. We use a commercial mixer with a 20-quart capacity and make 6-kg batches of diet. All the ingredients except the corn oil are added into the mixing bowl and blended slowly for 1–2 min. Then, the corn oil is added without stopping the mixer and mixing is continued for an additional 5 min. Overmixing can lead to destruction of vitamins.

It is our practice to mix selenium-deficient diets first and to remove them from the mixing room before using any selenium–sucrose. The diet is stored at 4° and can be kept for a month unless it is vitamin E-deficient. Vitamin E-deficient diets must be used within 1–2 weeks.

Animal Care and Feeding

The diet is the only substantial source of selenium to the animal. Water and air generally contain very small amounts of the element. It is our practice to supply tap water for drinking. Since, however, water has been found to have a significant selenium content in a few locations, it

---

[18] A. S. Csallany, L.-C. Su, and B. Z. Menken, *J. Nutr.* **114,** 1582 (1984).
[19] R. F. Burk and J. M. Lane, *Toxicol. Appl. Pharmacol.* **50,** 467 (1979).
[20] R. A. Sunde, G. E. Gutzke, and W. G. Hoekstra, *J. Nutr.* **111,** 76 (1981).

TABLE III
DEVELOPMENT OF SELENIUM DEFICIENCY IN RATS FED THE
SELENIUM-DEFICIENT DIET[a]

Weeks on diet	Percentage of control	
	Liver glutathione peroxidase activity[b]	Liver glutathione S-transferase activity[b]
1	17	99
2	6	91
3	2	114
4	1	133
5	1	131
6	1	131

[a] Diet described in Table II. Values are derived from means of three male Sprague–Dawley rats per diet at each time point. The rats were weanlings at the start of the experiment.

[b] Assays performed on 105,000 g supernatant fluid. The substrate used for the glutathione peroxidase assay was H_2O_2[22] and that for the glutathione S-transferase assay was 1-chloro-2,4-dinitrobenzene.[23]

should be tested before routine use. This is conveniently done by comparing liver glutathione peroxidase activity in rats fed a selenium-deficient diet and given tap water with that of similar rats given distilled water. When attempts are being made to produce very severe selenium deficiency, the use of highly purified water is advisable.

Many nutritional deficiencies decrease food intake, requiring pair-feeding for their study. Selenium deficiency causes a mild decrease in food intake[21] except in second generation animals in which the decrease is severe. Because of this, and the long period over which the diets must be fed, pair feeding is seldom used in studies of selenium deficiency. It should be used, however, when very severe deficiency is produced as in second generation selenium-deficient animals.

If possible, selenium-deficient and control rats should be housed in separate cage racks. We house animals in wire-bottom cages. When the animals are fed, the selenium-deficient diet should be fed first; the supply should then be returned to the refrigerator before the control diet is opened. This prevents contamination of the deficient diet.

[21] R. C. Ewan, *J. Nutr.* **106,** 702 (1976).

Development of Selenium Deficiency

Selenium deficiency is readily produced with the diet in Table II. Table III shows[22,23] results from a recent experiment in our laboratory in which this diet was used.[24] Liver glutathione peroxidase activity had fallen to 2% of control by 3 weeks, and a rise in glutathione S-transferase activity followed, indicating progression of the deficiency to the severe category (Table I).

Acknowledgments

The author is grateful to Mr. James M. Lane for help in developing the diet and procedures described here, and to Mrs. Rebecca E. Ortiz for typing the manuscript. This work has been supported by the National Institutes of Health, Grant ES 02497.

[22] R. A. Lawrence and R. F. Burk, *Biochem. Biophys. Res. Commun.* **71,** 952 (1976).
[23] W. H. Habig, M. J. Pabst, and W. B. Jakoby, *J. Biol. Chem.* **249,** 7130 (1974).
[24] K. E. Hill, R. F. Burk, and J. M. Lane, *J. Nutr.* **117** (1987).

[57] Intracellular Delivery of Cysteine

By MARY E. ANDERSON and ALTON MEISTER

Introduction

Intracellular cysteine is required for the synthesis of proteins, glutathione, and a number of other metabolites that are considered in this volume. Animals obtain cysteine from dietary peptide-bound cyst(e)ine and methionine. These amino acids are transported across the intestinal mucosa, both as the corresponding free amino acids and in the form of peptides.[1] Cysteine and cystine, which are present in the blood plasma, are transported into cells; the relative rates of uptake of cysteine and cystine vary in different cells. Conversion of methionine sulfur to cysteine sulfur by the cystathionine pathway constitutes a significant source of cysteine; this pathway functions chiefly in the liver but apparently also to some extent in other cells. Under normal conditions, these processes suffice to supply the amounts of cysteine moieties needed for protein synthesis and other metabolic purposes.

[1] Peptide Transport and Hydrolysis, *Ciba Found. Symp.* [N.S.] **50** (1977).

It is established that the dietary protein requirement of animals may be replaced by mixtures of free essential amino acids.[2] Administration of unbalanced amino acid mixtures may lead to decreased growth and to toxicity. Certain amino acids, including cysteine, are known to be toxic even when administered in moderate amounts. In contrast, the toxicity of many other amino acids is relatively low.[3] It has been reported that addition of L-cysteine to a basal essential amino acid diet for mice led to prompt loss of weight and death.[4] Necrosis of neurons in infant mouse retina and hypothalamus was reported after oral intake of L-cysteine; after subcutaneous administration of L-cysteine a disseminated pattern of neuronal degeneration in the brain rapidly appeared.[5] A single injection of L-cysteine (1.2 mg/g body weight) into 4-day-old rats led to brain atrophy which was well developed 27–32 days after administration.[6] In studies carried out by the present authors[7] it was found that mice given single intraperitoneal injections of L-cysteine (10 mmol/kg) became lethargic and some had convulsions.

It has been reported that L-cysteine added to culture media is toxic to cells, and it was therefore recommended that cysteine not be used as a constituent of culture media.[8] Toxicity of L-cysteine in such media was reported to be reduced significantly by storage of the media or by adding pyruvate to the media.[8]

The mechanisms by which cysteine exerts toxicity are not yet known, but there are several plausible possibilities. For example, it is well known that cysteine undergoes rapid spontaneous oxidation at neutral pH to form cystine and hydrogen peroxide. The formation of large amounts of hydrogen peroxide at certain extracellular and intracellular sites could have deleterious effects. Cysteine might also form mixed disulfides with essential protein thiol groups. Cysteine is known to form thiazolidine and hemithioketal derivatives with carbonyl-containing compounds. Cysteine inhibits a number of pyridoxal 5′-phosphate enzymes by forming relatively stable thiazolidine derivatives. It is also possible that high levels of cysteine can interfere with the transport or function of certain metal ions. Interestingly, the concentration of cysteine (and cystine) in tissues is

[2] W. C. Rose, *Fed. Proc., Fed. Am. Soc. Exp. Biol.* **8,** 546 (1949).
[3] J. P. Greenstein and M. Winitz, "Chemistry of the Amino Acids." Wiley, New York, 1961.
[4] S. M. Birnbaum, M. Winitz, and J. P. Greenstein, *Arch. Biochem. Biophys.* **72,** 428 (1957).
[5] J. W. Olney, O. L. Ho, and V. Rhee, *Brain Res.* **14,** 61 (1971).
[6] R. L. Karlsen, I. Grofova, D. Malthe-Sorenssen, and F. Fonnum, *Exp. Brain Res.* **208,** 167 (1981).
[7] M. E. Anderson and A. Meister, unpublished observations (1986).
[8] Y. Nishiuch, M. Sasaki, M. Nakayasu, and A. Oikawa, *In Vitro* **12,** 635 (1976).

regulated at a very low level, often in the range of 10–100 μM.[9] It seems relevant that important enzymes involved in the utilization of cysteine exhibit relatively low apparent K_m values; for example, the apparent K_m value for L-cysteine for human L-cysteinyl-tRNA synthetase (cysteine–tRNA ligase, EC 6.1.1.16) is equal to or less than 3 μM.[10] The apparent K_m value for L-cysteine in the reaction catalyzed by γ-glutamylcysteine synthetase (glutamate–cysteine ligase, EC 6.3.2.2) is about 0.35 mM.[11]

The observation that administered L-cysteine is followed by substantial toxicity whereas administration of L-2-oxothiazolidine 4-carboxylate in similar doses is nontoxic (see below) supports the belief that much of the toxicity of L-cysteine is related to its extracellular effects. The considerations reviewed above suggest that the administration of L-cysteine in experimental and therapeutic situations may be undesirable and that a means for delivering cysteine intracellularly would therefore be useful.

Cysteine Delivery Systems

On theoretical grounds, an ideal intracellular delivery system for cysteine would consist of a derivative of cysteine which is readily transported into cells and which, after transport, is effectively converted to cysteine. It is important that the other product (or products) formed intracellularly in the release of cysteine from its transport vehicle be nontoxic and/or readily metabolizable.

There are a number of ways in which cysteine might be delivered intracellularly. For example, various N-substituted cysteine derivatives including glutathione and other cysteine-containing peptides may be considered. N-Acetyl-L-cysteine, which is commonly used for the treatment of acetaminophen toxicity,[12–15] is apparently transported into cells. However, it is less effective than L-2-oxothiazolidine 4-carboxylate in increasing the glutathione levels in the livers of mice treated with acetaminophen[16] (see Fig. 1, p. 320). It seems likely that N-acetyl-L-cysteine

[9] M. K. Gaitonde, *Biochem. J.* **104**, 627 (1967).
[10] J. R. Waterson, W. P. Winter, and R. D. Schmickel, *J. Clin. Invest.* **54**, 182 (1974).
[11] P. G. Richman and A. Meister, *J. Biol. Chem.* **250**, 1422 (1975).
[12] J. R. Mitchell, D. L. Jollow, W. Z. Potter, D. C. Davis, J. R. Gillette, and B. B. Brodie, *J. Pharmacol. Exp. Ther.* **187**, 185 (1973).
[13] D. J. Jollow, J. R. Mitchell, W. Z. Potter, D. C. Davis, J. R. Gillette, and B. B. Brodie, *J. Pharmacol. Exp. Ther.* **187**, 195 (1973).
[14] W. Z. Potter, D. C. Davis, J. R. Mitchell, D. J. Jollow, J. R. Gillette, and B. B. Brodie, *J. Pharmacol. Exp. Ther.* **187**, 203 (1973).
[15] J. R. Mitchell, D. J. Jollow, W. Z. Potter, J. R. Gillette, and B. B. Brodie, *J. Pharmacol. Exp. Ther.* **187**, 211 (1973).
[16] J. M. Williamson, B. Boettcher, and A. Meister, *Proc. Natl. Acad. Sci. U.S.A.* **79**, 6246 (1982).

is deacetylated *in vivo,* but the site or sites at which this reaction takes place have not yet been determined. The kidney and to lesser extents liver and other tissues are known to exhibit amino acid *N*-acylase activity.[17]

Although glutathione is not transported into cells,[18] administration of glutathione may lead to increased cellular levels of cysteine. This may be explained by extracellular or membranous metabolism of glutathione to form products such as cysteinylglycine and γ-glutamylcysteine, which may be transported into certain cells and then cleaved, or by formation of extracellular cyst(e)ine followed by its transport. Thus, cysteine and often also glutathione levels may be increased after administration of glutathione. That glutathione administration protects to some extent against acetaminophen toxicity[19] is consistent with this interpretation. In contrast to glutathione, monoesters of glutathione (e.g., γ-glutamylcysteinylglycylethyl ester) are efficiently transported into cells.[19-21] However, although there is intracellular conversion of glutathione monoester to glutathione, there appear to be no intracellular mechanisms for the conversion of glutathione to cysteine. The observed increase in formation of cysteine and cysteine-containing products such as γ-glutamylcyst(e)ine and cyst(e)inylglycine after administration of glutathione monoesters[21] may probably be attributed to export of glutathione from cells followed by extracellular metabolism.

As noted above, peptides that contain cysteinyl residues such as cyst(e)inyl glycine and γ-glutamylcystine may serve to facilitate transport of cysteine moieties. Enzyme activity capable of hydrolyzing cyst(e)inylglycine occurs intracellularly. Administration of γ-glutamylcystine to mice leads to increased renal levels of glutathione. This result is explained by transport of γ-glutamylcystine followed by its reduction to γ-glutamylcysteine and cysteine and utilization of both of these compounds for glutathione synthesis.[22]

Other derivatives of cysteine that may be considered included those in which the carboxyl group is blocked in other ways (e.g., amides, esters). However, in general, cysteine derivatives that have free thiol groups

---

[17] S. M. Birnbaum, L. Levintow, R. B. Kingsley, and J. P. Greenstein, *J. Biol. Chem.* **194**, 455 (1952).

[18] A. Meister and M. E. Anderson, *Annu. Rev. Biochem.* **52**, 711 (1983).

[19] R. N. Puri and A. Meister, *Proc. Natl. Acad. Sci. U.S.A.* **80**, 5258 (1983).

[20] V. P. Wellner, M. E. Anderson, R. N. Puri, G. L. Jensen, and A. Meister, *Proc. Natl. Acad. Sci. U.S.A.* **81**, 4732 (1984).

[21] M. E. Anderson, F. Powrie, R. N. Puri, and A. Meister, *Arch. Biochem. Biophys.* **239**, 538 (1985).

[22] M. E. Anderson and A. Meister, *Proc. Natl. Acad. Sci. U.S.A.* **80**, 707 (1983).

might be undesirable because they form peroxide by spontaneous oxidation or react in other ways that are possible with cysteine itself. Cystine is not suitable as an intracellular delivery agent because of its marked insolubility. Various S-acyl, S-phospho, and other S-substituted derivatives of cysteine might also be considered.

Another category of compounds that might serve as useful intracellular delivery systems for cysteine are thiazolidines. Schubert[23] and Ratner and Clarke[24] discovered that L-thiazolidine-4-carboxylic acid is formed by reaction of formaldehyde with L-cysteine. Mackenzie and his colleagues[25,26] found that L-thiazolidine-4-carboxylic acid can replace dietary cysteine or cystine for growth of the rat. (However, they found that this thiazolidine did not completely replace either dietary methionine or homocystine.) These workers also found that L-thiazolidine-4-carboxylic acid is oxidized by liver mitochondria to form N-formylcysteine. The latter compound and N,N'-diformylcystine were found to be active in promoting the growth of rats on cyst(e)ine-deficient diets.

It is now recognized that the mitochondrial enzyme activity that acts on L-thiazolidine 4-carboxylate is proline oxidase; the N-formyl-L-cysteine formed is presumably hydrolyzed to L-cysteine. Mackenzie *et al.* noted that these reactions could serve "as a means of producing sulfhydryl groups within the body for detoxication or other purposes." Later workers concluded that ingested L-thiazolidine-4-carboxylic acid can act as an intracellular antioxidant and as a scavenger of free radicals.[27] L-Thiazolidine-4-carboxylic acid has been used as a protectant against various types of toxicity, but this compound has been reported to be toxic.[28]

2-Substituted thiazolidine-4-carboxylic acids have also been suggested as "prodrugs" of L-cysteine; certain of these compounds have been found to protect mice against acetaminophen toxicity.[29,30] Such compounds are reported to be more effective than L-thiazolidine-4-carboxylic acid. The 2-(R,S)-methyl-, 2-(R,S)-n-propyl-, and 2-(R,S)-n-pentylthiazolidine-4-(R)-carboxylic acids were nearly equipotent in protecting against acetaminophen toxicity, whereas the corresponding 2-(R,S)-ethyl,

---

[23] M. P. Schubert, *J. Biol. Chem.* **114**, 341 (1936).
[24] S. Ratner and H. T. Clarke, *J. Am. Chem. Soc.* **59**, 200 (1937).
[25] C. G. MacKenzie and J. Harris, *J. Biol. Chem.* **227**, 393 (1957).
[26] H. J. Debey, J. B. MacKenzie, and C. G. MacKenzie, *J. Nutr.* **66**, 607 (1958).
[27] H. U. Weber, J. F. Fleming, and J. Miquel, *Arch. Gerontol. Geriatr.* **1**, 299 (1982).
[28] See Weber *et al.*[27] and Nagasawa *et al.*[29] for additional literature citations.
[29] H. T. Nagasawa, D. J. W. Goon, R. T. Zera, and D. L. Yuzon, *J. Med. Chem.* **25**, 491 (1982).
[30] H. T. Nagasawa, D. J. W. Goon, W. P. Muldoon, and R. T. Zera, *J. Med. Chem.* **27**, 591 (1984).

2-(R,S)-phenyl, and 2-(R,S)-(4-pyridyl) derivatives were less protective. 2-(R,S)-Methylthiazolidine-4(S)-carboxylic acid (which has a configuration corresponding to D-cysteine) was totally ineffective in such protection. In contrast to L-thiazolidine-4-carboxylic acid, which is utilized by a pathway involving enzymatic activities, the 2-substituted thiazolidine-4-carboxylic acids liberate cysteine by nonenzymatic ring opening followed by hydrolysis.[30] Thus, the spontaneous liberation of L-cysteine from 2-(R,S)-methylthiazolidine-4(R)-carboxylic acid is accompanied by equimolar formation of acetaldehyde.

A novel procedure for the intracellular delivery of cysteine developed from studies on 5-oxoprolinase,[31,32] the enzyme that catalyzes the ATP-dependent conversion of 5-oxo-L-proline to L-glutamate according to reaction (1). The mechanism of this interesting reaction, in which the endergonic cleavage of 5-oxo-L-proline is coupled with the exergonic cleavage of ATP, involves phosphorylation of oxygen at the 5-carbonyl moiety of the substrate.[33] Studies of the specificity of this enzyme led to the finding that a number of 5-oxoproline analogs can interact with the enzyme.[34] Notably, the 5-oxoproline analog in which the 4-methylene moiety of 5-oxoproline is replaced by sulfur, i.e., L-2-oxothiazolidine 4-carboxylate is a good substrate[34,35]; thus, 5-oxoprolinase converts L-2-oxothiazolidine 4-carboxylate to cysteine according to reaction (2). In this reaction S-carboxycysteine (or an equivalent enzyme-bound intermediate) is formed and decarboxylated.

$$\text{5-Oxo-L-proline} + 2\ H_2O + ATP \rightarrow \text{L-glutamate} + ADP + P_i \quad (1)$$
$$\text{L-2-Oxothiazolidine-4-carboxylate} + 2\ H_2O + ATP \rightarrow$$
$$\text{L-cysteine} + CO_2 + ADP + P_i \quad (2)$$

Since L-2-oxothiazolidine 4-carboxylate, which can be readily synthesized,[36] is apparently nontoxic and is readily transportable into cells, its administration provides a useful intracellular cysteine delivery procedure. 5-Oxoprolinase is widely distributed in mammalian cells; it is, however, not present in the erythrocyte or in the ocular lens. As discussed below, administration of L-2-oxothiazolidine 4-carboxylate to mice is followed by the appearance of cysteine in many tissues.

[31] P. Van Der Werf, M. Orlowski, and A. Meister, *Proc. Natl. Acad. Sci. U.S.A.* **68**, 2982 (1971).
[32] A. Meister, O. W. Griffith, and J. M. Williamson, this series, Vol. 113, p. 445.
[33] A. P. Seddon and A. Meister, *J. Biol. Chem.* **261**, 11538 (1986).
[34] J. M. Williamson and A. Meister, *J. Biol. Chem.* **257**, 12039 (1982).
[35] J. M. Williamson and A. Meister, *Proc. Natl. Acad. Sci. U.S.A.* **78**, 936 (1981).
[36] B. Boettcher and A. Meister, this series, Vol. 113, p. 458 (1985).

Administration of L-2-Oxothiazolidine 4-Carboxylate as an Intracellular Cysteine Delivery System

The studies that led to this procedure included the following findings[16,33,34]: (1) L-2-Oxothiazolidine 4-carboxylate is a good substrate of 5-oxoprolinase. (The corresponding D-isomer is not a substrate). The apparent K_m value for the thiazolidine is 2 μM (K_m for 5-oxo-L-proline, 5 μM); the corresponding values for V_{max} are 0.73 and 1.34 μmol/min/mg, respectively. (2) Administration of 5-oxo-L-proline to mice leads to marked inhibition of the metabolism of 5-oxoproline but not of that of glutamate; in these studies metabolism was followed by the rate of formation of respiratory $^{14}CO_2$ after administration of the corresponding labeled materials. (3) Administration of L-2-oxothiazolidine 4-carboxylate to mice stimulated formation of glutathione in the liver.

These observations were interpreted to indicate that this thiazolidine is efficiently transported into cells and that it is a substrate *in vivo* of 5-oxoprolinase. It is therefore converted intracellularly to cysteine, which is used for the synthesis of glutathione. In these studies, mice fasted overnight were injected intraperitoneally with sodium L-2-oxothiazolidine 4-carboxylate (6.5 mmol/kg). The animals were sacrificed at intervals, and the level of glutathione in the liver was determined. The glutathione level of the kidney increased about 2-fold 4 hr after injection of the thiazolidine.[35]

In subsequent studies, it was found that administration of L-2-oxothiazolidine 4-carboxylate protected against acetaminophen toxicity in mice. This effect was correlated with an increase in the level of glutathione in the liver. Thus, as shown in Fig. 1,[16] when mice were injected intraperitoneally with a sublethal dose of acetaminophen (2.5 mmol/kg) and then with L-2-oxothiazolidine 4-carboxylate or N-acetyl-L-cysteine there was a considerable increase in glutathione level in the liver. This effect was substantially greater after giving the thiazolidine than after administration of N-acetyl-L-cysteine. Survival of all mice given a lethal (LD_{90}) dose of acetaminophen (9.5 mmol/kg, oral) was achieved by given an oral dose of the thiazolidine that was about twice that of the administered acetaminophen.[16]

The toxicity of acetaminophen is due to its conversion to a highly reactive intermediate that interacts with a number of cell constituents including glutathione.[12–15] Acetaminophen is also detoxified by pathways other than conjugation with glutathione; however, interaction with glutathione seems to be of major importance when very large doses of acetaminophen are administered. It is unlikely that cysteine itself protects significantly against acetaminophen since the tissue levels of cysteine are

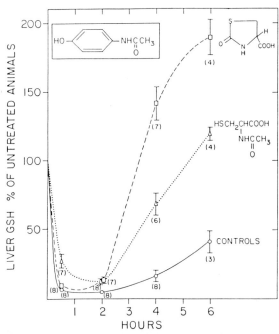

FIG. 1. Effects of L-2-oxothiazolidine 4-carboxylate and N-acetyl-L-cysteine on liver glutathione levels in mice treated with acetaminophen. Animals were injected intraperitoneally with acetaminophen (structure shown in inset) at 2.5 mmol/kg; 5 min later, one group was injected with the thiazolidine at 2 mmol/kg and another group was injected with N-acetyl-L-cysteine at 2 mmol/kg. The animals were sacrificed at the indicated intervals. Numbers in parentheses are numbers of animals used for the indicated time point. Results are means ± range; points given without bars had ranges of ±10%. Taken from Williamson et al.[16]

extremely low. Thus, it appears that a substantial level of glutathione is of major importance in reducing toxicity. Delivery of cysteine is required for protection because it is a necessary substrate for glutathione synthesis. Of relevance to the mechanism of protection, it may be noted that administration of buthionine sulfoximine, an inhibitor of glutathione synthesis[37] prevents the protective effect of L-2-oxothiazolidine 4-carboxylate against acetaminophen toxicity.[16]

Studies on the uptake of ^{35}S into various organs and body fluids of mice after administration of L-2-oxo[^{35}S]thiazolidine 4-carboxylate are summarized in Table I.[38] Significant amounts of radioactivity were found

[37] O. W. Griffith and A. Meister, J. Biol. Chem. 254, 7558 (1979).
[38] A. Meister, M. E. Anderson, and O. Hwang, J. Am. Coll. Nutr. 5, 137 (1986).

TABLE I
UPTAKE OF ^{35}S INTO VARIOUS ORGANS AND BODY FLUIDS
AFTER ADMINISTRATION TO MICE OF L-2-OXO
[^{35}S]THIAZOLIDINE 4-CARBOXYLATE[a]

Tissue or body fluid	Uptake (% of dose)	
	1 hr	4 hr
Kidney	2.77 ± 0.24	0.77 ± 0.02
Liver	4.19 ± 0.27	2.49 ± 0.27
Pancreas	0.52 ± 0.07	0.24 ± 0.02
Lung	0.43 ± 0.15	0.36 ± 0.02
Skeletal muscle[b]	25.8 ± 4.8	13.5 ± 1.1
Heart	0.33 ± 0.05	0.11 ± 0.08
Brain	0.94 ± 0.20	0.50 ± 0.02
Plasma[c]	9.40 ± 0.34[d]	11.7 ± 1[d]
Urine	18.2 ± 4.2[d]	47.7 ± 1.3[d]
Total recovery	63.2 ± 10.4	78.2 ± 3.9

[a] Mice (male, Rockefeller NCS, 20–25 g, fasted overnight) were injected intraperitoneally with a dose of 7.5 mmol/kg of sodium L-2-oxo[^{35}S]thiazolidine 4-carboxylate (0.188 M; 1 hr, 47.4 cpm/nmol; 4 hr, 117 cpm/nmol). At the times indicated, the animals were sacrificed, and the tissues were homogenized in 5 or 10 volumes of 5% 5-sulfosalicylic acid. Whole blood (about 2 ml) was collected in a 150-ml beaker that contained 50 μl of 0.5 M EDTA (pH 7) and centrifuged in 400-μl plastic tubes (1.5 min, 10,000 g; Beckman microfuge) to obtain plasma. Plasma and urine were deproteinized by adding 2 volumes of 10% 5-sulfosalicylic acid. After centrifugation (5 min, 10,000 g), the radioactivity present in the deproteinized solutions was determined by liquid scintillation counting. The data are the means of the values obtained in separate studies on mice. Taken from Meister et al.[38]
[b] A portion of thigh muscle was used. Total skeletal muscle was assumed to be 40% of body weight.
[c] Total plasma volume was assumed to be 2.5 ml.
[d] About 80% of the ^{35}S present in plasma and urine is accounted for as L-2-oxo[^{35}S]thiazolidine carboxylate.

in all of the tissues and body fluids examined. Increased levels of glutathione were found in liver, pancreas, and skeletal muscle. On the other hand, in this study smaller or no increases in glutathione levels were found in spleen, lung, heart, brain, and kidney. The level of glutathione in the kidney after administration of the thiazolidine was not increased by

TABLE II
LEVELS OF TISSUE CYSTEINE AFTER
ADMINISTRATION OF L-2-OXOTHIAZOLIDINE
4-CARBOXYLATE[a]

Tissue	Control (nmol/g)	Treated (nmol/g)
Kidney	280 ± 70	290 ± 90
Liver	180 ± 20	270 ± 30
Pancreas	490 ± 70	610 ± 50
Spleen	113 ± 15	129 ± 9
Lung	77 ± 19	82 ± 14[b]
Skeletal muscle	15 ± 0	60 ± 14
Heart	20 ± 3	45 ± 3
Brain	53 ± 2	161 ± 7.5

[a] Sodium L-2-oxothiazolidine 4-carboxylate (6.5 mmol/kg, 325 mM, pH 7) was administered subcutaneously to three mice (fasted). After 2 hr the animals were sacrificed, and the kidney, pancreas, and liver were immediately homogenized in 5% 5-sulfosalicylic acid (5–10 volumes per gram). The other tissues were frozen in a Dry Ice–ethanol bath; after 2–6 hr these were defrosted and homogenized as described above. The deproteinized homogenates were treated with monobromobimane, and the amounts of cysteine in the samples were determined by high-performance liquid chromatography.[38a] Taken from Meister et al.[38]

[b] An increase (50–70%) was found after 1 hr in another series of experiments.

simultaneous administration of glutamate and glycine. Under the same conditions, there was no increase in the level of glutathione in the liver after administration of cysteine or cysteine ethyl ester.

After administration of the thiazolidine, the tissue levels of cysteine increased substantially in liver, pancreas, skeletal muscle, heart, and brain (Table II).[38,38a] The specific radioactivities of tissue cysteine and glutathione found after administration of L-2-oxo[^{35}S]thiazolidine 4-carboxylate are recorded in Table III.[38,38a] It is of interest that the values for the specific radioactivity of cysteine were higher than those for glutathione in all organs examined. Notably, the differences between those for cysteine and for glutathione were smaller for liver and kidney than for

[38a] M. E. Anderson, this series, Vol. 113, p. 548 (1985).

TABLE III
RELATIVE SPECIFIC RADIOACTIVITY VALUES OF
CYSTEINE AND GLUTATHIONE AFTER
ADMINISTRATION OF L-2-OXO[^{35}S]THIAZOLIDINE
4-CARBOXYLATE[a]

Tissue	Cysteine	Glutathione
Kidney	41.7 ± 1.9	17.9 ± 0.57
Liver	18.3 ± 1.8	7.74 ± 1.2
Spleen	26.8 ± 0.32	1.19 ± 0.20
Pancreas	19.5 ± 1.6	3.63 ± 0.54
Lung	34.2 ± 2.4	3.15 ± 0.44
Skeletal muscle	47.2 ± 6.7	4.49 ± 0
Heart	31.3 ± 3.5	3.19 ± 0.65
Brain	44.6 ± 4.2	2.69 ± 0.09

[a] Mice were injected subcutaneously with a dose of 7.5 mmol/kg (0.188 M) of sodium L-2-oxo[^{35}S]-thiazolidine 4-carboxylate (47.4 cpm/nmol). The tissue samples were prepared as described in Table II. The levels of cysteine and glutathione in tissues were determined by high-performance liquid chromatography.[38a] The specific radioactivities of cysteine and glutathione were calculated from the radioactivity present in the corresponding fractions obtained by high-performance liquid chromatography and from the determined levels of cysteine and glutathione. The results represent means of values obtained on four mice. The data were normalized by dividing the observed specific radioactivities by those of the administered thiazolidine and multiplying by 100. Taken from Meister et al.[38]

the other tissues, a finding in accord with the more rapid synthesis of glutathione in these tissues.

It is notable that substantial amounts of label are found in the brain after administration of labeled thiazolidine (Table I) and also that cysteine levels are greatly increased in brain (Table II). In a study in which mice (fasted for 1 day) were injected intraperitoneally with the thiazolidine (10 mmol/kg), the values for cysteine in the brain increased from 0.07 µmol/g to 0.2 µmol/g after 3 hr; the level remained elevated for about 6 hr. In the same study, the level of glutathione increased about 15%.[7]

Methodological Considerations

In studies on rats and mice, doses of sodium L-2-oxothiazolidine 4-carboxylate of 2.5–15 mmol/kg body weight have been used intraperitoneally, subcutaneously, and orally. L-2-Oxothiazolidine-4-carboxylic acid is generally synthesized and available as the free acid. It should be dissolved in water and adjusted to pH 7.0–7.4 by addition of 1 equivalent of NaOH; 1 M solutions are easily obtained. The free acid and Na salt are stable in aqueous solution and may be stored at 5° for extended periods. L-2-Oxothiazolidine-4-carboxylic acid may be readily synthesized in about 1 day as described in a previous volume of this series.[36]

Some Applications

It is well established that glutathione is involved in the detoxication of many foreign compounds and in the protection of cells against the deleterious effects of radiation and of oxygen. Since cysteine is usually the limiting component for the synthesis of glutathione it is evident that a cysteine delivery agent such as L-2-oxothiazolidine 4-carboxylate has potential value in therapy. This subject has been considered elsewhere.[16,35,38–40] The possibility that intracellular cysteine itself might provide protection should also be considered.

In addition, the requirement of cysteine for protein synthesis cannot be ignored. This becomes of particular concern in situations in which the cystathionine pathway is decreased or absent, such as in certain inborn errors of metabolism[41] and in certain patients with liver disease.[42] The cystathionine pathway is deficient or absent in fetal tissues.[43,44] Therapy with amino acid mixtures is commonly employed in premature infants who may lack the cystathionine pathway. Most amino acid preparations currently used for such infants do not contain cysteine.

Addition of cysteine to amino acid solutions used for parenteral administration leads to complications because of the toxicity of cysteine and also because cysteine oxidizes spontaneously to yield cystine, which precipitates from solution. It thus appears that addition of L-2-oxothiazolidine 4-carboxylate to solutions used for parenteral amino acid administration may be useful. Possible uses of L-2-oxothiazolidine 4-carboxylate in agriculture have also been suggested.[16,45]

[39] A. Meister, this series, Vol. 113, p. 571 (1985).
[40] A. Meister, *Science* **220**, 472 (1983).
[41] D. Wellner and A. Meister, *Annu. Rev. Biochem.* **50**, 911 (1981).
[42] J. H. Horowitz, E. B. Rypins, J. M. Henderson, S. B. Heymsfield, S. D. Moffitt, R. P. Bain, R. K. Chawla, J. C. Bleier, and D. Rudman, *Gastroenterology* **81**, 668 (1981).
[43] J. A. Sturman, *in* "Natural Sulfur Compounds" (D. Cavallini, G. E. Gaull, and V. Zappia, eds.), p. 107. Plenum, New York, 1980.
[44] R. A. Burns and J. A. Milner, *J. Nutr.* **111**, 2117 (1981).
[45] J. L. Hilton and P. Pillai, *Weed Sci. Soc. Am., 1986 Meet.* Abstr. No. 208 (1986).

In addition to the potential value of a cysteine delivery system in therapy, there are a variety of applications in experimental work. For example, L-2-oxothiazolidine 4-carboxylate may be given, in place of L-cysteine, to experimental animals. In a recent study on mitochondrial and cytoplasmic glutathione, L-2-oxothiazolidine 4-carboxylate was given to mice previously treated with [^{35}S]cysteine to determine the rates at which the specific radioactivities of the two pools of [^{35}S]glutathione declined.[46] The use of this thiazolidine facilitates delivery of relatively high doses of cysteine. Similarly, inclusion of L-2-oxothiazolidine 4-carboxylate in tissue culture media is also of value for the growth of cells that are sensitive to the toxic effects of cysteine. For example, human lymphoid cell lines studied in this laboratory (line HSB) did not grow in media containing L-cysteine, but growth was optimal when the medium was supplemented with L-2-oxothiazolidine 4-carboxylate.[47] In cell cultures studies, levels in the range of 2–10 mM have been used.[47,48]

---

[46] O. W. Griffith and A. Meister, *Proc. Natl. Acad. Sci. U.S.A.* **77**, 3384 (1980).
[47] V. P. Wellner, unpublished data (1985).
[48] A. Russo and J. B. Mitchell, *Cancer Treat. Rep.* **69**, 1293 (1985).

Section III

Enzymes

A. Inorganic Sulfur and Sulfate
Articles 58 through 62

B. Sulfur Amino Acids: Mammalian Systems
Articles 63 through 70

C. Sulfur Amino Acids: Plants
Articles 71 through 77

D. Sulfur Amino Acids: Microbial Systems
Articles 78 through 88

E. Enzymes Active with Disulfides
Articles 89 through 91

[58] Overview: Inorganic Sulfur and Sulfate Activation

By JEROME A. SCHIFF and TEKCHAND SAIDHA

In nature sulfur exists in a wide range of oxidation states. The oxidized and reduced forms of sulfur can be interconverted by various organisms (Fig. 1). The metabolism of the most highly oxidized naturally occurring form of sulfur, sulfate, begins with the entrance of sulfate into the cell, which is accomplished through active transport mediated by a carrier-enzyme system.[1] The uptake of sulfate is followed by activation and by transfer and reduction reactions in appropriate organisms.

Reactions of Sulfate Activation and Their Distribution

Sulfate must be converted to an activated form before it can be utilized metabolically. There are two forms of activated sulfate, adenosine 5'-phosphosulfate (APS) and 3'-phosphoadenylylsulfate (adenosine 3'-phosphate 5'-phosphosulfate, PAPS). They are formed, as described in this volume [59], in two sequential enzyme-catalyzed reactions involving ATP-sulfurylase (sulfate adenylyltransferase, EC 2.7.7.4) and APS kinase (adenylylsulfate kinase, EC 2.7.1.25) (Fig. 2). The first step in the biosynthesis of PAPS, leading to the formation of APS, is catalyzed by the enzyme ATP-sulfurylase. This reaction is greatly favored energetically in the reverse direction as the free energy of hydrolysis of the sulfate group in APS is considerably higher than the free energy of the phosphate linkage of ATP. Therefore, the forward reaction proceeds to a reasonable extent only when the products of the reaction, APS and PP_i, are removed. PP_i is cleaved by the ubiquitous inorganic pyrophosphatase, and APS is removed by phosphorylation by APS-kinase and ATP in the second step or by utilization in other reactions. The resulting overall standard free energy change (Fig. 2)[2,3] helps explain accumulation of PAPS.

APS-kinase has a very high affinity for APS, giving the highest reaction rates at the lowest measurable concentration (5 μM). This characteristic of the enzyme helps to drive the first reaction in the forward direction by eliminating APS from the equilibrium. The three reactions together

---

[1] J. A. Schiff, *Encycl. Plant Physiol., New Ser.* **24**, 401 (1983).
[2] A. B. Roy and P. A. Trudinger, "The Biochemistry of Inorganic Compounds of Sulfur," p. 91. Cambridge Univ. Press, London and New York, 1970.
[3] T. W. Goodwin and E. I. Mercer, eds., "Introduction to Plant Biochemistry," p. 273. Pergamon, Oxford, 1982.

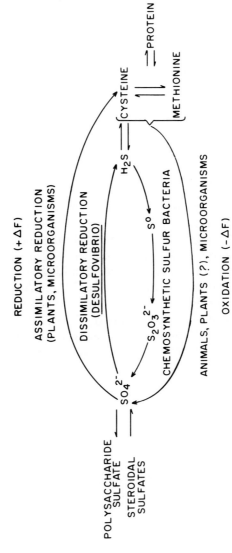

FIG. 1. Reactions of sulfur in the biosphere.

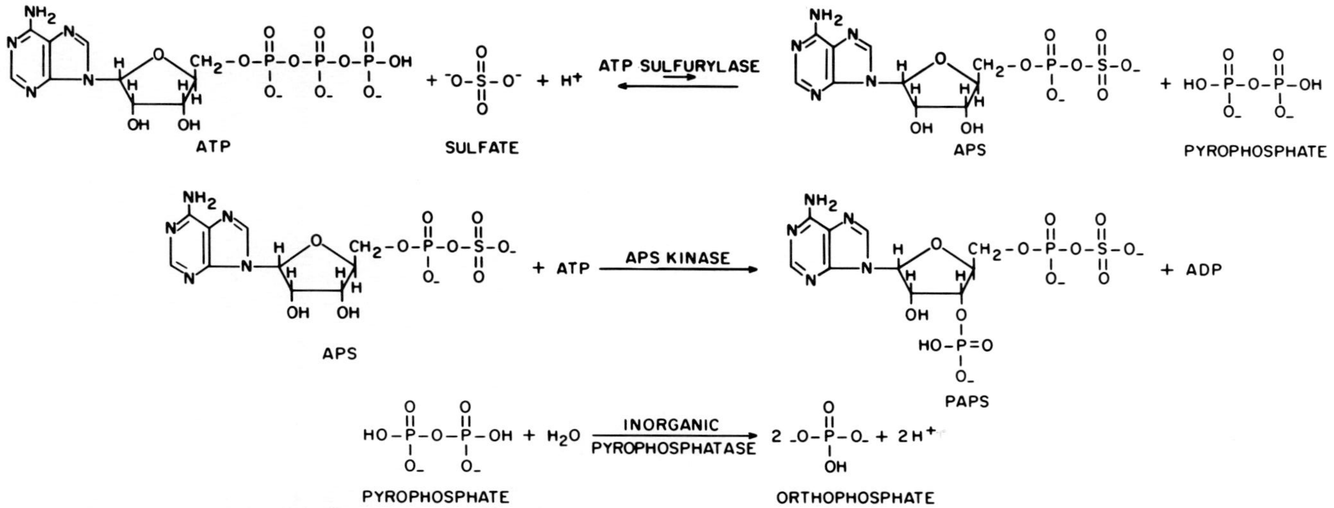

FIG. 2. Reactions of the sulfate-activating system. The $\Delta G^{\circ\prime}$ for the ATP-sulfurylase reaction is $+11$ kcal; for the APS kinase reaction, very negative; and for inorganic pyrophosphatase, -5 kcal.

comprise the sulfate activating system. A widely distributed 3'(2'),5'-diphosphonucleoside 3'(2')-phosphohydrolase (DPNPase) catalyzes the conversion of PAPS to APS. This enzyme may provide control of cellular levels of APS and PAPS which are required for different metabolic functions.[1]

The activating enzyme system that synthesizes PAPS catalyzes the following overall reaction: $2\ ATP + SO_4^{2-} \rightarrow ADP + PAPS + PP_i$, and is distributed widely in nature.[4] It is present in most mammalian tissues including those of adrenal glands, brain, cartilage, cornea, heart, ovaries, pancreas, placenta, retina, skin, spleen, and a variety of tumors. It has also been found in chick embryo, hen oviduct, and frog liver. Among invertebrates it occurs in snails, some sea urchins, and the clam *Spisula solidissima*. It is also present in higher plants, yeast, *Fusarium solani*, *Euglena gracilis*, and a number of other algae. The anaerobic sulfate-reducing bacterium, *Desulfovibrio*, appears to have only ATP-sulfurylase and to lack APS kinase.

Evidence relating to the cellular localization of the sulfate-activating system is scanty. The activating enzymes have been shown to be present in spinach chloroplasts[5] and on the outer surface of the mitochondrial inner membrane in *Euglena*.[6]

Sulfate Transfer and Reduction

Sulfate is used by living systems to form sulfate esters

$$R-O-\overset{\overset{O}{\|}}{\underset{\underset{O}{\|}}{S}}-O^-$$

of polysaccharides, phenols, steroids, and other organic compounds through transfer reactions in which PAPS is the sulfuryl group donor, catalyzed by sulfotransferases of various specificities.[4,7,8] APS or PAPS has also been suggested as a donor of the sulfonic acid group

---

[4] R. H. De Meio, in "Metabolic Pathways" (D. M. Greenberg, ed.), 3rd ed., Vol. 7, p. 287. Academic Press, New York, 1975.
[5] J. D. Schwenn and A. Trebst, in "The Intact Chloroplast" (J. Barber, ed.), p. 315. Elsevier, Amsterdam, 1976.
[6] T. Saidha, A. I. Stern, D.-H. Lee, and J. A. Schiff, *Biochem. J.* **232**, 357 (1985).
[7] J. A. Schiff and R. C. Hodson, *Annu. Rev. Plant Physiol.* **24**, 381 (1973).
[8] W. B. Jakoby, R. D. Sekura, E. S. Lyon, C. J. Marcus, and J.-L. Wang, in "Enzymatic Basis of Detoxication" (W. B. Jakoby, ed.), Vol. 2, p. 199. Academic Press, New York, 1980.

$$R-\overset{\overset{O}{\|}}{\underset{\underset{O}{\|}}{S}}-O^-$$

of the plant sulfolipid component, 6-sulfo-6-deoxy-D-glucose (6-sulfoquinovose); the sulfonic acid group is at the redox level of sulfite.[9] Other sulfonic acids of biological interest are cysteic acid, which will serve as a sulfur source for *Neurospora,* and taurine, which is present in animal systems and can be formed from oxidation of cysteine or from sulfate in chick embryos. Homocysteic acid, another sulfonic acid, is formed by Sat_2^-, a mutant of *Chlorella pyrenoidosa* which is blocked late in the sulfate reduction pathway.

Dissimilatory sulfate reduction occurs in certain anaerobic bacteria such as *Desulfovibrio* which use sulfate in place of molecular oxygen (Fig. 1). In this process APS is the nucleotide sulfate donor, and reduction results in the accumulation of hydrogen sulfide and in ATP synthesis through oxidative phosphorylation.[10] Assimilatory sulfate reduction, in which sulfate is reduced to form sulfur at the thiol level in sulfur-containing amino acids, coenzymes, and other organic compounds, now appears to exhibit two distinct patterns.[1] (1) In oxygen-evolving photosynthesizers including some blue-green algae (cyanobacteria), eukaryotic algae, higher plants, and spinach chloroplasts, the nucleotide sulfate donor is APS acting with a highly specific APS sulfotransferase (adenylyl sulfate : thiol sulfotransferase) which will not use PAPS. This is the "APS pathway." (2) PAPS is the nucleoside phosphosulfate donor for reduction by a specific PAPS sulfotransferase (3'-phosphoadenylyl : thiol sulfotransferase) which will not use APS. This is the "PAPS pathway" that operates in organisms which lack oxygen-evolving photosynthesis (besides animals which do not reduce sulfate to the thiol level), such as yeast and *Escherichia coli,* in other bacteria, and in a few oxygen-evolving blue-green algae. Detailed reactions of the two assimilatory pathways have been presented.[1]

Control of sulfate activation takes place in some of the microorganisms that reduce sulfate.[4] In *E. coli* the formation of PAPS is repressed by cysteine, and in *Bacillus subtilis* by both cysteine and glutathione. The two steps of activation seem to be repressed simultaneously, and there is an inverse relationship between the specific activity of the activating sys-

---

[9] J. L. Harwood, *in* "The Biochemistry of Plants" (P. K. Stumpf, ed.), Vol. 4, p. 301. Academic Press, New York, 1980.

[10] H. Bothe and A. Trebst, eds., "Biology of Inorganic Nitrogen and Sulfur." Springer-Verlag, Berlin and New York, 1981.

tem and the intracellular concentration of cysteine. This type of control is not present in all microorganisms. In *Desulfovibrio,* for instance, ATP-sulfurylase is not repressed by either cysteine or sulfite.[4] This could be explained by the fact that in this microorganism APS is the only activated form of sulfate that is produced; also, sulfate reduction is required constitutively for respiration and coupled phosphorylation. The sulfate-reducing pathway in yeast appears to be regulated at the ATP-sulfurylase step through feedback inhibition by sulfide and through methionine acting as a repressor.[4]

Most organisms, including higher animals and plants, oxidize reduced sulfur to sulfate, although the aerobic chemosynthetic bacteria (Fig. 1) are the only organisms which have been observed to couple the energy released, some 180 kcal/mol, to the reduction of carbon dioxide. The anoxygenic photosynthetic bacteria use reduced sulfur compounds as photosynthetic electron donors by oxidizing them to sulfur, thiosulfate, and sulfate.[10]

Acknowledgments

Supported by grants from the National Science Foundation.

[59] Sulfate-Activating Enzymes

By Irwin H. Segel, Franco Renosto, and Peter A. Seubert

Reactions

The two sulfate-activating enzymes, ATP-sulfurylase (ATP:sulfate adenylyltransferase; EC 2.7.7.4, sulfate adenylyltransferase), reaction (1), and APS kinase (ATP:adenylylsulfate 3'-phosphotransferase; EC 2.7.1.25, adenylylsulfate kinase), reaction (2), catalyze the activation of inorganic sulfate first to APS (adenosine- 5'-phosphosulfate or 5'-adenylylsulfate) and then to PAPS (adenosine 3'-phosphate 5'-phosphosulfate, or 3'-phospho-5'-adenylylsulfate, or 3'-phosphoadenosine 5'-phosphosulfate). Although the equilibrium of the sulfurylase reaction lies far to the left, the overall production of PAPS *in vivo* is promoted by the hydrolysis of the inorganic pyrophosphate, reaction (3), and the favorable APS kinase reaction.

$$ATP + SO_4^{2-} \xrightleftharpoons{\text{ATP-sulfurylase}} PP_i + APS \qquad (K_{eq} \sim 10^{-7}) \qquad (1)$$

$$ATP + APS \xrightleftharpoons{\text{APS kinase}} PAPS + ADP \qquad (K_{eq} \sim 10^{3}) \qquad (2)$$

$$PP_i + H_2O \xrightleftharpoons{\text{Pyrophosphatase}} 2 P_i \qquad (K_{eq} \sim 10^{3} M) \qquad (3)$$

$$2 ATP + SO_4^{2-} \rightleftharpoons PAPS + ADP + 2 P_i \qquad (K_{eq} \sim 10^{-1} M)$$

PAPS, the product of the two-step activation sequence, serves as the sulfate (sulfuryl) donor for the biosynthesis of sulfate esters. In aerobic sulfate assimilating microorganisms, PAPS is the substrate for an NADPH-linked reductase system, which yields sulfite. The sulfite (possibly enzyme bound) is further reduced to sulfide which then condenses with O-acetylserine to yield cysteine. APS rather than PAPS serves as the substrate for assimilatory sulfate reduction in higher plants and other oxygen-evolving eukaryotes. However, these organisms still possess APS kinase. (The PAPS formed may serve as a reserve of "active sulfate" and/or as the sulfuryl donor for sulfate ester biosynthesis.) In dissimilatory sulfate reducers (e.g., *Desulfovibrio*), APS plays an additional role, namely, that of terminal electron acceptor of anaerobic metabolism.

Sources of the Enzymes

ATP-sulfurylase has been highly purified from *Saccharomyces cerevisiae*,[1] *Penicillium chrysogenum*,[2,3] rat liver,[4] spinach,[5] and cabbage.[6] APS kinase has been purified from spinach,[7] *Chlamydomonas reinhardi*,[8] and *Penicillium chrysogenum*.[9]

Assay Methods for ATP-Sulfurylase

Colorimetric Molybdolysis (P_i Formation) Assay

The classical molybdolysis assay[10,11] is the easiest and least expensive method for monitoring column fractions during enzyme purification. The enzyme-catalyzed reaction can be formulated as shown by reaction (4) although there is no evidence for the actual formation of the nucleotide phosphomolybdate.

$$\text{MgATP} + \text{MoO}_4^{2-} \rightleftharpoons \text{MgPP}_i + \text{APMo} \tag{4}$$
$$\text{APMo} + \text{H}_2\text{O} \rightarrow \text{AMP} + \text{MoO}_4^{2-}$$

Reaction Mixture. The reaction mixture is prepared by mixing the following reagents: 8 ml purified (phosphate-free, sulfate-free) water;

[1] P. W. Robbins and F. Lipmann, *J. Biol. Chem.* **233**, 686 (1958).
[2] J. W. Tweedie and I. H. Segel, *J. Biol. Chem.* **246**, 2438 (1971).
[3] J. W. Tweedie and I. H. Segel, *Prep. Biochem.* **1**, 91 (1971).
[4] H. N. Burnell and A. B. Roy, *Biochim. Biophys. Acta* **527**, 239 (1978).
[5] W. H. Shaw and J. W. Anderson, *Biochem. J.* **127**, 237 (1972).
[6] T. Osslund, C. Chandler, and I. H. Segel, *Plant Physiol.* **70**, 39 (1982).
[7] J. N. Burnell and J. W. Anderson, *Biochem. J.* **134**, 565 (1973).
[8] H. G. Jender and J. D. Schwenn, *Arch. Microbiol.* **138**, 9 (1984).
[9] F. Renosto, P. A. Seubert, and I. H. Segel, *J. Biol. Chem.* **259**, 2113 (1984).
[10] R. S. Bandurski, L. G. Wilson, and C. L. Squires, *J. Am. Chem. Soc.* **78**, 6408 (1956).
[11] L. G. Wilson and R. S. Bandurski, *J. Biol. Chem.* **233**, 975 (1958).

4 ml 40 mM MgCl$_2 \cdot$ 6H$_2$O; 3 ml 0.4 M Tris buffer at pH 8.0 (all "Tris buffers" are prepared by titrating Tris, free base, with HCl); 40 mg Na$_2$MoO$_4 \cdot$ 2H$_2$O; 45 mg Na$_2$ATP; and 20 μl sulfate-free inorganic pyrophosphatase solution (Sigma; package of 500 units, dissolved in 2 ml water).

Stop and Developer Solutions. The stop solution is prepared by mixing 18 ml of 0.5 M sodium acetate with 100 ml of 0.5 M acetic acid. The final pH (glass electrode) is 4. The color developer is prepared by dissolving the following reagents in 10 ml of purified (P$_i$-free) water; 0.1 g (NH$_4$)$_6$Mo$_7$O$_{24} \cdot$ 4H$_2$O; 0.2 g sodium ascorbate; and 10–20 μl (one drop) concentrated H$_2$SO$_4$. The freshly prepared developer solution should be light yellow in color. Any greenish tinge indicates contamination with inorganic phosphate.

Procedure. The assay is started by adding 0.4 ml of the reaction mixture to 0.1 ml of appropriately diluted sample. (Sample dilutions are made with 40 mM Tris buffer, pH 8.0 at assay temperature.) After 10 min of incubation, the reaction is halted by adding 1 ml of 0.5 M sodium acetate at pH 4 and 0.2 ml of developer solution. After 20 min of development, the blue color is measured at 660 nm against a minus-enzyme blank. Blanks generally have an $A_{660\,nm}^{1\,cm}$ of 0.1 or less if new or acid (HCl)-washed test tubes and fresh (low P$_i$) ATP are used. A standard containing 0.1 μmol of PP$_i$ or 0.2 μmol of P$_i$ has a corrected $A_{660\,nm}$ of about 0.55. The assay is linear with time and enzyme concentration up to an $A_{660\,nm}$ of about 0.6. Because EDTA and thiols interfere with color development, it is best to omit these agents from elution and dilution buffers. "ATPase" activity can be distinguished from ATP-sulfurylase activity by omitting the molybdate from the reaction mixture (but not from the developer). The colorimetric molybdolysis assay may not be suitable for crude cell-free extracts because of high "ATPase" levels. (This includes endogenous protein kinase-phosphoprotein phosphatase reactions.) Additionally, a protein precipitate may form upon adding the acidic stop and developer solutions.

If sulfate is used in place of molybdate as the inorganic substrate, the reaction progress curve will be nonlinear because of the accumulation of strongly inhibitory APS. The inclusion of APS kinase in the reaction mixture will eliminate product inhibition. (A linear reaction curve in the absence of added APS kinase is an indication that the sulfurylase preparation already contains APS kinase.)

Continuous Spectrophotometric Molybdolysis (AMP Release) Assay

This assay is suitable for characterizing the activity of the enzyme on a variety of alternative "short-circuiting" inorganic substrates (MoO$_4^{2-}$,

WoO_4^{2-}, SeO_4^{2-}, CrO_4^{2-}).[12,13] It is also useful for scanning purification fractions that contain phosphate. (The rat liver enzyme is unstable in Tris buffer but stable in phosphate.) Each molecule of XO_4^{2-} that reacts with ATP leads to the oxidation of two molecules of NADH. The assay is not suitable for crude fractions that contain high "ATPase," "NADH oxidase," and "PEP hydrolase" activities. This assay (and all similar assays described below) requires a recording spectrophotometer capable of providing low-noise tracings at a full chart scale deflection equivalent to a $\Delta A_{340\,nm}$ of 0.1, or less, and an initial $A_{340\,nm}$ of 1.85. A continuously variable slit width is useful for readjusting the recorder pen after each full-scale excursion.

Stock Solutions

50 mM "MgATP" (prepared by dissolving 155 mg $Na_2ATP \cdot 3H_2O$ and 51 mg $MgCl_2 \cdot 6H_2O$ in 5 ml of 150 mM Tris, free base; alternatively, the reagents can be dissolved in Tris buffer and the solution readjusted to pH 8.0 with dilute NaOH).

100 mM $MgCl_2 \cdot 6H_2O$
100 mM Na_2MoO_4
8 mM Phosphoenolpyruvate (PEP)
20 mM KCl
6 mM NADH
} Each compound is dissolved in 50 mM Tris buffer, pH 8.0

150 units ml^{-1} Pyruvate kinase combined with
225 units ml^{-1} Lactate dehydrogenase
600 units ml^{-1} Adenylate kinase (myokinase)
} The enzymes are desalted by gel filtration through Sephadex G-25 with 40 mM Tris buffer, pH 8.0.

250 units ml^{-1} sulfate-free inorganic pyrophosphatase (enzyme plus accompanying buffer salts dissolved in water).

Procedure. The molybdate- and ATP-sulfurylase-dependent formation of AMP is monitored in a reaction mixture (1.0 ml total volume) containing the following final concentration of reagents: MgATP (5 mM, or varied as necessary for kinetics studies), $MgCl_2$ (5 mM in excess over MgATP), Na_2MoO_4 (10 mM, or varied as necessary), KCl (1 mM), PEP (0.4 mM), NADH (0.3 mM), inorganic pyrophosphatase (2.5 units ml^{-1}), adenylate kinase (30 units ml^{-1}), pyruvate kinase (40 units ml^{-1}), lactate dehydro-

---

[12] P. A. Seubert, L. Hoang, F. Renosto, and I. H. Segel, *Arch. Biochem. Biophys.* **225**, 679 (1983).

[13] P. A. Seubert, F. Renosto, P. Knudson, and I. H. Segel, *Arch. Biochem. Biophys.* **240**, 509 (1985).

genase (60 units ml^{-1}), ATP-sulfurylase (~0.1 μg of pure enzyme), and sufficient 50 mM Tris buffer, pH 8.0, to make a total volume of 1 ml. If available, APS kinase (0.25 unit ml^{-1}) is added to remove traces of sulfate which can give rise to inhibitory levels of APS. (PAPS is not a potent inhibitor of ATP-sulfurylase.) The reaction is started by adding the molybdate after any contaminating ADP and AMP have been depleted and the background rate of NADH oxidation has been established. The $\Delta A_{340\ nm}$ is measured as soon as the steady-state rate is achieved. Both primary substrates (MgATP and MoO$_4^{2-}$) are continuously regenerated, so that the limiting reagents are NADH and PEP.

$^{35}SO_4^{2-}$ Incorporation ($[^{35}S]$PAPS Synthesis) Assay

This assay is suitable for characterizing the activity of the purified enzyme on its natural inorganic substrate.[12] It can also be used to follow the fractionation of the enzyme in early purification steps. However, the presence of contaminating sulfohydrolases and various sulfotransferases will yield artifactually low ATP-sulfurylase levels. Purified APS kinase is required as a coupling enzyme.

Procedure. The reaction mixture, 1.5 ml total volume, contains 5 mM MgATP (or varied for kinetics studies), 5 mM excess (free) Mg^{2+} as MgCl$_2$, 10 mM ^{35}SO$_4^{2-}$ (or varied for kinetics; charcoal treated[12]; specific activity $> 4 \times 10^7$ cpm μmol^{-1}), 2–10 units ml^{-1} yeast inorganic pyrophosphatase, 0.3–3 units ml^{-1} purified APS kinase, and 0.1 ml of appropriately diluted ATP-sulfurylase (~1 μg of pure enzyme), all in 50 mM Tris buffer at pH 8.0 (standard buffer). After the desired incubation period, 0.2 ml of assay mixture is removed and added to 3 ml of ice-cold buffer containing 40 mg of acid-washed Norit charcoal (Pfanstiehl), 0.1 M unlabeled Na$_2$SO$_4$, and 0.3 mg bovine serum albumin. The mixture is centrifuged in a table-top clinical centrifuge, and the supernatant fluid is carefully removed by suction with a 22-gauge hypodermic needle. The charcoal pellet is vigorously suspended in 5 ml standard buffer containing 0.1 M unlabeled Na$_2$SO$_4$ and centrifuged. This washing procedure is repeated 5 or 6 times or until 0.5 ml of each successive supernatant solution contains a constant amount of ^{35}S. Finally the charcoal pellet is vigorously suspended in 2 ml 50% (v/v) ethanol containing 0.4% (w/v) NH$_3$. The charcoal is centrifuged, resuspended in the supernatant solution, and centrifuged once more. Then, 0.5 ml of the eluted ^{35}S-labeled nucleotide solution is carefully removed and counted in 5 ml of PCS scintillation fluid (Amersham). The specific activity of the ^{35}SO$_4^{2-}$ is determined by counting a suitable dilution of the stock solution in 0.5 ml of the above eluting solution. Blanks without enzyme are used at each ^{35}SO$_4^{2-}$ concentration.

If APS kinase is unavailable, the $^{35}SO_4^{2-}$ incorporation assay can still be used to follow the fractionation of the enzyme, although the reaction will not be linear with time. The assay mixture should contain as high as MgATP level as is feasible (in order to minimize the progressive product inhibition by APS); 10 mM MgATP and 2 mM $^{35}SO_4^{2-}$ are suitable concentrations. The separation of $^{35}SO_4^{2-}$ and [^{35}S]APS by ethanol precipitation of the former is the basis of another $^{35}SO_4^{2-}$-dependent assay for ATP sulfurylase.[14]

$^{32}PP_i$ Release Assay

The forward reaction can also be assayed by measuring the ATP-sulfurylase and sulfate-dependent release of charcoal-nonadsorbable $^{32}PP_i$ from [γ-^{32}P]MgATP.[13] This assay is quicker than the $^{35}SO_4^{2-}$ incorporation assay because the multiple charcoal washing steps are eliminated, but purified APS kinase is still required to pull the reaction to the right. APS kinase can be omitted if molybdate is used in place of sulfate. Contaminating "ATPase" activity will increase the minus-sulfate or minus-molybdate blanks. High quality [^{32}P]ATP is essential to minimize the blank values. (If more than 5–10% of the ^{32}P in the preparation is present as $^{32}P_i$ or $^{32}PP_i$, the labeled ATP should be purified.)

Procedure. The reaction mixture (0.5 ml total volume) contains 1 mM (or varied) [γ-^{32}P]MgATP (specific activity $> 4 \times 10^6$ cpm μmol^{-1}; obtained from a premixed solution of 50 mM Na$_2$ATP, 50 mM MgCl$_2$, and labeled ATP adjusted to pH 8.0 with either Tris, free base, or NaOH), 5 mM excess MgCl$_2$, 3 mM (or varied) Na$_2$SO$_4$, 4 units ml^{-1} inorganic pyrophosphatase, 0.6 units ml^{-1} APS kinase, and ATP-sulfurylase, all in 50 mM Tris buffer, pH 8.0. The reaction is started by adding ATP-sulfurylase. After 5 min of incubation, 3 ml of ice-cold buffer containing 60 mg charcoal is added to stop the reaction. The suspension is vortexed and centrifuged. The supernatant solution (1 ml) is counted in 3 ml of scintillation fluid. Blanks without enzyme are used at each ATP concentration. The specific activity of the labeled substrate is determined by adding 3 ml of buffer without charcoal to a reaction mixture and counting 1 ml of the resulting solution.

If the assay is used for kinetics studies, ATP utilization (two molecules per sulfate activated) must be kept to $<10\%$. If the contaminating $^{32}P_i$ level is 5%, the experimental counts will only be 2 to 3 times the blank values.

---

[14] Z. Reuveny and P. Filner, *Anal. Biochem.* **75,** 410 (1976).

Continuous Spectrophotometric (APS Kinase-Coupled) Assay

The continuous spectrophotometric assay is the most convenient assay for studying the kinetics of the forward reaction and determining the specific activity of the purified enzyme with its natural substrate. The procedure is identical to that of the continuous spectrophotometric molybdolysis assay except (1) excess APS kinase (3 units ml^{-1}) replaces adenylate kinase and (2) sulfate (up to 10 mM) replaces molybdate. One molecule of NADH is oxidized for each molecule of sulfate activated. Molybdate, tungstate, and chromate are inactive, but saturating selenate yields a V_{max} that is 24% of that with sulfate.

Reaction Progress (Average Velocity) Assays of the Forward Reaction with Sulfate in the Absence of APS Kinase

APS is a potent competitive product inhibitor of ATP-sulfurylase. Consequently, in the absence of the APS kinase as coupling enzyme, the reaction progress curves quickly depart from linearity at unsaturating substrate concentrations, making it impractical to study the initial velocity kinetics with sulfate as the inorganic substrate. However, because the K_i and K_m values of the substrates are about 1000-fold higher than the K_i and K_m values of the products, it is possible to exploit the progressive product inhibition as a diagnostic tool.[12,15] At the usual assay levels of substrates (0.3 to 3 K_m), with PP$_i$ continuously removed, the reaction slows for only one reason: progressive product inhibition by APS. Substrate depletion is negligible, and, therefore, [MgATP] and [SO$_4^{2-}$] can be considered to remain constant. The integrated rate equation for this situation is relatively simple and can be cast in the form of a linear Dixon plot of $1/\bar{v}$ versus [APS], where $\bar{v}$ is the average or apparent velocity at any time, i.e., [APS]/t.

The procedure is to monitor the appearance of [^{35}S]APS or P$_i$ (or ^{32}P$_i$) over time at several different initial (and essentially unchanging) concentrations of one of the substrates (e.g., MgATP) and a fixed concentration of the cosubstrate (SO$_4^{2-}$). Data points should be taken early in the reaction when changes in the concentration of the accumulating APS can be easily measured. A large excess of inorganic pyrophosphatase must be present to maintain [PP$_i$] close to 0. Data plotted as $1/\bar{v}$ versus [APS] intersect the $1/\bar{v}$-axis at $1/v_0$, the true initial velocity for [MgATP] at the constant [SO$_4^{2-}$]. The experiment is repeated at several different fixed sulfate concentrations, yielding v_0 values for several different MgATP concentrations at several different sulfate concentrations. These values

---

[15] N. D. Schmidt, J. J. Peschon, and I. H. Segel, *J. Theor. Biol.* **100,** 597 (1983).

can be analyzed in the usual way to obtain the relevant kinetic constants (V_{max_f}, K_{m_A}, K_{i_a}, K_{m_B}).

$^{35}SO_4^{2-}$ Release Assay

The MgPP$_i$-dependent release of $^{35}SO_4^{2-}$ from [^{35}S]APS is the most convenient assay for studying the reverse ATP-sulfurylase reaction.[13] The procedure is similar to the ^{32}PP$_i$ release assay: The reaction mixture (total volume, 1 ml) contains 0.02–0.5 μM [^{35}S]APS (specific activity $> 4 \times 10^8$ cpm μmol^{-1}), 0.33–10 μM total PP$_i$, 5 mM total (and essentially excess) MgCl$_2$, and ATP-sulfurylase, all in 50 mM Tris buffer, pH 8.0. The reaction is started by adding the enzyme. After 2.5 min of incubation, 2 ml of ice-cold buffer containing 40 mg charcoal and 0.1 M Na$_2$SO$_4$ are added to stop the reaction. The suspension is vortexed, centrifuged, and a 1-ml aliquot of the supernatant solution is counted in 3 ml of scintillation fluid. Minus-enzyme blanks are run at each APS concentration. The specific activity of the labeled substrate is determined by adding 2 ml of buffer–0.1 M Na$_2$SO$_4$ minus charcoal to a reaction mixture and counting 1.0 ml of the resulting solution. Enzyme concentration should be adjusted such that less than 10% of the [^{35}S]APS is consumed in the reaction. The success of the assay requires high purity [^{35}S]APS (<5% contamination with $^{35}SO_4^{2-}$). High specific radioactivity [^{35}S]APS, can be prepared by treating [^{35}S]PAPS with nuclease P$_1$. ([^{35}S]PAPS can be obtained commercially or prepared using the purified sulfate activating enzymes. Nuclease P$_1$ is available from Sigma.)

The extremely high K_{eq} of the reverse ATP-sulfurylase reaction can be exploited for measuring unknown concentrations of PP$_i$. If the reaction is allowed to go to completion, $^{35}SO_4^{2-}$ release is stoichiometric with the amount of PP$_i$ present.[15a]

$^{32}PP_i$ Incorporation and $^{32}PP_i$–ATP Exchange Assay

The APS-dependent conversion of ^{32}PP$_i$ to [^{32}P]ATP is the basis of another assay of the reverse ATP-sulfurylase reaction.[4,6] High specific radioactivity ^{32}PP$_i$ is required ($>10^9$ cpm μmol^{-1}). The labeled product is adsorbed onto charcoal, washed free of residual ^{32}PP$_i$, eluted with aqueous ethanol–NH$_3$, and counted (see $^{35}SO_4^{2-}$ Incorporation Assay). In the presence of sulfate (or selenate, but not molybdate), the enzyme catalyzes a ^{32}PP$_i$–ATP exchange reaction in the absence of added APS.[6,16]

[15a] L. A. Daley, F. Renosto, and I. H. Segel, *Anal. Biochem.* **157**, 385 (1986).
[16] W. H. Shaw and J. W. Anderson, *Plant Physiol.* **47**, 114 (1971).

Continuous Spectrophotometric Assay of the Reverse Reaction

MgATP formed in the reverse ATP-sulfurylase reaction can be monitored continuously in an assay coupled to hexokinase and glucose-6-phosphate dehydrogenase. The reaction mixture (1 ml total volume) contains 10–15 μM APS, 5 mM MgCl$_2$, 0.6 mM NADP$^+$, 1 mM glucose, 1 mM (or varied) sodium pyrophosphate, 2.5 units glucose-6-phosphate dehydrogenase (sulfate-free), 5 units hexokinase (sulfate-free), and ATP-sulfurylase, all in 50 mM Tris at pH 8.0. The reaction is started by adding APS after the background rate of NADPH formation has been established (the background is mainly the glucose dehydrogenase activity of glucose-6-phosphate dehydrogenase). This assay is useful for establishing V_{max_r} and the K_m for PP$_i$ at saturating concentrations of APS. Without a special light path-expanding modification, the usual recording spectrophotometer is not sufficiently sensitive to establish the K_i and K_m of APS. If allowed to go to completion in the presence of excess PP$_i$, the assay is useful for determining the APS concentration of stock solutions. (The commercial product is 80–90% pure, the major impurity being AMP.)

A bioluminescence assay for ATP has also been used to measure the reverse ATP-sulfurylase reaction.[17,18]

Assay Methods for APS Kinase

APS kinase is extremely difficult to assay accurately and may be difficult to detect in cell free extracts and early purification fractions. There are two major problems and the solutions to each are mutually cancelling. First, micromolar levels of APS are rapidly depleted by competing reactions of nucleotide sulfohydrolases and especially ATP-sulfurylase. ATP-sulfurylase will quantitatively convert micromolar APS to ATP and sulfate in the presence of micromolar PP$_i$. (The PP$_i$ can be present as a contaminant in the ATP or produced from the ATP by contaminating enzymes.) As little as 10 mM phosphate is sufficient to remove 10 μM APS if both ATP-sulfurylase and inorganic pyrophosphatase are present. The seemingly obvious solution, i.e., to use high levels of APS in the assay, is frustrated by the potent substrate inhibition of APS kinase by APS. At high ionic strength, the enzyme is inhibited by more than 6 μM APS; a dead-end enzyme · MgADP · APS complex is formed. At low ionic strength, inhibition starts at less than 1 μM APS.

Paper electrophoresis can be used to trace the distribution of APS kinase in early fractions. The incubation mixture in a total volume of 50 μl

---

[17] G. J. E. Balharry and D. J. D. Nicholas, *Biochim. Biophys. Acta* **220**, 513 (1970).
[18] B. C. Gerwick, S. B. Ku, and C. C. Black, *Science* **209**, 513 (1980).

contains 20 mM MgATP, 5 mM excess MgCl$_2$, 10 μM[^{35}S]APS (>10^8 cpm μmol^{-1}), and 20 μl of the APS kinase-containing fraction, all in 40 mM Tris–HCl at pH 8.0. After 10 min of incubation, 5–10 μl of the assay mixture is spotted onto a 50 × 2 cm strip of Whatman 3MM filter paper moistened with 50 mM potassium phosphate at pH 7.5. The strips are subjected to electrophoresis at 1000–2000 V until a yellow CrO$_4^{2-}$ marker (on a separate strip) reaches the end. (CrO$_4^{2-}$ migrates close to ^{35}SO$_4^{2-}$.) A separate yellow marker of DNP-cysteic acid migrates just behind PAPS. After drying, the strips are examined with a radiochromatogram scanner or are cut into small sections and counted in scintillation fluid. PAPS formation can also be measured by HPLC methods.[19–21]

After ATP-sulfurylase and ATPase activity have been removed or reduced to manageable levels, a continuous spectrophotometric assay can be used to follow the further purification of APS kinase and to characterize the kinetics of the pure enzyme. The reaction mixture is the same as that described earlier for the continuous molybdolysis assay of ATP-sulfurylase, except that 0.1 M Tris buffer is used, adenylate kinase and MoO$_4^{2-}$ are omitted, and APS (5.5 μM) and nuclease P$_1$ (25 units ml^{-1}) are included.[9] Nuclease P$_1$ continuously regenerates APS from the PAPS formed. (Rye grass 3′-nucleotidase is also suitable, but much more expensive than nuclease P$_1$. Alkaline phosphatase can not be used because the enzyme acts on ATP.) The commercial (Sigma) ammonium sulfate-containing pyruvate kinase–lactate dehydrogenase coupling enzymes are preferred. The high sulfate decreases the substrate inhibition of APS kinase by APS and also supresses the reverse ATP-sulfurylase reaction. Also, the background rate of NADH oxidation is lower than with desalted coupling enzymes. The above assay with nuclease P$_1$ omitted is useful for measuring the APS concentration of stock solutions.

The reverse APS kinase reaction can be studied with homogeneous enzyme by coupling the formation of APS to ATP-sulfurylase (1 molybdolysis unit ml^{-1}), hexokinase (5 units ml^{-1}), and low-sulfate glucose-6-phosphate dehydrogenase (2.5 units ml^{-1}) in the presence of the desired concentrations of PAPS and MgADP, MgCl$_2$ (5 mM excess), sodium pyrophosphate (1.0 mM), NADP (0.6 mM), glucose (1 mM), all in 0.1 M Tris buffer at pH 8.0.[9] The reaction is started by adding APS kinase after a brief preincubation to remove traces of APS in the PAPS. In this assay, two molecules of NADPH are formed for each molecule of PAPS and MgADP that react. Because the V_{max} of the reverse reaction is less than

---

[19] E. J. M. Pennings and M. J. Van Kempen, *J. Chromatogr.* **176**, 478 (1979).
[20] J. D. Schwenn and H. G. Jender *J. Chromatogr.* **193**, 285 (1980).
[21] D. M. Delfert and H. E. Conrad, *Anal. Biochem.* **148**, 303 (1985).

1% of V_{max_f}, the amount of APS kinase needed to achieve measurable rates is much greater than that required for measurements of the forward reaction.

Purification of the Sulfate-Activating Enzymes from *Penicillium chrysogenum*

The following procedure consistently provides high yields of homogeneous enzyme from mycelia of *P. chrysogenum* (wild type, ATCC 24791). With minor modifications, we have used the procedure to obtain ATP-sulfurylase from other fungi (*P. duponti, Aspergillus nidulans, Neurospora crassa*), plant leaves (cabbage, spinach), and rat liver and brain.

Growth Conditions. The organism is stored as a spore suspension in sterile soil. Spore plates are prepared periodically by aseptically sprinkling some soil onto the surface of solid medium (400 ml commercial tomato juice and 600 ml water, solidified with 30 g Bacto agar) contained in 11 × 7 cm flat culture bottles. Spores from one bottle are scraped from the surface into 30 ml sterile water and inoculated into 100 ml synthetic medium at pH 7.0 contained in a 500-ml Erlenmeyer flask.

The medium contains, per liter, 10 g dibasic ammonium citrate, 6 g dibasic ammonium phosphate, 16 g dibasic potassium phosphate, 1 g Na_2SO_4 (as sole sulfur source), 10 ml stock trace metals solution, and 40 g commercial cane sucrose [sterilized separately as a 40% (w/v) solution and added to the rest of the medium just before inoculation]. The trace metals solution contains, per liter, 5 g sodium citrate · $2H_2O$, 3 g $MnCl_2$, 2 g $ZnCl_2$, 2 g $FeCl_3$, 50 g $MgCl_2 \cdot 6H_2O$, 200 mg $CuCl_2 \cdot 2H_2O$, 750 mg $CaCl_2$, 200 mg $CoCl_2 \cdot 6H_2O$, 100 mg Na_2MoO_4, and 100 mg $Na_2B_4O_7$. The culture is incubated on a rotary shaker (250 rpm) at room temperature and transferred (10% inoculum) every 1–2 days until the mycelia grow as finely dispersed filaments (without pellets or lumps).

For enzyme production, two 3-liter Fernbach flasks, each containing 900 ml of medium, are inoculated with 100 ml of a 24-hr mycelial suspension. L-Cysteic acid (0.1 g liter^{-1}) is used as the sulfur source in this and subsequent cultures. After 24 hr of growth, 1–1.5 liters of the new culture is used to inoculate a New Brunswick Microferm fermentor jar containing 9 liters of medium. One or two jars are used in a preparation. The jars are incubated for 24 hr at 25° with an air flow of 9 liters min^{-1} and an agitator speed of 200 rpm.

Separation of ATP-Sulfurylase and APS Kinase. The mycelia from one or two jars are collected by suction filtration through four layers of cheesecloth and washed 3 times with cold deionized water. The washed and blotted mycelial pad, about 350 g per fermentor jar, is frozen in small batches with liquid nitrogen and then ground to a powder in a Waring blendor. The frozen powder is stirred into 1.5–2 volumes of 0.3 M Tris–HCl at pH 8.0 containing 5 mM EDTA, disodium salt. After thawing, the homogenate is centrifuged at 15,000 g for 10 min at 4°. All subsequent steps are carried out at 4°. To the supernatant solution, solid $(NH_4)_2SO_4$ is added in small portions to bring the solution to "30% saturation" (164 g of ammonium sulfate per liter added over a 1-hr period). After stirring for 15 min, the solution is centrifuged at 15,000 g for 10 min and the pellet discarded. The supernatant solution is brought to "55% saturation" with an additional 148 g of solid $(NH_4)_2SO_4$ per liter added over a 1-hr period. After stirring for 20 min, the precipitate is collected by centrifugation, dissolved in 200–400 ml of 40 mM Tris buffer at pH 8 (standard buffer), and dialyzed against 3–4 liters of the same buffer for 6 hr with at least two changes of buffer.

After dialysis, the protein solution is applied to a Blue Dextran–Sepharose[22] column (4 × 20 cm) connected in series to an Affi-Gel Blue (Bio-Rad) column (1.9 × 60 cm). Both columns are equilibrated with standard buffer. The columns (still in series) are washed with the standard buffer at a flow rate of 2 ml min^{-1} until the Affi-Gel Blue effluent has an A_{280} of 0.2 or less. ATP-sulfurylase is then eluted from the Blue Dextran column with a linear gradient of NaCl (0–1 M) in standard buffer (total volume of 800 ml). The flow rate is about 2 ml min^{-1}. Because freshly prepared Blue Dextran columns have a tendency to "bleed," the enzyme fractions may be contaminated with a small amount of eluted dye at this stage. APS kinase is eluted from the Affi-Gel Blue column with a linear gradient of 0–1.5 M NaCl in standard buffer (500 ml total volume). At this stage, ATP-sulfurylase is easily detected by the P_i colorimetric molybdolysis assay. APS kinase can be detected by the coupled spectrophotometric assay although the background rate of NADH oxidation remains high.

If only ATP-sulfurylase is sought, the Blue Dextran column can be replaced by a similar size Affi-Gel Blue column. (APS kinase, which will be eluted along with ATP-sulfurylase, does not interfere with the assay for the latter.)

Further Purification of ATP-Sulfurylase. The pooled active fractions from Blue Dextran (at 0.15–0.3 M NaCl) are dialyzed against two changes of 15 volumes each of standard buffer and applied to a DEAE-cellulose column (2.5 × 33 cm). The column is washed with standard buffer until the A_{280} is close to 0 and is eluted with a linear gradient of 0–1 M NaCl in standard buffer (800 ml total volume). ATP-sulfurylase elutes at about 0.15–0.3 M NaCl. Blue Dextran which bled from the previous column

[22] L. D. Ryan and C. S. Vestling, *Arch. Biochem. Biophys.* **160**, 279 (1974).

TABLE I
PURIFICATION OF ATP-SULFURYLASE FROM Penicillium chrysogenum[a]

Fraction	Volume (ml)	Protein (mg)	Activity[b] (units)	Specific activity[b] (units/mg protein)
Extract	1,430	13,800	(1,800)[c]	(0.13)[c]
Ammonium sulfate	300	5,900	(900)[c]	(0.16)[c]
Blue Dextran	170	248	458	1.85
DEAE-cellulose	50	33	308	9.3
BioGel A-1.5m	135	14.7	285	19.4

[a] From 637 g (wet wt., filtered and blotted) mycelia.
[b] Units in the molybdolysis assay at pH 8.0 and 30°.
[c] Not assayed. The parenthetical values are estimates based on a 50% recovery of activity in each of the first two steps.

remains bound to the DEAE. Active fractions are pooled and concentrated with a Diaflo PM30 membrane if necessary to about 0.5 mg protein per ml. Aliquots of 15 ml are passed through a BioGel A-1.5m gel filtration column (2.5 × 110 cm). The combined active fractions (about 65 ml total volume) are concentrated to about 0.5 mg ml^{-1} by filtration through a Diaflo PM30 membrane filter using a 50-ml Amicon stirred cell. At this stage, the preparation usually shows a single major Coomassie blue-staining band on sodium dodecyl sulfate gels. Trace protein impurities can be removed by a final adsorption–elution from a Matrex Gel Green A column (1.5 × 26 cm; elution with a 0–2 M NaCl gradient in standard buffer followed by dialysis). The final preparation is stored frozen at −20° in batches of 1 ml. Table I summarizes the purification of ATP-sulfurylase.

Properties of ATP-Sulfurylase. The *P. chrysogenum* enzyme has a subunit M_r of 69,000. The native enzyme appears to be a hexamer (M_r = 420,000). At pH 8.0 and 30° in the presence of 5 mM excess Mg^{2+}, the specific activities (V_{max}; units per mg protein) are as follows: 21 ± 3 (molybdolysis), 8 ± 1.5 (APS synthesis), 59 ± 9 (MgATP synthesis). The K_m for MgATP is 0.18 mM (with SO$_4^{2-}$) or 0.05 mM (with MoO$_4^{2-}$). The K_m values for SO$_4^{2-}$ and MoO$_4^{2-}$ are about 0.5 mM and 0.1 mM, respectively. In the reverse direction, the K_m values for APS and PP$_i$ (total of all species) are 0.3 and 6.5 μM, respectively. The K_i for APS is about 0.04 μM. Monovalent oxyanions (FSO$_3^-$, ClO$_3^-$, ClO$_4^-$, NO$_3^-$) are competitive with SO$_4^{2-}$ ($K_{iFSO_3^-}$ 3.4 μM) and uncompetitive with MgATP. Thiosulfate is also competitive with SO$_4^{2-}$ (K_i 0.35 mM) but noncompetitive with MgATP. Phenylglyoxal irreversibly inactivates the enzyme. Modification of SH groups with DTNB or tetranitromethane markedly

TABLE II
Purification of APS Kinase from *Penicillium chrysogenum*[a]

Fraction	Volume (ml)	Protein (mg)	Activity[b] (units)	Specific activity[b] (units/mg protein)
Extract	1,430	13,800	(1,680)[c]	(0.12)[c]
Ammonium sulfate	300	5,900	(840)[c]	(0.14)[c]
Affi-Gel Blue	96	67	420	6.3
Matrex Gel Green A	116	29	390	13.4
BioGel A-1.5m	66	15	360	24

[a] From 637 g (wet wt., filtered and blotted) mycelia.
[b] Units measured at 5 mM MgATP, 5.5 μM APS, 5 mM excess Mg^{2+}, ~0.15 M $(NH_4)_2SO_4$ (supplied by the coupling enzymes), and other factors indicated in the text, pH 8.0 and 30°.
[c] Not assayed. The parenthetical values are estimates based on a 50% recovery of activity in each of the first two steps.

increases the $[S]_{0.5}$ values for MgATP and SO_4^{2-} but has little effect on V_{max}.

Further Purification of APS Kinase. APS kinase elutes from Affi-Gel Blue between 0.5 and 0.7 M NaCl, just behind a peak of ATPase activity and partially overlapping a peak of yellow-pigmented material. The pooled active fractions are dialyzed for 3 hr against 3 liters of standard buffer with one change of buffer and then applied to a Matrex Gel green A column (2.5 × 27 cm) equilibrated with standard buffer. The column is washed with buffer until A_{280} indicates that no further protein is being eluted. A linear gradient of 0–1.5 M NaCl in standard buffer (total volume, 500 ml) is used to elute the enzyme at a flow rate of about 1 ml min^{-1}. Enzyme activity appears in the region of 0.5–0.7 M NaCl. The pooled active fractions are concentrated to about 10 ml by filtration through an Amicon PM30 Diaflo ultrafilter in a 50-ml stirred cell. After concentration, 5–7.5 ml of the solution is applied to a BioGel (Agarose) A-1.5m column (1.8 × 100 cm) equilibrated with standard buffer, and developed at a flow rate of 0.3 ml min^{-1}. The most active fractions are pooled and stored at −20° in batches of 1 ml. The preparation at this point generally contains 0.15–0.20 mg of protein per ml and is about 90% pure. Higher purity preparations can be obtained by collecting fractions more selectively. Table II summarizes the purification of the enzyme.

Properties of APS Kinase. The native *P. chrysogenum* enzyme is a dimer of two apparently identical 30,000-dalton subunits. At temperatures above 40°, the enzyme dissociates into inactive subunits. Upon cooling, the subunits reassociate and catalytic activity is restored; MgATP accelerates reassociation–reactivation. The enzyme is strongly inhibited by its substrate, APS. Neutral salts affect the v versus [APS] profile. Under standard assay conditions [pH 8.0, 30°, 5 mM MgATP, 5 mM excess Mg^{2+}, and 0.15 M $(NH_4)_2SO_4$], the highest activity (24 units per mg protein) is observed at 5.5 μM APS. (The extrapolated V_{max} is about 38 units per mg protein.) Activity at 5 mM MgATP is half-maximal at 1.3 and 27 μM APS. The K_m values for MgATP and APS are 1.5 mM and 1.4 μM, respectively. In the reverse direction (V_{max_r} 0.16 units per mg protein), the K_m for PAPS is 8 μM. APS kinase is irreversibly inactivated by trinitrobenzene sulfonate.

Preparation of "Carrier-Free" [^{35}S]PAPS and [^{35}S]APS

"Carrier-free" [^{35}S]PAPS is prepared by incubating the following at 30°: 2.5 mM MgATP, 5 mM $MgCl_2$, 10 units per ml sulfate-free inorganic pyrophosphatase, 0.1 APS synthesis units per ml (about 10 μg per ml) of purified ATP-sulfurylase, 1 unit per ml homogeneous APS kinase, and 10 mCi per ml of carrier-free $^{35}SO_4^{2-}$, all in 50 mM Tris–HCl at pH 8.0. The usual reaction volume is 1.0 ml. To check for the extent of conversion, 10-μl aliquots are removed periodically and diluted 50-fold. Ten microliters of the dilution are added to 2 ml of 2% (w/v) charcoal suspension. The mixture is vortexed and then centrifuged. An aliquot of 50 μl of the supernatant fluid is counted to determine residual $^{35}SO_4^{2-}$. Generally, greater than 90% conversion of $^{35}SO_4^{2-}$ to charcoal-adsorbable ^{35}S is achieved in 60 min. The labeled product is separated from ATP and any residual $^{35}SO_4^{2-}$ by flat-bed paper electrophoresis on Whatman 3MM sheets as described earlier. [^{35}S]PAPS is eluted from the paper with water. A convenient method is to place the cut-up paper into the barrel of a 3-ml syringe and force three or four 1.5-ml volumes of water through with the piston. The resulting solution is treated as a source of "carrier-free" [^{35}S]PAPS to obtain the desired specific activity of labeled substrate. The [^{35}S]PAPS will not really be "carrier-free" if any of the reagents used in the preparation contain contaminating sulfate.

"Carrier-free" [^{35}S]APS is prepared by treating "carrier-free" [^{35}S]PAPS with nuclease P_1. The reaction mixture, generally 2 ml containing 20 units of nuclease P_1, is incubated for 2 hr at 30°. Then, 0.4 ml of charcoal suspension (20 g per liter in 50 mM Tris–HCl of pH 8.0) is added with vigorous agitation (vortex). The suspension is centrifuged, and the supernatant liquid treated again with charcoal. The combined pellets are washed about 10 times with the buffer by resuspension and centrifugation. [^{35}S]APS is eluted with 0.3 ml of 0.4% triethanolamine in 50% aqueous ethanol. After centrifugation, the supernatant solution is filtered through a

Millipore Swinnex filter unit (0.45 μm pore size) to remove traces of charcoal and then concentrated to dryness with a Speed Vac. The [^{35}S]APS can be stored dry or dissolved in standard buffer and the solution stored at $-20°$. The solution is treated as a source of "carrier-free" [^{35}S]APS for mixing with solutions of commercial unlabeled APS (Sigma) to obtain the desired concentration and specific activity. The charcoal adsorption–elution procedure is performed periodically to rid the stock of $^{35}SO_4^{2-}$ that accumulates on prolonged storage.

Preparation of Unlabeled PAPS

If homogeneous sulfate-activating enzymes are unavailable, substrate amounts of unlabeled PAPS can be prepared using partially purified fractions from yeast,[23] rat liver,[24] *chlorella*,[25] chick embryo chondrocytes,[21] or *P. chrysogenum* (below), or can be prepared by organic synthesis.[26]

Procedure. The 30–55% ammonium sulfate fraction obtained from 100 g of mycelia is dissolved in about 25 ml of 50 mM Tris–HCl at pH 8.5 and dialyzed. MgATP (250 μmol from a 50 mM stock solution adjusted to pH 8), MgSO$_4$ (500 μmol), and inorganic pyrophosphatase (20 units) are added, and the mixture is adjusted to pH 8.5 by adding 0.15 M Tris, free base. The volume is brought to 50 ml with 50 mM Tris–HCl at pH 8.5. After 2 hr at 30°, the reaction vessel is placed in a boiling water bath, stirred for 2 min, and then cooled on ice. The precipitated protein is removed by centrifugation. PAPS is purified by the method of Singer.[24] Briefly, this involves adsorption to Dowex-1-Cl$^-$ (2.5 × 15 cm column; equilibrated with 1 mM Tris–HCl at pH 8.7) and stepwise elution, first with 3–5 liters of 0.3 M NaCl in 1 mM Tris–HCl at pH 8.7 (to elute ATP) and then with a 0.3–1.1 M NaCl gradient in 1 mM Tris–HCl at pH 8.7 (1 liter total volume). Those fractions having an $A_{260\,nm}$ value greater than 1 are pooled (about 300 ml total) and lyophilized. The dried material is suspended in about 30 ml of water, centrifuged to remove salt, and passed through a Sephadex G-10 column (2.5 × 95 cm, equilibrated and developed with the 1 mM Tris buffer at pH 8.7) in 15-ml portions. The fractions containing the bulk of the eluted PAPS (first peak) are pooled and lyophilized. (The second peak is PAP.) The dried PAPS is dissolved in 1–2 ml water. If analytical paper electrophoresis reveals contaminating PAP, a preparative run on large sheets of Whatman 3MM is included as the final step.

[23] P. W. Robbins, this series, Vol. 5, p. 970.
[24] S. S. Singer, *Anal. Biochem.* **96**, 34 (1979).
[25] M. L-S. Tsang, J. Lemieux, J. A. Schiff, and T. B. Bojarski, *Anal. Biochem.* **74**, 623 (1976).
[26] R. Cherniak and E. A. Davidson, *J. Biol. Chem.* **239**, 2986 (1964).

[60] Thiosulfate Reductase

By THOMAS R. CHAUNCEY, LAWRENCE C. UHTEG, and JOHN WESTLEY

$$SSO_3^{2-} + GSH \rightleftharpoons SO_3^{2-} + GSSH$$
$$(GSSH + GSH \rightleftharpoons H_2S + GSSG)$$

Thiol-dependent thiosulfate reductases occur in both prokaryotic and eukaryotic organisms of various phyla.[1-5] These enzymes use electrons from thiols to reduce the sulfane sulfur atoms of inorganic thiosulfate and organic thiosulfonate anions to the sulfide level. Because of problems with instability, only the yeast enzyme has been purified extensively.[6,7] Analysis of its reaction mechanism has shown that only the first of the reactions written above is enzyme catalyzed.[6,8] Spontaneous reaction of the persulfide (hydrodisulfide) product with excess thiol substrate releases the free inorganic sulfide.

Assay Method

Principle. Several different assay methods for this enzyme have been described. The one given here is historically the first of these, based on the reaction of the sulfite product with pararosaniline.[4,6,9] It is a discontinuous, practical method at the pH optimum, suitable for following the progress of a purification. The other procedures use either a coupled system with glutathione reductase and NADPH[8] or rely on the change in ultraviolet[10] or visible[11] absorbance when aromatic thiosulfonates are desulfurated to the corresponding sulfinates by thiosulfate reductase.[8] These latter methods, not given here, provide continuous monitoring of reaction progress and are thus especially useful for kinetic studies.

---

[1] A. Kaji and W. D. McElroy, *J. Bacteriol.* **77,** 630 (1959).
[2] H. D. Peck, Jr., and E. Fisher, Jr., *J. Biol. Chem.* **237,** 190 (1962).
[3] B. Sörbo, *Acta Chem. Scand.* **18,** 821 (1964).
[4] A. Koj, *Acta Biochim. Pol.* **15,** 161 (1968).
[5] B. Sido and A. Koj, *Acta Biol. Cracov., Ser. Zool.* **15,** 97 (1972).
[6] L. C. Uhteg and J. Westley, *Arch. Biochem. Biophys.* **195,** 211 (1979).
[7] T. R. Chauncey and J. Westley, *Biochim. Biophys. Acta* **744,** 304 (1983).
[8] T. R. Chauncey and J. Westley, *J. Biol. Chem.* **258,** 15037 (1983).
[9] B. Sörbo, *Acta Chem. Scand.* **12,** 1990 (1958).
[10] B. Sörbo, *Acta Chem. Scand.* **16,** 243 (1962).
[11] See this volume [42].

Reagents

Na_2SO_3, 0.1 M standard solution, fresh, deaerated
Tris–acetate buffer, 0.1 M, pH 9.0
$Na_2S_2O_3$, 80 mM
Glutathione, 80 mM, freshly prepared
NaOH, 1.0 M
$HgCl_2$, 0.23 M
Formaldehyde, 0.2% (w/v)
Pararosaniline, 0.04% (w/v) in 0.72 M HCl

Procedure. To small centrifuge tubes are added 0.5-ml aliquots of buffer and 0.25-ml aliquots of $Na_2S_2O_3$. These tubes and a separate vessel containing enough glutathione (GSH) to provide 0.25 ml per tube are equilibrated at 37°. Thiosulfate reductase samples, in 1–10 μl, are then added to the assay tubes. The pH of the GSH solution is adjusted to 9.0 by addition of NaOH, and the assay is started by addition of 0.25 ml of GSH to each tube, reserving an appropriate number of tubes for a reagent blank and sulfite standards.

After 5 min, the assay is terminated with 1 ml of $HgCl_2$, and the precipitate is removed by centrifugation. A 0.5-ml aliquot of the supernatant fluid is added to a mixture of 2 ml of the formaldehyde solution and 2 ml of pararosaniline, and the absorbance at 570 nm is measured. Absorbance is proportional to incubation time for at least 10 min and to enzyme content up to 0.1 unit. The unit is defined as that amount of enzyme which produces 1 μmol of sulfite per minute in the 1-ml incubation system.

Purification Procedure

The method given here for the isolation from dried yeast yields a mixture containing three active, charge-isomeric species of thiosulfate reductase but no detectable inactive proteins.[7] The enzyme isolated previously from this source by a mostly similar procedure was electrophoretically homogeneous but had a substantially lower specific activity.[6] The two isolates have otherwise indistinguishable properties.

With exception of the initial extraction, all purification steps are carried out at 0–5°. All centrifugations are at 26,000 ($\pm$1000) g, and all buffers contain $Na_2S_2O_3$ at 1.0 mM unless otherwise specified.

Step 1: Extraction. Commercial active dried yeast (400 g) is stirred into 1.2 liters of 1 M Tris–acetate at pH 9.5. After standing at room temperature for 3 hr with occasional stirring, the suspension is centrifuged for 1 hr at 5°.

Step 2: Salt Fractionation. To the extract is added 520 g of $(NH_4)_2SO_4$ per liter, slowly, with good stirring. The suspension is stirred for 1 hr after the salt addition has been completed, then centrifuged for 2 hr. The enzyme is extracted from the sedimented proteins by stirring for 1 hr in 600 ml of 0.10 M Tris–acetate at pH 8.0, in which 86 g of $(NH_4)_2SO_4$ has been dissolved, and then centrifuging for 1 hr. To the extracted enzyme is added 220 g of $(NH_4)_2SO_4$ per liter. After stirring for 1 hr, the precipitated enzyme is recovered by centrifugation for 1 hr and dissolved in 150 ml of 25 mM Tris–acetate at pH 8.0. The solution is dialyzed for 36 hr against 2 liters of the same buffer (4 changes) and then clarified by centrifugation if necessary.

Step 3: Polyethylene Glycol and QAE–Sephadex Fractionation. Solid polyethylene glycol (PEG 6000) is slowly added to the enzyme solution in a ratio of 200 g per liter. After 1 hr, precipitated inactive protein is removed by centrifugation for 30 min, and the preparation is slowly stirred into 1.6 liters of a 10% (w/v) suspension of QAE–Sephadex A-25 in 25 mM Tris–acetate at pH 8.0. After 3 hr of further stirring, the slurry is poured onto a 9 × 20 cm column and washed with 2 liters of 50 mM NaCl in the same buffer. The enzyme is eluted by passage of 4 liters of 200 mM NaCl in the same buffer, with the activity appearing between 0.6 and 2.4 liters of elution volume. The pooled active fractions are concentrated to less than 20 ml by ultrafiltration with an Amicon PM10 membrane.

Step 4: Sephadex G-100. For maximum resolution, the concentrated preparation is divided into four equal volumes that are applied separately to 2.5 × 90 cm columns of Sephadex G-100 in 0.1 M Tris–acetate at pH 8.0. Active fractions from all columns are pooled, concentrated by ultrafiltration, and dialyzed against 500-ml portions of 10 mM imidazole–acetate, pH 5.8, until the conductivity and pH of the dialyzate are the same as those of the buffer.

Step 5: DEAE–Sephadex at pH 5.8. The dialyzed preparation is applied to a 2.1 × 9 cm column of DEAE–Sephadex A-25 in the same imidazole buffer. The column does not retain the enzyme. Active fractions are pooled and the buffer is exchanged for 10 mM imidazole acetate at pH 6.45 during concentration (Amicon UM10 membrane).

Step 6: DEAE–Sephadex at pH 6.45. The preparation from the preceding step is applied to a 2.1 × 12 cm column of DEAE–Sephadex A-25 in the pH 6.45 buffer and washed with 150 ml of the buffer. The enzyme is eluted with a 350-ml linear gradient to 0.42 M NaCl in the same buffer. The recovered enzyme is concentrated as before, thoroughly dialyzed against 50 mM Tris–acetate at pH 8.0, containing $Na_2S_2O_3$ at 0.5 mM, in 50% glycerol, and stored at $-40°$.

The enzyme is isolated in 8% yield after a 900-fold purification (Table I). It is stable for many months under the specified conditions.

TABLE I
Purification of Thiosulfate Reductase from Baker's Yeast

Step	Volume (ml)	Protein (Mg)	Activity (units)	Specific activity (units/mg)
1. Extract	800	46,000	6800	0.15
2. Salt fractionation	370	25,000	5400	0.21
3. Polyethylene glycol/QAE–Sephadex	20	1,500	2150	1.4
4. Sephadex G-100	15	300	1460	4.9
5. DEAE–Sephadex, pH 5.8	35	15	900	61
6. DEAE–Sephadex, pH 6.45	2	4.0	536	134

Properties[6-8]

Molecular. The native yeast enzyme is a monomeric globular protein with an apparent molecular weight of 17,000 and a Stokes radius of 19 Å. It contains a single cysteinyl residue, but the free sulfhydryl group is not required for activity.

Specificity. The catalyzed reaction is the same as that of rhodanese (thiosulfate:cyanide sulfurtransferase; EC 2.8.1.1, thiosulfate sulfurtransferase), the transfer of a sulfane sulfur atom to a nucleophilic acceptor substrate. However, the two enzymes differ in both acceptor substrate specificity and mechanism (see below). Active donor substrates are $S_2O_3^{2-}$ and organic thiosulfonate anions with the structure $RS(O_2)S^-$. Acceptor substrates include GSH, cysteine, and homocysteine[1] but not CN^-.

Inhibition. Benzene sulfinate, SO_3^{2-}, and HS^- are competitive with both donor and acceptor substrates. GSSG is competitive with GSH, noncompetitive with respect to $S_2O_3^{2-}$. Organic thiosulfonate anions are substrates competitive with $S_2O_3^{2-}$. Double inhibition studies indicate that there are distinct but closely situated sites for the donor and acceptor substrates. Small inhibitors bind to the respective sites without mutual interference, but large inhibitors, benzene thiosulfonate and GSSG, for example, bind with mutual exclusion.

Mechanism and Constants. Kinetic analysis with $S_2O_3^{2-}$ and GSH as substrates at pH 9.0 indicated that the formal mechanism is a single displacement, either random or rapid equilibrium-ordered with a dead-end complex between the second substrate and the free enzyme (forms which are indistinguishable by steady-state kinetic analysis[12]). With benzene

[12] C. Frieden, *Biochem. Biophys. Res. Commun.* **68**, 914 (1976).

thiosulfonate and GSH as substrates at pH 8.1, the kinetic mechanism is unambiguously rapid equilibrium-ordered with GSH as the leading substrate. At pH 8.1 and 37°, the K_m for benzene thiosulfonate is 0.24 mM. GSH, as the leading substrate in a rapid equilibrium-ordered mechanism, has no K_m. The value of k_{cat} is 43 sec^{-1}. At pH 9.0 and 37°, the K_m for $S_2O_3^{2-}$ is 2.9 mM and the apparent K_m for GSH is 4.0 mM.

Acknowledgments

This work was supported by research grants from the National Science Foundation (DMB-8512836) and the National Institutes of Health (GM-30971).

[61] Adenylylsulfate–Ammonia Adenylyltransferase from *Chlorella*

By HEINZ FANKHAUSER, JEROME A. SCHIFF, LEONARD J. GARBER, and TEKCHAND SAIDHA

APA (adenosine 5′-phosphoramidate) has been used in the chemical synthesis of the phosphate anhydride bonds of many nucleotides including ADP and ATP.[1] Thus, the free energy of hydrolysis of the phosphoramidate bond of APA must be greater than that of the pyrophosphate linkage.[2] For this reason, the formation of APA requires the participation of a compound containing a bond with an even higher free energy of hydrolysis, the phosphosulfate bond of APS (adenosine 5′-phosphosulfate). The reaction of APS and ammonia to form APA (Scheme 1) catalyzed by adenylylsulfate–ammonia adenylyltransferase (EC 2.7.7.51) is widely distributed among various groups of organisms. In this article we describe the assay, purification, and properties of this enzyme from *Chlorella pyrenoidosa*.

Assay Methods

Principle. Two methods are used to measure enzyme activity. When both the adenylyltransferase and APS-sulfohydrolase are present in the enzyme preparation, an assay based on the conversion of [U-^{14}C]APS to [^{14}C]APA or 5′-[^{14}C]AMP is used. When highly purified enzymes are assayed, sulfate release from [^{35}S]APS is measured. Ammonia is added

---

[1] J. G. Moffatt and H. G. Khorana, *J. Am. Chem. Soc.* **83**, 649 (1961).
[2] H. Fankhauser, J. A. Schiff, and L. J. Garber, *Biochem. J.* **195**, 545 (1981).

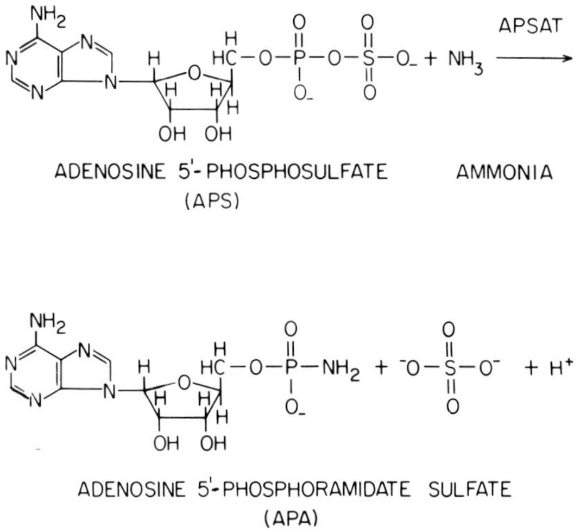

Scheme 1.

Adenosine 5'-phosphosulfate (APS) + NH₃ → Adenosine 5'-phosphoramidate (APA) + sulfate

when the transferase or transferase plus sulfohydrolase activity is to be measured; ammonia is omitted when only the sulfohydrolase activity is assayed.

Materials

[^{35}S]APS is prepared from adenosine 3'-phosphate 5'-phospho[^{35}S]-sulfate ([^{35}S]PAPS) by treatment with 3'-nucleotidase as described previously,[3,4] except that the final product is passed through 2 ml of Dowex 50W-X8 (100–200 mesh) cation-exchange resin in the sodium form in a Pasteur pipette to convert the products to the sodium salts. The resin, in the H$^+$ form, is converted to the Na$^+$ form by washing with 0.1 M NaOH and with water until neutral. The sample is applied in 5 ml of solution, and the column is washed with water until no more of the desired product appears in the effluent. The effluent is adjusted to pH 7.5 with Tris base.

[^{14}C]APS is prepared from [^{14}C]PAPS by the same methods previously used to obtain the ^{35}S-labeled compounds with some modifications.[5] The incubation mixture for PAPS formation contains the following in a total

---

[3] R. C. Hodson and J. A. Schiff, *Arch. Biochem. Biophys.* **132**, 151 (1969).
[4] M. L.-S. Tsang, J. Lemieux, and J. A. Schiff, *Anal. Biochem.* **74**, 623 (1976).
[5] We have recently found[6] that using 1 M triethylammonium bicarbonate at pH 7.5 instead of 0.4 M for elution of PAPS achieves more rapid elution of the Sephadex A-25 column with better resolution of the product.
[6] T. Saidha, A. I. Stern, D.-H. Lee, and J. A. Schiff, *Biochem. J.* **232**, 357 (1985).

volume of 1.3 ml: magnesium acetate (30 μmol), Tris–HCl at pH 9.0 (70 μmol), 2-mercaptoethanol (20 μmol), Na_2SO_4 (34 μmol), levamisole (20 μmol), disodium ATP (29.5 μmol), disodium [^{14}C]ATP (569 Ci/mol; 250 mCi in 200 μl of 0.1 M $KHCO_3$), and enzyme extract from *Chlorella* (1 ml). A smaller DEAE-cellulose column (1.0 × 35 cm) is used for separation of the products. Unused ATP is isolated from the column effluent and is used for another incubation. This is repeated twice; the four preparations are combined, yielding about 40% conversion of labeled ATP to labeled PAPS. For conversion of this product to [^{14}C]APS everything is scaled down proportionately by 10- to 15-fold. After isolation of [^{14}C]APS, it is converted to the sodium salt as described above for the ^{35}S-labeled material.

Procedures

Procedure (^{14}C-Based Assay).[2] The basic reaction mixture contains the following in a total volume of 1.0 ml: [^{14}C]APS (190,000 cpm/μmol, 2.35 μmol), NH_4Cl (500 μmol), Tris–HCl at pH 9.0 (100 μmol), and an appropriate amount of enzyme. The reaction is initiated by the addition of enzyme, and after incubation for 1 hr at 30° the tubes are chilled to 0° and acid-washed charcoal (15 mg/ml) is added. After a thorough agitation by vortex, the suspension is centrifuged at 5,000 g for 10 min at 4°. The charcoal is resuspended in the supernatant fluid, and the mixture is again centrifuged. The supernatant fluid is discarded and the charcoal is extracted 3 times with 2-ml portions of 1% (w/v) aqueous NH_3 in 50% (w/v) ethanol by centrifugation at 5,000 g for 10 min at 4°. The extracts are combined and concentrated under reduced pressure at 30° to about 0.5 ml. A sample is applied to Whatman 3MM filter paper (57 × 20 cm) and is subjected to electrophoresis (EC apparatus) in 0.1 M sodium borate at pH 8.0, for 1.5 hr at 1800 V and 0°. After the paper is dried, metabolites are located under 254 nm light; the spots are excised, cut into small pieces, placed in scintillation vials, and counted.[2]

Procedure (^{35}S-Based Assay).[2] [^{35}S]APS is used for the ^{35}S-based assay. The complete assay mixture contains the following in a total volume of 0.5 ml: [^{35}S]APS (195,000 cpm/μmol, 1.2 μmol), NH_4Cl (250 μmol), Tris–HCl at pH 9.0 (50 μmol), and an appropriate amount of enzyme. The incubation is for 1 hr at 30°. The analysis procedure is the same as for the ^{14}C-based assay up to and including adsorption of the nucleotides on charcoal. After adsorption, a sample of the supernatant fluid is added to a scintillation vial with a mixture containing 0.4% (w/v) 2,5-diphenyloxazole and 0.012% (w/v) 1,4-bis(5-phenyloxazol-2-yl) benzene in toluene–

Triton X-100 (2:1, v/v), and the radioactivity is determined to estimate the amount of $^{35}SO_4^{2-}$ released.

Without added enzyme, APA is not formed in either assay.

Organism and Enzyme Preparation. *Chlorella pyrenoidosa* Chick Emerson strain 3 from slants is grown phototrophically at 24° on the medium described previously[7] in 15-liter cultures in carboys or in 150-liter lots in barrels.[8] Unless otherwise indicated, all subsequent steps are performed at 4°. Cells are harvested in the late exponential phase of growth by centrifugation either with a Szent-Györgyi–Blum continuous-flow attachment in a Sorvall model RC2-B centrifuge or in a CEPA Schnell continuous centrifuge. The cells are washed with 0.1 M Tris–HCl at pH 7.0, containing 50 mM 2-mercaptoethanol (buffer I), by centrifugation at 10,000 g for 10 min and are resuspended in a minimum amount of the same buffer and stored at $-18°$.

After thawing the frozen cells, the suspension is diluted to a concentration of 1 g per 10 ml in the same buffer and is passed through a chilled French pressure cell at 41–48 Mpa (6000-7000 lb/in^2). The suspension is centrifuged at 10,000 g for 30 min, and the supernatant fluid is used as the crude enzyme preparation. This crude extract is fractionated further to purify the adenylyltransferase activity by procedures based on previous methods.[2] $(NH_4)_2SO_4$ (24.3 g) is added with stirring to the crude extract over the course of 1 hr to yield a concentration which is 40% of saturation as determined for 25° but performed at 4°. After being stirred for an additional hour the suspension is centrifuged at 16,000 g for 15 min, and the precipitate is discarded. An additional 13.2 g of $(NH_4)_2SO_4$ is added to the supernatant fluid in the same manner to yield a final concentration which is 60% of saturation. After centrifugation and isolation of the pellet, 14.3 g of $(NH_4)_2SO_4$ is added to the supernatant fluid to yield a concentration which is 80% of saturation, and the precipitate is collected by centrifugation. The pellets obtained from the 40–60% and 60–80% steps are each resuspended in the minimum amount of 20 mM Tris–HCl at pH 8.0, containing 50 mM 2-mercaptoethanol (buffer II) and are dialyzed with stirring against 2 liters of the same buffer overnight. The contents of each dialysis sac are centrifuged at 16,000 g for 15 min, the pellets are discarded, and the supernatant fluids are subjected, separately, to column chromatography on DEAE-cellulose (Whatman DE-52).

DEAE-cellulose chromatography of the 40–60% $(NH_4)_2SO_4$ fraction is carried out on a 24 × 2.2 cm column (bed volume 88 ml); a 10 × 1.0 cm

---

[7] J. A. Schiff and M. Levinthal, *Plant Physiol.* **43**, 547 (1968).
[8] H. Lyman and H. W. Siegelman, *J. Protozool.* **14**, 297 (1967).

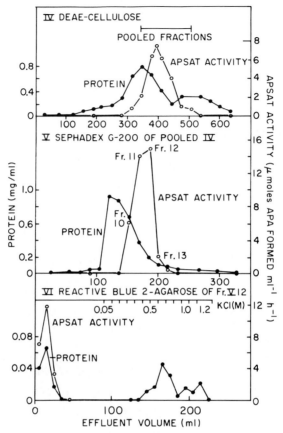

FIG. 1. Chromatography of adenylylsulfate–ammonia adenylyltransferase (APSAT) during various stages of purification. For experimental details see the text.

column (bed volume 9.0 ml) is used for the 60–80% $(NH_4)_2SO_4$ fraction. The DEAE-cellulose is washed with additional buffer II until the effluent reaches pH 8.0. The supernatant fluids from salt fractionation are then passed through their respective columns followed by 3 volumes of buffer II containing 50 mM NaCl, and the proteins are eluted with a linear gradient of 0.05–0.27 M NaCl in buffer II. After determination of enzyme activity directly in each fraction, the active fractions are pooled and dialyzed against buffer II in the same manner as before. The active fractions from the two columns are combined and chromatographed again on DE-52 DEAE-cellulose in the same manner as at first except for a 10 × 1.0 cm column. The active fractions (Fig. 1, step IV) are pooled and dialysed against buffer II, centrifuged at 16,000 g for 15 min, and the supernatant fluid used for further purification.

The supernatant fluid from the previous step is subjected to chromatography on Sephadex G-200 (Fig. 1, step V) which is equilibrated in 0.1 M Tris–HCl at pH 9.0, containing 50 mM 2-mercaptoethanol (buffer III), overnight at room temperature and poured as a 78 × 2.2 cm column. The supernatant fluid from the previous step is added to the column in a total volume of 7.2 ml, and the column developed in buffer III. The most active fractions (Fig. 1, step V) are isolated individually and further purified by Reactive 2-agarose (Sigma Chemical Co., St. Louis, MO) chromatography. The agarose column (5.5 × 1.5 cm) is equilibrated by passage of a large volume of buffer II. Each active fraction is processed separately. The active fraction (16 ml) is applied to the column, followed by 74 ml of buffer II which removes all of the adenylyltransferase activity (Fig. 1, step VI). For APS sulfohydrolase purification this is followed by 10 ml of 50 mM KCl in buffer II, 10 ml of 0.1 M KCl in buffer II, and 10 ml portions of buffer II containing successively increasing increments of KCl. Each increment increases the concentration by 0.1 M until 1.0 M KCl is reached; finally, two successive 10 ml volumes of 1.2 M KCl are applied in buffer II. APS-sulfohydrolase activity appears in the first 1.2 M KCl eluate. Selection of the appropriate fraction for reactive Blue 2-agarose chromatography (e.g., fraction 12 in Fig. 1, step V) yields a maximum purification of the enzyme of 1800-fold (Table I) with a recovery of 14% based on the adenylyltransferase activity in crude extracts, when all fractions from the Reactive Blue 2-agarose column containing the activity are pooled.[2]

Properties of Adenylylsulfate–Ammonia Adenylyltransferase[2]

The adenylylsulfate–ammonia adenylyltransferase activity has also been shown to be present in ammonium sulfate fractions obtained from crude extracts of *Euglena*, spinach, barley, *Dictyostelium discoideum*, and *Escherichia coli*.[2]

Polyacrylamide gel electrophoresis of the enzyme relative to proteins of known molecular weight indicates a molecular mass of 60,000–65,000 for the enzyme protein. SDS–polyacrylamide gel electrophoresis indicates the presence of three subunits of 26,000, 21,000 and 17,000 molecular mass consistent with a molecular mass of 64,000 for the native enzyme. Isoelectrofocusing indicates a single entity with a pI of 4.2.

The enzyme shows a very high specificity for its substrates, APS and ammonia. No detectable products are formed under standard assay conditions when enzyme is incubated with PAPS, ATP, ADP, or AMP in place of APS. Any of a large variety of amines, amino acids, or amides will not replace ammonia. The apparent K_m for APS is 0.82 mM; the apparent K_m

TABLE I
PURIFICATION OF ADENYLYLSULFATE–AMMONIA
ADENYLYLTRANSFERASE

Step	Total activity (units)[a]	Total protein (mg)	Specific activity (units/mg)
I. Extract	1580	15800	0.100
II. (NH$_4$)$_2$SO$_4$ fractionation			
40–60% saturation	562	2079	0.27
60–80% saturation	400	197	2.03
III. DEAE-cellulose I			
40–60% fraction	421	171	2.5
60–80% fraction	260	22	11.8
IV. DEAE-cellulose II	618	69	9.0
V. Sephadex G-200			
Fraction 10	98	11	9
Fraction 11	224	5.7	39
Fraction 12	239	3.0	80
Fraction 13	31	1.5	20
VI. Reactive Blue 2-agarose			
Fraction V 11	215	1.7	126
Fraction V 12	215	1.2	179
Fraction V 10 + 13	86	1.9	45

[a] A unit is defined as that amount of enzyme catalyzing the formation 1 μmol of adenosine 5'-phosphoramidate per hour under standard assay conditions.

for ammonia, assuming NH$_3$ to be the active species, is about 10 mM. Although the high K_m for ammonia leads to questions about the physiological role of the enzyme, the product of the reaction, APA, has been found among the soluble nucleotides of *Chlorella* cells indicating that it is normally formed under physiological conditions.[9]

The enzyme shows a broad pH optimum with a peak at 8.8. The enzymatic reaction at 30° forms only APA but the nonenzymatic reaction of APS and ammonia at elevated temperatures yields both this compound and 5'-AMP. Since the transferase does not catalyze any of the known enzymatic reactions of APS (ATP-sulfurylase, APS kinase, APS-sulfotransferase, or ADP-sulfurylase), it must be regarded as a unique enzyme. Its wide distribution among organisms indicates that the enzyme[2] and its product, APA,[9] may perform important cellular functions. APA will re-

[9] H. Fankhauser, G. A. Berkowitz, and J. A. Schiff, *Biochem. Biophys. Res. Commun.* **101**, 524 (1981).

place AMP as an activator of phosphorylase b,[10,11] threonine dehydratase,[12] and adenylate cyclase.[13]

Acknowledgments

This work was supported by grants from the National Science Foundation.

[10] T. Okazaki, A. Nakazawa, and O. Hayaishi, *J. Biol. Chem.* **243**, 5266 (1968).
[11] W. J. Black and J. H. C. Wang, *Biochim. Biophys. Acta* **212**, 257 (1970).
[12] A. Nakazawa, M. Tokushige, O. Haiashi, M. Ikehara, and Y. Mizuno, *J. Biol. Chem.* **242**, 3868 (1967).
[13] H. Hzuka, K. Adachi, K. Halprin, and V. Levine, *Biochim. Biophys. Acta* **444**, 685 (1976).

[62] Sulfatases from *Helix pomatia*

By ALEXANDER B. ROY

$$ROSO_3^- + H_2O \rightarrow ROH + H^+ + SO_4^{2-}$$
$$R_1R_2C{=}NOSO_3^- + H_2O \rightarrow R_1R_2C{=}NOH + H^+ + SO_4^{2-}$$

Sulfate esters are common metabolites, and their characterization can be aided by the use of a homogeneous, or highly purified, sulfatase of broad specificity as a hydrolyzing agent. The highly specific mammalian arylsulfatases are not suitable. The intestinal juice of the snail, *Helix pomatia*, which is commercially available, contains large amounts of aryl,[1,2] steroid,[2] and glucosinolate sulfatase[3] activities. The last is that of a true sulfatase which hydrolyzes the $\diagdown \!\!\! C{=}NOSO_3^-$ group in mustard oil glycosides and so is quite different from the classical myrosinase (thioglucosidase) of plants.[4] These sulfatase activities have been copurified and separated from the large amounts of β-glucuronidase in the starting material, but the product is not a homogeneous protein.

Assays

Arylsulfatase (EC 3.1.6.1) and Glucosinolate Sulfatase (EC 3.1.6.—)

Arylsulfatase and glucosinolate sulfatase are assayed in a pH-stat (titrant, 15 m*M* NaOH) with 4-nitrophenyl sulfate and sinigrin, respectively,

[1] K. S. Dodgson and G. M. Powell, *Biochem. J.* **73**, 672 (1959).
[2] P. Jarrige, J. Yon, and M. F. Jayle, *Bull. Soc. Chim. Biol.* **45**, 783 (1963).
[3] W. Thies, *Naturwissenschaften* **66**, 364 (1979).
[4] F. W. Challenger, "Aspects of the Organic Chemistry of Sulphur," p. 115. Butterworth, London, 1959.

TABLE I
CONDITIONS FOR THE ASSAY OF THE SULFATASES OF *Helix pomatia*[a]

Substrate	Buffer Type	pH	s (mM)	Enzyme (units)	Incubation (min)
4-Nitrophenyl sulfate	0.08 M KCl, 1.0 mM Tris[b]	7.4	20	0.5	—
	0.1 M Tris–HCl[c]	7.4	20	0.02	5
Sinigrin	0.09 M KCl, 1.0 mM Tris	7.4	10	1	—
Estrone sulfate	0.5 M acetate	5.5	0.5	0.1	5
Cortisone 21-sulfate	0.5 M acetate	4.6	0.6	1.5	5
Dehydroepiandrosterone sulfate	0.5 M acetate	4.6	0.6	3	60

[a] The amount of enzyme is given in arylsulfatase units, determined in the pH-stat (1 unit hydrolyzes 1 μmol min^{-1}).
[b] pH-stat.
[c] Spectrophotometric assay.

as substrates (Table I). Initial velocities are obtained[5] and, in the case of arylsulfatase (EC 3.1.6.1), corrected for the ionization of the liberated 4-nitrophenol (pK', 7.15; 64% ionized at pH 7.4). The velocity decreases with time ($t_{1/2}$ about 10 min).

A more sensitive spectrophotometric method was used to follow the chromatography of the arylsulfatase (Table I, see also this volume [37]). The arylsulfatase is highly dependent on the concentration of Tris, and, under the conditions in Table I, the specific activity determined spectrophotometrically is about 2.5 times that measured in the pH-stat.

Steroid Sulfatases (EC 3.1.6.2)

A slightly modified methylene blue method[6] has been used. Two sets of reaction mixtures are required. In the first, the assays, enzyme is added to start the reaction, and in the second, the controls, it is added only immediately before the reaction is stopped. The conditions[2] are given in Table I. The volume of the reaction mixture is 1 ml.

With cortisone 21-sulfate and dehydroepiandrosterone sulfate the reaction is stopped by heating for 5 min in a boiling water bath and, after

---

[5] A. B. Roy, *Biochim. Biophys. Acta* **526**, 489 (1978).
[6] A. B. Roy, *Biochem. J.* **62**, 41 (1956).

cooling, adding 1 ml of methylene blue reagent[6] (250 mg methylene blue, 50 g Na_2SO_4, 10 ml H_2SO_4 in 1 liter). With estrone sulfate, heating is omitted, otherwise extensive hydrolysis occurs, and the reaction is stopped by adding 1 ml of methylene blue reagent. The subsequent procedure is the same in all cases. After adding 10 ml of $CHCl_3$, the mixture was shaken vigorously for 30 sec and centrifuged for 5 min in a table-top clinical centrifuge. The aqueous phase together with any interface of denatured protein is removed, and a sample of the organic phase dried over Na_2SO_4 before its absorbance at 650 nm is measured (2-mm light path). The amount of steroid sulfate present is determined from the appropriate calibration curves, and the sulfatase activity obtained from the difference between the assay and the control determinations.

Recovery of the steroid sulfates in the controls are quantitative. Under the above conditions, 0.5 μmol of estrone sulfate, dehydroepiandrosterone sulfate, and cortisone 21-sulfate gave values of $A_{650\,nm}^{2\,mm}$ of 0.90, 0.79, and 0.68, respectively.

Purification

Only arylsulfatase activity was used to follow chromatography. Protein concentrations are based on an assumed $A_{280\,nm}^{1\%}$ of 1.0. Dialysis is avoided in the early stages because of the presence of a factor, presumably a cellulase, which attacked dialysis tubing.

Acetone Precipitation

Snail intestinal juice (Sigma H-2; 20 ml) is diluted, 1 g of sodium acetate added, the pH adjusted to 4.5 with acetic acid, and the volume made up to 250 ml. After cooling to about 1°, 203 ml of acetone is added during 45 min while the temperature is lowered to −10°. Following equilibration for 30 min, the precipitate is removed by centrifugation at −10° and discarded. To the supernatant fluid (405 ml) at −10° is slowly added an additional 126 ml of acetone; after equilibration, the precipitate is collected by centrifugation and dissolved in 50 ml of 50 mM sodium acetate–acetic acid at pH 4.6 held at 0°.

Ammonium Sulfate

The above solution is immediately made 55% saturated with ammonium sulfate (333 mg/ml) and, after equilibration for a few hours at 0°, centrifuged and the precipitate discarded. To the supernatant liquid is added additional ammonium sulfate (129 mg/ml) (75% of saturation), and the mixture maintained at 0° overnight. The precipitate was dissolved in starting buffer (75 mM NaCl, 10mM Tris–HCl at pH 7.5) 20°C to give about 10 ml of solution.

Sephadex G-200

The preparation is applied at once to a column (86 × 2.5 cm) of Sephadex G-200 in starting buffer and eluted at room temperature with the same buffer (rate, 0.25 ml/min; fraction volume, 4 ml). The arylsulfatase elutes as a single peak at about fraction 70. Fractions with a specific activity greater than about 50% of the maximum are combined and concentrated to about 5 ml with a UM10 or YM5 membrane (Amicon).

DEAE-Sephacel

The concentrate is applied to a column (53 × 1.5 cm) of DEAE-Sephacel in starting buffer and eluted at room temperature with the same buffer (rate, 0.3 ml/min; fraction volume, 4 ml). About 60% of the applied activity was eluted at about fraction 20 (sulfatase 1). After 160 ml of eluate has been collected, elution is started with a linear gradient formed from 200 g of starting buffer and 200 g of 0.5 M NaCl, 0.05 M Tris–HCl at pH 7.5. A further 25% of the activity (sulfatase 2) was eluted near fraction 70. The appropriate fractions are combined, and the sulfatase precipitated at 85% saturated ammonium sulfate, dissolved in 25 mM histidine–HCl at pH 6.3, and dialyzed against the same buffer to give about 3 ml of each enzyme.

The purification of the arylsulfatases is summarized in Table II. Table III shows the simultaneous purification of the other sulfatase activities.

TABLE II
PURIFICATION OF THE ARYLSULFATASE ACTIVITY FROM 20 ml OF INTESTINAL JUICE FROM *Helix pomatia*[a]

Step	Total protein (mg)	Total arylsulfatase (units)	Specific activity (units/mg)
Intestinal juice	5360	2780	0.52
1. Acetone fractionation	1640	2610	1.6
2. Ammonium sulfate	413	2730	6.6
3. Sephadex G-200	149	2173	15
4. DEAE-Sephacel			
Sulfatase 1	38	1110	29
Sulfatase 2	32	445	14

[a] Activity measured in pH-stat.

TABLE III
RELATIVE SULFATASE ACTIVITIES (ARYLSULFATASE = 1.0) DURING PURIFICATION OF
SULFATASES 1 AND 2 FROM *Helix pomatia*

Sulfatase	Intestinal juice	Sulfatase 1	Sulfatase 2
Arylsulfatase (pH-stat)	1.00	1.00	1.00
Glucosinolate sulfatase	0.66	0.61	0.58
Estrone sulfatase	0.20	0.27	0.27
Cortisone sulfatase	0.035	0.033	0.035
Dehydroepiandrosterone sulfatase	0.79×10^{-3}	1.1×10^{-3}	0.93×10^{-3}

Chromatofocusing

Sulfatases 1 and 2 have been applied to columns (16 × 0.9 cm) of Polybuffer Exchanger 94 in the above histidine buffer and eluted with Polybuffer 74 at pH 3.9, as described by the manufacturer. Both sulfatases yield two well-resolved peaks of activity which elute as single peaks, at the same pH, on rechromatography. The four fractions 1α, 1β, 2α, and 2β have apparent isoelectric points of 4.5, 4.3, 5.1, and 4.9, respectively. As no increase in specific activity was obtained, nor was there any separation of the different sulfatases, this step is usually omitted.

Properties

Both sulfatases 1 and 2 are stable and can be kept at 5° and pH 6 to 7 for many months without loss of activity. They are slightly acidic proteins (p*I* 4.3–5.1) with an apparent molecular weight (Sephadex G-75) of 79,000. These preparations are all devoid of β-glucuronidase activity ($<10^{-4}$ U/mg, see Table II).

The different sulfatase activities appear to be associated with a single protein. Previous work shows that this will have rather general arylsulfatase[1,2] and glucosinolate sulfatase[7] activities although cyclohexanone oxime *O*-sulfonate, which, like glucosinolates, contains the grouping $\diagup\hspace{-0.5em}^{\diagdown}C{=}NOSO_3^-$, was not hydrolyzed under a variety of conditions. The steroid sulfatase is more specific.[2,6] Estrogen 3-sulfates, corticosteroid 21-sulfates, and the 3β-sulfates of 5α- and Δ^5-steroids are hydrolyzed while the 3α-sulfates of 5α-steroids (e.g., androsterone sulfate), 17-sulfates

[7] D. Rakow, R. Gmelin, and W. Thies, *Z. Naturforsch., C. Biosci.* **36C**, 16 (1981).

(e.g., testosterone sulfate), and 20-sulfates are not. The 3-sulfates of 5β-steroids are hydrolyzed very slowly.[2]

Other possible sulfatase activities in the preparation have not been investigated. However, ascorbate 2-sulfate is hydrolyzed only slowly (<1% of the rate for 4-nitrophenyl sulfate), and early work[8] would suggest that any glycosulfatase will be slight.

Sulfate weakly inhibits the arylsulfatase[1,9] and glucosinolate[3] sulfatase activities, and much more powerfully the steroid sulfatase activities,[2] with a K_i of about 0.5 mM. The latter inhibition is noncompetitive.[2] Sulfite is a particularly powerful inhibitor.[1,9]

Sulfatases 1 and 2 are firmly bound to phosphocellulose and cannot be eluted from it, even by 0.5 M NaCl at pH 9. The bound enzyme is active and could be an economical means of characterizing metabolites of xenobiotics. Thies[3] has adopted the alternative technique of using glucosinolates bound to DEAE–Sephadex in his method of quantitatively determining these compounds in plant extracts. The technique is useful when gas–liquid chromatography is subsequently to be used because the sulfate formed on hydrolysis remains bound to the ion exchanger.

[8] K. S. Dodgson, *Biochem. J.* **78**, 324 (1961).
[9] P. Jarrige, *Bull. Soc. Chim. Biol.* **45**, 761 (1963).

[63] Mammalian Sulfur Amino Acid Metabolism: An Overview

By OWEN W. GRIFFITH

Introduction[1]

Methionine and cysteine, the two sulfur-containing protein amino acids, are metabolized by a variety of reactions and pathways to at least two dozen intermediates and products. Some of these metabolites serve functions that are essential for survival of the organism. Methionine and cysteine have, in addition, important catalytic roles in the active sites of many enzymes. Several of these metabolic processes and catalytic functions exploit unique aspects of sulfur chemistry to accomplish biochemical transformations that would be difficult or impossible to effect in the

[1] Studies from the author's laboratory were supported in part by the National Institutes of Health, Grant AM26912, and an Irma T. Hirsch Career-Scientist award.

absence of the sulfur amino acids. Methionine, for example, is metabolized primarily to S-adenosylmethionine (AdoMet), a sulfonium compound mediating most biochemical methylation reactions.[2] It is doubtful if other amino acid derivatives or other " 'onium" compounds could adequately serve this role; quarternary ammonium compounds are too thermodynamically stable to effectively methylate most acceptors, and oxonium compounds (e.g., a hypothetical oxygen analog of AdoMet) lack the kinetic stability to survive *in vivo*.

Cysteine similarly participates in a number of biochemical processes that depend directly on the particular reactivity of thiols. The high nucleophilicity of thiols facilitates the role of cysteine as an active site, covalent catalyst (e.g., in papain[3] and glyceraldehyde-3-phosphate dehydrogenase[4]) and allows the cysteine residue of glutathione to scavenge and detoxify electrophiles during mercapturic acid biosynthesis[5,6] and peroxide reduction.[7] The easy formation but low reactivity of sulfur free radicals permits glutathione to also capture and detoxify the more reactive free radicals of oxygen and carbon.[8] Coenzyme A and acyl carrier protein mediate a variety of acyl transfer reactions involving thioesters of the cysteamine residue (derived from cysteine) of these cofactors; the poor resonance stability of thioesters serves to maintain the high transfer potential of the acyl group.[9] Oxidized derivatives of cysteine have additional metabolic roles. The disulfide bonds of cystine residues stabilize the tertiary structure of proteins, for example. More extensive oxidation of cysteine yields sulfate and taurine, metabolites with important physiological roles in detoxification, bile acid formation, membrane stabilization and neurotransmission.[10]

[2] S. H. Mudd and H. L. Levy, *in* "The Metabolic Basis of Inherited Disease" (J. B. Stanbury, J. B. Wyngaarden, D. S. Fredrickson, J. L. Goldstein, and M. S. Brown, eds.), p. 522. McGraw-Hill, New York, 1983.
[3] A. N. Glazer and E. L. Smith, *in* "The Enzymes" (P. D. Boyer, ed.), 3rd ed., Vol. 3, p. 502. Academic Press, New York, 1971.
[4] J. I. Harris and M. Waters, *in* "The Enzymes" (P. D. Boyer, ed.), 3rd ed., Vol. 13, p. 1. Academic Press, New York, 1976.
[5] S. S. Tate, *in* "Enzymatic Basis of Detoxification" (W. B. Jakoby, ed.), Vol. 2, p. 95. Academic Press, New York, 1980.
[6] W. B. Jakoby, J. Stevens, M. W. Duffel, and R. A. Weisiger, *Rev. Biochem. Toxicol.* **6,** 97 (1984).
[7] L. Flohé, *Ciba Found. Symp.* [N.S.] **65,** 95 (1979).
[8] J. E. Packer, *in* "The Chemistry of the Thiol Group" (S. Patai, ed.), p. 481. Wiley, New York, 1974.
[9] W. P. Jencks, "Catalysis in Chemistry and Enzymology." McGraw-Hill, New York, 1969.
[10] S. W. Schaffer, S. I. Baskin, and J. J. Kocsis, eds., "The Effects of Taurine on Excitable Tissues." Spectrum Publications, New York, 1981.

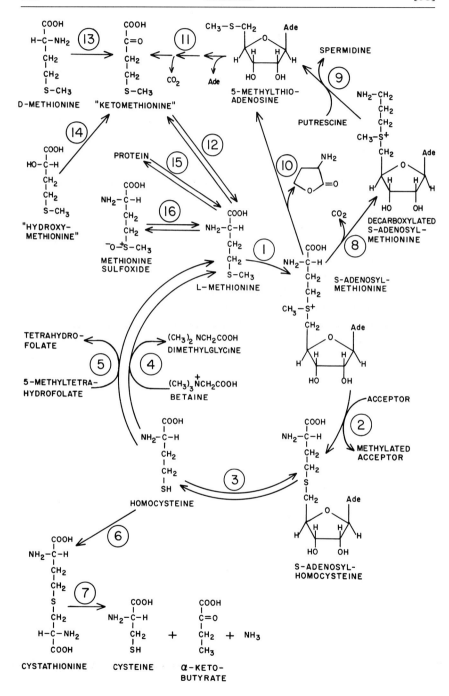

This chapter briefly summarizes the mammalian metabolism of methionine and cysteine with emphasis on the chemical transformations of sulfur; the metabolism of inorganic sulfur compounds is reviewed separately elsewhere in this volume.[11] Extensive reviews of the metabolism of sulfur in mammals are available.[2,12,13]

Methionine Metabolism

As noted, methionine metabolism is initiated by the formation of S-adenosylmethionine (AdoMet),[14] a sulfonium compound in which each of the carbons attached to sulfur is activated toward nucleophilic attack (reaction 1, Fig. 1). In most cases, AdoMet reacts by transfer of the S-methyl group to one of several possible acceptors including glycine (forming sarcosine), guanidinoacetate (forming creatine), phosphatidylethanolamine (forming lecithin), the lysine residues in specific proteins (forming ε-N,N,N-trimethyllysine, the precursor of carnitine), pyrimidine and purine bases of tRNA (leading to ribothymidine and other methylated nucleosides), and various xenobiotics bearing hydroxyl, amino, or sulfhydryl groups (reaction 2). It is estimated that methyl transfer reactions as a group consume about 95% of the AdoMet formed[15]; guanidinoacetate and glycine are believed to be the quantitatively most important acceptors.[15,16] In addition to the methylated acceptor, transmethylation yields S-adeno-

[11] J. A. Schiff and T. Saidha, this volume [58].
[12] T. P. Singer, in "Metabolic Pathways" (D. M. Greenberg, ed.), Vol. 7, p. 535. Academic Press, New York, 1975.
[13] A. J. L. Cooper, Annu. Rev. Biochem. **52**, 187 (1983).
[14] J. L. Hoffman, this series, Vol. 94, p. 223.
[15] S. H. Mudd and J. R. Poole, Metab., Clin. Exp. **24**, 721 (1975).
[16] H. Ogawa and M. Fujioka, Biochem. Biophys. Res. Commun. **108**, 227 (1982).

FIG. 1. Mammalian methionine metabolism. The circled numbers correspond to the following enzymes or metabolic processes (Enzyme Commission numbers are shown in parentheses): 1. Methionine adenosyltransferase (EC 2.5.1.6), an ATP-dependent reaction; 2. Various S-adenosylmethionine methyltransferases (EC 2.1.1.?); 3. Adenosylhomocysteinase (EC 3.3.1.1); 4. Betaine–homocysteine methyltransferase (EC 2.1.1.5); 5. 5-Methyltetrahydrofolate–homocysteine methyltransferase (methionine synthase) (EC 2.1.1.13); 6. Cystathionine β-synthase (EC 4.2.1.22); 7. Cystathionine γ-lyase (EC 4.4.1.1); 8. Adenosylmethionine decarboxylase (EC 4.1.1.50); 9. Spermidine synthase (EC 2.5.1.16); 10. Adenosylmethionine cyclotransferase (EC 2.5.1.4); 11. A multienzyme pathway including methylthioadenosine phosphorylase (EC 2.4.2.28) and probably two additional enzymes; 12. Various transaminases including particularly glutamine transaminase; 13. D-Amino-acid oxidase (EC 1.4.3.3); 14. Possibly (S)-2-Hydroxy-acid oxidase (EC 1.1.3.15) and D-2-Hydroxyacid dehydrogenase (EC 1.1.99.6); 15. Ribosomal protein synthesis and various proteases; 16. Methionine-S-oxide reductase (EC 1.8.4.?).

sylhomocysteine, a product reversibily hydrolyzed to adenosine and homocysteine (reaction 3). Since the equilibrium constant of reaction 3 favors S-adenosylhomocysteine synthesis, proper flux through the "methionine cycle"[2] depends on maintenance of a low *in vivo* concentration of homocysteine. Pathological conditions that cause homocysteine accumulation (e.g., cystathionine β-synthase deficiency or disorders of homocysteine methylation) result in S-adenosylhomocysteine (AdoHcy) accumulation; AdoHcy is a potent inhibitor of transmethylation reactions.[2,17]

As shown in Fig. 1, homocysteine occupies a branch point in methionine metabolism. In both the rat and human about half of the homocysteine formed is irreversibily converted by transsulfuration to cysteine, α-ketobutyrate, and ammonia (reactions 6 and 7) whereas the remainder is remethylated to methionine (reactions 4 and 5).[15,18] 5-Methyltetrahydrofolate is believed to be the quantitatively important methyl donor *in vivo* under normal nutritional conditions.[18–20] Methionine synthetase,[21,22] the enzyme catalyzing reaction 5, accounts for the final metabolic step in the *de novo* synthesis of methyl groups[2,12] and enjoys a wide tissue distribution. In contrast, activity of betaine–homocysteine methyltransferase (reaction 4) is confined to the liver in most species. Its activity is increased by high protein diets, and the enzyme probably serves mainly in betaine catabolism and, under conditions of high methionine intake, helps to dispose of excess homocysteine.[18–20] Transamination or oxidation of homocysteine to the corresponding α-keto acid is believed to be of limited importance under normal circumstances but may assume a more significant role in disorders of homocysteine metabolism.[2,15] The product, α-keto-γ-mercaptobutyrate, is a reactive and possibily toxic metabolite that *in vivo* is reduced in part to α-hydroxy-γ-mercaptobutyrate (reactions not shown).[23,24]

In humans polyamine biosynthesis[25] consumes 2–5% of the AdoMet pool[5] and is initiated by the decarboxylation of AdoMet (reaction 8).

[17] P. K. Chiang, *Methods Pharmacol.* **6**, 127 (1985).
[18] J. D. Finkelstein, *Metab., Clin. Exp.* **23**, 387 (1974).
[19] J. D. Finkelstein, W. E. Kyle, and B. J. Harris, *Arch. Biochem. Biophys.* **146**, 84 (1971).
[20] J. D. Finkelstein and J. J. Martin, *J. Biol. Chem.* **261**, 1582 (1986).
[21] C. S. Utley, P. D. Marcell, R. A. Allen, A. C. Antony, and J. F. Kolhouse, *J. Biol. Chem.* **260**, 13656 (1985).
[22] J. H. Mangum and J. A. North, *Biochemistry* **10**, 3765 (1971).
[23] A. J. L. Cooper and A. Meister, *Arch. Biochem. Biophys.* **239**, 556 (1985).
[24] H. Kodama, S. Ohmori, M. Susuki, S. Mizuhara, T. Oura, T. Isshiki, and I. Uemura, *Physiol. Chem. Phys.* **2**, 81 (1971).
[25] H. Tabor and C. W. Tabor, eds., this series, Vol. 94.

Subsequent transfer of the 3-aminopropyl moiety to putrescine, the product of ornithine decarboxylation, yields spermidine and 5-methylthioadenosine (reaction 9). Transfer of another 3-aminopropyl moiety to the unsubstituted putrescine nitrogen of spermidine yields spermine; in mammals this reaction is catalyzed by a distinct enzyme, spermine synthase.[26,27] 5-Methylthioadenosine is also formed by cyclization of AdoMet (reaction 10).[28] Although the pathway by which 5-methylthioadenosine is metabolized to methionine is not fully elucidated for mammals, the initial reaction appears to be phosphorylytic cleavage to adenine and 1-phospho-5-methylthioribose. The latter intermediate is metabolized to 2-oxo-4-methylthiobutyrate ("ketomethionine") and formate (from C-1 of the ribose) by an enzyme or enzymes that have not been purified (reaction 11).[29,30] "Ketomethionine" is transaminated to reform methionine (reaction 12). Although there does not appear to be a methionine-specific transaminase in mammalian cells, methionine formation is effectively catalyzed by several transaminases, most notably the glutamine transaminases isolated from liver and kidney.[31]

The transamination of methionine to form "ketomethionine" has been suggested as an important process *in vivo;* "ketomethionine" may be further metabolized via oxidative decarboxylation to 3-methylthiopropionate, methanethiol, and additional catabolites[32-34] (reactions not shown, see Refs. 2 and 13 for a contrasting view). Nutritional studies indicate that "ketomethionine" is also formed from both D-methionine (reaction 13) and DL-2-hydroxy-4-thiomethylbutyrate ("hydroxymethionine") (reaction 14)[35]; "ketomethionine" from these sources is converted in large part to L-methionine, thereby accounting for the observation that these compounds can replace L-methionine in the diet. The spontaneous formation and enzymatic reduction of methionine sulfoxide (reaction 16) have been reviewed.[36]

[26] P. Hannonen, J. Janne, and A. Raina, *Biochem. Biophys. Res. Commun.* **46**, 341 (1972).
[27] A. E. Pegg, K. Shuttleworth, and H. Hibasami, *Biochem. J.* **197**, 315 (1981).
[28] K. R. Swiatek, L. N. Simon, and K.-L. Chao, *Biochemistry* **12**, 4670 (1973).
[29] P. S. Backlund and R. A. Smith, *J. Biol. Chem.* **256**, 1533 (1981).
[30] P. C. Trackman and R. H. Abeles, *Biochem. Biophys. Res. Commun.* **103**, 1238 (1981).
[31] A. J. L. Cooper and A. Meister, "Chemical and Biological Aspects of Vitamin B_6 Catalysis: Part B," p. 3. Alan R. Liss, Inc., New York, 1984.
[32] N. J. Benevenga, *Adv. Nutr. Res.* **6**, 1 (1984).
[33] N. J. Benevenga and A. R. Egan, "Sulfur Amino Acids: Biochemical and Clinical Aspects," p. 327. Alan R. Liss, Inc., New York, 1985.
[34] G. L. Case and N. J. Benevenga, *J. Nutr.* **106**, 1721 (1976).
[35] D. H. Baker, this volume [55].
[36] N. Brot, H. Fliss, T. Coleman, and H. Weissbach, this series, Vol. 107, p. 352.

Cysteine Metabolism

In contrast to methionine, an essential amino acid, cysteine is synthesized by mammals; as shown in Fig. 1, cysteine sulfur is derived from methionine whereas the carbon and nitrogen of cysteine are derived from serine. It is notable that the dependence on methionine sulfur implies that cysteine remains a nonessential amino acid only if methionine intake is sufficient to meet the total sulfur amino acid requirement. Conversely, the dietary requirement for methionine is reduced, but not eliminated, when cyst(e)ine intake is adequate.[35]

As shown in Fig. 2, mammalian cysteine metabolism is complex. Cysteine participates conventionally in ribosomal protein synthesis (reaction 1), but posttranslational disulfide bond formation causes both cysteine and cystine to be released during protein degradation (reaction 2). As discussed below, cystine is reduced to cysteine by transhydrogenation with glutathione (reaction 18); direct reduction of cystine by NADH or NADPH has not been observed in mammals.

Several enzymes catalyzing cysteine transamination (reaction 3) have been isolated from mammalian tissues,[37-43] but the relative or absolute importance of the various activities *in vivo* is not established. Although much of the activity in liver and kidney homogenates is attributable to mitochondrial aspartate aminotransferase,[38-40] the contribution of this enzyme *in vivo* is unclear since the K_m for cysteine is 22 mM,[39] a value about 100-fold greater than its physiological concentration. β-Mercaptopyruvate, the product of cysteine transamination, is efficiently converted to pyruvate and a reduced sulfur species by β-mercaptopyruvate sulfurtransferase (reaction 4).[44] The sulfur product formed depends on the composition of the medium, but it is probable that any reduced sulfur species formed *in vivo* is ultimately converted to sulfate. In considering the partitioning of β-mercaptopyruvate between transamination (reaction 3) and conversion to sulfate and pyruvate (reaction 4) it is notable that rats given D-cysteine form sulfate but apparently do not form L-cysteine.[45]

[37] M. P. C. Ip, R. J. Thibert, and D. E. Schmidt, Jr., *Can. J. Biochem.* **55**, 958 (1977).
[38] T. Ubuka, Y. Ishimoto, and R. Akagi, *J. Inherited Metab. Dis.* **4**, 65 (1981).
[39] T. Ubuka, S. Umemura, S. Yuasa, M. Kinuta, and K. Watanabe, *Physiol. Chem. Phys.* **10**, 483 (1978).
[40] R. Akagi, *Acta Med. Okayama* **36**, 187 (1982).
[41] S.-M. Kuo, T. C. Lea, and M. Stipanuk, *Biol. Neonate* **43**, 23 (1983).
[42] M. Taniguchi, Y. Hosaki, and T. Ubuka, *Acta Med. Okayama* **38**, 375 (1984).
[43] A. J. L. Cooper, M. T. Haber, and A. Meister, *J. Biol. Chem.* **257**, 816 (1982).
[44] B. Sörbo, in "Metabolic Pathways" (D. M. Greenberg, ed.), Vol. 7, p. 433. Academic Press, New York, 1975.
[45] E. J. Glazenburg, I. M. C. Jekel-Halsema, A. Baranczyk-Kuzma, K. R. Krijgsheld, and G. J. Mulder, *Biochem. Pharmacol.* **33**, 625 (1984).

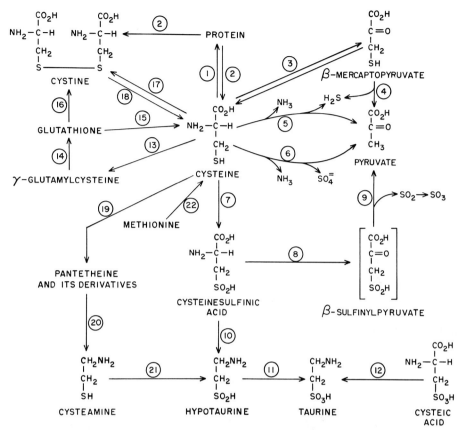

FIG. 2. Mammalian cysteine metabolism. The circled numbers correspond to the following enzymes or metabolic processes: 1. Ribosomal protein synthesis; 2. Formation of protein disulfide bonds; 3. Cysteine aminotransferase (EC 2.6.1.3) and probably other aminotransferases (see text); 4. 3-Mercaptopyruvate sulfurtransferase (EC 2.8.1.2); 5. Cystathionine γ-lyase (EC 4.4.1.1) (accounts for mammalian cysteine and cystine desulfhydrase activities, see text); 6. Poorly characterized mitochondrial activity (see text); 7. Cysteine dioxygenase (EC 1.13.11.20); 8. Aspartate aminotransferase (EC 2.6.1.1); 9. Spontaneous reaction; 10. Cysteinesulfinate decarboxylase (EC 4.1.1.29, sulfinoalanine decarboxylase); 11. "Hypotaurine oxidase" (mammalian enzyme not yet clearly identified); 12. Cysteinesulfinate decarboxylase (EC 4.1.1.29, sulfinoalanine decarboxylase); 13. γ-Glutamylcysteine synthetase (EC 6.3.2.2, glutamate–cysteine ligase); 14. Glutathione synthase (EC 6.3.2.3); 15. Combined activities of γ-glutamyltransferase (EC 2.3.2.2) and various dipeptidases acting on cysteinylglycine; 16. Enzymatic and nonenzymatic oxidation of glutathione to glutathione disulfide followed by the cleavage reactions listed as reaction 15 (see Ref. 75); 17. Various thiol oxidases; 18. Enzymatic and nonenzymatic transhydrogenations (disulfide interchange reactions) with glutathione coupled to glutathione reductase (EC 1.6.4.2); 19. Pantetheine and coenzyme A biosynthesis; 20. Pantetheine and coenzyme A catabolism; 21. Cysteamine dioxygenase (EC 1.13.11.19); 22. Transsulfuration pathway.

Assuming that the initial metabolism is to β-mercaptopyruvate (catalyzed by D-amino-acid oxidase), the findings suggest that the β-mercaptopyruvate sulfurtransferase reaction is greatly favored. The chemistry and biochemistry of β-mercaptopyruvate have been reviewed.[43]

Cysteine is converted to pyruvate by several additional pathways. Cysteine desulfhydrase (reaction 5) is an activity now attributed largely to cystathionine γ-lyase (γ-cystathionase), an enzyme which degrades, in addition to cystathionine, both cysteine and cystine. Whereas cysteine is converted directly to pyruvate, ammonia, and H_2S, cystine yields pyruvate, ammonia, and S-mercaptocysteine, a product reduced *in vivo* to cysteine and H_2S.[46–48] Although the intracellular concentration of cystine is low, it is notable that the K_m value for cystine is lower than that for cystathionine and much lower than that for cysteine.[49] Cystathionine β-synthase also catalyzes reaction 5 *in vitro*,[50,51] but its role in cysteine catabolism remains uncertain. The suggestion has been offered that reactions 3 and 4 together may account for cysteine desulfhydrase activity.[52] Although oxidative degradation of cysteine to pyruvate, ammonia, and sulfate (reaction 6) is reportedly catalyzed by an activity present in rat mitochondria,[53] a specific enzyme has not been identified; it is possible that the activity is due to the combined action of other enzymes of cysteine metabolism.

Oxidation of cysteine to cysteinesulfinate (reaction 7) is believed to be the major pathway of cysteine catabolism in mammals, particularly when cysteine availability is high.[54–56] The metabolic partitioning of cysteinesulfinate between transamination (reaction 8) and decarboxylation (reaction 10) shows considerable species variation. In mice, about 85% of parenterally administered cysteinesulfinate is converted to hypotaurine,[57]

---

[46] J. Loiselet and F. Chatagner, *Biochim. Biophys. Acta* **89**, 330 (1964).
[47] M. T. Costa, A. M. Wolf, and D. Giarnieri, *Enzymologia* **43**, 271 (1972).
[48] D. Cavallini, C. DeMarco, B. Mondovi, and B. G. Mori, *Enzymologia* **22**, 11 (1960).
[49] A. Kato, T. Matsuzawa, M. Suda, H. Nakagawa, and J. Ishizuka, *J. Biochem.* (*Tokyo*) **59**, 34 (1964).
[50] A. E. Braunstein, E. V. Goryachenkova, E. A. Tolosa, I. H. Willhardt, and L. L. Yefremova, *Biochim. Biophys. Acta* **242**, 247 (1971).
[51] M. H. Stipanuk and P. W. Beck, *Biochem. J.* **206**, 267 (1982).
[52] A. Meister, "Biochemistry of the Amino Acids." Academic Press, New York, 1965.
[53] A. Wainer, *Biochim. Biophys. Acta* **141**, 466 (1967).
[54] J. F. Wheldrake and C. A. Pasternak, *Biochem. J.* **102**, 45P (1967).
[55] K. Yamaguchi, S. Sakakibara, J. Asamizu, and I. Ueda, *Biochim. Biophys. Acta* **297**, 48 (1973).
[56] M. H. Stipanuk, *J. Nutr.* **109**, 2126 (1979).
[57] O. W. Griffith, *J. Biol. Chem.* **258**, 1591 (1983).

but the fraction decarboxylated in humans is much less.[58] The cat, which is cysteinesulfinate decarboxylase deficient,[59] forms little or no hypotaurine and in the absence of dietary taurine becomes blind.[60,61] β-Sulfinylpyruvate, the product of cysteinesulfinate transamination, spontaneously decomposes to pyruvate and sulfite (reaction 9); sulfite is oxidized by sulfite oxidase to sulfate. At least 90% of the hypotaurine formed from cysteinesulfinate is oxidized further to taurine (reaction 11); a small portion of the hypotaurine formed is transaminated to β-sulfinylacetaldehyde and catabolized, presumably through acetaldehyde, to CO_2 (not shown).[57,62] The mechanism by which hypotaurine is converted to taurine remains poorly understood, and both enzymatic[63] and nonenzymatic[64] pathways remain under consideration. Cysteic acid (cysteinesulfonic acid) is apparently not a mammalian metabolite but is contained in the diet as the result of nonenzymatic oxidation of cysteine and the action of enteric bacteria. As an alternative substrate of cysteinesulfinate decarboxylase, cysteinesulfonate is decarboxylated directly to taurine (reaction 12). Transamination of cysteinesulfonate, catalyzed by aspartate aminotransferase, yields β-sulfopyruvate; that product is reduced by malate dehydrogenase to β-sulfolactate (reactions not shown).[65]

Intracellular concentrations of cysteine are of the order of 30–200μM, values that are among the lowest observed for any of the protein amino acids.[66,67] Higher concentrations of cysteine are potentially dangerous to cells since cysteine forms a thiazolidine derivative with pyridoxal phosphate and thereby may deplete cells of that coenzyme. Cysteine is also easily oxidized nonenzymatically to cystine (reaction 17), an insoluble amino acid which is toxic if allowed to accumulate (c.f. cystinosis).[68] The multiple pathways of cysteine catabolism function efficiently to prevent

[58] J. G. Jacobsen, L. L. Thomas, and L. H. Smith, Jr., *Biochim. Biophys. Acta* **85**, 103 (1964).
[59] J. A. Worden and M. H. Stipanuk, *Comp. Biochem. Physiol. B* **B82**, 233 (1985).
[60] K. Knopf, J. A. Sturman, M. Armstrong, and K. C. Hayes, *J. Nutr.* **108**, 773 (1978).
[61] J. de la Rosa and M. H. Stipanuk, *Comp. Biochem. Physiol. B* **B81**, 565 (1985).
[62] J. H. Fellman and E. S. Roth, *Adv. Exp. Med. Biol.* **139**, 99 (1980).
[63] P. Kontro and S. S. Oja, *in* "Taurine: Biological Actions and Clinical Perspectives," p. 83. Alan R. Liss, Inc., New York, 1985.
[64] J. H. Fellman and E. S. Roth, *in* "Taurine: Biological Actions and Clinical Perspectives," p. 71. Alan R. Liss, Inc., New York, 1985.
[65] C. L. Weinstein and O. W. Griffith, *Anal. Biochem.* **156**, 154 (1986).
[66] J. D. Finkelstein, W. E. Kyle, B. J. Harris, and J. J. Martin, *J. Nutr.* **112**, 1011 (1982).
[67] N. Tateishi, T. Higashi, A. Naruse, K. Nakashima, H. Shiozaki, and Y. Sakamoto, *J. Nutr.* **107**, 51 (1977).
[68] J. A. Schneider and J. D. Schulman, *in* "The Metabolic Basis of Inherited Disease" (J. D. Stanbury, J. B. Wyngaarden, D. S. Fredrickson, J. L. Goldstein, and M. S. Brown, eds.), p. 1844. McGraw-Hill, New York, 1983.

toxic accumulations of cysteine. Glutathione biosynthesis (reactions 13 and 14) serves, in part, to preserve a pool of cysteine in a nontoxic, metabolically active form that can be easily transported between organs.[67,69,70] Thus, glutathione is actively synthesized in liver, a tissue rich in both dietary and biosynthetic cysteine, and is released into the plasma[70,71] and, to a lesser extent, bile.[72-74] Cysteine and cystine are then released from glutathione and its disulfide by enzymes localized on the external plasma membrane surfaces of cells throughout the organism (reactions 15 and 16).[67,75] Both cysteine and cystine are efficiently taken up by most cells. Cystine is reduced to cysteine intracellularly by transhydrogenation with glutathione (reaction 18). Since the glutathione pool is maintained in a highly reduced state by glutathione reductase, the cyst(e)ine pool is also maintained in a reduced state (>90% cysteine).

As shown in Fig. 2, cysteine is required for the biosynthesis of pantetheine and coenzyme A (pathway 19); the enzymes involved have been extensively studied and reviewed by Brown[76] and Abiko.[77] Catabolism of pantetheine and coenzyme A yield cysteamine, an intermediate metabolized to hypotaurine and taurine (reactions 20, 21, and 11). The possibility that this pathway accounts for quantitatively significant amounts of taurine biosynthesis has been considered.[78] Cysteamine and its disulfide, cystamine, may also play a role in mediating intracellular disulfide bond formation *in vivo*.[79]

---

[69] O. W. Griffith and A. Meister, *Proc. Natl. Acad. Sci. U.S.A.* **76**, 5606 (1979).

[70] M. E. Anderson, R. J. Bridges, and A. Meister, *Biochem. Biophys. Res. Commun.* **96**, 848 (1980).

[71] O. W. Griffith and A. Meister, *Proc. Natl. Acad. Sci. U.S.A.* **76**, 268 (1979).

[72] B. H. Lauterberg, C. V. Smith, H. Hughes, and J. R. Mitchell, *J. Clin. Invest.* **73**, 124 (1984).

[73] N. Kaplowitz, D. E. Eberle, J. Petrini, J. Tooloukian, M. C. Corvasce, and J. Kuhlenkamp, *J. Pharmacol. Exp. Ther.* **224**, 141 (1983).

[74] W. A. Abbot and A. Meister, *Proc. Natl. Acad. Sci. U.S.A.* **83**, 1246 (1986).

[75] O. W. Griffith, *J. Biol. Chem.* **256**, 12263 (1981).

[76] G. M. Brown, *J. Biol. Chem.* **234**, 370 (1959).

[77] Y. Abiko, in "Metabolic Pathways" (D. M. Greenberg, ed.), Vol. 7, p. 1. Academic Press, New York, 1975.

[78] D. Cavallini, R. Scandurra, S. DuprePh', G. Federici, L. Santoro, G. Ricci, and D. Barra, in "Taurine" (R. Huxtable and A. Barbeau, eds.), p. 59. Raven Press, New York, 1976.

[79] D. M. Ziegler and L. L. Poulsen, *Trends Biochem. Sci.* **2**, 79 (1977).

[64] Adenosylhomocysteinase (Bovine)

By PETER K. CHIANG

$$\text{Adenosine} + \text{L-homocysteine} \underset{+H_2O}{\overset{-H_2O}{\rightleftharpoons}} S\text{-adenosylhomocysteine} \qquad (1)$$

Adenosylhomocysteinase (S-adenosylhomocysteine hydrolase, EC 3.3.1.1) catalyzes the reversible reaction (1) first described by de la Haba and Cantoni.[1] The equilibrium of the reaction favors the synthesis of S-adenosylhomocysteine with a K_{eq} of 10^{-6} M. Discrete steps for the enzymatic reaction have been elucidated by Palmer and Abeles,[2] which involves the oxidation–reduction of enzyme-bound NAD. The enzyme has recently received much interest because of the many biological effects associated with its inhibition and modulation by adenosine or nucleosides.[3,4] The correlative biological effects have largely been attributed to the indirect inhibition of transmethylation reactions, by the accumulation of S-adenosylhomocysteine or its analogs as a result of the inhibition of the enzyme or the *de novo* synthesis of an analog of S-adenosylhomocysteine.[5]

Purification procedures have been described for the enzyme from mouse leukemia cells,[6] L1210 ascites tumor cells,[7] mouse mastocytoma P-815 cells,[8] hamster liver,[9] and *Dictyostelium discoideum*,[10] and rat livers.[11–14] This chapter describes the assay and purification procedures of

---

[1] G. de la Haba and G. L. Cantoni, *J. Biol. Chem.* **234**, 603 (1959).
[2] J. L. Palmer and R. H. Abeles, *J. Biol. Chem.* **254**, 1217 (1979).
[3] P. K. Chiang, *Methods Pharmacol.* **6**, 127 (1985).
[4] P. K. Chiang and G. L. Cantoni, *Biochem. Pharmacol.* **28**, 1897 (1979).
[5] A. Guranowski, J. A. Montgomery, G. L. Cantoni, and P. K. Chiang, *Biochemistry* **20**, 110 (1981).
[6] A. Merta, I. Votruba, J. Vesley, and A. Holý, *Collect. Czech. Chem. Commun.* **48**, 2701 (1983).
[7] E. L. White, S. C. Shaddix, R. W. Brockman, and L. L. Bennett, Jr., *Cancer Res.* **42**, 2260 (1982).
[8] A. Ichikawa, S. Sato, and K. Tomita, *J. Biochem. (Tokyo)* **97**, 189 (1985).
[9] I.-K. Kim, C.-Y. Zhang, P. K. Chiang, and G. L. Cantoni, *Arch. Biochem. Biophys.* **226**, 65 (1983).
[10] R. J. Hohman, M. C. Guitton, and M. Véron, *Arch. Biochem. Biophys.* **233**, 785 (1984).
[11] E. O. Kajander and A. M. Raina, *Biochem. J.* **193**, 503 (1981).
[12] M. Briske-Anderson and J. A. Duerre, *Can. J. Biochem.* **60**, 118 (1982).
[13] S. O. Døskeland and P. M. Ueland, *Biochim. Biophys. Acta* **708**, 185 (1982).
[14] B. Chabannes, L. Cronenberger, and H. Pachéco, *Experientia* **35**, 1014 (1979).

bovine liver adenosylhomocysteinase, of which much of the enzymology was described.[15–17]

Assay Method

Enzyme activity can be assayed both in the synthetic direction or in the hydrolytic direction.[3] The synthetic direction using ion-exchange column is the method of choice because it does not require laborious preparation of radioactive S-adenosylhomocysteine. However, inhibitors of adenosine deaminase must be included in the incubation mixture when assaying crude extracts. Furthermore, because the adenosylhomocysteinase is a very active enzyme, linearity with respect to protein concentrations should always be established.

Ion-Exchange Column

Synthetic Direction. The standard assay mixture contains in 0.5 ml: 50 mM potassium phosphate (pH 7.6), 2 mM dithiothreitol, 1 mM EDTA, 10% glycerol, 10 mM DL-homocysteine, and 20 μM [8-^{14}C]adenosine. When necessary, either erythro-9-(hydroxy-3-nonyl)adenine (Burroughs Wellcome Co., Research Triangle Park, NC) or 2'-deoxycoformycin (Developmental Therapeutics Program, National Cancer Institute, Bethesda, MD) is added at 10 μM to inhibit adenosine deaminase. The reaction is started by addition of enzyme and, after 10 min at 30°, 1 ml of 50 mM HCl is added to stop the reaction. The mixture is next poured onto a column (0.8 × 4 cm) of SP-Sephadex C-25 equilibrated with 10 mM HCl. After washing the column with 30 ml of 50 mM HCl, the S-[^{14}C]adenosylhomocysteine formed is eluted with 10 ml of 1 M HCl into a scintillation vial. Radioactivity is determined after the addition of 10 ml of scintillation fluor.

Hydrolytic Direction.[16] S-[8-^{14}C]Adenosylhomocysteine can be synthesized by the method of de la Haba and Cantoni,[1] and then purified by repeated crystallization. A faster alternative is to purify the synthesized product by high-pressure liquid chromatography.[18] Another enzymatic synthesis of radioactive S-adenosylhomocysteine labeled by L-[2(n)-^{3}H]homocysteine has been described.[19] It is based on the conversion of S-adenosyl-L-[2(n)-^{3}H]methionine to S-adenosyl-L-[2(n)-^{3}H]homocysteine

[15] P. K. Chiang, H. H. Richards, and G. L. Cantoni, *Mol. Pharmacol.* **13,** 939 (1977).
[16] H. H. Richards, P. K. Chiang, and G. L. Cantoni, *J. Biol. Chem.* **253,** 4476 (1978).
[17] R. H. Abeles, S. Fish, and B. Lapinskas, *Biochemistry* **21,** 5557 (1982).
[18] G. A. Miura, J. R. Santangelo, R. K. Gordon, and P. K. Chiang, *Anal. Biochem.* **141,** 161 (1984).
[19] R. A. Trewyn and S. J. Kerr, *Anal. Biochem.* **82,** 310 (1977).

by glycine N-methyltransferase using glycine as the acceptor. A convenient chemical synthesis of S-adenosylhomocysteine may also be applied to prepare its radioactive counterpart.[20]

The assay mixture contains in 0.5 ml: 50 mM potassium phosphate (pH 7.6), 2 mM dithiothreitol, 1 mM EDTA, 5 units[21] of calf intestinal adenosine deaminase, and 20 μM S-[^{14}C]adenosylhomocysteine. After 10 min of incubation at 30°, the reaction is stopped by the addition of 100 μl of 5 M formic acid. The mixture is poured onto a column of (0.8 × 2.5 cm) of SP-Sephadex C-25, equilibrated in 0.1 M formic acid. After rinsing each tube with 0.5 ml of 0.1 M formic acid, [^{14}C]inosine, a product of the deamination of [^{14}C]adenosine formed by the hydrolysis of S-[^{14}C]adenosylhomocysteine, is subsequently eluted by the addition of 3.5 ml of 0.1 M formic acid. Radioactivity is determined after the addition of 10 ml of scintillation fluor.

Thin-Layer Chromatography[5]

The incubation mixture contains in 50 μl: 50 mM Tris–HCl (pH 8.0), 2 mM dithiothreitol, 10 mM DL-homocysteine, and 20 μM [8-^{14}C]adenosine *or* [U-^{14}C]adenosine (Amersham Corporation). If the latter is used because of its high specific activity, care should be taken to ensure that it is in the β, and not the α, form since the enzyme is stereospecific for the β-pentose configuration. The enzyme is diluted in 50 mM Tris–HCl (pH 8.0) containing bovine serum albumin (1 mg/ml). After incubation at 30°, 15-μl aliquots are transferred at intervals to a small tube containing 5 μl of 0.15 M HCl to stop the reaction. An aliquot, 10 μl, is spotted on chromatographic plates coated with fluorescent indicator; 2 μl of 2 mM adenosine is added as carrier. The chromatographic plates can either be cellulose which is developed in 5% Na_2HPO_4 for 90 min, or silica gel which is developed for 60 min in 1-butanol/acetic acid/H_2O (12:3:5, v/v/v). The spots containing S-[^{14}C]adenosylhomocysteine are cut and counted after visualization under ultraviolet light.

To check the ability of various nucleosides to substitute adenosine as substrate, 10 mM DL-[^{35}S]homocysteine is incubated with 1 mM of the particular nucleoside of interest to generate the corresponding S-nucleosidylhomocysteine analog. DL-[^{35}S]Homocysteine is obtained by incubating DL-[^{35}S]homocystine (Amersham Corporation) in 10 mM Tris–HCl (pH 8.0) and a 6-fold excess of dithiothreitol for 30 min at 37°. Standards of S-nucleosidylhomocysteine analogs can be produced similarly by adding excess hydrolase to start the incubation at 30°; this is terminated 16 hr

[20] R. T. Borchardt, J. A. Huber, and Y. S. Wu, *J. Org. Chem.* **41**, 565 (1976).
[21] One unit is defined as 1 μmol of product formed per minute.

TABLE I
R_f Values of Nucleosides and Their Corresponding S-Nucleosidylhomocysteine (NucHcy) Analogs Synthesized by Bovine Liver Adenosylhomocysteinase[5]

	R_f values			
	Cellulose plate[a]		Silica gel plate[b]	
Nucleoside	Nucleoside	NucHcy	Nucleoside	NucHcy
Adenosine	0.43	0.53	0.53	0.23
Adenosine N^1-oxide	0.69	0.74	0.33	0.15
8-Aminoadenosine	0.26	0.33	0.48	0.14
(±)-Aristeromycin (carbocyclic adenosine)	0.40	0.40	0.43	0.20
8-Azaadenosine	0.58	0.65	0.65	0.35
2-Aza-3-deazaadenosine	0.45	0.52	0.40	0.12
3-Deazaadenosine	0.41	0.48	0.42	0.12
Formycin A (7-deaza-8-azaadenosine)	0.50	0.58	0.47	0.17
Inosine	0.69	0.73	0.42	0.17
Nebularine (purine ribonucleoside)	0.71	0.71	0.50	0.24
N^6-Methyladenosine	0.52	0.60	0.59	0.30
Pyrazomycin (pyrazofurin)	0.75	0.78	0.54	0.24

[a] Cellulose plates developed in 5% Na_2HPO_4; R_f for homocysteine is 0.9.
[b] Silica gel plates developed in 1-butanol/acetic acid/H_2O (12:3:5, v/v/v).

later by heating at 100°. The denatured protein is removed by filtration, and the S-nucleosidylhomocysteine analogs synthesized can be separated from their respective precursor nucleosides by thin-layer chromatography (Table I).

Purification Procedure

The purification procedure described here is based on the method of Richards *et al.*,[16] using bovine liver as starting material. One unit of enzyme activity is defined as 1 μmol of product formed per minute.

Crude Extract. Fresh bovine liver, 500 g, was homogenized in 1.5 liters of ice-cold 0.01 N acetic acid, and next centrifuged at 15,000 g for 20 min. Aftering filtering by glass wool, the pH of the supernatant fluid was adjusted to 7.6 by adding 1 N NaOH. EDTA and dithiothreitol were also added to final concentrations of 2 mM and 1 mM, respectively.

DEAE-Cellulose Batch. The supernatant fluid from the previous step was next stirred in an ice bath with 1.1 liters of DEAE-cellulose (Whatman DE52; calibrated by a suspension in a graduated beaker overnight), which was washed first with water, adjusted to pH 7.2 with HCl, washed extensively again with water and then equilibrated with 10 mM potassium phosphate (pH 7.2), 2 mM dithiothreitol, and 1 mM EDTA. After stirring for 15 min, the DEAE-cellulose was filtered with a coarse fritted disk funnel and washed with 2.2 liters of 20 mM potassium phosphate (pH 7.6), 2 mM dithiothreitol, and 1 mM EDTA. The enzyme was subsequently eluted by stirring the cellulose in 2 volumes of 750 ml each of 20 mM potassium phosphate (pH 7.6), 2 mM dithiothreitol, 1 mM EDTA, and 60 mM $(NH_4)_2SO_4$. The eluates were pooled, and $(NH_4)_2SO_4$ was added at 0° until saturation (69.7 g per 100 ml). After centrifugation, the precipitate was disolved in 150 ml of 8 mM potassium phosphate (pH 7.6), 5 mM dithiothreitol, and 1 mM EDTA. The enzyme was dialyzed overnight against 8 liters of the same buffer.

Hydroxyapatite. The dialyzed enzyme solution was charged onto a column of hydroxyapatite (5 × 28 cm) equilibrated with 8 mM potassium phosphate (pH 7.6), 5 mM dithiothreitol, and 1 mM EDTA. After washing with the buffer, elution was started by a linear gradient of 1.6 liters of potassium phosphate from 8 to 150 mM in the same buffer. Fractions containing the highest specific activity of the enzyme were pooled, and $(NH_4)SO_4$ powder was added at 0° until saturation. The precipitate was dissolved in a small volume of 50 mM Tris (pH 7.6), 20 mM KCl, 2 mM dithiothreitol, and 1 mM EDTA.

Sephadex G-100. The enzyme solution was subjected to column chromatography on Sephadex G-100 (3.5 × 90 cm) equilibrated with 50 mM Tris (pH 7.6), 2 mM dithiothreitol, and 1 mM EDTA. The peak fractions were pooled and diluted with 1.5 volumes of water. The enzyme solution was adsorbed onto a column of DEAE-cellulose (1 × 6 cm) equilibrated with 10 mM potassium phosphate (pH 7.6), 2 mM dithiothreitol, and 1 mM EDTA. The enzyme was eluted in a small concentrated volume by the same buffer containing 60 mM $(NH_4)_2SO_4$.

Crystallization. $(NH_4)_2SO_4$ powder was added at 0° until saturation and, after centrifugation, the precipitate was extracted with 10 ml of 10 mM potassium phosphate (pH 7.0), 42% $(NH_4)_2SO_4$, 5 mM dithiothreitol, and 1 mM EDTA. Next, the precipitate was extracted twice with 12 ml of 10 mM potassium phosphate (pH 7.0), 30% $(NH_4)_2SO_4$, 5 mM dithiothreitol, and 1 mM EDTA. The enzyme solution (between 60 and 90% in purity) was placed in a desiccator at 4° and allowed to crystallize. Crystallization should take place in 2–3 days. The crystals were collected and dissolved in 10 ml of 10 mM potassium phosphate (pH 7.0), 5 mM dithiothreitol, and 1 mM EDTA. $(NH_4)_2SO_4$ powder was added at 0° until saturation, and the precipitate was extracted again with the same buffer containing 40% $(NH_4)_2SO_4$, followed by 28% $(NH_4)_2SO_4$ solution, and allowed to crystallize as described.

TABLE II
NUCLEOSIDES AS SUBSTRATES AND COMPETITIVE INHIBITORS OF BOVINE
ADENOSYLHOMOCYSTEINASE[5,23,24]

Nucleoside	As substrate[a] (% adenosine)	K_i (M)
Adenosine	100	—
3-Deazaadenosine	165	4×10^{-6}
2-Aza-3-deazaadenosine	113	3×10^{-4}
Nebularine	34	1×10^{-3}
Formycin A	18	3×10^{-4}
N^6-Methyladenosine	16	2×10^{-4}
8-Azaadenosine	9	2×10^{-4}
Adenosine N^1-oxide	5	2×10^{-3}
Pyrazomycin	5	1×10^{-2}
8-Aminoadenosine	4	2×10^{-5}
Inosine	2	1×10^{-3}
(±)-Aristeromycin (carbocyclic adenosine)	0.1	5×10^{-9}
3-Deaza-(±)-aristeromycin (carbocyclic 3-deazaadenosine)	+	3×10^{-6}

[a] Assayed at 1 mM nucleoside and 10 mM DL-[^{35}S]homocysteine.

The specific activity of the purified enzyme varied between 0.5 and 0.65 units/mg, representing approximately 700-fold purification and about 20% yield. The enzyme at this stage appeared to be homogeneous by polyacrylamide gel electrophoresis and by analytical untracentrifugation. The purified enzyme could be lyophilized and stored at −70°.

Properties

Composition. The enzyme has four subunits, each with a molecular weight of about 5,000. There is 1 mol of NAD per mole of subunit, and the enzyme is a glycoprotein.

Specificities. The K_m for adenosine is 1.9 μM, and for L-homocysteine, 150 μM. Table II lists the substrate specificities of the enzyme for various adenosine analogs. The best nucleoside substrate is 3-deazaadenosine, with an efficiency about 1.5-fold that of adenosine. Next in line is 2-aza-3-deazaadenosine, which is slightly better than adenosine. Lacking the 6-amino group, nebularine (purine riboside) is 30% as active as adenosine. While formycin A and N^6-methyladenosine are poor substrates, 8-azaadenosine, adenosine N^1-oxide, pyrazomycin, and 8-aminoadenosine are marginally active (10% or less). Inosine can also function as a substrate with an efficiency only about 1.5% that of adenosine; it has a K_m of about 2 mM.

(±)-Aristeromycin (carbocyclic adenosine) is the worst substrate, and 3-deaza-(±)-aristeromycin (carbocyclic 3-deazaadenosine) can also serve as a substrate with an efficiency ranging from good to very poor.[22-24] In addition, N^6-hydroxyadenosine and 2-hydroxyadenosine are also substrates,[24] but their efficiency has not been determined. It has been reported that L-cysteine can substitute for L-homocysteine as the amino acid substrate.[24]

Inhibitors. There are three types of inhibitors for adenosylhomocysteinase: (1) competitive inhibitors; (2) irreversible inactivators; and (3) mixed type inhibitors that can either inhibit competitively and/or irreversibly. The K_i values for the competitive inhibition are listed in Table II. Among the most potent competitive inhibitors are 3-deazaadenosine, (±)-aristeromycin, and 3-deaza-(±)-aristeromycin.

The K_I values for the irreversible inactivators were calculated by the following equation[25]:

$$1/K_{app} = (K_I/k_2[I]) + (1/k_2)$$

where K_{app} is the pseudo-first-order rate constant, K_2 the rate constant of inactivation, and [I] the inactivator concentration. The K_I values are as follows[25]: 9-β-D-arabinofuranosyladenine, 2 μM; 9-β-D-arabinofuranosyl-3-deazaadenine, 41 μM; 2-chloro-3-deazaadenosine, 44 μM; 2-chloroadenosine, 57 μM; 9-β-D-arabinofuranosyl-2-fluoroadenine, 109 μM; 2'-deoxyadenosine 167 μM; nucleocidin, 204 μM; 9-β-D-arabinofuranosyl-2-chloroadenine, 353 μM. Mixed type inhibitors have not been determined for the bovine enzyme but have been found for the enzyme from other sources; they are 9-β-D-arabinofuranosyladenine, neplanocin A, and 9-β-D-arabinofuranosyl-3-deazaadenine.

---

[22] J. A. Montgomery, S. J. Clayton, H. J. Thomas, W. M. Shannon, G. Arnett, A. J. Bodner, I.-K. Kim, G. L. Cantoni, and P. K. Chiang, *J. Med. Chem.* **25,** 626 (1982).

[23] Garcia-Castro, I., J. M. Mato, G. I. Vasanthakumar, W. P. Weismann, E. Schiffmann, and P. K. Chiang, *J. Biol. Chem.* **258,** 4345 (1983).

[24] A. Guranowski and H. Jakubowski, *Biochim. Biophys. Acta* **742,** 250 (1983).

[25] I.-K. Kim, C.-Y. Zhang, G. L. Cantoni, J. A. Montgomery, and P. K. Chiang, *Biochim. Biophys. Acta* **829,** 150 (1985).

[65] Betaine–Homocysteine S-Methyltransferase (Human)

By WILLIAM E. SKIBA, MARILYN S. WELLS, JOHN H. MANGUM, and WILLIAM M. AWAD, JR.

Betaine–homocysteine S-methyltransferase (EC 2.1.1.5) catalyzes the transfer of a methyl group from betaine to L-homocysteine. It is confined to the liver in all vertebrates except for significant amounts in human and guinea pig kidney.[1-4] Identification in lower organisms has been noted but is rare.[1,5,6] Thetins can be substituted for betaine as the methyl donor.[7] Although a thetin–homocysteine S-methyltransferase (EC 2.1.1.3) has been described in the liver of several species, that enzyme is probably identical to the one described here.

Betaine + L-homocysteine → L-methionine + N,N-dimethylglycine

Assay Method

A modification of an earlier technique was utilized, based on the transfer of a [^{14}C]methyl group from betaine to homocysteine, yielding radioactive methionine.[8] Two radioactive products, methionine and dimethylglycine, are generated in equimolar amounts; methionine contributes one-third of the counts and dimethylglycine, two-thirds. Methyl-labeled [^{14}C]betaine (New England Nuclear, 1.49 mCi/mmol) was purified further by passage with water through a Dowex 1-X4 column (hydroxyl form, Bio-Rad) which retained radioactive impurities. Fresh homocysteine was generated from D,L-homocysteine thiolactone immediately before each experiment.[9] To 10 µl of 50 mM [^{14}C]betaine (0.149 mCi/mol) are added 20 µl 0.1 M homocysteine, 20 µl 1 M potassium phosphate (pH 7.4), 10–30 µl enzyme solution, and water for a final volume of 200 µl. Samples are

---

[1] L.-E. Ericson, *Acta Chem. Scand.* **14**, 2102 (1960).
[2] W. E. Skiba, M. P. Taylor, M. S. Wells, J. H. Mangum, and W. M. Awad, Jr., *J. Biol. Chem.* **257**, 14944 (1982).
[3] S. H. Mudd, H. L. Levy, and G. Morrow III, *Biochem. Med.* **4**, 193 (1970).
[4] G. E. Gaull, W. von Berg, N. C. R. Raiha, and J. A. Sturman, *Pediatr. Res.* **7**, 527 (1973).
[5] R. F. White, L. Kaplan, and J. Birnbaum, *J. Bacteriol.* **113**, 218 (1973).
[6] M. Balinska and A. Paszewski, *Biochem. Biophys. Res. Commun.* **91**, 1095 (1979).
[7] J. Durell, D. G. Anderson, and G. L. Cantoni, *Biochim. Biophys. Acta* **26**, 270 (1957).
[8] J. D. Finkelstein and S. H. Mudd, *J. Biol. Chem.* **242**, 873 (1967).
[9] S. H. Mudd, J. D. Finkelstein, F. Irreverre, and L. Laster, *J. Biol. Chem.* **240**, 4382 (1965).

incubated at 37° for 1 hr or more. The reaction was linear for at least 24 hr. if less than 10% of the betaine was utilized. Reactions are stopped by rapid freezing; methionine and dimethylglycine are separated from betaine by a 1 ml column (using a Pasteur pipette) of Dowex 1-X4 (hydroxyl form). After sample application, betaine is eluted with 4 ml of water. Thereafter, methionine and dimethylglycine are eluted with 4 ml of 1.5 M HCl into a scintillation vial to which is added 10 ml of Aquasol (New England Nuclear).

Purification

Cerebrally dead cadavers, with sustained cardiopulmonary function, served primarily as donors for renal transplantation. Hepatectomy was permitted by next of kin for investigational purposes. Following abdominal lymph node sampling for tissue typing and bilateral nephrectomy, aseptic total hepatectomy was done under continuous perfusion. The liver was sliced immediately and layered over powdered Dry Ice for storage at $-20°$. Tissue from a single subject was used in each preparation. All procedures were at 4° except where otherwise indicated, and all solutions described below contained 1 mM homocysteine and 1 mM dimethylglycine in order to stablize the enzyme.

Step 1: Extract. In the example presented, 90 g of liver is rinsed and thawed in 180 ml of 25 mM potassium phosphate at pH 7.4 and, after dicing to about 2-g cubes, is homogenized in a blender for 5 min. The homogenate is centrifuged 3 times at 27,000 g for 30 min; pellets and floating lipid layers are discarded.

Step 2: Heat. The cloudy supernatant liquid is placed in a 75° water bath; the temperature in the contained solution is allowed to rise to 70° over a period of 5 min and maintained for 5 min with continuous gentle stirring. Immediately thereafter, the solution is cooled in an ice bath, and particulate material removed by centrifugation for 20 min at 27,000 g.

Step 3: DEAE-Cellulose I. A clear supernatant fraction is obtained which is applied directly to a DEAE-cellulose (Whatman DE-52) column (6 × 13 cm) equilibrated in 25 mM potassium phosphate at pH 7.4. After washing with 600 ml of this solution, stepwise elution is done with 220 ml 0.3 M KCl and 430 ml 3.0 M KCl in the same buffer. A major component with 70-80% of the total activity eluted with the breakthrough peak during the initial wash; a minor peak with the remainder of the activity appears just before the application of the 3 M KCl. Breakthrough fractions containing the major component only of enzyme activity are pooled and diluted to 5 times the initial volume with 1 mM dimethylglycine–1 mM homocysteine (pH 7.4).

TABLE I
Purification of Betaine–Homocysteine S-Methyltransferase

Step	Volume (ml)	Protein (mg)	Activity (units[a] × 10⁻³)	Specific activity (units/mg)
Heat[b]	120	5220	323	62
DEAE-cellulose I[c]	95	487	144	295
DEAE-cellulose II	63	136	104	766
Sepharose 6B	42	45	71	1590
CM-cellulose I[d]	5	4.9	9.3	1900
CM-cellulose II	6.7	1.8	5.7	3140

[a] A unit is defined as the generation of 0.36 nmol methionine per hour.
[b] Representing 80% of activity in crude extract where accurate protein measurements could not be done.
[c] That portion recovered as the major form of the enzyme.
[d] Only 22% of the protein from the Sepharose column was applied.

Step 4: DEAE-Cellulose II. The solution is applied to a DE-52 column (5 × 15 cm) equilibrated with 5 mM potassium phosphate at pH 7.4. After washing with 700 ml of this buffer, the enzyme is eluted with 0.3 M KCl in the same buffer. Fractions with activity are concentrated by ultrafiltration to 5 ml.

Step 5: Sepharose 6B. The solution is applied to a Sepharose 6B column (2.6 × 90 cm) and eluted with 5 mM potassium phosphate at pH 7.4. Active fractions are concentrated by ultrafiltration through a PM10 membrane (Amicon Corp.). The retentate, 2 ml, is equilibrated with 5 mM potassium phosphate (pH 6.5) by passage through Sephadex G-25 (1.4 × 30 cm).

Step 6: CM-Cellulose I. The eluted enzyme fractions are pooled and applied to a CM-cellulose column (Whatman CM-52, 1.5 × 5 cm) equilibrated in the same buffer. The initial fractions containing the enzyme, which did not bind, are combined, concentrated by ultrafiltration, and stored at −20° in a solution of 50% glycerol.

Step 7: CM-Cellulose II. Portions of the preparation from Step 6 are utilized for further purification by equilibration with 5 mM potassium citrate (pH 5.7) and passage through a CM-52 column (1.5 × 3.2 cm) equilibrated in the same buffer. After an initial wash with 110 ml of this buffer the column is eluted with a linear gradient formed with 100 ml each of pH 5.7 buffer with and without 1 M KCl.

Table I provides a summary of the purification data. Chromatography through CM-52 at either pH 6.5 or 5.7 did not yield a homogeneous preparation. At pH 6.5 the protein did not bind to CM-cellulose. If chromatography through CM-cellulose at pH 5.7 was attempted directly after gel filtration through Sepharose 6B, components absorbing at 410 nm appeared to denature and distribute throughout the fractions containing the methyltransferase. This led to the procedure described above. A 250-fold purification was required to achieve homogeneity with a yield of approximately 15%. The protein was labile to lyophilization but could be stored in 50% (v/v) glycerol solutions at −20°C for over 1 year.

As an alternative to the Step 2 heating procedure, the supernatant fluid obtained after initial sedimentation at 27,000 g was placed in an ultracentrifuge and sedimented at 105,700 g for 90 min. The two components seen with the first DEAE-cellulose chromatography were not products of the heating step since the same elution pattern of two activities, in the same proportion, was noted when ultracentrifugation was substituted for heating at 70°. All liver specimens showed the two peaks and thus the results do not represent apparent genetic variation. When passed separately through the same column under the same buffer conditions, there was no change in the elution pattern of each of the two components, indicating that the separated transmethylases did not reflect column overloading or rapid generation of one component from the other. Two forms of the enzyme have been noted in horse liver also.[10]

Initial studies revealed that the enzyme was moderately labile as demonstrated by the substantial loss of activity with later steps of purification and by the generation of products of decreasing molecular weight, containing activity, during gel filtration steps. Neither nonspecific thiol-containing agents, such as mercaptoethanol, nor a series of divalent cations appeared to stablize the enzyme. The inclusion of one substrate, homocysteine, and one product, dimethylglycine, in all steps during the purification led to remarkable stabilization, preventing both the loss of activity and the generation of lower molecular weight products that were active.

Analysis for Purity

In the presence of 30 mM DL-homocysteine a single wide band was noted on gel electrophoresis at pH 9.5 using 5% acrylamide. In its absence, a series of discrete bands were generated which migrated more slowly than the major band. A plot of the relative migrations of absorption peaks of these bands against assumed integral aggregation numbers showed a straight line, confirming the assumption. A single polypeptide with an apparent molecular weight of 45,000 was found on electrophoresis in sodium dodecyl sulfate gels containing 1% 2-mercaptoethanol. Gel fil-

[10] W. A. Klee, H. H. Richards, and G. L. Cantoni, *Biochim. Biophys. Acta* **54**, 157 (1961).

tration with a calibrated Sephacryl S-300 column suggested an apparent molecular weight of 270,000 for the native enzyme. Thus, the protein appears to be a hexamer of identical subunits with the oligomer being able to form integral aggregates in the absence of homocysteine. The latter finding was noted earlier with the horse and pig enzymes.[7,11]

Other Studies

K_m values for homocysteine and betaine were 0.12 and 0.10 mM, respectively. Following the lead from studies with acetylcholine esterase, where it was observed that uncharged acetylcholine analogs bound well to that enzyme,[12-15] a similar series of studies were undertaken with the methyltransferase. 3,3-Dimethylbutyrate and isovalerate, the analogs, respectively, of betaine and dimethylglycine with nitrogens replaced by carbon atoms, were competitive inhibitors against betaine with K_i values of 0.45 and 0.30 mM, respectively. Interestingly, butyrate was found to inhibit significantly with a K_i value of 1.0 mM despite the fact that its isostere, sarcosine, did not show significant inhibition. These findings led to the generation of a dual-substrate inhibitor, S-(δ-carboxybutyl)-DL-homocysteine for which a K_i value of 6.5 μM was observed against betaine.[16] Since it is probable that only the L-isomer serves as the inhibitor, the real K_i value is approximately 3 μM. These studies provide a direction for the development of more potent inhibitors.

[11] L.-E. Ericson, *Acta Chem. Scand.* **14**, 2113 (1960).
[12] D. H. Adams, *Biochim. Biophys. Acta* **3**, 1 (1949).
[13] D. H. Adams and V. P. Whittaker, *Biochim. Biophys. Acta* **3**, 358 (1949).
[14] C. A. Mounter and R. M. Cheatham, *Enzymologia* **25**, 215 (1963).
[15] F. B. Hasan, S. G. Cohen, and J. B. Cohen, *J. Biol. Chem.* **255**, 3898 (1980).
[16] W. M. Awad, Jr., P. L. Whitney, W. E. Skiba, J. H. Mangum, and M. S. Wells, *J. Biol. Chem.* **258**, 12790 (1983).

[66] Cystathionine β-Synthase (Human)

By Jan P. Kraus

$$HOOC—CHNH_2—CH_2OH + HS—CH_2CH_2—CHNH_2—COOH \rightarrow$$
$$HOOC—CHNH_2—CH_2—S—CH_2CH_2—CHNH_2—COOH$$
$$\text{Cystathionine}$$

Cystathionine β-synthase (L-serine hydro-lyase [adding homocysteine], EC 4.2.1.22) is a pyridoxal 5′-phosphate-dependent enzyme in the transsulfuration pathway of higher eukaryotes which catalyzes the condensation of serine and homocysteine to form cystathionine. In rat liver, about 50% of sulfur present in the form of homocysteine is converted irreversibly to cystathionine and then to cysteine, whereas the other 50% is conserved by remethylation to methionine in the homocysteine-conserving cycle.[1] Marked deficiency of cystathionine β-synthase activity has been demonstrated in liver, cultured skin fibroblasts, and phytohemagglutinin-stimulated lymphocytes from patients with the most common form of inherited homocystinuria.[2] The primary translation product of the human and rat gene is a M_r 63,000 polypeptide. However, immunoprecipitation of synthase from fresh liver extracts with monospecific antisynthase antiserum yields two forms of the enzyme: a predominant form (~95%), a tetramer of M_r 63,000 subunits; and a minor form (~5%), a dimer of M_r 48,000 subunits.[3] It has been shown that the smaller form of the enzyme is derived from the larger form by the action of a specific protease, leading to approximately 60-fold increase in catalytic activity at physiological concentrations of homocysteine. Previous attempts at purification of the larger form of synthase from fresh liver extracts were largely unsuccessful due to the occurrence of multiple forms of the enzyme during purification, as well as to its tendency to form stable high molecular weight complexes with other proteins. We have, therefore, devised a purification procedure starting with an "aged" liver homogenate in which synthase molecules have been converted to the smaller, more active, and more stable form.[4]

Enzyme Assay

Enzyme Units

One unit of enzyme catalyzes the formation of 1 μmol of cystathionine in 1 hr at 37°. Specific activity is expressed as units per mg of protein. Protein is determined by the method of Lowry *et al.*[5] using bovine serum albumin as a standard.

---

[1] J. D. Finkelstein and J. J. Martin, *J. Biol. Chem.* **259**, 9508 (1984).
[2] S. H. Mudd and H. L. Levy, in "The Metabolic Basis of Inherited Disease" (J. B. Stanbury, J. B. Wyngaarden, D. S. Frederickson, J. L. Goldstein, and M. S. Brown, eds.), 5th ed., p. 522. McGraw-Hill, New York, 1983.
[3] F. Skovby, J. P. Kraus, and L. E. Rosenberg, *J. Biol. Chem.* **259**, 588 (1984).
[4] J. P. Kraus and L. E. Rosenberg, *Arch. Biochem. Biophys.* **222**, 44 (1983).
[5] O. H. Lowry, N. J. Rosebrough, A. L. Farr, and R. J. Randall, *J. Biol. Chem.* **193**, 265 (1951).

Preparation of L-Homocysteine and [^{14}C]Serine

One milliter of 200 mM L-homocysteine is prepared daily. L-Homocysteine thiolactone (30.72 mg, 0.2 mmol) is dissolved in 0.2 ml of 5 M NaOH, and, following incubation for 5 min at 37°, the pH is adjusted to 8.6 with 2 M HCl. Tris–HCl at pH 8.6 and dithiothreitol are added to final concentrations of 100 and 20 mM, respectively. All batches of commercial L-[U-^{14}C]serine are purified prior to use by paper chromatography in the solvent system used for the enzyme assay and stored at 4°.

Radioisotopic Assay

The assay is modified from that of Mudd et al.[6] and Fowler et al.[7] The final concentrations of the components of the assay in a total volume of 0.2 ml include the following: 100 mM Tris–HCl at pH 8.6, 250 μM pyridoxal 5′-phosphate, up to 100 μg extract protein, 0.5 mg/ml dialyzed bovine serum albumin, and 10 mM [^{14}C]serine (300 cpm/nmol). Following incubation for 5 min at 37°, the reaction is started by addition of L-homocysteine to 10 mM. After incubation for 30–60 min at 37°, an aliquot of the assay mixture is applied directly to Whatman No. 3 paper with drying in a hot air stream, and the [^{14}C]cystathionine formed is separated from the [^{14}C]serine by descending paper chromatography in 2-propanol/formic acid/H$_2$O (80 : 6 : 20, v/v) (≥16 hr). Radioactivity in the region of marker L-cystathionine (detected by staining the marker lane with ninhydrin, R_f ~0.15) is determined by cutting the chromatogram into strips and counting them in toluene-based scintillation fluid. Radioactivity from an enzyme-free blank is subtracted. Crude extracts are assayed in the presence of 1 mM L-cystathionine to protect the radioactive product from degradation by cystathionase.

In assays of cystathionine β-synthase from extracts of cultured human skin fibroblasts, the following concentrations of reagents are used: 1mM pyridoxal 5′-phosphate, up to 700 μg extract protein, 2.5 mM [^{14}C]serine (3,000 cpm/nmol), and 15 mM L-homocysteine. Incubations are at 37° for 4 hr.

Colorimetric Assay

The rapid colorimetric assay[8] is useful for monitoring purification Steps 4–8. The enzyme is assayed under conditions identical to the ra-

---

[6] S. H. Mudd, J. D. Finkelstein, F. Irreverre, and L. Laster, *J. Biol. Chem.* **240**, 4382 (1965).

[7] B. Fowler, J. Kraus, S. Packman, and L. E. Rosenberg, *J. Clin. Invest.* **61**, 645 (1978).

[8] S. Kashiwamata and D. M. Greenberg, *Biochim. Biophys. Acta* **212**, 488 (1970).

dioisotopic assay except for omission of the ^{14}C label. The reaction is stopped by addition of 3.3 ml of ninhydrin reagent. The mixture is heated for 5 min in a boiling water bath and then cooled in an ice bath for 2 min. After 20 min at room temperature, the absorbance at 453 nm is determined, and the enzyme-free blank is subtracted. The amount of cystathionine formed is determined relative to a standard reaction containing 0.5 μmol of L-cystathionine, without enzyme. Ninhydrin reagent is prepared by dissolving 6 g of ninhydrin in 600 ml of glacial acid, adding 200 ml of phosphoric acid, and mixing well; the reagent is stable for at least 1 month at 4°.

Enzyme Isolation

Materials

Cetyltrimethylammonium bromide (mixed alkyl chains, predominantly C_{14}) is obtainable from Sigma (Cat. No. M-7635). Hydroxylapatite is prepared by the method of Main et al.,[9] using ammonia for neutralization at boiling temperature. Postmortem liver, obtained at autopsy, was frozen within 3–6 hr after the death of the patient and stored at $-80°$.

Purification of Cystathionine β-Synthase

Step 1. Portions of frozen liver (300 g) are suspended in 4 volumes of 1 mM 2-mercaptoethanol and homogenized in a Waring Blendor at full speed for 1.5 min. The homogenate is adjusted to pH 6 with 1 M KH_2PO_4 (approximately 30 ml/liter of 20% homogenate) and stored for 1 week at 4° (fraction II, Table I).

Step 2. An equal volume of 0.1% (w/v) cetyltrimethylammonium bromide in 1 mM 2-mercaptoethanol is added to fraction II, and the mixture is stirred for 15 min at 4°, and centrifuged for 45 min at 16,000 g. The supernatant fluid is filtered through a plug of glass wool (fraction III).

Step 3. Fraction III is applied at 300 ml/hr to a column (5 × 15 cm) of DEAE-cellulose previously equilibrated with 15 mM potassium phosphate at pH 6.5, containing 1 mM 2-mercaptoethanol and 0.025% cetyltrimethylammonium bromide. The column is washed with this buffer, followed by 55 mM potassium phosphate–1 mM 2-mercaptoethanol, pH 6.5, until the A_{280} is less than 0.05 in each case. The enzyme is eluted with a 6-liter linear gradient from 55 to 170 mM potassium phosphate at pH 6.5 containing 1 mM 2-mercaptoethanol at a flow rate of 220 ml/hr. Fractions

[9] R. K. Main, M. J. Wilkins, and L. J. Cole, *J. Am. Chem. Soc.* **81**, 6490 (1959).

TABLE I
PURIFICATION OF CYSTATHIONINE β-SYNTHASE FROM HUMAN LIVER

Fraction	Purification step	Volume (ml)	Total activity (units)	Total protein (mg)	Specific activity (units/mg)
I	Fresh homogenate	1300	2526	25260	0.10
II	"Aged" homogenate (7 days, 4°, pH 6)	1300	4800	25260	0.19
III	14,000 g centrifugation in 0.05% CTB[a]	2365	5472	9600	0.57
IV	DEAE-cellulose	20.5	3936	219	18
V	Hydroxylapatite	5	2592	48	54
VI	Blue Sepharose	1	2496	31	81
VII	Preparative gel electrophoresis[b]	0.2	(1872)	(11.8)	220
VIII	Glycerol gradient[b]	0.4	(864)	(2.5)	340

[a] CTB, cetyltrimethylammonium bromide.
[b] One-tenth of fraction VI (0.1 ml, 250 units) was applied to the preparative gel. The numbers within parentheses correspond to the entire volume of fraction VI.

(22 ml) eluting between 95 and 125 mM phosphate contain the peak of activity. The enzyme is concentrated by ultrafiltration (Amicon, XM50 membrane) to about 20 ml, and dialyzed overnight against 15 mM potassium phosphate–1 mM 2-mercaptoethanol, pH 7.2 (fraction IV).

Step 4. Fraction IV is applied at 3 ml/hr to a column (2.8 × 8 cm) of hydroxylapatite equilibrated with the same buffer as used for dialysis. The column is washed with the equilibration buffer until A_{280} returns to baseline. The enzyme is eluted with a convex gradient (750 ml) from 15 to 80 mM potassium phosphate, 1 mM 2-mercaptoethanol, pH 7.2, using a fixed volume (230 ml) mixing chamber. Fractions (7.3 ml) with the highest activities, eluting between 34 and 54 mM phosphate, are pooled. The enzyme is concentrated to 5 ml by ultrafiltration (Amicon, XM50 membrane), diluted 10-fold with 1 mM 2-mercaptoethanol and reconcentrated to 5 ml (fraction V).

Step 5. Fraction V is applied to a column (2.7 × 6 cm) of Blue Sepharose (Pharmacia) at 13 ml/hr. The most active fractions (6.5 ml per fraction) in the breakthrough volume are collected and concentrated by ultrafiltration (Amicon, XM50 membrane) to a final volume of 1 ml (fraction VI).

Step 6. Up to 4 mg of protein (fraction VI) is adjusted to a final concentration of 35 mM Tris–HCl at pH 8.9 including 100 mM 2-mercaptoethanol and 15% glycerol containing bromophenol blue tracking dye, in a total volume of 0.2 ml. The sample is layered under the running buffer[10] onto a 7.5% polyacrylamide gel[10] (1.5 × 4 cm) in a Savant preparative gel cell (Model PAG 15WC, Hicksville, NY). A constant current of 2.5 mA is applied for 1 hr; current is then increased to 6.5 mA for electrophoresis overnight. The flow rate of the elution buffer (375 mM Tris–HCl, pH 8.9) should be 10 ml/hr. The gel cell is cooled to 4° using a cooling circulator (LKB 2209 Multi-temp). Fractions (1 ml) containing synthase activity are pooled, concentrated by ultrafiltration (Amicon, PM10 membrane) to 0.2 ml, diluted 100-fold with 25 mM potassium phosphate at pH 6.5 and 1 mM 2-mercaptoethanol, and concentrated to a final volume of 0.2 ml (fraction VII).

Step 7. Glycerol density gradient centrifugation is carried out at 50,000 rpm in a Beckman SW50 rotor for 20 hr at 4°. The gradient (5 ml) is from 10 to 30% glycerol in 10 mM potassium phosphate at pH 7.0 that includes 1 mM 2-mercaptoethanol to which 0.2 ml of sample (fraction VII) has been added. Fractions (0.2 ml) are collected using an ISCO gradient fractionator. Active fractions containing homogeneous enzyme, as determined by sodium dodecyl sulfate (SDS)–gel electrophoresis, are pooled and concentrated by ultrafiltration (Amicon YM10 membrane) to about 0.4 ml, diluted 10-fold with 25 mM potassium phosphate–1 mM 2-mercaptoethanol at pH 6.5, and concentrated. The enzyme is stored at −80° (fraction VIII).

Properties of Cystathionine β-Synthase

Estimation of Purity

Table I shows the results of a typical purification from 300 g of liver. The specific activity of the final product was about 3,400 and 1,800 times that of the fresh and "aged" homogenate, respectively. The purified enzyme was homogeneous[4,11] in its native and denatured forms as judged by polyacrylamide gel electrophoresis in four different systems: native gel, pH 3–10 electrofocusing gel, SDS gel, and 8 M urea gel.

pH Optimum and Isoelectric Focusing

The effect of pH on the formation of cystathionine was measured both in Tris–HCl and Ammediol–HCL buffers. A broad optimum from pH 8.4 to 9.0 was observed.[11] The isoelectric point was 5.2 at 2°.[4,11]

[10] B. J. Davis, *Ann. N.Y. Acad. Sci.* **21**, 404 (1964).
[11] J. Kraus, S. Packman, B. Fowler, and L. E. Rosenberg, *J. Biol. Chem.* **253**, 6523 (1978).

Absorption Spectrum

The homogeneous enzyme at pH 6.5 in 25 mM potassium phosphate had absorption maxima at 280, 363, and 427 nm.[4] The absorption peaks at 363 and 427 nm are characteristic of pyridoxal 5′-phosphate-containing enzymes, resulting from the nonprotonated and protonated Schiff bases, respectively, between the coenzyme and the apoenzyme.

Kinetic Analyses

Kinetic parameters of crude liver synthase composed of M_r 63,000 subunits were compared to those of purified synthase composed of M_r 48,000 subunits. No difference was noted in the apparent K_m for serine; a value of 4 mM was obtained for both preparations. The apparent K_m for homocysteine, however, was quite different for the two preparations. An apparent K_m of 25 mM was determined for the enzyme composed of M_r 63,000 subunits, a value 30 times higher than that (0.8 mM) determined for synthase composed of M_r 48,000 subunits.[3]

Antibody–Antigen Interactions

A monospecific antiserum against the purified enzyme was raised in rabbits. Rabbit antisera against human hepatic enzyme interacted readily with synthase in extracts from livers of other species.[4] About 1.3, 1.5, and 2 times as much antiserum was required to precipitate 1 unit of enzyme from mouse, rhesus monkey, and rat, respectively, as was required for 1 unit of human enzyme. In contrast, the bovine enzyme reacted only slightly with the anti-human synthase antibody. Finally, the antibody can be used to rapidly purify M_r 63,000 synthase subunits from crude extracts by immunoprecipitation followed by preparative SDS–polyacrylamide gel electrophoresis.[3]

Acknowledgment

I would like to acknowledge Dr. Leon E. Rosenberg in whose laboratory these studies were performed.

[67] Cysteine Dioxygenase

By KENJI YAMAGUCHI and YU HOSOKAWA

$$\text{L-Cysteine} + O_2 \rightleftharpoons \text{L-cysteinesulfinic acid}$$

General Introduction

Cysteine dioxygenase [EC 1.13.11.20] catalyzes the oxygenation of cysteine to cysteinesulfinic acid, a key intermediate of cysteine metabolism to hypotaurine, taurine, isethionic acid, pyruvate, and sulfate in mammalian tissues.[1-3] Attempts to purify the enzyme from rat liver[4-10] were unsuccessful because of its novel properties: rat liver cysteine dioxygenase is obtained in inactive form, which is activated by preincubation with L-cysteine under anaerobic conditions although the activated enzyme is rapidly and irreversibly inactivated during aerobic conditions. This inactivation can be prevented by a distinct cytoplasmic protein, called protein A.[11,12] On the basis of these findings the enzyme from cytosol of rat liver is highly purified.[13] On the other hand, a NAD-dependent cysteinesulfinic acid-forming enzyme activity[10-16] has been reported in some extrahepatic mammalian tissues at a much lower level than the

---

[1] J. G. Jacobsen and L. H. Smith, Jr., *Physiol. Rev.* **48**, 425 (1968).
[2] F. Chapeville and P. Fromageot, *Biochim. Biophys. Acta* **17**, 275 (1955).
[3] M. Tavachenik and H. Tarver, *Arch. Biochem. Biophys.* **156**, 115 (1955).
[4] B. Sőrbo and L. Ewetz, *Biochem. Biophys. Res. Commun.* **18**, 359 (1965).
[5] C. De Marco, R. Mosti, and D. Cavallini, *Boll. Soc. Ital. Biol. Sper.* **42**, 94 (1966).
[6] L. Ewetz and B. Sőrbo, *Biochim. Biophys. Acta* **128**, 296 (1966).
[7] J. B. Lombardini, P. Turini, D. R. Biggs, and T. P. Singer, *Physiol. Chem. Phys.* **1**, 1 (1969).
[8] J. B. Lombardini, T. P. Singer, and P. D. Boyer, *J. Biol. Chem.* **244**, 1172 (1969).
[9] K. Yamaguchi, S. Sakakibara, K. Koga, and I. Ueda, *Biochim. Biophys. Acta* **237**, 502 (1971).
[10] K. Yamaguchi, S. Sakakibara, J. Asamizu, and I. Ueda, *Biochim. Biophys. Acta* **297**, 48 (1973).
[11] S. Sakakibara, K. Yamaguchi, I. Ueda, and Y. Sakamoto, *Biochem. Biophys. Res. Commun.* **52**, 1093 (1973).
[12] S. Sakakibara, K. Yamaguchi, Y. Hosokawa, N. Kohashi, I. Ueda, and Y. Sakamoto, *Biochim. Biophys. Acta* **422**, 273 (1976).
[13] K. Yamaguchi, Y. Hosokawa, N. Kohashi, Y. Kori, S. Sakakibara, and I. Ueda, *J. Biochem. (Tokyo)* **83**, 479 (1978).
[14] G. H. Misra, *Brain Res.* **97**, 117 (1975).
[15] R. M. Di Giorgio, G. Tucci, and S. Macaione, *Life Sci.* **16**, 429 (1975).
[16] C. H. Misra, *Biochem. Pharmacol.* **28**, 1695 (1979).

NAD-independent cysteine dioxygenase activity that is exclusively found in liver. The NAD-dependent enzymes in extrahepatic tissues have not been purified.

Cysteine dioxygenase of rat liver is induced by the injection of hydrocortisone and sulfur amino acids such as cysteine or methionine,[8] and it is highly responsive to the dietary intake of sulfur amino acids[17-20] and protein,[17] as well as to the nutritional value of protein.[20] The urinary excretion of taurine[17,20] and hepatic glutathione[20] are proportionally changed with hepatic cysteine dioxygenase activity. It appears likely that hepatic cysteine dioxygenase activity regulates the cellular levels of cysteine, methionine, and glutathione and biosynthesis of taurine. Fetal and neonatal rat liver lack the enzyme, suggesting that taurine and sulfate are essential nutrients in these stages.[21,22] Malignant tumor cells also do not have the enzyme.[12]

Purification of Rat Liver Enzyme

Step 1: Extract. Rat liver (1 kg) is homogenized with 4 liters of 0.25 M sucrose and centrifuged at 15,000 g for 30 min. The supernatant fluid is next centrifuged at 105,000 g for 90 min to obtain a clear extract.

Step 2: Acetone Fraction. Acetone fractionation is carried out at $-15°$ in an ethanol–Dry Ice bath. Chilled acetone is added to the preparation to give 35% (v/v), and the precipitate is removed by centrifugation at 10,000 g for 10 min after standing for 15 min. To the clear supernatant liquid, chilled acetone is slowly added to a concentration of 50% (v/v), followed by standing for 10 min. The pellet obtained by centrifugation is dissolved in a small volume of 10 mM sodium phosphate at pH 6.8.

Step 3: DEAE-Cellulose. About 20% from Step 2 is applied to a DEAE-cellulose column (Whatman DE-52; 2.5 × 25 cm) equilibrated with same buffer at pH 6.8. After washing the column with 20 mM NaCl in the same buffer, to a point at which the absorbance at 280 nm is below 0.1, elution is carried out with a linear gradient of 600 ml of 20–250 mM NaCl in the same buffer at a flow rate of 30 ml/hr; fractions of 6 ml are collected.

[17] N. Kohashi, Y. Yamaguchi, Y. Hosokawa, Y. Kori, O. Fujii, and I. Ueda, *J. Biochem. (Tokyo)* **84**, 159 (1978).
[18] M. H. Stipnuk, *J. Nutr.* **109**, 2126 (1979).
[19] K. H. Daniels and M. H. Stipnuk, *J. Nutr.* **112**, 2130 (1982).
[20] K. Yamaguchi, Y. Hosokawa, S. Niizeki, H. Tojo, and I. Sato, *Prog. Clin. Biol. Res.* **179**, 23 (1985).
[21] Y. Hosokawa, K. Yamaguchi, N. Kohashi, Y. Kori, and I. Ueda, *J. Biochem. (Tokyo)* **88**, 389 (1978).
[22] C. Loriette and F. Chatagner, *Experientia* **34**, 981 (1978).

To the active fractions (tubes 46–54) are added chilled acetone at $-15°$ to 50% (v/v). After standing for 15 min at $-15°$, the resulting precipitates are collected by centrifugation and dissolved in a minimum volume of the same buffer containing 10% (v/v) ethanol. After this step, either ethanol, acetone, or glycerol is absolutely required to stabilize the enzyme.

Step 4: Sephadex G-100. The clear supernatant fluid is applied to a Sephadex G-100 (coarse) column (2.5 × 100 cm) equilibrated with buffer A (2.5 mM sodium phosphate buffer at pH 6.8, containing 10% ethanol). Elution is carried out with same buffer at a flow rate of 30 ml/hr; fractions of 5 ml are collected. The active fractions (58–64) are pooled.

Step 5: Hydroxylapatite. The enzyme solution is applied to a hydroxylapatite column (2.5 × 5 cm) equilibrated with buffer A. After washing with 500 ml of buffer, elution is carried out with a linear gradient of 200 ml changing from 2.5 to 150 mM sodium phosphate buffer at pH 6.8, both containing 10% ethanol, at a flow rate of 15 ml/hr; fractions of 3 ml are collected. The active fractions (24–29) are pooled.

Step 6: DEAE-Sephadex A-25. The fractions from Step 5 are dialyzed overnight against 2 liters of 20 mM NaCl in 10 mM sodium phosphate at pH 6.8, containing 10% ethanol, and are applied to a column (1 × 15 cm) of DEAE-Sephadex A-25 equilibrated with same buffer. Protein is eluted with 200 ml of a linear salt gradient, 20–250 mM NaCl in the same buffer at a flow rate of 15 ml/hr; fractions of 3 ml are collected. The active fractions (30–34) are concentrated with an Amicon microultrafiltration apparatus equipped with a PM10 membrane and stored $-20°$.

Step 7: Sephadex G-75. The concentrated fractions are applied to a Sephadex G-75 (superfine) column (1.6 × 70 cm) equilibrated with 10 mM sodium phosphate at pH 6.8, containing 10% ethanol. The column is developed with the same buffer at a flow rate of 8 ml/hr with fractions of 2 ml being collected. Activity is present in tubes 37 through 40, in which the enzyme appears to be homogeneous.

The purified enzyme can be stored at 0° without significant loss of activity for at least 1 month. After Step 3, no enzyme activity is detected without the anaerobic preactivation procedure with L-cysteine and the addition of protein A. The NAD-dependent enzyme activity is also not detected after Step 3. Table I summarizes the purification of cysteine dioxygenase.

Assay Procedure of Homogenate

Activation of Cysteine Dioxygenase Activity in Homogenate. Tissues are homogenized with 9 volumes of 0.25 M sucrose and centrifuged at

TABLE I
PURIFICATION OF CYSTEINE DIOXYGENASE FROM RAT LIVER

Step	Total volume (ml)	Total activity (units)	Total protein (mg)	Specific activity (units/mg protein)
1. Extract	3,900	107,000	65,500	1.6
2. Acetone	255	82,900	17,900	4.9
3. DEAE-cellulose	275	32,100	848	37.9
4. Sephadex G-100	140	17,500	21	854
5. Hydroxylapatite	18	8,180	5.4	1500
6. DEAE-Sephadex G-25	15.6	5,120	2.5	2070
7. Sephadex G-75	8.0	2,370	0.9	2650

105,000 g for 90 min; the residue is discarded since the cysteinesulfinic acid-degrading activity is present in the particulate fraction. Because only 10–30% of total cysteine dioxygenase activity is detectable at this stage, an activation procedure is required. In a Thunberg tube, 0.1 ml of enzyme solution, 10 μmol of L-cysteine adjusted at pH 6.8 just before use, and 100 μmol of glycine–NaOH buffer at pH 9.0 are added in a final volume of 1 ml. The Thunberg tube is connected with a nitrogen supply and a suction pump. The tube is subjected alternately to suction for 10 sec and filling with nitrogen through a three-way stopcock; the procedure is repeated 5 times. Maximum activation is obtained by incubation at 37° for 30–40 min under nitrogen.

The Assay of Cysteine Dioxygenase Activity. After cooling the Thunberg tubes in ice water for 3 min, the top of tube is removed, 0.1 ml L-[^{35}S]cysteine (0.2 μCi) is added, and the incubation is carried out for 10 min at 37° with shaking at 120 strokes/min under an atmosphere of air. The reaction is terminated by the addition of 0.5 ml 10% TCA. L-[^{35}S]Cystine is also available as tracer since labeled cystine is practically entirely reduced to cysteine by the disulfide-exchange reaction in the presence of a large excess of nonlabeled cysteine.

Determination of Product. A 1 ml-aliquot of supernatant fluid from a deproteinized reaction mixture is applied to a Dowex 50-X8 H$^+$ column (1 × 3 cm). The reaction product, cysteinesulfinic acid, is eluted with 9 ml water, and neutral amino acids, basic amino acids, and hypotaurine are absorbed on the resin. Sulfate, taurine, and cysteic acid are also eluted with cysteinesulfinic acid. Cysteinesulfinic acid, however, can be separated from these compounds with a Dowex 1-X2 (acetate form) column (1 × 5 cm). Thereby, taurine is first eluted with 30 ml of 0.1 M acetic acid; cysteinesulfinic acid and cysteic acid with 30 ml of 0.5 N HCl after washing with 40 ml of 2 M acetic acid; and sulfate and isethionic acid with 40 ml of 1 N HCl.

For the determination of cysteinesulfinic acid, either colorimetric assay with ninhydrin or a radiochemical assay using L-[^{35}S]cysteine is available. The details of determining cysteine sulfinic acid are as follows: 2-ml aliquots of the effluents are mixed with 2 ml of ninhydrin reagent[23] and heated for 15 min in a boiling water bath, followed by dilution with 2 ml of 50% ethanol after cooling. Absorbance is measured at 570 nm after 15 min at room temperature. The amount of cysteinesulfinic acid produced is calculated from the absorbance of authentic cysteinesulfinic acid solution treated in same manner as the samples. Aliquots (0.2–0.5 ml) of effluents are used for the radiochemical assay.

After Dowex 50 treatment, HPLC methods are also available for determination of cysteinesulfinic acid and related compounds. With an anion-exchange column (SA-121, Japan Spectroscopic Co. Ltd.) and 0.1 M KH_2PO_4 as the mobile phase at a flow rate of 1 ml/min, pressure of 40 kg/cm^2, and 50°, the retention time of hypotaurine, taurine, cysteinesulfinic acid, and cysteic acid were 3.8, 4.4, 9.7, and 12.9 min, respectively. Detection is fluorometric with o-phthalaldehyde.

Assay for Purified Cysteine Dioxygenase Activity. After purification Step 3, protein A is required to protect irreversible inactivation during assay. Assay conditions are the same for the crude homogenate except for the addition of protein A and 0.5 mg of bovine serum albumin. Partially purified protein A at 0.2 μg is sufficient for stabilizing 10 units of cysteine disoxygenase. Protein A does not participate in activation.

The Oxygen Electrode Method for Assay of Purified Enzyme Activity. Activity is easy to estimate with an oxygen electrode (model K-ICT-C, Gilson Co.) for purified enzyme.[24] With this method, the activation procedure should be performed separately. The activation mixture consists of 1 μmol L-cysteine, 10 μmol glycine–NaOH at pH 9.0, 50 μg bovine serum albumin and 10–30 units of enzyme in a total volume of 0.1 ml. Activation is performed as mentioned above. The reaction mixture includes 1.5 μmol of substrate, 150 μmol of the glycine buffer, an appropriate amount of protein A, and 0.5 mg of bovine serum albumin in total volume of 1.5 ml. An aliquot, 10 μl, of activated enzyme solution is added with a microsyringe to the electrode reaction cell mixture to initiate the reaction. A higher concentration of substrate interferes with the assay because of much higher oxygen consumption due to autoxidation of substrate. The oxygen consumption is determined in a closed system. Enzyme activity is

---

[23] J. R. Spies, this series, Vol. 3.
[24] Y. Hosokawa, O. Fujii, S. Nasu, and K. Yamaguchi, *Sulfur-Containing Amino Acids* **3**, 237 (1980).

calculated from the rate of oxygen consumption. This method is available for measuring protein A activity (see below).

Properties of Cysteine Dioxygenase

Homogeneity and Molecular Properties. The enzyme from Step 7 is homogeneous by polyacrylamide gel electrophoresis. Its isoelectric point, determined by isoelectrofocusing, is at pH 5.5. A single protein band is seen in sodium dodecyl sulfate (SDS)–gel electrophoresis for which an M_r of 22,500 has been calculated; an estimate from gel filtration is in good agreement, suggesting that the enzyme is a single peptide chain. The enzyme is inhibited by chelating agents having high affinity for iron but inhibition is not observed after activation. From atomic absorption spectrophotometric analysis, 0.8–0.9 g atom of iron are present per 22,500 g of purified protein. There is no heme prosthestic group.

Catalytic Properties. The enzyme is highly specific for L-cysteine which has a K_m of 0.45 mM. The stoichiometric relation between oxygen consumed and cysteine sulfinate produced has been obtained using an oxygen electrode. Neither L-cystine, D-cysteine, carboxymethyl-L-cysteine, carboxyethyl-L-cysteine, S-methyl-L-cysteine, N-acetyl-L-cysteine, DL-homocysteine, nor cysteamine serves as substrate. These compounds, however, exhibit activation and inhibition on cysteine dioxygenase, as shown in Table II. In the activation and inhibition experiments, preactivation is carried out separately. An aliquot of the activated enzyme solution is added to the assay mixture, followed by the aerobic incubation under standard conditions. The activator and inhibitor to be

TABLE II
ACTIVATION AND INHIBITION OF CYSTEINE DIOXYGENASE ACTIVITY BY L-CYSTEINE-RELATED COMPOUNDS

Substance	Activation (relative activity, %)		Inhibition (%)	
	10 mM	100 mM	1 mM	10 mM
L-Cysteine	100	—	—	0
Carboxymethyl-L-cysteine	89	—	37	—
Carboxyethyl-L-cysteine	92	—	53	—
S-Methyl-L-cysteine	35	—	34	—
D-Cysteine	14	—	47	—
Cysteamine	—	53	—	39
N-Acetyl-L-cysteine	—	24	—	35
DL-Homocysteine	—	16	—	87

examined are added to the activation mixture or to the assay mixture, respectively.

Purification of Protein A

Step 1: Extract. The preparation obtained in Step 1 of the enzyme is used as starting material for protein A.

Step 2: Acetone. The extract is treated with acetone at $-20°$ to give a concentration of 42% (v/v). After standing for 15 min, the precipitate is collected by centrifugation and the pellet dissolved in a minimum volume of 10 mM sodium acetate at pH 5.5. The pH of solution is adjusted at pH 5.5 with 20% acetic acid. After standing for 10 min, denatured protein is removed by centrifugation.

Step 3: CM-Cellulose. The clear supernatant is applied to a CM-cellulose column (Whatman CM-52; 2.5 × 20 cm) equilibrated with the same acetate buffer. After washing with 2 liters of 10 mM sodium acetate containing 0.05 M NaCl at pH 5.5, elution with a linear salt gradient of 600 ml of 0.05–0.8 M NaCl in the equilibration buffer is carried out at a flow rate of 30 ml/hr, collecting 6-ml fractions. Those fractions with stabilizing activity (46–54) are collected. After the addition of 2% of the volume of 0.5 M sodium phosphate at pH 6.8, the pH of the solution is adjusted to 6.8 with 2 N NaOH.

Step 4: Sephacryl S-300. The solution is concentrated to about 3 ml by ultrafiltration with an Amicon PM30 membrane, and the denatured protein is removed by centrifugation. The enzyme solution is applied to a Sephacryl S-300 column (2.5 × 100 cm) equilibrated with 5 mM sodium phosphate at pH 6.8, and the column is eluted with same buffer at a flow rate of 25 ml/hr. Fractions of 5 ml are collected. Active fractions (54–58) are pooled.

Step 5: Hydroxylapatite. The solution from Step 4 is applied to a hydroxylapatite column (1 × 5 cm) equilibrated with 5 mM sodium phosphate at pH 6.8. After washing with 500 ml of the phosphate buffer, elution with a linear gradient of 100 ml of 5–250 mM of sodium phosphate at pH 6.8, is carried out at a flow rate of 12 ml/hr, collecting fractions of 3 ml. The active fractions appear in tubes 41 through 45. The preparation can be stored for at least 2 months in the presence of 1 mg/ml bovine serum albumin at $-45°$ without significant loss of stabilizing activity.

Molecular Properties and Specificity of Protein A

The preparation obtained in Step 5 is not homogeneous by acrylamide gel electrophoresis, and small amounts of nonactive proteins contaminate. An M_r of 78,000 has been estimated with a Sephadex G-200 column,

TABLE III
PURIFICATION OF PROTEIN A

Step	Total volume (ml)	Total protein (mg)	Specific activity: $t_{1/2}$ prolonged per µg protein (sec)
1. Extract	785	13,000	9.5
2. Acetone	50	3,120	23.6
3. CM-cellulose	55	52.1	472
4. Sephacryl S-300	25	3.23	2670
5. Hydroxylapatite	15	0.95	7100

and 25,000 by SDS–gel electrophoresis. The isoelectric point, determined by the electrofocusing, is 7.30. Protein A cannot be substituted for other proteins which have been examined.

Assay of Protein A Activity[24]

The oxygen electrode method, noted above, is most suitable for assay of the stabilizing activity of protein A, since the reaction velocity is readily calculable from the slope of oxygen consumption versus the reaction time. If protein A is omitted from the assay mixture, the slope of cumulative oxygen consumption (P) is nonlinear from the beginning of the reaction. Thus, the instantaneous reaction velocity ($\Delta P/\text{sec} = V_t$) at the given time ($t$, seconds) decreases during the course of assay. The instantaneous reaction velocity should obey an equation of the form $V_t = V_0 e^{-kt}$. The true initial velocity (V_0) and the time to reach half of V_0 ($t_{1/2}$) can be obtained from the equation. The $t_{1/2}$ value (in seconds) and V_0 are proportionally increased as a function of protein A and enzyme volume, respectively. However, $t_{1/2}$ and V_0 values are not changed by the concentrations of enzyme and protein A, respectively. Accordingly, the specific stabilizing activity of protein A can be expressed as the $t_{1/2}$ value prolonged per microgram of protein A. The $t_{1/2}$ value in the absence of protein A is about 75 sec. A summary of purification of protein A is given in Table III, which uses this parameter.

Distribution of Cysteine Dioxygenase and Protein A

Cysteine dioxygenase activity is found in the liver of many mammals in the following order of decreasing activity: mouse, rat, pig, cow, dog, rabbit.[25] In contrast, protein A is widely distributed in tissues such as liver, brain, heart, kidney, and spleen of rat.[25-27]

NAD-Dependent Cysteine Dioxygenase Activity in Extrahepatic Tissues

No cysteine sulfinate-forming enzyme having the novel properties of purified cysteine dioxygenase of mammalian liver, noted above, has been found in extrahepatic tissues. NAD-dependent cysteinesulfinic acid-forming enzyme activity, however, has been found not only in liver but also in some extrahepatic tissues such as brain,[10,14] retina,[15] reproductive tissues,[16] spleen, stomach, kidney, heart, and intestine,[10] although NAD may play a role as an activator or stabilizer of enzyme but not as hydrogen carrier[6-9] and these activities are much lower than the hepatic NAD-independent cysteine dioxygenase activity. The enzyme activities of these tissues are subjected to neither activation by anaerobic incubation with L-cysteine nor stabilization by protein A, and the subcellular localization of extrahepatic tissue is quite different; hepatic cysteine dioxygenase is found exclusively in the cytosol, but the brain enzyme is associated with a particulate fraction.[14]

Assay of NAD-Dependent Enzyme Activity in Extrahepatic Tissue. Extracts are prepared as noted in Assay Procedure of Homogenate. The standard assay mixture consisted of 20 μmol (0.4–0.8 μCi) L-[^{35}S]cysteine, 4 μmol NAD, 0.5 μmol $Fe_2(NH_4)_2(SO_4)_2$, 10 μmol hydroxylamine, 100 μmol of the phosphate buffer at pH 6.8, and enzyme in a final volume of 2 ml. The solutions of cysteine hydrochloride and hydroxylamine are neutralized with NaOH just before use. The mixture is incubated without substrate for 5 min at room temperature before the reaction is initiated by addition of substrate and terminated by the addition of 1 ml of 8% TCA. Incubation is carried out in small 20-ml flasks using air as the gas phase with shaking at 120 strokes/min for 30–60 min at 37°. Controls are carried out in the same manner except for the enzyme solution which is added after TCA. The determination of product is carried out as noted in Assay Procedure of Homogenate. Substrate concentration for the assay of the brain enzyme should be higher than that for other tissues.[14]

---

[25] Y. Hosokawa, N. Kohashi, and K. Yamaguchi, *Sulfur-Containing Amino Acids* **1**, 251 (1980).
[26] K. Yamaguchi, in "The Effects of Taurine on Excitable Tissues" (S. W. Schaffer, S. L. Baskin, and J. J. Kocsis, eds.), p. 5. Spectrum Publications, New York, 1981.
[27] K. Yamaguchi, in "Natural Sulfur Compounds" (D. Cavallini, G. E. Gaull, and V. Zappia, eds.), p. 175. Plenum, New York, 1980.

[68] Cysteinesulfinate Decarboxylase

By CATHERINE L. WEINSTEIN and OWEN W. GRIFFITH

Introduction[1]

$$\begin{array}{c} COO^- \\ |\\ ^+NH_3-C-H \\ |\\ CH_2 \\ |\\ SO_2^- \end{array} \rightarrow \begin{array}{c} ^+NH_3 \\ |\\ CH_2 \\ |\\ CH_2 \\ |\\ SO_2^- \end{array} + CO_2$$

L-Cysteinesulfinate decarboxylase 3-sulfino-L-alanine carboxylyase, EC 4.1.1.29 catalyzes the decarboxylation of cysteinesulfinate to form hypotaurine, a product efficiently oxidized to taurine *in vivo*. Cysteinesulfinate, the major oxidative metabolite of cysteine in most mammalian species, is also transaminated by aspartate aminotransferase to form β-sulfinylpyruvate, an unstable intermediate that breaks down to pyruvate and sulfite. Sulfite is oxidized to sulfate by sulfite oxidase. The *in vivo* partitioning of cysteine sulfur between taurine and sulfate formation is controlled by the relative activities of cysteinesulfinate decarboxylase and aspartate aminotransferase and by the access of cysteinesulfinate to those enzymes.[2-4] Cysteinesulfinate decarboxylase been purified to homogeneity from bovine brain by Wu,[5] and has been purified extensively from rat liver by several investigators.[3,6,7] The procedure described here yields enzyme with a specific activity of 10–12 μmol min^{-1} mg^{-1}, values at least 3-fold higher than previously reported for the liver enzyme.[3,6,7]

---

[1] Studies from the authors' laboratory were supported by the National Institutes of Health, Grant AM26912. O.W.G. is a recipient of an Irma T. Hirschl Career-Scientist award. We thank Michael Hayward, Ernest Campbell, and Bryan Wong for excellent technical assistance.

[2] O. W. Griffith, this volume [63].

[3] O. W. Griffith, *J. Biol. Chem.* **258**, 1591 (1983).

[4] K. Yamaguchi, S. Sakakibara, J. Asamizu, and I. Ueda, *Biochim. Biophys. Acta* **297**, 48 (1973).

[5] J. Y. Wu, *Proc. Natl. Acad. Sci. U.S.A.* **79**, 4270 (1982).

[6] Y. C. Lin, R. H. Demeio, and R. M. Metrione, *Biochim. Biophys. Acta* **250**, 558 (1971).

[7] M. C. Gion-Rain, C. Portemer, and F. Chatagner, *Biochim. Biophys. Acta* **384**, 265 (1975).

Assay

Cysteinesulfinate decarboxylase activity has been determined by quantitating CO_2 formation manometrically[6,8,9] or radiometrically[10,11] and by quantitating hypotaurine formation radiometrically.[3,12] The last method, using [^{35}S]cysteinesulfinate, is most convenient and is described below. Procedures for the synthesis of radiolabeled cysteinesulfinate are given elsewhere.[3,13–15]

Reagents

Buffer: 1.0 M potassium phosphate at pH 6.8
Pyridoxal 5′-phosphate, 0.1 M, adjusted to pH 6.8 with KOH
Dithiothreitol, 1.0 M
L-[^{35}S]Cysteinesulfinic acid, 50 mM, adjusted to pH 6.8 with dilute KOH, 0.04 μCi/μmol

Procedure. A working solution consisting of 3 ml of buffer, 50 μl of pyridoxal 5′-phosphate, 125 μl of dithiothreitol, and water to 25 ml is prepared daily and stored on ice. In a 1.5-ml capped, conical, plastic vial is placed 100 μl of working solution and up to 50 μl of the enzyme solution to be assayed. The volume is brought to 150 μl with water, and the mixed solution is kept at 37° for 10–30 min (typically 15 min) to activate the enzyme. L-[^{35}S]Cysteinesulfinate (50 μl) is added to initiate the reaction and incubation is continued at 37°. After 15 min, the reaction is stopped by placing the vials in a boiling water bath for 2 min, after which the vials are chilled on ice and centrifuged. A portion of the supernatant fluid (150 μl) is applied to small columns (0.5 × 7 cm) of Dowex 1-X8 (200–400 mesh, acetate form). The resin is washed with 5.0 ml of 0.1 M acetic acid, and the eluent, which contains [^{35}S]hypotaurine, is collected and mixed.

A 2-ml portion of the eluent is taken for liquid scintillation counting. Radioactivity due to background radiation and unbound [^{35}S]cysteinesulfinate (<0.02%) is determined from reaction mixtures lacking enzyme. Product formation is calculated on the basis of the initial cysteinesulfinate specific activity, which is determined by counting an aliquot of the cysteinesulfinate stock solution in liquid scintillation cocktail containing 2 ml of 0.1 M acetic acid. A unit of cysteinesulfinate decarboxylase activity is

---

[8] D. B. Hope, *Biochemistry* **59**, 497 (1955).
[9] M. C. Gion-Rain and F. Chatagner, *Biochim. Biophys Acta* **276**, 272 (1972).
[10] J. G. Jacobsen, L. Thomas, and L. H. Smith, *Biochim. Biophys. Acta* **85**, 103 (1964).
[11] K. M. Daniels and M. H. Stipanuk, *J. Nutr.* **112**, 2130 (1982).
[12] J. Y. Wu, L. G. Moss, and M. S. Chen, *Neurochem. Res.* **4**, 201 (1979).
[13] R. Emiliozzi and L. Pichart, *Bull. Soc. Chim. Fr.* **11–12**, 1887 (1959).
[14] R. M. Spears and D. L. Martin, *J. Labelled Compd., Radiopharm.* **18**, 1055 (1980).
[15] O. W. Griffith and C. L. Weinstein, this volume [49].

the amount of enzyme that catalyzes the formation of 1 μmol of hypotaurine per minute under the conditions given.

The assay is linear with time and enzyme concentration up to at least 50% substrate conversion. The initial incubation, without substrate in the presence of dithiothreitol, has little effect on freshly isolated enzyme but increases the activity of enzyme aged several weeks by as much as 3-fold. Although activity is reportedly increased by assaying under nitrogen,[9] we have not observed this effect; initial incubation with dithiothreitol may accomplish the same activation. Solutions of [^{35}S]cysteinesulfinate slowly oxidize to [^{35}S]cysteinesulfonate during storage at 4° or −20°. The purity of the reagent can be checked by thin-layer chromatography on cellulose plates in *tert*-butanol : formic acid : water (75 : 10 : 15) (R_f cysteinesulfinate, 0.2; R_f cysteinesulfonate, 0.1). Since cysteinesulfonate is an alternative substrate of the decarboxylase, an assay similar to that described using commercially available [^{35}S]cysteinesulfonate is possible; activity with cysteinesulfonate is about 14% of that with cysteinesulfinate under the conditions given.

Purification Procedure

General. Hepatic cysteinesulfinate decarboxylase activity is reported to be higher in male rats than female rats[16,17] and, in contrast to that of cysteine dioxygenase, is decreased by increased sulfur amino acid intake.[11] In the procedure described, fresh livers were obtained from male Sprague–Dawley rats maintained on Purina laboratory chow; frozen livers from commercial suppliers yield less activity per weight and lower final specific activities. All procedures are carried out at 0–5°; centrifugations are at 16,000 g. Protein is monitored by its absorption at 280 nm during chromatographic procedures and by the Hartree modified Lowry procedure[18] in bulk or pooled fractions. The results of a typical isolation are summarized in Table I. The method is based in part on previous work.[3,6,19]

Step 1: Homogenization. Livers from 30 decapitated and exsanguinated rats (480 g wet weight) are minced with scissors and homogenized in two volumes (w/v) of isolation medium (70 mM sucrose, 220 mM mannitol, 2 mM potassium HEPES at pH 7.4) with four passes of a motor driven Potter–Elvehjem tissue grinder. The resulting mixture is diluted with iso-

---

[16] J. A. Worden and M. H. Stipanuk, *Comp. Biochem. Physiol. B* **82B,** 233 (1985).
[17] C. Loriette and F. Chatagner, *Experientia* **34,** 981 (1978).
[18] E. F. Hartree, *Anal. Biochem.* **48,** 422 (1972).
[19] B. Sörbo and T. Heyman, *Biochim. Biophys. Acta* **23,** 624 (1957).

TABLE I
Purification of Cysteinesulfinate Decarboxylase

Step	Volume (ml)	Total protein (mg)	Total enzyme (units)[a]	Specific activity (units/mg)
Crude homogenate	4,000	71,600	1,250	0.017
7000 g Supernatant	3,600	36,000	1,120	0.029
pH 5.6 Supernatant	3,600	13,000	961	0.074
50% Ammonium sulfate pellet	107	2,560	877	0.343
Phenyl-Sepharose pool	244	317	789	2.5
Chromatofocusing pool				
Major	104	104	489	4.7
Minor	45	189	32	0.17
Ultrogel AcA 44 pool				
Major	44	35	351	10.0
Minor[b]	50	4	12	3.0

[a] A unit is defined as 1 μmol product formed per minute.
[b] The yield and specific activity of the minor isozyme is 2- to 3-fold higher immediately following chromatography. Specific activities greater than 3 have, however, been difficult to maintain in the pooled and concentrated fractions.

lation medium to give a 10% (w/v) homogenate. The crude homogenate is centrifuged for 15 min, and the pellet discarded.

Step 2: Acid Precipitation. The supernatant liquid is cooled to 2° in an ice–salt bath and, with good magnetic stirring, is slowly adjusted to pH 5.6 by addition of 2 M acetic acid. Care is taken to frequently wash the pH electrode since protein precipitating on the tip will otherwise invalidate the readings. The solution is centrifuged for 1 hr, and the pellet is discarded.

Step 3: Ammonium Sulfate. Solid ammonium sulfate (20 g/100 ml) is added to the supernatant solution to yield 35% saturation. As the salt dissolves, the pH is adjusted to 6.8 with 1 M ammonium hydroxide. The resulting mixture is stirred for 30 min and centrifuged for 30 min. Additional ammonium sulfate (9 g/100 ml) is added to the supernatant fluid to yield 50% saturation. After stirring and centrifuging as before, the pellet is dissolved in 100 ml of buffer I (10 mM potassium phosphate buffer at pH 6.8, containing 0.1 mM EDTA, 0.1 mM pyridoxal phosphate, and 1 mM dithiothreitol). The resulting solution is clarified by brief centrifugation.

Step 4: Phenyl-Sepharose. The solution from Step 3 is mixed with 80 ml of Phenyl-Sepharose CL-4B (Pharmacia) that had been equilibrated with buffer I containing 11.4 g/100 ml of ammonium sulfate. The resulting brown slurry is added to the top of a column (2.5 cm i.d.) containing 320 ml of similarly equilibrated resin. The resin is washed sequentially with 1 liter of buffer I containing 11.4 g/100 ml of ammonium sulfate, 2 liters of buffer I containing 2.8 g/100 ml of ammonium sulfate, and 1.5 liters of buffer I containing 1.4 g/100 ml of ammonium sulfate. The enzyme is eluted by the last buffer. Appropriate fractions are pooled, and solid ammonium sulfate is added to 60% saturation (39 g/100 ml). After stirring and centrifuging, the pellet is dissolved in 100 ml of buffer I and dialyzed overnight against that buffer. The Phenyl-Sepharose resin is regenerated using sequential washes with water, ethanol, and butanol in various combinations as recommended by the manufacturer. Although the resin remains discolored, it may be reused without difficulty.

Step 5: Chromatofocusing. The dialyzed enzyme solution is adjusted to pH 7.2 with KOH and is applied to a column (1.5 × 25 cm) of PBE 94 chromatofocusing resin (Pharmacia) previously equilibrated with buffer II (25 mM imidazole·HCl at pH 7.4 containing 1 mM dithiothreitol). The resin is washed with 100 ml of buffer II followed by 800 ml of a 1–8 dilution of Polybuffer 74 (Pharmacia) adjusted to pH 4 with HCl. Fractions of 10 ml are collected at a flow rate of 80 ml/hr. A major and a minor peak of enzyme elute at pH values of 6.1 and 5.6, respectively; they are pooled separately.

Step 6: Gel Filtration. Fractions containing the major cysteinesulfinate decarboxylase chromatofocusing isozyme are concentrated by adding ammonium sulfate to 60% saturation as described above; the resulting pellet is redissolved in 4 ml of buffer I. That solution is applied to a column (2.5 × 120 cm) of Ultrogel AcA 44 (Pharmacia) which is equilibrated and eluted by reverse flow with buffer I. The enzyme activity coelutes with the major protein peak. Fractions having substantial activity are pooled and concentrated by addition of ammonium sulfate to 60% saturation; the resulting pellet is dissolved in 0.5–1.0 ml of buffer I containing 5 mM dithiothreitol and is dialyzed overnight against that buffer. The minor chromatofocusing isozyme is purified similarly. It elutes from the gel filtration column at the same position as the major isozyme. The major and minor isozymes obtained after gel filtration represent 30% of the enzyme activity initially present. The major isozyme is purified 590-fold and is homogeneous by sodium dodecyl sulfate (SDS)–gel electrophoresis.

Enzyme Properties

Stability. At concentrations above 5 mg/ml, purified cysteinesulfinate decarboxylase loses about 30% of its initial activity in 2 months at 4°. In dilute solution it is significantly less stable. The apparent activity loss is greater if the incubation with dithiothreitol is omitted from the assay protocol. The enzyme appears to be stable for months when stored at $-20°$; freezing in small aliquots avoids losses due to repeated freezing and thawing.

Physical Properties. SDS–gel electrophoresis of the purified enzyme indicates a subunit molecular weight of 52,000, which is in agreement with the value previously reported.[7] The M_r of the native enzyme has been the subject of considerable debate. Whereas gel filtration chromatography using various resins indicates values of 60,000–80,000,[6,7] sedimentation velocity experiments, using a Stokes radius of 3.4 determined by gel filtration, indicates an apparent M_r of 96,000.[7] More recently, samples of either the major or minor chromatofocusing isozymes have been cross-linked by treatment with dimethyl suberimidate and subjected to SDS–gel electrophoresis; both isozymes yield dimers of 52,000 subunits.

Although the purified major isozyme is homogeneous by SDS–gel electrophoresis, it exhibits considerable heterogeneity by gel electrofocusing. Three major and two minor bands are visualized by Coomassie Blue staining with p*I* values ranging between 5.0 and 5.6. All of the major bands represent active enzyme; the minor bands may also represent enzyme but were too faint or poorly resolved to be definitively assayed. The purified minor chromatofocusing isozyme contains 8–10 electrofocusing bands with p*I* values ranging from 4.3 to 5.6; at least 2 of the bands near pH 4.9 have cysteinesulfinate decarboxylase activity. The relationship of the liver isozymes to the brain isozyme(s) is not clear. Preliminary studies using crude homogenate from rat brain indicate 1 to 2 bands of cysteinesulfinate decarboxylase activity with p*I* values in the range of 5.0–5.6. It is notable that the homogeneous brain enzyme has a specific activity of 0.072 units mg^{-1},[5] a value much lower than that observed for the liver isozyme.

Cysteinesulfinate decarboxylase contains pyridoxal 5′-phosphate as a cofactor[8] and exhibits an absorbance maximum at 435 nm characteristic of a protein–pyridoxal phosphate Schiff base.[7] Enzyme purified by the procedure given here but using buffers not containing added coenzyme contains about 0.85 mol of pyridoxal phosphate per 52,000 subunit.

Catalytic Properties. Cysteinesulfinate decarboxylase has a relatively flat pH rate profile with optimal activity between pH 6.8 and 7.6 in phosphate buffers. Activity decreases to 50% at pH values of about 6.4 and 8.2.[10] The major chromatofocusing isozyme exhibits a K_m for cysteinesulfinate of 0.12 mM, a value in close agreement with those reported previously for total rat liver enzyme.[6,7,10] The purified minor isozyme exhibits the same K_m within experimental error. Both chromatofocusing isozymes also decarboxylate cysteinesulfonate (K_m 0.63 mM) with a V_{max}

that is 9–10% of that observed with cysteinesulfinate. At concentrations of 25 mM, L-aspartate, L-glutamate, DL-2-amino-3-phosphonopropionic acid, and DL-homocysteic acid are decarboxylated by the major isozyme at rates which are 0.3%, 0.01%, 0.06%, and 0.001%, respectively, of the rate of cysteinesulfinate decarboxylation. The substrate specificity of the minor isozyme is similar. The enzyme is not inhibited by hypotaurine or taurine at concentrations up to at least 25 mM.

Species and Tissue Distribution. There is a large variation in cysteinesulfinate decarboxylase activity among species[10,16]; activity is highest in rat, mouse, and dog liver, is low in adult human liver, and is nearly absent in cat liver. Within some species enzyme activity also varies with the age and sex of the animal.[16,17,20] Activity is, for example, very low in newborn human liver, a finding suggesting that the human infant may have a dietary requirement for taurine. It is notable in this respect that taurine is present in high concentration in human milk but is low or absent in cow's milk and some formula diets.[21] Cysteinesulfinate decarboxylase activity is highest in liver in all species examined. Moderate enzyme activity is found in kidney; lesser amounts occur in brain, skeletal muscle, heart, spinal cord, lung, and spleen.[10,12]

[20] R. M. DiGiorgio, S. Macaione, and G. DeLuca, *Ital. J. Biochem.* **27**, 83 (1978).
[21] D. K. Rassin, J. A. Sturman, and G. E. Gaull, *Early Hum. Dev.* **2**, 1 (1978).

[69] Cysteamine Dioxygenase

By RUSSELL B. RICHERSON and DANIEL M. ZIEGLER

$$H_2NCH_2CH_2SH + O_2 \rightarrow H_2NCH_2CH_2SO_2^- + H^+$$

Cysteamine dioxygenase (EC 1.13.11.19) catalyzes the oxygenation of the aminothiol, cysteamine, to hypotaurine. The enzyme is present in several tissues and was first purified to homogeneity from equine kidney by Cavallini and associates.[1] The kidney dioxygenase is a metalloprotein consisting of two subunits similar in size (M_r 52,000).[2] Changes in the EPR signal at $g = 4.3$ in the presence of substrate suggest that enzyme-bound

[1] D. Cavallini, R. Scandurra, and S. Dupré, this series, Vol. 17, p. 479.
[2] D. Cavallini, C. Cannella, G. Federici, S. Dupré, A. Fiore, and E. Del Grosso, *Eur. J. Biochem.* **16**, 537 (1970).

iron has a functional role in catalysis.[3] During the course of purification, activity becomes dependent on catalytic amounts of hydroxylamine or other reductants with an E_0' higher than +0.011 V.[4]

Assay Method

Principle. The activity in liver cytosol is readily determined by measuring cysteamine-dependent oxygen uptake in essentially metal-free media. The autoxidation of cysteamine below pH 8 is quite slow, and the dioxygenase is the only enzyme present in the cytosol fraction that appears capable of catalyzing oxygenation of cysteamine.

Reagents

1.0 M Potassium phosphate at pH 7.8
20 mM Ethylenediaminetetraacetic acid (EDTA)
5 mM Hydroxylamine hydrochloride
10 mM Cysteamine hydrochloride

Procedure. Oxygen uptake is measured in a 2-ml, thermostatted (37°) vessel fitted with a Clark-type electrode. The basic assay medium contains 100 mM phosphate, 1 mM EDTA, 0.25 mM hydroxylamine, 0.25 mM cysteamine, and enzyme. All of the compounds except cysteamine are added to the reaction vessel and after 3–5 min of temperature equilibration, the reaction is started by adding 10 μl of the cysteamine solution through the capillary access port. Oxygen uptake is recorded for an additional 2–3 min. The measurement is repeated without enzyme to test for nonenzymatic oxidation of cysteamine which, in the presence of EDTA, should be essentially zero. Activity is expressed as micromoles cysteamine-dependent oxygen uptake per minute per milligram protein.

Purification Procedure

All operations are carried out at 0–5° unless specified otherwise.

Step 1: Preparation of Liver Cytosol. Hog livers obtained from a local slaughterhouse are collected and transported, on ice, to the laboratory as described previously.[5] Upon arrival in the laboratory the liver is minced with an electric meat grinder and 800 g is homogenized in 2 liters of 0.25 M sucrose. The homogenate is centrifuged at 2,000 rpm for 20 min, and the

---

[3] G. Rotilio, G. Federici, L. Calabrese, M. Costa, and D. Cavallini, *J. Biol. Chem.* **245**, 6235 (1970).
[4] D. Cavallini, C. De Marco, R. Scandurra, S. Dupré, and T. Graziani, *J. Biol. Chem.* **241**, 3189 (1966).
[5] D. M. Ziegler and L. L. Poulsen, this series, Vol. 52, p. 142.

turbid supernatant fraction is decanted through cheese cloth. The pH of the supernatant is lowered to 5.5 by adding 1 M phosphoric or 1 M acetic acid and insoluble material removed by centrifugation at 14,000 rpm for 20 min.

Step 2: PEG Fractionation. The clear supernatant fluid, about 1700 ml, is fractionated with polyethylene glycol (PEG 8000) by adding 50% (w/v) PEG in water. The fraction precipitating between 5 and 8% PEG is collected by centrifugation and dissolved in 10 mM potassium phosphate at pH 7.5. This fraction can be stored overnight on ice with little or no loss in activity.

Step 3: DEAE–Cellulose Chromatography. The PEG fraction is transferred to a column (2.5 × 25 cm) of DEAE-cellulose (Whatman DE-52) equilibrated with 10 mM potassium phosphate at pH 7.5. After washing the loaded column with about 2 liters of the same buffer (or until absorbance of the effluent at 280 nm is no greater than 0.05 cm^{-1}), the dioxygenase is eluted by increasing the salt concentration with a nonlinear gradient. The gradient is formed by dropwise addition of 20 mM potassium phosphate at pH 7.5 containing 0.1 M KCl, into a closed mixing chamber containing 300 ml of 10 mM phosphate at pH 7.5. Fractions of 15 ml are collected and assayed for activity. The dioxygenase is usually recovered in fractions eluting between 310 and 400 ml after starting the gradient.

Step 4: Salt Fractionation. Active fractions from the preceding step are pooled and fractionated by adding solid, "enzyme grade" ammonium sulfate. The fraction precipitating between 35 and 55% saturation is collected by centrifugation and redissolved in 10 mM potassium phosphate at pH 7.5.

Step 5: Gel-Filtration Chromatography. The ammonium sulfate fraction is transferred to a Sephadex G-150 column (2.5 × 55 cm) equilibrated with 10 mM potassium phosphate at pH 7.5, and the dioxygenase is then eluted with the same buffer at a flow rate of 0.5 ml/min. The fractions between 18 and 30 ml, containing the highest activity, are collected and concentrated by pressure dialysis to about 6–10 mg protein/ml.

Step 6: Isoelectric Focusing. An aliquot (1.5 ml) of the preceding fraction, mixed with IEF-Sephadex containing 2% ampholytes (pH range 3–10) is loaded onto a focusing plate (3.5 × 22.5 × 0.3 cm). The plate is developed at a constant power of 2 W for 8 hr in a Pharmacia FBE-3000 unit. The dioxygenase focuses at pH 4.7, and 3-mm segments of the gel in this region are removed and suspended in 10 mM of the phosphate buffer. The active fractions from 5 to 8 electrofocusing runs are combined, ampholytes removed by gel filtration on Sephadex G-10, and the purified enzyme concentrated by pressure dialysis. The purified dioxygenase is

TABLE I
Purification of Cysteamine Dioxygenase[a]

Step	Volume	Total activity (μmol/min)	Total protein (mg)	Specific activity (μmol min^{-1} mg^{-1})
1. Cytosol	1700	232	58,500	0.004
2. PEG fraction	650	140	18,200	0.008
3. DEAE–Cellulose	365	70	1,100	0.064
4. (NH$_4$)$_2$SO$_4$ (35–55%)	15	55	600	0.091
5. Sephadex G-150	33	50	190	0.25
6. Isoelectric focusing	36	46	16	2.80

[a] Prepared from 800 g hog liver.

quite stable and can be stored at $-20°$ for up to 3 years with little or no loss in activity. The yield and purification at each step are summarized in Table I.

Properties

The active fractions after isoelectric focusing are of greater than 93% purity as judged by electrophoresis on polyacrylamide gels in the presence of sodium dodecyl sulfate (SDS). The hog liver enzyme, like the equine kidney dioxygenase,[2] appears to be a dimer of similar subunits, although the subunit M_r of the liver enzyme may be somewhat larger (64,000 versus 52,000). However, in all other respects the hog liver enzyme appears similar to the equine dioxygenase,[1] and only those properties relevant to the use of the enzyme for measuring cysteamine (this volume [25]) will be considered.

Substrate Specificity and Inhibitors. Of the physiological compounds tested, only cysteamine is oxidized at a significant rate, although, below 10 μM, pantetheine exhibits slight substrate activity (less than 3% of the cysteamine-dependent rate). However, the pantetheine-dependent rate decreases with increasing pantetheine concentration and drops essentially to zero above 100 μM. At this concentration, pantetheine also inhibits cysteamine oxidation; pantetheine appears to be a better inhibitor than substrate for the dioxygenase. Other thiols—cysteine, glutathione, coenzyme A, and lipoic acid—or the disulfides—pantethine, cystamine, glutathione disulfide, and cystine—are not substrates nor do they inhibit the oxidation of cysteamine at concentrations approaching saturation in the assay medium. Taurine and hypotaurine also do not affect activity, and it would appear that physiological compounds do not interfere in the use of

TABLE II
SUBSTRATE SPECIFICITY

Compound	Activity[a]
Cysteamine	2.8
Piperazinylcysteamine	2.1
N,N-Dimethylcysteamine	1.3
Trimethyl(2-mercaptoethyl)ammonium chloride	0.95
2-Mercaptoethanol	0.22
Cysteine methyl ester	0.48

[a] Expressed as micromoles substrate-dependent oxygen uptake per minute per milligram enzyme at 37°, pH 7.8, and at saturating substrate.

the dioxygenase to measure cysteamine in body fluids of animals treated with this aminothiol.

On the other hand, the dioxygenase catalyzes oxidation of many of the synthetic analogs of cysteamine that are used to protect animals against ionizing radiation[6] (Table II). Those that are oxidized at a significant rate would interfere in the estimation of cysteamine in serum from animals pretreated with combinations of cysteamine and synthetic radioprotective agents.

While many of the radioprotective agents are substrates, one of the more commonly used derivatives of cysteamine, mercaptoethylguanidine, is a potent inhibitor of the dioxygenase with a K_i of approximately 7 μM. The inhibitor is noncompetitive with cysteamine and this compound may be a useful, selective inhibitor for this enzyme. The dioxygenase is also quite sensitive to some dithiols. For example, 1,3-dithiopropane and 1,4-dithiobutane inhibit the enzyme-catalyzed oxidation of cysteamine with K_i values close to 20 μM. Whereas both dithiols show limited substrate activity, they are not competitive with cysteamine and appear to inhibit by combining with a site or sites other than the substrate binding site. The more polar dithiothreitol also inhibits but at a much higher concentrations. However, unlike the lipophilic alkyl dithiols, dithiolthreitol is not a substrate for the dioxygenase.

Reaction Stoichiometry and Kinetic Constants. Within the limits of detection, the dioxygenase catalyzes only dioxygenation of cysteamine. The ratio of cysteamine oxidized to oxygen consumed is consistently 1 : 1, and hypotaurine is the only product detected in the enzyme-catalyzed

[6] A. M. Bacq, *in* "Sulfur-Containing Radioprotective Agents" (A. M. Bacq, ed.), p. 1. Pergamon, Oxford, 1975.

reaction. Constants for the reaction, calculated from kinetic data collected at pH 7.8 and 37°, indicate that the K_m for oxygen is near 0.1 mM and that the K_m for cysteamine is somewhat less than 1 μM. However, the latter value was calculated from data collected only at substrate concentrations well above K_m, and the actual constant may be considerably less than 1 μM.

[70] Selenocysteine β-Lyase (Porcine)

By NOBUYOSHI ESAKI and KENJI SODA

$$HSeCH_2CHNH_2COOH \rightarrow CH_3CHNH_2COOH + Se$$

Selenocysteine β-lyase (EC 4.99.1.—) catalyzes specifically the β-elimination reaction of L-selenocysteine to form L-alanine and elemental selenium. The enzyme occurs widely in various mammalian tissues[1] and in the cells of aerobic bacteria.[2] The enzyme has been purified to homogeneity from pig liver[3] and from *Citrobacter freundii*.[4]

Assay Method

Principle. Elemental selenium, inherently produced from selenocysteine, is reduced spontaneously to H$_2$Se with dithiothreitol that is included in the reaction mixture. H$_2$Se is determined with lead acetate.

Reagents

4 mM DL-Selenocystine dissolved in 0.2 M Tricine–NaOH at pH 8.5
100 mM Dithiothreitol
0.2 mM Pyridoxal 5'-phosphate
0.4 mg/ml Bovine serum albumin
5 mM Lead acetate dissolved in 0.1 N HCl

Procedures. The standard assay system consists of 0.25 ml dithiothreitol/Tricine–NaOH solution, 50 μl dithiothreitol, 50 μl pyridoxal 5'-phosphate, 50 μl bovine serum albumin, and enzyme in a final

---

[1] N. Esaki, T. Nakamura, H. Tanaka, T. Suzuki, Y. Morino, and K. Soda, *Biochemistry* **20**, 4492 (1981).
[2] P. Chocat, N. Esaki, T. Nakamura, H. Tanaka, and K. Soda, *J. Bacteriol.* **156**, 455 (1983).
[3] N. Esaki, T. Nakamura, H. Tanaka, and K. Soda, *J. Biol. Chem.* **257**, 4386 (1983).
[4] P. Chocat, N. Esaki, K. Tanizawa, K. Nakamura, H. Tanaka, and K. Soda, *J. Bacteriol.* **163**, 669 (1985); see this volume [86].

volume of 0.5 ml. After incubation at 37° for 20 min, 3.5 ml of lead acetate solution is added to the reaction mixture. In a blank tube, lead acetate is added to the reaction mixture prior to enzyme. Turbidity of the resultant yellowish-brown PbSe colloid is measured at 400 nm within 15 min; turbidity diminishes at a rate of 0.3% per minute. An apparent molar turbidity coefficient of colloidal PbSe at 400 nm is 2.36×10^4 on the basis of determination of H_2Se with 5,5'-dithiobis(2-nitrobenzoic acid) ($\varepsilon = 2.80 \times 10^4$ at pH 7.4).

Definition of Units. One unit of selenocysteine β-lyase is defined as the amount of enzyme that catalyzes the appearance of 1 μmol of H_2Se per minute. Specific activity is expressed as units per milligram protein. Protein is determined by the method of Lowry *et al.*[5] with bovine serum albumin as a standard. For most column fractions, the protein elution patterns are estimated by their absorption at 280 nm.

Pig Liver Selenocysteine β-Lyase

Purification Procedure

All steps are carried out at 4° unless otherwise stated. Potassium phosphate at pH 7.4, containing 20 μM pyridoxal 5'-phosphate and 0.01% 2-mercaptoethanol, is used as the standard buffer. Hydroxyapatite is prepared according to the method of Tiselius *et al.*[6]

Step 1: Extract. Fresh pig livers (6.5 kg) are minced with a meat mincer and subsequently homogenized at full speed for 2 min with a Waring Blendor in 2 volumes of 50 mM standard buffer. The homogenate is centrifuged at 10,000 g for 1 hr.

Step 2: Heat Treatment. The supernatant solution is kept at 50° for 30 min and centrifuged after cooling in ice.

Step 3: Ammonium Sulfate. The supernatant solution (8,030 ml) is brought to 25% saturation with solid ammonium sulfate, and after standing for 20 min the precipitate is removed by centrifugation. Solid ammonium sulfate is slowly added to the supernatant solution to 45% saturation. The precipitate collected by centrifugation is dissolved in 900 ml of 10 mM standard buffer and dialyzed against 4 changes of 10 liters each of the same buffer.

Step 4: DEAE-Cellulose. The dialyzed solution is applied to a DEAE-cellulose column (9.5 × 60 cm), equilibrated with 10 mM standard buffer.

[5] O. H. Lowry, N. J. Rosebrough, A. L. Farr, and R. J. Randall, *J. Biol. Chem.* **193**, 265 (1951).
[6] A. Tiselius, S. Hjerten, and O. Levin, *Arch. Biochem. Biophys.* **65**, 132 (1956).

TABLE I
PURIFICATION OF SELENOCYSTEINE β-LYASE FROM PIG LIVER

Step	Volume (ml)	Total protein (mg)	Total activity (units)	Specific activity (units/mg)
1. Extract	12,400	651,000	13,600	0.021
2. Heat treatment	8,030	271,000	7,300	0.027
3. Ammonium sulfate	1,410	103,000	6,290	0.061
4. DEAE-cellulose	154	5,640	2,980	0.53
5. First hydroxyapatite	28	425	1,360	3.2
6. Second hydroxyapatite	3.8	240	1,210	5.0
7. Sephadex G-200	5.2	27.2	635	23
8. Third hydroxyapatite	2.3	7.6	278	37

After application of protein, the column is washed with 5 liters of the same buffer and then with 5.1 liters of the buffer containing 0.1 M KCl. Fractions containing the enzyme are concentrated by the addition of ammonium sulfate (50% saturation) and dialyzed against 10 mM standard buffer.

Step 5: First Hydroxyapatite. The dialyzed solution is applied to a hydroxyapatite column (8 × 45 cm) equilibrated with 10 mM standard buffer. The enzyme is eluted stepwise with 0.05 and 0.1 M of the buffer at a flow rate of 60 ml/hr. Active fractions are combined and concentrated with ammonium sulfate (50% saturation) and dialyzed against 10 mM buffer.

Step 6: Second Hydroxyapatite. The dialyzed enzyme is rechromatographed on a hydroxyapatite column (2.7 × 25 cm) equilibrated with 50 mM buffer. The enzyme is eluted with 70 mM buffer at a flow rate of 10 ml/hr. The active fractions are combined and concentrated with ammonium sulfate (50% saturation) followed by dialysis against 10 mM buffer.

Step 7: Sephadex G-200. The enzyme solution is applied to a Sephadex G-200 column (2.2 × 130 cm) equilibrated with 10 mM buffer and eluted at a flow rate of 12 ml/hr. Active fractions are combined and concentrated with an Amicon 202 ultrafiltration unit.

Step 8: Third Hydroxyapatite. The enzyme solution is applied to a hydroxyapatite column (1.2 × 13 cm) equilibrated with 10 mM buffer. The enzyme is eluted with 40 mM buffer at a flow rate of 0.4 ml/hr. The active fractions are collected and concentrated by ultrafiltration.

A summary of the purification is presented in Table I.

Properties

Stability. The purified enzyme can be stored at $-20°$ for a few weeks without loss of activity when the protein concentration is greater than 0.5 mg/ml, whereas it is inactivated significantly by freezing in dilute solutions (<60 μg/ml). However, the enzyme is fully stable in the presence of 20% sucrose, 20% glycerol, or 1% crystalline bovine serum albumin under the same conditions. Therefore, the enzyme is routinely stored in deep-freeze ($-20°$) in the presence of 20% sucrose.

Physicochemical Properties. The purified enzyme is homogeneous by the criteria of disk gel electrophoresis and ultracentrifugation. The sedimentation coefficient ($s_{0,w}$) is calculated to be 5.5 S ($20°$; 10 mM potassium phosphate at pH 7.4, containing 20 μM pyridoxal 5'-phosphate, 0.1 M KCl, and 0.01% 2-mercaptoethanol; protein concentration, 2.0 mg). The molecular weight of the enzyme is estimated to be 93,000 (by Sephadex G-200 gel filtration method) and 85,000 $\pm$ 3,000 (by sedimentation equilibrium method, assuming a partial specific volume of 0.74). The enzyme consists of two subunits of identical M_r (48,000). The enzyme exhibits absorption maxima at 278 and 420 nm. At 278 nm ($A_{1\,cm}^{1\%}$), ε was calculated to be 4.80. After incubation with 10 mM hydroxylamine (pH 7.2), followed by dialysis against 3 changes of the standard buffer for 12 hr, the enzyme required pyridoxal 5'-phosphate and did not have an absorption maximum at 420 nm. Activity is about 94% restored by addition of 20 μM pyridoxal 5'-phosphate. The K_m value for pyridoxal 5'-phosphate is estimated to be 0.33 μM.

Effect of pH. Maximum activity is at about pH 9.0 when measured in Tricine–NaOH and glycine–NaOH buffers.

Substrate Specificity. The enzyme acts specifically on selenocysteine to produce alanine and H_2Se. An apparent K_m for L-selenocysteine is 0.83 mM. L-Cysteine behaves as a competitive inhibitor against DL-selenocysteine (K_i of 1 mM). None of the following compounds inhibit the enzyme reaction with 4 mM DL-selenocysteine: 5 mM L-serine, L-alanine, L-homocysteine, L-selenohomocysteine, and H_2Se, and 10 mM glutathione.

[71] Sulfur Amino Acids of Plants: An Overview

By JOHN GIOVANELLI

Role of Sulfur Amino Acid Synthesis by Plants in the Sulfur Cycle

Plants are the most important producers of sulfur amino acids, assimilating an estimated 4.6×10^{11} kg of inorganic sulfur per annum.[1] The crucial role of plants in Nature is illustrated in Fig. 1. Man and other nonruminants require a dietary source of methionine, which they convert ultimately to inorganic sulfate. Plants complete this cycle by reductive assimilation of inorganic sulfate to cysteine and methionine, and thus provide the ultimate source of methionine in most animal diets.

Sulfur Amino Acid Constituents of Plants

This overview focuses primarily on cysteine and methionine, the major end products of sulfate assimilation. These amino acids comprise up to 90% of the total sulfur of most plants,[2,3] and are present predominantly (99% or more) in protein.[3,4] The nonprotein fraction commonly contains glutathione as a major constituent, together with much smaller amounts of sulfur amino acid intermediates involved in protein cysteine and protein methionine biosynthesis (e.g., cysteine, cystathionine, homocysteine, and methionine), and in polyamine synthesis and methyl transfer reactions (e.g., AdoMet,[5] S-adenosylhomocysteine and 5'-methylthioadenosine). Plants are unusual in that the nonprotein fraction may also contain a variety of sulfur amino acids whose metabolic function and biochemistry have not been clearly defined.[6,7] Examples include S-methylcysteine and its γ-glutamyl and sulfoxide derivatives, S-methylmethioninesulfonium,

---

[1] J. W. Anderson, "Sulphur in Biology," p. 23. University Park Press, Baltimore, Maryland, 1978.
[2] W. H. Allaway and J. F. Thompson, *Soil Sci.* **101**, 240 (1966).
[3] J. Giovanelli, S. H. Mudd, and A. H. Datko, *in* "The Biochemistry of Plants" (B. J. Miflin, ed.), Vol. 5, p. 454. Academic Press, New York, 1980.
[4] A. H. Datko and S. H. Mudd, *Plant Physiol.* **75**, 474 (1984).
[5] Abbreviations: AdoMet, S-adenosylmethionine; ACC, 1-aminocyclopropane-1-carboxylic acid.
[6] G. A. Rosenthal, "Plant Nonprotein Amino and Imino Acids," p. 236. Academic Press, New York, 1982.
[7] S. Hunt, *in* "Chemistry and Biochemistry of the Amino Acids." (G. C. Barrett, ed.), p. 111. Chapman & Hall, New York, 1985.

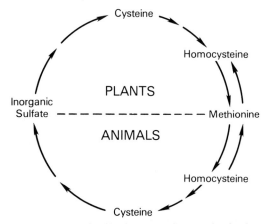

FIG. 1. Cycle of sulfur between plants and animals.

homomethionine, and djenkolic acid. In a number of plants, these "unusual" amino acids make an appreciable contribution to the total nonprotein amino acid fraction. For example, γ-glutamyl-S-methylcysteine accounts for over 40% of the free amino nitrogen of kidney beans.[8] Detailed analyses of sulfur amino acids have been reported for the lower plant *Chlorella*[3] and the higher plant *Lemna*.[3,4]

Cysteine

Synthesis of cysteine by Eqs. (1) and (2) provides the major portal for organic reduced sulfur compounds in plants (Fig. 1). Serine acetyltransferase (EC 2.3.1.30) has been partially purified from plants, and was isolated free of cysteine synthase.[9] By contrast, serine acetyltransferase from *Salmonella typhimurium* was isolated predominantly as a complex with cysteine synthase.[10]

$$\text{L-Serine} + \text{acetyl-CoA} \xrightarrow{\text{serine acetyltransferase}} O\text{-acetylserine} + \text{CoA} \quad (1)$$

$$O\text{-Acetylserine} + H_2S \xrightarrow{\text{cysteine synthase}} \text{L-cysteine} + \text{acetate} \quad (2)$$

Cysteine synthase (*O*-acetylserine(thiol)-lyase, EC 4.2.99.8; *O*-acetylserine sulfhydrase) has been highly purified from a variety of plant sources and its properties determined.[3,11] Although definitive proof is lacking, available evidence suggests that *O*-acetylserine is the physiological substrate for cysteine synthase in plants.[3] The original report by Brug-

[8] J. F. Thompson, C. J. Morris, W. A. Arnold, and D. H. Turner, *in* "Amino Acid Pools" (J. T. Holden, ed.), p. 54. Am. Elsevier, New York, 1962.
[9] I. K. Smith and J. F. Thompson, *Biochim. Biophys. Acta* **227**, 288 (1971).
[10] N. M. Kredich, M. A. Becker, and G. M. Tomkins, *J. Biol. Chem.* **244**, 2428 (1969).
[11] Bacterial enzymes are described in this volume [81] and [82].

gemann et al.[12] that serine is the substrate for plant cysteine synthase can probably be explained by the small activity, less than 10% of that with O-acetylserine, of this enzyme with serine.[3] To date, only H_2S has been demonstrated directly as a sulfur substrate for cysteine synthase. The physiological substrate for the enzyme is not clear, since inorganic sulfate can be reduced, after initial activation to adenosine 5′-sulfatophosphate, by two pathways in plants.[13] One pathway involves free sulfate as an intermediate, which is reduced by sulfite reductase (EC 1.8.1.2) to H_2S. The other pathway proceeds *via* carrier-bound sulfite (carrier—S—SO_3^-), which is reduced by thiosulfonate reductase to carrier-bound sulfide (carrier—S—S^-). In contrast to *Escherichia coli* in which thioredoxin is thought to be the carrier for sulfate reduction,[14] studies with *Chlorella* suggest that glutathione may serve this function in plants.[15,16] The relative physiological roles of the two pathways of sulfate assimilation in plants remain to be determined. Studies with *Chlorella* mutants show that thiosulfonate reductase is indispensable for sulfate reduction, even in organisms containing sulfite reductase, indicating the physiological importance of the bound pathway in this organism.[17]

Both negative and positive control have been implicated in the regulation of cysteine synthesis.[3] Negative control is effected by one or more products of the sulfate assimilation pathway, the most likely of which is cysteine. The sites for negative control include sulfate uptake[18,19] and the steps catalyzed by adenosine 5′-sulfatophosphate sulfotransferase[20] and sulfate adenylyltransferase (EC 2.7.7.4).[21,22] The latter enzyme is also subject to positive control by an unidentified nitrogen-containing compound.[21,22] Cysteine levels may further be controlled by H_2S emission from cyst(e)ine or its precursors, or by the availability of O-acetylserine.[23–25]

[12] J. Bruggemann, K. Schlossmann, M. Merkenschlager, and M. Waldschmidt, *Biochem. Z.* **335**, 392 (1962).
[13] J. W. Anderson, in "The Biochemistry of Plants" (B. J. Miflin, ed.), Vol. 5, p. 203. Academic Press, New York, 1980.
[14] M. L.-S. Tsang and J. A. Schiff, *J. Bacteriol.* **125**, 923 (1976).
[15] M. L.-S. Tsang and J. A. Schiff, *Plant Cell Physiol.* **17**, 1209 (1976).
[16] M. L.-S. Tsang and J. A. Schiff, *Plant Sci. Lett.* **11**, 177 (1978).
[17] A. Schmidt, W. R. Abrams, and J. A. Schiff, *Eur. J. Biochem.* **47**, 423 (1974).
[18] A. H. Datko and S. H. Mudd, *Plant Physiol.* **75**, 466 (1984).
[19] I. K. Smith, *Plant Physiol.* **66**, 877 (1980).
[20] C. Brunold, in "Biology of Inorganic Nitrogen and Sulfur" (H. Bothe and A. Trebst, eds.), p. 352. Springer-Verlag, Berlin and New York, 1981.
[21] Z. Reuveny and P. Filner, *J. Biol. Chem.* **252**, 1858 (1977).
[22] M. W. Zink, *Can. J. Bot.* **62**, 2107 (1984).
[23] H. Rennenberg, *Annu. Rev. Plant Physiol.* **35**, 121 (1984).
[24] This volume [75].
[25] This volume [77].

In addition to incorporation into protein, cysteine is metabolized predominantly to methionine (see below) and glutathione. Synthesis of the latter compound in plants appears to proceed by the two-step sequence demonstrated in animals and microorganisms. The enzymes catalyzing these steps in plants [γ-glutamylcysteine synthetase (EC 6.3.2.2, glutamate–cysteine ligase) and glutathione synthetase (EC 6.3.2.3)] have been studied only in crude preparations.[26]

Methionine

Biosynthesis of homocysteine, the immediate precursor of methionine, proceeds predominantly, if not exclusively, by transsulfuration[3]:

$$\text{Cysteine} + O\text{-phosphohomoserine} \xrightarrow{\text{Cystathionine } \gamma\text{-synthase}} \text{Cystathionine} + P_i \quad (3)$$

$$\text{Cystathionine} + \text{water} \xrightarrow{\text{Cystathionine } \beta\text{-lyase}} \text{homocysteine} + \text{pyruvate} + \text{ammonia} \quad (4)$$

At present, studies of plant cystathionine γ-synthase[27] have been restricted to crude preparations.[28–30] Purification and properties of cystathionine β-lyase (EC 4.4.1.8) from spinach are described elsewhere in this volume.[31] Transsulfuration in plants [Eqs. (3) plus (4)] resembles that in bacteria in proceeding only in the direction of cysteine to homocysteine.[3] No transsulfuration from homocysteine to cysteine, a process characteristic of animals and fungi, could be detected. Plants are unique in using O-phosphohomoserine as the physiological donor of the 4-carbon moiety of homocysteine.[3] All other organisms studied use either O-succinylhomoserine (bacteria) or O-acetylhomoserine (bacteria and fungi).

Plant cystathionine γ-synthase also catalyzes reaction (5) *in vitro*.[32] However, this reaction has little, if any, significance *in vivo*. This conclusion is supported by (1) *in vivo* labeling patterns of $^{35}SO_4^{2-}$ assimilation in plants[33,34] and (2) the isolation of a mutant of *Nicotiana plumbaginifolia* that lacks cystathionine β-lyase and requires methionine for growth.[35]

[26] H. Rennenberg, *Phytochemistry* **21**, 2771 (1982).
[27] EC 4.2.99.9 specifies O-succinylhomoserine as substrate for cystathionine γ-synthase and is therefore not an appropriate designation of the plant enzyme.
[28] H. Aarnes, *Plant Sci. Lett.* **19**, 81 (1980).
[29] A. H. Datko, J. Giovanelli, and S. H. Mudd, *J. Biol. Chem.* **249**, 1139 (1974).
[30] G. A. Thompson, A. H. Datko, S. H. Mudd, and J. Giovanelli, *Plant Physiol.* **69**, 1077 (1982).
[31] This volume [76].
[32] A. H. Datko, S. H. Mudd, and J. Giovanelli, *J. Biol. Chem.* **252**, 3436 (1977).
[33] J. Giovanelli, S. H. Mudd, and A. H. Datko, *J. Biol. Chem.* **253**, 5665 (1978).
[34] P. K. Macnicol, A. H. Datko, J. Giovanelli, and S. H. Mudd, *Plant Physiol.* **68**, 619 (1981).
[35] I. Negrutiu, D.De Brouwer, R. Dirks, and M. Jacobs, *Mol. Gen. Genet.* **199**, 330 (1985).

$$O\text{-Phosphohomoserine} + H_2S \rightarrow \text{Homocysteine} + P_i \tag{5}$$

Absence of significant synthesis of homocysteine by direct sulfhydration [Eq. (5)] *in vivo* may be explained by the relative values of K_m and V_{max} for cysteine synthase and cystathionine γ-synthase.[30] Thus, the V_{max} for cysteine synthesis is three orders of magnitude greater than that for homocysteine synthesis in Eq. (5), whereas the K_m for sulfide in cysteine synthesis is two orders of magnitude less than that for homocysteine synthesis. These combined factors would cause sulfide to pass almost exclusively into cysteine, rather than directly into homocysteine.[30]

Methylation of homocysteine to methionine in plants is catalyzed by tetrahydropteroyltriglutamate methyltransferase (EC 2.1.1.14, 5-methyltetrahydropteroyltriglutamate–homocysteine methyltransferase). The plant enzyme does not require cobalamin or AdoMet for activity, and uses the triglutamyl derivative of N^5-methyltetrahydrofolic acid at a faster rate than the monoglutamyl derivative.[36] By contrast, the corresponding enzyme in animals and certain bacteria requires cobalamin and catalytic amounts of AdoMet, and uses the monoglutamyl derivative at least as rapidly as the triglutamyl.[36]

Because of its unique role in plants, O-phosphohomoserine is the last intermediate common to the methionine and threonine biosynthetic branches. In all other organisms, the "branch point" is at homoserine. The regulatory patterns for synthesis of methionine (and threonine) might therefore be expected to be different in plants. *In vivo* studies of the patterns of $^{35}SO_4^{2-}$ assimilation in *Lemna* have clearly established that methionine feedback regulates its own *de novo* synthesis and that cystathionine synthesis is an important regulatory site.[37] Cystathionine γ-synthase is not inhibited by potential effectors such as methionine, AdoMet, or S-methylmethionesulfonium, but it is down-regulated in *Lemna* growing with concentrations of methionine sufficient to regulate methionine biosynthesis *in vivo*.[30] Inhibition of cystathionine γ-synthase *in vivo* with the suicide inhibitor propargylglycine showed that down-regulation of the enzyme, by itself, was not sufficient for regulation of methionine biosynthesis and that additional factors must be involved. Two such factors that have been proposed[30] are cooperative inhibition by AdoMet of the lysine-sensitive aspartate kinase,[38] and stimulation by AdoMet of threonine synthase.[39,40]

---

[36] E. A. Cossins, in "The Biochemistry of Plants" (D. D. Davies, ed.), Vol. 2, p. 365. Academic Press, New York, 1980.
[37] J. Giovanelli, S. H. Mudd, and A. H. Datko, *Plant Physiol.* **77**, 450 (1985).
[38] S. E. Rognes, P. J. Lea, and B. J. Miflin, *Nature (London)* **287**, 357 (1980).
[39] J. T. Madison and J. F. Thompson, *Biochem. Biophys. Res. Commun.* **71**, 684 (1976).
[40] J. Giovanelli, K. Veluthambi, G. A. Thompson, S. H. Mudd, and A. H. Datko, *Plant Physiol.* **76**, 285 (1984).

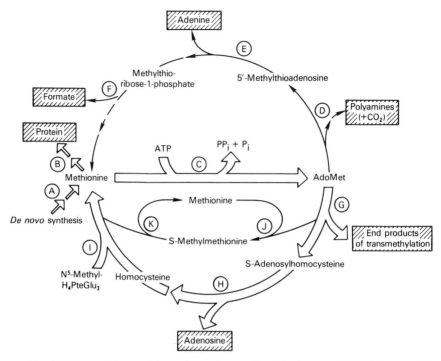

FIG. 2. Estimated fluxes of the major pathways of methionine metabolism in *Lemna* [J. Giovanelli, S. H. Mudd, and A. H. Datko, *Plant Physiol.* **78,** 555 (1985); S. H. Mudd and A. H. Datko, *ibid.* **81,** 103 (1986).] The width of the arrows approximates the magnitude of fluxes. Fluxes indicated by solid lines are about one-third that into protein. Shaded boxes indicate major end products of methionine metabolism. Reactions (enzymes) are as follows: A, net synthesis of methionine by transsulfuration (see text); B, incorporation of methionine into protein; C, methionine adenosyltransferase (EC 2.5.1.6) [J. Giovanelli, S. H. Mudd, and A. H. Datko, in "The Biochemistry of Plants" (B. J. Miflin, ed.), Vol. 5, p. 454. Academic Press, New York, 1980]; D, polyamine synthesis via AdoMet decarboxylase (EC 4.1.1.50) and spermidine synthase (EC 2.5.1.16) [J. Giovanelli, S. H. Mudd, and A. H. Datko, in "The Biochemistry of Plants" (B. J. Miflin, ed.), Vol. 5, p. 454. Academic Press, New York, 1980]; E, F, metabolism of 5'-methylthioadenosine to methionine by a series of reactions in which the methylthio moiety and 4 of the 5 ribose carbons of 5'-methylthioadenosine are converted, respectively, to the methylthio and 4-carbon moieties of methionine [J. Giovanelli, A. H. Datko, S. H. Mudd, and G. A. Thompson, *Plant Physiol.* **71,** 319 (1983); S. F. Yang and N. E. Hoffman, *Annu. Rev. Plant Physiol.* **35,** 155 (1984)]; G, transmethylation; H, adenosylhomocysteinase (EC 3.3.1.1) (this volume [73]); I, 5-methyltetrahydropteroyltriglutamate–homocysteine methyltransferase (EC 2.1.1.14); enzymatic catalysis of J or K has been described in a variety of plants [J. Giovanelli, S. H. Mudd, and A. H. Datko, in "The Biochemistry of Plants" (B. J. Miflin, ed.), Vol. 5, p. 454. Academic Press, New York, 1980], and the combined presence of these enzymes demonstrated in *Lemna* [S. H. Mudd and A. H. Datko, *Plant Physiol.* **81,** 103 (1986).

A comprehensive picture (Fig. 2) of *in vivo* fluxes through the major pathways of methionine metabolism has recently been provided by studies with *Lemna*.[41,42] The salient points of this scheme are as follows: (1) Of the pathways illustrated, incorporation of methionine into protein (reaction B) is the only pathway consuming the entire methionine molecule, i.e., methyl, sulfur, and 4-carbon moieties. In the other pathways, methionine acts essentially as a catalyst via AdoMet. For example, utilization of the 4-carbon moiety of methionine for synthesis of polyamines (reaction D) is accompanied by recycling of the methylthio moiety and regeneration of methionine (reactions E and F). Ultilization of the methyl group of methionine in transmethylation (reaction G) is accompanied by recycling of the homocysteinyl moiety and regeneration of methionine via reactions H and I. As explained below, transmethylation to form *S*-methylmethioninesulfonium (reaction J) is exceptional in that it comprises part of a cycle, the net effect of which is hydrolysis of ATP. (2) Synthesis of AdoMet (reaction C) is clearly the major pathway for methionine metabolism, with over 4-fold as much methionine being metabolized through AdoMet as accumulates in protein. (3) AdoMet is used predominantly for transmethylation. The major end product of transmethylation is phosphatidylcholine, with progressively smaller amounts directed to synthesis of methyl groups of pectin methyl esters, chlorophyll methyl esters, unidentified neutral lipids, nucleic acid derivatives, and methylated basic amino acids.[42] Little, if any, oxidation of the methyl group of methionine occurs. The flux of methyl groups into *S*-methylmethioninesulfonium, tentatively estimated at 15–20% of that into phosphatidylcholine,[43] is not accompanied by a comparable accumulation of *S*-methylmethioninesulfonium, indicating extensive turnover of the methyl groups of this amino acid.[42] The net effect of turnover of *S*-methylmethioninesulfonium in the cycle proposed in Fig. 2 is conversion of ATP to adenosine + PP_i + P_i.[42]

The metabolic patterns outlined in Fig. 2 for *Lemna*, a rapidly growing, vegetative tissue, are probably typical of most plant tissues.[41,42] An interesting departure occurs in specific plants, such as ripening apples and tomatoes, that are unusual in evolving massive amounts of ethylene.[44] In these tissues, methionine is metabolized predominantly to ethylene[45] by the following reactions[44]:

$$\text{Methionine} \xrightarrow{(C)} \text{AdoMet} \xrightarrow{(L)} \text{ACC} \xrightarrow{(M)} \text{ethylene} (+ CO_2 + HCN)$$
$$\text{5'-methylthioadenosine}$$

---

[41] J. Giovanelli, S. H. Mudd, and A. H. Datko, *Plant Physiol.* **78**, 555 (1985).
[42] S. H. Mudd and A. H. Datko, *Plant Physiol.* **81**, 103 (1986).
[43] S. H. Mudd, personal communication.
[44] S. F. Yang and N. E. Hoffman, *Annu. Rev. Plant Physiol.* **35**, 155 (1984).
[45] S. P. Burg and C. O. Clagett, *Biochem. Biophys. Res. Commun.* **27**, 125 (1967).

Reaction C is described in Fig. 2. Synthesis of ACC is catalyzed by 1-aminocyclopropane-1-carboxylate synthase (EC 4.4.1.14),[46] with the by-product, 5'-methylthioadenosine, recycled by reactions E and F (Fig. 2). Conversion of ACC to ethylene proceeds via a series of reactions that remain to be characterized.[44]

[46] This volume [72].

[72] 1-Aminocyclopropane-1-carboxylate Synthase

By Douglas O. Adams and Shang Fa Yang

The gaseous plant hormone ethylene is biosynthesized in plant tissue by the following sequence[1,2]: methionine → S-adenosyl-L-methionine (AdoMet) → 1-aminocyclopropane-1-carboxylic acid (ACC) → ethylene. 1-Aminocyclopropane-1-carboxylate synthase (EC 4.4.1.14, S-adenosyl-L-methionine methylthioadenosine-lyase) catalyzes an α,γ-elimination of the methylthioadenosine (MTA) group of AdoMet, leading to the formation of ACC [Eq. (1)].

$$\overset{+}{\underset{\underset{Ado}{|}}{CH_3-S}}-CH_2-CH_2-\underset{\underset{NH_2}{|}}{CH}-COOH \longrightarrow \underset{H_2C}{\overset{H_2C}{\diagdown}}\underset{NH_2}{\overset{COOH}{\diagup}}C + MTA \qquad (1)$$

Assay Method

Principle. Enzymatic activity is assayed by following the appearance of ACC in the presence of AdoMet. The amount of ACC thus formed is quantitated by oxidizing ACC to ethylene with subsequent quantitation of ethylene by gas chromatography.[3] Since the method is specific, sensitive, convenient, and has been used in nearly all studies, it will be described here.

Procedure. Usually, about 400 µl of the dialyzed extract described below is placed in a 13 × 100 mm test tube. After adding 50 µl of 0.5 M HEPPS[3a] at pH 8.5, the reaction is started by adding 50 µl of 0.5 mM

[1] D. O. Adams and S. F. Yang, *Proc. Natl. Acad. Sci. U.S.A.* **76**, 170 (1979).
[2] N. E. Hoffman and S. F. Yang, *Annu. Rev. Plant Physiol.* **35**, 155 (1984).
[3] M. C. C. Lizada and S. F. Yang, *Anal. Biochem.* **100**, 140 (1979).
[3a] Abbreviations: HEPES, 4-(2-hydroxyethyl)-1-piperazineethanesulfonic acid; HEPPS, N-2-hydroxyethylpiperazine-N'-3-propanesulfonic acid.

AdoMet and incubating at 30°.[4,5] The appropriate incubation time should be determined experimentally but is usually in the range of 0.5–3.0 hr. At the end of the incubation, 50 µl of 10 mM $HgCl_2$ is added to the reaction mixture, and the tube is placed in an ice bath and sealed with a serum cap. The $HgCl_2$ serves to stop the reaction and is essential for the degradation of ACC to ethylene with the NaOCl reagent.[3] ACC is degraded to ethylene by injecting 0.1 ml (about 10 drops) of a cold mixture of 5% NaOCl (commercial bleach solution) and saturated NaOH (2 : 1, v/v) through the serum cap with a 1-ml syringe. After vortexing the cold reaction tubes, ethylene released into the gas phase is determined by gas chromatography.[3,5] Since ACC is converted to ethylene in less than a minute by this procedure,[3] analysis of ethylene can be carried out with little delay.

To determine the efficiency of the conversion of ACC to ethylene during the degradation, a known amount of authentic ACC is added as an internal standard to another similarly incubated reaction mixture and degraded as described above. The amount of ACC formed in the sample is calculated as the quotient (ethylene released)/(conversion efficiency). The conversion efficiency is usually about 70%. Numerous columns and a variety of detectors have been used to assay ethylene by gas chromatography. The reader is referred to Cowper and DeRose for a detailed discussion.[6] As little as 1 pmol of ethylene can be quantitated by routine gas chromatographic techniques.

Very recently, Miura and Chiang[7] have developed a radioisotopic method for assaying the synthase. The procedure utilizes a cation-exchange column to separate unreacted S-adenosyl-L-[1-^{14}C]methionine from labeled ACC formed in the reaction. The assay is conducted as described above except that 15,000 dpm of [^{14}C]AdoMet is included. The reaction is stopped by adding 1.2 ml of ice-cold water. The diluted reaction mixture is loaded onto a 0.9 × 2.5 cm column of ion-exchange resin AG50W-X4 (100–200 mesh) in the ammonium form. The reaction vessel is then washed with three 1-ml portions of water which are added to the column. The unreacted AdoMet remains bound whereas labeled ACC in the eluate is determined by scintillation counting. This method might be preferred by those who have access to a liquid scintillation counter but not a gas chromatograph. However, [^{14}C]AdoMet should be purified before use to assure that there is little or no contamination with labeled degradation products.

[4] T. Boller, R. C. Herner, and H. Kende, *Planta* **145**, 293 (1979).
[5] Y. B. Yu, D. O. Adams, and S. F. Yang, *Arch. Biochem. Biophys.* **198**, 280 (1979).
[6] C. J. Cowper and A. J. DeRose, "The Analysis of Gases by Chromatography." Pergamon, Oxford, 1983.
[7] G. A. Miura and P. K. Chiang, *Anal. Biochem.* **147**, 217 (1985).

Extraction and Purification Procedures

Extraction. Since wounding greatly increases the amount of the synthase in tomato pericarp, tissue is chopped into small pieces and allowed to stand for 3 hr at 25° before homogenization. The tissue is homogenized in 100 mM HEPPS at pH 8.5 containing 4 mM dithiothreitol and 0.5 μM pyridoxal phosphate in a ratio of 1 ml of buffer per gram fresh weight of tissue.[8] The homogenate is filtered through four layers of cheesecloth and centrifuged at 10,000 g for 20 min.

Since the ACC synthase assay is based on measurement of ACC formation from AdoMet, and tissues from which the enzyme is extracted frequently contain large amounts of ACC, ACC must be removed prior to performing the assay. This is readily accomplished by dialyzing the extract against 2 mM HEPPS at pH 8.5 containing 0.1 mM dithiothreitol and 0.2 μM pyridoxal phosphate, or by passing the extract through a Sephadex G-50 or G-25 column using the same buffer as for dialysis. Gel filtration offers the advantage of quickly obtaining an extract for assay.

The enzyme is seemingly quite labile,[8] and in some cases quick assay of the enzyme has been crucial to obtaining detectable activity.[9] Care must be taken during extraction to avoid severe pH changes as this causes rapid loss of enzyme activity.[10]

Purification. The synthase has not been highly purified, largely because of its instability and low concentration in tissue.[10] The most extensive purification, about 60- to 70-fold, is from tomato pericarp.[8] An extract of tomato pericarp, prepared essentially as described above, was brought to 30% saturation with ammonium sulfate (164 g/liter) and centrifuged for 20 min at 20,000 g. The supernatant fluid was taken to 95% saturation (additional 445 g/liter) and centrifuged. The precipitate was suspended in 10 mM HEPES[3a] at pH 8.0 containing 0.1 mM dithiothreitol and 0.2 μM pyridoxal phosphate and taken to 10% saturation with ammonium sulfate. This mixture was applied to a column of phenyl-Sepharose CL-4B and the column was eluted with 48 ml of the same buffer used for suspension, i.e., containing 10% ammonium sulfate (w/v). A stepwise elution with NaCl in the same buffer was used to elute ACC synthase. The first step of 0.5 M NaCl (48 ml) removed a large amount of protein but no ACC synthase activity. Most of the ACC synthase was eluted with 48 ml of buffer containing 0.1 M NaCl. A 60- to 70-fold purification was achieved.

---

[8] M. A. Acaster and H. Kende, *Plant Physiol.* **72**, 139 (1983).
[9] C. Y. Wang and D. O. Adams, *Plant Physiol.* **69**, 424 (1982).
[10] H. Kende, M. A. Acaster, and M. Guy, in "Ethylene and Plant Development" (J. A. Roberts and G. A. Tucker, eds.), p. 23. Butterworths, London, 1985.

Mattoo et al.[11] reported that ACC synthase will also bind to alkyl-agarose and aminoalkyl-agarose gels and that the activity can be recovered by elution with 0.5 M NaCl.

Properties

Molecular Weight. ACC synthase has been estimated to have a molecular weight of 57,000 by gel filtration.[8]

Kinetic Properties and Substrate Specificity. The enzyme from tomato (*Lycopersicon esculentum* Mill.) has been found to have a K_m for AdoMet of about 20 μM.[4,5,7,12] The K_m for the mung bean [*Vigna radiata* (L.) Wilczek] enzyme has been estimated as 60 μM[13] while that for sliced winter squash (*Cucurbita maxima* Duch.) as 120 μM[14] and 13.3 μM.[15] ACC synthase can use only (−)-Ado-L-Met,[12] which can convert to (+)-Ado-L-Met in solution.[16] The recent report[17] that commercial AdoMet preparations contain only 17.8–69% of the 260 nm absorbing material as (−)-AdoMet suggests that when very precise kinetic studies are undertaken, workers should consider purification of (−)-AdoMet by the HPLC procedure described by Creason et al.[17]

Coenzyme and Inhibition. Partially purified preparations require low concentrations of pyridoxal phosphate for maximum activity.[5] The enzyme is strongly inhibited by well-known inhibitors of pyridoxal enzymes such as aminoethoxyvinylglycine, a competitive inhibitor of the tomato enzyme with a K_i of 0.2 μM[4]; methoxyvinylglycine is much less active, and vinylglycine is inactive.[5] Aminooxyacetic acid is a competitive inhibitor of the tomato enzyme with a K_i of 0.8 μM.[5] Sinefungin is an effective inhibitor with an I_{50} of 2.5 μM whereas S-adenosylhomocysteine and its sulfoxide show I_{50} values of 150 and 240 μM, respectively.[7] Three of the stereoisomers of AdoMet have recently been shown to inhibit ACC synthase.[12] Whereas only the naturally occurring (−)-Ado-L-Met can serve as substrate, (+)-Ado-L-Met and (±)-Ado-D-Met act as inhibitors having K_i values of 15 and 70 μM, respectively.

---

[11] A. K. Mattoo, D. O. Adams, G. W. Patterson, and M. Lieberman, *Plant Sci. Lett.* **28,** 173 (1982–1983).
[12] S. Khani-Oskouee, J. P. Jones, and R. W. Woodard, *Biochem. Biophys. Res. Commun.* **121,** 181 (1984).
[13] H. Yoshii and H. Imaseki, *Plant Cell Physiol.* **22,** 369 (1981).
[14] H. Hyodo, K. Tanaka, and K. Watanabe, *Plant Cell Physiol.* **24,** 963 (1983).
[15] H. Hyodo, K. Tanaka, and J. Yoshisaka, *Plant Cell Physiol.* **26,** 161 (1985).
[16] S. Wu, W. P. Huskey, R. T. Borchardt, and R. L. Schowen, *Biochemistry* **22,** 2828 (1983).
[17] G. L. Creason, J. T. Madison, and J. F. Thompson, *Phytochemistry* **24,** 1151 (1985).

[73] Adenosylhomocysteinase from Yellow Lupine

By ANDRZEJ GURANOWSKI and HIERONIM JAKUBOWSKI

Adenosylhomocysteinase (EC 3.3.1.1) catalyzes the following reversible reaction:

$$S\text{-Adenosyl-L-homocysteine} + H_2O \rightleftharpoons \text{adenosine} + \text{L-homocysteine}$$

Equilibrium favors synthesis of AdoHcy.[1-3] Adenosylhomocysteinase seems to occur in all eukaryotes[4] and was recently demonstrated in some bacteria.[5] Yellow lupine seeds are suitable for preparation of homogeneous enzyme.

Assay Method

Principle. Adenosylhomocysteinase activity can be determined by measurement of radioactive AdoHcy formation from Hcy[1] and labeled Ado.[1] The product is separated from substrates by thin-layer chromatography.

Chromatographic Supports. Aluminium or plastic sheets precoated with silica gel 60 F_{254} or cellulose F_{254} were purchased from E. Merck, Darmstadt. Aminohexyl-Sepharose was prepared by the procedure of Cuatrecasas.[6] Hydroxyapatite (BioGel HTP) was from Bio-Rad Laboratories (Richmond, CA).

Buffers. Buffers A, B, and C are prepared from 10% (v/v) glycerol and 1 mM 2-mercaptoethanol in 10, 60, and 20 mM potassium phosphate at pH 6.8, respectively.

Procedure. The standard incubation mixture contains the following in a final volume of 50 μl: 50 mM Tris–HCl at pH 8.5, 2 mM dithiothreitol, 10 mM DL-homocysteine, 1 mM radioactive adenosine, and the enzyme fraction. Samples are incubated at 37° for the appropriate time, depending on enzyme activity, and the reaction is stopped by transferring 5 μl aliquots onto aluminum sheets precoated with silica gel 60 F_{254}. The chro-

---

[1] Abbreviations: AdoHcy, S-adenosylhomocysteine; Ado, adenosine; Hcy, L-homocysteine; AdoCys, S-adenosylcysteine; Cys, cysteine.
[2] G. De La Haba and G. L. Cantoni, *J. Biol. Chem.* **234,** 603 (1959).
[3] A. Guranowski and J. Pawelkiewicz, *Eur. J. Biochem.* **80,** 517 (1977).
[4] R. D. Walker and J. A. Duerre, *Can. J. Biochem.* **53,** 312 (1975).
[5] S. Shimizu, S. Shiozaki, T. Ohshiro, and H. Yamada, *Eur. J. Biochem.* **141,** 385 (1984).
[6] P. Cuatrecasas, *J. Biol. Chem.* **245,** 3059 (1970).

matograms are developed for 60 min in ethyl acetate/propan-2-ol/ammonia/water (27:23:4:4, v/v). The spots containing radioactive AdoHcy are visualized under ultraviolet light, cut out, and counted.

Comments on the Procedure. A stock solution of radioactive [8-^{14}C]- or [U-^{14}C]adenosine is diluted with the unlabeled nucleoside to give about 50,000 counts/min per sample. The solution of Hcy should be fresh. To slow down the oxidation of Hcy, the stock solutions of Hcy and dithiothreitol should be combined. To speed up enzyme purification, column eluates can be assayed, but only qualitatively, in the absence of radioactivity. In that case, because the difference in the R_f values for AdoHcy and Ado is large (0.35 and 0.80, respectively[3]), development of the chromatograms can be shortened to 15 min.

Alternatively, two other chromatographic systems can be used for separation of labeled Ado and AdoHcy: (1) cellulose plates developed in acetone/water (5:2, v/v), R_f values 0.61 and 0.22, respectively,[3] and (2) silica gel plates developed in 1-butanol/acetic acid/water (12:3:5, v/v), R_f values 0.53 and 0.23, respectively.[7] If it is necessary to work with radioactive [^{35}S]Hcy and unlabeled Ado, good separation of [^{35}S]AdoHcy from [^{35}S]Hcy can be obtained on cellulose plates developed in 5% Na_2HPO_4 (R_f values of the compounds are 0.53 and 0.90, respectively).[7]

The enzyme assay described above is fully applicable to crude extracts from lupine seeds since there are no competing activities with adenosylhomocysteinase for substrates in this material. However, beginning from the second day of lupine seed germination, activity of adenosine nucleosidase appears in the cotyledons; the enzyme cleaves adenosine to adenine and ribose.[8] For the assay adenosylhomocysteinase under such circumstances, the hydrolysis of AdoHcy, rather than AdoHcy synthesis, should be measured.

Definition of Unit. One unit of adenosylhomocysteinase catalyzes the synthesis of 1 μmol of AdoHcy during 1 min at 37° under the conditions defined above.

Purification

General Remarks. The higher the ionic strength of extracting buffer, the more lupine seed proteins can be solubilized.[9] Since adenosylhomocysteinase is effectively extracted with buffer of low ionic strength, it can be separated from the bulk of other seed proteins that are insoluble under

---

[7] A. Guranowski, J. A. Montgomery, G. L. Cantoni, and P. K. Chiang, *Biochemistry* **20**, 110 (1981).

[8] A. Guranowski and J. Pawelkiewicz, *Planta* **139**, 245 (1978).

[9] H. Jakubowski and J. Pawelkiewicz, *Acta Biochim. Pol.* **21**, 271 (1974).

such conditions. After selective extraction, the four subsequent conventional purification steps allow homogeneous adenosylhomocysteinase with good (26%) yield. All the following purification steps are carried out at 4°.

Crude Extract. Extract the meal of 1 kg of yellow lupine seeds (*Lupinus luteus*) with 3 liters of buffer A for 30 min. Pass the extract through cheesecloth and remove cellular debris by centrifugation at 30,000 g for 30 min.

Ammonium Sulfate. Bring the extract to 30% ammonium sulfate saturation by adding solid ammonium sulfate (17 g/100 ml). Centrifuge the mixture at 30,000 g for 10 min, and discard the pellet. From the resultant supernatant fluid precipitate adenosylhomocysteinase by adding ammonium sulfate to 50% saturation (additional 12 g/100 ml). Collect the precipitate by centrifugation, dissolve in about 250 ml of buffer B, dialyze against this buffer, and remove insoluble particles by centrifugation.

Aminohexyl-Sepharose. The protein solution (270 ml) is applied to an aminohexyl-Sepharose column (5 × 33 cm) equilibrated with buffer B. Wash the column with 3 liters of the same buffer and apply a linear salt gradient of 4 liters (0–0.3 M KCl in buffer B); the enzyme emerges at 150–180 mM KCl. Pool active fractions (530 ml) and concentrate the enzyme by ammonium sulfate precipitation (60% saturation, i.e., 37 g/100 ml). Collect the precipitate by centrifugation and dissolve in a small volume of buffer B.

Sephadex G-200. Apply the solution from the previous step onto a Sephadex G-200 column equilibrated with buffer B. Pool active fractions (48 ml) which appear at $V_e/V_o = 1.4$ and precipitate adenosylhomocysteinase with ammonium sulfate (37 g/100 ml). Collect the precipitate by centrifugation, dissolve in buffer C, dialyze extensively against this buffer, and discard insoluble material by centrifugation.

Hydroxyapatite. Apply the clear solution to a column (1.5 × 15 cm) of hydroxyapatite equilibrated with buffer C. Adenosylhomocysteinase does not adsorb under these conditions and emerges in the break-through fraction. Store this fraction at −20°. The enzyme tolerates repeated freeze–thawing.

A summary of the purification procedure is presented in Table I.

Properties

Purity.[3] Only one distinct band of M_r 55,000 appears on the polyacrylamide gel after electrophoresis of the purified lupine adenosylhomocysteinase in the presence of sodium dodecyl sulfate and 2-mercaptoethanol.

TABLE I
PURIFICATION OF ADENOSYLHOMOCYSTEINASE FROM YELLOW LUPINE SEEDS

Fraction	Volume (ml)	Total protein (mg)	Total activity (units)	Specific activity (units/mg)
Crude extract	1,710	42,800	132.5	0.0031
30–50% Ammonium sulfate	270	16,200	118.2	0.0073
Aminohexyl-Sepharose	530	646	69.7	0.094
Sephadex G-200	48	190	65.5	0.345
Hydroxyapatite	26	9.3	34.4	3.70

Only one band, with pI of 4.9 ± 0.1, was found upon gel isoelectric focusing.

Physical Properties.[3] When estimated by Sephadex G-200 gel filtration, the active enzyme is an α_2 dimer of M_r 110,000. The purified enzyme forms multimers which are estimated from pore gradient gel electrophoresis as M_r 55,000, 110,000, 165,000, and about 200,000 for monomer, dimer, trimer, and tetramer, respectively.

Miscellaneous.[3] The optimum activity of the lupine enzyme is between pH 8.5 and 9.0, and sulfhydryl groups are essential for the catalyzed reactions. The energy of activation computed from the Arrhenius plot for the synthesis of AdoHcy was 14.4 kcal/mol (60.5 kJ/mol).

Substrate Specificity. Specificity with respect to the amino acid is very narrow. Only cysteine but not homoserine, serine, penicillamine, glutathione, or 2-mercaptoethanol will substitute for homocysteine.[10] Specificity with respect to the nucleoside substrate is less strict. In addition to adenosine, the following nucleoside condense with Hcy and form corresponding S-nucleosidylhomocysteines[3,7,10]: 3-deazaadenosine (slightly better substrate than Ado), 2-aza-3-deazaadenosine, nebularine (a purine riboside), formycin A, N^6-methyladenosine, 8-azaadenosine, adenosine N^1-oxide, pyrazomycin, 8-aminoadenosine, inosine, aristeromycin, 2-aminoadenosine, N^6-hydroxyadenosine, 2-hydroxyadenosine, 2-chloroadenosine, and 3-deazaaristeromycin.

Kinetic Constants. In the reaction of AdoHcy synthesis, K'_m values are 2.3 μM[3] and 0.13 mM[10] for adenosine and homocysteine, respectively. In the analogous system, K'_m for Cys is 35 mM.[10] The k_{cat} values for synthesis of AdoHcy and AdoCys are 4.3 and 0.11 sec^{-1}, respectively.[10] For AdoHcy hydrolysis, the estimated K'_m for AdoHcy is in the range of 12–

[10] A. Guranowski and H. Jakubowski, *Biochim. Biophys. Acta* **742**, 250 (1983).

20 μM.[3,10] Specificity (k_{cat}/K_m) exhibited by the lupine adenosylhomocysteinase toward AdoHcy and AdoCys is 60,000 and 0.9 M^{-1} sec^{-1}, respectively.[10] The equilibrium constant for hydrolysis of AdoHcy is 0.5–0.7 μM.[3]

Inhibition Studies. Certain Ado analogs act as competitive inhibitors in the synthesis of AdoHcy.[3] Aristeromycin was found to be the strongest inhibitor (K_i 5 nM). The K_i values for the other analogs tested were as follows: 3-deazaadenosine (1 μM), 8-aminoadenosine (14 μM), 8-azaadenosine (0.2 mM), N^6-methyladenosine (0.25 mM), 2-aza-3-deazaadenosine (0.34 mM), formycin A (0.39 mM), inosine (1.1 mM), adenosine N^1-oxide (1.5 mM), nebularine (1.5 mM), and pyrazomycin (11 mM).

Formation of Complex with Adenosine.[11,12] The lupine adenosylhomocysteinase, an α_2 dimer, forms a specific 1 : 2 enzyme–adenosine complex. Binding of the first molecule of Ado is fast ($k > 7 \times 10^5$ M^{-1} sec^{-1}) whereas the second molecule of Ado binds in a slow process with a half-life of 5 min. Adenosine in the 1 : 1 and 1 : 2 enzyme–substrate complexes reacts slowly (k = 0.05 min^{-1}) to eventually yield free active enzyme, adenine, and ribose.

[11] H. Jakubowski and A. Guranowski, *Biochem. Biophys. Res. Commun.* **84,** 1060 (1978).
[12] H. Jakubowski and A. Guranowski, *Biochemistry* **20,** 6877 (1981).

[74] S-Substituted L-Cysteine Sulfoxide Lyase from Shiitake Mushroom[1]

By KYODEN YASUMOTO and KIMIKAZU IWAMI

$$2\ R\overset{O}{\overset{\uparrow}{S}}CH_2CH(NH_2)COOH + H_2O \rightarrow R\overset{O}{\overset{\uparrow}{S}}SR + 2\ NH_3 + 2\ CH_3COCOOH$$

Fruiting bodies of shiitake mushrooms (*Lentinus edodes*) give a characteristic flavor on maceration of the fresh mushroom and on rehydration of the dried mushroom. Previous studies[2,3] have demonstrated that the principal flavor component "lenthionine (1,2,3,5,6-pentathiepane)" arises enzymatically from the nonvolatile precursor "lentinic acid" which was

[1] See this series, Vol. 17 [198] and [199], for *S*-alkyl-L-cysteine lyase from *Pseudomonas* and *S*-alkyl-L-cysteine sulfoxide lyase from *Allium cepa* (onion).
[2] K. Yasumoto, K. Iwami, and H. Mitsuda, *Agric. Biol. Chem.* **35,** 2059 (1971).
[3] K. Yasumoto, K. Iwami, and H. Mitsuda, *Agric. Biol. Chem.* **35,** 2070 (1971).

identified as 2-(γ-glutamylamino)-4,6,8,9,10-pentaoxo-4,6,8,10-tetrathioundecanoic acid[4]:

$$CH_3-\underset{\underset{O}{\|}}{\overset{\overset{O}{\|}}{S}}-CH_2-\overset{\overset{O}{\uparrow}}{S}-CH_2-\overset{\overset{O}{\uparrow}}{S}-CH_2-\overset{\overset{O}{\uparrow}}{S}-CH_2\underset{\underset{O}{\|}}{CHNHCCH_2}CH_2\overset{\overset{COOH}{|}}{CHNH_2}$$

At least two distinct enzymes are involved in the sequence of action; one is γ-glutamyltransferase (EC 2.3.2.2) which acts on lentinic acid to liberate the γ-glutamyl moiety[5] and the other is the S-substituted L-cysteine sulfoxide lyase (EC 4.4.1.—) which acts on the primary product, deglutamyllentinic acid, to produce ammonia, pyruvate, and a thiosulfinate by catalyzing a β-elimination reaction[6] similar to that catalyzed by *Allium* alliinases. The thiosulfinate produces lenthionine, probably spontaneously.

Assay Method

Principle. The enzyme can be assayed by measuring the rate of pyruvate formation. Pyruvate is conveniently estimated by the color intensity of its 2,4-dinitrophenylhyrazone in alkali, or more precisely, by NADH oxidation in a coupled reaction with lactate dehydrogenase.

Reagents

(±)-S-Alkyl-L-cysteine sulfoxide, 50 mM, in distilled water
Pyridoxal phosphate, 0.25 mM, in distilled water
Tris–HCl, 0.2 M, at pH 8.4
2,4-Dinitrophenylhydrazine, 0.1% in 2 N HCl

Procedure. The reaction mixture (1.0 ml) containing 10 mM substrate, 50 μM pyridoxal phosphate, and an appropriate amount of enzyme in 0.1 M Tris buffer is incubated at 37° for the desired time, usually 5 min, and the reaction is terminated by adding an equal volume of the 2,4-dinitrophenylhydrazine solution. The mixture is allowed to stand for 30 min at 37°, after which 0.2 ml is removed. To the aliquot are added 2.8 ml water and 1 ml of 2 N NaOH, followed by measurement of absorbance at 420 nm.

Definition of Unit. One unit of enzyme activity is defined as the amount capable of catalyzing release of 1 μmol pyruvate per minute under

---

[4] G. Höfle, R. Gmelin, H. H. Luxa, M. N'Galamulume-Treves, and S. Hatanaka, *Tetrahedron Lett.* p. 3126 (1976).
[5] K. Iwami, K. Yasumoto, K. Nakamura, and H. Mitsuda, *Agric. Biol. Chem.* **39**, 1933 (1975).
[6] K. Iwami, K. Yasumoto, and H. Mitsuda, *Agric. Biol. Chem.* **39**, 1947 (1975).

the assay conditions described. Specific activity is expressed in units per milligram of protein.

Enzyme Preparation[7]

Unless otherwise stated, all operations are carried out at 0–4° and centrifugation at 8,000 g for 20 min.

Step 1: Extract. Fresh caps of fruiting bodies of *Lentinus edodes* (1 kg) are ground in a large Waring Blendor with 3 liters of Dry Ice-cooled ethanol. The homogenate is centrifuged, and the supernatant liquid is discarded. The sediment is suspended in 10 liters of 50 mM potassium phosphate at pH 6.8 with continuous stirring for 30 min. The suspension is centrifuged.

Step 2: Ammonium Sulfate. The solution is saturated to 40% with solid ammonium sulfate (24.3 g/100 ml) and the mixture is centrifuged after standing for 30 min. The precipitate is discarded, and the supernatant liquid is brought to 75% of saturation by adding 27.3 g/100 ml of solid ammonium sulfate. After standing for 60 min, the precipitate is separated by centrifugation, dissolved in a small volume of the phosphate buffer, and dialyzed against several changes of the same buffer. Insoluble material is removed by centrifugation.

Step 3: DEAE–Sephadex A-25 Chromatography. The clarified dialysate is applied in two portions to a DEAE-Sephadex A-25 column (2.6 × 40 cm), previously equilibrated with the phosphate buffer. Protein is eluted with a linear concentration gradient of NaCl (0–1 M) in the phosphate buffer and collected in fractions of 3 ml. Fractions containing the bulk of the activity are combined and brought to 75% saturation with solid ammonium sulfate (51.6 g/100 ml). The precipitate formed is separated by centrifugation and dissolved in about 12 ml.

Step 4: Sephacryl S-200 Chromatography. The enzyme solution is applied in portions of 1.5 ml to a Sephacryl S-200 column (2.6 × 70 cm) equilibrated with the phosphate buffer and eluted with the same buffer. The eluate is collected in fractions of 3 ml. The enzyme is eluted in a single major peak at the effluent volume expected for a molecular weight of 80,000. Active fractions are combined and concentrated with a Diaflo YM10 membrane.

Step 5: Preparative Electrophoresis. A 0.5-ml portion of the concentrate is placed on the top of a 7.5% polyacrylamide slab gel (0.2 × 13 × 13 cm) in the glycine buffer system of Davis.[8] After overnight electrophoresis at a constant current of 3 mA/cm^2, the gel is cut into a series of

[7] K. Iwami and K. Yasumoto, *Agric. Biol. Chem.* **44**, 3003 (1980).
[8] B. J. Davis, *Ann. N.Y. Acad. Sci.* **121**, 404 (1964).

TABLE I
PURIFICATION OF S-SUBSTITUTED L-CYSTEINE SULFOXIDE LYASE FROM
FRUITING BODIES OF *Lentinus edodes*

Step	Total volume (ml)	Total protein (mg)	Total activity (units)	Specific activity (units/mg)
1. Extract	3,200	11,900	20,800	1.7
2. Ammonium sulfate	120	3,550	18,700	5.3
3. DEAE–Sephadex A-25	12	433	11,200	26
4. Sephacryl S-200	7	217	8,640	40
5. Electrophoresis				
Enzyme I	5	10.3	3,020	293
Enzyme II	5	0.35	107	306

segments of 0.5 cm width. The gel segments are extracted with 2 volumes of potassium phosphate at pH 6.5 by gentle homogenization followed by centrifugation. Activities are found located in one major and one minor peak. The extracts from the two peaks are stored at −20°.

A summary of the purification procedure is shown in Table I.

Properties

Stability. The final preparations, in contrast to those of *Allium* alliinases,[9,10] can be stored in the presence of pyridoxal phosphate at −25° for a few months with less than 10% loss of activity.

Purity. The purified enzyme is homogeneous on electrophoresis in polyacrylamide gel. On isoelectric focusing in 10% polyacrylamide gel containing carrier ampholyte of pH 3–5, two isoenzymes are observed. Enzymes I and II give single bands corresponding to isoelectric points of pH 4.15 and 4.65, respectively.

pH Optimum. Maximal activity is found at about pH 8.7 in 0.1 M Tris–HCl containing 50 μM pyridoxal phosphate for both isoenzymes.

Kinetic Properties. Both enzymes give K_m values of 50–60 μM for deglutamyllentinic acid, whereas the K_m values for S-methyl-, S-ethyl-, and S-propyl-L-cysteine sulfoxides are in the range of several millimolar, similar to those reported for *Allium* alliinases.[9,10]

Cofactor and Inhibitors. The enzymes are activated by pyridoxal phosphate and inhibited by carbonyl reagents such as hydroxylamine and

[9] M. Mazelis and L. Crews, *Biochem. J.* **108**, 725 (1964).
[10] S. Schwimmer, C. A. Ryan, and F. F. Wong, *J. Biol. Chem.* **239**, 777 (1964).

aminooxyacetic acid. The inhibition is completely reversed by pyridoxal phosphate, but not by other vitamin B_6 vitamers. Sulfhydryl reagents such as p-chloromercuribenzoate, N-ethylmaleimide, and iodoacetamide are not inhibitory.

Specificity. The preferred substrate of the *Lentinus* enzymes is deglutamyllentinic acid, although various S-alkyl-L-cysteines and their sulfoxides can serve as substrates. The relative rates of pyruvate formation from different substrates with enzyme I are as follows, with deglutamyllentinic acid as 100 under otherwise standard assay conditions at pH 8.4: S-methyl-L-cysteine (0.9), S-ethyl-L-cysteine sulfoxide (5.4), S-ethyl-L-cysteine (2.3), S-ethyl-L-cysteine sulfoxide (5.1), S-propyl-L-cysteine (3,2), S-propyl-L-cysteine sulfoxide (5.2), S-allyl-L-cysteine (4.3), cystathionine (3.1), L-cysteine (0.4), L-cystine (4.5 at 2 mM), L-alanine (0) and β-chloro-L-alanine (95.4). L-Methionine and L-serine are not substrates. Enzyme II shows entirely similar substrate specificity. This spectrum of activity differs markedly from those of garlic and onion alliinases, but resembles that from *Albizzia lophanta*[11] and *Acasia farneisana*.[12] Deglutamyllentinic acid, the natural substrate, and commercially available β-chloro-L-alanine are both useful for rapid assay.

Preparation of Deglutamyllentinic Acid[2,13]

About 10 kg of fresh fruiting bodies of *Lentinus edodes* are ground in a large Waring Blendor with 20 liters of ice-cold ethanol. The supernatant solution after centrifugation in a basket-type centrifuge is evaporated under reduced pressure to below 40° and the resulting concentrate is washed with ethyl ether to remove fatty materials, followed by ion-exchange chromatography on an Amberlite CG-120 (H^+ form) column (2 × 70 cm). The column is eluted with 0.5 N NaOH, and the eluate is applied directly onto another Amberlite CG-120 column (2 × 50 cm). The column eluate is assayed for lentinic acid by the following procedures: an aliquot of the eluate is incubated with the crude extract of mushroom (the solution obtained after Step 1 in Enzyme preparation), and the amount of lenthionine formed is estimated by observing the organoleptic change (or more quantitatively by measuring the wave height at around −0.6 V in AC polarography).[3] Active fractions are combined and evaporated under reduced pressure. The concentrate is applied to a Dowex 2-X8 (acetate) column (2 × 50 cm), which is subsequently eluted with 0.5 N formic acid.

[11] S. Schwimmer and A. Kjaer, *Biochim. Biophys. Acta* **42**, 316 (1960).
[12] M. Mazelis and R. K. Creveling, *Biochem. J.* **147**, 485 (1975).
[13] K. Yasumoto, K. Iwami, H. Mizusawa, and H. Mitsuda, *Nippon Nogei Kagaku Kaishi* **50**, 563 (1976).

White precipitates form after standing for several days at 0–4°. The precipitate, lentinic acid, is collected by filtration, dissolved in hot water, and recrystallized from methanol–water. The yield was 1.5 g (m.p. 157°, dec.).

To 1 g of lentinic acid, dissolved in 100 ml of 0.1 M glycine–NaOH at pH 8.6, is added a partially purified γ-glutamyltransferase (2.8 mg protein, 28 units).[14] After incubation for 30 min at 37°, the reaction mixture is concentrated to a volume of less than 20 ml and allowed to stand overnight in the refrigerator. The precipitate is collected by filtration and crystallized from hot water. The yield of deglutamyllentinic acid is 0.45 g (m.p. 194°, dec.).

[14] K. Yasumoto, K. Iwami, T. Yonezawa, and H. Mitsuda, *Phytochemistry* **16,** 1351 (1977).

[75] Cystine Lyase from Cabbage

By IVAN K. SMITH and DAVID I. HALL

Cystine + H_2O → Cysteine persulfide + pyruvate + ammonia

Cystine lyase has been isolated from the roots and leaves of several members of the Cruciferae belonging to the genera *Brassica*.[1–4] The trivial name of the enzyme is probably a misnomer, in the sense that it implies a specificity and physiological function which has not been demonstrated.

Assay Method

Principle. Although quantitation of cysteine persulfide is not difficult,[5] most investigators prefer to assay enzyme activity by measuring the initial rate of pyruvate formation. This is done either by following the oxidation of NADH in the presence of lactate dehydrogenase, or by preparing the dinitrophenylhydrazone, treating with base to develop color, and measuring absorbance at 520 nm.

[1] N. W. Anderson and J. F. Thompson, *Phytochemistry* **18,** 1953 (1979).
[2] M. Mazelis, K. Scott, and D. Gallie, *Phytochemistry* **20,** 991 (1982).
[3] D. I. Hall and I. K. Smith, *Plant Physiol.* **72,** 654 (1983).
[4] M. Mazelis, N. Beimer, and R. K. Creveling, *Arch. Biochem. Biophys.* **120,** 371 (1967).
[5] M. Flavin, *J. Biol. Chem.* **237,** 768 (1962).

Reagents

Substrate, L-cystine, 1 mM dissolved in 0.4 M Tris–HCl at pH 8.5 (For higher concentrations, dissolve cystine in 50 mM NaOH, add an equal volume of 0.8 M Tris–HCl, and adjust the pH to 8.5 with 10 M HCl.)
Pyridoxal phosphate, 0.1 mM
Dinitrophenylhydrazine, 0.1% solution in 2 M HCl
NaOH, 1 M

Procedure. The following are mixed in a tube: 0.5 ml of the L-cystine solution in buffer, 0.1 ml pyridoxal phosphate, 0.3 ml of water, and 0.1 ml of enzyme extract. Reaction mixtures are incubated at 30° for 15 min, and the reaction is stopped by the addition of 0.3 ml of 0.1% dinitrophenylhydrazine in 2 M HCl. The tube is incubated for a further 15 min at 30°, to allow formation of the dinitrophenylhydrazone. Color is developed by adding 2.5 ml of 1 M NaOH; if necessary, the solution is clarified by centrifugation. Absorbance at 520 nm is measured.

The A_{520} is converted into nanomoles of pyruvate by using a calibration curve, derived by incubating standard amounts of pyruvate (from 20 to 200 nmol) with dinitrophenylhydrazine and adding base. Routinely, two controls are used: one lacks enzyme and the second lacks substrate. Initially, the linearity of the reaction is checked, because cysteine, produced by the spontaneous breakdown of cysteine persulfide, inhibits. Linearity is obtained when the incubation times are short (10–15 min) or when 20 μl of 4 mM Ellman's reagent [5,5'-dithiobis(2-nitrobenzoic acid)] is included in the reaction mixture to remove cysteine.

Purification Procedure[3]

All purification steps are conducted at 0–4°.

Extract. Commercially available head cabbage (*Brassica oleracea* var. *capitata*) leaves (1 kg) are deveined, washed, and homogenized in 1 liter of ice-cold 0.1 M sodium phosphate at pH 7.2, containing 0.5 mM pyridoxal phosphate, in a Waring Blendor at top speed for 3 min. The debris is removed by straining through cheesecloth, and the filtrate is centrifuged at 10,000 g for 20 min.

Ammonium Sulfate. Ammonium sulfate (243 g/liter) is added to the crude homogenate, the solution is slowly stirred for 20 min, and the precipitated protein is sedimented by centrifuging at 10,000 g for 30 min. The pellet is discarded, and the supernatant liquid brought to 60% saturation by adding an additional 132 g of $(NH_4)_2SO_4$. The resulting precipitate is collected by centrifugation, dissolved in 50 ml of 10 mM sodium phosphate at pH 7.5, containing 10 μM pyridoxal phosphate, and dialyzed for 12 hr in a 30-fold excess of the same buffer, with changes at 1, 4, and 8 hr.

DEAE-Cellulose. DEAE-cellulose (Whatman DE-52) is stirred for 1 hr in 10 volumes of 0.5 M HCl, followed by treatment with 10 volumes of 0.5 M NaOH for 1 hr, and finally is washed with distilled water to lower the pH to 8. The precycled DEAE-cellulose is suspended in a 4-fold excess of buffer A (10 mM sodium phosphate at pH 8.5, 10 μM pyridoxal phosphate), containing 0.5 M NaCl, and equilibrated overnight. The DEAE-cellulose is poured into a prechilled column to yield a packed volume of 8 × 2.5 cm, and is washed overnight with buffer A, containing 30 mM NaCl.

The dialyzed $(NH_4)_2SO_4$ fraction is clarified by centrifugation at 10,000 g for 30 min and applied to the column. Nonabsorbed protein is washed from the column (1 ml/min) with 250 ml of buffer A containing 30 mM NaCl. Protein is eluted (0.5 ml/min) with a linear salt gradient, prepared by mixing 250 ml of 50 mM NaCl in buffer A and 250 ml of 500 mM NaCl in buffer A. Two peaks of cystine lyase activity are resolved; one is eluted by 100 ± 50 mM NaCl and is designated isozyme I, and a second is eluted by 300 ± 50 mM NaCl and is designated isozyme II. The fractions with the greatest enzyme activity are pooled and brought to 60% saturation with $(NH_4)_2SO_4$. Precipitated protein is sedimented by centrifugation, and the resultant pellet is dissolved in 5 ml of buffer A containing 200 mM NaCl.

The relative amounts of isozyme I and II varies from one preparation to another, with isozyme I representing from 10 to 50% of the total activity. When isozyme II, eluted from DEAE-cellulose, is incubated for 30 min in 1 M NaCl, dialyzed, and chromatographed on DEAE-cellulose, between 15 and 25% of the cystine lyase activity is eluted at the position of isozyme I. This observation led to the suggestion that isozyme I is a dissociated form of isozyme II.

BioGel A-0.5M. Two milliliters of the DEAE fraction is applied to a BioGel A-0.5M column (90 × 1.5 cm), preequilibrated with buffer A containing 200 mM NaCl. Protein is eluted with the same solution at a rate of 8 ml/hr. Molecular sieving of isozyme I yields a single peak of enzyme activity corresponding to a molecular weight of about 150 kDa. Molecular sieving of isozyme II resolves two peaks of activity, corresponding to molecular weights of 240 kDa and greater than 500 kDa.

A summary of the results of purification is presented in Table I.

Properties[1-4]

General. The pH optimum of cystine degradation is between 8.5 and 9, using either isozyme. Both isozymes have a K_m for cystine of 0.3 mM. The enzyme is stabilized during purification by pyridoxal phosphate, and a requirement for this cofactor is demonstrable for preparations from some plants. The molecular weight of isozyme I is about 150 kDa,

TABLE I
PURIFICATION OF CYSTINE LYASE I AND II FROM *Brassica oleracea* var. *capitata*

Fraction	Total volume (ml)	Total protein (mg)	Total activity[a] (units)	Specific activity (units/mg)
Homogenate	1500	3210	5140	1.6
Ammonium sulfate 40–60%	80	480	3600	7.5
Isozyme I				
DEAE-Cellulose peak I	20	12.2	575	47.1
BioGel A-0.5M	6	0.5	43	87.5
Isozyme II				
DEAE-Cellulose peak II	20	7.0	377	53.9
BioGel A-0.5M	8	0.08	28	350

[a] Micromoles pyruvate per hour.

whereas isozyme II is resolved by BioGel into two high molecular weight forms, 240 kDa and greater than 500 kDa.

Inhibitors. The enzyme is totally inhibited by either 0.5 mM NaCN or 0.5 mM hydroxylamine but not by similar concentrations of *p*-chloromercuribenzoate, iodoacetate, or *N*-ethylmaleimide. Activity is reduced 40%

TABLE II
SPECIFICITY OF CYSTINE LYASES ISOLATED FROM TURNIP (*Brassica rapa*) ROOTS AND CABBAGE (*Brassica oleracea*) LEAVES

	% Activity with L-cystine			
Substrate	Isozyme I cabbage[3]	Turnip[1]	Isozyme II cabbage[3]	Turnip[2]
L-Cystine	100	100	100	100
D-Cystine	N.A.[a]	0	N.A.	0
L-Cysteine	41	14	21	N.A.
D-Cysteine	1	N.A.	0	N.A.
S-Methyl-L-cysteine sulfoxide	38	46	38	25
S-Methyl-L-cysteine	14	32	0	0
L-Djenkolic acid sulfoxide	N.A.	N.A.	N.A.	116
L-Djenkolic acid	16	16	7	0
S-Sulfo-L-cysteine	N.A.	86	N.A.	129

[a] N.A., not assayed

by 0.5 mM mercaptoethanol, probably because of the conversion of cystine to the less active substrate, cysteine.

Specificity. The substrate specificity of the two isozymes differs (Table II). Isozyme I is less specific, catalyzing an α,β-elimination reaction with a range of S-substituted cysteines, L-cysteine, and O-acetylserine. The substrate specificity is similar to that of the S-alkylcysteine lyase (EC 4.4.1.6) purified from *Acacia farnesiana* Willd seedlings.[6] In contrast, isozyme II degrades only S-substituted cysteines when the sulfur atom is oxidized to the sulfoxide or is attached to an oxidized sulfur atom, as in S-sulfocysteine. The alternate substrates are competitive inhibitors, meaning that total pyruvate production is lower in mixtures of L-cystine and poor substrates (cysteine or cystathionine) than it is when only L-cystine is present.

Physiological Role. The physiological role of the enzyme is debatable. The characteristics discussed above are of enzymes isolated from plants containing S-methylcysteine sulfoxide. Assuming that cystine and S-methylcysteine sulfoxide are substrates of a single protein, the physiological function of the enzyme in these plants is probably degradation of the latter substrate, because the concentration of cystine in healthy tissue is negligible. In contrast, L-cysteine desulfhydrase (EC 4.4.1.1, cystathionine γ-lyase) activity has been demonstrated in a variety of plants not known to contain S-substituted cysteines, and its function is thought to be regulation of the cysteine pool size[7]; unfortunately, this enzyme has not been characterized, and therefore it is not known whether or not L-cystine is an alternate substrate.

[6] M. Mazelis and R. K. Creveling, *Biochem. J.* **147**, 485 (1975).
[7] H. Rennenberg, *Annu. Rev. Plant Physiol.* **35**, 121 (1984).

[76] Cystathionine β-Lyase from Spinach

By JOHN GIOVANELLI

L-Cystathionine + H_2O → L-homocysteine + pyruvic acid + NH_3

Cystathionine β-lyase (EC 4.4.1.8), the enzyme catalyzing the above reaction, is widely distributed in plants. In none of the plant tissues examined was γ-cleavage of cystathionine, i.e., the formation of cysteine, α-ketobutyrate, and NH_3, detected.[1]

[1] J. Giovanelli and S. H. Mudd, *Biochim. Biophys. Acta* **227**, 654 (1971).

Assay Procedures

Principle. Enzyme activity is determined by the rate of formation of pyruvate, measured either isotopically or spectrophotometrically. In the isotopic assay, enzyme is incubated with cystathionine labeled with ^{14}C specifically in the 3-carbon moiety. The incubation is terminated with trichloroacetic acid, and ^{14}C compounds fractionated on a column of Dowex 50 (H^+). Unreacted [^{14}C]cystathionine is retained by the column, while the radioactive product of β-cleavage of cystathionine (pyruvate) is recovered in the effluent and provides a measure of cystathionine β-lyase activity. The isotopic assay is suitable for crude extracts. It is specific for β-cleavage of cystathionine, since the keto acid formed by γ-cleavage (α-ketobutyrate) is derived from the 4-carbon moiety of cystathionine, and is therefore not radioactive. Removal of NADH oxidase during purification allows use of a spectrophotometric assay in which pyruvate formation is coupled to the oxidation of NADH in the presence of lactic dehydrogenase.

Isotopic Assay

Reagents

Tris–HCl, 1 M, pH 8.65
Sodium pyruvate, 0.1 M, freshly prepared
Trichloroacetic acid, 5%
Dowex 50 (H^+): AG 50W-X4, 200–400 mesh (Bio-Rad) is converted to the H^+ form and used in columns of 0.9 × 2.9 cm.
L[3-^{14}C]Cystathionine (radioactivity located in the β-carbon of the 3-carbon moiety) is prepared by incubation of L-[3-^{14}C]serine and L-homocysteine with partially purified cystathionine β-synthase.[2] L-Homocysteine is freshly prepared from the thiolactone,[2] and L-[3-^{14}C]serine (Amersham) is purified by column chromatography on Dowex 50 (H^+).[2] The reaction mixture contains the following in a final volume of 1.3 ml: L-[3-^{14}C]serine, 0.6 mCi, 78 μmol; L-homocysteine, 150 μmol; Tris–HCl at pH 8.3, 60 μmol; EDTA, 2 μmol; pyridoxal phosphate, 0.025 μmol; and partially purified rat liver cystathionine synthase sufficient to form 13 μmol of cystathionine in 135 min under the standard assay conditions described by Mudd *et al.*[2] L-[3-^{14}C]Cystathionine formed in the reaction is isolated and purified as described.[2] Oxidation to the sulfoxide is minimized by addition of 10 mM 2-mercaptoethanol to the solvent used

---

[2] S. H. Mudd, J. D. Finkelstein, F. Irreverre, and L. Laster, *J. Biol. Chem.* **240,** 4382 (1965).

for preparative chromatography. Carrier L-cystathionine (Calbiochem) is added to this preparation to obtain the L-[3-^{14}C]cystathionine of lower specific activity that is used in the assay.

Procedure.[1] The reaction mixture contains, in a final volume of 0.5 ml, Tris–HCl at pH 8.65, 100 μmol; L-[3-^{14}C]cystathionine, 9.8 nCi, 0.0475 μmol; sodium pyruvate, 5 μmol; and cystathionine β-lyase (up to 0.5 units). The reaction is started by addition of enzyme, and allowed to proceed under an atmosphere of N_2 for 30 min at 30°. The reaction is stopped by addition of 1 ml of 5% trichloroacetic acid, and the precipitate removed by centrifugation. The supernatant solution is applied to a column of Dowex 50 (H^+) which is washed with 3 ml of water. Radioactivity in the combined effluent and wash fractions (α-keto acid fraction) is used to determine the amount of pyruvate formation. Correction is made for a relatively small rate of appearance of ^{14}C in the α-keto acid fraction in the presence of heat-inactivated enzyme.

Spectrophotometric Assay

Reagents

Tris–HCl, 1 M, pH 8.65

L-Cystathionine (Calbiochem), 0.1 M: The compound (11.2 mg) is dissolved in a minimal amount of 1 N KOH and diluted to 5.0 ml with water.

NADH, 1.7 mM

Lactate dehydrogenase, crystalline preparation from heart in 2.2 M ammonium sulfate (Calbiochem), containing 9.6 mg protein/ml with a specific activity of 4090 international units/ml

Procedure. Reactions are carried out in cuvettes of 1-cm light path in a thermoregulated (30°) Cary Model 15 spectrophotometer. The reaction contains, in a final volume of 1 ml, Tris–HCl at pH 8.65, 200 μmol; L-cystathionine, 10 μmol; NADH, 0.175 μmol; lactic dehydrogenase, 30 μg; and cystathionine β-lyase. The assay is linear up to at least 2 units of cystathionine β-lyase. The reaction is started by addition of cystathionine β-lyase, and the absorbance at 340 nm is recorded continuously with reference to a cuvette containing identical components except for omission of cystathionine β-lyase. The assay, used chiefly for enzyme purified from Steps 4 and 5, is dependent on addition of the enzyme, cystathionine, NADH, and lactic dehydrogenase. Enzyme from Step 3 can also be assayed spectrophotometrically, provided a correction (approximately 15%) is made for the cystathionine β-lyase-independent rate of NADH oxidation. Rates of cystathionine β-lyase determined with the spectrophotometric assay are approximately 2.2-fold those determined with the isotopic assay, as expected from the K_m for cystathionine and the respective concentrations of cystathionine in the two assays.

Definition of Unit and Specific Activity. A unit of enzyme is defined as that catalyzing the cleavage of 1 nmol of cystathionine per minute. Specific activity is the number of units of enzyme per milligram protein. Protein was determined by the Lowry method,[3] using bovine serum albumin as a standard.

Preparation of Enzyme

Reagents

Pyridoxal phosphate: A stock solution (0.25 M) is prepared from 165 mg of pyridoxal phosphate (Calbiochem, A grade) in a mixture of 0.6 ml 1 M KOH and 0.25 ml 2 M Tris–HCl at pH 8.65, followed by dilution with water to a final volume of 2.5 ml. The solution remains stable for many months when stored in the dark at $-65°$. Solutions of pyridoxal phosphate described in the text are prepared by fresh dilution of this stock solution.

Buffer A: 0.1 M potassium phosphate, 0.1 mM EDTA, and 0.14 mM 2-mercaptoethanol, final pH 7.25

Buffer B: 5 mM potassium phosphate, 0.1 mM EDTA, 0.1 mM pyridoxal phosphate, and 0.14 mM 2-mercaptoethanol, final pH 7.25

Buffer C is identical to Buffer B, with omission of pyridoxal phosphate.

Aluminum oxide (Merck No. 71707) was triturated with 2 N HCl, the slurry thoroughly washed with water on a Büchner funnel, and then dried in an oven.

$(NH_4)_2SO_4$, Mann, special enzyme grade

Hydroxyapatite, prepared by the method of Tiselius *et al.*,[4] was equilibrated with 5 mM K_2HPO_4 adjusted to pH 5.6 with acetic acid.

Bentonite, Fisher, U.S.P.

Procedure. Unless stated otherwise, all operations are carried out at $4°$, and centrifugations are at 9,000 g in a Sorvall RC-2B centrifuge. Sephadex G-25 (coarse) was used for all gel filtrations.

Step 1: Acetone Powder. Spinach (*Spinacia oleracea* L.) is grown in a greenhouse maintained at approximately $24°$ or purchased from a local market. Leaves are deveined, washed thoroughly in distilled water, and cut into strips approximately 1 cm wide. The shredded leaves (160 g) are

[3] E. Layne, this series, Vol. 3 [73].
[4] A. Tiselius, S. Hjertén, and O. Levin, *Arch. Biochem. Biophys.* **65,** 132 (1956).

homogenized in a Waring Blendor (1-gallon capacity) with 2.4 liters of acetone at $-40°$. Excess acetone is removed on a Büchner funnel, and the protein paste suspended in 2 liters of ether at room temperature. The suspension is again filtered with suction on a Büchner funnel, and residual solvent removed in a vacuum desiccator containing paraffin flakes and magnesium perchlorate. The dry powder (9 g) retains activity for at least 1 year when stored at 4°.

Step 2: Extraction. Six grams of acetone powder is ground in a mortar and pestle with 6 g of aluminum oxide and 120 ml of Buffer A. The slurry is centrifuged for 40 min. The supernatant solution is decanted through muslin and centrifuged for an additional 2 hr to yield 88 ml of clear supernatant solution.

Step 3: Ammonium Sulfate Fractionation. The supernatant solution is brought to 24% saturation by addition of $(NH_4)_2SO_4$, 13.45g/100 ml. The precipitate is removed by centrifugation, followed by addition of more $(NH_4)_2SO_4$, 5.5 g/100 ml, to the supernatant solution to obtain 34% saturation. The precipitate is collected by centrifugation, dissolved in a minimal volume of Buffer A, and gel filtered through Sephadex G-25 equilibrated with Buffer B.

Step 4: Negative Adsorption with Bentonite. The preparation from Step 3 is diluted with Buffer B to a concentration of 7.9 mg protein/ml. To each milliliter of this solution is added a slurry of 24 mg bentonite and 0.4 ml of 0.1 M acetic acid. After stirring for 15 min, the mixture is centrifuged at 20,000 g to remove bentonite. Enzyme activity, recovered in the supernatant fraction, is unstable (see below) and is therefore subjected immediately to the next purification step.

Step 5: Hydroxyapatite Chromatography. The supernatant solution (30 ml) is acidified with 6 ml of 25 mM acetic acid and aliquots of 12 ml applied to each of three columns (0.9 × 2.2 cm) of hydroxyapatite. Multiple columns are preferable to a single large column, because of the slow flow rates of the latter. Each column is washed with 2 ml of 0.1 M potassium acetate, pH 5.5, and eluted with 4 ml of 0.5 M potassium phosphate at pH 6.0. Eluates from the columns are combined, concentrated approximately 10-fold by "dialysis" against Aquacide II (Calbiochem), and gel filtered through Sephadex G-25 equilibrated with Buffer C.

A summary of the purification is given in Table I.

Properties

Stability. Activities of crude extracts and the 24–34% $(NH_4)_2SO_4$ fraction were stable at $-80°$ for at least 5 months. More purified fractions

TABLE I
PURIFICATION OF CYSTATHIONINE β-LYASE FROM SPINACH

Purification step	Volume (ml)	Total protein (mg)	Total activity[a] (units)	Specific activity[a] (units/mg)
2. Crude extract	88	1090	284	0.26
3. Ammonium sulfate	12	136	258	1.9
4. Bentonite treatment	30	6.2	114	18.4
5. Hydroxyapatite column	1.5	0.43	48	111

[a] Activities of the crude extract were determined by the isotopic assay and were multiplied by 2.2 to yield the reported values equivalent to those determined by the spectrophotometric assay. All other fractions were assayed spectrophotometrically. Permission of Elsevier Science Publishers to reproduce previously published[1] portions of this table is gratefully acknowledged.

(Steps 4 and 5) lose up to 70% of their activity after storage for 5 days at 4° and at −80°, thereby making it impractical to purify the enzyme further.

Cofactor. Addition of pyridoxal phosphate does not significantly increase the activity of spinach cystathionine β-lyase at any stage of purification. Enzyme activity is abolished by treatment with hydroxylamine and is restored by incubation with pyridoxal phosphate. These observations suggest that spinach cystathionine β-lyase is a pyridoxal phosphate enzyme which is isolated as the holoenzyme.[1] Studies on the inhibition of the spinach enzyme by rhizobitoxine (see below) provide additional evidence for pyridoxal phosphate as a prosthetic group.

pH Optimum. The pH optimum in Tris–HCl buffer is 8.8.

Substrate Specificity and Kinetic Properties. The enzyme is most active with L-cystathionine, L-djenkolate, and the mixed disulfide of L-cysteine and L-homocysteine. No γ-cleavage of cystathionine (to α-ketobutyrate, cysteine, and NH_3) can be detected. K_m values for cystathionine and djenkolate are 0.13 and 0.25 mM, respectively; the V_{max} for cystathionine is approximately 90% that for djenkolate.[1] The substrate specificity of spinach cystathionine β-lyase[1] resembles that of the bacterial[5] enzyme in being most active with cystathionine and djenkolate and far less active with cystine, S-methylcysteine, and serine; it contrasts with that of the fungal[6] enzyme which exhibits activities with cystine, serine, and lanthionine approximating that with cystathionine.

[5] C. Delavier-Klutchko and M. Flavin, *Biochim. Biophys. Acta* **99**, 375 (1965).
[6] M. Flavin and C. Slaughter, *J. Biol. Chem.* **239**, 2212 (1964).

Inhibitors. β-Cyanoalanine is a strong competitive inhibitor, with a K_i of 40 μM. The enzyme is inhibited also by 5,5'-dithiobis(2-nitrobenzoic acid) and N-ethylmaleimide. With respect to sensitivity to the latter two compounds, the plant enzyme resembles fungal cystathionine β-lyase[6] and differs from bacterial cystathionine β-lyase which is not sensitive to either of these inhibitors.[7] Rhizobitoxine, a bacterial toxin similar in structure to cystathionine,[8] is a potent inhibitor of plant cystathionine β-lyase, both *in vivo*[9] and *in vitro*.[10] Inhibition of the purified enzyme from spinach is of the active-site-directed irreversible type, probably involving covalent linkage of an enzyme-catalyzed cleavage product of rhizobitoxine with the pyridoxal phosphate prosthetic group[10]:

$$EP + R \rightleftharpoons EPR \rightarrow EPR_1$$

where EP is the holoenzyme form of cystathionine β-lyase (P = pyridoxal phosphate), R rhizobitoxine, EPR a dissociable enzyme–rhizobitoxine complex, and EPR_1 an inactivated enzyme. The dissociation constant (K_i) of EPR is 80 μM. The rate of inactivation at infinite rhizobitoxine concentration is equivalent to a minimum half-time of inactivation of 0.2 min. The kinetics of protection of the enzyme by substrates (cystathionine, djenkolate) or the competitive inhibitor β-cyanoalanine are consistent with competition between these amino acids and rhizobitoxine for the active enzyme.

[7] C. Delavier-Klutchko and M. Flavin, *J. Biol. Chem.* **240**, 2537 (1965).
[8] L. D. Owens, J. F. Thompson, R. G. Pitcher, and T. Williams, *J. Chem. Soc., Chem. Commun.* p. 714 (1972).
[9] J. Giovanelli, L. D. Owens, and S. H. Mudd, *Plant Physiol.* **51**, 492 (1973).
[10] J. Giovanelli, L. D. Owens, and S. H. Mudd, *Biochim. Biophys. Acta* **227**, 671 (1971).

[77] D-Cysteine Desulfhydrase from Spinach

By AHLERT SCHMIDT

Introduction

Cysteine and its derivatives are important intermediates in plant sulfur metabolism. It is well documented that, in general, the amino acids are used in the L-form and enzymes involved in their metabolism are stereospecific for the L-enantiomers. Recent findings have shown, however, that green algae and cyanobacteria are able to grow on either L- or D-cysteine as the sole sulfur source,[1] and that higher plants are able to emit volatile hydrogen sulfide after feeding of L- or D-cysteine. These findings indicate that enzymes capable of metabolizing D-cysteine or D-cysteine derivatives should be present in plant and algal cells, and enzymes specific for the D-enantiomer of cysteine have now been demonstrated in plants and algae.[2,3] Presented here is a procedure for the separation of the stereospecific D-cysteine desulfhydrase from spinach (*Spinacia oleracea* L.) leaves.

Assay Method

Principle. D-Cysteine desulfhydrase activity is assayed by measuring the production of hydrogen sulfide from D-cysteine. Hydrogen sulfide is determined by a reaction that forms methylene blue.[4]

Reagents

Tris–HCl, 1 M, pH 9.0
D-Cysteine, 8 mM
Dithioerythritol, 100 mM
N',N'-Dimethyl-p-phenylenediamine dihydrochloride, 20 mM in 7.2 N HCl (Solution I)
FeCl$_3$, 30 mM in 1.2 M HCl (Solution II)

Procedure. The reaction mixtures contain the following in a final volume of 1.0 ml: 0.1 ml Tris–HCl buffer (100 μmol), 25 μl dithioerythritol (2.5 μmol), and enzyme. After 30 min at 37°, the reaction is terminated by addition of 0.1 ml of solution I and 0.1 ml of Solution II. After 30 min, formation of methylene blue is determined spectrophotometrically at 670 nm using a molar extinction coefficient of 28.5×10^6.[2,3,7] It should be noted that the conditions given are optimal for the assay of D-cysteine desulfhydrase; higher concentrations of D-cysteine, dithioerythritol, or other thiols inhibit methylene blue formation.

Step 1: Extract. Spinach leaves, 4 kg, are homogenized in a Waring Blendor in 20 mM Tris–HCl at pH 8.0 (buffer I). The crude extract is clarified by centrifugation for 10 min at 10,000 g.

Step 2: Polymin Treatment. The supernatant fluid is mixed with

---

[1] K. F. Krauss, Thesis, München (1984).
[2] A. Schmidt, *Z. Pflanzenphysiol.* **107**, 301 (1982).
[3] A. Schmidt and I. Erdel, *Z. Naturforsch., C. Biosci.* **38C**, 428 (1983).
[4] F. Pachmayr, Thesis, München (1960).

polymin P (20 μl/ml), and the precipitate formed is removed by centrifugation as noted above.

Step 3: Ammonium Sulfate. Solid ammonium sulfate is added, and the precipitate between 35 and 80% of saturation is collected by centrifugation and dissolved in a small volume of buffer I.

Step 4: Gel Chromatography. The protein solution is clarified by centrifugation and further fractionated on an AcA 54 (Pharmacia) gel column (2.6 × 70 cm) equilibrated and eluted with 0.1 M Tris–HCl at pH 8.0, containing 0.1 M KCl. Batches of 26 ml are applied for each run, and fractions of 4 ml are collected. The D-cysteine desulfhydrase elutes between 180 and 210 μl. These fractions are combined and concentrated using an Amico diaflow system with a 10,000 M_r cut-off membrane.

Step 5: DEAE-Cellulose. The concentrated enzyme is loaded onto a DEAE-cellulose column (2 × 8 cm) previously equilibrated with 20 mM Tris–HCl at pH 8.0. After washing with 50 ml of this buffer, the column is developed with a linear gradient (400 ml) from 0 to 0.5 M NaCl in the same buffer. The enzyme is found between about 120 and 150 ml. Active fractions are pooled, concentrated, and rechromatographed using identical conditions. Active fractions are pooled. DEAE-Cellulose chromatography separates the D-cysteine desulfhydrase from L-cysteine-specific activities.

Step 6: Biogel A-1.5M. The final step is based on molecular size using a BioGel A-1.5M matrix (1.5 × 75 cm). The enzyme is collected between 98 and 110 ml; combined fractions are stored at $-18°$.

By the procedure outlined, a 110-fold purification with a yield of 34% was achieved. The purified enzyme had a specific activity of 5.7 μmol H_2S formed per milligram protein per hour using standard assay conditions. A summary of the results of the purification procedure are presented in Table I.

Properties of the Enzyme

Substrate Specificity. The purified enzyme is specific for D-cysteine; no activity was found with the following analogs: D-cysteine, L-cysteine, L-cystine, N-acetyl-L-cysteine, L-cysteine methyl ester, L-cysteine ethyl ester, DL-homocysteine, mercaptoacetic acid, mercaptoethanol, mercaptolactic acid, and dithioerythritol. Addition of dithiothreitol enhances sulfide production, probably by preventing autoxidation of D-cysteine.[2,3]

Influence of Ions and Inhibitors. The following compounds do not inhibit at concentrations of 2 mM: $MgCl_2$, $CaCl_2$, KCl, NaCl, $AlCl_3$,

TABLE I
Purification of d-Cysteine Desulfhydrase from Spinach Leaves

Step	Volume (ml)	Activity (units[a]/ml)	Protein (mg/ml)	Specific activity (units[a]/mg)
1. Crude extract	3,540	0.66	12.8	0.052
2. Polymin P supernatant	3,610	0.73	9.1	0.08
3. Ammonium sulfate	239	18.7	35.3	0.53
4. Gel chromatography	80	36.9	27.5	1.34
5. DEAE–Cellulose chromatography	34	55.1	21.8	2.52
6. BioGel A-1.5M gel chromatography	15	53.6	9.3	5.74

[a] Units are defined in terms of micromoles of product formed per hour.

EDTA, and trisodium citrate. Inhibition, probably due to complex formation with d-cysteine, was observed with $NiCl_2$ and $CoCl_2$.[2,3] Hydroxylamine and aminooxyacetic acid inhibit enzymatic activity 80% at 2 mM,[1] indicating the involvement of pyridoxal 5′-phosphate. However, addition of pyridoxal 5′-phosphate did not enhance the activity, suggesting that d-cysteine desulfhydrase has a high affinity for the coenzyme.

Products. Analysis of the products of the d-cysteine desulfhydrase reaction has shown H_2S to be the major one; its formation is the basis of the assay used. Pyruvic acid and ammonia are also produced, but the amounts detected are not stoichiometric with sulfide,[2,3] suggesting that another compound(s) is formed as well. We have evidence that a 4-methylthiazolidine-1,4-dicarboxylic acid could be formed; that product was also detected with the l-cysteine desulfhydrase from *Salmonella typhimurium*.[5] In addition, a thiol-exchange reaction leading to a dithiothreitol cysteine adduct and free sulfide has not been excluded.

Comments

An enzyme specific for the d-enantiomer of cysteine can be isolated from spinach and may be used for the analysis of d-cysteine and for d-cystine if the later is reduced prior to analysis. The same procedure can be applied to isolate an enzyme from the green alga *Chlorella fusca* having identical properties.[3] The *Chlorella* enzyme, however, is not stable when stored frozen for longer periods. No function for the d-cysteine-specific enzyme is known. One might speculate, however, about different routes

[5] H. M. Kredich, L. J. Foote, and B. S. Keenan, *J. Biol. Chem.* **248**, 6187 (1981).

for synthesis and degradation, a "compartmentation" of thiol pools without morphologically separate compartments, a possible signal theory for regulation, or specific biosynthetic routes.[2,6,8] Such possibilities would appear to require a cysteine racemase, which has not yet been detected in green plants or algae.

[6] K. R. Krijsheld, E. J. Glazenburg, E. Scholtens, and G. J. Mulde, *Biochim. Biophys. Acta* **677,** 7 (1981).
[7] L. M. Siegel, *Anal. Biochem.* **11,** 126 (1965).
[8] T. Nagasawa, T. Ishii, H. Kumagai, H. Yamada, *Eur. J. Biochem.* **153,** 541 (1985).

[78] Microbial Sulfur Amino Acids: An Overview

By KENJI SODA

Incorporation of Inorganic Sulfur into Amino Acids in Microorganisms

Various microorganisms assimilate inorganic sulfur to synthesize sulfur amino acids. Hydrogen sulfide, the final product of bacterial sulfate reduction, reacts microbially with serine and homoserine to form cysteine and homocysteine, respectively. O-Acetylserine is derived from serine and acetyl-CoA, and serves as a better β-substituent acceptor of H_2S for the key enzyme, O-acetylserine (thiol)-lyase (O-acetylserine sulfhydrylase, EC 4.2.99.8) than serine in *Salmonella*,[1] other bacteria,[2] and yeast,[3] whereas both amino acids are converted to cysteine at similar rates in *Neurospora*. Cystathionine β-synthase (serine sulfhydrylase, EC 4.2.1.22), which catalyzes the reaction of serine and H_2S to yield cysteine and H_2O, is found in a variety of bacteria, yeasts, and fungi,[4] but its metabolic importance in the assimilation of inorganic sulfur has not been shown.

Alternatively, hydrogen sulfide is incorporated into homocysteine by the γ-replacement reaction of O-acylhomoserines. O-Acetylhomoserine is the preferred γ-substituent acceptor in fungi,[5] yeast,[6] and in such bacteria

[1] N. M. Kredich and G. M. Tomkins, *J. Biol. Chem.* **241,** 4955 (1966).
[2] L. A. Chambers and P. A. Trudinger, *Arch. Microbiol.* **77,** 164 (1971).
[3] J. L. Wiebers and H. R. Garner, *J. Biol. Chem.* **241,** 5644 (1967).
[4] J. Bruggemann, K. Schlossmann, M. Merckenschlager, and M. Waldschmidt, *Biochem. Z.* **335,** 392 (1962).
[5] J. Giovanelli and S. H. Mudd, *Biochem. Biophys. Res. Commun.* **27,** 150 (1967).
[6] S. Nagai and M. Flavin, *J. Biol. Chem.* **241,** 4463 (1966).

as *Brevibacterium*,[7] *Bacillus*,[8] and others. O-Succinylhomoserine is the predominant substrate in *Escherichia coli*[4] and *Salmonella typhimurium*.[9] O-Acylhomoserine enzymatically reacts also with thiols other than H_2S, e.g., CH_3SH, to form the corresponding S-substituted homocysteines, e.g., methionine. Homoserine is only a poor substrate of O-acetylhomoserine (thiol)-lyase (EC 4.2.99.10) and, therefore, is probably not a direct precursor of sulfur amino acids.[10]

Metabolism of Cysteine and Its Derivatives

Cysteine, homocysteine, and their derivatives synthesized from inorganic sulfur are converted to a number of other sulfur amino acids in microorganisms. Methionine and cysteine are regarded as the two metabolic centers of sulfur amino acids. Figure 1 depicts the metabolic pathways of cysteine and its derivatives. Homocysteine, another key intermediate of sulfur amino acid metabolism, is synthesized from O-succinylhomoserine and cysteine through cystathionine by coupled reactions of O-succinylhomoserine (thiol)-lyase (cystathionine γ-synthase, EC 4.2.99.9) and cystathionine β-lyase (EC 4.4.1.8) (transsulfurylation) in *Salmonella*[11] and *E. coli*. O-Acetylhomoserine is substituted for O-succinylhomoserine in *Neurospora*, yeasts, and some bacteria. The reverse transsulfurylation, i.e., the formation of cysteine from homocysteine catalyzed by cystathionine β-synthase (EC 4.2.1.22) and cystathionine γ-lyase (EC 4.4.1.1), also occurs in *Neurospora* as it does in mammalian systems. Cystathionine γ-lyase is also found in *Streptomyces* and other actinomycetes.[12]

Cysteine is deaminated oxidatively to pyruvate by a pyridoxal enzyme, cysteine desulfhydrase (EC 4.4.1.1, cystathionine γ-lyase), and oxidized to cysteine sulfinate by cysteine dioxygenase (EC 1.13.11.20), which is found in yeasts.[13] Cysteine sulfinate is also an important metabolite of cysteine, resembling aspartate in its chemical properties. It undergoes enzymatic transamination with α-ketoglutarate and several other keto acids to form the corresponding amino acids and β-sulfinopyruvate,

---

[7] R. Miyajima and I. Shiio, *J. Biochem. (Tokyo)* **73**, 1061, (1973); see also this volume [81].
[8] A. H. Datko, J. Giovanelli, and S. Mudd, *J. Biol. Chem.* **249**, 1139 (1974); see also this volume [82].
[9] M. M. Kaplan and M. Flavin, *J. Biol. Chem.* **242**, 3884 (1967).
[10] S. Yamagata and K. Takeshima, *J. Biochem. (Tokyo)* **80**, 777 (1976).
[11] M. Flavin, this series, Vol. 17, p. 416.
[12] T. Nagasawa, H. Kanzaki, and H. Yamada, *J. Biol. Chem.* **259**, 10393 (1984); see also this volume [85].
[13] V. Kumar, *Biochemistry* **22**, 762 (1983); see also this volume [67].

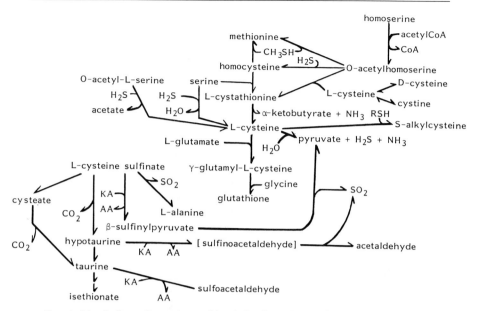

FIG. 1. Metabolism of cysteine and its derivatives. KA, amino acceptors; AA, amino donors.

which spontaneously degrades to pyruvate and SO_3^{2-}; thus, this transamination is essentially irreversible.[14] Cysteine sulfinate is desulfinated to alanine by bacterial aspartate β-decarboxylases (EC 4.1.1.12), and may be metabolized through hypotaurine, although the metabolic route in microbes has not been studied in detail. Taurine is transaminated with α-ketoglutarate to produce β-sulfoacetaldehyde by bacterial taurine aminotransferase (EC 2.6.1.55).[15] Hypotaurine also is transaminated with pyruvate by this enzyme and by bacterial ω-amino-acid aminotransferase (EC 2.6.1.—) to form β-sulfinoacetaldehyde, which is spontaneously degraded to acetaldehyde and sulfite.[16] Taurine is metabolized to isethionate in bacteria with taurine dehydrogenase (EC 1.4.99.2) catalyzing the first metabolic step.[17] Cysteine undergoes enzymatic β-replacement reactions with a number of thiols to form the corresponding β-substituted cysteines, although the physiological function of the reactions remains unknown.

[14] A. Meister, "Biochemistry of the Amino Acids." Academic Press, New York, 1965.
[15] S. Toyama and K. Soda, *J. Bacteriol.* **109,** 533 (1972).
[16] K. Yonaha, S. Toyama, M. Yasuda, and K. Soda, *Agric. Biol. Chem.* **41,** 1701 (1977); see also this volume [88].
[17] H. Kondo, K. Kagotani, M. Oshima, and M. Ishimoto, *J. Biochem.* (Tokyo) **73,** 1269 (1973); see also this volume [87].

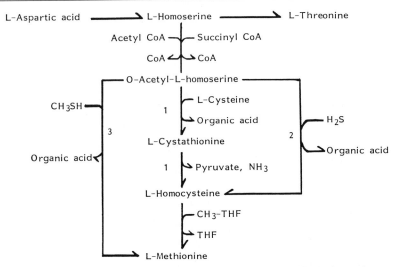

FIG. 2. Synthesis of methionine and its derivatives. (1) Transsulfurylation pathway, (2) Sulfhydrylation pathway, (3) methylsulfhydrylation pathway.

Bacterial methionine γ-lyase (EC 4.4.1.11),[18] which also catalyzes α,β-elimination of cysteine and S-substituted cysteines, participates in the replacement reaction.

Synthesis of Methionine and Its Derivatives

Methionine can be synthesized from O-acylhomoserine derived from homoserine in microorganisms (Fig. 2). Homocysteine is produced from O-acylhomoserine through cystathionine as described above, or directly by γ-replacement with H_2S. This pathway is the main route in many microorganisms but not in *Brevibacterium flavum,* in which direct production of homocysteine from O-acetylhomoserine and H_2S is predominant.[19] Methionine can be synthesized directly from O-acylhomoserine and CH_3SH through methylsulfhydration as catalyzed by O-acetylhomoserine (thiol)-lyase, i.e., methionine synthase. Homocysteine is S-methylated by two distinct pathways in microbial methionine synthesis. One is the vitamin B_{12}-dependent homocysteine transmethylation reaction found in *E. coli,*[20] *Aerobacter aerogenes,*[21] and some other species. This re-

[18] K. Soda, H. Tanaka, and N. Esaki, *Trends Biochem. Sci.* **8**, 214 (1983); see also this volume [79].
[19] H. Ozaki and J. Shiio, *J. Biochem. (Tokyo)* **91**, 1163 (1982); see also this volume [81].
[20] M. A. Foster, G. Tejerina, J. R. Guest, and D. D. Woods, *Biochem. J.* **92**, 476 (1964).
[21] D. Karr, J. Tweto, and P. Albersheim, *Arch. Biochem. Biophys.* **121**, 732 (1967).

quires a reduction system such as $FADH_2$, and a trace amount of S-adenosyl methionine [Eq. (1): CH_3—THF, N^5-methyltetrahydrofolate;

$$CH_3\text{—THF} + \text{Homocysteine} \xrightarrow{B_{12} \text{ coenzyme}} \text{methionine} + \text{THF} \quad (1)$$

THF, tetrahydrofolate]. The vitamin B_{12}-coenzyme is required to form the holo form of N^5-methyltetrahydrofolate–homocysteine methyltransferase (EC 2.1.1.13). Methionine is probably synthesized as shown in Eqs. (2) and (3), although the role of S-adenosylmethionine (AdoMet) has

$$N^5\text{-}CH_3\text{—THF} + B_{12} \text{ enzyme} \xrightarrow[\text{AdoMet}]{\text{NADH}} CH_3\text{—}B_{12} \text{ enzyme} + \text{THF} \quad (2)$$

$$CH_3\text{—}B_{12} \text{ enzyme} + \text{Homocysteine} \longrightarrow \text{methionine} + B_{12} \text{ enzyme} \quad (3)$$

not been fully elucidated.[22] The vitamin B_{12}-independent methylation is catalyzed by N^5-methyltetrahydrofolate–homocysteine methyltransferase (EC 2.1.1.14) [Eq. (4)]; the reaction occurs in bacteria, yeasts, and fungi that cannot produce vitamin B_{12}.[23]

$$N^5\text{-}CH_3\text{—THF} + \text{Homocysteine} \xrightarrow{Mg^{2+}, P_i} \text{methionine} + \text{THF} \quad (4)$$

Catabolism of Methionine and Its Derivatives

Methionine plays a central role in the catabolism of sulfur amino acids as shown in Fig. 3. It is metabolized through α-ketobutyrate by three possible pathways: (1) the conversion of methionine to cystathionine through S-adenosylmethionine and homocysteine, and then to α-ketobutyrate and NH_3; (2) the deamination to α-keto-γ-thiomethylbutyrate by L-amino-acid oxidase and aminotransferases, or by coupling of amino-acid racemase[24] and D-amino-acid oxidase reactions with subsequent dethiomethylation to α-ketobutyrate, the dethiomethylation not having been clearly shown in microorganisms; or (3) the simultaneous deamination and dethiomethylation to α-ketobutyrate as catalyzed by bacterial methionine γ-lyase. Methionine is decarboxylated to the corresponding amine, β-thiomethylpropylamine, by bacterial methionine decarboxylase.[25] Adenosylhomocysteinase (S-adenosylhomocysteine hydrolase, EC 3.3.1.1) participates in transsulfurylation in bacteria, and is useful in the enzymatic production of S-adenosylhomocysteine.[26] S-Adenosylmethionine is decarboxylated by S-adenosylmethionine decarboxylase

[22] D. M. Greenberg, ed. "Metabolic Pathways," Vol. 7. Academic Press, New York, 1975.
[23] R. T. Taylor and H. Weissbach, in "The Enzymes" (P. D. Boyer, ed.), 3rd ed., Vol. 9, p. 121. Academic Press, New York, 1973.
[24] K. Soda and T. Osumi, this series, Vol. 17, p. 629.
[25] H. Misono, Y. Kawabata, M. Toyosato, T. Yamamoto, and K. Soda, *Bull. Inst. Chem. Res., Kyoto Univ.* **58**, 323 (1980).
[26] S. Shimizu, S. Shiozaki, T. Ohshiro, and H. Yamada, *Eur. J. Biochem.* **141**, 385 (1984).

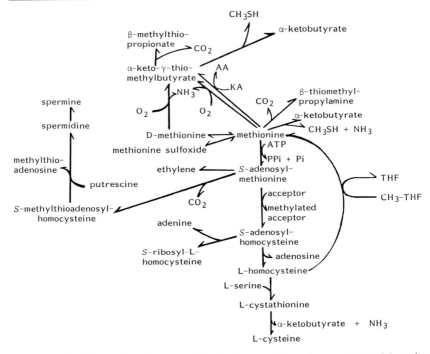

FIG. 3. Catabolism of methionine and its derivatives. KA, amino acceptors; AA, amino donors.

(EC 4.1.1.50), which occurs in E. coli,[27] Saccharomyces cerevisiae,[28] and several other microorganisms. The product, S-methylthioadenosylhomocysteamine, is metabolized to polyamines. Ethylene, a plant hormone promoting ripening in fruits, is derived from methionine through 1-aminocyclopropane 1-carboxylate.[29] Ethylene production from methionine occurs in Saccharomyces cerevisiae as well.[30] Methionine is oxidized chemically with active oxygen to methionine sulfoxide and methionine sulfone, although the microbial oxidation has not been examined. Nutritionally, methionine sulfoxide can be substituted for methionine, but methionine sulfone cannot. Methionine sulfoxide is reduced to methionine by the coupled reaction of methionine sulfoxide reductase and thioredoxin reductase found in E. coli[31] as shown in the scheme below. This reductase system does not act on methionine sulfoxide residues in protein.

[27] R. B. Wickner, C. W. Tabor, and H. Tabor, J. Biol. Chem. **245**, 2132 (1970).
[28] M. S. Cohn, C. W. Tabor, and H. Tabor, J. Biol. Chem. **252**, 8212 (1977).
[29] D. O. Adams and S. F. Yang, Trends Biochem. Sci. **6**, 161 (1981).
[30] K. C. Thomas and M. Spencer, Can. J. Microbiol. **23**, 1669 (1977).
[31] S. Ejiri, H. Weissbach, and N. Brot, Anal. Biochem. **102**, 393 (1980).

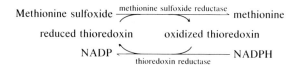

Microbial Metabolism of Selenium Amino Acids

Selenium is located between sulfur and tellurium in the VIb group of the periodic table, and its properties closely resemble those of sulfur. Selenium amino acids are metabolized essentially according to the metabolic pathways of sulfur amino acids. Enzymes participating in the metabolism of sulfur amino acids act on the selenium counterparts in a similar manner. For example, methionine γ-lyase also catalyzes the α,γ-elimination and γ-replacement of selenomethionine, as well as the γ-replacement of methionine with various selenols. However, selenocysteine β-lyase of *Citrobacter freundii* distinguishes selenocysteine from cysteine, acting only on the former.[32]

The mechanism of incorporation of selenium into amino acid residues of selenium-containing enzymes, e.g., glycine reductase and formate dehydrogenase of *Clostridium,* remains to be elucidated.

The general metabolism of selenium amino acids[33] and sulfur amino acids[34] have been reviewed in detail recently.

[32] P. Chocat, N. Esaki, K. Tanizawa, K. Nakamura, H. Tanaka, and K. Soda, *J. Bacteriol.* **163,** 669 (1985); see also this volume [86].
[33] R. J. Shamberger, "Biochemistry of Selenium," p. 1. Plenum, New York, 1983.
[34] A. J. L. Cooper, *Annu. Rev. Biochem.* **52,** 187 (1983).

[79] L-Methionine γ-Lyase from *Pseudomonas putida* and *Aeromonas*

By Nobuyoshi Esaki and Kenji Soda

$$CH_3SCH_2CH_2CH(NH_2)COOH + H_2O \rightarrow CH_3SH + CH_3CH_2COCOOH + NH_3$$

L-Methionine γ-lyase (EC 4.4.1.11) catalyzes the α,γ-elimination of L-methionine to produce α-ketobutyrate, methanethiol, and ammonia. The enzyme has been purified from *Clostridium sporogenes,*[1] *Pseudomonas*

[1] W. Kreis and C. Hession, *Cancer Res.* **33,** 1862 (1973).

putida,[2,3] and Aeromonas sp.[4] The present chapter describes the assay method, purification procedures, and general properties of L-methionine γ-lyase from *P. putida* and *Aeromonas* sp.

Assay Method

Principle. The enzymatic α,γ-elimination of L-methionine can be routinely followed by spectrophotometric determination of α-ketobutyrate with 3-methyl-2-benzothiazolone hydrazone hydrochloride.[5]

Reagents

L-Methionine, 0.1 M
Pyridoxal 5'-phosphate, 0.1 mM
Potassium phosphate, 0.5 M, at pH 8.0
Trichloroacetic acid, 50% (w/v)
Sodium acetate, 1.0 M, at pH 5.0
3-Methyl-2-benzothiazolone hydrazone hydrochloride, 0.1%

Procedure. The standard assay system consists of 0.4 ml potassium phosphate buffer, 0.5 ml L-methionine, 0.2 ml pyridoxal 5'-phosphate, and enzyme in a final volume of 2.0 ml. Enzyme is replaced by water in a blank. After the mixture is incubated at 37° for 10 min, the reaction is terminated by addition of 250 μl of 50% trichloroacetic acid. After centrifugation, 1 ml of the supernatant solution is mixed with 2 ml of the acetate buffer and 0.8 ml of 3-methyl-2-benzothiazolone hydrazone hydrochloride. After 30 min at 50°, absorbance at 320 nm is measured against the blank.

Definition of Unit. One unit of enzyme is defined as the amount that catalyzes the formation of 1 μmol of α-ketobutyrate. Specific activity is expressed as units per milligram protein. The protein concentration is determined from the absorbance coefficient of the purified enzyme: *Pseudomonas* enzyme, $A^{1\%}_{278\,nm} = 7.72$; *Aeromonas* enzyme, $A^{1\%}_{278\,nm} = 9.98$.

Purification Procedures

Preparation from Pseudomonas putida

Pseudomonas putida ICR 3460 is grown in a medium containing 0.25% L-methionine, 0.1% polypeptone, 0.1% glycerol, 0.1% KH_2PO_4, 0.1%

---

[2] S. Ito, T. Nakamura, and Y. Eguchi, *J. Biochem. (Tokyo)* **78,** 1105 (1975).
[3] T. Nakayama, N. Esaki, K. Sugie, T. T. Berezov, H. Tanaka, and K. Soda, *Anal. Biochem.* **138,** 421 (1984).
[4] T. Nakayama, N. Esaki, W.-J. Lee, I. Tanaka, H. Tanaka, and K. Soda, *Agric. Biol. Chem.* **48,** 2367 (1984).
[5] K. Soda, *Agric. Biol. Chem.* **31,** 1054 (1967).

K_2HPO_4, 0.01% $MgSO_4 \cdot 7H_2O$, and 0.025% yeast extract at 28° for 18 hr with shaking. The cells, harvested by centrifugation, are washed with 0.85% NaCl and subsequently with 10 mM potassium phosphate at pH 7.2, containing the following ingredients: 1 mM phenylmethylsulfonyl fluoride, 50 µM N-p-tosyl-L-phenylalanine chloromethyl ketone, 1 mM ethylenediaminetetraacetic acid, 20 µM pyridoxal 5'-phosphate, 0.01% 2-mercaptoethanol, and 2% ethanol. All subsequent operations are performed at 0–5° unless otherwise stated. The buffers used throughout contain the same ingredients in the phosphate buffer described in this paragraph.

Step 1: Extract. Washed cells, about 13 g (wet weight), are ground in a mortar chilled on ice with 13 g of levigated aluminum oxide for 10 min, suspended in 25 ml of 10 mM potassium phosphate at pH 7.2, and centrifuged. Deoxyribonuclease I of bovine pancreas (Sigma), 30 µg, is added to the supernatant solution, and the mixture is stirred at room temperature for 1 hr, followed by dialysis against 1,000 volumes of the above buffer for 24 hr. The precipitate formed during dialysis is removed by centrifugation.

Step 2: DEAE-Toyopearl 650 M. The supernatant solution is applied to a DEAE-Toyopearl 650 M column (6 × 7 cm; bed volume, 60 ml) equilibrated with the dialysis buffer. The column is washed with the same buffer containing 40 mM KCl until the absorbance of the eluate at 280 nm decreases to less than 0.03. Enzyme is eluted with the same buffer containing 0.08 M KCl. The eluate is pooled (250 ml) and brought to pH 8.2 ± 0.1 by addition of 63 ml of 0.1 M sodium pyrophosphate at pH 8.3 containing 0.12 M KCl.

Step 3: DEAE-Sephadex A-50. The enzyme solution is applied to a column of DEAE-Sephadex A-50 (4 × 5 cm; bed volume, 20 ml) equilibrated with 20 mM sodium pyrophosphate at pH 8.3 containing 0.12 M KCl. The column is washed with the same buffer to reduce the absorbance of the eluate at 280 nm to less than 0.01. The enzyme is eluted with the same buffer containing 0.25 M KCl.

Crystallization. The enzyme is crystallized by the vapor diffusion method of McPherson.[6] The crystallization is carried out at 4° in a sealed plastic sandwich box (320 ml), in which an concave slide glass is set over a petri dish reservoir containing 20 ml of 16% polyethylene glycol 6000 in 10 mM potassium phosphate (pH 7.2). The enzyme solution (1.65 mg/45 µl) is mixed with 15 µl of the reservoir solution in the concave of the slide glass. Yellow pyramidal crystals appear after 24 hr.

A summary of a typical purification is presented in Table I. The procedure is generally accomplished within 2 days with an average yield of 75%. Larger scale purification can also be achieved in a similar manner

---

[6] A. McPherson, Jr., *J. Biol. Chem.* **251**, 6300 (1976).

TABLE I
PURIFICATION OF L-METHIONINE γ-LYASE FROM *Pseudomonas putida*

Step	Volume (ml)	Total protein (mg)	Total activity (units)	Specific activity (units/mg)
1. Extract	27	229	65.8	0.29
2. DEAE-Toyopearl 650 M	250	77	53.3	0.69
3. DEAE-Sephadex A-50	66	2.4	48.4	20.4

with a proportional increase in the amount of ion-exchange matrices; e.g., 460 mg of the homogeneous enzyme is obtained from 1.3 kg cells (wet weight), which are disrupted with a Dyno-Mill (Willy A, Basel, Switzerland), with a yield of 70% within 4 days. The specific activity of the final preparation is 20.4 units/mg.

Preparation from Aeromonas sp.

L-Methionine γ-lyase is purified from *Aeromonas* sp. ICR 3470 essentially by the same method as described above, using DEAE-Toyopearl 650 M and DEAE-Sephadex A-50 column chromatography. About 30 mg of a homogeneous preparation (specific activity, 31 units/mg) is obtained from a crude extract of 204 g of wet cells (total protein, 4.6 g; specific activity, 1.04 units/mg), with an overall yield of 20%.

Properties

Stability. The *Pseudomonas putida* enzyme can be stored at $-20°$ for several months without loss of activity when 0.1 M or higher concentrations of potassium phosphate at pH 7.2 is used as a solvent; it is inactivated significantly by freezing and thawing in a dilute (less than 10 mM) buffer; about 70% of the original activity is lost after storage for 6 months. However, activity can subsequently be recovered to a level of 80% of the original by incubation with 50 mM dithiothreitol in 10 mM potassium phosphate (pH 7.2) at 0° for 1 hr. The enzyme is stable up to 50°, when heated for 30 min in 0.1 M potassium phosphate buffer (pH 7.2). The *Aeromonas* sp. enzyme retains 82% of the original activity after incubation at 60° for 5 min in 0.2 M potassium phosphate buffer (pH 7.2). Both the enzyme preparations are stable in the pH range of 6–9 for at least 3 hr at 0°.

TABLE II
PROPERTIES OF L-METHIONINE γ-LYASE

Properties	Source	
	P. putida	Aeromonas sp.
$s^0_{20, w}$	8.3 S	—
M_r Sedimentation equilibrium	165,000	—
M_r Gel sieving	—	149,000
M_r Light scattering	174,000	159,000
Absorption maxima	278 nm (ε = 134,000)	278 nm (ε = 159,000)
	420 nm (ε = 38,900)	423 nm (ε = 31,300)
No. subunits	4	4
Subunit M_r	43,000	41,000
Pyridoxal phosphate (mol/mol of enzyme)	4	4

TABLE III
SUBSTRATE SPECIFICITY OF THE P. putida ENZYME

	Relative activity (%)	
Substrate	α,γ-Elimination	γ-Replacement with 2-mercaptoethanol
L-Methionine	100	100
D-Methionine	0	0
L-Ethionine	55	57
DL-Methionine sulfone	55	39
DL-Methionine DL-sulfoxide	21	43
L-Methionine DL-sulfoximine	1	5
S-Methyl-L-methionine	4	0
L-Homocysteine	180	176
O-Acetyl-L-homoserine	140	202
O-Ethyl-L-homoserine	99	—
DL-Selenomethionine	282	114
L-Norleucine	0	0
L-Norvaline	0	0
L-Cysteine	10	59
D-Cysteine	0	0
L-Cystathionine	0	—
S-Methyl-L-cysteine	9	125
O-Acetyl-L-serine	15	—

Physicochemical Properties. Table II summarizes the physicochemical properties of the enzyme purified from both species. The absorption spectrum of the *Pseudomonas* enzyme with λ_{max} at 278 and 420 nm is not influenced by variation of the pH. The formyl group of pyridoxal 5'-phosphate, 4 mol of which are contained per mole of enzyme, is bound in an aldimine linkage to an ε-amino group of a lysine residue of the enzyme; reduction of the enzyme with $NaBH_4$ results in irreversible inactivation. The K_m value for pyridoxal phosphate is 1.4 μM. The enzyme activity is strongly inhibited by both carbonyl and sulfhydryl reagents, e.g., hydroxylamine, penicillamine, L-cycloserine, and N-ethylmaleimide. The properties of the *Aeromonas* enzyme are similar to those of the *Pseudomonas* enzyme, except that the M_r of the latter is higher than that of the former.

Catalytic Properties. The enzyme from *Pseudomonas putida* shows multiple catalytic functions. It catalyzes α,γ-elimination and γ-replacement reactions of L-homocysteine and its S-substituted derivatives, O-substituted L-homoserines, and DL-selenomethionine, as well as α,β-elimination and β-replacement reactions of L-cysteine, S-methyl-L-cysteine, and O-acetyl-L-serine (Table III). The K_m value for L-methionine in the α,γ-elimination reaction was determined to be 1.0 mM. The enzyme also catalyzes deamination and γ-addition reactions of L-vinylglycine (K_m 8.6 mM, V_{max} 21 units/mg in deamination). A number of S-substituted L-homocysteines and S-substituted L-cysteines, and their selenium analogs, can be prepared by the β- and γ-replacement reactions and γ-addition reactions catalyzed by L-methionine γ-lyase as described previously.[7] The proton-exchange reactions of various straight-chain L-amino acids, with deuterium as solvent, are also catalyzed by the enzyme. Thus, L-[α-^{2}H,β-^{2}H]methionine, L-[α-^{2}H,β-^{2}H]norleucine, L-[α-^{2}H,β-^{2}H]norvaline, L-[α-^{2}H,β-^{2}H]-α-aminobutyrate, and L-[α-^{2}H,β-^{2}H]alanine can be prepared enzymatically.

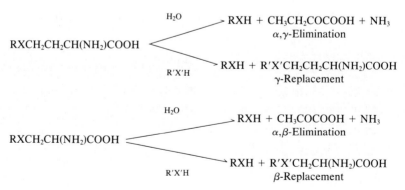

[7] H. Tanaka, N. Esaki, and K. Soda, *Enzyme Microb. Technol.* **7**, 530 (1985).

$$CH_2=CHCH(NH_2)COOH \begin{array}{c} \xrightarrow{H_2O} CH_3CH_2COCOOH + NH_3 \\ \text{Deamination} \\ \xrightarrow{R'X'H} R'X'CH_2CH_2CH(NH_2)COOH \\ \gamma\text{-Addition} \end{array}$$

X = S, O, or Se
X' = S or Se

$CH_3SCH_2CH_2CH(NH_2)COOH + {}^2H_2O \rightarrow CH_3SCH_2C^2H_2C^2H(NH_2)COOH$
$CH_3SCH_2CH(NH_2)COOH + {}^2H_2O \rightarrow CH_3SC^2H_2C^2H(NH_2)COOH$
$CH_3CH_2CH_2CH_2CH(NH_2)COOH + {}^2H_2O \rightarrow CH_3CH_2CH_2C^2H_2C^2H(NH_2)COOH$
$CH_3CH_2CH_2CH(NH_2)COOH + {}^2H_2O \rightarrow CH_3CH_2C^2H_2C^2H(NH_2)COOH$
$CH_3CH_2CH(NH_2)COOH + {}^2H_2O \rightarrow CH_3C^2H_2C^2H(NH_2)COOH$
$CH_3CH(NH_2)COOH + {}^2H_2O \rightarrow C^2H_3C^2H(NH_2)COOH$
$CH_2(NH_2)COOH + {}^2H_2O \rightarrow HC^2H(NH_2)COOH$
(R)-Glycine-d_1

The *Aeromonas* enzyme is similar to the *Pseudomonas* enzyme with respect to substrate specificity except that the *Aeromonas* enzyme catalyzes the elimination of L-cystathionine, a reaction not carried out by the *Pseudomonas* enzyme (Table III).

[80] O-Acetylhomoserine Sulfhydrylase from *Schizosaccharomyces pombe*

By SHUZO YAMAGATA

O-Acetyl-L-homoserine + H_2S → L-homocysteine + acetic acid	(1)
O-Succinyl-L-homoserine + H_2S → L-homocysteine + succinic acid	(2)
L-Homoserine + H_2S → L-homocysteine + H_2O	(3)

O-Acetylhomoserine sulfhydrylase is the first sulfhydrylase that has been demonstrated to form L-homocysteine with any of three forms of L-homoserine.[1] The enzyme appears to be physiologically functional with O-acetyl-L-homoserine as substrate, since no cystathionine γ-synthase activity is detectable.

Assay Method

Principle. The colorimetric determination of L-homocysteine produced by reactions (1), (2), and (3) is essentially the same as that de-

---
[1] S. Yamagata, *J. Biochem. (Tokyo)* **96**, 1511 (1984).

scribed for the O-acetylserine–O-acetylhomoserine sulfhydrylase of *Saccharomyces cerevisiae*.[2] The major differences are in the higher concentration of amino acid substrates and the longer reaction times that are required to determine the activity accurately. Since the enzyme is very unstable, sucrose or glycerol is added in the reaction mixture as a "stabilizer," particularly when a longer incubation period is required for the assay of a preparation with very low activity.

Procedure. In a total volume of 1.0 ml are mixed 0.1 M potassium phosphate at pH 7.8, 1 mM EDTA, 0.2 mM pyridoxal phosphate, 25% sucrose, an appropriate amount of enzyme (0.01–0.10 units), 10 mM O-acetyl-L-homoserine (O-succinyl-L-homoserine, or L-homoserine), and 1 mM Tris–sulfide. Solutions of the two substrates are added last, after a preincubation of 5 min at 30° for the other components; the two final substrates are unstable at pH 7.8.[3] The incubation itself is carried out for 20 min at 30°. Other conditions and the colorimetric determination of the product are the same as described for the enzyme from *Saccharomyces cerevisiae*.[2] When accurate determination of the enzyme activity in a crude extract is required, utilization of radioactive sulfide, $H_2^{35}S$, is recommended.[4]

Definition of Unit and Specific Activity. One unit is defined as the amount of enzyme catalyzing the production of 1 μmol of homocysteine per minute by sulfhydrylation of any of the three forms of L-homoserine. Specific activity is expressed as units per milligram protein. Protein is determined according to Lowry *et al.*[5]

Purification Procedure

Organism and Culture. A wild-type strain of the fission yeast *Schizosaccharomyces pombe* (IFO-0363) is grown in a rich medium containing, per liter, the following: peptone, 3 g; KH_2PO_4, 3 g; $MgSO_4 \cdot 7H_2O$, 1 g; $(NH_4)_2SO_4$, 2 g; yeast extract, 2 g; glucose, 40 g; and $CaCl_2 \cdot 2H_2O$, 0.25 g. The medium, 1.5 liters in a 3-liter flask, is inoculated with 20 ml of a fully grown seed culture and shaken vigorously with 30°. Cells are harvested in the early stationary phase by centrifugation to yield about 15 g wet weight per flask. After washing once with 5 volumes of cold distilled water, the cells are safely maintained for at least 2 months at −50° until use.

[2] This volume [83].
[3] S. Yamagata, K. Takeshima, and N. Naiki, *J. Biochem. (Tokyo)* **75**, 1221 (1974).
[4] D. Kerr, this series, Vol. 17, Part B, p. 446.
[5] O. H. Lowry, N. J. Rosebrough, A. L. Farr, and R. J. Randall, *J. Biol. Chem.* **193**, 265 (1951).

Extraction. Cells of *S. pombe* (700 g wet weight) are suspended in 1.2 liters of PEDP buffer (50 mM potassium phosphate at pH 7.0, containing 1 mM EDTA, 0.2 mM dithiothreitol, and 0.2 mM pyridoxal phosphate). The suspension is subjected to agitation in a Dyno Mill[6] at 3,000 rpm and at a flow rate of 5–6 liters/hr. The treated suspension is diluted with 1.2 liters of the same buffer and centrifuged at 10,000 g for 30 min.

Step 1: Salt Fractionation. The supernatant fraction thus obtained (2.3 liters) is fractionated with ammonium sulfate. The precipitate formed between 35% (200 g/liter) and 65% saturation (an additional 195 g/liter) are collected by centrifugation. This fraction is dissolved in a small volume of PEDP buffer and dialyzed against 3 liters of the same buffer overnight. The clear supernatant fraction obtained by centrifugation is further fractionated with ammonium sulfate between 35% (200 g/liter) and 45% saturation (additional 60 g/liter) after dilution with 1.8 liters of PEDP buffer. The proteins are dissolved in a small volume of the buffer and dialyzed as above.

Step 2: DEAE-Cellulose Chromatography 1. The dialysate is diluted to 1.5 liters with PEDP buffer containing 25% sucrose (PEDP–sucrose buffer) and applied to a DEAE-cellulose (DE-52, Whatman) column (4.2 × 10 cm) equilibrated with the same buffer without pyridoxal phosphate. After washing with 500 ml of the equilibration buffer, proteins are eluted with 800 ml of a linear (50–250 mM) potassium phosphate (pH 7.0) gradient, which is supplemented to 1 mM EDTA, 0.2 mM dithiothreitol, and 25% sucrose. Fractions (7.5 ml each) having high *O*-acetylhomoserine sulfhydrylase activity, which elutes at a phosphate concentration of about 120 mM, are combined.

Step 3: DEAE-Cellulose Chromatography 2. The protein solution is diluted to 320 ml with PEDP–sucrose buffer and subjected to chromatography on a DEAE-cellulose column (2.7 × 11 cm). Chromatography is carried out in the same manner as above, using 400 ml of a linear phosphate concentration gradient from 50 to 200 mM as the elution buffer. Active fractions (5 ml each) are combined.

Step 4: Concentration. The preparation is subjected to concentration to approximately 10 ml with a collodion bag (Sartorius membrane filter, SM 13200) following which it is dialyzed against 200 ml of PEDP–sucrose buffer overnight to adjust the sucrose concentration to 25% (10 ml).

Step 5: Sephadex G-150 Chromatography. A portion, 8 ml, of the

---

[6] The cell suspension is continuously sent into a glass container of 600 ml in a Dyno Mill (W. A. Bachofen Maschinenfabrik, type KDL). The container holds about 500 ml of glass beads (0.25–0.5 mm in diameter) which are agitated by 4 disks while cooling with running ice water.

TABLE I
PARTIAL PURIFICATION OF O-ACETYLHOMOSERINE SULFHYDRYLASE FROM
S. pombe CELLS[a]

Step	Volume (ml)	Total protein (mg)	Enzyme amount (units)	Specific activity (units/mg protein)
1. Ammonium sulfate	170	7,510	107	0.014
2. DEAE-cellulose 1	80	189	96.0	0.51
3. DEAE-cellulose 2	89	111	82.3	0.74
4. Concentration	9.9	123	72.3	0.59
5. Sephadex G-150				
Fraction No. 32	4.9	2.6	7.4	2.8
33	4.9	2.6	10.5	4.0
34	4.9	2.8	12.4	4.4
35	4.9	3.3	10.5	3.2
36	4.9	4.0	6.4	1.6

[a] 700 g (wet weight).

dialysate is subjected to gel filtration on a column (2.6 × 79 cm) of Sephadex G-150 equilibrated with PEDP–sucrose buffer. Fractions of 5 ml are collected at a flow rate of 5 ml/hr.

Comments. A summary of the purification is presented in Table I. The specific enzyme activity, about 4 units/mg of protein in the fractions obtained in the final step, are approximately 13% of those for the homogeneous O-acetylserine–O-acetylhomoserine sulfhydrylase from *Saccharomyces cerevisiae*.[7]

Properties

Stability. The enzyme quickly loses catalytic activity, particularly during storage at −20°. However, sucrose or glycerol at a concentration of 25% is effective in protecting against inactivation induced by either freezing or incubation with the reaction mixture. The enzyme is most stable at pH 7.0 and loses activity more readily at an alkaline pH.

Optimum pH for Activity. The activity curve is almost symmetric, with optimal pH at around 8.0 when the reaction is carried out in either 0.1 M Tris–HCl or potassium phosphate buffers. The former give slightly higher activity than the latter. Half of maximal activity is at pH 6.7 and 9.0.

[7] S. Yamagata and K. Takeshima, *J. Biochem.* (*Tokyo*) **80**, 777 (1976).

Inhibitors. The following amino acids inhibit the enzyme at a concentration of 10 mM: L-methionine (74%), S-adenosyl-L-methionine (52%), S-adenosyl-L-homocysteine (46%), O-acetyl-L-serine (46%), O-phospho-L-serine (46%), DL-ethionine (46%), and L-serine O-sulfate (17%). Preincubation of the enzyme with the following carbonyl reagents at a concentration of 10 mM and at 30° for 15 min also inhibits: hydroxylamine · HCl (98%), semicarbazide · HCl (92%), phenylhydrazine · HCl (88%), and KCN (77%). The enzyme is not sensitive to such sulfhydryl reagents as p-chloromercuribenzoic acid (1 mM), 5,5′-dithiobis(2-nitrobenzoic acid) (1 mM) or monoiodoacetate (5 mM).

Repression. L-Methionine, added at concentrations from 0.1–3 mM in a synthetic medium,[8] has no repressive effect on the synthesis of the enzyme.

Substrate Specificity. The enzyme reacts with O-acetyl-L-homoserine, O-succinyl-L-homoserine, and L-homoserine, but does not react with O-acetyl-L-serine, O-acetyl-L-threonine, O-acetyl-L-tyrosine, O-acetyl-L-hydroxyproline, O-phospho-DL-threonine, O-phospho-L-serine, or L-serine O-sulfate.

Kinetic Properties. In the three sulfhydrylase reactions, the following apparent K_m values are found: O-acetyl-L-homoserine, 13 mM; O-succinyl-L-homoserine, 11 mM; and L-homoserine, 10 mM. The respective V_{max} values were 15.5, 6.2, and 2.5 units/mg protein. The K_m for H_2S is 0.053 mM in the O-acetylhomoserine sulfhydrylase reactions.

Cofactor. Apoenzyme prepared by incubation of a holoenzyme preparation with 10 mM phenylhydrazine hydrochloride shows a K_m value of 83 nM for pyridoxal phosphate.

Molecular Weight and pI. Gel filtration on a Sephadex G-150 column and centrifugation in a sucrose concentration gradient suggest M_r of 186,000 and 170,000, respectively. These values do not change in the absence of sucrose or in the presence of 10 mM L-methionine or 1 M CsCl. The apoenzyme has a similar value for M_r. Electrofocusing of the enzyme, with the carrier Ampholite (pH 3–10, LKB-Produkter), shows an isoelectric point 4.1.

---

[8] S. Yamagata, K. Takeshima, and N. Naiki, *J. Biochem. (Tokyo)* **77**, 1029 (1975).

[81] O-Acetylhomoserine Sulfhydrylase from *Brevibacterium flavum*

By ISAMU SHIIO and HACHIRO OZAKI

O-Acetyl-L-homoserine + sulfide → L-homocysteine + acetate

In *Brevibacterium flavum*, L-methionine is synthesized from L-homoserine via O-acetylhomoserine and L-homocysteine. The formation of L-homocysteine from O-acetyl-L-homoserine is catalyzed by O-acetylhomoserine sulfhydrylase, in contrast to enteric bacteria and *Neurospora*, in which L-homocysteine is synthesized through cystathionine formation.[1]

Assay Method

Principle. The assay procedure, which uses the nitroprusside reaction to determine homocysteine in the presence of sulfide, is a slight modification of that for O-acetylserine sulfhydrylase described by Kredich and Becker.[2] O-Acetyl-L-homoserine and sulfide are incubated with bacterial extract. The homocysteine formed is converted to a stable S-nitrosothiol derivative by adding nitrous acid, which also stops the reaction. Excess nitrous acid is removed by the addition of ammonium sulfamate, and then mercuric chloride, sulfanilamide, and N-1-naphthylethylenediamine are added. The S-nitrosothiol derivative of homocysteine decomposes in the presence of mercuric ion to give nitrous acid which diazotizes the sulfanilamide. Finally, the diazotized sufanilamide couples with naphthylethylenediamine to give the chromophore, an azo dye (see schematic presentation of the assay by Kredich and Becker[2]).

Reagents

Tris–HCl, 0.2 M, containing 1 mM EDTA, disodium salt, at a final pH 7.4
Na_2S, 0.1 M
O-Acetyl-L-homoserine, 0.1 M[3]
$NaNO_2$, 0.10 M
H_2SO_4, 0.4 N
Ammonium sulfamate, 2% in water

---

[1] H. Ozaki and I. Shiio, *J. Biochem. (Tokyo)* **91**, 1163 (1982).
[2] N. M. Kredich and M. A. Becker, this series Vol. 17B [197].
[3] S. Nagai and M. Flavin, *J. Biol. Chem.* **241**, 3861 (1966).

HgCl$_2$, 2% in 0.4 N HCl
Sulfanilamide, 6.88% in 0.4 N HCl
N-1-Naphthylethylenediamine dihydrochloride, 0.2% in 0.4 N HCl
Enzyme solution containing 100–1000 units/ml, 50 mM potassium phosphate at pH 7.3, 50 µM pyridoxal phosphate, and 0.1 mM β-mercaptopropionic acid

Procedure. The reaction is carried out at room temperature in capped 12 × 105 mm glass test tubes. A fresh solution of Na$_2$S is prepared daily and stored in an ice bath. Immediately prior to assay, the Na$_2$S solution is diluted 1:25 with the Tris–HCl–EDTA buffer. O-Acetyl-L-homoserine solution, 20 µl, and 20 µl of enzyme are mixed, and the reaction started by the addition of 0.16 ml of the Tris–HCl–EDTA–sulfide mixture. O-Acetyl-L-homoserine is omitted from blanks. After incubation for 4 min, the test tube is uncapped, and 1 ml of freshly prepared 1 mM nitrous acid (1 part 0.1 M NaNO$_2$ to 99 parts 0.4 N H$_2$SO$_4$) is added to the reaction mixture, which is then vigorously agitated. Six minutes later, 0.1 ml of 2% ammonium sulfamate is added, and the solution is mixed. After 10 min (shorter incubation leads to higher blank value), 1.6 ml of the following mixture is added with mixing: 1 part HgCl$_2$, 4 parts sulfanilamide, and 2 parts N-1-naphthylethylenediamine. Full color develops in 5 min and is stable for at least 15 min; the absorbance at 540 nm is measured. The 0.2 ml of 0.1 mM homocysteine yields an A_{540} of 0.28, as reported by Kredich and Becker[2] for 0.1 mM cysteine.

Definition of Unit and Specific Activity. One unit of O-acetylhomoserine sulfhydrylase is defined as the amount of enzyme that catalyzes the formation of 1 nmol of homocysteine per minute. Specific activities are expressed as units per milligram protein as determined by the method of Lowry et al.[4] with bovine serum albumin as standard.

Purification Procedure

Growth of the Organism. A homoserine auxotroph of *Brevibacterium flavum* No. 2247 (ATCC 14067), strain H1013[5] (FERM-P6465) is cultured aerobically with shaking at 30° for 24 hr in 500-ml flasks, each containing 50 ml of medium-5[6] (a glucose–vitamin–salts medium) supplemented with 20 mg of L-methionine and 500 mg of L-threonine per liter, which are sterilized at 115° for 10 min. Medium-5 contains the following (amounts per liter): 36 g glucose, 10 g urea, 1 g KH$_2$PO$_4$, 0.4 g MgSO$_4$·7H$_2$O, 10 mg

[4] O. H. Lowry, N. J. Rosebrough, A. L. Farr, and R. J. Randall, *J. Biol. Chem.* **193**, 265 (1951); see also this series, Vol. 3 [73].
[5] K. Sano and I. Shiio, *J. Gen. Appl. Microbiol.* **13**, 349 (1967).
[6] S. Sugimoto and I. Shiio, *J. Biochem. (Tokyo)* **81**, 823 (1977).

$FeSO_4 \cdot 7H_2O$, 8 mg $MnSO_4 \cdot 4H_2O$, 100 μg thiamin·HCl, 30 μg d-biotin, and 7 ml 6 N HCl. Cell harvests and all subsequent procedures are performed at 0–4°.

Step 1: Preparation of Cell Extracts. Fresh cells are harvested from 1.5 liters of culture medium and are washed twice by suspension in and centrifugation from 200 ml of 0.2% KCl. After washing, the cells (14 g wet weight) are suspended in 40 ml of 50 mM potassium phosphate at pH 7.3 containing 50 μM pyridoxal phosphate and 0.1 mM β-mercaptopropionic acid (buffer A). Cells are broken by treatment in 10-ml batches for 10 min in a sonic oscillator (Toyoriko Type N-50-3, 10 kHz) with cooling under tap water. Cell debris is removed by centrifugation at 30,000 g for 30 min. The supernatant fluid is used as cell extract.

Step 2: Ammonium Sulfate 1. Solid ammonium sulfate is added to the cell extract to 0.4 of saturation (243 mg/ml). After centrifugation at 17,000 g for 10 min, the supernatant solution is made to 0.55 saturation with ammonium sulfate by the addition of 97 mg/ml. The mixture is centrifuged, and the resulting precipitate is dissolved in a small volume of buffer A (8.9 ml, final volume).

Step 3: Sephadex G-200. The enzyme solution is applied to a Sephadex G-200 column (2.4 × 40 cm) equilibrated and used with buffer A at a flow rate of 0.5 ml/min. Fractions with enzyme activity are pooled (38 ml).

Step 4: DEAE-Cellulose. The pool fraction is charged onto a DEAE-cellulose (Midorijuji Co.) column (2.5 × 17 cm), previously equilibrated with 50 mM Tris–HCl at pH 7.5 containing 0.1 mM β-mercaptopropionic acid, 50 μM pyridoxal phosphate, and 0.2 M NaCl (buffer B). After the column is washed with 100 ml of buffer B, the enzyme is eluted with a NaCl gradient formed from 400 ml of buffer B and 400 ml of buffer B containing 0.5 M NaCl (final concentration). Fractions showing enzyme activity are pooled and dialyzed overnight against a 50-fold volume of 10 mM N-tris(hydroxymethyl)methyl-2-aminoethanesulfonic acid (TES)–NaOH at pH 7.5 containing 0.1 mM β-mercaptopropionic acid and 50 μM pyridoxal phosphate (buffer C).

Step 5: Hydroxylapatite. The dialyzed enzyme (40 ml) is placed on a hydroxylapatite column (2.5 × 17 cm, BioGel HT, a product of Bio-Rad Laboratories) equilibrated with buffer C. After the column is washed with 80 ml of buffer C, the enzyme is eluted with a phosphate gradient formed from 40 ml of buffer C and 400 ml of buffer C containing 0.2 M potassium phosphate. Fractions showing enzyme activity are pooled.

Step 6: Ammonium Sulfate 2. The solution is mixed with solid ammonium sulfate to give 90% saturation. The precipitate formed is dissolved in a small volume of buffer A and used as the partially purified enzyme.

TABLE I
PURIFICATION OF O-ACETYLHOMOSERINE SULFHYDRYLASE FROM Brevibacterium flavum H1013

Step	Total volume (ml)	Total protein (mg)	Total activity (units)	Specific activity (units/mg protein)
1. Cell extract	38	1,070	209,000	200
2. Ammonium sulfate 1	8.9	353	100,500	285
3. Sephadex G-200	38	102	90,000	880
4. DEAE-Cellulose	90	21	65,600	3,170
5. Hydroxylapatite	100	11	53,400	4,880
6. Ammonium sulfate 2	10	10.3	48,600	4,720

The total activity of the partially purified enzyme corresponds to a recovery of 23%; the specific activity indicates that a 24-fold purification was obtained. A typical purification is summarized in Table I.

Notes on the Purification Procedure

1. Synthesis of the enzyme is strongly repressed by L-methionine; addition of 500 mg/liter of L-methionine to the culture medium leads to a decrease in specific activity to the level of 20 units/mg protein from the derepressed level of 200 units/mg protein obtained by the addition of 20 mg of methionine per liter.

2. Polyacrylamide disk–gel electrophoresis of the final preparation shows two minor protein bands and one major band that has the enzyme. The enzyme is estimated to be about 70% pure.

3. The enzyme preparation appears free of homoserine O-acetyltransferase and β-cystathioninase activity, but has slight O-acetylserine sulfhydrylase activity. Additional chromatography on DEAE-cellulose of the final preparation shows two peaks of O-acetylserine sulfhydrylase activity; one is eluted in the same fraction as O-acetylhomoserine sulfhydrylase activity, suggesting that the enzyme has slight O-acetylserine sulfhydrylase activity. On the other hand, the behavior of the cystathionine γ-synthase activity is similar to that of O-acetylhomoserine sulfhydrylase during the course of purification and upon rechromatography on DEAE-cellulose. Therefore, both activities are attributable to the same enzyme.

Properties

The molecular weight of the enzyme is 360,000, determined by gel filtration on Sephadex G-200. The pH optimum is 8.7. Phosphate buffer is superior to Tris–HCl and TES buffers. The enzyme is inhibited by hydroxylamine and reactivated by pyridoxal phosphate, indicating that pyridoxal phosphate is a coenzyme for the enzyme. The enzyme reacts specifically with O-acetyl-L-homoserine. Activity with O-acetyl-L-serine is only one-hundredth that observed with O-acetyl-L-homoserine; no activity is observed with O-succinyl-L-homoserine, L-homoserine, or L-serine. The K_m values for sulfide and O-acetylhomoserine are 83 μM and 2.0 mM, respectively. The enzyme is inhibited 50, 23, and 29% by L-methionine, L-homoserine, and O-acetyl-L-serine, respectively, each at 10 mM. No inhibition is observed with L-serine, glycine, cystathionine, or S-adenosyl-L-methionine.

[82] O-Acetylserine Sulfhydrylase from *Bacillus sphaericus*

By TORU NAGASAWA and HIDEAKI YAMADA

$$CH_3COOCH_2CH(NH_2)COOH + H_2S \rightarrow HSCH_2CH(NH_2)COOH + CH_3COOH$$

O-Acetylserine sulfhydrylase activity, an enzyme of importance in cysteine biosynthesis,[1,2] was found in high concentration in β-chloro-L-alanine-resistant *Bacillus sphaericus*.[3] The formation of the enzyme is enhanced by β-chloro-L-alanine, and this effect is counteracted by a corepressor, L-cysteine.[4] O-Acetylserine sulfhydrylase (EC 4.2.99.8) was purified and crystallized for the first time from this strain.[3]

Assay Procedure

Principle. The enzyme is assayed by measuring the rate of formation of L-cysteine from O-acetyl-L-serine and hydrogen sulfide. To repress the inhibitory effect of the cysteine formed, the reaction is carried out in the presence of acetone; acetone reacts with cysteine to form 2,2-dimethyl-4-thiazolidinecarboxylic acid, from which cysteine can be regenerated by heating in the presence of HCl.[5]

[1] D. M. Greenberg, *in* "Metabolic Pathways" (D. M. Greenberg, ed.), Vol. 7, p. 505. Academic Press, New York, 1975.
[2] N. M. Kredich and M. A. Becker, this series, Vol. 17A, p. 459.
[3] G. S. Dhillon, T. Nagasawa, and H. Yamada, *J. Biotechnol.* **1**, 47 (1984).
[4] T. Nagasawa, G. S. Dhillon, T. Ishii, and H. Yamada, *J. Biotechnol.* **2**, 365 (1984).
[5] J. P. Greenstein and M. Winitz, *in* "Chemistry of the Amino Acids" (J. P. Greenstein and M. Winitz, eds.), p. 1879. Wiley, New York, 1961.

Reagents

Potassium phosphate, 1 M, pH 7.5
O-Acetyl-L-serine–HCl, 0.1 mM
Sodium hydrosulfide, 0.5 mM
Trichloroacetic acid, 1.84 M

Procedure. To a centrifuge tube are added 0.1 ml potassium phosphate, 0.2 ml O-acetyl-L-serine, 0.2 ml sodium hydrosulfide, 20 μl acetone, from 0.003 to 0.08 units of enzyme, and water to make the total volume 1 ml. A control tube is prepared without the enzyme and O-acetyl-L-serine. Each mixture is incubated for 30 min at 30°, during which the rate should be linear. The reaction is stopped by adding 0.2 ml of trichloroacetic acid. The precipitate is removed by centrifugation and cysteine in the supernatant fluid is determined by the method of Gaitonde.[6]

Definition of Enzyme Unit and Specific Activity. One unit is defined as the amount of enzyme catalyzing the formation of 1 μmol of cysteine in 1 min. Specific activity is expressed in terms of enzyme units per milligram protein, as determined from the absorption at 280 nm. The absorption coefficient for the purified protein was calculated to be 0.693 mg^{-1} ml cm^{-1} by absorbance and dry weight determination.[3]

Purification Procedure

Source and Growth. Bacillus sphaericus L-118, which is resistant to 50 mM 3-chloro-L-alanine, was isolated from soil by an enrichment culture technique.[3] The formation of O-acetylserine sulfhydrylase by this strain is highly enhanced by the addition of β-chloro-L-alanine–HCl to the basal nutrient medium.[3,4] The basal medium consists of 10 g glucose, 5 g meat extract, 5 g yeast extract, 0.1 g pyridoxine · HCl, 2 g KH$_2$PO$_4$, 1 g MgSO$_4$ · 7H$_2$O, 40 mg CaCl$_2$ · 2H$_2$O, 20 mg FeSO$_4$ · 7H$_2$O, 20 mg MnSO$_4$, and 20 mg of CuSO$_4$ · 5H$_2$O per liter of tap water (pH 7.0). *Bacillus sphaericus* L-118 is cultivated in 2-liter flasks containing 500 ml of the basal medium supplemented with 0.5% β-chloro-L-alanine–HCl at 28° for 24 hr with reciprocal shaking. The cells are harvested by centrifugation at 10,000 g at 4° and washed with 0.15 M NaCl solution.

All the purification steps (1–7) are performed at 0–5°C. Tris–HCl buffer at pH 7.5 or potassium phosphate buffer at pH 7.5 each containing 0.1 mM pyridoxal phosphate, 0.1 mM EDTA, disodium salt, and 0.2 mM dithiothreitol, are used unless otherwise specified.

Step 1: Cell-Free Extract. Cells (about 45 g, wet weight) are harvested from 7 liters of culture broth, washed, and suspended in 500 ml of 0.1 M

---

[6] M. Gaitonde, *Biochem. J.* **104**, 626 (1967).

Tris buffer, and then disrupted with a Dyno-Mill (W. A. Bachofen, Switzerland) for 30 min at below 10°. Cell debris is removed by centrifugation at 10,000 g for 30 min.

Step 2: Salt Fractionation. The extract is diluted to 2 liters with 0.1 M Tris buffer, and then solid ammonium sulfate is added to 40% saturation (24.3 g/100 ml). The pH is adjusted to 7.0 with 2 N HCl. After stirring overnight, the resulting precipitate is removed by centrifugation at 10,000 g, and the supernatant fluid is further saturated with ammonium sulfate to 65% saturation (additional 16.8 g/100 ml). After 2 hr of mixing at 5°, the precipitate is recovered, dissolved in 10 mM Tris buffer, and dialyzed overnight against the same buffer for 18 hr.

Step 3: DEAE-Cellulose. The dialyzed solution is applied to a DEAE-cellulose column (4.0 × 40 cm) equilibrated with 0.1 M Tris buffer, washed with the same buffer, and eluted with the buffer supplemented with 0.2 M KCl. Active fractions are pooled and precipitated with ammonium sulfate, 43 g/100 ml. The precipitate is collected by centrifugation, dissolved in 0.1 M Tris, and dialyzed overnight against the same buffer.

Step 4: Octyl-Sepharose CL-4B. The enzyme solution from Step 3 at 5° is treated with ammonium sulfate, 11.4 g/100 ml, in small portions with stirring. This preparation is charged onto a column (1.8 × 15 cm) of octyl-Sepharose CL-4B, previously equilibrated with a 20% saturated ammonium sulfate solution (11.4 g/100 ml) containing 0.1 M potassium phosphate at pH 7.5. The enzyme is eluted by lowering the ionic strength of ammonium sulfate in a linear manner (20–0%, 200 ml in each container) in the same buffer. Active fractions are combined, precipitated with solid ammonium sulfate, 43 g/100 ml, and the precipitate dissolved in 6.9 ml of 0.1 M Tris buffer to a protein concentration of about 35 mg/ml.

Step 5: Heat Treatment. The solution is placed in a water bath at 60° and maintained there for 5 min, followed by immediate cooling in an ice bath. The resultant suspension is clarified by centrifugation at 10,000 g for 30 min, treated with solid ammonium sulfate, 43 g/100 ml, and dissolved in 10 ml of 10 mM Tris buffer.

Step 6: Hydroxyapatite. The enzyme is applied to a hydroxyapatite column (1.8 × 15 cm) previously equilibrated with 10 mM Tris buffer without EDTA. After washing the column with 100 ml of 0.1 M Tris buffer containing 0.2 M KCl, the enzyme is eluted with 0.1 M Tris buffer containing 0.5 M ammonium sulfate. Ammonium sulfate, 43 g/100 ml, is added to active enzyme fractions. After 12 hr, the suspension is centrifuged at 12,000 g, and then the precipitated enzyme is dissolved in 0.1 M potassium phosphate buffer and dialyzed against the same buffer overnight.

Step 7: Sephadex G-100. The precipitate is dissolved in 1 ml of 1 M potassium phosphate buffer containing 0.1 M KCl and then applied to a

TABLE I
PURIFICATION OF O-ACETYLSERINE SULFHYDRYLASE FROM
B. sphaericus L-118

Step	Total volume (ml)	Total protein (mg)	Total activity (units)	Specific activity (units/mg)
1. Extract	535	30,000	19,100	0.64
2. Salt fractionation	52	8,970	13,700	1.53
3. DEAE-Cellulose	98	1,880	13,100	6.97
4. Octyl-Sepharose CL-4B	6.9	244	7,230	29.6
5. Heat treatment	10	127	6,130	48.3
6. Hydroxyapatite	1.8	34.8	5,340	153
7. Sephadex G-100	9.7	21.0	5,010	239

column (2.0 × 82 cm) of Sephadex G-100 previously equilibrated with the same buffer. The column is developed at a flow rate of 3 ml/hr with the aid of a peristaltic pump. The active fractions are combined and precipitated with solid ammonium sulfate (43 g/100 ml). The precipitated enzyme is dissolved in 10 mM potassium phosphate buffer and dialyzed against the same buffer.

The overall purification achieved is about 380-fold with a yield of 25%. The purified enzyme catalyzes the β-replacement reaction of O-acetyl-L-serine in the presence of sodium hydrosulfide, yielding 239 μmol min^{-1} mg^{-1} of protein under the standard conditions. The results of a typical purification are summarized in Table I.

Crystallization. The purified enzyme solution is diluted with 10 mM potassium phosphate buffer at pH 7.5 containing 0.1 mM pyridoxal phosphate and 0.2 mM dithiothreitol to obtain a protein concentration of about 20 mg/ml. Ammonium sulfate, 24.3 g/100 ml, is added to this solution at 0° under gentle stirring with a glass rod. Addition of ammonium sulfate is continued to the point at which induced turbidity ceases to disappear from a silky sheen which appears upon stirring the mixture. After incubation for 2 hr, the enzyme begins to crystallize and is complete after 18 hr. The crystalline enzyme is collected by centrifugation at 8,000 g for 15 min. It is fully active in the absence of pyridoxal phosphate.

Properties

Stability. The purified enzyme can be stored in 10 mM potassium phosphate at pH 7.0 containing 0.1 mM pyridoxal phosphate and 0.1 mM dithiothreitol at 2° for 4 weeks without loss of activity. Heat treatment at

50, 60, 70, 75, and 80° for 5 min causes a 0, 7.5, 15, 49, 96, and 98% loss of the initial activity, respectively.

M_r and Cofactor. The enzyme has a molecular weight of about 68,000 ± 2,000 and consists of two subunits identical in molecular weight. The holoenzyme exhibits absorption maxima at 278 and 412 nm, with a $A_{278}:A_{412}$ ratio of approximately 2.92, and contains 2 mol of pyridoxal phosphate per mole of enzyme.

Substrates. β-Chloro-L-alanine (V_{max} 38 μmol min^{-1} mg^{-1}, K_m 15 mM) can be substituted for O-acetyl-L-serine (V_{max} 239 μmol^{-1} mg^{-1}, K_m 7.8 mM) as a substrate for the enzyme. L-Serine, D-serine, L-homoserine, O-succinyl-L-homoserine, and O-acetyl-L-homosereine are not substrates. Hydrogen sulfide cannot be replaced by methyl mercaptan, ethyl mercaptan, isopropyl mercaptan, isobutyl mercaptan, *n*-propyl mercaptan, *n*-butyl mercaptan, *sec*-butyl mercaptan, *tert*-butyl mercaptan, thiophenol, benzyl mercaptan, *meta*-thiocresol, *para*-thiocresol, or *ortho*-thiocresol.

Inhibitors. The enzyme is sensitive to $HgCl_2$, $CuSO_4$, phenylhydrazine, and hydroxylamine (inhibition at 10 mM: 97, 95, 88, and 88%, respectively).

[83] O-Acetyl-L-Serine–O-Acetyl-L-Homoserine Sulfhydrylase from *Saccharomyces cerevisiae*

By Shuzo Yamagata

$$O\text{-Acetyl-L-serine} + H_2S \rightarrow \text{L-cysteine} + \text{acetic acid} \quad (1)$$
$$O\text{-Acetyl-L-homoserine} + H_2S \rightarrow \text{L-homocysteine} + \text{acetic acid} \quad (2)$$

The physiological function of O-acetylserine–O-acetylhomoserine sulfhydrylase of *Saccharomyces cerevisiae* is the biosynthesis of both cysteine and homocysteine.[1,2] Reaction (1) is also catalyzed in this organism by another enzyme, L-serine sulfhydrylase,[3,4] which was first observed by Schlossman and Lynen in 1957.[5] The latter enzyme, however, does not seem to function as a cysteine synthase in this organism.[1] Reactions (1) and (2) have both been observed in extracts of many other microorgan-

---

[1] S. Yamagata, T. Kawai, and K. Takeshima, *Biochim. Biophys. Acta* **701**, 334 (1982).
[2] S. Yamagata, K. Takeshima, and N. Naiki, *J. Biochem. (Tokyo)* **77**, 1029 (1975).
[3] S. Yamagata, *J. Biochem. (Tokyo)* **88**, 1419 (1980).
[4] S. Yamagata, *J. Bacteriol.* **147**, 688 (1981).
[5] K. Schlossmann and F. Lynén, *Biochem. Z.* **328**, 591 (1957).

isms and plants[6] but are catalyzed by separate enzymes. Only a few other organisms have been shown to have enzymes catalyzing both the reactions.[7]

Assay Method

Principle. The products of these reactions, cysteine and homocysteine, are determined after removing sulfide at an acidic pH by reaction with nitroprusside.[8] Other spectrophotometric determinations of cysteine[9,10] and homocysteine[9] are also convenient. Use of radioactive sulfide (^{35}S)[11] is recommended for an accurate determination of the activity in crude extracts.

Reagents

Potassium phosphate, 0.5 M, at pH 7.8
Tris–HCl, 1 M, at pH 7.8
Pyridoxal 5′-phosphate, 10 mM
Ethylenediaminetetraacetic acid (EDTA), 0.2 M
O-Acetyl-L-serine, 50 mM
O-Acetyl-L-homoserine, 50 mM
Tris–sulfide,[11] 10 mM
Trichloroacetic acid (TCA), 30%
NaCl, saturated solution
Na_2CO_3, 1.5 M, containing 67 mM KCN
Sodium pentacyanonitrosylferrate (nitroprusside), 2%

Procedure. In a glass-stoppered centrifuge tube is prepared, in a total volume of 0.8 ml, a mixture containing 0.2 ml of the potassium phosphate buffer, 20 μl pyridoxal phosphate, 5 μl EDTA, an appropriate amount of enzyme (0.01–0.10 units), and water. After preincubation at 30° for 5 min, the reaction is started by adding 0.1 ml of O-acetylserine or O-acetylhomoserine and, finally, 0.1 ml sulfide. The tube is stoppered and incubated for 10 min at 30°. The reaction is stopped by the addition of 0.1 ml TCA, and the tube is centrifuged at 3,000 rpm for 5 min. The supernatant liquid is transferred to a test tube by decantation and gently bubbled with nitrogen gas for 5 min to remove remaining sulfide. An aliquot of 0.5 ml of the

---

[6] P. A. Trudinger and R. E. Loughlin, *in* "Comprehensive Biochemistry" (A. Neuberger, ed.) Vol. 19A, p. 165. Elsevier/North-Holland Biomedical Press, Amsterdam, 1981.
[7] A. Paszewski and J. Grabski, *Acta Biochim. Pol.* **23,** 321 (1976).
[8] R. R. Grunert and P. H. Phillips, *Arch. Biochem.* **30,** 217 (1951).
[9] N. M. Kredich and M. A. Becker, this series, Vol. 17, Part B, p. 459.
[10] M. K. Gaitonde, *Biochem. J.* **104,** 627 (1967).
[11] D. Kerr, this series, Vol. 17, Part B, p. 446.

treated solution is transferred to a colorimeter cell, to which is then added 0.5 ml distilled water, 3 ml NaCl, 0.5 ml Na_2CO_3, and 0.5 ml nitroprusside. After 1 min, absorbance at 520 nm may be measured with a Bausch and Lomb spectrophotometer (Model SP 20A). Cysteine or homocysteine, 0.5 μmol/0.5 ml of the reaction mixture, yields an A_{520} of 0.74.

Comment. When O-acetyl-L-homoserine sulfhydrylase activity is being determined, phosphate buffer can be replaced by 0.1 ml of Tris–HCl at the same pH in the incubation mixture, yielding a 10% greater activity than with phosphate. But Tris–HCl buffer should not be used in the O-acetylserine sulfhydrylase reaction since this amino acid appears to convert to N-acetylserine more easily in this buffer.[12]

Definition of Unit and Specific Activity. One unit of enzyme catalyzes the sulfhydrylation of 1 μmol of O-acetylserine or O-acetylhomoserine per minute. Specific activity is expressed as units per milligram protein. Protein is determined colorimetrically.[13]

Purification

Material. The purification procedure given here is essentially that described by Yamagata et al.[12,14] Pressed baker's yeast, *Saccharomyces cerevisiae,* was purchased from Oriental Yeast Company, Osaka. Other sources may also be employed.

Extraction by Autolysis. Pressed yeast cake, 5 kg, is crumbled, spread to a thickness of 2–3 cm over stainless steel trays, and mixed with an equal quantity of powdered Dry Ice. Thus frozen the cells are thawed at 30° with the aid of an electric fan. The freezing–thawing procedure is repeated once. The resulting thick cell suspension is dried at room temperature in front of an electric fan and subsequently kept under reduced pressure over pellets of NaOH. The dried material (1.6 kg) is suspended in 5.5 liters of 20% saturated ammonium sulfate (115 g/liter) solution, and the pH is adjusted to 10.0 with 28% ammonia in water. The suspension is maintained at 30° overnight (about 17 hr).

Step 1: Salt Fractionation I. The suspension is diluted with 5.5 liters of cold (4°) 20% saturated ammonium sulfate solution. Proteins are henceforth handled at 4°, unless otherwise stated. The ammonium sulfate concentration is adjusted to 45% saturation by adding 1.27 kg of the salt; the precipitate formed, including cell debris, is removed by centrifugation at 10,000 g for 30 min. The supernatant liquid is mixed with ammonium

[12] S. Yamagata, K. Takeshima, and N. Naiki, *J. Biochem. (Tokyo)* **75,** 1221 (1974).
[13] O. H. Lowry, N. J. Rosebrough, A. L. Farr, and R. J. Randall, *J. Biol. Chem.* **193,** 265 (1951).
[14] S. Yamagata and K. Takeshima, *J. Biochem. (Tokyo)* **80,** 777 (1976).

sulfate (135 g/liter, 65% saturation), and the proteins precipitated are collected by centrifugation, dissolved in 400 ml distilled water, and subsequently subjected to dialysis overnight against 4.5 liters of 20 mM Tris–HCl at pH 7.8, containing 1 mM EDTA and 0.2 mM pyridoxal phosphate. The dialysate is centrifuged at 10,000 g for 30 min.

Step 2: Salt Fractionation II. The supernatant fluid is fractionated with ammonium sulfate between 50% (315 g/liter) and 60% saturation (additional 65 g/liter). The precipitate is dialyzed overnight against two changes of 2 liters of the same buffer as in Step 1.

Step 3: Heat Treatment. The supernatant liquid after centrifugation of the dialyzed solution is diluted to a protein concentration of 30 mg/ml with the same buffer, and the concentration of pyridoxal phosphate is adjusted to 0.6 mM. After adjusting the pH precisely to 6.0 with 1 N and 0.1 N HCl, the solution is heated in a water bath at 80° and stirred vigorously until the temperature of the solution reaches 55°. It is then immediately chilled to 4° with ice water and centrifuged at 10,000 g for 20 min.

Step 4: Salt Fractionation III. The resulting supernatant fraction is subjected to another ammonium sulfate fractionation, and the precipitate obtained between 50% (315 g/liter) and 65% saturation (additional 95 g/liter) is dialyzed as in Step 2.

Step 5: DEAE-Cellulose I. This fraction (193 ml) is diluted 10-fold and adsorbed onto a DEAE-cellulose (DE-52, Whatman) column (3.2 × 14 cm) equilibrated with the same buffer. Enzyme activity is eluted with this buffer containing 0.2 M NaCl (43 ml).

Step 6: DEAE-Cellulose II. The eluate is diluted 4-fold as above and applied to a DEAE-cellulose column (3.2 × 14 cm) equilibrated with the same buffer containing 0.05 M NaCl. After washing the column with 350 ml of the equilibrating buffer, proteins are eluted with 1 liter of a linear NaCl concentration gradient from 50 to 200 mM formed in the same buffer. The eluate is collected in fractions of 10 ml with active fractions being combined.

Step 7: DEAE-Cellulose III. The enzyme solution obtained above is adsorbed on a DEAE-cellulose column (3.2 × 2.5 cm) equilibrated with the same buffer, and the column is washed with 60 ml of the buffer. The enzyme is eluted with 20 mM potassium phosphate at pH 6.5, containing 1 mM EDTA.

Comments. The purification summarized in Table I has been found to be reproducible and yields one major protein band on disk gel electrophoresis. Electrophoresis in the presence of sodium dodecyl sulfate also yields a single band. Extraction of the enzyme can also be effectively carried out by agitating the cell suspension with glass beads in a Dyno Mill as described previously for extraction of low molecular weight *O*-ace-

TABLE I
PURIFICATION OF O-ACETYLSERINE–O-ACETYLHOMOSERINE SULFHYDRYLASE FROM S. cerevisiae[a]

Step	Volume (ml)	Total protein (mg)	Enzyme amount[b] (units)	Specific activity[b] (units/mg protein)
1. Salt fractionation I	1,920	107,000	26,700	0.250
2. Salt fractionation II	569	53,400	22,000	0.412
3. Heat treatment	1,560	19,800	14,800	0.747
4. Salt fractionation III	193	15,900	17,600	1.11
5. DEAE-Cellulose 1	43	2,340	8,770	3.75
6. DEAE-Cellulose 2	118	400	3,810	9.55
7. DEAE-Cellulose 3	20.8	49.6	1,490	30.0

[a] Purified from 5 kg of pressed baker's yeast.
[b] O-Acetylhomoserine is used as the substrate.

tylserine sulfhydrylase[1,3,4] from a mutant-type strain of the same organism.[2]

Properties

Stability. No loss in activity is observed after storage at $-20°$ for 6 months when dissolved in 20 mM Tris–HCl at pH 7.8, supplemented with 1 mM EDTA and 0.2 mM pyridoxal phosphate.

Specificity. The enzyme reacts with both O-acetyl-L-serine and O-acetyl-L-homoserine, giving rise to L-cysteine [reaction (1)] and L-homocysteine [reaction (2)], respectively, when sulfide is used as a cosubstrate. The rate of reaction with the former amino acid is about 14% as much as with the latter. No other amino acid serves as substrate; O-acetyl-L-threonine, O-acetyl-L-tyrosine, O-acetyl-L-hydroxyproline, and O-succinyl-DL-homoserine are inactive.[15] Methyl mercaptan and ethyl mercaptan are also substrates with O-acetyl-L-homoserine as the primary substrate, producing methionine and ethionine, respectively; cysteine and aminoethyl mercaptan are not substrates.[15]

Kinetic Constants. For O-acetylserine the K_m is 5.1 mM [reaction (1)] and 6.7 mM for O-acetylhomoserine [reaction (2)].[12] A mutant-type enzyme shows similar values.[2]

Requirement and Inhibitors. Pyridoxal phosphate is always required in the assay. The association constant of the enzyme for the cofactor is

[15] S. Yamagata, *J. Biochem. (Tokyo)* **70**, 1035 (1971).

$6.1 \times 10^4\ M^{-1}$, which is considerably lower than the values reported for other pyridoxal enzymes, e.g., $7 \times 10^5\ M^{-1}$ for rat liver cystathionase.[16] The K_m values for the cofactor are 18.2 μM [reaction (1)] and 11.8 μM [reaction (2)], but the latter value is greater, 44 μM, in a mutant enzyme.[2] O-Succinyl-DL-homoserine, L-homoserine, and L-methionine competitively inhibit the enzyme, at concentrations of 30 mM, by 31, 37, and 73%, respectively.[15]

Molecular Properties.[17] The enzyme has an M_r of approximately 200,000 as obtained by gel filtration and by centrifugation in a sucrose density gradient ($s_{20,w}$ 9.0 S). Sodium dodecyl sulfate (1%) or urea (6 M) dissociates the enzyme into four subunits calculated to have M_r of 51,000 or 57,000, respectively. The subunits appear identical, each bearing one sulfhydryl group "buried" inside and binding 1 mol of pyridoxal phosphate. Partial restoration (30%) of the catalytic activity is possible when urea-denatured enzyme is dialyzed in the presence of dithiothreitol against urea free buffer. The isoelectric point of the enzyme is at pH 4.4; there are more acidic residues than basic residues found on amino acid analysis.[17]

[16] K. Oh and J. E. Churchichi, *J. Biol. Chem.* **248**, 7370 (1973).
[17] S. Yamagata, *J. Biochem. (Tokyo)* **80**, 787 (1976).

[84] Cystathionine β-Lyase from *Escherichia coli*

By JACK R. UREN

Cystathionine + H$_2$O → homocysteine + pyruvate + ammonia

Cystathionine β-lyase (β-cystathionase, EC 4.4.1.8) is an enzyme in methionine biosynthesis present in enteric microorganisms. The enzyme has broad substrate specificity capably of degrading L-cystine, L-cysteine, L-djenkolic acid, L-homolanthionine, and L-*meso*-lanthionine in addition to L-cystathionine. This chapter updates a previous source for the enzyme from a *Salmonella metA-15* mutant which overproduced the enzyme 7-fold over wild-type levels.[1] By cloning the *Esherichia coli metC* gene on a ColE1 plasmid, and by transferring this plasmid into a regulation-deficient *metJ* mutant, 100-fold overproduction of β-cystathionase above the wild-

[1] S. Guggenheim, this series, Vol. 17, Part B [195f].

type *E. coli* level could be obtained.[2] This strain (ATCC 37384) now represents the most enriched source for the enzyme that is available.

Assay Method

Principle. The enzyme was assayed by measuring the rate of formation of free thiol groups by spontaneous disulfide interchange with 5,5'-dithiobis(2-nitrobenzoic acid) (DTNB), similar to the described procedure for γ-cystathionase.[3]

Reagents

Tris–HCl, 0.1 M, pH 9.0
L-Cystathionine, 10 mM, in 0.01 N HCl
5,5'-Dithiobis(2-nitrobenzoic acid), 10 mM, in 0.1 M potassium phosphate at pH 7.0
Enzyme solution containing between 1 and 50 units/ml

Procedure. Store all reagents at 4°. Add, 0.78 ml of the Tris buffer, 0.2 ml L-cystathionine, and 20 μl DTNB to a 1-ml cuvette. Add 1–10 μl of the enzyme solution and record the increase in absorbance at 412 nm with time. A molar extinction coefficient for the aryl mercaptide of 13,200 was used to calculate the results. One enzyme unit represents the formation of 1 μmol of mercaptide per minute at 37°.

Purification

Procedure. The RG 100/pLC4-14 cells were grown overnight at 37° in Fernbach flasks containing 2 liters of 0.8% Bacto nutrient broth. Cells were harvested by centrifugation at 8,000 g and washed twice with 10 mM sodium phosphate at pH 7.0, containing 0.1 mM dithioerythritol and 10 μM pyridoxal phosphate (buffer A). The cells were stored frozen until used. Cells, 18 g, were suspended in 360 ml of buffer A and broken in a French press at 20,000 p.s.i. The extract was centrifuged at 10,000 g, and the supernatant fluid loaded onto a column (20 × 2.5 cm) of DE-52 cellulose that had been equilibrated with buffer A. The column was washed with 250 ml of buffer A and eluted with a linear gradient of 500 ml of 0.05–0.35 M NaCl in buffer A. Fractions of 10 ml each were collected, and those with more than 7.5 units/ml of enzyme were pooled.

Pooled fractions were diluted with buffer A to a conductivity of 5 mmho/cm and treated with 25 g of BioGel HTP that had been equilibrated

---

[2] C. Dwivedi, R. Ragin, and J. Uren, *Biochemistry* **21**, 3064 (1982).
[3] M. Flavin and C. Slaughter, this series, Vol. 17, Part B [195e].

TABLE I
PURIFICATION SCHEME FOR *E. coli* β-CYSTATHIONASE[a]

Step	Volume (ml)	Activity (units)	Protein (mg)	Specific activity (units/mg)
Crude extract	365	4220	3500	1.2
DE-52 cellulose	150	3490	480	7.3
Hydroxyapatite	250	3030	200	15
Blue Sepharose	105	2260	10.5	215

[a] Taken from Dwivedi *et al.*[2]

with buffer A. After 30 min the suspension was centrifuged, and the supernatant fluid was discarded. The pellet was suspended in 500 ml of 60 mM sodium phosphate at pH 7.0, containing 0.1 mM dithioerythritol and 10 μM pyridoxal phosphate. After 30 min the suspension was again centrifuged, and the supernatant liquid collected. The supernatant fluid was loaded onto a column (75 × 1.7 cm) of Blue Sepharose CL-6B equilibrated with buffer A. The column was washed and eluted under conditions identical to the DE-52 column. Fractions of 10 ml were collected, and those having specific activities over 200 units/mg protein were pooled. For concentration purposes the pooled fractions were diluted with buffer A to a conductivity of 5 mmho/cm and charged onto a column (5 ml) of DE-52 cellulose that was eluted with buffer A containing 0.3 M NaCl. All purification steps were carried out at 4–10°.

Table I summarizes the results of purification. Compared to the previous purification from the *Salmonella typhimurium metA-15* mutant,[1] the specific activity of the crude extract was 13-fold higher, the overall yield was 5-fold higher, and the final purified specific activity was 27-fold higher.

TABLE II
APPARENT KINETIC CONSTANTS FOR
β-CYSTATHIONASE FROM *E. coli*[a]

Substrate	K_m (mM)	V_{max} (units/mg)
L-Cystathionine	0.04	249
L-Cystine	0.25	263
L-Homolanthionine	4.5	255
L-*meso*-Lanthionine	0.83	262
L-Djenkolic acid	0.36	198

[a] Taken from Dwivedi *et al.*[2]

Properties

Stability. Following sterile filtration, the enzyme was stable when stored at 4° for up to 4 months. The enzyme lost activity when frozen or lyophilized but remained stable at −20° in the presence of 50% glycerol for 4 months. Unless grown in the presence of colicin, the plasmid may be unstable, resulting in a loss of the overproduction phenotype.

pH Optimum. The enzyme showed a bell-shaped pH-rate profile with an optimum between pH 8 and 9. At pH 7.2 the enzyme had 36% of its optimal activity.

Substrate Specificity. The enzyme followed Michaelis–Menten kinetics and showed broad substrate specificity. Table II presents several kinetic properties. Using the coupled enzyme assay[3] in the presence of 10 mM dithioerythritol, L-cysteine was a poor substrate with a maximal velocity only 1% of that with L-cystathionine. L-Cysteine also showed excess substrate inhibition.

Inhibitors. Adenosine (1 mM), S-adenosylmethionine (1 mM), S-adenosylhomocysteine (1 mM), allylglycine (16 mM), diaminopimelic acid (16 mM), 2,7-diaminooct-4-ynedioic acid (16 mM), and propargylglycine (1 mM) did not significantly inhibit the enzyme. β-Cyanoalanine inhibited to only 43% at 16 mM. L-Cysteine showed competitive inhibition with a K_i of 0.4 mM in the presence of 10 mM dithioerythritol. 3,3,3-Trifluoroalanine showed time-dependent irreversible inhibition with a binding constant of 0.55 mM and a saturating inactivation half-time of 1.7 min.

Physical Properties. The enzyme had an aggregate size of 280,000 as determined by gel filtration, with 6 subunits of 45,000 determined by sodium dodecyl sulfate gel electrophoresis. It bound 6 mol of pyridoxal phosphate with a K_m of 5 μM. All 36 cysteine residues could react with DTNB, indicating that there were no disulfide bonds.

[85] Cystathionine γ-Lyase from *Streptomyces phaeochromogenes*

By Toru Nagasawa, Hiroshi Kanzaki, and Hideaki Yamada

L-Cystathionine + H_2O → L-cysteine + H_2O + α-ketobutyric acid

Cystathionine γ-lyase (EC 4.4.1.1) was first purified and crystallized from rat liver,[1] but is also widely distributed among fungi. It has been

[1] Y. Matsuo and D. M. Greenberg, *J. Biol. Chem.* **230**, 545 (1958).

shown to be present in *Saccharomyces*,[2,3] *Neurospora*,[2] and *Aspergillus*.[4] Although previous studies all indicated that prokaryotic organisms lack cystathionine γ-lyase activity,[2,5] the enzyme is now known to be widely distributed in actinomycetes: *Streptomyces, Micromonospora, Micropolyspora, Mycobacterium, Nocardia, Streptosporangium*, and *Streptoverticillium*.[6] Strains belonging to the genus *Streptomyces* show relatively strong cystathionine γ-lyase activity. The microbial enzyme has been purified and crystallized from *Streptomyces phaeochromogenes* (IFO 3105).[6] Cystathionine β-synthase and cystathionine γ-synthase activities were also detected in crude extracts of *S. phaeochromogenes*, but cystathionine β-lyase was not. Consequently, the reverse transsulfuration pathway in actinomycetes appears to be similar to that in yeast and molds.[6]

Assay Procedures

Principle. The enzyme is assayed by measuring the rate of formation of either α-ketobutyric acid or cysteine from L-cystathionine, or the formation of α-ketobutyric acid from L-homoserine. Gaitonde's acid/ninhydrin assay[7] is highly specific, there being essentially no reaction with homocysteine. The lack of color development with homocysteine means that the assay is specific for γ-lyase activity and that β-lyase activity does not interfere. The assay is suitable, therefore, for crude extracts with cystathionine as substrate.

Assay I: Measurement of α-Ketobutyric Acid Formed from L-Homoserine

Reagents

Potassium phosphate, 0.2 M, pH 8.0
L-Homoserine, 0.2 M
Pyridoxal phosphate, 1 mM
Trichloroacetic acid, 1.84 M

Procedure. To a centrifuge tube are added 0.2 ml potassium phosphate, 0.25 ml L-homoserine, 40 µl pyridoxal phosphate, 0.1–0.5 units of

---

[2] C. Delavier-Klutchko and M. Flavin, *J. Biol. Chem.* **240**, 2537 (1965).
[3] E. Morzycka and A. Paszewski, *FEBS Lett.* **101**, 97 (1979).
[4] A. Paszewski and J. Grabski, *Acta Biochim. Pol.* **20**, 159 (1973).
[5] M. Flavin, in "Metabolic Pathways (D. M. Greenberg, ed.), Vol. 7, p. 457. Academic Press, New York, 1975.
[6] T. Nagasawa, H. Kanzaki, and H. Yamada, *J. Biol. Chem.* **259**, 10393 (1984).
[7] M. Gaitonde, *Biochem. J.* **104**, 626 (1967).

enzyme, and water to a total volume of 1.0 ml. For each assay tube, a control tube is prepared from which L-homoserine is omitted. The mixture is incubated for 30 min at 30°; during this period a linear rate should be attained. Incubation is terminated by adding 0.2 ml of trichloroacetic acid. The precipitate is removed by centrifugation, and α-ketobutyric acid in the supernatant fluid is determined by the method of Friedemann and Haugen.[8]

Assay II: Measurement of L-Cysteine Formed from L-Cystathionine

Reagents

Potassium phosphate, 0.2 M, pH 8.0
L-Cystathionine, 8.3 mM
Pyridoxal phosphate, 1 mM
Trichloroacetic acid, 1.84 M

Procedure. The reaction is carried out as described for Assay I except that the L-homoserine is replaced by 0.25 ml of L-cystathionine. The incubation is stopped by adding 0.2 ml of trichloroacetic acid. The precipitate is removed by centrifugation, and cysteine is determined by the method of Gaitonde.[7]

Definition of Unit and Specific Activity. One unit of enzyme is defined as the amount of enzyme catalyzing the formation of 1 μmol of α-ketobutyric acid from L-homoserine in 1 min. Specific activity is expressed in terms of enzyme units per milligram protein. Protein is determined from its absorption at 280 nm. The absorption coefficient was found to be 0.839 mg^{-1} ml cm^{-1} by absorbance and by dry weight determination.[6]

Purification Procedure

Source and Growth. Streptomyces phaeochromogenes (IFO 3105) obtained from a stock culture of the Institute of Fermentation, Osaka (IFO Type Culture Collection), was the source of the enzyme. The basal medium for cultivation consists of 2 g yeast extract, 0.3 g glucose, 0.2 g K_2HPO_4, 0.1 g $MgCl_2 \cdot 2H_2O$, 4 mg $CaCl_2 \cdot 2H_2O$, 2 mg $FeSO_4 \cdot 7H_2O$, 2 mg $MnSO_4 \cdot 4$–$6H_2O$, 2 mg $CuSO_4 \cdot 5H_2O$, and 1 mg $ZnSO_4 \cdot 7H_2O$ per 100 ml of tap water. The pH of the medium is adjusted to 7.0 with 4 M NaOH. *Streptomyces phaeochromogenes* is collected from an agar slant of the basal medium and inoculated into subculture flask. The subculture (3.5 liters) is shaken reciprocally at 28° for 48 hr and then inoculated into a 100-liter jar fermentor containing 70 liters of the basal medium supplemented

---

[8] T. E. Friedemann, this series, Vol. 3, p. 414.

with 35 g of antifoam (AF-emulsion). Incubation is carried out at 28° for 20 hr with aeration (37 liters/min). The cells from 33.5 liters of broth are harvested by continuous-flow centrifugation and washed with 0.15 M NaCl containing 0.1 mM EDTA. The yield of wet cells is approximately 20 g/liter of medium.

All operations in steps 1 through 9 are carried out at 4°, and potassium phosphate buffer containing 20 μM pyridoxal phosphate, 0.12 mM EDTA, and 0.2 mM dithiothreitol is used throughout, unless otherwise specified.

Step 1: Cell-Free Extract. Washed cells (670 g) are suspended in 2 liters of 0.1 M potassium phosphate at pH 7.5 and disrupted by ultrasonic oscillation (19 kHz). Cell debris is removed by centrifugation at 12,000 g for 30 min. The supernatant fluid is dialyzed overnight against 10 mM potassium phosphate at pH 7.5 at 5°. The precipitate that appears during dialysis is removed by centrifugation at 12,000 g for 30 min and discarded.

Step 2: Salt Fractionation I. Solid ammonium sulfate, 17.6 g, is added per 100 ml of cell-free extract (30% saturation) at 0°. The pH is maintained at 7.5 with 6.3 M NH$_4$OH. After stirring for 2 hr or more, the precipitate is removed by centrifugation at 12,000 g, and 19.8 g of ammonium sulfate is added per 100 ml of fluid (60% saturation). The suspension is centrifuged at 12,000 g, and the pellet dissolved in 0.1 M potassium phosphate at pH 7.5. The solution is dialyzed for 36 hr against three changes of 10 liters each of 10 mM potassium phosphate at pH 7.5.

Step 3: DEAE-Cellulose. The enzyme solution is applied to a DEAE-cellulose column (3.5 × 60 cm), previously equilibrated with 0.1 M potassium phosphate at pH 7.5. The column is washed with 0.1 M potassium phosphate buffer (pH 7.5) containing 0.1 M KCl until the absorbance of the effluent at 280 nm is reduced to 0.2 or less. The enzyme is eluted with the same buffer containing 0.2 M KCl. Fractions of 17 ml are collected, the active eluates are combined, and 39 g of solid ammonium sulfate are added per 100 ml with stirring. The precipitate is collected by centrifugation, dissolved in a small amount of the phosphate buffer, and dialyzed against three changes of 10 liters each of 10 mM potassium phosphate at pH 7.5.

Step 4: Salt Fractionation II. The dialyzed enzyme preparation from Step 3 is again subjected to fractionations with ammonium sulfate (finely ground). The precipitate appearing between 50 and 60% of saturation (31.1 g and 39 g per 100 ml, respectively) is dissolved in the 0.1 M phosphate buffer and dialyzed against three changes of 5 liters each of the 10 mM phosphate buffer.

Step 5: DEAE-Sephacel. The enzyme solution from Step 4 is applied to a column (2.7 × 27 cm) of DEAE-Sephacel previously equilibrated with

0.1 M phosphate buffer. After the column has been washed with the same buffer, the enzyme is eluted with a linear gradient of KCl (0–0.5 M, 400 ml in each container) in the same buffer at a flow rate of 80 ml/hr. Fractions of 3.3 ml are collected, and the active fractions are pooled. Solid ammonium sulfate is added to the enzyme solution to 39 g/100 ml of solution. After centrifugation of the suspension at 12,000 g, the precipitate is dissolved in the 0.1 M buffer and dialyzed against 5 liters of 10 mM buffer.

Step 6: Hydroxylapatite. The enzyme solution is applied to a column (2.8 × 14 cm) of hydroxylapatite previously equilibrated with the 10 mM phosphate buffer. The column is washed fully with 10 mM phosphate until the absorbance of the effluent at 280 nm is reduced to 0.1 or less. The enzyme is then eluted with 200 ml of 50 mM potassium phosphate. Fractions of 3 ml are collected, and the active fractions are combined.

Step 7: Phenyl-Sepharose CL-4B. After cooling to 0°, 7.67 g/100 ml of ammonium sulfate is added in small portions with stirring (10% of saturation). The enzyme solution is applied to a column (1.0 × 26 cm) of phenyl-Sepharose CL-4B previously equilibrated with 50 mM potassium phosphate at pH 7.5 containing 10% saturated ammonium sulfate. The column is washed with 50 ml of 50 mM potassium phosphate at pH 7.5 containing sequentially 10% and 5% saturated ammonium sulfate. The enzyme is eluted by lowering linearly the ionic strength of ammonium sulfate (from 5 to 0%, 200 ml in each container) in the same buffer. Fractions of 2.3 ml are collected, and the active fractions are combined.

Step 8: Octyl-Sepharose CL-4B. Ammonium sulfate is added in small portions with stirring to bring the enzyme solution from Step 7 to 10% saturation (5.6 g/100 ml). After application to a column (0.8 × 10 cm) of octyl-Sepharose CL-4B, previously equilibrated with 50 mM phosphate buffer containing 10% ammonium sulfate, the enzyme passes through the column without loss of activity.

Step 9: Sephadex G-200. Ammonium sulfate, 32.6 g/100 ml, is added in small portions with stirring to the solution from Step 8. The precipitate is collected at 12,000 g and dissolved in 10 mM potassium phosphate buffer. The enzyme solution is concentrated to about 2 ml with a Diaflo YM30 membrane and charged onto a column (2.3 × 93 cm) of Sephadex G-200 previously equilibrated with 10 mM potassium phosphate at pH 7.5 containing 0.1 M KCl. The rate of sample loading and column elution is maintained at 4 ml/hr with a peristaltic pump. Protein is eluted with the same buffer, and fractions containing cystathionine γ-lyase activity are combined and concentrated by ultrafiltration.

Purification of approximately 3,000-fold can be achieved, with a yield of about 30%. After the last step, the enzyme appears to be homogeneous

TABLE I
PURIFICATION OF CYSTATHIONINE γ-LYASE FROM S. phaeochromogenes[a]

Step	Total volume (ml)	Total protein (mg)	Total activity (units)	Specific activity (units/mg)
1. Cell-free extract	2,210	85,800	73.3	0.0009
2. Salt fractionation I	800	48,500	70.6	0.0015
3. DEAE-Cellulose	435	1,970	69.7	0.0483
4. Salt fractionation II	43.5	582	43.2	0.0741
5. DEAE-Sephacel	106	279	41.2	0.148
6. Hydroxylapatite	64.8	114	32.1	0.383
7. Phenyl-Sepharose CL-4B	9.5	21.3	22.3	1.05
8. Octyl-Sepharose CL-4B	18	9.54	21.8	2.28
9. Sephadex G-200	48.2	8.15	21.2	2.60

[a] The reaction was carried out under the conditions of Assay II.

by the criteria of polyacrylamide gel electrophoresis, analytical centrifugation, and double diffusion in agarose.

The purified enzyme catalyzes the α,γ-elimination reaction of L-homoserine (Assay I) and L-cystathionine (Assay II) at 2.60 and 1.90 μmol/min/mg of protein, respectively. The purification procedure is summarized in Table I.

Crystallization. The purified enzyme solution is diluted with 10 mM of the potassium phosphate buffer containing 20 μM pyridoxal phosphate and 0.2 mM dithiothreitol to a protein concentration of about 30 mg/ml. Ammonium sulfate is added to this solution to 31.3 g/100 ml at 0° with gentle stirring with a glass rod. Additional ammonium sulfate is added until the induced turbidity ceases to disappear upon stirring. The presence of cystallized materials is evident from the silky sheen seen when the mixture is stirred. If denatured protein strands become visible, the solution is centrifuged to remove them before the enzyme begins to crystallize. The cystallized enzyme is in the apo form. Full activity is restored by the addition of 20 μM pyridoxal phosphate.

Properties

Stability. The highly purified enzyme can be stored for 2 weeks without loss of activity in 10 mM potassium phosphate at pH 7.5 containing 20 μM pyridoxal phosphate, 0.12 mM EDTA, and 0.2 mM dithiothreitol at 4°. The enzyme remains stable for 2 or more months at −20° in the above buffer containing 45% (v/v) glycerol.

pH Optimum. In two buffer systems, Tris–HCl and NH$_4$Cl–NH$_4$OH (final concentration, 42 mM), the optima for the α,γ-elimination reaction of L-cystathionine and L-homoserine are at approximately pH 9.0.

M_r and Cofactor. The enzyme has an M_r of about 166,000 and consists of four subunits of identical size. The holoenzyme exhibits absorption maxima at 278 and 421 nm with an $A_{278} : A_{421}$ ratio of approximately 3.97, and contains 4 mol of pyridoxal phosphate per mole of enzyme.

Substrate Specificity and Catalytic and Kinetic Properties. L-Cystathionine, L-homoserine, DL-lanthionine, L-djenkolic acid, and L-cystine are cleaved by the *Streptomyces* enzyme. L-Cysteine is also cleaved by the enzyme, although at concentrations greater than 1 mM, substrate inhibition is seen. For L-cystathionine and L-homoserine the apparent K_m values are 0.2 and 13 mM, respectively. The α,β-elimination reaction of L-cystathionine is also catalyzed by the enzyme at a rate about 15% of that of the α,γ-elimination reaction. The enzyme can also catalyze the γ-replacement reaction of L-homoserine in the presence of L-cysteine to form L-cystathionine.[9] D-Cysteine, L- and D-homocysteine, and 3- and 2-mercaptopropionate can replace L-cysteine as substrates for γ-replacement.[10]

Inhibitors. The enzyme is strongly inhibited by carbonyl reagents such hydroxylamine, D- and L-penicillamine, semicarbazide, phenylhydrazine, 3-methyl-2-benzothiazolinone hydrazone, and D-cycloserine. The enzyme displays high sensitivity to some thiol reagents (at 1 mM), e.g., *p*-chloromercuribenzoate, HgCl$_2$, AgNO$_3$, and ZnCl$_2$, yielding from 60 to 100% inhibition after 30 min at 30° with each reagent. Sodium cyanide causes 82% inhibition at 1 mM. DL-Propargylglycine and β-cyano-L-alanine strongly inhibit (80–100% inhibition at 1 mM).

Immunological Properties. Antiserum prepared against the purified cystathionine γ-lyase of *S. phaeochromogenes* cross-reacts with the cystathionine γ-lyase from *S. olivochromogenes* (IFO 3178), *S. lydicus* (AKU 2502), *S. lactamdurans* (IFO 13305), *Streptoverticillium kentuchense* (IFO 12880), *Streptosporangium roseum* (IFO 3776), *Micromonospora chalcea* (IFO 12135), *Micropolyspora angispora* (IFO 13155), and *Microellobosporia violacea* (IFO 12517). Spurs are formed when the enzymes from these actinomycetes are placed in neighboring wells on double-immunodiffusion analysis.

---

[9] H. Yamada, H. Kanzaki, and T. Nagasawa, *J. Biotechnol.* **1**, 205 (1984).
[10] H. Kanzaki, M. Kobayashi, T. Nagasawa, and H. Yamada, *Agric. Biol. Chem.* **50**, 391 (1986).

[86] Selenocysteine β-Lyase from *Citrobacter freundii*

By NOBUYOSHI ESAKI and KENJI SODA

$$HSeCH_2CH_2NH_2COOH \rightarrow CH_3CH_2NH_2COOH + Se$$

Selenocysteine β-lyase (EC 4.99.1.—) catalyzes the β-elimination reaction in which L-alanine and elemental selenium are formed from L-selenocysteine. The enzyme is specific for L-selenocysteine and is found widely distributed.[1-3] Assay procedures for the enzyme are detailed elsewhere in this volume [70], describing the preparation of a similar enzyme from pig liver.[3]

Purification Procedure[4]

All operations are performed at 0–5° unless otherwise stated. Potassium phosphate, 10 mM at pH 7.2, containing 0.1 mM phenylmethylsulfonyl fluoride, 20 μM pyridoxal 5′-phosphate, and 0.1% 2-mercaptoethanol, is used as the standard buffer (buffer A). Buffer B is identical to buffer A, except that 1 mM dithiothreitol is substituted for 2-mercaptoethanol and 1 M sucrose is added. After each step, the enzyme solution is concentrated with an Amicon 200 ultrafiltration unit.

Step 1: Extract. *Citrobacter freundii* (ICR 0070) is grown in a medium containing 1.5% polypeptone, 0.1% glycerol, 0.1% KH_2PO_4, 0.1% K_2HPO_4, 0.025% yeast extract, 0.01% $MgSO_4 \cdot 7H_2O$, 0.01% meat extract, and 0.5% NaCl (pH 7.2). Cells are harvested by centrifugation at the end of the log phase (about 12 hr), and washed twice with 0.85% NaCl. Washed cells (wet weight, 2.5 kg) are suspended in 2 liters of buffer A containing 1 mM phenylmethylsulfonyl fluoride and disrupted with a Dyno Mill (Willy A., Bachofen, Switzerland) at a flow rate of 1 liter/hr. After centrifugation at 10,000 g for 30 min, the supernatant solution is treated with ribonuclease and deoxyribonuclease (0.1 mg/liter each) at room temperature for 1 hr, followed by centrifugation at 10,000 g for 1 hr. The clear supernatant solution is dialyzed against four changes of 20 liters of buffer A.

[1] N. Esaki, T. Nakamura, H. Tanaka, T. Suzuki, Y. Morino, and K. Soda, *Biochemistry* **20**, 4492 (1981).
[2] P. Chocat, N. Esaki, T. Nakamura, H. Tanaka, and K. Soda, *J. Bacteriol.* **156**, 455 (1983).
[3] N. Esaki, T. Nakamura, H. Tanaka, and K. Soda, *J. Biol. Chem.* **257**, 4386 (1983).
[4] P. Chocat, N. Esaki, K. Tanizawa, K. Nakamura, H. Tanaka, and K. Soda, *J. Bacteriol.* **163**, 669 (1985).

Step 3: Phenyl-Sepharose. The enzyme solution is applied to two phenyl-Sepharose columns (5 × 30 cm) equilibrated with buffer A containing 0.5 M potassium phosphate at the same pH. After successive washing with buffer A containing 0.5, 0.3, 0.2, and 0.15 M phosphate, the enzyme is eluted with buffer A containing 0.1 M phosphate at a flow rate of 0.2 liter/hr. Active fractions are concentrated and dialyzed against 100 volumes of buffer A.

Step 4: DEAE-Sephadex A-50. The enzyme solution is applied to a DEAE-Sephadex A-50 column (4 × 25 cm), equilibrated with buffer A in which potassium phosphate buffer is replaced by Tris–HCl at pH 8.0. The enzyme does not bind to the resin, and the unadsorbed fractions are concentrated and dialyzed as described above.

Step 5: Hydroxyapatite. The enzyme solution is applied to a column (4 × 10 cm) of hydroxyapatite[5] equilibrated with buffer A. After the column is washed with buffer A containing 20 mM phosphate, the enzyme is eluted with buffer A containing 30 mM phosphate at a flow rate of 20 ml/hr and dialyzed against 100 volumes of buffer B.

Step 6: Sephadex G-150. The enzyme solution is applied to a Sephadex G-150 column (2 × 85 cm) equilibrated with buffer B and eluted with the same buffer at a flow rate of 20 ml/hr. The active fractions are pooled and concentrated.

Step 7: Toyo Soda TSK DEAE-3W. The resulting enzyme solution is chromatographed on a Toyo Soda TSK DEAE-3SW prepacked column equilibrated with buffer B. After washing with buffer B containing KCl (up to 220 mM), the enzyme is eluted with buffer B containing 240 mM KCl and dialyzed against 100 volumes of buffer B.

Step 8: FPLC MonoQ NR 5/5. The dialyzed enzyme is chromatographed on a MonoQ HR 5/5 anion-exchange column of the Pharmacia Fast Protein Liquid Chromatography (FPLC) system. At this step, Tris–HCl at pH 7.4 is substituted for potassium phosphate in buffer B. The elution is carried out with a linear gradient of KCl (0.1–0.12 M) in the Tris buffer B at a flow rate of 1 ml/min. Active fractions are pooled and dialyzed against 100 volumes of buffer B.

Table I shows a typical result of purification of the enzyme.

Properties

Stability. The presence of 0.1% 2-mercaptoethanol or 1 mM dithiothreitol is essential to prevent enzyme inactivation. A high concentration of sucrose, typically 1 M, is also effective. However, even with a buffer containing 1 M sucrose, 1 mM dithiothreitol, 50 μM pyridoxal 5'-

---

[5] A. Tiselius, S. Hjerten, and O. Levin, *Arch. Biochem. Biophys.* **65**, 132 (1956).

TABLE I
PURIFICATION OF SELENOCYSTEINE β-LYASE FROM *Citrobacter freundii*

Step	Volume (ml)	Total activity (units)	Total protein (mg)	Specific activity (units/mg)
1. Extract	3,700	860	102,000	0.008
2. DEAE-Toyopearl 650M	1,800	315	13,500	0.023
3. Phenyl-Sepharose	680	172	2,800	0.061
4. DEAE-Sephadex A-50	140	122	1,070	0.114
5. Hydroxyapatite	1.2	27	48	0.56
6. Sephadex G-150	1.0	15	17	0.97
7. Toyo Soda TSK DEAE-3W	0.8	16	4	4.0
8. FPLC MonoQ NR 5/5	0.3	0.39	0.06	6.47

phosphate, and 0.1 mM ethylenediaminetetraacetic acid, enzyme activity gradually decreases during storage. More than 60% of the original activity is lost after storage at $-20°$ for 1 month.

Physicochemical and Catalytic Properties. The M_r of the enzyme is estimated to be 63,000 ± 3,000 (high-performance gel permeation method) and 64,000 ± 1,000 (polyacrylamide gel electrophoresis in sodium lauryl sulfate). The enzyme is composed of a single polypeptide chain. The isoelectric point of the protein is estimated to be 6.6 ± 0.1. The enzyme shows maximum reactivity at pH 7.0 when assayed in potassium phosphate and sodium borate buffers. The absorption spectrum of the enzyme (pH 7.2) exhibits mamima at 278 and 420 nm. When enzyme is treated with 10 mM hydroxylamine hydrochloride, activity decreases to less than 1% but is 79% restored by the addition of 50 μM pyridoxal 5'-phosphate. The K_m value for pyridoxal 5'-phosphate is 0.25 μM. An average pyridoxal 5'-phosphate content of 0.8 mol/64,000 g of enzyme is obtained; 1 mol of pyridoxal 5'-phosphate is bound per mole of enzyme.

The enzyme catalyzes exclusively the β-elimination of L-selenocysteine to produce L-alanine and elemental selenium. The apparent K_m for L-selenocysteine is 0.95 mM. The enzyme is strongly inhibited by 1 mM of the following metallic dications: Zn^{2+}, Ni^{2+}, Pb^{2+}, and Hg^{2+}. It is completely inactivated by incubation at 37° for 3 hr with 1 mM thiol reagents such as iodoacetic acid, iodoacetamide, and N-ethylmaleimide. L-Cysteine and L-cysteine methyl ester behave as competitive inhibitors against L-selenocysteine (K_i, 0.65 and 0.6 mM, respectively), even in the presence of a large excess (0.1 mM) of pyridoxal 5'-phosphate. None of the following compounds inhibited the enzyme reaction with 4 mM DL-selenocysteine: 4 mM L-selenocysteine ethyl ester, L-aspartate, glycine, L-alanine, L-norleucine, DL-serine, and hydrogen selenide. Neither D-cysteine nor cysteamine inhibited the enzyme reaction at 5 mM. The enzyme catalyzes the α,β-elimination of β-chloro-L-alanine to form ammonia, pyruvate, and chloride and is irreversibly inactivated during the reaction. An apparent K_{inact} value and a maximum inactivation rate constant are calculated to be 3.1 mM and 0.23 min^{-1}, respectively. Based on an enzyme molecular weight of 64,000, the partition ratio between α,β-elimination and inactivation is 825.

[87] Taurine Dehydrogenase

By HIROYUKI KONDO and MAKOTO ISHIMOTO

$$^-O_3SCH_2CH_2NH_3^+ + H_2O + \text{Acceptor} \rightarrow {}^-O_3SCH_2CHO + NH_4^+ + \text{Acceptor (reduced)}$$

Taurine dehydrogenase (EC 1.4.99.2) catalyzes the dehydrogenation of taurine (aminoethanesulfonic acid) with the formation of sulfoacetaldehyde, ammonia, and a reduced electron acceptor. The nature of the acceptor is not known.

Assay Method

Principle. The assay is based on the determination of oxygen consumption with taurine as substrate in the presence of phenazine methosulfate (PMS). The determination can be made either with Clark oxygen electrodes or with a Warburg manometer.

Reagents

Tris–HCl, 50 mM, pH 8.5
PMS, 20 mM, stored in a dark bottle
Potassium cyanide, 20 mM, neutralized with dilute HCl to pH 8.5
Enzyme dissolved in 50 mM Tris–HCl at pH 8.5
Taurine, 1 M

Procedure. A mixture of 2.1 ml of the Tris buffer, 0.15 ml PMS, 0.15 ml KCN, and 0.3 ml of enzyme solution is incubated at 30° in a 3-ml vessel in which oxygen electrodes are inserted. The reaction is started by adding 0.3 ml taurine. Oxygen concentration is measured with the oxygen electrodes equipped with a recorder. The total volume of the reaction mixture depends only on the size of the oxygen electrode vessel.

Definition of Unit and Specific Activity. One unit of enzyme is defined as the amount that catalyzes the consumption of 1 μmol of oxygen per minute under the above conditions. Specific activity is expressed as units per milligram protein.

Organism and Cultivation

Taurine-decomposing bacteria can be isolated from the natural environment by the enrichment culture technique according to Ikeda et al.[1] The organism used here is available from the authors on request. The bacterial cells are grown at 30° for 10 hr in a medium containing, per liter, 5 g taurine, 4.5 g K_2HPO_4, 1.5 g KH_2PO_4, 0.2 g $MgCl_2 \cdot 6H_2O$, and 0.2 g yeast extract. A long-necked 500-ml flask, a 5-liter conical flask, or a 60-liter jar fermentor is used for the cultivation depending on the culture size. The volume of the liquid medium is 40% of the total capacity of the vessel. The inoculum size is 5% in each case. Shaking at 120 strokes/min is required for the cultivation in a flask and aeration of 25 liters/min for that in a jar fermentor.

Purification Procedure

Taurine dehydrogenase is a membrane-bound enzyme. When the enzyme preparation of the membrane-bound form is not needed, Step 1 may be skipped over and the enzyme is solubilized directly from the cells as described in Step 2. All operations are carried out at 0–4° unless stated otherwise.

Step 1: Preparation of Crude Extract and Membrane-Bound Enzyme. The cells (40 g wet weight) are suspended in 80 ml of 0.05 M Tris–HCl at pH 8.5, and 0.001% DNase is added to the suspension. The cells are disrupted by treatment with a sonic disintegrator (20 kHz) at 133 W at 1-min intervals for a total of 10 min to keep the temperature low. Instead of sonic treatment, passage through a French pressure cell at 400 kg/cm^2 can be employed. The extract is centrifuged at 5,000 g for 15 min, and the precipitate is discarded. The membrane-bound enzyme is obtained by further centrifugation at 100,000 g for 60 min. The residue is washed with 20 mM Tris–HCl at pH 8.5, suspended in 52 ml of the same buffer, and dialyzed against the same buffer overnight.

Step 2: Solubilization of the Enzyme. The enzyme is solubilized by treatment with a detergent, 0.1% Emulgen 810 (Kao Atlas), for 30 min. The enzyme may also be obtained in solubilized form by adding Emulgen 810 at 0.1% directly to the cell suspension at the time of cell disruption in Step 1. In either instance, the suspension is centrifuged at 20,000 g for 30

---

[1] K. Ikeda, H. Yamada, and S. Tanaka, *J. Biochem. (Tokyo)* **54**, 312 (1963).

TABLE I
PURIFICATION OF TAURINE DEHYDROGENASE[a]

Step	Activity (units)	Protein (mg)	Specific activity (units/mg protein)
Extract	268	7,300	0.037
Supernatant at 20,000 g	177	2,070	0.085
Supernatant at 100,000 g	103	1,350	0.076
Sepharose gel filtration	69	145	0.476

[a] Typical data obtained with 40 g (wet weight) of bacteria.[2]

min, and the supernatant liquid obtained is centrifuged at 100,000 g for 60 min. About 40% of the enzyme activity in the crude extract is recovered in the resultant supernatant fluid.

Step 3: Ammonium Sulfate Precipitation and Gel Filtration on Sepharose 4B. The solution is saturated with ammonium sulfate to 40% (24.3 g/100 ml) by adding the solid salt with stirring. After standing for 1 hr, the precipitate is centrifuged and dissolved in 10 ml of 50 mM Tris–HCl at pH 8.5, insoluble materials being removed by centrifugation at 20,000 g for 20 min. A portion, 6 ml, of the supernatant fluid is applied to a column (2 × 90 cm) of Sepharose 4B (Pharmacia) which had been equilibrated with the same buffer. The column is developed with the same buffer. Enzyme activity appears in the effluent a little after the appearance of blue dextran. Active fractions are combined and dialyzed overnight against 50 mM Tris–HCl at pH 8.5. A typical purification is summarized in Table I.[2] Approximately 13-fold purification is achieved with an overall yield of 26%.

Properties

Size. The molecular weight of the enzyme is considered to be in the range of several hundred thousands on the basis of the gel filtration on Sepharose 4B.

Stability. The enzyme is stable to freeze-thawing and to storage at −20° at a protein concentration of 1.5 mg/ml for at least 1 month. It is,

[2] H. Kondo, K. Kagotani, M, Oshima, and M. Ishimoto, *J. Biochem.* (*Tokyo*) **73**, 1269 (1973).

however, unstable to heat, losing 50% of its activity after incubation at 40° for 20 min.

pH Optimum. The optimum pH for taurine dehydrogenation is 8.4.

K'_m for Taurine and PMS. K'_m values are 20 mM for taurine and 56 μM for PMS.

Substrate Specificity. Hypotaurine is oxidized only slowly. Ethanolamine, ethylamine, *n*-propylamine, *n*-butylamine, benzylamine, histamine, cadaverine, cysteic acid, DL-alanine, β-alanine, and aminoethanephosphonic acid are not oxidized in the presence of PMS.

Specificity for Electron Acceptor. Taurine is not oxidized by the soluble, purified enzyme in the absence of PMS and cyanide but is oxidized by the membrane-bound preparation without PMS. NAD^+, $NADP^+$, 2,6-dichlorophenolindophenol, or ferricyanide do not replace PMS, although the latter two are active in the crude extract prior to solubilization.

Inhibitors. Because of the complex reactivity of PMS with some of the compounds examined, effects of those compounds should be determined with the membrane-bound enzyme. The most active inhibitors are isonicotinic acid 2-isopropylhydrazine (iproniazid), phenelzine, and *p*-chloromercuribenzoate. Neocuproine, 2-heptyl-4-hydroxyquinoline *N*-oxide, atebrin, and methylene blue are inhibitory to some extent. Iproniazid and *p*-chloromercuribenzoate require preincubation with the enzyme for maximum inhibition.

Reaction Products and Stoichiometry of the Oxidation of Taurine. Reaction products of taurine dehydrogenation by this enzyme are sulfoacetaldehyde and ammonia.[2,3] The stoichiometry and reaction equations are as follows:

$$^-O_3SCH_2CH_2NH_3^+ + 2\ PMS + H_2O \rightarrow {}^-O_3SCH_2CHO + 2\ PMS\ (reduced) + NH_4^+$$
$$2\ PMS\ (reduced) + O_2 \xrightarrow{nonenzymatic} 2\ PMS + H_2O_2$$

Comments

The enzyme is inducible by taurine. The enzyme should be called taurine dehydrogenase or taurine : (acceptor) oxidoreductase (deaminating) (EC class 1.4.99.—). Sulfoacetaldehyde, the reaction product, is decomposed to sulfite and acetate by sulfoacetaldehyde lyase (EC 4.4.1.12), which is dependent on thiamin pyrophosphate and magnesium for activity.[4-6]

[3] H. Kondo, H. Anada, K. Ohsawa, and M. Ishimoto, *J. Biochem. (Tokyo)* **69**, 621 (1971).
[4] H. Kondo and M. Ishimoto, *J. Biochem. (Tokyo)* **72**, 487 (1972).
[5] H. Kondo and M. Ishimoto, *J. Biochem. (Tokyo)* **76**, 229 (1974).
[6] H. Kondo and M. Ishimoto, *J. Biochem. (Tokyo)* **78**, 317 (1975).

[88] ω-Amino Acid–Pyruvate Aminotransferase

By KAZUO YONAHA, SEIZEN TOYAMA, and KENJI SODA

ω-Amino acid + pyruvate ⇌ aldehydic acid + L-alanine

Assay Method

Principle. ω-Amino acid–pyruvate aminotransferase (EC 2.6.1.18; β-alanine–pyruvate aminotransferase) was originally designated taurine–pyruvate aminotransferase[1] and may correspond to β-alanine–pyruvate aminotransferase (EC 2.6.1.18). The enzyme catalyzes the reversible transamination of various sulfur-containing ω-amino acids and ω-amino carboxylic acids with pyruvate to yield the corresponding aldehydic acids and L-alanine. The assay method is based on the measurement of aldehydic acid or L-alanine.

Reagents

β-Alanine (or another ω-amino acid), 0.2 M
Sodium pyruvate, 0.2 M
Pyridoxal 5'-phosphate, 1 mM
Potassium phosphate, 0.2 M, pH 8.0
Trichloroacetic acid, 25% (w/v)
o-Aminobenzaldehyde, 25 mM in 10% ethanol
Potassium hydroxide, 0.8 M
Glycine–KCl–KOH buffer, 1 M pH 8.0

Procedures. The standard assay system consists of 20 μmol of β-alanine or other ω-amino acid, 20 μmol sodium pyruvate, 0.1 μmol pyridoxal 5'-phosphate, 100 μmol potassium phosphate at pH 8.0, and enzyme in a final volume of 1.0 ml. In a control tube, β-alanine is replaced by water. After incubation at 37° for 30 min, the reaction is terminated by addition of 0.1 ml of 25% trichloroacetic acid followed by centrifugation at 17,000 g for 10 min. The enzyme is assayed by determining L-alanine (Method A) or malonic semialdehyde (or aldehydic acid) (Method B).

METHOD A: DETERMINATION OF L-ALANINE. Alanine in the supernatant fluid is determined with a conventional amino acid analyzer.

METHOD B: DETERMINATION OF MALONIC SEMIALDEHYDE (OR ALDEHYDIC ACID). The procedure is described in detail in a previous volume of this series (Vol. 113 [20], p. 102).

[1] S. Toyama, K. Miyazato, M. Yasuda, and K. Soda, *Agric. Biol. Chem.* **37**, 2939 (1973).

Definition of Unit and Specific Activity. One unit of enzyme is defined as that amount which catalyzes the formation of 1 μmol of alanine. Specific activity is expressed as units per milligram protein. Protein is determined by the method of Lowry et al.,[2] or estimated from the absorbance at 280 nm ($E_{1\,cm}^{1\%} = 9.84$).[3]

Purification Procedure[4]

Culture Conditions.[1] *Pseudomonas* sp. F-126 (this organism can be obtained from Department of Agricultural Chemistry, University of The Ryukyus, Nishihara, Okinawa 903-01, Japan) is grown in a medium composed of 0.3% β-alanine (as an inducer), 0.2% polypeptone, 0.1% glycerol, 0.1% KH_2PO_4, 0.2% K_2HPO_4, 0.1% NaCl, 0.01% $MgSO_4 \cdot 7H_2O$, and 0.01% yeast extract. The pH of the medium is adjusted to 7.0 with sodium hydroxide. Growth is carried out in a 50-liter fermenter at 30° for 20–22 hr under aeration. Cells, harvested by centrifugation, are washed twice with 0.85% NaCl, and subsequently with 20 mM potassium phosphate at pH 8.0, containing 0.1 mM pyridoxal 5′-phosphate and 0.01% 2-mercaptoethanol.

All buffers used for the purification contain 20 μM pyridoxal 5′-phosphate and 0.01% 2-mercaptoethanol unless otherwise specified. All subsequent procedures are carried out at 4°.

Step 1: Extract. Washed cells (1 kg, wet weight) are suspended in 20 mM potassium phosphate at pH 7.4, containing 0.01% 2-mercaptoethanol and 0.1 mM pyridoxal 5′-phosphate, and subjected to sonication for 20 min in a 19-kHz oscillator followed by centrifugation.

Step 2: Polyethyleneimine Treatment. To the cell-free extract is added 0.1 ml of 10% polyethyleneimine solution (pH 7.4) per milligram protein with stirring at 4°. After 10 min, the bulky precipitate is removed by centrifugation at 10,000 g for 20 min. The supernatant solution is brought to 70% saturation with ammonium sulfate (47.2 g/100 ml). The precipitate is dissolved in 10 mM potassium phosphate at pH 7.4 and dialyzed overnight against 100 volumes of the same buffer. The insoluble materials formed during dialysis are removed by centrifugation.

Step 3: DEAE-Cellulose I. The enzyme solution is placed on a column (11 × 100 cm) of DEAE-cellulose equilibrated with 10 mM potassium phosphate at pH 7.4. After the column is washed thoroughly with the same buffer, the enzyme is eluted with the buffer supplemented with 50

---

[2] O. H. Lowry, N. J. Rosebrough, A. L. Farr, and R. J. Randal, *J. Biol. Chem.* **193**, 265 (1951).
[3] K. Yonaha, S. Toyama, M. Yasuda, and K. Soda, *Agric. Biol. Chem.* **41**, 1701 (1977).
[4] K. Yonaha, S. Toyama, M. Yasuda, and K. Soda, *FEBS Lett.* **71**, 21 (1976).

TABLE I
PURIFICATION OF ω-AMINO ACID–PYRUVATE AMINOTRANSFERASE FROM *Pseudomonas* sp. F-126

Step	Volume (ml)	Total protein (mg)	Total activity (units)	Specific activity (units/mg)
1. Extract	2500	140,000	9800	0.07
2. Polyethyleneimine	2000	102,000	7240	0.07
3. DEAE-Cellulose I	1500	6,210	5030	0.81
4. DEAE-Cellulose II	700	2,760	4310	1.56
5. Hydroxyapatite	100	992	3200	3.23
6. Sephadex G-150	50	736	2590	3.52
7. Crystallization	—	660	2310	3.50

mM NaCl. Active fractions are pooled and concentrated by addition of ammonium sulfate (70% saturation). The precipitate is dissolved in the same phosphate buffer and dialyzed against 100 volumes of it.

Step 4: DEAE-Cellulose II. The enzyme solution is subjected to a column (6 × 70 cm) of DEAE-cellulose in the same manner as in Step 3. The active fractions are concentrated by ammonium sulfate (70% saturation) and dissolved in the phosphate buffer followed by dialysis against 100 volumes of the same buffer.

Step 5: Hydroxyapatite. The enzyme solution is applied to a column (4 × 45 cm) of hydroxyapatite equilibrated with the same buffer, washed with it, and eluted with 0.3 M potassium phosphate at pH 7.4. The active fractions are concentrated by ammonium sulfate (70% saturation) and dissolved in a small volume of 20 mM potassium phosphate at pH 7.4.

Step 6: Sephadex G-150. The enzyme is applied to a Sephadex G-150 column (2.5 × 100 cm) equilibrated with 20 mM potassium phosphate at pH 7.4 and eluted with the same buffer. The active fractions are pooled.

Step 7: Crystallization. To the enzyme solution is gradually added ammonium sulfate until a faint turbidity is obtained. The pH of the solution is maintained at about 7.4 with 1 N NaOH. On standing overnight, crystals with a rectangular form are obtained. Purification of the enzyme is summarized in Table I. About 50-fold purification is achieved with an overall yield of about 25%.

Properties

Stability. The crystalline enzyme can be stored with little loss of activity at 0–5° for several months in 10 mM potassium phosphate at pH 7.4

containing ammonium sulfate (40% saturation). The enzyme is stable in the pH range of 8.0–10.0 and up to 65° at pH 8.0 (10 mM potassium phosphate) for 15 min.

Physicochemical Properties.[3,4] The purified enzyme is homogenous by the criteria of disk gel electrophoresis and ultracentrifugation. The M_r is estimated as 172,000 by Sephadex G-150 gel filtration. Assuming a partial specific volume of 0.74, an M_r of 167,000 ± 3000 is obtained. The enzyme consists of four identical subunits (M_r 44,000).

Spectrometric Properties and Coenzyme.[5] The absorption spectrum of the enzyme is atypical for a pyridoxal 5'-phosphate-dependent transaminase. There is no distinct maximum in the 410–420 nm region at neutral pH. The absorbance around 400 nm is increased at an acidic pH with a decrease in 340 nm absorption. An isosbestic point is obtained at 378 nm. The enzyme contains 1 mol pyridoxal 5'-phosphate and 1 mol vitamin B_6 compound (structure unknown) per enzyme tetramer. Pyridoxal 5'-phosphate, absorbing at 400 nm, forms a Schiff base with ω-amino groups of the enzyme and participates directly in catalysis. The function of the unknown vitamin B_6 compound (absorption peak at 340 nm) has not been settled.

The sequence around the active site lysine residue is as follows: Cys(cm)-Ile-Ala-Lys(Pxy)-Gln-Val-Thr-Asn, where Cys(cm) represents carboxymethylcysteine and Lys(Pxy) represents phosphopyridoxyllysine.

The enzyme fluoresces with maxima at 330 and 380 nm upon excitation at 280 nm. The maximum is at 465 nm when excited at 400 nm. The excitation spectrum of the enzyme, analyzed at 380 nm, shows maxima at 280 and 340 nm, and at 400 nm by analysis at 465 nm.

Circular dichroic spectrum of the enzyme shows positive peaks at 340 and 415 nm with molecular elipticity of 8.4×10^4 and 2.0×10^4, respectively. α-Helix content of the enzyme is calculated to be 50%.

Substrate Specificity.[3,6,7] The enzyme catalyzes transamination with several ω-amino acids and pyruvate. 2-Aminoethane sulfinate (hypotaurine) (relative activity, 132), 3-aminopropane sulfonate (80), β-alanine (100), γ-aminobutyrate (40), 7-aminoheptanoate (35), 8-aminooctanoate (77), DL-β-amino-*n*-butyrate (96.4), and DL-β-aminoisobutyrate (54.2) are good amino donors, whereas 2-aminoethane sulfonate (taurine) (3.6), 5-aminopentanoate (6.0), and 6-aminohexanoate (8.2) are not as active. In addition to ω-amino acids, mono- and diamines also have amino donor

---

[5] K. Yonaha, S. Toyama, and H. Kagamiyama, *J. Biol. Chem.* **258**, 2260 (1983).
[6] K. Yonaha and S. Toyama, *Agric. Biol. Chem.* **42**, 2263 (1978).
[7] K. Yonaha and S. Toyama, *Agric. Biol. Chem.* **43**, 1043 (1979).

activity for the enzyme: n-butylamine (55), n-amylamine (80), n-hexylamine (65), n-heptylamine (60), 1,4-diaminobutane (55), 1,6-diaminohexane (87), and 1,7-diaminoheptane (92) are good amino donors.

In contrast to broad specificity for the amino donor, the enzyme is specific for pyruvate as the exclusive amino acceptor. α-Ketoglutarate, oxaloacetate, phenylpyruvate, α-ketoisocaproate, and α-ketoisovalerate are inactive.

Kinetics.[3,6,7] Michaelis constants have been calculated: taurine, 19 mM; β-alanine, 3.3 mM; n-butylamine, 66.6 mM; putrescine, 76.9 mM; and pyruvate, 3.3 mM. The pH optimum is at 8.5–10.5. Maximal activity is obtained at 60°.

Inhibitors.[3] The enzyme is inhibited by carbonyl reagents such as phenylhydrazine, 3-methyl-2-benzothiazolone hydrazone hydrochloride, D-cycloserine, hydroxylamine, and aminooxyacetate; inactivation by these reagents is reversed by incubation with pyridoxal 5'-phosphate. Gabaculine and ethanolamine O-sulfate are suicide substrates of the enzyme.[8,9] Sulfhydryl reagents and chelating reagents have no effect on the enzyme reaction.

Distribution.[10] The ω-amino acid transaminase is found in many strains of bacteria and yeast grown in a medium containing β-alanine, but it has not been observed in actinomycetes or fungi.

---

[8] G. Burnett, K. Yonaha, S. Toyama, K. Soda, and C. Walsh, *J. Biol. Chem.* **255,** 428 (1980).
[9] K. Yonaha and S. Toyama, *Agric. Biol. Chem.* **49,** 229 (1985).
[10] K. Yonaha, K. Suzuki, H. Minei, and S. Toyama, *Agric. Biol. Chem.* **47,** 2257 (1983).

[89] Sulfhydryl Oxidase from Milk

By HAROLD E. SWAISGOOD and H. ROBERT HORTON

Sulfhydryl oxidase was first obtained in substantially purified form from bovine milk by Janolino and Swaisgood.[1] This enzyme has been shown to catalyze *de novo* synthesis of disulfide bonds with cysteine,[2] cysteamine,[2] glutathione,[1] and reduced forms of ribonuclease,[1] xanthine

---

[1] V. G. Janolino and H. E. Swaisgood, *J. Biol. Chem.* **250,** 2532 (1975).
[2] M. X. Sliwkowski, H. E. Swaisgood, D. A. Clare, and H. R. Horton, *Biochem. J.* **220,** 51 (1984).

oxidase,[3] and chymotrypsinogen[4] as substrates. The enzyme is a constituent of the plasma membrane of mammary secretory cells[5] and occurs in milk within skim milk membrane vesicles.[6] Solubilization of the enzyme, yielding an increase in specific activity, can be achieved by addition of nonionic detergents, such as polyoxyethylene-9-lauryl ether ($C_{12}E_9$) or β-octylglucoside, to the vesicles from whey. Sulfhydryl oxidase can be isolated from these membrane vesicles by several procedures; however, a recently developed transient covalent affinity chromatographic procedure[7] is the most rapid and involves the fewest steps. This procedure, which will be described in detail here, is based on the catalytic mechanism of the enzyme.[2]

Preparation of Cysteinylsuccinamidopropyl-Controlled-Pore Glass

The size of the solubilized enzyme–detergent complex is rather large so that penetration of pores smaller than 200 nm is quite limited. Therefore, controlled-pore glass with a diameter of 200 nm or larger is used to prepare the affinity matrix. The procedure[7] for preparation of the affinity matrix consists of three principal steps: (1) silanization, (2) succinylation, and (3) covalent attachment of cysteine. The reactions are shown schematically in Fig. 1.

Step 1: Silanization. The controlled-pore glass surface is cleaned, generating a maximum number of silanol groups, by incubation in concentrated nitric acid at 100° for 1 hr. The beads are washed with distilled water until the pH of the washings is neutral. A 10% (v/v) aqueous solution of 3-aminopropyltriethoxysilane is prepared, and the pH is adjusted to 8 with 6 N HCl. The cleaned CPG beads[8] are added to three volumes of the 10% 3-aminopropyltriethoxysilane, degassed at reduced pressure to ensure that all of the pore volume is filled with reagent, and incubated with occasional mixing at 80° for 4 hr. Following reaction, excess reagent is decanted and the amino beads are dried overnight at 85–110° under reduced pressure. We have found that silanization at pH 8 results in better

---

[3] D. A. Clare, B. A. Blakistone, H. E. Swaisgood, and H. R. Horton, *Arch. Biochem. Biophys.* **211**, 44 (1981).

[4] V. G. Janolino, M. X. Sliwkowski, H. E. Swaisgood, and H. R. Horton, *Arch. Biochem. Biophys.* **191**, 269 (1978).

[5] D. A. Clare, H. R. Horton, T. J. Stabel, H. E. Swaisgood, and J. G. Lecce, *Arch. Biochem. Biophys.* **230**, 138 (1984).

[6] M. B. Sliwkowski, H. E. Swaisgood, and H. R. Horton, *J. Dairy Sci.* **65**, 1681 (1982).

[7] M. X. Sliwkowski, M. B. Sliwkowski, H. R. Horton, and H. E. Swaisgood, *Biochem. J.* **209**, 731 (1983).

[8] The following abbreviations are used: CPG, controlled pore glass; DTNB, 5,5'-dithiobis(2-nitrobenzoic acid); EDC, 1-ethyl-3-(3-dimethylaminopropyl)carbodiimide.

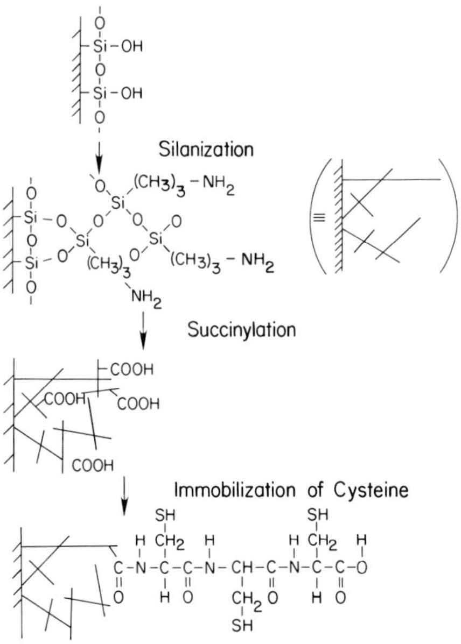

FIG. 1. Schematic representation of the preparation of cysteinylsuccinamidopropyl–glass matrix.

performance for the affinity matrix than does similar treatment at pH 4. Furthermore, aqueous silanization, as opposed to derivatization in an organic phase, produces a much better matrix for isolation of sulfhydryl oxidase.[9]

Step 2: Succinylation. Aminopropyl-CPG beads can be succinylated most conveniently by reaction with succinic anhydride dissolved in acetone. The dry beads are rinsed several times with acetone, and a solution containing 1% (v/v) triethylamine and 10% (v/v) succinic anhydride in acetone is added in a ratio of approximately three volumes of solution per volume of beads. The mixture can be degassed briefly under reduced pressure to ensure complete exposure of the pore surfaces. Reaction is complete within 10 min at room temperature. The succinamidopropyl-glass beads are thoroughly rinsed with distilled water prior to reaction with carbodiimide.

Step 3: Immobilization of Cysteine. A sequential surface activation/immobilization procedure developed by Janolino and Swaisgood[10] has been used to covalently attach the cysteine. The sequential activation/

[9] G. W. Koszalka, Ph.D. Dissertation, North Carolina State Univ. Raleigh (1985).
[10] V. G. Janolino and H. E. Swaisgood, *Biotechnol. Bioeng.* **24**, 1069 (1982).

immobilization procedure consists of (1) formation of the O-acylisourea derivative of the matrix carboxyl groups by incubating the succinamidopropyl-beads with 0.1 M EDC[8] at pH 4.75 and room temperature for 20 min, (2) rapid rinsing (<2 min) of the beads with a cold (4°) neutral salt solution to remove excess EDC, and (3) immobilization of cysteine by addition and recirculation of a 0.1 M solution (2.5 ml/g of beads) of cysteine in 50 mM sodium phosphate, pH 7, at 4° for 20 hr. To optimize the functioning of this matrix for subsequent isolation of sulfhydryl oxidase, it was found necessary to repeat the activation/immobilization procedure 3–5 times.[9] Presumably, formation of cysteine peptides increases the accessibility of these residues to the enzyme. It should be noted that other thiol matrices not incorporating cysteine, which is a substrate for the enzyme, are not effective as affinity materials.

The activated carboxyl groups also can react with the thiol group of cysteine to give a thiol ester; however, since this bond is not very stable, the number of such bonds in the final derivatized matrix may be small. In any case, the cysteinylsuccinamidopropyl-glass matrices typically contain 2–3 μmol SH/g or about 0.22–0.33 μmol SH/m^2 of surface area.

Preparation of Solubilized Skim Milk Membrane Vesicles

Fresh raw milk is skimmed by centrifugation at 4100 g for 30 min at 30°. It is important to remove all lipid material, otherwise fat globule membrane proteins will contaminate the preparation.[1] This may be accomplished by carefully pouring the skim milk from the centrifuge bottle without disrupting the cream layer. Casein micelles are coagulated by addition of 4 units chymosin/100 ml of skim milk and incubation for 30 min at 30°. Under these conditions the Phe$_{105}$—Met$_{106}$ bond of κ-casein is specifically hydrolyzed and micelles are destabilized. The micelles are removed by centrifuging at 16,300 g for 45 min at 30°, which yields a clear supernatant whey fraction.

Membrane vesicles can be isolated from the whey fraction by chromatography using a column (2.5 × 200 cm) of glycerolpropyl-CPG 3000 glass (pore diameter 300 nm).[11] Prior to chromatography, the whey is subjected to an Amicon concentrator–dialyzer equipped with a 100,000 M_r exclusion limit hollow fiber membrane against 6 volumes of 47 mM sodium phosphate at pH 7.0 and concentrated 10-fold. The concentrated whey fraction, 80 ml, is applied to the CPG column and eluted with the phosphate buffer at 4°. The membrane vesicles are eluted in the mobile phase volume of the CPG-3000 column.

[11] V. G. Janolino and H. E. Swaisgood, *J. Dairy Sci.* **67**, 1161 (1984).

TABLE I
PURIFICATION OF SULFHYDRYL OXIDASE FROM BOVINE MILK WHEY BY
TRANSIENT COVALENT AFFINITY CHROMATOGRAPHY WITH
CYSTEINYL-CPG-GLASS[a]

Fraction	Volume (ml)	Total protein (mg)	Total[b] units	Specific activity (units/mg)
Whey	910	8190	40	0.005
Solubilized membrane	156	565	17	0.031
Covalent chromatography	99	0.545	13	24

[a] Adapted from data presented by Sliwkowski et al.[7]
[b] One unit of activity is defined as the oxidation of 1 μmol GSH per minute at 35°, measuring thiol depletion by reaction with DTNB.

Alternatively, a crude fraction of skim milk membrane vesicles can be obtained by removing the smaller whey proteins by diafiltration through a 100,000 M_r exclusion limit membrane. For example 1 liter of whey can be rapidly diafiltered at room temperature against 6 liters of sodium phosphate at pH 7.0 and concentrated to 150 ml with a Millipore Pellicon-Casette system.

In preparation for transient covalent affinity chromatography, the membrane vesicles are solubilized by addition of 1% (w/v) polyoxyethylene-9-lauryl ether ($C_{12}E_9$). After incubation with the nonionic detergent for 30 min at 4°, the mixture is centrifuged at 100,000 g for 1 hr at 4°, and the supernatant fraction is carefully removed with a syringe.

Purification of Sulfhydryl Oxidase[7]

Prior to application of the solubilized crude enzyme preparation, the cysteinylsuccinamidopropyl-glass matrix is completely reduced by washing with 8 M urea containing 0.1 M dithiothreitol, 0.1 M acetic acid, and 1 M NaCl. The reduced matrix is equilibrated with 47 mM sodium phosphate at pH 7.0. Enzyme is bound to the reduced matrix by recirculation of the solubilized preparation through a column of the beads for 4 hr at 35°. (We have also used 18 hr at 25°.) Typically, with 60 ml of matrix, 75–85% of the activity will be bound from a solubilized preparation obtained from 1 liter of whey (e.g., see Table I). Noncovalently bound proteins are removed by washing the column with a 1% (w/v) solution of nonionic detergent ($C_{12}E_9$) in 50 mM sodium phosphate at pH 7.0 (1 liter). Sulfhydryl oxidase is recovered by recirculation of substrate solution at 35° with 2 mM GSH in the phosphate buffer, until no thiol groups are detectable by

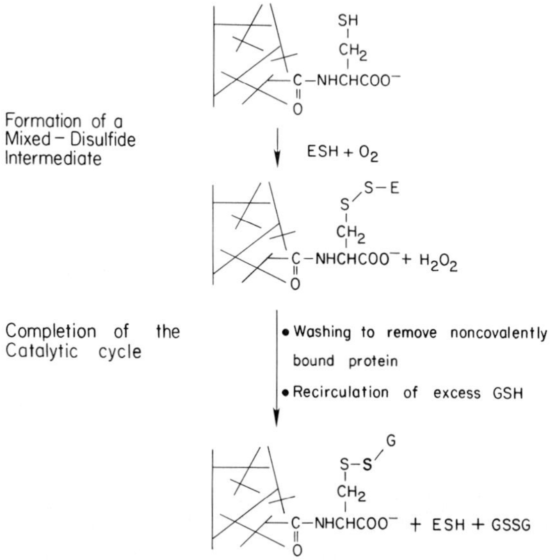

Formation of a
Mixed–Disulfide
Intermediate

$$\text{Matrix}-C(=O)-NHCHCOO^-\text{ with }CH_2-SH$$

$$\downarrow ESH + O_2$$

$$\text{Matrix}-C(=O)-NHCHCOO^- \text{ with } CH_2-S-S-E + H_2O_2$$

Completion of the
Catalytic cycle

- Washing to remove noncovalently bound protein
- Recirculation of excess GSH

$$\text{Matrix}-C(=O)-NHCHCOO^- \text{ with } CH_2-S-S-G + ESH + GSSG$$

FIG. 2. Schematic illustration depicting the attachment and release of sulfhydryl oxidase using a cysteinylsuccinamidopropyl–glass matrix for transient covalent affinity chromatography.

reaction of an aliquot with DTNB.[8] Essentially complete recovery of the bound enzyme has been observed using from 25 ml to 1 liter of substrate solution for about 60 ml of matrix. The proposed reactions responsible for transient covalent binding to the matrix are shown schematically in Fig. 2.

The released enzyme solution is adjusted to 0.5 M sucrose and stored at 4° to stabilize the activity. All of the above solutions should be filter-sterilized to minimize microbial activity.

The transient covalent affinity chromatographic step typically gives 600- to 800-fold purification (Table I). The specific activity of these preparations range from 15 to 25 units/mg. This matrix also has been used in a 2500-fold purification of sulfhydryl oxidase from bovine kidney.[12]

Assay of Sulfhydryl Oxidase Activity

Catalytic activities of sulfhydryl oxidase have been assayed by measuring O_2 consumption, production of glutathione disulfide, and production of H_2O_2, as well as depletion of thiol groups.[1,7] Several of these

[12] C. H. Schmelzer, H. E. Swaisgood, and H. R. Horton, *Biochem. Biophys. Res. Commun.* **107**, 196 (1982).

methods are presented elsewhere in this volume.[13] The most commonly used method for following oxidation of thiol groups will be presented here.

A typical assay[1,3,6,7] consists of 3.0 ml of 0.8 mM glutathione (GSH) in 47 mM sodium phosphate (0.1 ionic strength) at pH 7.0. Enzyme solution (250 μl) is added at zero time to this solution equilibrated at 35°. At intervals, 250-μl aliquots are removed and added to 5.0 ml of 0.1 mM DTNB in 0.1 M sodium phosphate at pH 8.0, containing 10 mM EDTA, at room temperature. The absorbance of this solution is measured at 412 nm after 2 min of reaction. Sufficient enzyme activity should be added to the assay solution to permit the assay to be completed within 10 min. From the linear portion of the progress curve, the moles SH oxidized per minute are calculated using a molar absorptivity of $\varepsilon_{412} = 13{,}600\ M^{-1}\ \text{cm}^{-1}$. Thus for the volumes used above, the activity per milliliter enzyme solution, units/ml, is given in Eq. (1). With some crude enzyme preparations, inter-

$$\text{units/ml} = \frac{21}{0.25}\left(\frac{\Delta A_{412/\text{min}}}{\varepsilon}\right)(3.25 \times 10^{-3})(10^6) = 20.1(\Delta A_{412/\text{min}}) \quad (1)$$

ference from γ-glutamyltransferase-induced SH autoxidation may be observed due to hydrolysis of GSH yielding cysteinylglycine. Such interference can be eliminated by substituting 1.4 mM GlyGlyCys for GSH as a substrate for sulfhydryl oxidase.[14]

[13] M. X. Sliwkowski and H. E. Swaisgood, this volume [19].
[14] C. H. Schmelzer, H. E. Swaisgood, and H. R. Horton, *Biochim. Biophys. Acta* **827**, 140 (1985).

[90] Sulfhydryl Oxidase from Rat Skin

By LOWELL A. GOLDSMITH

Disulfide bonds are essential in the maturation of normal epidermis and normal hair. In the living layers of the epidermis the keratins, the bulk of the epidermal proteins, can be solubilized without disulfide reduction. In contrast, the keratins in the finally cornified stratum corneum require reduction of their disulfides for solubilization. In hair the α-helical fibrous keratins are intimately related with the nonhelical high-sulfur cysteine-rich matrix proteins. In hair there is evidence that the formation of disulfide bonds is controlled. In copper-deficient sheep[1] there are decreased disulfide bonds, and in Menkes disease[2] (an X-linked disorder of metallothionein metabolism with abnormal copper storage and utilization) there is a decrease in the concentration of disulfide groups in hair and a reciprocal increase in free sulfhydryl groups in the hair. This background led us to analyze skin for enzymes which could oxidize sulfhydryl groups and form disulfide bonds.

Assay of Sulfhydryl Oxidase Activity[3]

Enzyme activity was assayed by using two different procedures: the ability of enzymatically active preparations to oxidize Ellman's reagent and the ability of active fractions to oxidize denatured ribonuclease to form enzymatically active ribonuclease. The ease and simplicity of the oxidation of Ellman's reagent made it the method of choice used to follow enzyme purification.

Assay with Ellman's Reagent

Principle. Unreduced Ellman's reagent, 5,5'-dithiobis(2-nitrobenzoic acid) (DTNB), has a molar extinction coefficient of 13,000 at 412 nm while oxidized Ellman's reagent does not absorb at this wavelength.

Reagents

Diluent buffer: Sodium phosphate, 50 mM, with 1 mM EDTA, pH 7.6
Substrate solution: Freshly prepared 2 mM dithiothreitol in diluent buffer
DTNB stock solution: 5,5'-dithiobis (2-nitrobenzoic acid), 40 μg, in 10 ml diluent buffer

Procedure. A typical reaction mixture contains 0.1 ml dithiothreitol substrate, 0.1–0.4 ml enzyme mixture, and diluent buffer to make a total volume of 1.2 ml. The mixture is incubated at 37°, and aliquots of 0.3 ml are removed at 0, 30 and 60 min and added to 3.0 ml diluent buffer. DTNB, 50 μl are added, and absorbance at 412 is measured. Using the molar absorbance of DTNB, data are expressed as nanomoles of oxidized DTNB per hour per microgram protein.

Assay of Renaturation of Ribonuclease A

Principle. Reduced denatured pancreatic ribonuclease A is devoid of enzymatic activity but slowly renatures spontaneously when reducing and

---

[1] R. W. Burley, *Nature (London)* **174**, 1019 (1954).
[2] D. M. Danks, in "The Biochemistry and the Physiology of the Skin" (L. A. Goldsmith, ed.), p. 1102. Oxford Univ. Press, London and New York, 1983.
[3] K. Takamori, J. M. Thorpe, and L. A. Goldsmith, *Biochim. Biophys. Acta* **615**, 309 (1980).

denaturing agents are removed. The rate of this renaturation is increased in the present of enzymes that form its disulfide bonds.

Reagents

Denatured ribonuclease: Thirty micrograms of ribonuclease A (64 Kunitz units/mg, Sigma) are dissolved in 3.0 ml of 8.0 M urea which has been passed over a mixed-bed ion-exchange column. The pH is adjusted to 8.6 with 5% methylamine. 2-Mercaptoethanol, 1 μl/mg ribonuclease, is added. The tube containing this reaction mixture is flushed with nitrogen and incubated at 25° for 4.5 hr. The pH is then adjusted to 4.5 with glacial acetic acid, and the entire solution chromatographed on a Sephadex G-10 column (3 × 25 cm) in 0.1 M acetic acid. Fractions with ribonuclease activity are pooled and diluted with 0.1 M acetic acid to contain 600–660 μg/ml (absorption at 280 nm of 0.35–0.40). Fractions of 3 ml are stored frozen at −20° and kept on ice when being prepared for use.

Procedure. Enzyme is incubated at 37° with 1 ml reduced ribonuclease in a final volume of 2 ml, with the volume adjusted with diluent buffer. A reagent blank is always included. Samples are removed at 0, 30, 60, and 90 min. Ribonuclease is assayed by measuring the hydrolysis of yeast RNA according to the procedure of Kalnitsky et al.[4]

Enzyme Purification[3]

Table I summarizes the purification procedure which is described below. All biochemical operations are performed at 4° unless otherwise noted.

Preparation of Skin Homogenate. Male Lewis rats (150–200 g) are sacrificed with ether, and the hair is removed from the lower back using forceps. The subcutaneous tissues are removed with a scalpel, and the tissue is weighed and is minced with fine scissors. Three hundred grams of minced skin are suspended in 2700 ml of 50 mM sodium phosphate at pH 7.6 containing 1 mM EDTA.

The suspension is homogenized with a Polytron (Brinkmann Instruments) for three 30-sec periods, with intervening 30-sec cooling periods. The resulting suspension is centrifuged at 27,000 g for 30 min. The supernatant fluid is decanted, defatted, and centrifuged as above. The final supernatant liquid is frozen at −20° until used. Enzyme activity is stable in this preparation for at least 2 months. After thawing at room temperature the preparation is centrifuged and concentrated 8.3-fold using an

[4] G. Kalnitsky, J. P. Hummel, and C. Diersk, *J. Biol. Chem.* **234**, 1512 (1959).

TABLE I
Purification of Sulfhydryl Oxidase from Rat Skin[a]

Step	Volume (ml)	Total protein (mg)	Total enzyme activity (units)[b]	Specific activity
Concentrated extract	360	4710	76300	16
Ammonium sulfate	40	1500	36600	24
DEAE-Cellulose	15	92.3	27500	291
CM-Sephadex	12.5	8.8	15300	1540
CM-Cellulose	8.0	3.2	9160	2580
Sephadex G-75	6.2	0.4	4580	9510

[a] Obtained from 300 g of rat skin.
[b] Enzyme activity: manomoles of reduced DTNB per 1 ml for 60 min at 37°.

Amicon PM10 filter. Solid ammonium sulfate is added over 1 hr to a 30% saturation (164 mg/ml). After stirring for 1 hr the suspension is centrifuged at 12,000 g for 10 min, and additional ammonium sulfate (181 mg/ml) is added. This suspension is centrifuged at 27,000 g for 30 min, and the residue dissolved is in 1 mM sodium phosphate at pH 7.6 and dialyzed against several changes of the same buffer for 48 hr. A small precipitate after dialysis is removed by centrifugation at 27,000 g for 30 min. Although there is significant loss of enzyme activity due to ammonium sulfate fractionation, the step is essential for a preparation of high specific activity.

Chromatography. The preparation is chromatographed on a column (2.5 × 14 cm) of DEAE-cellulose which has been equilibrated with 1 mM sodium phosphate at pH 7.6. Elution is with the same buffer. The enzyme does not bind under these conditions, and tubes containing enzyme activity are pooled and concentrated with ultrafiltration (Amicon, PM10). The concentrated enzyme is dialyzed overnight against 5 mM sodium phosphate with 1 mM EDTA, pH 7.0. After centrifugation at 27,000 g for 30 min, the supernatant liquid is applied to a column (1.0 × 15 cm) of CM-Sephadex in 5 mM sodium phosphate with 1 mM EDTA, pH 7.0. The column is washed with 200 ml of this equilibrating buffer and eluted at 15 ml/hr with a 400-ml linear gradient of sodium phosphate (5–200 mM) at pH 7.0 containing 1 mM EDTA. Enzyme activity elutes between 35 and 70 mM phosphate. Tubes with enzyme activity are pooled and concentrated on an Amicon PM10 filter.

The concentrated solution is dialyzed overnight against 5 mM sodium phosphate containing 1 mM EDTA, pH 6.0, and centrifuged at 27,000 g.

The supernatant fluid is applied to a column (1.0 × 20 cm) of CM-cellulose (Whatman, CM-52) equilibrated with 5 mM sodium phosphate containing 1 mM EDTA, pH 6.0. The column is washed with 160 ml of equilibrating buffer and eluted with a 400-ml linear gradient of sodium phosphate (5–200 mM) containing 1 mM EDTA, pH 6.0. Enzyme activity elutes between 60 and 74 mM phosphate, and active fractions are concentrated to 2 ml on an Amicon PM10 filter. The preparation is clarified by centrifugation at 27,000 g for 30 min.

The concentrated enzyme solution is placed onto a column of Sephadex G-75 (2.5 × 80 cm) in 50 mM sodium phosphate with 1 mM EDTA, pH 7.6. The resulting peak of enzyme activity is homogenous upon electrophoresis in sodium dodecylsulfate (SDS) gels.

Properties of Rat Skin Sulfhydryl Oxidase[3]

Molecular Properties. The apparent molecular mass is 66,000 ± 2000 by both SDS–gel electrophoresis and gel filtration in Sephadex G-75. The pI of the enzyme is 4.65 as determined by isoelectric focusing in 1% ampholine. The most purified forms of the enzyme have no color or spectrophotometric absorption at visible wavelengths.

Enzymatic Activity. In general, crude preparations of skin can be used for determining many of the catalytic properties of the enzyme. Enzyme activity, as determined by oxidation of Ellman's reagent, is thermostable and increases with temperature up to 45°. The pH optimum for both the oxidation of Ellman's reagent and the reactivation of ribonuclease is between 8.0 and 8.2. However, dialysis of the crude homogenate against phosphate buffers between pH 5.5 and 7.5 causes a 50% increase in total activity, suggesting the possibility of a dialyzable low molecular weight inhibitor.

The renaturation of ribonuclease and the oxidation of Ellman's reagent occur in a time-dependent manner, proportional to the amount of enzyme in the reaction mixture. Since it has not been possible to obtain saturation of enzyme activity with any of the substrates used, kinetic parameters for the enzyme are not presented. High concentrations of dithiothreitol are inhibitory, as discussed below.

Thiol Substrate Specificity. Substrate specificity was evaluated by substituting a number of thiols for dithiothreitol in the standard assay. With dithiothreitol and dithioerythritol both having a relative activity of 100%, the activity of D-penicillamine was 33%, L-cysteine 17%, N-acetylcysteine 4.6%, 2-mercaptoethanol 3.7%, and reduced glutathione 0.7%.

Preincubation with those thiols which are poor substrates leads to inhibition of enzyme activity, even after removal of these compounds by dialysis. The permanent loss of enzyme activity is related to the ability of the thiol to function effectively as a substrate for the enzyme. Iodoacetamide and iodoacetic acid decrease enzyme activity, but N-ethylmaleimide has almost no effect. The degree of inhibition by these alkylating agents is increased if the enzyme is preincubated with thiols which are good substrates, but unaffected when the enzyme is preincubated with glutathione (GSH). This is a property similar to that of other sulfhydryl–disulfide exchange enzymes.

Metal Cofactors. Enzyme activity is not affected by preincubation with EDTA, EGTA, α,α'-dipyridyl or o-phenanthroline, azide, catalase, or cyanide. The similar enzyme isolated from cow snout is inhibited in a concentration-dependent fashion by diethyldithiocarbamate. Native enzyme activity is increased 2-fold by incubating it with copper.[5] This is the strongest evidence to date that the skin enzyme may be a metal (copper)-dependent sulfhydryl oxidase.

Heat Stability. The crude enzyme is very heat stable, but the purified enzyme when incubated at 60° has only 15% of its initial activity after 15 min and no activity after 30 min. Dithiothreitol (1 mM) increases the stability to heating, but the same concentration of GSH is ineffective.

Tissue Localization. When adult skin is separated by preincubation in 2 M KBr, and the epidermis and dermis assayed directly, the specific activity of the enzyme in the epidermal and dermal fractions are similar. Since the dermis is a much larger fraction of the entire skin, the total enzyme activity in the dermis is greater than in the epidermis. All of our initial studies showed the enzyme to be soluble in the 100,000 g supernatant of the skin as prepared in our scheme. The epidermal enzyme, studied extensively in the hair-free cow snout, has the same properties as that reported for the rat skin enzyme, the latter representing enzyme predominantly from the hair follicle. In the cow snout most of the enzyme activity is localized in the granular layer, a reasonable location for an enzyme which functions in the formation of disulfide bonds.[5] Measurable activity is present in the epidermis and dermis of the newborn rat. Preliminary data from sheep skin show enzyme activity with both the Ellman's reagent and the ribonuclease renaturation assays.

Acknowledgments

Much of the initial work on this enzyme was performed by K. Takamori and Judy Thorpe-Summerhays in Durham, North Carolina, at Duke University. Dr. Takamori continues studies of this enzyme in skin at Juntendo University, Tokyo, Japan.

[5] H. Yamada, K. Usui, K. Takamori, and H. Ogawa, *J. Invest. Dermatol.* **84,** 317 (abstr.) (1985).

[91] Asparagusate Reductase

By HIROSHI YANAGAWA

Asparagusate + NADH + H$^+$ ⇌ dihydroasparagusate + NAD$^+$

Asparagusate reductase (EC 1.6.4.7) is an essential component of the pyruvate dehydrogenase complex of asparagus,[1] which catalyzes a CoA- and NAD-linked oxidative decarboxylation of pyruvate.[2] The reductase is a flavoprotein and catalyzes the oxidoreduction of asparagusate[3] and lipoate.[4]

Synthesis of Asparagusic Acid[5]

Ethyl malonate (**I**) (1.12 kg, 7.0 mol) and 40% formalin (2 kg, 26.7 mol) are mixed. Ten percent NaOH is added dropwise to the mixture in an ice bath with stirring and maintained at pH 8–9. After 2 hr, the ice bath is removed, and the mixture is further stirred at room temperature for 2 days. The reaction mixture is extracted with petroleum ether and dried with anhydrous magnesium sulfate. Evaporation of the extract gives colorless ethyl bis (hydroxymethyl)malonate[6] (**II**); yield: 1.2 kg (79%).

A mixture of (**II**) (88 g, 0.4 mol) and 57% hydriodic acid (270 g, 1.2 mol) is refluxed for 5 hr, during which period volatile materials are removed in a hood to produce a final volume about 65% of the starting volume. Upon cooling overnight at room temperature, crude β,β'-diiodoisobutyric acid (**III**) precipitates. It is recrystallized from carbon tetrachloride to afford pure **III** (colorless plate); yield: 68 g (50%); m.p. 127–129.5°.

Aqueous sodium trithiocarbonate[7] (4.2 g, 8.9 mmol) is added dropwise to **III** at 25° with stirring under an atmosphere of nitrogen; the reaction mixture is heated at 50° for 5 hr. After acidification with 6 N sulfuric acid, the aqueous layer is extracted 3 times with ethyl acetate. The ethyl acetate extracts are dried with anhydrous magnesium sulfate and concen-

---

[1] H. Yanagawa, Ph.D. Dissertation, University of Tohoku (1973).
[2] H. Yanagawa, T. Kato, and Y. Kitahara, *Plant Cell Physiol.* **14**, 1213 (1973).
[3] H. Yanagawa, T. Kato, Y. Kitahara, N. Takahashi, and Y. Kato, *Tetrahedron Lett.* p. 2549 (1972).
[4] U. Schmidt, P. Grafen, K. Atland, and W. Goedde, *Adv. Enzymol.* **32**, 423 (1969).
[5] H. Yanagawa, T. Kato, H. Sagami, and Y. Kirtahara, *Synthesis* p. 607 (1973).
[6] K. N. Welch, *J. Chem. Soc.* p. 257 (1930).
[7] D. J. Martin and C. C. Greco, *J. Org. Chem.* **33**, 1275 (1968).

$$\underset{(I)}{\underset{COOC_2H_5}{\overset{COOC_2H_5}{CH_2}}} \xrightarrow{HCHO} \underset{(II)}{\underset{HOCH_2 \quad COOC_2H_5}{\overset{HOCH_2 \quad COOC_2H_5}{C}}} \xrightarrow{HI} \underset{(III)}{\underset{ICH_2}{\overset{ICH_2}{CHCOOH}}} \xrightarrow{Na_2CS_3, H^+}$$

$$\underset{(IV)}{\underset{HSCH_2}{\overset{HSCH_2}{CHCOOH}}} \xrightarrow{DMSO} \underset{(V)}{\underset{S-CH_2}{\overset{S-CH_2}{CHCOOH}}}$$

trated to give crystals of dihydroasparagusic acid (**IV**), which are recrystallized from benzene–cyclohexane; yield 0.38 g (88% from **III**; m.p. 59.5–60.5°).

Compound **IV** (1 g, 6.6 mmol) is dissolved in dimethyl sulfoxide (10 ml) and stirred at 70–75° under nitrogen. The reaction is monitored by silica gel thin-layer chromatography[8] to prevent further oxidation. The reaction is usually completed within 5 hr. The mixture is poured into cold water, and extracted 3 times with benzene. The benzene layer is washed with water and dried over anhydrous magnesium sulfate. Evaporation of the extract gives crude asparagusic acid (**V**), which is recrystallized from benzene–cyclohexane to yield yellowish prismatic crystals; yield: 0.65 g (70%); m.p. 76–77°.

Enzyme Assay

Principle. Asparagusate reductase and lipoyl reductase activities are measured spectrophotometrically by the decrease in absorption at 340 nm due to NADH oxidation in the presence of asparagusate or lipoate.

Reagents

Sodium phosphate at pH 5.9, 0.6 M
EDTA, 7.5 mM
Asparagusic acid[3,5] in 10% ethanol, 24 mM, prepared fresh daily
DL-Lipoic acid (Tokyo Kasei Co., Ltd.) in ethanol, 48 mM, prepared fresh daily
NADH, 3 mM
NAD, 3 mM

---

[8] R_f 0.65 on thin-layer plates of silica gel developed in toluene–ethylformate–formic acid (5:1:1).

Procedure. Assays are carried out in cuvettes of 10 mm light path maintained at 25° and containing 50 μl each of sodium phosphate buffer and EDTA and 20 μl each of NADH, NAD, and asparagusic acid or DL-lipoic acid. The reaction is started by adding enzyme. Absorbance at 340 nm is measured with a spectrophotometer. A linear relationship is maintained for the first 10 min.

Units. A unit of activity is defined as that amount of enzyme which oxidizes 1 nmol of NADH per minute under standard assay conditions. Specific activity is defined as units per milligram protein; protein is determined by the method of Lowry *et al.*,[9] using crystalline bovine serum albumin as a standard.

Enzyme Purification[10]

All operations are carried out at 4°. Centrifugation is performed in a Kubota KR-200A centrifuge, and an Amicon PM10 filter is used for concentration. Sodium phosphate serves as buffer throughout purification because the enzymes are stabilized by it.

Preparation of Mitochondrial Fraction.[2] Fresh etiolated or green asparagus (*Asparagus officinalis* L.) shoots are washed with distilled water, and moisture is removed by pressing them gently between sheets of absorbent paper. In a typical preparation, 2 kg of the tissues are chopped and then homogenized in a Waring Blendor in cold 0.30 M sucrose–70 mM sodium phosphate at pH 7.4. The homogenate is squeezed through gauze and centrifuged at 2700 g for 15 min to remove particulates. The supernatant liquid is centrifuged at 15,000 g for 15 min to sediment mitochondrial particles, which are resuspended in the homogenizing buffer and again sedimented at 15,000 g for 15 min. The mitochondrial pellet thus obtained is suspended in 67 mM sodium phosphate at pH 7.0, containing 10 mM EDTA to yield a protein concentration of about 30 mg/ml.

Solubilization of Asparagusate Reductases. The mitochondrial suspension is first treated by freezing and thawing to solubilize lipoyl dehydrogenase. Freezing at −50° and thawing at 37° is repeated at least twice over a period of 20 min. The treated mitochondrial suspension is then centrifuged at 36,000 g for 30 min. The precipitates are subsequently treated with sodium dodecyl sulfate (SDS) to solubilize asparagusate reductase. The SDS concentration used is 0.5% in the ratio of 0.5 mg SDS per milligram of mitochondrial protein. Solubilization is performed at 37°

---

[9] O. H. Lowry, N. J. Rosebrough, A. L. Farr, and R. J. Randall, *J. Biol. Chem.* **193**, 265 (1951).

[10] H. Yanagawa and F. Egami, *Biochim. Biophys. Acta* **384**, 342 (1975).

in a shaking water bath for 20 min. The solubilized enzyme is separated by centrifugation at 36,000 g for 30 min.

Purification of Asparagusate Reductases

Gel Filtration I. A fraction (50 ml) of the enzymes from the SDS treatment is concentrated to 4 ml and applied to a Sephadex G-200 column (3 × 80 cm) which is eluted with the 67 mM phosphate buffer containing 10 mM EDTA.

Gel Filtration II. Fractions from the preceding step are combined and concentrated to 3 ml. The resulting solution is passed through a Sephadex G-200 column (3 × 80 cm) using the 67 mM phosphate buffer.

DEAE-Cellulose I. The active fractions from the preceding step are combined and concentrated to 8 ml. The concentrate is applied to a DEAE-cellulose column (1.2 × 37 cm) previously equilibrated with 10 mM sodium phosphate at pH 7.2. A few milliliters of the buffer are added to wash the column after the sample has been adsorbed, and elution is begun with 600 ml of a linear gradient in a concentration range from 10 to 500 mM of sodium phosphate at pH 6.95. Asparagusate reductase activity is eluted at phosphate concentrations of 150–160 mM (fraction I) and 165–190 mM (fraction II). Lipoyl reductase activity is eluted coincident with both peaks of asparagusate reductase.

DEAE-Cellulose II. Fractions I and II from the preceding step are combined, dialyzed against 10 mM sodium phosphate at pH 7.2, and concentrated to 10 ml. The resulting solution is adsorbed on a DEAE-cellulose column (1.4 × 37 cm), and the enzymes are eluted with a linear gradient of the phosphate buffer used in step 3 from 10 to 500 mM. Two fractions are again obtained. Results obtained after each step of the preparation leading to the two fractions are summarized in Table I. In both fractions I and II, the ratio of specific activity with asparagusic acid to that with DL-lipoic acid is 2.55 : 1. Fractions I and II were named asparagusate reductases I and II.

Properties[11]

Criteria of Purity. Ultracentrifugation studies show that asparagusate reductases I and II move as a single boundary during the entire run at 60,000 rpm. Each enzyme, examined by polyacrylamide disk gel electrophoresis in Tris–glycine at pH 8.3, yields a single band of protein. The sedimentation coefficients ($s_{20,\,w}$) of asparagusate reductases I and II were calculated to be 6.22 and 6.39 S, respectively.

[11] H. Yanagawa and F. Egami, *J. Biol. Chem.* **251**, 3637 (1976).

TABLE I
PURIFICATION OF ASPARAGUSATE REDUCTASE OF ASPARAGUS MITOCHONDRIA

Purification step	Volume (ml)	Total protein (mg)	Asparagusate reductase		Lipoyl dehydrogenase	
			Total activity (units)	Specific activity (units/mg)	Total activity (units)	Specific activity (units/mg)
0.5% Sodium dodecyl sulfate extract	50	315	58.80	0.19	85.4	0.27
Sephadex G-200 I	135	129	98.40[a]	0.76	117[a]	0.91
Sephadex G-200 II	60	50	70.30[a]	1.95	340[a]	4.73
DEAE-Cellulose I						
Enzyme I	24	0.282	12.5	44.1	32.0	113
Enzyme II	30	0.308	18.5	60.2	47.2	153
DEAE-Cellulose II						
Enzyme I	5	0.024	10.1	425	26.0	1080
Enzyme II	5	0.071	15.3	215	38.8	546

[a] Anomalous excess yield of the activity in these steps is probably due to removal of SDS.

The molecular weights of asparagusate reductases I and II, estimated from sedimentation equilibrium data, were 111,000 and 110,000, respectively, and from gel filtration on Sephadex G-200, 112,000 each. Partial specific volumes of asparagusate reductases I and II were calculated to be 0.719 and 0.715 ml/g from the amino acid composition. Asparagusate reductases I and II are slightly acidic proteins with isoelectric points of 6.75 and 5.75, respectively.

Enzyme in the oxidized form displays a characteristic flavoprotein absorption spectrum, with maxima at 272, 356, and 453 nm for both asparagusate reductases. The absorption at 450 nm is significantly decreased by the addition of sodium dithionite. FAD is released from asparagusate reductases by boiling the enzyme solution. The flavin contents of enzymes I and II were calculated[12] to be 10 and 12 nmol/mg protein, respectively, corresponding to approximately 1 mol of FAD per mole of protein.

Substrate and Cofactor Specificities. Asparagusate reductases I and II are active with either asparagusic or lipoic acids but cannot act on other disulfides tested; the relative rates of NADH oxidation are in the ratio 1:2.5 for asparagusic acid and lipoic acid. NADPH is inactive for both enzymes. The K_m values (~3 mM) for lipoic acid are lower than those

[12] H. Beinert and E. Page, *J. Biol. Chem.* **225**, 479 (1957).

(~20 mM) for asparagusic acid. The optimum concentration of NADH is 0.1 mM; higher concentrations are inhibitory. The pH optimum for both substrates and both enzymes is 5.9.

With $K_3Fe(CN)_6$ as an electron acceptor for NADH oxidation at the optimum pH of 5.25, asparagusate reductases I and II reduce $K_3Fe(CN)_6$ with NADH. K_m values for the acceptor in the $K_3Fe(CN)_6$–NADH activity of asparagusate reductases I and II are 0.9 and 0.8 mM.

Effect of Phospholipids and Surfactants on Asparagusate Reductase. Natural and synthetic lecithins and Tween 80 cause a 2- to 4-fold increase in lipoyl reductase activity but have no effect on asparagusate reductase activity. Triton X-100, SDS, and deoxycholate do not appreciably affect either activity, but cetyltrimethylammonium bromide inhibits lipoyl reductase activity.

Stability. Both asparagusate reductases in 67 mM sodium phosphate at pH 7.0 are very unstable at 4° but do not lose activity for at least 1 month at −80°. EDTA and phosphate ion stabilize the enzymes. After heating at 50° for 5 min, both asparagusate and lipoyl reductase activities are completely retained. Moreover, the asparagusate reductase activity is activated 1.4-fold by heating at 60–70° for 5 min, whereas lipoyl reductase activity gradually decreases above 50° and is completely lost at 90°.

Appendix: Index of Related Articles in Other Volumes of Methods in Enzymology

Related articles	Volume and page or [article] number
N-Acetylcysteamine thioesters	**14**: 696
O-Acetylhomoserine sulfhydrylase	**17B**: 446
O-Acetylserine sulfhydrylase	**17B**: 459
Acid-labile sulfide	**53**: 273
Adenosine 3'-phosphate 5'-phosphosulfate	**77**: 413
S-Adenosyl-L-homocysteine cleaving enzyme	**5**: 752
Adenosyl-3-methylthiopropylamine	**94**: 73
S-Adenosyl-3-thiopropylamine	**94**: 463
S-Adenosyl-L-methionine assay	**17B**: 397; **94**: 57, 66
S-Adenosyl-L-methionine decarboxylation	**94**: 68
in *E. coli*	**94**: 228
in *S. cerevisiae*	**5**: 756; **94**: 231
inhibition	**94**: 239
S-Adenosyl-L-methionine preparation	**3**: 600; **6**: 566
S-Adenosylmethionine synthetase	**94**: 219
Adenosylsulfur compounds, HPLC	**94**: 57
Alcohol sulfotransferase	**77**: 206
S-Alkylation	**11**: 204
S-Alkyl-L-cysteine lyase	**17B**: 470; **77**: 253; **113**: 510
S-Alkyl-L-cysteine sulfoxide lyase	**17B**: 475
Amine N-methyltransferases	**77**: 263; **142**: [77]
δ(α-Aminoadipyl)cysteinylvaline synthetase	**43**: 471
S-Aminoethylation	**11**: 315
Arylsulfatases	**86**: 17
Arylsulfotransferase	**77**: 197
ATP : L-methionine S-adenosyltransferase	**17B**: 393
Bile salt sulfotransferase	**77**: 213
S-(Bilin)cysteine	**106**: 359
Binding proteins for sulfatide	**138**: [38]
Calmoduline-activating methyltransferases	**139**: [50]
S-Carboxymethylation	**11**: 199
Catechol O-methyltransferase	**5**: 748; **15**: 722
Chondroitin sulfatase	**8**: 663; **28**: 917
Chondroitin sulfate, biosynthesis	**28**: 638
Cystathionases	**2**: 311; **5**: 936; **17B**: 425, 439

(*continued*)

Related articles	Volume and page or [article] number
β-Cystathionase	**17B**: 439
γ-Cystathionase	**17B**: 433
Cystathionine synthase	**17B**: 454
Cystathione γ-synthase	**17B**: 425
Cysteamine oxygenase	**17B**: 479
Cysteine alkylation	**47**: 116
Cysteine, assay	**11**: 59; **25**: 55
cyanylation and cleavage	**47**: 129
Cysteine S-conjugate β-lyase	**17B**: 470; **77**: 253; **113**: 510
Cysteine S-conjugate N-acetyltransferase	**113**: 516
Cysteine peptides	**11**: 390, 475
Cysteine, as S-sulfocysteine	**25**: 55
Cysteine desulfhydrase	**2**: 315
Cysteine protease	**45**: 3
Cysteine protease inhibitors	**45**: 740
Cysteinesulfinic acid	**17A**: 681, 689
Cysteinylglycine peptidase	**113**: 471
Cystine, isotopic	**4**: 772
Cystine, as S-sulfocysteine	**25**: 55
Dehydroascorbate reductase	**122**: 10
Dedimethylamino 4-aminoanhydrotetracycline N-methyltransferase	**43**: 603
O-Demethylpuromycin O-methyltransferase	**43**: 508
Desulfhydrases: *see* specific enzyme	
Dimethylthetin–homocysteine methyltransferases	**5**: 743
Disulfides, assay	**77**: 353
Disulfide, asymetrical, preparation	**77**: 420
Disulfide formation, proteins	**25**: 387; **107**: 301, 305
Ergothionine	**4**: 776
Erythromycin C O-methyltransferase	**43**: 487
Exocystine desulfhydrase	**2**: 318
Flavin, sulfite interaction	**18**: 468
S-Formylglutathione hydrolase	**77**: 320
γ-Glutamylcysteine synthetase	**17B**: 484, 495; **113**: 379, 390
γ-Glutamyltranspeptidase	**2**: 263, 267; **19**: 782, **77**: 237; **86**: 38; **113**: 400, 419
Glutaredoxin	**113**: 525
Glutathione	
assay	**3**: 606; **4**: 331; **77**: 353, 373; **113**: 548
depletion	**77**: 50, 59
isotopic labeling	**17B**: 509; **113**: 567

Related articles	Volume and page or [article] number
metabolism	**113**: 571
purification	**3**: 603; **17B**: 509
transport	**113**: 571
Glutathione dehydrogenase (ascorbate)	**122**: 10
Glutathione disulfide	
assay	**77**: 353, 373; **113**: 548
efflux	**105**: 445
Glutathione peroxidase	**52**: 506; **77**: 325; **105**: 114; **107**: 593, 602; **113**: 490
Glutathione reductase	**17B**: 500, 503; **113**: 484
Glutathione synthetase	**2**: 343; **17B**: 488; **113**: 393
Glutathione thioesters	**77**: 424
Glutathione transferase	**113**: 495
assay	**77**: 353
preparation	**77**: 218, 231, 235; **113**: 499, 504, 507
Glutathionylspermidine	**94**: 434
Glycosulfatases	**81**: 670, **138**: [67]
Glyoxalase I	**77**: 297
Glyoxalase II	**77**: 320
Guanidinoacetate methyltransferase	**2**: 260
Histamine N-methyltransferase	**17B**: 767
S^{β}-(2-Histidyl)cysteine	**106**: 355
Homocysteine desulfhydrase	**2**: 318
Homocysteine S-methyltransferase	**17B**: 400
Hydroxyindole O-methyltransferase	**17B**: 764, **142**: [68]
Hydroxysteroid sulfotransferase	**77**: 206
S-(2-Hystidyl)cysteine	**106**: 355
Iduronate sulfatase	**50**: 150; **83**: 573
Indoleamine N-methyltransferase	**142**: [77]
Indolepyruvate 3-methyltransferase	**43**: 498
Inorganic sulfur compounds, assay	**3**: 995
Iron–sulfur proteins	**53**: 259; **126**: 360
Iron–sulfur centers	**67**: 238; **69**: 792
Iron–sulfur clusters	**53**: 268
Mercaptopyruvate : cyanide transsulfurase	**5**: 987
2-Mercaptopyruvate sulfurtransferase	**77**: 291
Mixed disulfides, synthesis	**77**: 420
β-Mercaptopyruvate : sulfite transsulfurase	**5**: 990
Metallothioneins, radioimmunoassay	**84**: 121
Methionine adenosyltransferase	**94**: 223
Methionine, isotopic labeling	**4**: 772
Methionine biosynthesis	**17B**: 371, 388

(*continued*)

Related articles	Volume and page or [article] number
Methionine peptides	**11**: 390
Methionine sulfoxide, assay	**11**: 487
Methionine sulfoxide in protein	**107**: 352
Methyltetrahydrofolate–homocysteine methyltransferase	**17B**: 379
5'-Methylthioadenosine nucleosidase	**94**: 365
5'-Methylthioadenosine phosphorylase	**94**: 355
5'-Methylthioribose kinase	**94**: 361
Methyltransferases (see also specific methyltransferases)	**4**: 764
calmodulin activation	**139**: [50]
Mucopolysaccharides, sulfation	**8**: 496
Nicotinamide methyltransferase	**2**: 257
Norepinephrine N-methyltransferase	**142**: [76]
Oligosaccharyl sulfatase	**138**: [67]
L-2-Oxathiazolidine 4-carboxylate	**113**: 458
PAPS, synthesis	**77**: 413
Penicillin, isotope labeled	**4**: 776
Penicillin acylase	**43**: 698, 705, 721
Phenol sulfotransferase	**77**: 197
Phenylethanolamine N-methyltransferase	**17B**: 761
Phosphatidylethanolamine methyltransferase	**14**: 125
3'-Phosphoadenosyl 5-phosphosulfate reductase	**17B**: 546
3'-Phosphoadenosyl 5-phosphosulfate, synthesis	**77**: 413
N-methyltransferase protein (arginine)	**106**: 268
Protein–carboxyl O-methyltransferase	**106**: 295, 310, 321, 330
assay	**106**: 265
Protein N-methyltransferase	**106**: 268, 274
assay	**106**: 265
Protein disulfide isomerase	**107**: 281, 301
Protein disulfides	**25**: 387; **107**: 301, 305; **131**: [5]
Protein methylation and demethylation	**106**: 274
Protein disulfide isomerase	**107**: 381
Protoporphyrin methyltransferase	**17**: 222
Radioactive sulfur compounds (see also specific compound)	**4**: 771
Reduction	
disulfides	**5**: 992; **11**: 481; **25**: 185, 387; **47**: 111
sulfoxide	**5**: 992
sulfoxide, protein	**91**: 549
sulfo group	**107**: 620
Rhodanese	**2**: 334; **77**: 285
Rhodopsin, sulfhydryl	**81**: 223
Riboflavin 5'-monosulfate	**18**: 465
Selenomethionine proteins	**107**: 620
Selenocysteine proteins	**107**: 576, 582, 602
Steroid sulfate esters	**15**: 351, 684

APPENDIX 527

Related articles	Volume and page or [article] number
Sulfatases	**2**: 325
chondroitin sulfatase	**8**: 663; **28**: 917
glycosulfatases	**8**: 670; **83**: 573; **138**: [67]
iduronate sulfatase	**50**: 150
oligosaccharyl sulfatase	**138**: [67]
steroid	**15**: 684
Sulfatide	**83**: 191
Sulfatide-binding protein	**138**: [38]
Sulfate-activating enzymes	**5**: 964
Sulfhydryl group assay	**25**: 457; **77**: 373
acid labile (*see also* glutathione)	**10**: 634
Sulfhydryl peptides	**91**: 392
Sulfide, radioactive	**4**: 771
6-Sulfido–peptide leukotrienes	**86**: 252
Sulfite reductase	**5**: 983; **17B**: 520, 528, 539
Sulfitolysis	**11**: 206
S-Sulfo group reduction	**47**: 123
Sulfonylation	**11**: 7066
Sulfotransferases (*see also* specific enzyme)	**5**: 977
Sulfoxide reduction	**5**: 992
Sulfoxides (protein), reduction	**91**: 549
Sulfur amino acids, preparation and assay	**3**: 578
Sulfur, radioactive	**4**: 771
Sulfuryl adenylates	**6**: 766
Taurine, isotope labeled	**4**: 775
Taurocholic acid, isotope labeled	**4**: 776
Taurine–glutamate aminotransferase	**113**: 102
Thiazolidine carboxylate dehydrogenase	**5**: 736
Thiocyanate, radioactive	**4**: 771
Thioesters	
N-acetylcysteamine	**14**: 696
Coenzyme A	**77**: 430
Glutathione	**77**: 424
Thioester hydrolase	**71**: 181, 188, 200, 230
Thioglycoside polyacrylamide gels	**83**: 299
Thiolation	**25**: 541
Thiol–disulfide interchange	**34**: 531; **107**: 330
Thiol-disulfide transhydrogenase	**17B**: 510, 515
Thiol group reactions	**4**: 256
Thiolhistidine, isotopic	**4**: 776
Thiol *S*-methyltransferase	**77**: 257
Thiol–protein disulfide exchange	**17B**: 510; **107**: 330; **113**: 520, 541
Thiol transferase	**77**: 281; **113**: 520, 541

(*continued*)

Related articles	Volume and page or [article] number
Thioredoxin	**107**: 295
Thioredoxin reductase	**69**: 382
Thiosulfate : cyanide sulfurtransferase	**7**: 285
γ-Tocopherol methyltransferase	**111**: 544
Transsulfuration	**17B**: 416, 450
Tryptophan C-methyltransferase	**142**: [30]
Tyrosine O-sulfate in proteins	**107**: 200

Author Index

Numbers in parentheses are footnote reference numbers and indicate that an author's work is referred to although the name is not cited in the text.

A

Aarnes, H., 422
Abbot, W. A., 376
Abdolrasulinia, R., 26
Abeles, R. H., 233, 371, 377, 378
Abiko, Y., 376
Abounassif, M. A., 187
Abrams, W. R., 421
Acaster, M. A., 428, 429(8)
Ackerman, R. J., 56
Adachi, K., 361
Adachi, O., 293
Adams, D. H., 388
Adams, D. O., 426, 427, 428, 429, 458
Adams, J. B., 202
Adams, R. N., 110, 114(2)
Adler, H., 235
Aherne, G. W., 76
Akagi, R., 17, 18, 22(4), 372
Akerboom, T. P. M., 102, 124, 125(2), 128(2), 269
Akerfeld, S., 60, 67(60)
Albersheim, P., 456
Albert, A., 133
Albertini, D. F., 269
Aley, S. B., 76(23), 77
Allen, F., 216
Allaway, W. H., 419
Allen, N. P., 269
Allen, R. A., 370
Allison, L. A., 110
Allison, W. S., 57
Ames, G. F.-L., 145
Ampola, M. G., 178
Anada, H., 499
Anders, M. W., 231
Andersen, K. K., 272, 285
Anderson, B. M., 65, 66
Anderson, C., 7
Anderson, D. G., 384
Anderson, G. H., 299
Anderson, J. W., 335, 341, 419, 421

Anderson, M. E., 89, 124, 287, 290(14), 314, 316, 320, 322, 323(7, 38, 38a), 376
Anderson, N. W., 439, 441(1)
Anderson, P. A., 299, 302
Anderson, W. L., 115, 256
Ando, Y., 50, 53(o), 54, 248, 254
Andreen, J. H., 242
Andreo, C. S., 51, 52(ee), 55
Andrews, J. L., 67
Anfinsen, C. B., 251
Antonaccio, M. J., 258
Antony, A. C., 370
Apontoweil, P., 269
Applegarth, D. A., 173
Aprahamian, S., 297
Ariki, M., 52(aa), 55
Ariko, M., 51
Armstrong, M., 299, 375
Arnett, G., 383
Arnold, W. A., 420
Aronson, N. N., Jr., 199
Arrick, B., 76(23), 77
Asamizu, J., 374, 395, 403(10), 404
Ashley, D. V. M., 299
Askelof, P., 48
Atland, K., 516
Aull, J. L., 48, 56
Avedovech, N. A., 173, 184, 224, 225(4)
Avi-Dor, Y., 66
Awad, W. M., Jr., 384, 388
Awapara, J., 173
Axelsson, K., 48, 151
Axén, R., 56, 97
Ayling, P. D., 166
Azam, H., 63
Azumi, M., 17, 18(4), 22(4)

B

Baba, A., 163, 164
Babson, J. R., 101, 105(1), 106(1)
Baccanari, D. P., 45

Backlund, P. S., 371
Bacq, A. M., 414
Bacq, Z. M., 149
Baddiley, J., 245
Bader, J., 192
Bafundo, K. W., 298
Baggenstoss, A. H., 186
Bailey, J. L., 115, 116(4), 118(4)
Bain, R. P., 324
Baines, B. S., 56, 62
Baker, D. H., 297, 298, 299, 300, 301, 302, 303, 304, 305, 371, 372(35)
Baker, W. L., 49
Balharry, G. J. E., 342
Balinska, M., 384
Bandurski, R. S., 335
Bank, B., 167
Baranczyk-Kuzma, A., 204, 205(18), 372
Barber, M., 216
Barboni, E., 25
Barclay, R. K., 45
Bard, A. J., 42
Barr, R., 65
Barra, D., 149, 376
Bartkowiak, F., 51
Basford, R. E., 58
Baskin, S. I., 367
Baum, H., 48, 208
Baumeister, R., 235
Bayer, R. J., 132
Beales, D., 188
Beatty, P. W., 101, 105(1), 106(1)
Beck, P. W., 374
Becker, M. A., 420, 470, 474, 479
Beilstein, M. A., 307
Beimer, N., 439, 441(4)
Beinert, H., 520
Beister, H. E., 232
Belcher, R., 7
Bell, J. M., 305
Bellomo, G., 63
Bender, F., 299
Benesch, R. E., 44, 59
Benevenga, N. J., 371
Benisek, W. F., 60
Benkouka, F., 52(i), 54
Bennett, C. D., 52(d), 54
Bennett, L. L., Jr., 377
Bennett, M. A., 280
Ben-Shoshan, M., 76

Benson, E. S., 51
Benson, J. R., 278
Bentley, H. R., 286, 287
Benya, P. D., 248
Berends, W., 269
Berezof, T. T., 244, 292, 460
Berger, E. A., 144, 146(1)
Berglund, F., 4, 6(12)
Bergstrom, R. F., 187
Berkowitz, G. A., 360, 361(9)
Berlin, R. D., 269
Bernacchi, A. S., 149
Berni, R., 236
Bernstein, J., 76
Bertagnolli, B. L., 30
Beukhof, J. R., 3
Beutler, E., 45
Beutler, H. O., 11
Beuzard, Y., 149
Bhat, R., 66
Bhattacharya, S. K., 257, 259(4), 261(4), 264(4)
Biggs, D. R., 395, 403(7)
Bigwood, E. J., 279
Bilzer, M., 269
Bir, K., 178
Birkett, D. J., 59, 66(49), 188
Birnbaum, J., 384
Birnbaum, S. M., 314, 316
Bixby, E. M., 178
Black, C. C., 342
Black, W. J., 361
Blakeley, R. L., 47, 53(cc), 55, 117, 130
Blakistone, B. A., 505, 510(3)
Bleier, J., 324
Bleszynski, W., 214
Blitenthal, R. C., 298
Bloch, K., 86
Bloch-Tardy, M., 185
Bloxham, D. P., 53(ii), 55
Blumenfeld, O. O., 65
Boatmen, S., 58, 125
Bodner, A. J., 383
Boebel, K. P., 297, 298
Boelens, H., 135
Boettcher, B., 315, 318, 319(16), 320(16), 324(16)
Bogdanski, S. L., 7
Boggs, R. W., 298
Bojarski, T. B., 349

Boller, T., 427, 429
Bond, J. S., 52(y), 55
Bonicel, J. J., 52(i), 54
Bonting, S. C., 65
Borch, R. F., 231
Borchardt, R. T., 191, 204, 205(18), 379, 429
Borochov-Neori, H., 76, 83(13)
Bostrom, H., 300
Bothe, H., 333, 334(10)
Boyer, P. D., 60, 395
Boyne, A. F., 49
Bradley, K. H., 248
Bradshaw, R. A., 249
Brandsma, L., 135
Bratcher, S. C., 59, 66(49a)
Braunstein, A. E., 374
Breimer, D. D., 240
Brenner, D. G., 276
Brewer, C. F., 65
Bridges, R. J., 376
Briske-Anderson, M., 377
Brocklehurst, K., 46, 48, 49, 56, 58, 62, 65, 132
Brockman, R. W., 377
Brodie, A. E., 101, 105(1), 106(1)
Brodie, B. B., 315, 319(12, 13, 14, 15)
Brooksbank, B. W. L., 216
Brosemer, R. W., 155
Brot, N., 371, 458
Brown, C., 227
Brown, G. M., 376
Brown, J. R., 256
Brown, M. K., 107, 108(4)
Brown, N. R., 192
Brown, W. B., 252
Bruggemann, J., 421, 453, 454(4)
Brunold, C., 421
Bryant, R. G., 47, 49(15)
Buck, J. S., 302
Budna, K. W., 188
Burg, S. P., 425
Burggraaf, M., 7
Burk, R. F., 307, 308, 310, 311, 313
Burley, R. W., 511
Burnell, H. N., 335, 341(4)
Burnett, G., 504
Burns, R. A., 324
Burnworth, L., 143, 155
Burrous, M. R., 237, 238(23)
Burt, R. J., 58

Burton, N. K., 76
Buschbacher, R. M., 95
Bussolati, O., 271, 279, 280(4)
Butler, J., 248
Butler, M., 187
Buys, W., 230
Byard, J. L., 196, 197(9)

C

Cadle, L. S., 155
Calabrese, L., 411
Calabro-Jones, P. M., 94
Calloway, D. H., 297
Calvin, M., 249
Campbell, C. W., 65
Campbell, J. A., 3
Campbell, R. E., 232
Cannella, C., 26, 53(oo), 55, 236, 410, 413(2)
Cannon, G. N., 242
Cantoni, G. L., 192, 377, 378, 379(5), 380(5, 16), 382(5), 383, 384, 387, 430, 431, 433(7)
Carlberg, I., 48
Carlen, P. L., 269
Carlsen, J. B., 61
Carlsson, J., 48, 56, 97
Carmack, C., 216
Carne, T., 52(11), 55
Carnegie, P. R., 94
Carney, M. M., 29
Carrillo, N., 57
Carruthers, G., 187
Carter, D. E., 186
Cary, E., 196, 309
Case, G. L., 371
Castro, J. A., 149
Catsimpoolas, N., 25, 26(4)
Cavallini, D., 25, 26, 53(oo), 55, 115, 149, 179, 180, 226, 252, 374, 376, 395, 410, 411, 413(1, 2)
Caygill, C. P. J., 32, 34(1), 35(1), 36(1)
Cecil, R., 115, 254, 256
Cennerazzo, M. J., 49, 134, 251
Chabannes, B., 377
Challenger, F., 227, 361
Chambers, L. A., 453
Chan, J. C. M., 4

Chan, W. W. C., 262
Chandler, C., 335, 341(6)
Chang, T. S. K., 269
Chao, K.-L., 371
Chapeville, F., 395
Charles, A. M., 11
Charton, M., 280
Chatagner, F., 374, 396, 404, 405, 406, 409(7)
Chaudhry, A. G., 52(a), 54
Chauncey, T. R., 53(dd), 55, 235, 350
Chawla, R. K., 324
Cheatham, R. M., 388
Cheftel, C., 299
Chemla, M., 259
Chen, L. J., 202
Chen, M. S., 405, 410(12)
Chen, Z., 52(pp), 55
Chendler, J. P., 227
Cheung, D. T., 248
Chiang, P. K., 370, 377, 378, 382(23), 383, 427, 429(7), 431, 433(7)
Chikamatsu, K., 245
Chinali, G., 52(e), 54
Ching, T.-Y., 284, 286(5)
Chinn, P. C., 76, 77(12)
Chinard, F. P., 44
Chishti, S. B., 276
Chocat, P., 240, 242, 243, 244(3), 415, 459, 493
Chovan, J., 173
Chow, S. F., 236
Christen, P., 64
Christensen, H. V., 271, 279, 280(4)
Christensen, B. W., 291
Christensen, J. J., 133
Christomanou, H., 208, 217(6)
Churchichi, J. E., 483
Chyan, O., 128
Cirakoglu, C., 144, 145
Clagett, C. O., 425
Clare, D. A., 120, 504, 505, 510(3)
Clark, A. G., 214
Clark, D. G., 173
Clark, S., 255
Clarke, H. T., 280, 281(17), 317
Clayton, S. J., 383
Clelland, W. W., 59, 115, 132, 138, 139(24), 246

Cockle, S. A., 252
Coffey, D. S., 58
Cohen, G., 95
Cohen, H. J., 11
Cohen, J. B., 276, 388
Cohen, P. P., 47, 53(k), 54
Cohen, S. G., 276, 388
Cohn, M. S., 458
Cole, D. E. C., 3, 4, 7
Cole, L. J., 391
Cole, R. D., 115, 116(4), 118(4), 141, 255
Coleman, T., 371
Combs, G. F., 307
Conan, J. W., 3
Conrad, H. E., 343
Cook, P. F., 30
Cooper, A. J. L., 178, 179, 369, 370, 371, 372, 374(43), 459
Cooper, J. D. H., 141, 142, 155
Cooper, J. M., 120
Corbin, J. E., 297, 299, 302, 305
Cordasco, M. G., 286
Corkins, H. G., 288
Correa, W. S., 264, 265
Corvasce, M. C., 376
Cosby, E. L., 28
Cossins, E. A., 423
Cosstick, R., 76
Cossu, P., 185
Costa, M., 179, 180, 374, 411
Cowan, J. C., 232
Cowper, C. J., 427
Cox, D. W. G., 280
Crane, F. L., 65
Cranston, R. W., 251, 253
Crawford, L. V., 50
Crawford, M. A., 84, 93, 99
Crawhall, J. C., 149, 178, 186, 248
Creason, G. L., 429
Creighton, T. E., 119, 132, 248
Crestfield, A. M., 115
Creveling, C. R., 191, 438, 439, 443
Crews, L., 437
Criddle, R. S., 242
Cronenberger, L., 377
Crozet, N., 81, 84(33)
Csallany, A. S., 311
Cserpan, I., 53(n), 54
Cuatrecasas, P., 430

Cuq, J. L., 299
Cummins, J. M., 81, 84(33)
Czarnecki, G. L., 298, 300

D

Dabrowski, J., 276
Dailey, H. A., 65
D'Alessio, G., 134
Dall'Asta, V., 271, 279, 280(4)
Damodaran, S., 116
Danehy, J. P., 25
Daniels, A. L., 300
Daniels, K. H., 396, 405
Danks, D. M., 511
Darling, S., 221
Daron, H. H., 48, 56
Das Sarma, B., 45, 49(9)
Datko, A. H., 419, 420(3), 421, 422, 423, 425, 454
Daub, E., 130
Dautrevaux, M., 63
David, N. M., 253
Davidson, B., 236
Davis, B. J., 393, 436
Davis, D. C., 315, 319(12, 13, 14, 15)
Davis, P., 186
Davison, A. N., 299
Debey, H. J., 317
De Brouwer, D., 422
De Carlo, J. D., 52(i), 54
Decastro, C. R., 149
Decker, R. S., 83
Deeprose, R. D., 192
DeFarreyra, F. C., 149
DeFenos, O. M., 149
Deffner, G. G. J., 280
DeGrip, W. J., 53(mm), 55
DeGrood, R. M., 130, 131(9), 154
DeGroot, A. P., 298
de la Haba, G., 377, 378, 430
De La Rosa, J., 155, 375
Delavier-Klutchko, C., 448, 449, 487
Delfert, D. M., 343, 349(21)
Del Grosso, E., 410
Dell'Anna, L., 4
DeLuca, G., 410
DeMarco, C., 26, 185, 226, 255, 374, 395, 411

DeMaster, E. G., 110, 111, 112(3), 113(3), 114
De Meio, R. H., 3, 201, 332, 333(4), 334(4), 404, 405(6), 409(6)
Depont, J. J., 65
DeRose, A. J., 427
Deshmukh, M. N., 285
Desnuelle, P. A., 52(i), 54
Deuticke, B., 52(f), 54
Dhillon, G. S., 474, 475(3, 4)
DiCesare, P., 248
Dickson, E. R., 186
Dickman, S. R., 60
Diersk, C., 512
DiGiorgio, R. M., 395, 403(15), 410
DiMonte, D., 63
Dinovo, E. C., 53(j), 54
Diplock, A. T., 32, 34(1), 35(1), 36(1)
Dirks, R., 422
Dittmer, J. C., 3
Dixon, G. H., 118
Dodgson, K. S., 207, 208, 209, 214, 215, 216, 217, 361, 366
Dodiuk, H., 76
Dohan, J. S., 257, 259(2)
Dohn, D. R., 231
Dollar, A. M., 251
Dolly, J. O., 208
Doremus, M., 264
Dorian, R., 76, 77, 85, 268
Døskeland, S. O., 377
Drent, G., 235
Drescher, D. G., 141
Dreven, H., 56, 97
Dreyfuss, J., 15
Drobnica, L., 66
Dubler, R. E., 65
Duerre, J. A., 377, 430
Duffel, M. W., 201, 202(2), 228, 367
Dunach, E., 285
Duncan, G. S., 192
Dunlap, R. B., 48, 53(t), 54
Dunn, J. T., 56, 255
Dunn, W. A., 199
Dupré, S., 115, 149, 252, 376, 410, 411, 413(1, 2)
Durell, J., 384, 388(7)
D'Urso, M., 269
Durst, T., 275, 282, 283(1), 284(1), 285(1)

du Vigneaud, V., 227, 243, 298
Dwivedi, C., 484, 485(2)
Dyer, H. M., 298

E

Eady, R. R., 50
Eberle, D. E., 376
Eckstein, F., 53(u), 54, 76
Edgren, M., 124, 125(3), 128(3)
Efrom, M., 178
Egami, F., 518, 519
Egan, A. R., 371
Egli, P., 258
Eguchi, Y., 460
Eisenberg, D., 84
Ejiri, S., 458
Eklow, L., 63
Elberling, J. A., 111
Eldridge, A. D., 232
Elfarra, A. E., 231
Elkind, J. L., 276
Ellis, R. J., 65
Ellis, W. W., 101, 105(1), 106(1)
Ellman, G. L., 46, 49, 86, 97, 116, 117(23), 130, 252, 257, 268, 272
Elsami, M., 155
Emiliozzi, R., 272, 274(12), 405
Engel, M. H., 167
Engel, P. C., 56
Epand, R. M., 252
Epstein, F. H., 7
Erdle, I., 166, 450, 451(3), 452(3)
Ericson, G. R., 248
Ericson, L.-E., 384, 388
Erikkson, S., 48
Eriksson, B., 18
Eriksson, S. A., 18, 56
Emst-DeVries, S. E., 65
Esaki, N., 35, 240, 242, 243, 244, 291, 292, 415, 456, 459, 460, 464, 493
Esmann, M., 52(hh), 55
Evans, D. R., 62
Ewetz, L., 395

F

Fahey, R. C., 76, 77, 85, 89, 94, 95, 96, 268
Fair, T. W., 286

Faltin, Z., 76, 81, 268, 269
Fankhauser, H., 354, 356(2), 357(2), 359(2), 360
Fariss, M. W., 107, 108(4)
Farooqui, A. A., 208
Farr, A. L., 389, 416, 466, 471, 480, 501, 518
Farrington, G. K., 275
Faulkner, L. R., 42
Faust, D., 76
Favre, G., 63
Fearnley, I., 188
Federici, G., 25, 26, 149, 376
Feigle, F., 221, 222(3)
Feil, V. J., 234
Fellman, J. H., 173, 183, 184, 185(3), 224, 225(4), 226, 271, 375
Feng, S. Y., 155
Fenton, S. S., 96
Ferro, A. J., 166
Field, D., 49, 134, 251
Filner, P., 339, 421
Finazzi-Agrò, A., 26
Finch, R., 188
Finkelstein, J. D., 370, 375, 384, 389, 390, 444
Finley, J. W., 124, 298, 301
Finn, J. M., 167
Fiore, A., 410, 413(2)
Fish, S., 378
Fisher, E., Jr., 350
Flavin, M., 26, 28(12), 439, 448, 449, 453, 454, 470, 484, 486(3), 487
Fleming, A. D., 81, 84(33), 269
Fleming, C. R., 186
Fleming, J. F., 317
Fletcher, J. C., 27
Fliss, H., 371
Flohé, L., 367
Florence, T. M., 64
Floris, P., 185
Foley, J. W., 285
Fonnum, F., 314
Foote, C. S., 284, 286(5)
Foote, L. J., 452
Forsch, R. A., 280
Foss, O., 26, 28(11)
Foster, M. A., 456
Foster, S. J., 33, 37, 196, 197, 198, 199(2), 200(11)
Fothergill, J. E., 75

Fothergill, L. A., 75
Fowler, B., 390, 393
Fowler, L. R., 215, 216(29)
Fox, H., 297
Franceschin, A., 4
Frank, H., 310
Frank, J., 114, 187
Frankel, M., 242
Fraser, P. E., 26, 221, 279, 280(4)
Frederici, G., 410, 411, 413(2)
Fredga, A., 240
Freemann, D., 187
Freisheim, J. H., 280
Frendo, J., 235
Fridovitch, I., 62, 120
Friedemann, T. E., 488
Frieden, C., 353
Frieden, E., 60
Friedman, E., 280
Friedman, O. M., 66
Friedmann, M., 252, 300
Fromageot, C., 163
Fromageot, P., 395
Fry, E. G., 60, 252
Fujii, O., 396, 399
Fujioka, M., 369
Fujita, T. S., 173
Furlong, C. E., 144, 145, 146(2)

G

Gaffield, W., 285
Gainer, H., 76, 81(9), 82(9)
Gaitonde, M. K., 18, 19(14), 148, 225, 248, 315, 475, 479, 487, 488(7)
Galante, Y., 52(b), 54
Gallie, D., 439, 441(2)
Ganther, H. E., 32, 33, 35, 37, 195, 196, 197, 198, 199(2), 200(10, 11)
Garber, L. J., 354, 356(2), 357(2), 359(2)
Garcia-Castro, I., 192, 383
Garen, L., 56, 255
Garner, H. R., 453
Garrido, L., 4
Garvin, J., 155
Gatlin, D. M., 307
Gaull, G. E., 384, 410
Gautheron, D. C., 52(b), 54
Gavilanes, F., 53(r), 54

Gazzola, G. C., 271, 279, 280(4)
Geall, M. C., 186
Gehring, H., 64
Geokas, M. C., 216
Geraci, G., 134
Gerber, N. H., 144, 145(4)
Gerwick, B. C., 342
Ghosh, P., 67
Giarnieri, D., 374
Gibson, Q. H., 61
Gil-Av, E., 167, 241
Gilbert, D. D., 115
Gilbert, H. F., 129
Gillette, J. R., 315, 319(12, 13, 14, 15)
Gilman, A., 276
Gion-Rain, M. C., 404, 405, 409(7)
Giovagnoli, C., 26
Giovanelli, J., 419, 420, 421(3), 422, 423, 425, 443, 448(1), 449, 453, 454
Girard, M. L., 259
Giudice, L. C., 247
Glazenburg, E. J., 372, 453
Glazer, A. N., 367
Glick, D., 197
Gmelin, R., 365, 435
Go, J. G., 3
Godinot, C., 52(b), 54
Goedde, W., 516
Goldberg, I., 76, 84
Goldberger, R. F., 256
Goldsmith, L. A., 511, 512(3), 514(3)
Goldstein, B. J., 269
Goldstein, E., 76
Goldstein, F., 27
Gonias, S. L., 60, 67(59)
Gonnard, P., 185
Goodman, L. S., 276
Goodwin, T. W., 329
Goon, D. J. W., 110, 112(3), 113(3), 317
Gordon, R. K., 192, 378
Gore, M. G., 52(a), 54
Goreck, M., 251
Gori, B., 26
Gorin, G., 25
Goryachenkova, E. V., 374
Götzendörfer, C., 276
Graber, G., 297
Grabski, J., 479, 487
Grafen, P., 516
Grant, W. M., 15, 16(1)

Grassetti, D. R., 46, 97
Graziani, M. T., 115, 252, 255, 411
Gray, P. N., 280
Greco, C. C., 516
Greenberg, D. M., 390, 457, 474, 486
Greenstein, J. P., 169, 170(15), 233, 280, 314, 316, 474
Greer, L., 66
Gregory, P. E., 307
Griffin, M., 141
Griffin, S. T., 240
Griffith, O. W., 167, 170, 171(16), 270, 271, 274, 276, 278, 281, 287, 288(11, 16), 290(13, 14), 318, 320, 325, 374, 375, 376, 404, 405, 406(3)
Grimshaw, C. E., 132
Grofova, I., 314
Gronow, M., 63
Gruber, B., 53(j), 54
Grunbaum, B. W., 4
Grunerts, R. R., 59, 479
Gubitz, G., 167
Guest, J. R., 456
Guggenheim, S., 483
Guidoni, A. A., 52(i), 54
Guilleux, F., 299
Guitton, M. C., 377
Gunther, W. H. H., 248
Guranowski, A., 192, 377, 379(5), 380(5), 382(5), 383, 430, 431, 432(3), 433, 434
Gurusiddaaiah, S., 155
Gutfreund, H., 208
Gutman, A., 22, 27, 28(19)
Guttenberg, C., 48
Gutzke, G. E., 311
Guy, M., 428
Guzman-Barrón, E. S., 44

H

Habeeb, A. F., 44, 49(6), 115, 124, 128(4), 181
Haber, M. T., 179, 372, 374(43)
Habig, W. H., 313
Hablitz, J. J., 280
Haest, C. W., 52(f), 54
Hafeman, D. G., 308
Hafter, R. E., 280
Haguenauer-Tsapis, R., 62, 65(72)

Haiashi, O., 361
Hall, D. I., 439, 441(3)
Halloway, B. E., 51
Halpin, K. M., 297, 298
Halprin, K., 361
Hamilton, P. B., 127
Han, L.-P. B., 130
Hancher, C. W., 144
Handler, P., 62
Hannestad, U., 178, 179
Hannonen, P., 371
Hannum, C. H., 166
Hansen, L. D., 133
Hansen, S., 155
Hanson, J. K., 259, 264(12)
Harada, M., 58
Hare, P. E., 167, 241, 278
Harinath, B. C., 217
Harnish, D. P., 25
Harrap, K. R., 252
Harris, B. J., 370, 375
Harris, E. J., 48
Harris, J. I., 367
Harris, J. W., 269, 317
Harrison, J. H., 58, 125
Harrison, L. C., 255
Harter, J. M., 298, 299, 304
Harth, M., 187
Hartree, E. F., 406
Harwood, J. L., 333
Hassan, W., 149
Hata, T., 257, 259(3), 264(3)
Hatanaka, S., 435
Hatefi, Y., 52(b), 54
Hausinger, R. P., 49, 50(23)
Haverback, B. J., 216
Hayaishi, O., 61, 361
Hayden, P., 231
Hayenga, K. J., 118
Hayes, K. C., 299
Headley, M. H., 280
Hedlund, B. E., 51
Heil, B. M., 80
Heinrikson, R. L., 235
Helfferich, F., 8
Helland, P., 173
Hellerman, L., 44, 58
Helmerhorst, E., 256
Henderson, J. M., 324
Henkin, J., 56

AUTHOR INDEX

Henrichs, A. M. A., 186, 189(8)
Heppel, L. A., 144, 146(1)
Herbert, J. A., 56
Herner, R. C., 427, 429(4)
Herrington, K. A., 66
Hershkowitz, E., 76, 85, 268
Hession, C., 459
Hewitt, J., 269
Heyman, T., 406
Heymsfield, S. B., 324
Hibasami, H., 371
Higashi, T., 63, 375, 376(67)
Hill, B. T., 252
Hill, D., 143, 155
Hill, K. E., 313
Hill, L., 300
Hille, R., 53(v), 55, 254
Hilton, J. L., 324
Hirakawa, D. A., 302
Hiramatsu, K., 48
Hird, F. J. R., 251
Hirs, C. H. W., 127
Hirschberger, L. L., 155
Hiskey, R. G., 58, 125
Hjerten, S., 416, 446, 494
Ho, O. L., 314
Hoang, L., 337, 338(12), 340(12)
Hodson, R. C., 332, 355
Hoekstra, W. G., 33, 196, 308, 311
Hoffman, H. P., 249
Hoffman, J. L., 369
Hoffman, N. E., 425, 426
Höfle, G., 435
Hogan, D. L., 141
Hohman, R. J., 377
Hol, W. G. J., 235
Holý, A., 377
Hook, G. E. R., 208
Hope, D. B., 180, 271, 276, 405
Horowitz, J. H., 324
Horowitz, P. M., 236
Horton, H. R., 120, 123(2, 4), 504, 505, 509, 510
Hosaki, Y., 372
Hosokawa, Y., 395, 396, 399, 403
Houk, J., 130, 132(8), 133(8)
House, D. W., 167
Howard, J. B., 49, 50(23)
Howell, B. A., 63
Howorth, P. N. J., 186

Hsieh, H. S., 35
Hsiung, M., 149
Huang, T. T. F., 76
Huber, J. A., 379
Huber, R. E., 242
Huennekens, F. M., 58
Hughes, H., 376
Hulbert, P. B., 84
Hummel, J. P., 512
Humphrey, R. E., 253
Humphries, B. A., 58, 125
Hunt, S., 419
Hupe, D. J., 51, 130, 131(9), 132, 154
Hurt, H. D., 309
Huskey, W. P., 429
Hwang, O., 320, 322(38), 323(38)
Hyun, M. H., 167
Hzuka, H., 361
Hyodo, H., 429

I

Iacobazzi, V., 271, 280
Iavarone, C., 180
Ibata, Y., 164
Iborra, F., 65
Ichikawa, A., 377
Ida, S., 149, 164
Ikeda, K., 497
Ikeda, M., 56
Ikegami, T., 270
Ikehara, M., 361
Imai, K., 67, 68, 69(6), 70(3), 71, 72(4), 73(8), 74
Imaseki, H., 429
Inoue, K., 74
Inouye, M., 36
Ip, M. P. C., 372
Irreverre, F., 384, 390, 444
Isenberg, J. I., 141
Ishii, T., 474, 475(4)
Ishimoto, M., 455, 498, 499
Ishimoto, Y., 179, 372
Ishizuka, J., 374
Isles, T. E., 61
Isshiki, G., 270, 370
Itiaba, K., 149
Ito, S., 460
Iwami, K., 434, 436, 438, 439

Iwanaga, S., 68, 69(6)
Iwasaki, I., 28
Iwata, H., 163, 164
Iyer, K. S., 115, 248
Izatt, R. M., 133

J

Jackson, R. C., 252
Jackson, S. G., 4
Jacobs, M., 422
Jacobsen, J. G., 375, 395, 405, 409(10), 410(10)
Jacobson, G. R., 255
Jacobson, K. B., 45, 49(9)
Jaffe, I. J., 186
Jakobson, I., 231
Jakoby, W. B., 201, 202, 204(7, 8), 205(14), 228, 313, 332, 367
Jakubowski, H., 383, 431, 433, 434
Jamieson, G. A., 245
Janne, J., 371
Janolino, V. G., 504, 505, 506, 507, 509(1), 510(1)
Jansen, G., 3
Jarabak, R., 236
Jarrige, P., 361, 362(2), 366
Jarvis, D., 253
Javor, B., 76(22), 77, 97, 100(2)
Jayle, M. F., 361, 362(2), 366(2)
Jefferies, T. M., 187
Jeffery, E. H., 32, 34(1), 35(1), 36(1)
Jeffery, H. J., 209, 210(15), 214(15)
Jekel-Halsema, I. M. C., 372
Jellenz, W., 167
Jencks, W. P., 52(c), 54, 367
Jender, H. G., 335, 343
Jensen, G. L., 316
Jensen, L. S., 309
Jerfy, A., 208, 214(7)
Jocelyn, P. C., 44, 49, 61, 102, 108, 129, 251, 272
John, A., 305
Johnson, C. R., 283, 284(3), 285, 288
Johnson, D. F., 141
Johnson, J. L., 11
Johnson, R. B., 52(g), 54
Jollès-Bergeret, B., 280
Jollow, D. L., 315, 319(12, 13, 14, 15)

Jonas, A. J., 149
Jones, A. J., 256
Jones, J. D., 299
Jones, J. F., 201
Jones, J. P., 429
Jong, P., 7
Jordan, P. M., 52(a), 54
Jowett, D. A., 214
Jutisz, M., 163

K

Kaczmarek, L. K., 299
Kadin, H., 99, 257, 258(1a), 261
Kagamiyama, H., 503
Kagan, H. B., 285
Kågedal, B., 179
Kagotani, K., 455, 498
Kaiser, E. T., 56
Kajander, E. O., 377
Kaji, A., 350, 353(1)
Kalk, K. H., 235
Källberg, M., 179
Kalnitsky, G., 512
Kamin, H., 11
Kamp, D., 52(f), 54
Kandrach, M., 145
Kanety, H., 76, 265, 270
Kanzaki, H., 454, 487, 488(6), 492
Kaplan, A. P., 56, 255
Kaplan, E., 114
Kaplan, L., 384
Kaplan, M. M., 454
Kaplan, N. O., 86, 149
Kaplowitz, N., 376
Karalis, A. J., 49, 134
Karasawa, K., 74
Karlsen, R. L., 314
Kasahara, K., 179
Kashiwamata, S., 390
Kato, A., 374
Kato, T., 516, 518(2)
Kawabata, Y., 457
Kawahara, Y., 68
Kawai, T., 478, 482(1)
Kay, D. R., 187
Kearney, E. B., 59, 164, 271, 279, 280(4)
Keenan, B. S., 452

AUTHOR INDEX

Keim, P. S., 235
Kellmers, A. D., 144
Kende, H., 427, 428, 429(8)
Kenyon, G. L., 171
Kepes, A., 62, 65(72)
Kerr, D., 466, 479
Kerr, L. M. H., 209
Kerr, M. A., 53(rr), 55
Kerr, S. J., 378
Kessler, D. L., 11, 14
Khani-Oskouee, S., 429
Khorana, H. G., 354
Kidby, D. K., 59
Kierstan, M. P. J., 48
Kies, C., 297
Kiguchi, S., 17, 18(4), 22(4)
Kim, I.-K., 377, 383
Kimori, M., 164
Kimura, T., 56
King, J. C., 307
Kingsley, R. B., 316
Kinon, B. J., 265
Kinuta, M., 17, 18, 22(4), 372
Kirchhoff, R. A., 288
Kirkeland, J. J., 10
Kirkpatrick, A., 253
Kirowa-Eisner, E., 84, 93, 99
Kissinger, P. T., 114
Kitahara, Y., 516, 518(2)
Kitson, T. M., 52(ss), 55
Kizer, D. E., 63
Kjaer, A., 291, 438
Klayman, D. L., 240
Klee, W. A., 115, 248, 387
Kleeman, C. R., 7
Klouwen, H. M., 57
Knapp, F. F., Jr., 245
Knight, C. G., 83
Knight, R. H., 230
Knobler, Y., 242
Knopf, K., 299, 375
Knowles, J. R., 276
Knox, J. H., 10
Knudson, P., 337, 339(13), 341(13)
Kobashi, K., 121
Kobayashi, T., 74
Kocsis, J. J., 367
Kodama, H., 270, 370
Koechlin, B. A., 172, 173(1)
Koga, K., 395, 403(9)

Koh, T., 28
Kohashi, N., 395, 396, 403
Kohn, L. D., 293
Koj, A., 235, 350
Kolhouse, J. F., 370
Kolthoff, I. M., 44, 116
Kondo, H., 455, 498, 499
Konigsberg, W., 65
Konishi, Y., 56, 115, 116(18), 117(18)
Kontro, P., 375
Koopman, B. J., 3
Koppel, R. L., 267, 268(6)
Kori, Y., 395, 396
Kosower, E. M., 76, 77(10, 11), 79, 80, 81(6, 8, 9), 82(9), 84, 85, 93, 99, 101, 264, 265, 267, 268, 269, 270, 272
Kosower, N. S., 76, 80, 81, 82(9), 84, 85, 93, 99, 101, 264, 265, 267, 268, 269, 270, 272
Koszalka, G. W., 506
Kotaki, A., 58
Kraemer, K. L., 141
Kraus, J. P., 389, 390, 393, 394(3)
Kraus, R. J., 32, 33, 35(3), 36(3), 37, 196, 197, 198, 199(2), 200(11)
Krauss, K. F., 450
Kredich, H. M., 452
Kredich, N. M., 420, 453, 470, 474, 479
Kreis, W., 459
Kress, L. F., 252
Kreuzig, F., 114, 187
Krijgsheld, K. R., 372, 453
Kripalani, K. J., 258
Krohn, J. A., 130
Kronman, M. J., 59, 66(49a)
Krull, L. H., 252
Ku, S. B., 342
Kudo, I., 74
Kuehl, G. V., 53(w), 55, 252
Kuehl, T. J., 81, 84(33)
Kuhlenkamp, J., 376
Kulka, M., 237
Kumagai, H., 166
Kumar, A., 275
Kumar, V., 454
Kun, E., 179
Kuo, S.-M., 372
Kuper, S. W. A., 216
Kuriyama, K., 149, 164
Kushnir, M., 269

Kustu, S., 145
Kyle, W. E., 370, 375

L

Labardie, J., 199
Labia, R., 276
Labouesse, B., 65
Lambeth, D. O., 248
Lan, S. J., 258
Landry, D. A., 7
Lane, J., 239
Lane, J. M., 307, 311, 313
Lange, J., 3
Lapinskas, B., 378
Larsen, G. L., 234
Larson, K., 151
Lascelles, J., 236
Laskowski, M., 252
Laster, L., 384, 390, 444
Lauterberg, B. H., 376
Lavine, T. F., 17, 18, 224
Lawlis, V. B., 118
Lawrence, R. A., 307, 313
Layne, E., 446
Lea, P. J., 423
Lea, T. C., 372
Leach, S. J., 115, 116(6)
Lecavalier, D. R., 186
Lederer, E., 163
Lee, D.-H., 332, 355
Lee, K. S., 141
Lee, M. L., 53(w), 54, 252
Lee, W.-J., 460
Le Gaillard, F., 63
Legler, G., 51
Leibach, F. H., 269
Leinweber, F.-J., 15, 160, 161(1)
Lelievre, V., 276
Lemieux, J., 349, 355
Leonard, N. J., 283, 285(4)
Leslie, J., 57
Lesnicki, A., 214
Letonoff, T. V., 7
Letts, E. A., 227
Levander, O. A., 307
Levin, O., 416, 446, 494
Levine, V., 361
Levinthal, M., 357
Levintow, L., 316

Levy, H. L., 367, 369(2), 370(2), 384, 389
Levy, L., 270
Levy, M. H., 185
Lewis, S. D., 67, 134
Li, G. S. K., 299
Libaw, E., 248
Lieberman, M., 429
Light, A., 252
Lilburn, J. E., 130, 132(7), 133(7)
Limb, G. R., 251
Lin, Y. C., 404, 405(6), 406(6), 409(6)
Linde, O., 237
Lindley, H., 251, 253
Lipmann, F., 335
Lippi, U., 4
Lipscomb, W. N., 62
Lipton, S. H., 280, 281(19)
Little, G., 46
Liu, C. H., 307
Liu, T. Y., 129
Livesey, J. C., 108
Livingston, J. N., 269
Lizada, M. C. C., 426, 427(3)
Lloyd, A. G., 3, 4
Loiselet, J., 374
Lombardini, J. B., 395, 396(8), 403(7, 8)
Loriette, C., 396, 406, 410(17)
Lou, M. J., 128
Loughlin, R. E., 479
Lovgren, G., 60, 67(60)
Lowe, O. G., 280
Lowry, O. H., 389, 416, 466, 471, 480, 501, 518
Lu, Z. H., 307
Luise, M., 11
Lumper, L., 76
Lundquist, P., 4, 5(19), 6(19)
Lunte, S. M., 114
Luthra, N. P., 48
Luxa, H. H., 435
Lyman, H., 357
Lynén, F., 478
Lyon, E. S., 201, 202(2, 3), 204(7, 8), 332
Lysko, H., 252, 257

M

Ma, R. S. N., 4
Macaione, S., 395, 410

McCall, J. T., 186
McCandless, E. L., 4
McCants, D., Jr., 283, 284(3), 285(3)
McCarthy, K. D., 173, 184
McCarty, R. E., 65
McClaren, J. A., 253
McCoy, K. E. M., 309
McCully, K. S., 299
McDermott, E. E., 286
McDonald, D., 202
McDonald, R. M., 249
McElroy, W. D., 350, 353(1)
McIntosh, J., 141
McKay, M. J., 52(y), 55
McKeller, R., 128
MacKenzie, C. G., 317
MacKenzie, J. B., 317
McKeown-Longo, P. J., 65
McKinney, L. L., 232
McKinstry, D. N., 258
McLaughlin, L. W., 76
McLean, A. E. M., 188
McMurray, C. H., 62
Macnicol, P. K., 422
McPhee, J. R., 115, 254, 256
McPherson, A., Jr., 461
Macquarrie, R., 52(gg), 55
McWherter, C. A., 118
Madison, J. T., 423, 429
Mager, J., 66
Maggio, E. T., 171
Main, R. K., 391
Mainka, L., 80
Makino, N., 53(m), 54
Makkear, H. P., 250
Makowske, M., 271, 279, 280(4)
Maloney, C. M., 4
Malthe-Sorenssen, D., 314
Malthouse, J. P., 49, 56
Manabe, H., 36
Mandell, R., 29
Mangum, J. H., 370, 384, 388
Mann, J., 190
Mannervik, B., 48, 56, 151
Manning, J. M., 287, 291(10)
Mantsala, P., 53(ff), 55
Manz, F., 4
Marcell, P. D., 370
March, J., 36
Marcus, C. J., 201, 202(3), 204(8), 332
Mareni, C., 269
Marks, V., 76
Marsh, M. W., 52(y), 55
Marsh, R. L., 52(e), 54
Marshall, M., 47, 53(k), 54
Mårtensson, J., 4, 5(19), 6(19), 178, 179(5)
Martin, D. J., 516
Martin, D. L., 274, 405
Martin, J. J., 370, 375, 389
Martin, W. G., 155, 300
Mason, H. L., 59
Massey, V., 53(v), 55, 254
Masuoka, N., 17, 18(5), 20(5), 21(5), 22(5), 224
Mato, J. M., 382(23), 383
Matsueda, R., 56
Matsuo, Y., 486
Matsuzawa, T., 374
Matthews, I. T. W., 83
Mattock, P., 201
Mattoo, A. K., 429
May, P. C., 280
Maybury, R. H., 118
Mazelis, M., 437, 438, 439, 441(2), 443
Mefford, I., 110, 114(2)
Meguro, H., 257
Meister, A., 17, 26, 52(x), 53(1), 54, 55, 66, 124, 167, 179, 221, 271, 275, 279, 287, 290(13, 14), 291(10), 314, 315, 316, 318, 320, 322(38), 323(7, 38), 324, 325, 370, 371, 372, 374, 376, 455
Meites, L., 42
Melcher, M., 196
Melis, A., 84, 93, 99
Melville, T. H., 217
Menken, B. Z., 311
Mercer, E. I., 329
Meredith, M., 109
Merkenschlager, M., 421, 453, 454(4)
Merrifield, B., 134
Merta, A., 377
Metrione, R. M., 404, 405(6), 406(6), 409(6)
Michalk, D., 4
Miflin, B. J., 423
Migdalof, B. H., 258
Mikami, H., 17, 18(5), 20(5), 21(5), 22(5), 224
Miles, E. W., 293
Miller, D. M., 62
Miller, S. A., 299
Milner, J. A., 324
Milsom, D. W., 207, 215(3)

Minchinton, A. I., 76
Minei, H., 504
Miners, J. O., 188
Mintel, R., 23, 26, 237
Miquel, J., 317
Miraglia, R. J., 300
Misani, F., 286, 287
Misono, H., 457
Misra, D. C., 134
Misra, G. H., 395, 403(14)
Misra, H. P., 119
Mitchell, J. R., 315, 319(12, 13, 14, 15, 16), 325, 376
Mitchell, P. D., 190
Mitsuda, H., 434, 436, 438, 439
Miura, G. A., 192, 378, 427, 429(7)
Miyajima, R., 454
Miyata, T., 68, 69(6)
Miyazato, K., 500
Mizuhara, S., 270, 370
Mizuno, Y., 361
Mizuo, H., 163, 164
Mizusawa, H., 438
Mizushima, H., 74
Modig, H. G., 124, 125(3), 128(3), 252
Moffatt, J. G., 354
Moffitt, S. D., 324
Mohyaddin, F., 4, 7
Moir, A. J. G., 75
Mondovi, B., 26, 226, 374
Montal, M., 76
Montgomery, J. A., 192, 377, 379(5), 380(5), 382(5), 383, 431, 433(7)
Monty, K. J., 15, 160, 161(1)
Moore, S., 115, 127, 141, 275, 279, 287, 291(10), 301
Morales, M. F., 80
Moran, R. G., 280
Moran, T., 286
Morgan, C. D., 271, 276
Mori, B. G., 226, 374
Morimoto, N., 164
Morino, Y., 415, 493
Mornet, D., 80
Moroney, J. V., 65
Morris, C. J., 420
Morris, J. G., 305
Morris, R. G., 144, 145
Morris, V. C., 307
Morrison, A. R., 216

Morrow, G., 384
Morrow, P. M. F., 214
Mortensen, E., 59
Moscarello, M. A., 252
Mosher, D. F., 52(g), 54, 65
Moss, L. G., 405, 410(12)
Mosti, R., 255, 395
Motiu-De Grood, R., 51
Mountain, I. M., 45
Mounter, C. A., 388
Moyer, A. M., 227
Mudd, S. H., 367, 369(2), 370(2), 384, 389, 390, 419, 420(3), 421, 422, 423, 425, 443, 444, 448(1), 449, 453, 454
Muench, K. H., 53(w), 54, 252
Muijsers, A. O., 186, 189(8)
Mulde, G. J., 453
Mulder, G. J., 372
Muldoon, W. P., 317
Muramoto, K., 66
Murphy, J. B., 45, 49(9)
Murphy, M. J., 11
Murray, J. F., 46, 97
Myohanen, T., 53(ff), 55

N

Nagai, S., 453, 470
Nagainis, M. P., 52(jj), 55
Nagasawa, H. T., 110, 111, 112(3), 113(3), 114, 317
Nagasawa, T., 454, 474, 475(3, 4), 487, 488(6), 492
Naiki, N., 295, 466, 469, 478, 480, 482(12), 483(12)
Nakagawa, H., 374
Nakagawa, Y., 56
Nakamura, K., 436, 459
Nakamura, M., 45
Nakamura, T., 244, 415, 460, 493
Nakamuro, T., 35
Nakashima, K., 375
Nakayama, T., 244, 292, 460
Nakayasu, M., 314
Nakazawa, A., 361
Nakazawa, H., 167
Nara, Y., 257
Naruse, A., 63, 375, 376(67)

Nasu, S., 399
Neal, R. A., 309
Negi, S. S., 250
Negrutiu, I., 422
Neidle, S., 291
Neumann, H., 256
Neumann, N. P., 275
Neurath, H., 118
Newcomer, M. E., 62
Newton, G. L., 76, 77, 80(6, 7), 81(6), 85, 89, 94, 95, 96, 97, 101, 268
N'Galamulume-Treves, M., 435
Ngo, T. T., 30
Nicholas, D. J. D., 342
Niizeki, S., 396
Nilsson, L., 178
Nimmi, M. E., 248
Nisbet, A. D., 75
Nishida, C., 235
Nishikawa, D., 166
Nishiuch, Y., 314
Nitta, K., 59, 66(49a)
Njaa, L. R., 299
Noble, E. P., 53(j), 54
Nojima, S., 74
Norris, R., 56
North, J. A., 370
Novelli, G. D., 144, 149
Novogrodsky, A., 271

O

Oae, S., 282, 284(2), 285(2)
Odom, J. D., 48
Offerman, M. K., 52(y), 55
Offermanns, H., 186
Ogawa, H., 369, 515
Ogden, G., 141
Ogura, F., 245
Oh, K., 483
Oh, S.-H., 33, 196
Ohkishi, H., 166
Ohkuma, S., 149, 164
Öhman, S., 4, 5, 6(19), 6(22), 29
Ohmori, S., 270, 370
Ohsawa, K., 499
Ohshiro, T., 430, 457
Oikawa, A., 314
Oja, S. S., 375

Okazaki, T., 361
Olafsdottir, K., 108
Oldfield, J. E., 307
Oliver, J. M., 269
Olney, J. W., 314
Olson, B. R., 249
Olson, D. L., 149
Olson, L. M., 302
Olson, O. E., 196
Orlowski, M., 287, 318
Orrenius, S., 63
Osborn, M., 249
Oshima, M., 455, 498
Oshima, R. G., 144, 146(2), 149
Osslund, T., 335
Osumi, T., 457
Oton, J., 66
Otsubo, T., 245
Oura, T., 270, 370
Overdank-Bogart, T., 76(23), 77
Owens, L. D., 449
Ozaki, H., 456, 470

P

Pabst, M. J., 313
Pachéco, H., 377
Pace, J., 286
Pace, N., 4
Pachmayr, F., 450
Packer, J. E., 367
Packman, S., 390, 393
Page, E., 520
Palau, J., 52(kk), 55
Palek, J., 76
Palmer, I. S., 196
Palmer, J. L., 377
Palmer, T., 141
Palmieri, F., 271, 280
Pande, C., 76, 83(16)
Papanikolaou, N. E., 285
Parente, A., 134
Parker, D. J., 57
Parker, M. G., 259, 264(12)
Parker, R., 178
Parmeggiai, A. P., 52(e), 54
Pascoe, P. J., 4, 6
Pasternak, C. A., 374
Paszewski, A., 384, 479, 487

Patchornik, A., 251
Patrick, H., 300
Patterson, G. W., 429
Patterson, M. A. K., 130, 132(8), 133(8)
Patterson, W. I., 243
Pawelkiewicz, J., 430, 431, 432(2), 433(2)
Pazhenchevsky, B., 76, 85, 268
Peake, M. J., 4, 6
Pearson, W. N., 310
Pechère, J. F., 118
Peck, E. J., Jr., 173
Peck, H. D., Jr., 350
Peduzzi, J., 276
Pegg, A. E., 371
Pekas, J. C., 234
Pennings, E. J. M., 343
Pensa, B., 53(oo), 55, 179, 180
Perkins, R. I., 285
Perreault, S. D., 269
Perrett, D., 190
Perrin, C. L., 42
Perry, T. L., 155
Peschon, J. J., 340
Peters, C. F., 242
Peterson, D., 53(r), 54
Peterson, R. P., 309
Petrini, J., 376
Pfeifer, R., 143, 155
Phares, E. F., 144
Phillips, P. H., 59, 479
Pichat, L., 272, 274(12), 405
Picken, J. C., 232
Pierce, J. G., 247
Pigiet, V., 76, 77(12)
Pillai, P., 324
Pillion, D., 269
Pinnick, C. L., 204, 205(18)
Pirkle, W. H., 167
Piscopo, M., 269
Pitchen, P., 285
Pitcher, R. G., 449
Pizzo, S. V., 60, 67(59)
Pleiderer, G., 51
Plese, P. C., 53(t), 54
Ploegman, J. H., 235
Poet, R. B., 99, 257, 258(1a), 261
Pointer, K., 66
Poland, J., 166
Poole, J. R., 369, 370(15)
Portemer, C., 404, 409(7)

Potter, D. W., 101, 105(1), 106(1)
Potter, J. L., 253
Potter, W. Z., 315, 319(12, 13, 14, 15)
Potts, K. T., 276
Poulos, A., 202
Poulsen, L. L., 376, 411
Powell, G. M., 214, 361, 366
Powrie, F., 316
Pozzo, A. D., 66
Prasad, A. R., 53(nn), 55
Price, C. A., 65, 94
Price, N. C., 59, 66(49)
Price, S. J., 141
Pringle, J. R., 249
Prockop, D. J., 249
Prody, C., 145
Prosser, C. I., 202
Provansal, M., 299
Puri, R. N., 316
Purkiss, P., 178

Q

Quebbeman, A. J., 231
Quinn, D., 269
Quiocho, F. A., 62

R

Rabenstein, D. L., 110, 114, 139, 186, 189(11), 258
Radda, G. K., 59, 66(49)
Radkowsky, A. E., 79, 80(28), 84
Raferty, M. A., 141
Ragin, R., 484, 485(2)
Raiha, N. C. R., 384
Raina, A., 371, 377
Rajagopalan, K. V., 11, 14
Rakow, D., 365
Ramachandran, J., 66
Ramaswamy, S., 201, 202, 204(14), 205(14)
Rammler, D. H., 215, 216(29)
Randall, L. O., 122
Randall, R. J., 389, 416, 466, 471, 480, 501, 518
Ranney, H. M., 76, 80(6, 7), 81(6), 85, 101, 268
Rao, G. S., 25

AUTHOR INDEX

Rassin, D. K., 410
Rathod, P. K., 226
Ratner, S., 317
Ravizzini, R. A., 51, 52(ee), 55
Ray, P. D., 248
Read, W. O., 173
Reddy, M. K., 134, 139(20), 251
Reddy, V. A., 53(bb), 55
Redfern, B., 110, 112(3), 113(3)
Reed, D. J., 101, 105(1), 106(1), 107, 108, 109, 228
Reese, H., 276
Reeve, J. R., 247
Reiner, L., 286, 287
Reinhold, J. G., 7
Reiter, R., 308
Reith, J. F., 11
Remtulla, M. A., 173
Rennenberg, H., 421, 422, 443
Renosto, F., 335, 337, 338(12), 339(13), 340(12), 341(13), 343(9)
Reuveny, Z., 339, 421
Revesz, L., 124, 125(3), 128(3)
Rhee, V., 314
Rhodes, W. G., 58, 125
Ricci, G., 149, 236, 376
Rich, J. J., 300
Richards, H. H., 192, 378, 380(16), 387
Riches, P. G., 252
Richman, P. G., 287, 315
Riddles, P. W., 47, 53(cc), 55, 117, 130
Ridge, B., 58
Riehm, J. P., 65
Rigau, J. J., 285
Rinaldi, A., 185
Rinderknecht, H., 216
Riordan, J. F., 44, 64(5)
Ristow, S. S., 246
Rix, I. H. B., 7
Robbins, K. R., 298, 302
Robbins, P. W., 335, 349
Roberts, M. F., 50, 53(p), 54, 60
Robeson, B. L., 155
Robins, E., 217
Robinson, D., 214
Robinson, J., 120
Robison, R., 264
Robson, A., 27
Robson, J. S., 257, 259(4), 261(4), 264(4)
Robyt, J. F., 56

Rogers, D., 291
Rogers, W. I., 59
Rogers, Q. R., 305
Rognes, S. E., 423
Rohrbach, M. S., 58, 125
Rolland, B., 185
Ronzio, R. A., 275, 287
Rose, F. A., 207, 208, 215(3)
Rose, W. C., 314
Rosebrough, N. J., 389, 416, 466, 471, 480, 501, 518
Rosen, B. P., 145
Rosenberg, L. E., 389, 390, 393, 394(3)
Rosenthal, G. A., 419
Rosenthal, S. M., 153
Rosowsky, A., 280
Ross, D., 63
Rossa, J., 149
Roth, E. S., 173, 184, 185(3), 375
Roth, M., 141, 143(4), 157
Rothfus, J. A., 262
Rothwell, R., 187
Rotilio, G., 411
Rotruck, J. T., 298
Rousselet, F., 259
Rovery, M., 52(i), 54
Rowe, W. B., 275, 287
Roy, A. B., 17, 202, 207, 208, 209, 210, 214(7, 15), 216, 217, 329, 335, 341(4), 362, 363(6), 365
Rudinger, J., 253, 257
Rudman, D., 324
Rüegg, U. T., 118, 253, 257
Russell, A. S., 186
Russell, J., 235
Russell, J. T., 155
Russo, A., 325
Ruth, G. R., 307
Ryan, C. A., 437
Ryan, L. D., 345
Rydon, H. N., 58
Rypins, E. B., 324

S

Sabry, Z. I., 3
Saetre, R., 110, 114, 186, 189(11), 257, 258, 259(5)
Sagami, H., 516

Saidha, T., 332, 355
Sakakibara, S., 374, 395, 396(12), 403(10), 404
Sakamoto, Y., 63, 375, 395, 396(12)
Salmon, A. G., 59, 66(49)
Sandhoff, K., 208, 217(6)
Sano, K., 471
Santangelo, J. R., 192, 378
Santi, W., 167
Santoro, L., 149, 376
Sasaki, M., 314
Sass, N. L., 300
Sasse, C. E., 299
Sato, I., 396
Sato, S., 377
Satterwhite, H. G., 45
Saundry, R. H., 75
Saville, B., 63
Scandurra, R., 149, 376, 410, 411, 413(1)
Schachman, H. K., 62
Schaeffer, M. C., 305
Schaffer, M. H., 255
Schaffer, S. W., 173, 367
Scheibe, R., 53(z), 55
Schellenberg, G. D., 144, 145(4)
Scheraga, H. A., 56, 115, 116(18), 117(18), 118, 119(29)
Scherberich, P., 186
Schiewelbein, H., 235
Schiff, J. A., 329, 332, 333(1), 349, 354, 355, 356(2), 357, 359(2), 360, 361(9), 369, 421
Schiffmann, E., 192, 382(23), 383
Schirch, L., 53(r), 54
Schlenk, F., 166
Schlesinger, P., 235
Schlessinger, D., 269
Schlossmann, K., 421, 453, 454(4), 478
Schmelzer, C. H., 509, 510
Schmetz, F. J., 149
Schmickel, R. D., 315
Schmidt, A., 166, 421, 450, 451(2, 3), 452(2, 3), 453(2)
Schmidt, D. E., Jr., 372
Schmidt, N. D., 340
Schmidt, U., 516
Schmidt, V., 51
Schmitz, J. A., 107, 108(4), 307
Schneider, J. A., 144, 149, 248, 375
Schneider, J. F., 24, 25
Scholtens, E., 453

Schonbrunn, A., 233
Schoot, B. M., 65
Schowen, R. L., 429
Schram, E., 279
Schubert, M. P., 317
Schubert, K. R., 50, 53(p), 54, 60
Schulert, A. R., 309
Schulman, J. D., 248, 375
Schütte, I., 11
Schwarz, M., 84, 93, 99
Schwenn, J. D., 332, 343
Schwimmer, S., 437, 438
Scioscia-Santoro, A., 226
Scleem, M., 10
Scott, K., 439, 441(2)
Scowen, E. F., 186
Scriver, C. R., 3, 4, 7
Seaver, N., 94
Seddon, A. P., 318, 319(33)
Seegmiller, J. E., 248
Seehra, J. S., 52(a), 54
Seelig, G. F., 52(x), 55
Segal, S., 248
Segel, I. H., 335, 337, 338(12), 339(13), 340, 341(13)
Seibles, T. S., 249
Seitz, A. W., 299
Sekura, R. D., 201, 202, 203(15), 204(15), 206(15), 287, 332
Sela, M., 251, 256
Sement, E. G., 259
Seon, B. K., 252
Serjeant, E. P., 133
Seubert, P. A., 335, 337, 338(12), 339(13), 340(12), 341(13), 343(9)
Sevem, A., 173
Shaddix, S. C., 377
Shaderevian, S. B., 3
Shafai, J., 7
Shafer, J. A., 67, 134
Shaked, Z., 133, 137(17), 140(17)
Shamberger, R. J., 459
Shamoo, A. E., 51, 52(aa), 55
Shannon, W. M., 383
Shargool, P. D., 30
Sharma, O. P., 250
Shaw, E. H., Jr., 173
Shaw, W. H., 335, 341
Sherman, W. R., 216
Sherry, P. A., 302

Shih, V. E., 29
Shiio, I., 454, 456, 470, 471
Shimizu, S., 430, 457
Shiozaka, H., 375
Shiozaki, S., 430, 457
Shipton, M., 56, 58, 132
Shirota, F. N., 110, 112(3), 113(3), 114
Shomrat, R., 76, 81, 268
Shoup, R. E., 110
Shuster, L., 86
Shuttleworth, K., 371
Siami, G., 309
Sido, B., 350
Siegel, L. M., 11, 17, 453
Siegelman, H. W., 357
Sies, H., 102, 124, 125(2), 128(2), 269
Silverberg, M., 56, 255
Silverman, R. B., 185
Simmons, S. S., Jr., 141
Simon, R. A., 371
Simpson, M. I., 227
Sinet, P. M., 95
Singer, S. S., 349
Singer, T. P., 59, 164, 271, 279, 280(4), 369, 395, 396(8), 403(7, 8)
Singhvi, S. M., 258
Sinha, N. K., 252
Sjodahl, R., 178, 179(5)
Skea, W., 143, 155
Skiba, W. E., 384, 388
Skovby, F., 389, 394(3)
Slaughter, C., 448, 449(6), 484, 486(3)
Sliwkowski, M. B., 120, 123(2), 504, 505, 508(7), 510
Sliwkowski, M. X., 120, 123(2)
Sloboda, R. D., 83
Slump, P., 298
Sluyterman, L. A. A. E., 180
Smalley, K. A., 305
Smiley, K. L., 49
Smith, A. J., 236
Smith, C. H., 252
Smith, C. V., 376
Smith, D. E., 52(h), 54
Smith, D. J., 171
Smith, D. K., 76
Smith, E. L., 367
Smith, I., 33, 420, 421
Smith, I. K., 439, 441(3)
Smith, J. N., 214

Smith, K. J., 188
Smith, L. H., Jr., 375, 395, 405, 409(10), 410(10)
Smith, R. A., 371
Smith, R. E., 52(gg), 55
Smoluk, G. D., 94
Smyth, D. G., 65
Sneddon, W., 178, 190
Snedecor, B. R., 118
Snell, C. T., 22
Snell, F. D., 22
Snyder, G. H., 49, 134
Snyder, H. R., 242
Snyder, L. R., 10
Soda, K., 35, 240, 242, 243, 244, 271, 291, 292, 415, 455, 456, 457, 459, 460, 464, 493, 500, 501, 503(3, 4), 504
Solney, E. M., 45
Solomon, F., 52(c), 54
Sörbo, B., 4, 5, 6(12, 19, 22), 22, 23, 24(7), 26, 29, 178, 179, 235, 236, 350, 372, 395, 403(6), 406
Soucek, D. A., 269
Southern, L. L., 298
Spaeth, D. G., 300
Spears, R. M., 274, 405
Spencer, B., 7, 207, 208, 214, 215, 216, 217
Spencer, M., 458
Spero, L., 249
Spetling, R., 253
Spielberg, S. P., 248
Spies, J. R., 399
Squires, C. L., 335
Srivastava, S., 45
Stabel, T. J., 505
Stanbury, J. B., 178
Stanfield, E. F., 216
Stan-Lotter, H., 52(qq), 55
Stark, G. R., 50, 255
Start, J. D., 155
Stedman, D., 155
Stein, W. H., 115, 275
Steinberg, I. Z., 253, 256
Steiner, M., 50, 53(o), 54, 248, 254
Steiner, R. F., 66
Stephens, A. D., 190
Stern, A. I., 332, 355
Stern, R. B., 186
Stevens, J. L., 228, 231, 367
Stevenson, K. J., 50, 53(s), 54

Stewart, C. P., 257, 259(4), 261(4), 264(4)
Stinshoff, K., 209
Stinson, R. A., 62
Stipani, I., 271, 280
Stipanuk, M. H., 155, 372, 374, 375, 396, 405, 406, 410(16)
Stokes, G. B., 256
Stokke, O., 173
Stollery, J. G., 252
Strickland, R. D., 4
Stricks, W., 44, 116
Strumeyer, D., 86
Stuart, J. D., 155
Stuchbury, T., 56
Sturdik, E., 66
Sturman, J. A., 299, 375, 384, 410
Su, L.-C., 311
Su, Q., 307
Subramani, S., 62
Suda, M., 374
Sueyoshi, T., 68, 69(6)
Sugie, K., 244, 292, 460
Sugimoto, S., 471
Sugita, Y., 53(m), 54
Sumner, A. T., 82
Sumner, H. W., 114
Sumner, J. B., 28
Sunde, R. A., 308, 311
Suschitzky, H., 56
Susuki, M., 370
Sutherland, B., 307
Sutherland, I. T., 3
Suzuki, H., 36
Suzuki, I., 11
Suzuki, K., 504
Suzuki, T., 415, 493
Swaisgood, H. E., 120, 123(2, 4), 504, 505, 506, 507, 509, 510
Swan, J. M., 118
Swanepoel, O. A., 254
Swaroop, A., 7
Sweetman, B. J., 18
Swiatek, K. R., 371
Switzer, R. L., 50, 53(p), 54, 60
Sysak, P. K., 284, 286(5)
Szajewski, R. P., 130, 132(7, 10), 133, 137(17), 139(10), 140(10, 17)
Szczepkowski, T. W., 224, 226
Szyper, M., 262

T

Tabor, C. W., 127, 153, 370, 458
Tabor, H., 127, 370, 457
Taborsky, E., 7
Takahashi, H., 257
Takahashi, N., 516, 517(3)
Takamori, K., 511, 512(3), 514(3), 515
Takeshima, K., 295, 454, 466, 468, 469, 478, 480, 482(1, 2, 12), 483(12)
Tallgren, G., 7
Tan, K. K., 250
Tanaka, H., 35, 240, 242, 243, 244, 291, 292, 415, 456, 459, 460, 464, 493
Tanaka, I., 460
Tanaka, K., 429
Tanaka, S., 497
Tanaka, Y., 149, 164
Tanchoco, M. L., 66
Taniguchi, M., 17, 18(5), 20(5), 21(5), 22(5), 224, 372
Tanizawa, K., 415, 459
Tannenbaum, S. R., 299
Tarbell, D. S., 25
Tarka, S. M., 300
Tarnowski, G. S., 45
Tarver, H., 3, 395
Tate, S. S., 367
Tateishi, N., 63, 375, 376(67)
Tavachenik, M., 395
Taylor, G., 231
Taylor, M. P., 384
Taylor, R. T., 457
Teeter, R. G., 297, 305
Tejerina, G., 456
Teng, S. S., 269
Ternai, B., 67
Thannhauser, T. W., 56, 115, 116(18), 117(18), 118, 119(29)
Theriault, Y., 139
Thibert, R. J., 372
Thies, W., 361, 365, 366(3)
Thoene, J. G., 149
Thomas, C., 32, 34(1), 35(1), 36(1)
Thomas, H. J., 383
Thomas, J. H., 4
Thomas, K. C., 458
Thomas, L. L., 375, 405, 409(10), 410(10)
Thompson, G. A., 422, 423

Thompson, J. F., 419, 420, 423, 429, 439, 441(1), 449
Thorneley, R. N., 50
Thorpe, J. M., 511
Tice, S. V., 26, 221, 279, 280(4)
Tietze, F., 96, 129
Tiselius, A., 416, 446, 494
Tishbee, A., 167, 241
Tkachuk, R., 71
Todd, P., 63
Toennies, G., 17, 18(6), 224, 280
Tojo, H., 396
Tokushige, M., 361
Tolosa, E. A., 374
Tomita, K., 377
Tomkins, G. M., 420, 453
Toniolo, D., 269
Tooloukian, J., 376
Townshend, A., 7
Toyama, S., 455, 500, 501, 503, 504
Toyo'oka, T., 67, 68, 69(6), 70(3), 71, 72(4), 73(8), 74
Toyosato, M., 457
Trackman, P. C., 371
Traeger, J., 237
Trebst, A., 332, 333, 334(10)
Trentham, D. R., 62
Trewyn, R. A., 378
Trudinger, P. A., 17, 453, 479
Truex, R.C., 300
Tsang, M. L.-S., 349, 355, 421
Tudball, N., 4
Tucci, G., 395, 403(15)
Tuppy, H., 65
Turini, P., 395, 403(7)
Turnell, D. C., 141, 142, 155
Turner, J., 208
Tuzimura, K., 257
Tweedie, J. W., 335
Tweto, J., 456

U

Ubuka, T., 17, 18, 22(4), 178, 179, 224, 372
Ue, K., 80
Ueda, I., 374, 395, 396, 403(9, 10), 404
Ueland, P. M., 377
Uemura, I., 270, 370
Uhteg, L. C., 235, 350, 351(6)
Umemura, S., 372
Uren, J., 484, 485(2)
Usdin, E., 191
Usui, K., 515
Utley, J. F., 94
Utley, C. S., 370

V

Valle, E. M., 57
Vallee, B. L., 44, 64(5)
Vallejos, R. H., 51, 52(ee), 55, 57
Vanaman, T. C., 255
van Bladderen, P. J., 230
van der Gen, A., 135, 230
Van der Henn, G. K., 3
Vander Jagt, D. L., 130
van der Korst, J. K., 186, 189(8)
Van Der Vies, S. M., 50, 53(s), 54
Van der Werf, P., 318
van de Stadt, R. J., 186, 189(8)
Van Etten, R. L., 7, 208
Van Haard, P. M., 65
Van Kempen, M. J., 343
Van Rensburg, N. J., 254
Van Vleet, J. F., 307
Varrichio, F., 57
Vas, M., 53(n), 54
Vasanthakumar, G., 192, 382(23), 383
Veeger, C., 50, 53(s), 54
Veluthambi, K., 423
Véron, M., 377
Vesleý, J., 377
Vestling, C. S., 345
Visek, W. J., 309
Vogel, F., 76
Vogel, R., 235
Volini, M., 235, 236(1)
von Berg, W., 384
Voordouw, G., 50, 53(s), 54
Votruba, I., 377

W

Wadsworth, P., 83
Wagner, J. G., 187

Waheed, A., 7, 208
Wainer, A., 374
Wake, R. G., 254
Waldschmidt, M., 421, 453, 454(4)
Walker, R. D., 430
Wall, J. S., 141
Walmsley, R. W., 4, 6
Walsh, C., 233, 504
Walsh, K. A., 249
Walshe, J. M., 186, 248
Walters, F. H., 155
Walther, L., 11
Wälti, M., 180, 271, 276
Wang, C. Y., 428
Wang, J. H. C., 361
Wang, J.-L., 201, 202(3), 204(8), 332
Wang, T. P., 86
Ward, J. F., 94
Warr, J. R., 269
Watanabe, K., 372, 429
Watanabe, Y., 67
Waters, M., 367
Waterson, J. R., 315
Watkins, J. C., 280
Watts, R. W. E., 186
Weakely, R. B., 232
Webb, E. C., 214
Weber, H. U., 317
Weber, K., 249
Wedding, R. T., 30
Wedler, F. C., 275
Weigert, W. M., 186
Weil, L., 249
Weinmann, W., 11
Weinstein, C. L., 375, 405
Weinstein, S., 167
Weir, D., 187
Weisiger, R. A., 228, 367
Weismann, W. P., 383
Weiss, P., 286
Weissbach, H., 371, 457, 458
Weitzman, P. D. J., 115, 259, 264(12)
Welch, K. N., 516
Wellner, V. P., 316, 324, 325
Wells, M. A., 3
Wells, M. S., 384, 388
Welty, J. D., 173
Welz, B., 196
Wendel, A., 308
Weng, L., 235

Werman, R., 269
Wertheim, B., 264
Westley, A., 238
Westley, J., 23, 24, 25, 26, 53(dd), 55, 235, 236, 237, 238, 239, 350, 351(6)
Westley, L., 235
Weswig, P. H., 307, 309
Wetlaufer, D. B., 115, 246, 256
Wetzel, R., 53(u), 54
Whanger, P. D., 307
Wheeler, E. L., 124
Wheldrake, J. F., 374
Wheler, G. H. T., 155
Whistler, R. L., 132
White, E. L., 377
White, F. H., 251
White, H., 52(c), 54
White, R. F., 384
Whitehead, J. E. M., 216
Whitehead, J. K., 286, 287
Whitehouse, M. W., 67
Whitesides, G. M., 130, 132(7, 8, 10), 133, 137(17), 139(10), 140(10, 17)
Whitney, P. L., 388
Whitney, R., 310
Whittaker, V. P., 388
Wick, M., 280
Wiebers, J. L., 453
Wiedner, H., 53(u), 54
Wieland, T., 276
Wiesman, W. P., 192
Wijdenes, J., 180
Wijers, H. E., 135
Wilkins, M. J., 391
Wilkinson, S. P., 186
Willems, J. J. L., 11
Willhardt, I. H., 374
Williams, I. H., 173
Williams, M. N., 52(d), 54
Williams, R., 186
Williams, R. T., 214
Williams, T., 449
Williams, W. J., 7
Williamson, G., 56
Williamson, J. M., 53(1), 54, 315, 318, 319(16, 34, 35), 320(16), 324(16)
Willis, G. M., 298
Willis, R. C., 144, 146(2)
Wilson, J. M., 51, 130, 131(9), 132, 154
Wilson, L. G., 335

Wilson, R. B., 299
Wilson, R. P., 307
Wilson, T. D., 155
Winitz, M., 169, 170(15), 233, 280, 314, 474
Winter, W. P., 315
Wishnia, A., 76, 83(16)
Witt, S. C., 124
Witter, A., 65
Woitczak, L., 235
Wolberg, G., 192
Wolf, A. M., 374
Wolff, H., 288
Wolthers, B. G., 3
Wong, F. F., 437
Wood, H. G., 53(q), 54
Wood, J. L., 22, 25, 26, 226
Woodard, R. W., 429
Woods, D. D., 456
Woodward, G. E., 60, 252, 257, 259(2)
Worden, J. A., 375, 406, 410(16)
Worwood, M., 208
Wu, D., 51, 130, 131(9), 154
Wu, J. Y., 404, 405, 409(5), 410(12)
Wu, S., 429
Wu, Y. S., 379
Wynn, C. H., 209, 214(16), 216

Y

Yagi, K., 58
Yakubu, S. I., 84
Yamada, H., 166, 430, 454, 457, 474, 475(3, 4), 487, 488(6), 492, 497, 515
Yamagami, S., 163, 164
Yamagata, S., 295, 454, 465, 466, 468, 469, 478, 480, 482, 483
Yamaguchi, H., 245
Yamaguchi, K., 374, 395, 396, 399, 403, 404
Yamamoto, S., 61
Yamamoto, T., 457
Yamuchi, T., 61
Yanagawa, H., 516, 517(3, 5), 518, 519

Yanagimachi, R., 76, 81, 84(33), 269
Yang, S. F., 299, 425, 426, 427, 458
Yao, K., 178
Yasuda, M., 455, 500, 501, 503(3, 4), 504(3)
Yasumoto, K., 434, 436, 438, 439
Yefremova, L. L., 374
Yeh, L. S., 301
Yeo, Y. Y., 149
Yon, J., 361, 362(2), 366(2)
Yonaha, K., 455, 501, 503, 504
Yoneda, H., 167
Yonezawa, T., 439
Yorek, M. A., 248
Yoshida, H., 53(q.), 54
Yoshii, H., 429
Yoshida, T., 257
Yost, E. J., Jr., 125
Yost, F., 58
Young, E. P., 178
Young, L., 216, 230
Yu, Y. B., 427, 429(5)
Yuasa, S., 18, 372
Yuzon, D. L., 317

Z

Zahler, W. L., 115, 246
Zdansky, G., 240
Zera, R. T., 317
Zerner, B., 47, 53(cc), 55, 117, 130
Zezulka, A. Y., 297
Zhang, C.-Y., 377, 383
Ziegler, D. M., 124, 125(1), 129, 376, 411
Zimmer, G., 80
Zimmer, L. T., 186
Zimmerman, T. P., 192
Zink, M. W., 421
Zipser, J., 76, 81
Zipser, Y., 268
Zirkin, B. R., 269
Zuman, P., 42

Subject Index

A

ABD-F. See 4-(Aminosulfonyl)-7-fluoro-2,1,3-bonzoxadiazole
Acacia farnesiana, 438, 443
ACC, formation, 426
ACESC. See S-(2-Amino-2-carboxyethylsulfonyl)-L-cysteine
Acetaminophen, toxicity, 315–317, 319–320
Acetothetin bromide, chemical synthesis, 227
  synthesis, 170–171
N-Acetyl-S-(methanethio)-DL-[1-^{14}C]cysteine
  resolution, 171–172
Acetylcholine esterase, 388
N-Acetylcysteamine, 293
N-Acetylcysteine
  derivatized with ABD-F
    detection limit, 74
    fluorescence, 73
    HPLC, 74
  derivatized with SBD-F
    detection limit, 69
    HPLC separation, 71
  S-1,2-dichlorovinyl conjugate, 234
  mBBr derivative
    HPLC chromatography, 90–92
    retention time, 88
N-Acetyl-L-cysteine
  effect on liver glutathione levels, 320
  transport into cells, 315–316
O-Acetylhomocystine sulfhydrylase, 242
O-Acetylhomoserine, 453
O-Acetylhomoserine (thiol)-lyase, 454, 456
O-Acetylhomoserine sulfhydrylase
  from B. flavum, 470–474
    assay, 470–471
    inhibition, 474
    properties, 473–474
    purification, 471–473
    specificity, 474
  in preparation of L-selenocystine and L-selenohomocystine, 295–296
  from S. pombe, 465–469
    assay, 465–466
    inhibition, 469
    kinetic properties, 469
    properties, 468–469
    purification, 466–468
    repression, 469
    stability, 468
    substrate specificity, 469
N-Acetyl-D-methionine, biological activity, 298
N-Acetyl-L-methionine, bioefficacy, 298
Acetylphenylhydrazine, 264
4-Acetylphenyl sulfate
  structure, 210
  substrate for arylsulfatase, 211, 213, 215
O-Acetylserine, 453
O-Acetylserine–O-acetylhomoserine sulfhydrylase, from S. cerevisiae, 466, 468, 478–483
  assay, 479–480
  inhibition, 482–483
  kinetic constants, 482
  properties, 482–483
  purification, 480–482
  requirements, 482–483
  specificity, 482
O-Acetylserine sulfhydrase. See Cysteine synthase
O-Acetylserine sulfhydrylase. See also O-Acetylserine (thiol)-lyase
  from B. sphaericus, 474–478
    assay, 474–475
    inhibition, 478
    properties, 477–478
    purification, 475–476
O-Acetyl-L-serine sulfhydrylase, assay, using sulfide electrode, 30–31
O-Acetylserine (thiol)-lyase, 453. See also Cysteine synthase
Actinomycetes, reverse transulfuration pathway, 487
Acyl carrier protein, in acyl transfer reactions, 367
O-Acylhomoserine, 453–454
Adenosine

R_f values, 380
  as substrate and inhibitor, 382
Adenosine N^1-oxide
  R_f values, 380
  as substrate and inhibitor, 382
Adenosine 3'-phosphate 5'-phosphosulfate.
  See PAPS
Adenosine 5'-phosphoramidate sulfate. See
  APA
Adenosine 5'-phosphosulfate. See APS
Adenosine 5'-sulfatophosphate sulfotransferase, 421
S-Adenosylcysteine, 192
Adenosylhomocysteinase, 369, 377–383, 424, 457. See also S-Adenosylhomocysteine hydrolase
  assay, 378–380
  inhibitors, 377, 382–383
  modulation, biological effects, 377
  properties, 382–383
  purification, 377, 380–382
  specificities, 382–383
  thin-layer chromatography, 379–380
  from yellow lupine, 430–434
    assay, 430–431
    definition of unit, 431
    energy of activation, 433
    formation of complex with adenosine, 434
    inhibition, 434
    kinetic constants, 433–434
    properties, 432–434
    purification, 431–433
    substrates, 433
  yield, 382
S-Adenosylhomocysteine, 191–192, 369–370, 457
  biological effects, 192
S-Adenosylhomocysteine hydrolase, 191–192. See also Adenosylhomocysteinase
S-Adenosylmethionine, 191–195, 457
  decarboxylated, 191
  formation, 367, 369
  function, 367
  HPLC, 193–194
    applications, 195
    automated, 195
  metabolites, 191–192
    HPLC, 193–194
    preparation of samples, 193–194

Adenosylmethionine cyclotransferase, 369
Adenosylmethionine decarboxylase, 369
S-Adenosylmethionine decarboxylase, 457–458
S-Adenosyl-L-methionine methyladenosine-lyase. See 1-Aminocyclopropane-1-carboxylate synthase
S-Adenosylmethionine methyltransferase, 369
Adenylate kinase
  activation, 135
  bisthiomethyl derivative, 135–136
  reduction by LMW thiol, measurement, 136–137
5'-Adenylylsulfate. See APS
Adenylyl sulfate:thiol sulfotransferase. See
  APS sulfotransferase
Adenylylsulfate–ammonia adenylyltransferase
  from Chlorella, 354–361
    assay, 354–357
    preparation, 357–360
    properties, 359–361
    ^{35}S-based assay, 354, 356–357
    specificity, 359–360
  distribution, 361
Adenylylsulfate kinase. See APS kinase
AdoMet. See S-Adenosylmethionine
AdoMet decarboxylase, 424
Aeromonas, L-methionine γ-lyase, 459–465
Agarose, as stabilizing agent for sulfate, 4
Alanine
  derivatized with ABD-F, 73
  derivatized with SBD-F, detection limit, 69
β-Alanine–pyruvate aminotransferase, 500
L-Alanine 3-sulfinic acid, transamination, determination of sulfite formed by, 22
Alanine sulfodisulfane, 223–227
  analytical properties, 224–227
  anion-exchange chromatography, 225
  biosynthetics, 225–226
  chemical properties, 225
  synthesis, 223–224
    from cystine and thiosulfate, 224
    from cystine disulfoxide and thiosulfate, 224
L-Alanine sulfodisulfane, acidic ninhydrin reaction of, 19–21
Alanyl-tRNA synthetase, SH groups, assay by aromatic disulfides, 52

SUBJECT INDEX

Albizzia lophanta, 438
Albumin, disulfide bond reduction, 252
Alcohol sulfotransferase, 201
Aldehyde dehydrogenase, SH groups, assay by aromatic disulfides, 52
S-Alkylcysteine lyase, 443
Alkyl thiosulfonate, cyanolysis, 27
Alliinase, 435, 438
Amanitin, sulfone analogs, 276
Amine sulfotransferase, 202
  assay, with labeled aniline, 205–206
  ion-pair extraction assay, 203
ω-Amino-acid aminotransferase, bacterial, 455
D-Amino-acid oxidase, 166, 369
ω-Amino acid–pyruvate aminotransferase, 500–504
  assay, 500
  inhibitors, 504
  kinetics, 504
  properties, 502–503
  purification, 501–502
  spectrophotometric properties, 503
  substrates, 503–504
Amino acids
  HPLC separation, 159, 165–166
  radiolabeled, 272
  toxicity, 314
Amino acid sulfoximines, 286–291
  synthesis, 288
8-Aminoadenosine
  R_f values, 380
  as substrate and competitive inhibitor of adenosylhomocysteinase, 382
S-(2-Amino-2-carboxyethylsulfonyl)-L-cysteine
  preparation, 18–19
  reactivity with thiol compounds, 18
  as reagent for converting inorganic sulfite and thiosulfite into organic cysteine derivatives, 17–22
  procedure, 19–21
  sensitivity of assay, 21
  structure, 17–18
1-Aminocyclopropane-1-carboxylate synthase, 426–429
  assay, 426–427
  inhibition, 429
  properties, 429
  purification, 428–429
  substrates, 429

1-Aminocyclopropane-1-carboxylic acid. *See* ACC
5-Aminolevulate dehydratase, SH groups, assay by aromatic disulfides, 52
4-(Aminosulfonyl)-7-fluoro-2,1,3-benzoxadiazole
  thiol adducts, properties, 72
  thiol determination with, 71–75
Ammonium 7-fluoro-2,1,3-benzoxadiazole 4-sulfonate
  determination of thiols with, 67–71
  fluorescent thiol adducts, properties, 68
APA
  cellular functions, 361
  formation, 354
APS
  APA formation from, 354–355
  [^{14}C], preparation, 355–356
  in dissimilatory sulfate reduction, 335
  formation, 334
  [^{35}S]
    carrier-free, preparation, 348–349
    preparation, 355
    as substrate for assimilatory sulfate reduction, 335
APS kinase, 329–332, 335
  assay methods, 342–344
  from *P. chrysogenum*
    properties, 347–348
    purification, 347
  paper electrophoresis, 342–343
  reverse reaction, assay, 343–344
  separation from ATP-sulfurylase, 344–345
APS pathway, 333
APS sulfotransferase, 333
(±)-Aristeromycin
  R_f values, 380
  as substrate and inhibitor, 382–383
Aromatic disulfides
  oxidation of thiols, spectrophotometric assay of thiols based on, 45 –451
  reagent for thiol assay, factors affecting choice of, 48–49
ARS—R reagents, 56
Ar—SS—R, for assaying SH groups, 50–51
Arylamine sulfotransferase, 202
Arylsulfatase, 207–217
  assay, 207
  from *H. pomatia*

assay, 361–362
properties, 366
purification, 363–365
lysosomal, 207
microsomal, 207
physiological roles, 208
specificity, 207
spectrophotometric assay, 208
tissue distribution, 207
Arylsulfotransferase, 201
Ascorbate, in food, removal, 12, 14
L-Ascorbate oxidase, to remove ascorbate from food, 12, 14
Asparagus, pyruvate dehydrogenase complex, 516
Asparagusate reductase, 516–521
assay, 517–518
cofactor specificity, 520–521
criteria of purity, 519
effect of phospholipid, 521
effect of surfactants, 521
properties, 519–521
purification, 518–520
substrates, 520–521
Asparagusic acid, synthesis, 516–517
Aspartate, HPLC measurement, 106
Aspartate aminotransferase, 372, 373, 375, 404
source, 161
L-Aspartate aminotransferase, 160
Aspartate β-decarboxylase, bacterial, 455
Aspergillus, cystathionine γ-lyase, 487
Aspergillus nidulans, ATP-sulfurylase, 344
Atherosclerosis, 299
ATP:adenylylsulfate 3′-phosphotransferase. *See* APS kinase
ATP:sulfate adenylyltransferase. *See* ATP-sulfurylase
ATPase
reduction, by mercaptoethanol, 250
SH groups, assay by aromatic disulfides, 52
ATP-sulfurylase, 329–332, 334
assay, 335–342
colorimetric molybdolysis (P_i formation) assay, 335–336
continuous spectrophotometric (APS kinase-coupled) assay, 340
continuous spectrophotometric molybdolysis (AMP release) assay, 336–338

forward reaction, reaction progress (average velocity) assay of, with sulfate in absence of APS kinase, 340–341
from *P. chrysogenum*
properties, 346–347
purification, 345–346
$^{32}PP_i$ release assay, 339
reverse reaction
bioluminescence assay for ATP used to measure, 342
$^{32}PP_i$ incorporation and $^{32}PP_i$–ATP exchange assay, 341
$^{35}SO_4^{2-}$ release assay, 341
separation from APS kinase, 344–345
$^{35}SO_4^{2-}$ incorporation ([^{35}S]PAPS synthesis) assay, 338–339
8-Azaadenosine
R_f values, 380
as substrate and inhibitor, 382
2-Aza-3-deazaadenosine
R_f values, 380
as substrate and inhibitor, 382

B

Bacillus, sulfur amino acid metabolism, 454
Bacillus sphaericus
O-acetylserine sulfhydrylase, 474–478
culture, 475
Bacillus subtilis, PAPS formation, 333
Bacteria
anoxygenic photosynthetic, use of reduced sulfur compounds as photosynthetic electron donors, 334
diamide treatment, 269
Barley, adenylylsulfate–ammonia adenylyltransferase, 359
bBBr
absorbance, 78
biological properties, 79
commercial name, 77
commercial source, 77
fluorescence, 78
formal name, 77
short form name, 77
solution, 80
uses, 80

SUBJECT INDEX

Bence Jones protein, reduction, by mercaptoethanol, 250
Benzenethiol
  reactivity in synthesis of S-substituted cysteines from L-serine, 294
  in synthesis of S-substituted homocysteines by L-methionine γ-lyase, 293
Benzidine, 3, 7
Benzofuroxan, reagent for thiols, 58
S-Benzyl amino acids, synthesis, 276
Benzyl chloride, 276–277
S-Benzyl-L-cysteine
  preparation, 293–294
  synthesis, 230–231
S-Benzyl-α-methylcysteine, 275
S-Benzyl-DL-α-methylcysteine, synthesis, 276–278
S-Benzyl-α-methylcysteine sulfone, 275
S-Benzyl-DL-α-methylcysteine sulfone
  cleavage of, to sulfinic acid, 278–279
  synthesis of, 276–278
Se-Benzyl-L-selenocysteine, preparation, 294–295
S-Benzyl sulfones, 276
  cleavage to sulfinic acids, 278–279
Betaine–homocysteine methyltransferase, 369, 370
Betaine–homocysteine S-methyltransferase
  distribution, 384
  human
    analysis for purity, 387–388
    assay, 384–385
    competitive inhibitors, 388
    dual-substrate inhibitor, 388
    inhibitors, 388
    K_m values for homocysteine and betaine, 388
    purification, 385–387
Bile-salt sulfotransferase, 202
Bimane-labeled products, spectroscopic data for, 82
Bimanes, 76
  anti series, 78
  syn series, 78
4,6-Bis(bromomethyl)-3,7-dimethyl-1,5-diazabicyclo[3.3.0]octa-3,6-diene-2,8-dione. See bBBr
Bisbromobimane. See bBBr
4,4'-Bisdimethylaminodiphenylcarbinol, reagent for thiols, 58

Borohydride, disulfide reduction, 251–252, 257
Bovine serum albumin
  derivatized with SBD-F, 68
  detection limit, 69
  disulfide bonds, reduction, 252
Brassica oleracea, cystathionine β-lyase, 439–443
Brassica rapa, cystine lyase, 442
Brevibacterium, sulfur amino acid metabolism, 454
Brevibacterium flavum
  O-acetylhomoserine sulfhydrylase, 470–474
  culture, 471
  methionine synthesis, 456
Bromine, for conversion of sulfides to sulfoxides, 284
Bromine complex with diazobicyclo[2.2.2]octane, for conversion of sulfides to sulfoxides, 284
Bromobimanes, 76–77, 85
  absorbance, 78
  biological properties, 79–80
  fluorescence, 78
  for fluorescent labeling of biological systems, 77
  labeled products, fluorescence microscopy, 82
  labeling procedures, 80–82
    reaction in solution, 81
    reactions in cells and tissues, 81–82
  materials labeled with
    absorbance, 78
    biological properties, 80
    fluorescence, 78
    photochemistry, 79
    stability, 79
  reaction with thiolate, 79
  reaction with tripeptide thiol, 79
  solutions, 80–81
  spectroscopic data for, 82
syn-(1-Bromoethyl,methyl)bimane, formation of sulfide from two thiols via reaction with, 84
4-Bromomethyl-3,7-dimethyl-6-trimethylammoniomethyl-1,5-diazabicyclo-[3.3.0]octa-3,6-diene-2,8-dione. See qBBr
4-Bromomethyl-3,6,7-trimethyl-1,5-diazabi-

cyclo[3.3.0]octa-3,6-diene-2,8-dione. See mBBr
N-Bromosuccinimide, for conversion of sulfides to sulfoxides, 284
1-Butanethiol
  reactivity in synthesis of S-substituted cysteines from L-serine, 294
  in synthesis of S-substituted homocysteines by L-methionine γ-lyase, 293
2-Butanethiol, reactivity in synthesis of S-substituted cysteines from L-serine, 294
DL-Buthionine (S,R)-sulfoximine, 287–288
L-Buthionine (S,R)-sulfoximine, 288
Buthionine sulfoximine
  inhibition of γ-glutamylcysteine synthetase, 287
  structure, 287
tert-Butyl hypochlorite, for conversion of sulfides to sulfoxides, 284–286

C

C1-esterase inhibitor, reduction, by DTT, 247
Cabbage. See also Brassica oleracea
  ATP-sulfurylase, 335, 344
Captopril
  derivatized with ABD-F, fluorescence intensity and maxima, 73
  derivatized with SBD-F, 68
  detection limit, 69
  thiol equilibrium constant determination, 139
  urinary, selective analysis, 257–258
Captopril disulfide, in urine, electrochemical reduction, 261–262
Carboxylic acids
  analogs of, 276
  compared to sulfinic acids, 271
2-Carboxy-2-oxoethanesulfonic acid. See β-Sulfopyruvate
Casein, dietary, in sulfur studies, 304–305
Cell extracts
  acidified, stability, 90
  preparation, for analysis of LMW thiols, 87
  reagent-derived products, retention times, 87–88
Cell fractions, bromobimane-labeled thiols,
  absorption and fluorescence measurements, 82–83
Cells
  bromobimane-labeled thiols, absorption and fluorescence measurements, 82–83
  labeling of thiols in, with bromobimanes, 81
CH_3Se—DNP
  HPLC elution times, 33
  molecular ion, 36
  R_f values, 33
Chlamydomonas, diamide treatment, 269
Chlamydomonas reinhardi, APS kinase, 335
Chlorella, sulfate reduction in, 421
Chlorella fusca, D-cysteine-specific enzyme, 452
Chlorella pyrenoidosa
  adenylylsulfate–ammonia adenylyltransferase, 354–361
  growth, 357
  sulfonic acid formation, 333
p-Chloromercuribenzenesulfonic acid, for colorimetric assay of thiols, 60
p-Chloromercuribenzoic acid, for colorimetric assay of thiols, 60–62
m-Chloroperbenzoic acid, for conversion of sulfides to sulfoxides, 283–284
Chloroplast, photosystem II, effect of mBBr, 84
Chloroplast CF_1, SH groups, assay by aromatic disulfides, 52
N-Chlorosuccinimide, for conversion of sulfides to sulfoxides, 284
2-Chloro-1,1,2 trifluoroethylene, cysteine conjugate of, 231
Choline, sulfone analogs, 276
Cholinesterase, reduction, by DTT, 247
Chromic acid, for conversion of sulfides to sulfoxides, 285
Citrobacter freundii
  metabolism of selenium amino acids, 459
  selenocysteine β-lyase, 493–496
Clostridium, metabolism of selenium amino acids, 459
Clostridium sporogenes, L-methionine γ-lyase, 459
CoA transferase, SH groups, assay by aromatic disulfides, 52

Coenzyme A
  in acyl transfer reactions, 367
  biosynthesis, 373, 376
  catabolism, 373, 376
  derivatized with ABD-F, fluorescence intensity and maxima, 73
  derivatized with SBD-F, 68
    detection limit, 69
  mBBr derivative
    half-life in 1% acetic acid, 90
    HPLC chromatography, 90–92
    retention time, 88
    quantitative determination, 95
    reaction with diamide, 265
Coenzyme M, mBBr derivative
  HPLC chromatography, 90–92
  retention time, 88
Collagen III, reduction, by DTT, 247
Complement C2, SH groups, assay by aromatic disulfides, 52
Cortisone sulfatase, from *H. pomatia*, purification, 365
CSA. *See* Cysteine sulfinate
Cucurbita maxima, 1-aminocyclopropane-1-carboxylate synthase, 429
Cupric copper, oxidant for thiols, 60
Cyanide, disulfide reduction, 257
Cyanolysis,
  cold, 26
    in presence of cupric ion, 26, 28–29
    procedure, 27
    reagents, 27
  of disulfides, 255
  hot, 26, 27–28
    procedure, 28
    reagents, 28
  principle, 26
  reaction conditions, 27
  sulfane sulfurs identified by susceptibility to, 26–29
Cyclohexanethiol, in synthesis of S-substituted homocysteines by L-methionine γ-lyase, 293
Cystamine
  in mediating intracellular disulfide bond formation, 376
  in plasma, enzymatic determination, 150–152
  tissue level, colorimetric assay, 152–154

β-Cystathionase. *See* Cystathionine β-lyase
γ-Cystathionase. *See also* Cystathionine γ-lyase
  persulfides formed by, 26
Cystathionine, HPLC measurement, 106
Cystathionine β-lyase, 422, 454
  bacterial, 448
  from cabbage, 439–443
    assay, 439–440
    inhibitors, 442–443
    properties, 441–442
    purification, 440–442
    specificity, 443
  distribution in plants, 443
  from *E. coli*, 483–486
    assay, 484
    inhibitors, 486
    purification, 484–485
    substrate, 483, 485–486
  fungal, 448
  from *S. metA-15* mutant, 483, 485
  from spinach, 443–449
    assay, 444–446
    inhibitors, 449
    isotopic assay, 444–445
    preparation, 446–448
    properties, 447–449
    protein determination, 446
    spectrophotometric assay, 445–446
    stability, 447–448
    substrate, 448
  from turnip, 442
Cystathionine γ-lyase, 369, 373, 374, 454. *See also* Cysteine desulfhydrase; γ-Cystathionase
  distribution among fungi, 486–487
  metabolism of cystine, 225–226
  from *S. phaeochromogenes*, 486–492
    absorption coefficient, 488
    assay, 487–488
    catalytic properties, 492
    immunological properties, 492
    inhibitors, 492
    properties, 491–492
    purification, 488–491
    substrates, 492
Cystathionine pathway, 313, 324
Cystathionine β-synthase, 369, 374, 388–394, 454, 487

absorption spectrum, 394
antibody–antigen interactions, 394
assay, 389–391
colorimetric assay, 390–391
deficiency, 389
estimation of purity, 393
forms, 389
human, isolation, 391–393
isoelectric focusing, 393
kinetic analyses, 394
microbial, 453
properties, 393–394
radioisotopic assay, 390
Cystathionine γ-synthase, 422, 423, 487. See also O-Succinylhomoserine (thiol)-lyase
inhibition, 423
Cysteamine
clinical applications, 149
derivatized with ABD-F
detection limit, 74
fluorescence intensity and maxima, 73
HPLC, 74
derivatized with SBD-F, 68
detection limit, 69
HPLC separation, 71
formation, 149
mBBr derivative
HPLC chromatography, 90–92
retention time, 88
in mediating intracellular disulfide bond formation, 376
in plasma, enzymatic determination, 150–152
in synthesis of S-substituted homocysteines by L-methionine γ-lyase, 293
tissue level, colorimetric assay, 152–154
Cysteamine dioxygenase, 373, 410–415
assay, 411
inhibitors, 413–414
kinetic constants, 414–415
properties, 413–415
purification, 411–413
reaction stoichiometry, 414–415
substrates, 413–414
tissue distribution, 410
Cysteic acid, 279, 333, 375
bioactivity, 299
HPLC, 164–166

HPLC measurement, 106
Cyst(e)ine
dietary, quantification, 300–301
in foods, 298–299
D-isomer, biological efficacy, 298
Cyst(e)inyl glycine, in cysteine delivery to cells, 316
Cysteine, 453
as active site covalent catalyst, 367
in biosynthesis of pantetheine and coenzyme A, 376
DL-[^{14}C]-, preparative resolution of, 169–172
catabolism, 375–376
in mammals, 374–375
catalytic functions, 366–367
cellular uptake, 313, 376
delivery systems, 315–318
applications to experimental work, 325
potential therapeutic uses, 324–325
derivatized with ABD-F
detection limit, 74
fluorescence intensity and maxima, 73
HPLC, 74
derivatized with SBD-F, 68
detection limit, 69
determination in solution, 68–69
HPLC separation, 71, 72
S-1,2-dichlorovinyl conjugate, 234
D-enantiomer, plant enzymes specific for, 450, 452
enantiomers, resolution, 166–172
hepatic, 112–114
HPLC measurement, 106
HPLC with electrochemical detection, 110
intracellular concentration, 375
intracellular delivery, 313–325
mammalian metabolism, 372–376, 395
mBBr derivative
HPLC chromatography, 90–92
retention time, 88
metabolism, 166, 366
metabolites, reversed-phase HPLC, 155–160
apparatus, 156
chromatographic procedure, 157–158
linearity, 158–159
reproducibility, 158–159
sample preparation, 156–157

sensitivity, 158–159
standard solutions, 157
uses, 155
microbial metabolism, 454–456
oxidation, to cysteine sulfinate, 374
oxidized derivatives, metabolic roles, 367
o-phthalaldehyde derivatives, HPLC, 141–143
physiological requirement for, 297
of plants, 419–422
quantitative determination, 94
reaction with diamide, 265
requirement, for protein synthesis, 324
synthesis, by mammals, 372
toxicity, 314
in culture media, 314
mechanisms, 314–315
transamination, 372
transamination pathway, 178
transaminative metabolites, determination, 178–182
Cysteine aminotransferase, 373
Cysteine conjugates, of toxic gases, 231
Cysteine desulfhydrase, 374, 454
in sulfide assay, 31
D-Cysteine desulfhydrase, from spinach, 449–453
assay, 450–451
effect of ions, 451–452
inhibitors, 451–452
products, 452
properties, 451–453
purification, 450–452
substrates, 451
L-Cysteine desulfhydrase, 443
from *S. typhimurium*, 452
Cysteine dioxygenase, 373, 395–403, 454–455
activation and inhibition by L-cysteine-related compounds, 400
activation of activity in liver homogenate, 397–398
catalytic properties, 400–401
conversion of cysteine into cysteine sulfinate, 226
distribution, 402–403
determination of product, 398–399
hepatic activity, function, 396
homogeneity, 400

NAD-dependent activity in extrahepatic tissues, 395–396, 403
assay, 403
properties, 400–401
purified
assay for, 399
oxygen electrode method for assay of, 399–400
rat liver, 395
assay, 397–400
purification, 396–398
storage, 397
L-Cysteines, S-substituted, preparation with tryptophan synthase, 291–295
Cysteine S-conjugates, 228–234
labeled
desalting on Amberlite XAD-2, 234
LH-20 Sephadex chromatography, 234
purification, 233–234
reversed-phase HPLC separation, 233–234
silica gel chromatography, 234
metabolism, 228
separation of unwanted N-conjugates from, 229
synthesis, 228
beginning with *N*-acetyl-L-cysteine, 232–233
beginning with β-chloro-L-alanine, 233
from *N*-carbobenzyloxy-*O*-tosylserine benzyl ester, 233
choice of reaction medium, 229
electrophile to cysteine stoichiometry in, 228–229
improving radiochemical yield of, 228–229
thin-layer chromatography, 229–230
trials with unlabeled reactants, 229
using liquid ammonia and sodium amide, 231–232
using methanol and sodium methoxide as base, 230–231
Cysteine sulfinate, 270–271, 404, 454–455
separation from other compounds, by HPLC, 159–160
sulfite production from, 15
unlabeled, synthesis, 274
yield, 273
L-[^{35}S]Cysteine sulfinate
characterization, 274

purification, 273
storage, 273
synthesis, 272-274
  cleavage of thiosulfinate intermediate, 273
  formation of thiosulfinate intermediate, 272-273
L-[1-^{14}C]Cysteine sulfinate, synthesis, 274
L-[3-^{14}C]Cysteine sulfinate, synthesis, 274
Cysteine-sulfinate decarboxylase, 373, 375
L-Cysteine-sulfinate decarboxylase. *See* Sulfinoalanine decarboxylase
Cysteinesulfinic acid, 395
  determination of, 399
    fuchsin method, 161-163
      interfering compounds, 162-163
      sensitivity, 163
      specificity, 162
    HPLC, 164-166
      preparation of samples, 164-165
      system, 165
  relationship to aspartate, 160
  reversed-phase HPLC, 155-160
  R_f value, 274
  as sole source of sulfur in bacteria, 160
Cysteine sulfonate, 279
Cysteinesulfonic acid, 280
  R_f value, 274
L-Cysteine sulfoxide lyase, S-substituted, from shiitake mushroom, 434-439
  assay, 435-436
  inhibition, 437-438
  kinetic properties, 437
  preparation, 436-437
  properties, 437-438
  purity, 437
  specificity, 438
Cysteine synthase, 420, 423
  assay, using sulfide electrode, 30-31
Cysteine thiosulfate. *See* Alanine sulfodisulfane
Cysteine-tRNA ligase. *See* L-Cysteinyl-tRNA synthetase
Cysteinyl-glutathione disulfide
  formation, 108
  HPLC measurement, 104, 106
Cysteinylglycine
  S-1,2-dichlorovinyl conjugate, 234
  HPLC measurement, 106
  mBBr derivative

    HPLC chromatography, 90-92
    retention time, 88
    standard solution, for HPLC, 86
Cysteinylsuccinamidopropyl-controlled-pore glass
  attachment and release of sulfhydryl oxidase, 509
  preparation, 505-507
L-Cysteinyl-tRNA synthetase, K_m for L-cysteine, 315
Cystine
  cellular uptake, 313, 376
  derivatized with ABD-F, fluorescence intensity and maxima, 73
  derivatized with SBD-F, detection limit, 69
  dietary, excess, 304
  dietary requirement, for laboratory animals, 303, 304
  electrochemical reduction, 262
  HPLC measurement, 106
  o-phthalaldehyde derivatives, HPLC, 141-143
  quantitative assay, using cystine-binding protein, 146-148
  radiolabeled, 272
  reduction, 250
  R_f value, 274
  toxicity, 375
Cystine-binding protein
  of *E. coli*, 144
  purification, 144-145
  in quantitative assay of cystine in biological samples, 146-148
Cystine lyase. *See* Cystathionine β-lyase
Cystinosis, 149, 375

D

3-Deazaadenosine
  R_f values, 380
  as substrate and competitive inhibitor of adenosylhomocysteinase, 382-383
3-Deaza-(±)-aristeromycin, as substrate and competitive inhibitor of adenosylhomocysteinase, 382-383
1-Decanethiol
  reactivity in synthesis of S-substituted cysteines from L-serine, 294-295
  in synthesis of S-substituted homocys-

teines by L-methionine γ-lyase, 293
Deglutamyllentinic acid, from *L. edodes*, preparation, 438–439
Dehydroepiandrosterone sulfatase, from *H. pomatia*, purification, 365
Dephosphocoenzyme A, mBBr derivative
　HPLC chromatography, 90–92
　retention time, 88
Desulfovibrio
　sulfate-activating enzyme system, 332
　sulfate reduction, 333–335
Dextran, as stabilizing agent for sulfate, 4
Diamide
　disulfide formation with, 264–270, 272
　　in cell suspensions, 269
　　chemical background, 265
　　reaction in solutions, 267
　　reactions in RBC, 267–268
　　in tissues, 269
　　treatment procedures, 267–268
　formula, 264
　oxidation of thiols to disulfides, functional consequences, 269–270
　properties of, 266
　protein response to, 265
　reaction with nonprotein thiols, 265
1,5-Diazabicyclo[3.3.0]octadienediones. *See* Bimanes
Diazenedicarboxylic acid bis(N,N-dimethylamide). *See* Diamide
Dichlorophenolindophenol, reagent for thiols, 58
S-1,2-Dichlorovinyl-L-cysteine, synthesis, 231–232
Dictyostelium discoideum, adenylylsulfate–ammonia adenylyltransferase, 359
Di-DNP-diselenide
　HPLC elution times, 33
　R_f values, 33
　TLC, 36
Di-DNP-disulfide
　HPLC elution times, 33
　R_f values, 33
Di-DNP-monoselenide
　HPLC elution times, 33
　molecular ion, 36
　R_f values, 33
Di-DNP-monosulfide
　HPLC elution times, 33

R_f values, 33
Diet
　amino acid, 301
　　for cats, 302
　　　composition of amino acid mixtures, 303
　　for chicks, 302
　　　composition of amino acid mixtures, 303
　　cost, 303–304
　　for dogs, 302
　　　composition of amino acid mixtures, 303
　　for rats/mice, 302
　　　composition of amino acid mixtures, 303
　　texture, 302
　crystalline amino acid, 301
　intact protein
　　for rodents, sulfur amino-acid deficient, 305–306
　　in sulfur studies, 304
　nitrogen- or methionine-free, methionine and cyst(e)ine depletion with, 304
　purified amino acid, carbohydrate source in, 301–302
　selenium-deficient, 307
　sulfur amino acid assay, for adult laboratory animals, 306–307
　for sulfur amino acid studies, 301–307
Dietary sulfur compounds, biological activity of, 297–300
Dihydrofolate reductase, SH groups, assay by aromatic disulfides, 52
Dihydrolipoate, carboxyl-bound polymer, disulfide reduction, 250–251
4-(Dimethylamino)-4'-azobenzene sulfinate
　characterization, 238
　substrate for sulfurtransferases, 236–239
　synthesis, 237
4-(Dimethylamino)-4'-azobenzene thiosulfonate anion
　characterization, 238
　substrate for sulfurtransferases, 236–239
　synthesis, 237
5-Dimethylamino-1-naphthalene sulfinate
　substrate for sulfurtransferase, 239
　synthesis, 239
5-Dimethylamino-1-naphthalene thiosulfonate anion

substrate for sulfurtransferase, 239
  synthesis, 239
2,2-Dimethylpropane-1-thiol, reactivity in synthesis of S-substituted cysteines from L-serine, 294–295
Dimethyl-β-propiothetin chloride, 227
Dimethyl selenide, trapping, 37
Dinitrogen tetraoxide, for conversion of sulfides to sulfoxides, 285
2,4-Dinitrophenol
  HPLC elution times, 33
  R_f values, 33
Dinitrophenol derivatives
  derivatization, 103–104
  HPLC, 101–109
    applications, 107–109
    limit of detection, 107
    procedure, 104–106
    sample preparation, 102
    standard procedure, 102–103
DIP, 270
DIP + 2, 270
Diphenylpicrylphenylhydrazine, reagent for thiols, 57
3′(2′),5′-Diphosphonucleoside. See DPNPase
Disulfide bonds
  cleavage with sodium sulfite, 116
  concentration, determination, 115
  formation, enzyme-catalyzed oxygen-dependent, assay of, 119–123
  NTSB assay
    advantages, 115
    applications, 118–119
    automated, 118–119
    procedure, 117
    reactions, 115–117
  pairings in proteins, 2-D reversed-phase HPLC, 118
  in peptides and proteins, 115–119
Disulfides
  chemical reduction, 246–256
  cyanolysis, 255
  derivatized with SBD-F, determination in solution, 68–70
  dinitrophenol derivatives, HPLC, 101–109
  direct reduction, 251–253
  electrochemical reduction, 257–264
    agar bridge substitutes, 262

  electrical characteristics of cell, 261
  electrochemical reduction cell, 258–259
  electrochemical reduction of sample, 260–261
  method, 258–261
  porous glass-bridged electrochemical reduction cell, 262–263
  preparation of agar salt bridge, 260
  procedure, 261–264
  reagents, 258
  reduction media, alternatives, 264
  simpler systems, 263–264
  smaller cells, 264
  formation, with diamide, 264–270, 272
  hepatocyte, HPLC, 107–108
  hydrolytic cleavage, 256
  mBBr derivatives, 96
  protein–glutathione mixed, HPLC, 108–109
  quantitative determination, 96
  reduction
    with borohydride, 251–252, 257
    by carboxyl-bound polymer of dihydrolipoate, 250–251
    with cyanide, 257
    by dithioerythritol, 246–249
    by dithiothreitol, 246–249, 257
    by 2-mercaptoethanol, 249–250, 257
    with phosphorothioate, 256
    with substitution, 253–256
    by thiols, 246–251
    with thiophenol, 251
    with tributylphosphine, 257
    with zinc–acid, 252, 257
  sulfitolysis, 254–255, 257
2,2′-Dithiodipyridine, 46
4,4′-Dithiodipyridine, 46
Dithioerythritol, disulfide reduction, 246–249
Dithiothreitol
  disulfide reduction, 246–249, 257
  mBBr derivative
    HPLC chromatography, 90–92
    retention time, 88
  reaction with diamide, 265
Djenkolic acid, 243, 420
  enzymatic degradation, 244
DNase

cysteine thiol group, determining pK_a value of, 134
  reduction by LMW thiols, measurement, 136–137
DPNPase, 332
DTNB, 46
2,2'-DTP. See 2,2'-Dithiodipyridine
4,4'-DTP. See 4,4'-Dithiodipyridine

E

Egg albumin, derivatized with ABD-F, 74–75
  α-chymotryptic digestion of, 75
  isolation of ABD-labeled peptides from, 75
Eggs, diamide treatment, 269
Ehrlich ascites tumor cells, diamide treatment, 269
Ellman's reagent, for analysis of thiols in diamide-treated RBC, 268
Elongation factor G, SH groups, assay by aromatic disulfides, 52
Enterokinase, reduction, by DTE, 247
Enterotoxins, reduction, by mercaptoethanol, 250
Ergothioneine, mBBr derivative
  HPLC chromatography, 90–92
  retention time, 88
Erythrocyte membrane, SH groups, assay by aromatic disulfides, 52
Escherichia coli
  adenylylsulfate–ammonia adenylyltransferase, 359
  cysteine metabolism, 454
  cystine-binding protein, 144
  methionine catabolism, 458
  osmotic shock-sensitive transport system, 144
  PAPS formation, 333
  sulfate reduction in, 421
  sulfur amino acid metabolism, 454
  vitamin B_{12}-dependent homocysteine transmethylation reaction, 456–457
Estrone sulfatase, from *H. pomatia*, purification, 365
Estrone sulfotransferase, 202
Ethanethiol, in synthesis of S-substituted homocysteines by L-methionine γ-lyase, 293

S-Ethyl-L-cysteine, preparation, 294
Ethylene
  biosynthesis, 426
  production, 458
Ethyl O-mesitylenesulfonyl acethydroxyamate, 288
DL-α-Ethylmethionine, preparation, 288–290
α-Ethylmethionine sulfoximine
  inhibition of glutamine synthetase, 287
  structure, 287
DL-α-Ethylmethionine (S,R)-sulfoximine, synthesis, 288–291
N-Ethylmorpholine, retention time, 88
N-Ethymaleimide, for colorimetric assay of thiols, 64–65
Euglena, adenylylsulfate–ammonia adenylyltransferase, 359
Euglena gracilis, sulfate-activating enzyme system, 332

F

(Fab)$_2$, conversion to thiol-containing derivative, labeling with mBBr, and use as probe, 83
Fat cells, diamide treatment, 269
FDNB. See 1-Fluoro-2,4-dinitrobenzene
Ferricyanide, oxidant for thiols, 59
Fibrinogen, reduction, by mercaptoethanol, 250
Fibronectin
  reduction, by DTT, 247
  SH groups, assay by aromatic disulfides, 52
Fluorescent labeling, 77
1-Fluoro-2,4-dinitrobenzene, 32
  HPLC elution times, 33
  R_f values, 33
Food, sulfite measurement in, 11–14
Formate dehydrogenase, 459
Formycin A
  R_f values, 380
  as substrate and inhibitor adenosylhomocysteinase, 382
Fructose-bisphosphate aldolase, SH groups, assay by aromatic disulfides, 52
Fuchsin method, for sulfite determination, 15–17

color stability, 16
expanded sensitivity, 16
reagent, 15
specificity, 16
standard procedure, 15
sulfate standard, preparation, 16–17
Fumarylacetoacetase, SH groups, assay by aromatic disulfides, 52
Fungi, sulfur amino acid metabolism, 453
Fusarium solani, sulfate-activating enzyme system, 332

G

Gelatin, as stabilizing agent for sulfate, 4
Glucosinolate sulfatase, from *H. pomatia*
  assay, 361–362
  properties, 365, 366
  purification, 365
Glutamate, HPLC measurement, 106
Glutamate–ammonia ligase. *See* Glutamine synthetase
Glutamate–cysteine ligase. *See* γ-Glutamylcysteine synthetase
Glutamic-oxaloacetic transaminase. *See* L-Aspartate aminotransferase
Glutamine synthetase, inactivation, by sulfoximines, 287
Glutamine transaminase, 369, 371
α-Glutamylcysteine, mBBr derivative, retention time, 88
γ-Glutamylcysteine, 97, 100
  HPLC measurement, 106
  mBBr derivative
    HPLC chromatography, 90–92
    retention time, 88
Glutamylcysteine synthetase, SH groups, assay by aromatic disulfides, 52
γ-Glutamylcysteine synthetase, 373, 422
  inactivation, by sulfoximines, 287
  K_m for L-cysteine, 315
γ-Glutamylcystine, in cysteine delivery to cells, 316
γ-Glutamyl glutamate, HPLC measurement, 106
γ-Glutamyl-S-methylcyteine, 420
γ-Glutamyltransferase, 373, 435
Glutathione, 99, 251
  biosynthesis, 376
  cysteine residue, function, 367
  derivatized with ABD-F
    detection limit, 74
    fluorescence intensity and maxima, 73
    HPLC, 74
  derivatized with SBD-F, 68
    detection limit, 69
    HPLC separation, 71, 72
  S-1,2-dichlorovinyl conjugate, 234
  effect on cellular levels of cysteine, 316
  in food, 298
  hepatic, 112–114
  HPLC measurement, 104, 106
  HPLC with electrochemical detection, 110
  in human RBC, oxidation with diamide, 267
  mBBr derivative
    HPLC chromatography, 90–92
    retention time, 88
  of plants, 419
  quantitative determination, 94–95
  reaction with bromobimane, 79
  reaction with diamide, 265
  in red cells, 264
  response to diamide, 266
  synthesis, in plants, 422
Glutathione disulfide
  electrochemical reduction, 262
  HPLC measurement, 102–109
  hydrolysis, 256
  reduction, 250
  thiol equilibrium constant determination, 139
Glutathione peroxidase
  activity, in selenium deficiency, 308, 313
  decomposition of selenium in, 37–38
Glutathione sulfide, production, from glutathione, via reaction with bromobimane, 84
Glutathione synthetase, 422
Glutathione transferase
  mBBr as substrate, 84
  SH groups, assay by aromatic disulfides, 52
Glutathione S-transferase, activity, in selenium deficiency, 313
Glycerol, as stabilizing agent for sulfate, 4
Glycerol-3-phosphate dehydrogenase, SH groups, assay by aromatic disulfides, 52

Glycine reductase, 459
Gonadotropin, reduction, by DTE, 247

H

Hair, disulfide bond formation, 510
Halides, for colorimetric assay of thiols, 66–67
Halobacterium halobium, thiol
 purification, 97–99
 purified, electrolytic reduction, 99
Helix pomatia, sulfatases, 361–365
Hemoglobin SH groups, diamide-reactive, 267
Hepatocytes
 dibutyl phthalate separation method, 107–108
 mitochondrial thiols and disulfides, HPLC, 109
HEPPS, retention time, 88
1-Heptanethiol
 reactivity in synthesis of S-substituted cysteines from L-serine, 294
 in synthesis of S-substituted homocysteines by L-methionine γ-lyase, 293
Heptathionine sulfoximine, inhibition of γ-glutamyl-histamine synthetase, 287
1-Hexanethiol
 reactivity in synthesis of S-substituted cysteines from L-serine, 294
 in synthesis of S-substituted homocysteines by L-methionine γ-lyase, 293
Hexathionine sulfoximine, inhibition of γ-glutamyl-histamine synthetase, 287
Histone H3, SH groups, assay by aromatic disulfides, 52
Homocyst(e)ine, diet supplement, 299–300
Homocysteic acid, 333
 HPLC measurement, 106
Homocysteine, 370, 453
 derivatized with ABD-F
  detection limit, 74
  fluorescence intensity and maxima, 73
  HPLC, 74
 derivatized with SBD-F
  detection limit, 69
  determination in solution, 68–69
  HPLC separation, 71
 hepatic, 112–114

HPLC measurement, 106
HPLC with electrochemical detection, 110
mBBr derivative
 HPLC chromatography, 90–92
 retention time, 88
synthesis, 454
 in plants, 422
 transsulfuration to cysteine, 422
L-Homocysteines, S-substituted, preparation with L-methionine γ-lyase, 291–292
Homocysteine sulfinate, 270–271
Homocysteine sulfonate, 279–280
L-Homocysteinesulfonic acid, synthesis, 280–281
Homocysteinyl-glutathione disulfide, HPLC measurement, 106
Homocystine, HPLC measurement, 106
Homocystinuria, 299, 389
Homoglutathione, mBBr derivative
 HPLC chromatography, 90–92
 retention time, 88
Homomethionine, 420
HPEA, derivatized with SBD-F, detection limit, 69
Hydrogen peroxide, for conversion of sulfides to sulfoxides, 284–286
Hydrogen selenide
 identification, volatilization procedure, 32–38
 volatilization procedure
  advantages, 35
  materials, 32–33
  methods, 32–35
DL-2-Hydroxy-4-(methylthiobutyrate), bioefficacy, 298
D-2-Hydroxy-acid dehydrogenase, 369
(S)-2-Hydroxy-acid oxidase, 369
N^6-Hydroxyadenosine, as substrate of adenosylhomocysteinase, 383
2-Hydroxyadenosine, as substrate of adenosylhomocysteinase, 383
2-Hydroxyethanesulfonic acid. *See* Isethionic acid
α-Hydroxy-γ-mercaptobutyrate, 370
p-Hydroxymercuribenzoic acid, for colorimetric assay of thiols, 60
D-Hydroxymethionine, biological activity, 298

L-Hydroxymethionine, biological activity, 298
Hypotaurine, 149, 270, 376, 410, 455
  conversion to taurine, 375
  [2-^{3}H]-
    as substrate for hypotaurine aminotransferase, 183–185
    synthesis, 183–184
  reversed-phase HPLC, 155–160
  separation from other compounds, by HPLC, 159–160
Hypotaurine aminotransferase
  assay, 183–185
  methodology, 184–185
Hypotaurine oxidase, 373

I

Inorganic pyrophosphatase, 331
Inorganic sulfate
  contribution to useful sulfur pool in diet, 300
  reduction, in plants, 421
Inorganic sulfur, incorporation into amino acids in microorganisms, 453–454
Inosine
  R_f values, 380
  as substrate and inhibitor, 382
S-Inosylhomocysteine, 192
Insulin
  disulfide bonds, sulfitolysis, 254
  hydrolysis, 256
Insulin receptor, reduction, by DTT, 247
Iodine, oxidant for thiols, 60
Iodobenzene dichloride, for conversion of sulfides to sulfoxides, 284
Iodosobenzene, for conversion of sulfides to sulfoxides, 285
Iodosobenzoic acid, reagent for thiols, 51–57
Ion chromatography
  apparatus, 9
  data evaluation, 9–11
  materials, 9
  principle, 7–9
Isethionic acid, 172–177, 280
  chloroethylsulfonylchloride derivatives, GLC, 176–177
  formula, 172
  gas–liquid chromatography, 173–177
  isolation
    from animal tissue, 176
    from squid axon, 176
  methyl ether/methyl ester derivatives, GLC, 174–175
  paper electrophoresis, 177
  relationship to taurine, 173
  silyl derivatives, GLC, 175–176
  species distribution, 172–173

K

Keratins, 510
α-Ketobutyrate, 370
α-Keto-γ-mercaptobutyrate, 370
Ketomethionine, 371
Kosower's reagent, 272

L

Lactalbumin, reduction, by DTE, 247
Lanthionine, dietary, 298
  quantification, 301
Lemna, methionine metabolism, 423–425
Lenthionine (1,2,3,5,6-pentathiepane), 434
Lentinic acid, 434
Lentinus edodes, L-cysteine sulfoxide lyase, 434–439
Lipase, SH groups, assay by aromatic disulfides, 52
Lipoamide disulfide, reduction by thiols, equilibrium constants for, 139
Lipoic acid, reaction with diamide, 265
Lipoyl reductase, assay, 517
Lithium β-sulfopyruvate, synthesis, 221–222
Liver, rat, ATP-sulfurylase, 335, 344
Lycopersicon esculentum, 1-aminocyclopropane-1-carboxylate synthase, 429
Lymphocytes, diamide treatment, 269
Lymphoid cell lines, growth, in medium supplemented with L-2-oxothiazolidine 4-carboxylate, 325
Lysophospholipase L$_2$, reaction with ABD-F, 74–75
Lysozyme
  cysteine thiol group, determining pK_a value of, 134
  reduction
    by DTT, 247
    by mercaptoethanol, 250

M

Malate dehydrogenase, SH groups, assay by aromatic disulfides, 53
Maleimides, for colorimetric assay of thiols, 64–66
Malic enzyme, SH groups, assay by aromatic disulfides, 53
mBBr
  absorbance, 78
  for analysis of LMW thiols, 85
  for analysis of thiols in diamide-treated RBC, 268
  biological properties, 79
  commercial name, 77
  commercial source, 77
  fluorescence, 78
  formal name, 77
  for histochemical detection of disulfides, 82
  labeling of thiols, 101
  labeling of thiols in tissues, 81–82
  short form name, 77
  solution, 80
  uses, 80
Menkes disease, 511
Menthyl sulfinates, reaction of Grignard reagents with, synthesis of optically pure sulfoxides by, 285
Mercaptoacetate, 178
  gas chromatography, 179–182
Mercaptoethanol
  derivatized with ABD-F, fluorescence intensity and maxima, 73
  derivatized with SBD-F, 68
    detection limit, 69
  disulfide reduction, 249–250, 257
2-Mercaptoethanol
  mBBr derivative
    HPLC chromatography, 90–92
    retention time, 88
  in synthesis of S-substituted homocysteines by L-methionine γ-lyase, 293
3-Mercaptolactate, 178
  gas chromatography, 179–182
3-Mercaptolactate cysteine disulfiduria, 178
(S,S)-1-(3-Mercapto-2-D-methyl-1-oxopropyl)-L-proline. See Captopril
5-Mercapto-2-nitrobenzoate, 47
α-Mercaptopropionylglycine
  derivatized with ABD-F, fluorescence intensity and maxima, 73
  derivatized with SBD-F, detection limit, 69
2-Mercaptopyridine
  mBBr derivative
    HPLC chromatography, 90–92
    retention time, 88
  standard solution, for HPLC, 86
3-Mercaptopyruvate, 178
  gas chromatography, 179–182
β-Mercaptopyruvate sulfurtransferase, 372, 374
3-Mercaptopyruvate sulfurtransferase, 373
Mercurials
  for colorimetric assay of thiols, 60–63
  phenolic, for colorimetric assay of thiols, 62
Mercurous acetate, disulfide reduction with, 253
O-Mesitylsulfonylhydroxylamine, in preparation of sulfoximines, 288
Methanethiol
  mBBr derivative
    HPLC chromatography, 90–92
    retention time, 88
  reactivity in synthesis of S-substituted cysteines from L-serine, 294
  standard solution, for HPLC, 86
Methionine, 454
  catalytic functions, 366–367
  dietary
    excess, 304
    quantification, 300–301
  dietary requirement, for laboratory animals, 303, 304
  enantiomers, resolution, 166–172
  in foods, 298–299
  formation, 371
  mammalian metabolism of, 369–372
  metabolism, 166, 366, 367
  microbial catabolism, 457–459
  microbial synthesis, 456–457
  mixed DL-isomer, bioefficacy, 297
  oxidation to methione sulfoxide, with hydrogen peroxide, 286
  physiological requirement for, 297
  of plants, 419
  synthesis, in plants, 422–426
D-Methionine, bioefficacy, 297

L-Methionine
  bioefficacy, 297
  deuterated and tritiated, preparation with L-methionine γ-lyase, 296–297
Methionine adenosyltransferase, 369, 424
Methionine analogs, α-, β-, and γ-substituted, synthesis from αβ-unsaturated aldehydes and ketones, 290
Methionine decarboxylase, bacterial, 457
Methionine γ-lyase, 459
  bacterial, 456
  decomposition of selenodjenkolic acid, 244
L-Methionine γ-lyase
  from *Aeromonas*
    catalytic properties, 465
    inhibition, 464
    substrate, 465
  assay, 460
  definition of unit, 460
  distribution, 459–460
  from *P. putida*
    catalytic properties, 464–465
    inhibition, 464
    substrate, 463–464
  in preparation of deuterated and tritiated L-methionine and S-methyl-L-cysteine, 296–297
  in preparation of S-substituted L-homocysteines, 291–292
  properties, 462–465
  protein concentration, 460
  purification
    from *Aeromonas*, 462
    from *P. putida*, 460–462
  storage, 462
Methionine S-oxide reductase, 369
Methionine sulfone, 275
  bioactivity, 299
Methionine sulfoxide
  bioactivity, 299
  reduction, 458
Methionine sulfoxide reductase, 458
Methionine sulfoximine, 275, 286–287
  structure, 287
  as suicide substrate, 287
Methionine synthase, 456
Methionine synthetase, 370
4-Methumbelliferone sulfate
  structure, 210
  substrate for arylsulfatase, 211
N^6-Methyladenosine
  R_f values, 380
  as substrate and inhibitor, 382–383
2-Methyl-1-butanethiol, in synthesis of S-substituted homocysteines by L-methionine γ-lyase, 293
2-Methyl-2-butanethiol, in synthesis of S-substituted homocysteines by L-methionine γ-lyase, 293
3-Methyl-1-butanethiol, in synthesis of S-substituted homocysteines by L-methionine γ-lyase, 293
S-Methylcysteine, 419
S-Methyl-L-cysteine, deuterated and tritiated, preparation with L-methionine γ-lyase, 296–297
α-Methyl-DL-cysteine sulfinate, unlabeled, synthesis, 274
α-Methylcysteinesulfinic acid, 275
DL-α-Methylcysteinesulfinic acid, synthesis, 278–279
3,3'-Methylenediselenobis(2-aminopropionic acid). See Selenodjenkolic acid
2-Methyl-2-heptanethiol, in synthesis of S-substituted homocysteines by L-methionine γ-lyase, 293
2-Methyl-2-hexanethiol, in synthesis of S-substituted homocysteines by L-methionine γ-lyase, 293
Methyl methanethiolsulfonate, 135–136
S-Methylmethioninesulfonium, 419
Methylnitrocatechol sulfate
  structure, 210
  substrate for arylsulfatase, 211–214
2-Methyl-2-octanethiol, in synthesis of S-substituted homocysteines by L-methionine γ-lyase, 293
2-Methyl-2-pentanethiol, in synthesis of S-substituted homocysteines by L-methionine γ-lyase, 293
2-Methylpropane-1-thiol, reactivity in synthesis of S-substituted cysteines from L-serine, 294
1-Methyl-1-propanethiol, in synthesis of S-substituted homocysteines by L-methionine γ-lyase, 293
2-Methyl-1-propanethiol, in synthesis of S-substituted homocysteines by L-methionine γ-lyase, 293

2-Methyl-2-propanethiol, in synthesis of S-substituted homocysteines by L-methionine γ-lyase, 293
Se-Methyl-L-selenocysteine, preparation, 295
Methylselenol, identification, 32
Methylselenyl sulfides, identification, 36
Methylsulfhydrylation, 456
5-Methyltetrahydrofolate–homocysteine methyltransferase, 369
N^5-Methyltetrahydrofolate–homocysteine methyltransferase, 457
5-Methyltetrahydropteroyltriglutamate–homocysteine methyltransferase, 424. See also Tetrahydropteroyltriglutamate methyltransferase
Methylthioadenosine, 191
5-Methylthioadenosine, 371
Methylthioadenosine phosphorylase, 369
4-Methylumbelliferone sulfate, substrate for arylsulfatase, 213, 216
MHPEA, derivatized with SBD-F, detection limit, 69
Microellobosporia violacea, cystathionine γ-lyase, 492
Micromonospora, cystathionine γ-lyase, 487
Micromonospora chalcea, cystathionine γ-lyase, 492
Micropolyspora, cystathionine γ-lyase, 487
Micropolyspora angispora, cystathionine γ-lyase, 492
Microsomes, brain, SH groups, assay by aromatic disulfides, 53
Microtubules, labeling with qBBr, 83
Mixed disulfides, reagents for assay of thiols, 50–51
Moffatt oxidation, 282
Monobromobimane. See mBBr
MSH. See O-Mesitylsulfonylhydroxylamine
Mucus, bronchial, reduction, by mercaptoethanol, 250
Mung bean. See also *Vigna radiata*
  extraction procedure, for LMW thiol analysis, using acetonitrile to denature proteins, 89–90
  LMW thiols, analysis, by HPLC, 92–94
Mycobacterium, cystathionine γ-lyase, 487

Myosin, SH groups, assay by aromatic disulfides, 53
Myrosinase, 361

N

Na^+,K^+-ATPase, SH groups, assay by aromatic disulfides, 52
NADH:hydrogen-peroxide oxidoreductase, 12
β-Naphthalenethiol, in synthesis of S-substituted homocysteines by L-methionine γ-lyase, 293
Nebularine
  R_f values, 380
  as substrate and competitive inhibitor of adenosylhomocysteinase, 382
Neurospora
  cystathionine γ-lyase, 487
  cysteine metabolism, 454
  sulfur amino acid metabolism, 453
  sulfur source, 333
Neurospora crassa, ATP-sulfurylase, 344
Neutrophiles, diamide treatment, 269
Nicotiana plumbaginifolia, mutant lacking cystathionine β-lyase, 422
Nitric acid, for conversion of sulfides to sulfoxides, 285
Nitrocatechol sulfate
  structure, 210
  substrate for arylsulfatase, 209–212
2-Nitro-5-mercaptobenzoic acid, mBBr derivative, retention time, 88
Nitrophenol, mercury derivatives, for colorimetric assay of thiols, 62
4-Nitrophenyl sulfate
  structure, 210
  substrate for arylsulfatase, 211, 213–215
Nitroprusside, oxidant for thiols, 59
Nitroquinol sulfate
  structure, 210
  substrate for arylsulfatase, 210–214
2-Nitro-5-thiosulfobenzoate
  assay of disulfide bonds with, 115–119
  preparation, 117
Nitrous acid, for colorimetric assay of thiols, 63–64
Nocardia, cystathionine γ-lyase, 487
1-Nonanethiol, reactivity in synthesis of S-

substituted cysteines from L-serine, 294
NTSB. See 2-Nitro-5-thiosulfobenzoate
NTSB–protein reaction, 117–118

O

1-Octanethiol, reactivity in synthesis of S-substituted cysteines from L-serine, 294
Ornithine transcarbamylase, SH groups, assay by aromatic disulfides, 53
5-Oxoprolinase
  reaction catalyzed, 318
  SH groups, assay by aromatic disulfides, 53
L-2-Oxothiazolidine 4-carboxylate
  in cell culture studies, 325
  doses, for rodents, 323–324
  as intracellular cysteine delivery system, 318–324
  in parenteral amino acid administration, 324
  potential value in therapy, 324–325
  [^{35}S]
    relative specific radioactivity values of cysteine and glutathione after administration of, 322–323
    uptake of ^{35}S into organs and body fluids after administration to mice, 320–322
  tissue levels of cysteine after administration of, 322
Oxyhemoglobin, SH groups, assay by aromatic disulfides, 53
Oxytocin, hydrolysis, 256
Ozone, for conversion of sulfides to sulfoxides, 285

P

Palladium salts, for colorimetric assay of thiols, 67
Pantetheine, 149, 413
  biosynthesis, 373, 376
  catabolism, 373, 376
  mBBr derivative
    HPLC chromatography, 90–92
    retention time, 88

Pantoate dehydrogenase, SH groups, assay by aromatic disulfides, 53
PAP, formation, 201
Papain
  activation, 135
  cysteine thiol group, determining pK_a value of, 134
  reduction, by mercaptoethanol, 250
  S-thioalkyl derivative, 135–136
    reduction by LMW thiols, determination, 136–137
PAPS
  activating enzyme system that synthesizes, 329–332
  formation, 334
  [^{35}S], carrier-free, preparation, 348
  as sulfate (sulfuryl) donor for biosynthesis of sulfate esters, 335
  unlabeled, preparation, 349
PAPS pathway, 333
PAPS sulfotransferase, 333
Penicillamine
  assay methods, 186
  colorimetric assay, 190–191
  electrochemical reduction, 262
  forms, 186
  HPLC measurement, 106
  liquid chromatography, 186–190
    with determination by amino acid autoanalysis, 189–190
    with determination by derivatization, 188–189
    with electrochemical detection, 186–188
  mBBr derivative, retention time, 88
  therapeutic applications, 186
  thiol equilibrium constant determination, 139
Penicillanic acid sulfone, 275–276
Penicillium chrysogenum
  APS kinase, 335
  ATP-sulfurylase, 335
  growth, 344
  sulfate-activating enzymes, purification, 344–348
Penicillium duponti, ATP-sulfurylase, 344
1-Pentanethiol
  reactivity in synthesis of S-substituted cysteines from L-serine, 294

in synthesis of S-substituted homocysteines by L-methionine γ-lyase, 293
Pepstatin, conversion to thiol-containing derivative, labeling with mBBr, and use as probe, 83
Permanganate, oxidant for thiols, 60
Persulfide, 25
  cyanolysis, 27
Persulfide groups
  absorbance spectrum, 26
  formation, 25–26
Phenolphthalein bissulfate
  structure, 210
  substrate for arylsulfatase, 211, 213, 215–216
Phenol sulfotransferase, ion-pair extraction assay, 202–203
Te-Phenyltellurocysteine
  preparation, 245
Te-Phenyltellurohomocysteine
  preparation, 245
3′-Phosphoadenosine 5′-phosphosulfate, sulfite production from, 15
3′-Phosphoadenylyl:thiol sulfotransferase. See PAPS sulfotransferase
3′-Phosphoadenylylsulfate. See PAPS
3-Phosphoglycerate kinase, SH groups, assay by aromatic disulfides, 53
O-Phosphohomoserine, in plants, 423
N-(Phosphonoacetyl)-L-aspartate, sulfone analogs, 275
4′-Phosphopantetheine, mBBr derivative, retention time, 88
Phosphorothioate, disulfide reduction, 256
o-Phthalaldehyde
  preparation, 141
  reaction with amino acids, 141
Piperazine, diazenecarbonyl reagents derived from, 270
Plants, nonprotein fraction, sulfur amino acids, 419
Plasma
  SBD-thiols in, HPLC, 70–72
  sulfate, analysis, 4–6
Platelet membranes
  disulfide bonds, sulfitolysis, 254
  SH groups, assay by aromatic disulfides, 53
Platinum salts, for colorimetric assay of thiols, 67

Polarogram, 40
  half-wave potential, 40
  wave, 40
Polarograph
  components, 38–39
  detector of current flow, 39
  direct current source of electrical potential, 38–39
  electrode, 39
Polarography, 254
  applications, 39–42
  diffusion current, 41
  interface effects, 44
  of mercury ion-reactive species, 41
  principles, 39–42
  procedure, 42–43
  of redox compounds, 39–41
Polyethylene glycol, as stabilizing agent for sulfate, 4
Polysulfide, 25
  cyanolysis, 27
Polythionate, 25
  cyanolysis, 28
Polythionate thiosulfate, cyanolysis, 27
Potassium chloride, retention time, 88
Potassium phosphate, retention time, 88
Proline
  derivatized with ABD-F, fluorescence intensity and maxima, 73
  derivatized with SBD-F, detection limit, 69
Proline oxidase, activity, 317
1-Propanethiol
  reactivity in synthesis of S-substituted cysteines from L-serine, 294–295
  in synthesis of S-substituted homocysteines by L-methionine γ-lyase, 293
2-Propanethiol, in synthesis of S-substituted homocysteines by L-methionine γ-lyase, 293
Propiothetin bromide, chemical synthesis, 227
Protamines, reduction, by mercaptoethanol, 250
Protein
  dietary requirement, 314
  response to diamide, 265
  SH groups, ionization behavior, 134
Protein A, 395
  activity, assay, 402

purification, 401–402
specificity, 401–402
Protein-mixed disulfides
  cellular, estimation of, 124
  determination, 125
    preparation of samples, 125–126
    procedure, 126
    reagents, 126
Protein—SS—Ar, for assaying SH groups, 50
Prothionine sulfoximine, inhibition of γ-glutamylcysteine synthetase, 287
PRPP-synthetase, SH groups, assay by aromatic disulfides, 53
Pseudomonas putida, L-methionine γ-lyase, 459–465
Pummerer rearrangement, 282
Pyrazomycin
  R_f values, 380
  as substrate and inhibitor, 382
Pyridine–bromine complex, for conversion of sulfides to sulfoxides, 284
Pyridoxal 5′-phosphate-dependent enzymes, in preparation of sulfur- and selenium-bearing amino acids, reactions, 291
Pyridoxal 5′-phosphate enzymes, inhibition by cysteine, 314
Pyruvate diphosphate kinase, SH groups, assay by aromatic disulfides, 53
Pyruvate kinase, SH groups, assay by aromatic disulfides, 53

Q

qBBr
  absorbance, 78
  biological properties, 79
  commercial name, 77
  commercial source, 77
  fluorescence, 78
  formal name, 77
  short form name, 77
  solution, 80
  uses, 80
Quinone, reagent for thiols, 58

R

Red blood cells
  diamide-treated, analytical procedures for thiols in, 268
  labeling of thiols in, with bromobimanes, 81
  thiol oxidation with diamide, 267–268
  thiol status, 265
Rhodanese. *See also* Sulfurtransferase
  assay, DAB substrate, 238–239
  assay methods, 236
  persulfides formed by, 26
  SH groups, assay by aromatic disulfides, 53
  for sulfur transfer from thiosulfate to cyanide, 24
Rhodopsin
  energy transfer studies, bromobimane labeling used in, 83
  SH groups, assay by aromatic disulfides, 53
Ribonuclease
  protein disulfide groups, reduction with thioglycolic acid, 251
  reduction, by mercaptoethanol, 250
Ribonuclease A, refolding, chemical heterogeneity in, 119
Ribosomal protein, synthesis, 373

S

Saccharomyces cerevisiae
  O-acetylserine–O-acetylhomoserine sulfhydrylase, 466, 468, 478–483
  ATP-sulfurylase, 335
  ethylene production, 458
  methionine catabolism, 458
Saetre–Rabenstein agar-bridged electrochemical reduction cell, 258–259
Salmonella
  O-acetylserine (thiol)-lyase, 453
  cysteine metabolism, 454
Salmonella typhimurium
  L-cysteine desulfhydrase, 452
  sulfur amino acid metabolism, 454
Sanger's reagent, 32
  derivatization of thiols and disulfides with, 103
SBD-F. *See* Ammonium 7-fluoro-2,1,3-benzoxadiazole 4-sulfonate
Schizosaccharomyces pombe
  O-acetylhomoserine sulfhydrylase, 465–469
  culture, 466
Schmidt reaction, 288

Sea urchin eggs, diamide treatment, 269
Selenate, 36
Selenide, identification, 36
Selenite, identification, 36
Selenium
  biochemical functions, 308
  from urine, identification, 38
  volatilization apparatus, 34
  volatilization procedure, 34–35
Selenium amino acids
  microbial metabolism, 459
  preparation with microbial pyridoxal phosphate enzymes, 291–297
Selenium deficiency, in rats, 307–313
  animal care and feeding, 311–312
  assessing severity of, 307–308
  classification, 308
  development, 312–313
  diet for producing, 309–311
  strategies for producing, 308–309
Selenocysteine, 240–243
  determination, Gaitonde's method for, 148–149
  properties, 240
  Se-alkyl and Se-aryl derivatives, synthesis, 242
  synthesis, 240
    procedures, 240–241
L-Selenocysteine, Se-substituted, preparation with tryptophan synthase, 291–295
Selenocysteine β-lyase, 415–418, 459
  assay, 415
  from *C. freundii*, 493–496
    absorption spectrum, 495
    catalytic properties, 495–496
    properties, 494–496
    purification, 493–495
    specificity, 493
    stability, 494–495
  distribution, 415, 493
  effect of pH, 418
  pig liver, 415
    purification, 416–417
  properties, 418
  protein determination, 416
  stability, 418
  storage, 418
  substrates, 418
Selenocystine
  ^{1}H-NMR spectrum, 242

chemical and physical properties, 242–243
optically active, preparation, 242
L-Selenocystine, preparation, with O-acetylhomoserine sulfhydrylase, 295–296
Selenodjenkolic acid, 243–244
  enzymatic decomposition, 244
  properties, 244
  synthesis, procedure, 243
Selenoenzymes, 308
Selenohomocysteine, Se-alkyl and Se-aryl derivatives, synthesis, 242
L-Selenohomocysteine, synthesis, 242
Selenohomocystine
  chemical and physical properties, 242–243
  optically active, preparation, 242
  synthesis, 242
L-Selenohomocystine, preparation, with O-acetylhomoserine sulfhydrylase, 295–296
Selenolanthionine, as byproduct of selenocysteine synthesis, 241
Selenoprotein, determination of volatile selenium released from, 37
Selenotrisulfides, identification, 36
Serine, derivatized with SBD-F, detection limit, 69
Serine acyltransferase, 420
  complex with cysteine synthase, 420
  from *S. typhimurium*, 420
Serine hydroxymethyltransferase, SH groups, assay by aromatic disulfides, 53
L-Serine sulfhydrylase, 478
Serine sulfhydrylase. *See* Cystathionine β-synthase
Serum, sulfate, analysis, 4
Shiitake mushrooms. *See Lentinus edodes*
Sickle cell anemia, 149
Singlet oxygen, for conversion of sulfides to sulfoxides, 284
Skim milk membrane vesicles, solubilized, preparation, 507–508
Sodium acetate, retention time, 88
Sodium borate, retention time, 88
Sodium hydride, in anhydrous dimethyl sulfoxide, disulfide reduction with, 252
Sodium mercury amalgam, disulfide reduction with, 252

Sodium metaperiodate, for conversion of sulfides to sulfoxides, 283, 285
Sodium methane sulfonate, retention time, 88
Spermatozoa
  bromobimane labeling, and motility, 84
  diamide treatment, 269
  labeling of thiols in, with bromobimanes, 81
Spermidine, 371
Spermidine synthase, 369
Spermine, 371
Spermine synthase, 371
Spinach
  adenylylsulfate–ammonia adenylyltransferase, 359
  APS kinase, 335
  ATP-sulfurylase, 335, 344
Spisula solidissima, sulfate-activating enzyme system, 332
Steroid sulfatase, from *H. pomatia*
  assay, 362–363
  properties, 366
Steroid sulfotransferase, 202
  assay, 206
Streptomyces
  cystathionine γ-lyase, 487
  cysteine metabolism, 454
Streptomyces lactamdurans, cystathionine γ-lyase, 492
Streptomyces lydicus, cystathionine γ-lyase, 492
Streptomyces olivochromogenes, cystathionine γ-lyase, 492
Streptomyces phaeochromogenes
  cystathionine γ-lyase, 486–492
  growth, 488
Streptosporangium, cystathionine γ-lyase, 487
Streptosporangium roseum, cystathionine γ-lyase, 492
Streptoverticillium, cystathionine γ-lyase, 487
Streptoverticillium kentuchense, cystathionine γ-lyase, 492
Succinyl-CoA synthetase, SH groups, assay by aromatic disulfides, 53
O-Succinylhomoserine, 454
O-Succinylhomoserine (thiol)-lyase, 454
Suicide substrate, 287

Sulfane
  atoms, stability, 25
  definition of, 25
Sulfane sulfur
  determination, 25–29
  identification, by susceptibility to cyanolysis, 26–29
  reactivity, toward cyanolysis, 27
Sulfatase, from *H. pomatia*, 361–365
  properties, 365–366
Sulfatase A, 207–208
Sulfatase B, 207–208
Sulfatase C, 207
Sulfate, 396
  activated, forms of, 329
  activation
    control of, 333–334
    reactions of, 329–332
  assimilation, in plants, 421
  in biological materials
    colorimetry, 3
    estimation, 3
  carrier-enzyme system, 329
  determination, in urine, 3
  dietary, 300
  formation, 367, 404
  ion chromatography, 7–11
    analytical procedure, 9
    apparatus, 9
    column selectivity, 10
    data evaluation, 9–11
    materials, 9
    principle, 7–9
    system capacity factor, 10
    system resolution, 10–11
    theoretical plate number, 10
  metabolism, 329
  nephelometry, 3–6
    advantages, 3–4
    precautions, 4
    procedure, 5–6
    stabilizing agents, 4
  precipitation
    with barium ions, 3
    with benzidine, 3, 7
    with chloranilate, 3, 7
    with rhodizonate, 7
  reduction
    assimilatory, 333, 335
    dissimilatory, 333, 335

selective barium sulfate precipitation, 4, 7
serum, automated laser nephelometry, after deproteinization with trichloroacetic acid, 4
transfer, 332–333
turbidimetry, 3–6
  advantages, 3–4
  precautions, 4
  procedure, 5
  stabilizing agents, 4
uptake, 329
Sulfate-activating enzymes, 334–349
  from *P. chrysogenum*, purification, 344–348
Sulfate-activating system, 329–332
  cellular localization, 332
  species distribution, 332
Sulfate adenylyltransferase, 421. *See also* ATP-sulfurylase
Sulfate esters, characterization, 361
Sulfenic acids, 279
Sulfhydrylation, 456
Sulfhydryl oxidase
  assay, 120
    comparison of methods, 123
    horseradish peroxidase system, 122–123
    oxygen electrode–catalase-coupled system, 120–121
  in cow snout, 515
  in dermis, 515
  epidermal, 515
  from milk, 504–510
    assay, 509–510
    distribution, 505
    isolation, 505
    purification, 508–509
    solubilization, 505
  purification, by covalent affinity chromatography, 123
  from rat skin, 510–515
    assay, 511
    enzymatic activity, 514
    heat stability, 515
    metal cofactors, 515
    properties, 514–515
    purification, 512–514
    ribonuclease renaturation assay, 511–512

thiol substrate specificity, 514–515
tissue localization, 515
Sulfide
  conversion to sulfoxides, choice of oxidants, 283
  determination, with ion-specific electrode, 29–31
  ion-specific electrode
    equipment, 29–30
    principle, 29
  mBBr derivative
    HPLC chromatography, 90–92
    retention time, 88
  oxidation, 281
    asymmetric induction methods, 285
    to sulfones, 278
  preparation of sulfoxides from, 281–282
Sulfinic acids, 276, 279
  formation, 271–272
  properties, 270–271
  synthesis, 272–274
β-Sulfinoacetaldehyde, 270
3-Sulfino-L-alanine carboxylyase. *See* Sulfinoalanine decarboxylase
Sulfinoalanine decarboxylase, 404–410
  assay, 405–406
  catalytic properties, 409–410
  properties, 408–410
  purification, 404, 406–408
  pyridoxal phosphate content, 409
  species and tissue distribution, 410
  yield, 404
β-Sulfinopyruvate, 270
β-Sulfinyl-α-aminopropionic acid. *See* Cysteinesulfinic acid
β-Sulfinyl pyruvate, 221
  formation, 404
Sulfite
  and ACESC, reaction of, 18
  as cleavage reagent for proteins, 254
  determination
    fuchsin method, 15–17
    methods, 11
    by reaction with ACESC, 20–21
    by sulfite oxidase assay, 11–14
    as S-sulfo-L-cysteine, 19–20
    using S-(2-amino-2-carboxyethyl-sulfonyl)-L-cysteine, 17–22
  disulfide reduction, 257
  mBBr derivative

HPLC chromatography, 90–92
    retention time, 88
  oxidation to sulfate, 404
  uses, 14
Sulfite dehydrogenase, 11
Sulfite oxidase, 404
  assay of sulfites in food, 11–14
    calculation, 13
    general method, 12–13
    interferences, 14
    optimal reaction conditions, 14
    principle, 12
    procedure, 13
    reaction conditions, 13–14
    reagents, 12
    specificity, 14
    statistical results, 14
  inhibition, 14
Sulfite reductase, 421
Sulfite reductase (NADPH), 11
Sulfitolysis, of disulfides, 254–255
Sulfoacetaldehyde lyase, 499
S-Sulfo-L-cysteine, acidic ninhydrin reaction of, 19–21
Sulfones, 282
  properties, 274–275
Sulfonic acids, 274
  of biological interest, 333
  in nature, 280
  preparation, 280
    oxidizing agents, 280
  properties, 279
Sulfoproteins, stability, 119
S-Sulfoproteins
  preparation, 118
  properties, 118
β-Sulfopyruvate, 280
  assay, 222–223
  formation, 221
  metabolism, 221
  NADH oxidation assay, 222–223
  solutions
    assay, 222
    stability, 222
Sulfosalicylic acid, retention time, 88
Sulfotransferase
  acting on hydroxysteroids, binding assay, 206
  assay, 201–207
    using radioactive sulfuryl group acceptors, 205

assay methods, comparisons, 206–207
    Ecteola-cellulose chromatographic assay, 204–205
    thin-layer chromatography, 203–204
Sulfoxides, 274–275, 281–286
  chemistry, 282
  enantiomeric, conversion to sulfoximines, 288
  as oxidizing reagents, 282–283
  preparation
    nonoxidative methods for, 285
    from sulfides, 281–282
  reduced to sulfides, 282
  synthesis of sulfoximines from, 291
Sulfoximides. See also Sulfoximines
Sulfoximines, 275
  structures, 287
  synthesis, 288
    from sulfoxides, 291
Sulfur
  in chicken feathers, identification, 38
  cycle between animals and plants, 419–420
  elemental, 25
    cyanolysis, 27
  reactions of, in biosphere, 329–330
Sulfur amino acids
  analytical resolution of, 167–169
  enantiomers, resolution, 170
  mammalian metabolism, 366–376
  microbial, 453–459
  of plants, 419–426
    constituents, 419–426
  preparation with microbial pyridoxal phosphate enzymes, 291–297
  synthesis, by plants, role in sulfur cycle, 419
Sulfur amino acid studies, assay diets for, 301–307
Sulfur compounds
  identification, volatilization procedure, 32, 37
  polarography, 38–44
    effect of pyridine on half-wave potentials, 44
    half-wave potentials, 43–44
Sulfur dioxide, colorimetric determination, Grant's method, 15
Sulfurtransferase. See also Rhodanese; Thiosulfate reductase
  assay, DAB thiosulfonate and sulfinate

substrates, 238–239
assay methods, 236
chromogenic substrates, 236–239
distributions, 235
fluorigenic substrates, 239
Swern oxidation, 282

T

Taka-amylase A, disulfide reduction, with borohydride, 252
Taurine, 280, 333, 375, 376, 396, 455
  conversion to isethionate, 173
  in food, 299
  formation, 367, 404
  oxidation, 499
  reversed-phase HPLC, 155–160
  separation from other compounds, by HPLC, 159–160
Taurine:(acceptor) oxidoreductase (deaminating), 499
Taurine aminotransferase, bacterial, 455
Taurine-decomposing bacteria, 497
Taurine dehydrogenase, 496–499
  assay, 496–497
  bacterial, 455
  inhibitors, 499
  K_m for taurine and PMS, 499
  properties, 498–499
  purification, 497–498
  reaction products, 499
  size, 498
  specific activity, 496–497
  specificity for electron acceptor, 499
  stoichiometry of oxidation of taurine, 499
  substrates, 499
Taurine transaminase. See ω-Amino acid–pyruvate aminotransferase
Taurocholic acid, 280
Tellurium, 245
Tellurium compounds, identification, volatilization procedure, 37–38
Telluroamino acids, 245
Tetrahydropteroyltriglutamate methyltransferase, 423
Thetin, 227
Thetin–homocysteine S-methyltransferase, 384
L-Thiazolidine-4-carboxylic acid, in intracellular delivery of cysteine, 317
Thiazolidine-4-carboxylic acid, 2-substituted, 317–318
Thiazolidines, in intracellular delivery of cysteine, 317
Thimidylate synthase, SH groups, assay by aromatic disulfides, 53
Thiobenzyl alcohol
  reactivity in synthesis of S-substituted cysteines from L-serine, 294
  in synthesis of S-substituted homocysteines by L-methionine γ-lyase, 293
Thiocyanate
  determination, methods, 22
  ferric complex ions, 22–24, 26
    principle, 23
    procedure, 24
    reagents, 23–24
  ion-selective electrode for, 22
Thioesters
  mBBr derivatives, 96
  quantitative determination, 96
Thioglycolate, disulfide reduction, 251
Thioglycolic acid
  mBBr derivative
    HPLC chromatography, 90–92
    retention time, 88
  in synthesis of S-substituted homocysteines by L-methionine γ-lyase, 293
Thioglycolic acid ethyl ester, in synthesis of S-substituted homocysteines by L-methionine γ-lyase, 293
Thiolactic acid, in synthesis of S-substituted homocysteines by L-methionine γ-lyase, 293
Thiol–disulfide exchange, 246
Thiol–disulfide interchange reactions, 129–130
  equilibrium constants, 138–140
  rates of, 130–131
    Brønsted correlation, 132
Thiol equilibrium constants, determination of values of, 138–140
Thiol oxidases, 373
Thiol pK_a values, determination, 132–138
  in proteins, 134–138
  for small molecules, 132–134
  for thiols containing groups with proton affinities similar to that of SH, 134
Thiolproteinase inhibitor, reduction, by DTT, 247

Thiols
  carbon-based electrode, 110
  derivatized with ABD-F
    determination, 71–75
    determination in solution, 72–73
    HPLC, 73–74
  derivatized with SBD-F
    determination in solution, 68–70
    HPLC separation, 70–72
  dinitrophenol derivatives, HPLC, 101–109
  disulfide reduction, 246–251
  enzyme-catalyzed oxidation, 119
  fluorometric assay, with fluorobenzadiazoles, 67–75
  hepatic, 110, 112–114
  hepatocyte, HPLC, 107–108
  HPLC with electrochemical detection, 110–114
    calculations, 112–114
    chromatography, 111–112
    principle, 110
    procedure, 111–112
    reagents, 110–111
    recoveries, 112
    sample preparation, 111–112
    sample storage, 112
    standard curves, 112
    system, 111
  inorganic oxidants, 59–60
  labeling, with bromobimanes, 76–84
  low-molecular-weight, determination using mBBr labeling and HPLC, 85–96
    chromatography, 90–92
    materials, 85–86
    preparation of samples, 86–90
    quantitative determinations, 94–96
  mBBr derivatives
    characteristics, 100–101
    electrolytic reduction, 99–100
    isolation, 96
  mercury-based electrode, 110
  metal ion-catalyzed oxidation, 118
  nonprotein
    DTNB-based assays, 47
    labeling with bromobimanes, 79
  organic oxidants, 51–58
  oxidation, 127
    in extraction, 95
    protection against, 104, 124–125
  spectrophotometric assays based on, 45–60
  oxidized, derivatized with SBD-F, determination in solution, 69
  oxidizing agents for, 45
  protein, spectrophotometric assay, 49–50
  protein-bound, liberation, 127
  purification
    from biological samples, 96–101
    materials, 97
    procedure, 96–99
    quantitative determinations, 94–96
  reaction with diazenecarbonyl derivatives, 265
  reaction with SBD-F, 69–70
  reduced, derivatized with SBD-F, determination in solution, 68–69
  reduction, 128–129
  spectrophotometric assay, 44–67
    based on thiol oxidation, 45–60
    based on thiol substitution, 60–67
    choice of buffer, 45
    preliminary protein precipitation, 45
    standard samples, HPLC chromatography, 90–92
  standard solutions
    for HPLC, 86
    storage, 87
  stock solutions, for HPLC, 86
  sulfonation, as preparative technique, 118–119
Thiol status, of biological systems, 265–266
Thioredoxin, reduction, by DTT, 247
Thioredoxin reductase, 458
Thiosulfate, 25
  cyanolysis, 27, 28
  determination, methods, 22
  mBBr derivative
    HPLC chromatography, 90–92
    retention time, 88
  quantitative cyanolysis in presence of cupric ions, 23, 24
  quantitative determination, 94–95
  sulfite production from, 15
  thiocyanate formed from, colorimetric determination as ferric thiocyanate complex ion, 24–25
  in urine, determination by cyanolysis, 29
Thiosulfate:cyanide sulfurtransferase, 24

Thiosulfate reductase, 350–354. *See also* Sulfurtransferase
  assay, 236, 350–351
  DAB substrate, 239
  inhibition, 353
  kinetic constants, 353–354
  mechanism, 353–354
  properties, 353–354
  purification, 351–353
  SH groups, assay by aromatic disulfides, 53
  specificity, 353
Thiosulfate sulfurtransferase, 24
  assay, with ACESC, 22
Thiosulfite
  and ACESC, reaction of, 19
  determination
    as L-alanine sulfodisulfane, 20
    by reaction with ACESC, 20–21
    using S-(2-amino-2-carboxyethylsulfonyl)-L-cysteine, 17–22
Thiosulfonate, cyanolysis, 27
Thiosulfonate ions, 25
Thiosulfonate reductase, 421
Thyroid-stimulating hormone, reduction, by DTT, 247
Tissue
  animal, extraction for LMW thiol analysis, using methanesulfonic acid to denature enzymes, 89
  diamide treatment, 269
  labeling of thiols in, with bromobimanes, 81–82
Tomato. *See Lycopersicon esculentum*
Torula yeast, rat diet based on, 309–311
Transaminase, 371
Transhydrogenase, SH groups, assay by aromatic disulfides, 53
Transsulfuration, in plants, 422
Transsulfurylation, 456
  in bacteria, 457
  in microorganisms, 454
Tributylphosphine, disulfide reduction, 253, 257
Trimethylselenonium ion, urinary, quantification, 195–200
  applications, 200
  cation HPLC using gradient elution, 198–200
  general procedures, 196
  Reineckate precipitation, 197–198, 200
Trinitrobenzenesulfonic acid, reagent for thiols, 58
Tris–methane sulfonate, retention time, 88
Trithionate, cyanolysis, 27
tRNA-$_f$Met, conversion to thiol-containing derivative, labeling with mBBr, and use as probe, 83
Trypsin, disulfide reduction, with borohydride, 252
Trypsin inhibitor, disulfide reduction, with borohydride, 252
Trypsinogen, disulfide reduction, with borohydride, 252
Tryptophan synthase, in preparation of S-substituted L-cysteines and Se-substituted L-selenocysteines, 291–295
Tryptophanyl-RNA ligase, disulfide reduction, with borohydride, 252
Tryptophanyl-RNA synthetase, SH groups, assay by aromatic disulfides, 53
Turnip. *See Brassica rapa*
Tween 80, as stabilizing agent for sulfate, 4
Tyrosine, derivatized with SBD-F, detection limit, 69
Tyrosine-ester sulfotransferase, 201

U

Urease, SH groups, assay by aromatic disulfides, 53
Urine
  minimization of free captopril oxidation during storage, 258
  selenium in, identification, 38
  sulfate, analysis, 3–4
  thiosulfate, determination by cyanolysis, 29

V

Vigna radiata, 1-aminocyclopropane-1-carboxylate synthase, 429
Volatile selenols, identification, 32

W

Water—mBBr, retention time, 88
Winter squash. *See Cucurbita maxima*
Wool, disulfide bonds in, reduction, 253

WR-1065
  mBBr derivative
    HPLC chromatography, 90–92
    retention time, 88
    quantitative determination, 94

X

Xanthine oxidase
  disulfide bonds, sulfitolysis, 254
  SH groups, assay by aromatic disulfides, 53

Y

Yeast, sulfur amino acid metabolism, 453

Z

Zinc–acid, disulfide reduction, 252, 257

227265